STUDENT SOLUTIONS MANUAL

Dennis Kletzing

INTRODUCTORY

LINEAR ALGEBRA

An Applied First Course

8/E

Bernard Kolman ■ David R. Hill

PEARSON

Prentice
Hall

Upper Saddle River, New Jersey 07458

Editor in Chief: *Sally Yagan*
Acquisitions Editor: *George Lobell*
Supplements Editor: *Jennifer Urban*
VP of Production and Manufacturing: *David W. Riccardi*
Executive Managing Editor: *Kathleen Shiaparelli*
Managing Editor: *Nicole Jackson*
Production Editor: *Donna Crilly*
Supplement Cover Manager: *Daniel Sandin*
Manufacturing Buyer: *Ilene Kahn*

© 2005 by Pearson Education, Inc.
Pearson Prentice Hall
Pearson Education, Inc.
Upper Saddle River, NJ 07458

The author and publisher of this book have used their best efforts in preparing this book. These efforts include the development, research, and testing of the theories and programs to determine their effectiveness. The author and publisher make no warranty of any kind, expressed or implied, with regard to these programs or the documentation contained in this book. The author and publisher shall not be liable in any event for incidental or consequential damages in connection with, or arising out of, the furnishing, performance, or use of these programs.

Pearson Prentice Hall® is a trademark of Pearson Education, Inc.

Printed in the United States of America

ISBN 0-13-143742-9

Pearson Education Ltd., *London*
Pearson Education Australia Pty. Ltd., *Sydney*
Pearson Education Singapore, Pte. Ltd.
Pearson Education North Asia Ltd., *Hong Kong*
Pearson Education Canada, Inc., *Toronto*
Pearson Educación de Mexico, S.A. de C.V.
Pearson Education—Japan, *Tokyo*
Pearson Education Malaysia, Pte. Ltd.

Contents

Preface

This manual is to accompany the Eighth Edition of Bernard Kolman's *Introductory Linear Algebra with Applications*. Detailed solutions to all odd numbered exercises are included. It was prepared by Dennis Kletzing, Stetson University, and Nina Edelman and Kathy O'Hara, Temple University. It contains many of the solutions found in the Fifth Edition, prepared by David R. Hill, Temple University, as well as solutions to new exercises included in the Eighth Edition of the text.

Chapter 1

Linear Equations and Matrices

Section 1.1, p. 8

1. $x + 2y = 8$
$3x - 4y = 4$

   Add 2 times the first equation to the second one to obtain $5x = 20$. Therefore $x = 4$. Substituting this value into the first equation, we obtain:

$$4 + 2y = 8 \implies y = 2.$$

   The solution is $x = 4$, $y = 2$.

3. $3x + 2y + z = 2$
$4x + 2y + 2z = 8$
$x - y + z = 4$

   To eliminate x, add (-3) times the third equation to the first one and (-4) times the third equation to the second one to obtain the system

$$
\begin{aligned}
5y - 2z &= -10 \\
6y - 2z &= {-8}.
\end{aligned}
\tag{1.1}
$$

   To eliminate z from (1.1), add (-1) times the first equation to the second to obtain

$$y = 2.$$

   Substitute $y = 2$ into the first equation of (1.1) to obtain $5(2) - 2z = -10$. Therefore $z = 10$. Finally, substitute $y = 2$, $z = 10$ into the first equation of the given system to get $3x + 2(2) + 10 = 2$. Therefore $x = -4$. The solution is $x = -4$, $y = 2$, $z = 10$.

5. $2x + 4y + 6z = -12$
$2x - 3y - 4z = 15$
$3x + 4y + 5z = {-8}$

   To simplify the system, multiply the first equation by $\left(\frac{1}{2}\right)$ to obtain the system

$$
\begin{aligned}
x + 2y + 3z &= -6 \\
2x - 3y - 4z &= 15 \\
3x + 4y + 5z &= -8.
\end{aligned}
\tag{1.2}
$$

To eliminate x, add (-2) times the first equation to the second one and (-3) times the first equation to the third one to obtain

$$-7y - 10z = 27$$
$$-2y - 4z = 10.$$

(1.3)

Multiply the second equation by $\left(\frac{1}{2}\right)$:

$$-7y - 10z = 27$$
$$-y - 2z = 5.$$

(1.4)

Add (-5) times the second equation in (1.4) to the first one to obtain

$$-2y = 2 \implies y = -1.$$

Substitute $y = -1$ into the second equation in (1.4) to obtain

$$+1 - 2z = 5 \implies z = -2.$$

Substituting the values $y = -1$, $z = -2$ into the first equation in (1.2), we obtain

$$x + 2(-1) + 3(-2) = -6 \implies x = 2.$$

The solution is $x = 2$, $y = -1$, $z = -2$.

7. $x + 4y - z = 12$
     $3x + 8y - 2z = 4$

To eliminate x, add (-3) times the first equation to the second one to obtain

$$-4y + z = -32.$$

We now solve this equation for y in terms of z:

$$y = \tfrac{1}{4}z + 8.$$

The variable z can be chosen to be any real number. To find x in terms of z, substitute the expression for y into the first equation:

$$x + 4\left(\tfrac{1}{4}z + 8\right) - z = 12 \implies x = -20.$$

The solution is $x = -20$, $y = \tfrac{1}{4}r + 8$, $z = r$, where r is any real number.

9. $x + y + 3z = 12$
     $2x + 2y + 6z = 6$

To eliminate x, add (-2) times the first equation to the second one to obtain $0 = -18$. This makes no sense. Therefore the linear system has no solution. It is inconsistent.

11. $2x + 3y = 13$
     $x - 2y = 3$
     $5x + 2y = 27$

To eliminate x, add (-2) times the second equation to the first equation and (-5) times the second equation to the third one to obtain

$$7y = 7$$
$$12y = 12.$$

(1.5)

Therefore $y = 1$. Substituting the value $y = 1$ into the second equation, we obtain

$$x - 2(1) = 3 \implies x = 5.$$

The solution is $x = 5$, $y = 1$.

13. $x + 3y = -4$
    $2x + 5y = -8$
    $x + 3y = -5$

The first and third equations have the same left side but different right sides. Therefore the system has no solution. It is inconsistent.

15. $2x - y = 5$
    $4x - 2y = t$

(a) We first eliminate y from this system by adding (-2) times the first equation to the second to obtain $0 = -10 + t$. Therefore, to be a consistent system we must have $t = 10$.

(b) The system is inconsistent if t has any value other than 10. For example, if $t = 3$, we are led to the conclusion that $0 = 10 - t = 7$, which makes no sense.

(c) There are an infinite number of choices for t that result in an inconsistent system.

17. $2x + y - 2z = -5$
    $3y + z = 7$
    $z = 4$

Substituting the value $z = 4$ from the last equation into the second equation, we obtain

$$3y + 4 = 7 \implies y = 1.$$

Substituting $z = 4$ and $y = 1$ into the first equation gives

$$2x + 1 - 2(4) = -5 \implies x = 1.$$

Thus the solution is $x = 1$, $y = 1$, $z = 4$.

19. $2x + 3y - z = 11$
    $x - y + 2z = -7$
    $4x + y - 2z = 12$

If such a value exists, then the values $x = 1$, $y = 2$, $z = r$ must satisfy each of the equations. Substituting these values into each of the equations, we obtain:

$$\begin{aligned}
\text{First Equation:} \quad & 2(1) + 3(2) - r = 11 && \implies && r = -3 \\
\text{Second Equation:} \quad & 1 - 2 + 2r = -7 && \implies && r = -3 \\
\text{Third Equation:} \quad & 4(1) + 2 - 2r = 12 && \implies && r = -3.
\end{aligned}$$

Thus, for $r = -3$ we have that $x = 1$, $y = 2$, $z = r$ is a solution.

21. (a) A unique point.

(b) No points simultaneously lie in all three planes.

(c) There are infinitely many points.

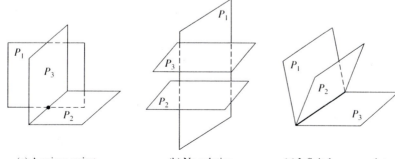

(a) A unique poimt   (b) No solution   (c) Infinitely many points

23. Let the amount of low-sulfur fuel be denoted by x_1 and the amount of high-sulfur fuel be denoted by x_2. Convert all times to minutes. Then,

$$\begin{aligned}\text{blending plant requirements:} \quad & 5x_1 + 4x_2 = 180 \\ \text{refining plant requirements:} \quad & 4x_1 + 2x_2 = 120.\end{aligned}$$

To eliminate x_1 we multiply the first equation by 4 and add (-5) times the second equation to it to obtain

$$6x_2 = 120 \quad \Longrightarrow \quad x_2 = 20.$$

To solve for x_1, substitute $x_2 = 20$ into either of the original equations. Using the blending plant equation, we have

$$5x_1 + 4(20) = 180 \quad \Longrightarrow \quad 5x_1 = 100 \quad \Longrightarrow \quad x_1 = 20.$$

Thus, 20 tons of low-sulfur fuel and 20 tons of high-sulfur fuel should be manufactured so that the plants are fully utilized.

25. Let

$$\begin{aligned} x_1 &= \text{the number of ounces of food A per meal,} \\ x_2 &= \text{the number of ounces of food B per meal,} \\ x_3 &= \text{the number of ounces of food C per meal.} \end{aligned}$$

Then,

$$\begin{array}{lrcrcrcl}\text{protein requirements:} & 2x_1 & + & 3x_2 & + & 3x_3 & = & 24 \\ \text{fat requirements:} & 3x_1 & + & 2x_2 & + & 3x_3 & = & 24 \\ \text{carbohydrate requirements:} & 4x_1 & + & 1x_2 & + & 2x_3 & = & 21.\end{array}$$

To eliminate x_1 from the first two equations, we multiply the first equation by 3 and add (-2) times the second equation to it to obtain $5x_2 + 3x_3 = 27$. To eliminate x_1 from the second and third equations, we multiply the second equation equation by 4 and add (-3) times the third equation to it to obtain $5x_2 + 6x_3 = 33$. Thus we have the pair of equations

$$\begin{aligned} 5x_2 + 3x_3 &= 27 \\ 5x_2 + 6x_3 &= 33 \end{aligned}$$

to solve for x_2 and x_3. To eliminate x_2, we add (-1) times the first equation to the second equation to obtain

$$3x_3 = 6 \quad \Longrightarrow \quad x_3 = 2.$$

Substituting $x_3 = 2$ into the equation $5x_2 + 3x_3 = 27$ and solving for x_2, we obtain

$$5x_2 + 3(2) = 27 \quad \Longrightarrow \quad 5x_2 = 21 \quad \Longrightarrow \quad x_2 = \tfrac{21}{5} = 4.2.$$

Finally, substitute $x_2 = 4.2$ and $x_3 = 2$ into any of the original equations and solve for the remaining variable x_1. Using the equation for carbohydrate requirements, we obtain

$$4x_1 + 1(4.2) + 2(2) = 21 \quad \Longrightarrow \quad 4x_1 = 12.8 \quad \Longrightarrow \quad x_1 = 3.2.$$

Thus, each meal should contain 3.2 ounces of food A, 4.2 ounces of food B, and 2 ounces of food C.

27. (a) $\begin{aligned} p(1) &= a(1)^2 + b(1) + c = a + b + c = -5 \\ p(-1) &= a(-1)^2 + b(-1) + c = a - b + c = 1 \\ p(2) &= a(2)^2 + b(2) + c = 4a + 2b + c = 7. \end{aligned}$

(b) Subtract the second equation from the first to obtain $2b = -6$. Therefore, $b = -3$. Substituting $b = -3$ into the first and third equations, we obtain

$$a - 3 + c = -5 \qquad\qquad a + c = -2$$
$$\text{or}$$
$$4a - 6 + c = 7 \qquad\qquad 4a + c = 13.$$

Subtract the second equation from the first to obtain $-3a = -15$ or $a = 5$. Finally, substitute $a = 5$ into the first equation to obtain $5 + c = -2$ or $c = -7$. Thus, the solution is $a = 5$, $b = -3$, $c = -7$.

T.1. Let a solution to the system of equations in (2) be given by

$$x_1 = s_1, \quad x_2 = s_2, \ldots, \quad x_n = s_n.$$

Interchanging the position of two of the equations in (2) gives a system in which $x_j = s_j$, $j = 1, 2, \ldots, n$, still satisfies each equation.

T.3. If $s_1, s_2, \ldots, s_n$ is a solution to (2), then the pth and qth equations are satisfied:

$$a_{p1}s_1 + \cdots + a_{pn}s_n = b_p$$
$$a_{q1}s_1 + \cdots + a_{qn}s_n = b_q.$$

Thus, for any real number r,

$$(a_{p1} + ra_{q1})s_1 + \cdots + (a_{pn} + ra_{qn})s_n = b_p + rb_q$$

and so $s_1, \ldots, s_n$ is a solution to the new system.

Conversely, any solution to the new system is also a solution to the original system (2).

Section 1.2, p. 19

1. (a) $a_{12} = -3$, $a_{22} = -5$, $a_{23} = 4$.
   (b) $b_{11} = 4$, $b_{31} = 5$.
   (c) $c_{13} = 2$, $c_{31} = 6$, $c_{33} = -1$.

3. Two matrices are equal if corresponding entries agree. Hence,

$$\begin{bmatrix} a + 2b & 2a - b \\ 2c + d & c - 2d \end{bmatrix} = \begin{bmatrix} 4 & -2 \\ 4 & -3 \end{bmatrix}$$

implies that we have two systems of equations:

$$a + 2b = 4 \qquad\qquad 2c + d = 4$$
$$\text{and}$$
$$2a - b = -2 \qquad\qquad c - 2d = -3.$$

We solve the system on the left by multiplying the first equation by (-2) and adding it to the second equation to obtain

$$-5b = -10 \quad \Longrightarrow \quad b = 2.$$

Substituting $b = 2$ into the first equation, we find that

$$a + 2(2) = 4 \quad \Longrightarrow \quad a = 0.$$

Similarly, to solve the system on the right for c, we multiply the first equation by 2 and add it to the second equation to obtain

$$5c = 5 \quad \Longrightarrow \quad c = 1.$$

Substituting $c = 1$ into the first equation, we obtain

$$2(1) + d = 4 \quad \Longrightarrow \quad d = 2.$$

Thus, $a = 0$, $b = 2$, $c = 1$, and $d = 2$.

5. (a) $3D + 2F = 3\begin{bmatrix} 3 & -2 \\ 2 & 4 \end{bmatrix} + 2\begin{bmatrix} -4 & 5 \\ 2 & 3 \end{bmatrix} = \begin{bmatrix} 9 & -6 \\ 6 & 12 \end{bmatrix} + \begin{bmatrix} -8 & 10 \\ 4 & 6 \end{bmatrix} = \begin{bmatrix} 1 & 4 \\ 10 & 18 \end{bmatrix}.$

   (b) $3(2A) = 3\left(2\begin{bmatrix} 1 & 2 & 3 \\ 2 & 1 & 4 \end{bmatrix}\right) = 3\begin{bmatrix} 2 & 4 & 6 \\ 4 & 2 & 8 \end{bmatrix} = \begin{bmatrix} 6 & 12 & 18 \\ 12 & 6 & 24 \end{bmatrix}.$

   $6A = 6\begin{bmatrix} 1 & 2 & 3 \\ 2 & 1 & 4 \end{bmatrix} = \begin{bmatrix} 6 & 12 & 18 \\ 12 & 6 & 24 \end{bmatrix}.$

   Note: $3(2A) = 6A$.

   (c) $3A + 2A = 3\begin{bmatrix} 1 & 2 & 3 \\ 2 & 1 & 4 \end{bmatrix} + 2\begin{bmatrix} 1 & 2 & 3 \\ 2 & 1 & 4 \end{bmatrix} = \begin{bmatrix} 5 & 10 & 15 \\ 10 & 5 & 20 \end{bmatrix}.$

   $5A = 5\begin{bmatrix} 1 & 2 & 3 \\ 2 & 1 & 4 \end{bmatrix} = \begin{bmatrix} 5 & 10 & 15 \\ 10 & 5 & 20 \end{bmatrix}.$

   Note: $3A + 2A = 5A$.

   (d) $2(D + F) = 2\left(\begin{bmatrix} 3 & -2 \\ 2 & 4 \end{bmatrix} + \begin{bmatrix} -4 & 5 \\ 2 & 3 \end{bmatrix}\right) = 2\begin{bmatrix} -1 & 3 \\ 4 & 7 \end{bmatrix} = \begin{bmatrix} -2 & 6 \\ 8 & 14 \end{bmatrix}.$

   $2D + 2F = 2\begin{bmatrix} 3 & -2 \\ 2 & 4 \end{bmatrix} + 2\begin{bmatrix} -4 & 5 \\ 2 & 3 \end{bmatrix} = \begin{bmatrix} 6 & -4 \\ 4 & 8 \end{bmatrix} + \begin{bmatrix} -8 & 10 \\ 4 & 6 \end{bmatrix} = \begin{bmatrix} -2 & 6 \\ 8 & 14 \end{bmatrix}.$

   Note: $2(D + F) = 2D + 2F$.

   (e) $(2 + 3)D = 5\begin{bmatrix} 3 & -2 \\ 2 & 4 \end{bmatrix} = \begin{bmatrix} 15 & -10 \\ 10 & 20 \end{bmatrix}.$

   $2D + 3D = 2\begin{bmatrix} 3 & -2 \\ 2 & 4 \end{bmatrix} + 3\begin{bmatrix} 3 & -2 \\ 2 & 4 \end{bmatrix} = \begin{bmatrix} 6 & -4 \\ 4 & 8 \end{bmatrix} + \begin{bmatrix} 9 & -6 \\ 6 & 12 \end{bmatrix} = \begin{bmatrix} 15 & -10 \\ 10 & 20 \end{bmatrix}.$

   Note: $(2 + 3)D = 2D + 3D$.

   (f) $3(B + D)$ is undefined since B is 3×2 and D is 2×2.

7. (a) $(2A)^T = \left(2\begin{bmatrix} 1 & 2 & 3 \\ 2 & 1 & 4 \end{bmatrix}\right)^T = \begin{bmatrix} 2 & 4 & 6 \\ 4 & 2 & 8 \end{bmatrix}^T = \begin{bmatrix} 2 & 4 \\ 4 & 2 \\ 6 & 8 \end{bmatrix}.$

   Note: $(2A)^T = 2A^T$.

   (b) $(A - B)^T$ is undefined since A is 2×3 and B is 3×2.

   (c) $(3B^T - 2A)^T = \left(3\begin{bmatrix} 1 & 0 \\ 2 & 1 \\ 3 & 2 \end{bmatrix}^T - 2\begin{bmatrix} 1 & 2 & 3 \\ 2 & 1 & 4 \end{bmatrix}\right)^T$

   $= \left(3\begin{bmatrix} 1 & 2 & 3 \\ 0 & 1 & 2 \end{bmatrix} - 2\begin{bmatrix} 1 & 2 & 3 \\ 2 & 1 & 4 \end{bmatrix}\right)^T = \begin{bmatrix} 1 & 2 & 3 \\ -4 & 1 & -2 \end{bmatrix}^T = \begin{bmatrix} 1 & -4 \\ 2 & 1 \\ 3 & -2 \end{bmatrix}.$

   (d) $(3A^T - 5B^T)^T$ is undefined since A^T is 3×2 and B^T is 2×3.

   (e) $(-A)^T = \begin{bmatrix} -1 & -2 & -3 \\ -2 & -1 & -4 \end{bmatrix}^T = \begin{bmatrix} -1 & -2 \\ -2 & -1 \\ -3 & -4 \end{bmatrix};\ -(A^T) = -\begin{bmatrix} 1 & 2 \\ 2 & 1 \\ 3 & 4 \end{bmatrix} = \begin{bmatrix} -1 & -2 \\ -2 & -1 \\ -3 & -4 \end{bmatrix}.$

   Note: $(-A)^T = -(A^T)$.

   (f) $(C + E + F^T)^T$ is undefined since $C + E$ is 3×3 and F^T is 2×2.

9. No. If there are scalars x_1 and x_2 such that

$$x_1\begin{bmatrix} 1 & 0 \\ 0 & 1 \end{bmatrix} + x_2\begin{bmatrix} 1 & 0 \\ 0 & 0 \end{bmatrix} = \begin{bmatrix} 4 & 1 \\ 0 & -3 \end{bmatrix},$$

then equating the (1,2) entries gives $0 = 1$, a contradiction.

11. Use the definition of matrix addition and Table 1.1 for problems involving bit matrices.

(a) $A + B = \begin{bmatrix} 1 & 0 & 1 \\ 1 & 1 & 0 \\ 0 & 1 & 1 \end{bmatrix} + \begin{bmatrix} 0 & 1 & 1 \\ 1 & 0 & 1 \\ 1 & 1 & 0 \end{bmatrix} = \begin{bmatrix} 1+0 & 0+1 & 1+1 \\ 1+1 & 1+0 & 0+1 \\ 0+1 & 1+1 & 1+0 \end{bmatrix} = \begin{bmatrix} 1 & 1 & 0 \\ 0 & 1 & 1 \\ 1 & 0 & 1 \end{bmatrix}.$

(b) $B + C = \begin{bmatrix} 0 & 1 & 1 \\ 1 & 0 & 1 \\ 1 & 1 & 0 \end{bmatrix} + \begin{bmatrix} 1 & 1 & 0 \\ 0 & 1 & 1 \\ 1 & 0 & 1 \end{bmatrix} = \begin{bmatrix} 1 & 0 & 1 \\ 1 & 1 & 0 \\ 0 & 1 & 1 \end{bmatrix}.$

(c) $A + B + C = (A + B) + C = \begin{bmatrix} 1 & 1 & 0 \\ 0 & 1 & 1 \\ 1 & 0 & 1 \end{bmatrix} + \begin{bmatrix} 1 & 1 & 0 \\ 0 & 1 & 1 \\ 1 & 0 & 1 \end{bmatrix} = \begin{bmatrix} 0 & 0 & 0 \\ 0 & 0 & 0 \\ 0 & 0 & 0 \end{bmatrix}.$

   Note: $(A + B) + C = A + (B + C)$.

(d) $A + C^T = \begin{bmatrix} 1 & 0 & 1 \\ 1 & 1 & 0 \\ 0 & 1 & 1 \end{bmatrix} + \begin{bmatrix} 1 & 0 & 1 \\ 1 & 1 & 0 \\ 0 & 1 & 1 \end{bmatrix} = \begin{bmatrix} 0 & 0 & 0 \\ 0 & 0 & 0 \\ 0 & 0 & 0 \end{bmatrix}.$

(e) $B - C = B + C = \begin{bmatrix} 0 & 1 & 1 \\ 1 & 0 & 1 \\ 1 & 1 & 0 \end{bmatrix} + \begin{bmatrix} 1 & 1 & 0 \\ 0 & 1 & 1 \\ 1 & 0 & 1 \end{bmatrix} = \begin{bmatrix} 1 & 0 & 1 \\ 1 & 1 & 0 \\ 0 & 1 & 1 \end{bmatrix}.$

   Note: See 11(b).

13. (a) Let $B = \begin{bmatrix} b_{11} & b_{12} \\ b_{21} & b_{22} \end{bmatrix}$. Then

$$A + B = \begin{bmatrix} 1 & 0 \\ 0 & 0 \end{bmatrix} + \begin{bmatrix} b_{11} & b_{12} \\ b_{21} & b_{22} \end{bmatrix} = \begin{bmatrix} 0 & 0 \\ 0 & 0 \end{bmatrix}.$$

Equate corresponding elements and solve each equation:

$$\begin{aligned} 1 + b_{11} = 0 &\implies b_{11} = 1 \\ 0 + b_{12} = 0 &\implies b_{12} = 0 \\ 0 + b_{21} = 0 &\implies b_{21} = 0 \\ 0 + b_{22} = 0 &\implies b_{22} = 0 \end{aligned}$$

Thus $B = \begin{bmatrix} 1 & 0 \\ 0 & 0 \end{bmatrix}.$

(b) Let $C = \begin{bmatrix} c_{11} & c_{12} \\ c_{21} & c_{22} \end{bmatrix}$. Then

$$A + C = \begin{bmatrix} 1 & 0 \\ 0 & 0 \end{bmatrix} + \begin{bmatrix} c_{11} & c_{12} \\ c_{21} & c_{22} \end{bmatrix} = \begin{bmatrix} 1 & 1 \\ 1 & 1 \end{bmatrix}.$$

Equate corresponding elements and solve each equation:

$$\begin{aligned} 1 + c_{11} = 1 &\implies c_{11} = 0 \\ 0 + c_{12} = 1 &\implies c_{12} = 1 \\ 0 + c_{21} = 1 &\implies c_{21} = 1 \\ 0 + c_{22} = 1 &\implies c_{22} = 1 \end{aligned}$$

Thus $C = \begin{bmatrix} 0 & 1 \\ 1 & 1 \end{bmatrix}.$

15. Let $\mathbf{v} = \begin{bmatrix} a & b & c & d \end{bmatrix}$. Then

$$\begin{bmatrix} 0 & 1 & 0 & 1 \end{bmatrix} + \begin{bmatrix} a & b & c & d \end{bmatrix} = \begin{bmatrix} 1 & 1 & 1 & 1 \end{bmatrix}.$$

Equate corresponding elements and solve each equation:

$$\begin{aligned} 0 + a = 1 &\implies a = 1 \\ 1 + b = 1 &\implies b = 0 \\ 0 + c = 1 &\implies c = 1 \\ 1 + d = 1 &\implies d = 0 \end{aligned}$$

Thus $\mathbf{v} = \begin{bmatrix} 1 & 0 & 1 & 0 \end{bmatrix}$.

T.1. Let A and B be diagonal $n \times n$ matrices. Let $C = A + B$, $D = A - B$, $c_{ij} = a_{ij} + b_{ij}$, and $d_{ij} = a_{ij} - b_{ij}$. For $i \neq j$, a_{ij} and b_{ij} are each equal to 0, so $c_{ij} = 0$ and $d_{ij} = 0$. Thus C and D are diagonal.

T.3. (a) $A - A^T = \begin{bmatrix} 0 & b-c & c-e \\ c-b & 0 & 0 \\ e-c & 0 & 0 \end{bmatrix}$. (b) $A + A^T = \begin{bmatrix} 2a & b+c & c+e \\ c+b & 2d & 2e \\ e+c & 2e & 2f \end{bmatrix}$.

(c) Same as (b) since $(A + A^T)^T = A^T + A = A + A^T$.

T.5. (a) Let $A = \begin{bmatrix} a_{ij} \end{bmatrix}$ and $B = \begin{bmatrix} b_{ij} \end{bmatrix}$ be upper triangular matrices, and let $C = A + B$. Then for $i > j$, $c_{ij} = a_{ij} + b_{ij} = 0 + 0 = 0$, and thus C is upper triangular. Similarly, if $D = A - B$, then for $i > j$, $d_{ij} = a_{ij} - b_{ij} = 0 - 0 = 0$, so D is upper triangular.

(b) Proof similar to that for (a).

(c) Let $A = \begin{bmatrix} a_{ij} \end{bmatrix}$ be both upper and lower triangular. Then $a_{ij} = 0$ for $i > j$ and for $i < j$. Thus, A is a diagonal matrix.

T.7. Let $A = \begin{bmatrix} a_{ij} \end{bmatrix}$. Then the diagonal entries of both A and A^T are a_{ii} ($i = 1, 2, \ldots, n$). So their difference is 0.

T.9. There are four possibilities: two choices in the first position and two choices in the second.

$$\begin{bmatrix} 0 \\ 0 \end{bmatrix}, \quad \begin{bmatrix} 0 \\ 1 \end{bmatrix}, \quad \begin{bmatrix} 1 \\ 0 \end{bmatrix}, \quad \begin{bmatrix} 1 \\ 1 \end{bmatrix}.$$

T.11. There are 2^4 possibilities: two choices for each position.

$$\begin{bmatrix} 0 \\ 0 \\ 0 \\ 0 \end{bmatrix}, \begin{bmatrix} 1 \\ 0 \\ 0 \\ 0 \end{bmatrix}, \begin{bmatrix} 1 \\ 1 \\ 1 \\ 1 \end{bmatrix}, \begin{bmatrix} 0 \\ 1 \\ 1 \\ 1 \end{bmatrix}, \begin{bmatrix} 0 \\ 1 \\ 0 \\ 0 \end{bmatrix}, \begin{bmatrix} 1 \\ 1 \\ 0 \\ 0 \end{bmatrix}, \begin{bmatrix} 1 \\ 0 \\ 1 \\ 1 \end{bmatrix}, \begin{bmatrix} 0 \\ 0 \\ 1 \\ 1 \end{bmatrix},$$

$$\begin{bmatrix} 0 \\ 0 \\ 1 \\ 0 \end{bmatrix}, \begin{bmatrix} 1 \\ 0 \\ 1 \\ 0 \end{bmatrix}, \begin{bmatrix} 1 \\ 1 \\ 0 \\ 1 \end{bmatrix}, \begin{bmatrix} 0 \\ 1 \\ 0 \\ 1 \end{bmatrix}, \begin{bmatrix} 0 \\ 0 \\ 0 \\ 1 \end{bmatrix}, \begin{bmatrix} 1 \\ 0 \\ 0 \\ 1 \end{bmatrix}, \begin{bmatrix} 1 \\ 1 \\ 1 \\ 0 \end{bmatrix}, \begin{bmatrix} 0 \\ 1 \\ 1 \\ 0 \end{bmatrix}.$$

T.13. There are 2^4 possibilities: two choices for each position in the 2×2 matrix.

$$\begin{bmatrix} 0 & 0 \\ 0 & 0 \end{bmatrix}, \begin{bmatrix} 0 & 0 \\ 0 & 1 \end{bmatrix}, \begin{bmatrix} 0 & 0 \\ 1 & 0 \end{bmatrix}, \begin{bmatrix} 0 & 0 \\ 1 & 1 \end{bmatrix}, \begin{bmatrix} 0 & 1 \\ 0 & 0 \end{bmatrix}, \begin{bmatrix} 0 & 1 \\ 0 & 1 \end{bmatrix}, \begin{bmatrix} 0 & 1 \\ 1 & 0 \end{bmatrix}, \begin{bmatrix} 0 & 1 \\ 1 & 1 \end{bmatrix},$$

$$\begin{bmatrix} 1 & 0 \\ 0 & 0 \end{bmatrix}, \begin{bmatrix} 1 & 0 \\ 0 & 1 \end{bmatrix}, \begin{bmatrix} 1 & 0 \\ 1 & 0 \end{bmatrix}, \begin{bmatrix} 1 & 0 \\ 1 & 1 \end{bmatrix}, \begin{bmatrix} 1 & 1 \\ 0 & 0 \end{bmatrix}, \begin{bmatrix} 1 & 1 \\ 0 & 1 \end{bmatrix}, \begin{bmatrix} 1 & 1 \\ 1 & 0 \end{bmatrix}, \begin{bmatrix} 1 & 1 \\ 1 & 1 \end{bmatrix}.$$

T.15. There are 2^{n^2} possibilities: two choices for each position in the $n \times n$ matrix.

T.17. Rewrite A as

$$A = \begin{bmatrix} 1 & 1 & 0 \\ 0 & 1 & 0 \\ 0 & 1 & 1 \end{bmatrix}.$$

We want

$$A + B = \begin{bmatrix} 1 & 1 & 1 \\ 1 & 1 & 1 \\ 1 & 1 & 1 \end{bmatrix}.$$

Let $B = \begin{bmatrix} 0 & 0 & 1 \\ 1 & 0 & 1 \\ 1 & 0 & 0 \end{bmatrix}.$

ML.1. Once you have entered matrices A and B you can use the commands given below to see the items requested in parts (a) and (b).

(a) Commands: **A(2,3)**, **B(3,2)**, **B(1,2)**

(b) Use command **A(1,:)** for $\text{row}_1(A)$
Use command **A(:,3)** for $\text{col}_3(A)$
Use command **B(2,:)** for $\text{row}_2(B)$
(In this context the colon means 'all'.)

(c) Matrix **B** in **format long** is

8.00000000000000	0.66666666666667
0.00497512437811	-3.20000000000000
0.00001000000000	4.33333333333333

ML.3. (a) Use the command **bingen(0,7,3)** to view all possible 3-bit vectors. Each vector corresponds to a column.

(b) Use the command **bingen(0,15,4)** to view all possible 4-bit vectors. Each vector corresponds to a column.

ML.5. Let

$$\mathbf{A} = [1 \ \ 0 \ \ 1; 1 \ \ 1 \ \ 0; 0 \ \ 1 \ \ 1]; \quad \mathbf{B} = [0 \ \ 1 \ \ 1; 1 \ \ 0 \ \ 1; 1 \ \ 1 \ \ 0];$$
$$\text{and} \quad \mathbf{C} = [1 \ \ 1 \ \ 0; 0 \ \ 1 \ \ 1; 1 \ \ 0 \ \ 1]$$

(a) **binadd(A,B)**

```
ans =
    1   1   0
    0   1   1
    1   0   1
```

(b) **binadd(B,C)**

```
ans =
    1   0   1
    1   1   0
    0   1   1
```

(c) **F = binadd(A,B)**

F =

    1 1 0
    0 1 1
    1 0 1

binadd(F,C)

ans =

    0 0 0
    0 0 0
    0 0 0

(d) **binadd(A,C′)**

ans =

    0 0 0
    0 0 0
    0 0 0

(e) **binadd(B,C)** ($= B - C$ since in binary arithmetic a matrix is its own additive inverse.)

ans =

    1 0 1
    1 1 0
    0 1 1

Section 1.3, p. 34

1. (a) $\mathbf{a} \cdot \mathbf{b} = (1)(4) + (2)(-1) = 2.$

    (b) $\mathbf{a} \cdot \mathbf{b} = (-3)(1) + (-2)(-2) = 1.$

    (c) $\mathbf{a} \cdot \mathbf{b} = (4)(1) + (2)(3) + (-1)(6) = 4.$

    (d) $\mathbf{a} \cdot \mathbf{b} = (1)(1) + (1)(0) + (0)(1) = 1.$

3. We have $\mathbf{a} \cdot \mathbf{b} = (-3)(-3) + (2)(2) + (x)(x) = 13 + x^2 = 17.$ It follows that $x^2 = 4.$ Hence $x = 2$ or $x = -2.$

5. We have

$$\mathbf{v} \cdot \mathbf{v} = \begin{bmatrix} \frac{1}{2} \\ -\frac{1}{2} \\ x \end{bmatrix} \cdot \begin{bmatrix} \frac{1}{2} \\ -\frac{1}{2} \\ x \end{bmatrix} = \frac{1}{4} + \frac{1}{4} + x^2 = 1 \implies x^2 = \frac{1}{2} \implies x = \pm\frac{\sqrt{2}}{2}.$$

Hence $x = \frac{\sqrt{2}}{2}$ or $x = -\frac{\sqrt{2}}{2}.$

7. (a) $AB = \begin{bmatrix} 1 & 2 & -3 \\ 4 & 0 & -2 \end{bmatrix} \begin{bmatrix} 3 & 1 \\ 2 & 4 \\ -1 & 5 \end{bmatrix} = \begin{bmatrix} 3+4+3 & 1+8-15 \\ 12+0+2 & 4+0-10 \end{bmatrix} = \begin{bmatrix} 10 & -6 \\ 14 & -6 \end{bmatrix}.$

    (b) $BA = \begin{bmatrix} 3 & 1 \\ 2 & 4 \\ -1 & 5 \end{bmatrix} \begin{bmatrix} 1 & 2 & -3 \\ 4 & 0 & -2 \end{bmatrix} = \begin{bmatrix} 3+4 & 6+0 & -9-2 \\ 2+16 & 4+0 & -6-8 \\ -1+20 & -2+0 & 3-10 \end{bmatrix} = \begin{bmatrix} 7 & 6 & -11 \\ 18 & 4 & -14 \\ 19 & -2 & -7 \end{bmatrix}.$

    (c) $CB + D$ is undefined since CB is 3×2 and D is $2 \times 2.$

(d) $AB + DF = \begin{bmatrix} 1 & 2 & -3 \\ 4 & 0 & -2 \end{bmatrix} \begin{bmatrix} 3 & 1 \\ 2 & 4 \\ -1 & 5 \end{bmatrix} + \begin{bmatrix} 2 & 3 \\ -1 & -2 \end{bmatrix} \begin{bmatrix} 2 & -3 \\ 4 & 1 \end{bmatrix}$

$\qquad\qquad = \begin{bmatrix} 10 & -6 \\ 14 & -6 \end{bmatrix} + \begin{bmatrix} 16 & -3 \\ -10 & 1 \end{bmatrix} = \begin{bmatrix} 26 & -9 \\ 4 & -5 \end{bmatrix}.$

(e) $BA + FD$ is undefined since BA is 3×3 and FD is 2×2.

9. (a) $(AB)_{12} = \text{row}_1(A) \cdot \text{col}_2(B) = \begin{bmatrix} 2 & 3 \end{bmatrix} \begin{bmatrix} -1 \\ 2 \end{bmatrix} = (2)(-1) + (3)(2) = 4.$

(b) $(AB)_{23} = \text{row}_2(A) \cdot \text{col}_3(B) = \begin{bmatrix} -1 & 4 \end{bmatrix} \begin{bmatrix} 3 \\ 4 \end{bmatrix} = (-1)(3) + (4)(4) = 13.$

(c) $(AB)_{31} = \text{row}_3(A) \cdot \text{col}_1(B) = \begin{bmatrix} 0 & 3 \end{bmatrix} \begin{bmatrix} 3 \\ 1 \end{bmatrix} = (0)(3) + (3)(1) = 3.$

(d) $(AB)_{33} = \text{row}_3(A) \cdot \text{col}_3(B) = \begin{bmatrix} 0 & 3 \end{bmatrix} \begin{bmatrix} 3 \\ 4 \end{bmatrix} = (0)(3) + (3)(4) = 12.$

11. We find that

$$AB = \begin{bmatrix} 1 & 2 \\ 3 & 2 \end{bmatrix} \begin{bmatrix} 2 & -1 \\ -3 & 4 \end{bmatrix} = \begin{bmatrix} -4 & 7 \\ 0 & 5 \end{bmatrix} \quad \text{and} \quad BA = \begin{bmatrix} 2 & -1 \\ -3 & 4 \end{bmatrix} \begin{bmatrix} 1 & 2 \\ 3 & 2 \end{bmatrix} = \begin{bmatrix} -1 & 2 \\ 9 & 2 \end{bmatrix}.$$

Therefore $AB \neq BA$.

13. (a) The first column of AB is

$$A \cdot \text{col}_1(B) = \begin{bmatrix} 1 & -1 & 2 \\ 3 & 2 & 4 \\ 4 & -2 & 3 \\ 2 & 1 & 5 \end{bmatrix} \begin{bmatrix} 1 \\ 3 \\ 4 \end{bmatrix} = \begin{bmatrix} 6 \\ 25 \\ 10 \\ 25 \end{bmatrix}.$$

(b) The third column of AB is

$$A \cdot \text{col}_3(B) = \begin{bmatrix} 1 & -1 & 2 \\ 3 & 2 & 4 \\ 4 & -2 & 3 \\ 2 & 1 & 5 \end{bmatrix} \begin{bmatrix} -1 \\ -3 \\ 5 \end{bmatrix} = \begin{bmatrix} 12 \\ 11 \\ 17 \\ 20 \end{bmatrix}.$$

15. $AC = \begin{bmatrix} 2 & -3 & 4 \\ -1 & 2 & 3 \\ 5 & -1 & -2 \end{bmatrix} \begin{bmatrix} 2 \\ 1 \\ 4 \end{bmatrix} = \begin{bmatrix} (2)(2) + (-3)(1) + (4)(4) \\ (-1)(2) + (2)(1) + (3)(4) \\ (5)(2) + (-1)(1) + (-2)(4) \end{bmatrix}$

$\qquad\qquad = \begin{bmatrix} (2)(2) \\ (-1)(2) \\ (5)(2) \end{bmatrix} + \begin{bmatrix} (-3)(1) \\ (2)(1) \\ (-1)(1) \end{bmatrix} + \begin{bmatrix} (4)(4) \\ (3)(4) \\ (-2)(4) \end{bmatrix}$

$\qquad\qquad = 2 \begin{bmatrix} 2 \\ -1 \\ 5 \end{bmatrix} + 1 \begin{bmatrix} -3 \\ 2 \\ -1 \end{bmatrix} + 4 \begin{bmatrix} 4 \\ 3 \\ -2 \end{bmatrix}.$

17. (a) We find

$$AB = \begin{bmatrix} 2 & -3 & 1 \\ 1 & 2 & 4 \end{bmatrix} \begin{bmatrix} 3 \\ 5 \\ 2 \end{bmatrix} = \begin{bmatrix} -7 \\ 21 \end{bmatrix}$$

and

$$3\mathbf{a}_1 + 5\mathbf{a}_2 + 2\mathbf{a}_3 = 3 \begin{bmatrix} 2 \\ 1 \end{bmatrix} + 5 \begin{bmatrix} -3 \\ 2 \end{bmatrix} + 2 \begin{bmatrix} 1 \\ 4 \end{bmatrix} = \begin{bmatrix} -7 \\ 21 \end{bmatrix}.$$

Therefore $AB = 3\mathbf{a}_1 + 5\mathbf{a}_2 + 2\mathbf{a}_3$.

(b) We find

$$AB = \begin{bmatrix} [2 \ -3 \ 1] \begin{bmatrix} 3 \\ 5 \\ 2 \end{bmatrix} \\ \\ [1 \ 2 \ 4] \begin{bmatrix} 3 \\ 5 \\ 2 \end{bmatrix} \end{bmatrix} = \begin{bmatrix} -7 \\ 21 \end{bmatrix}.$$

19. (a) $\begin{bmatrix} 2 & 0 & 0 & 1 \\ 3 & 2 & 3 & 0 \\ 2 & 3 & -4 & 0 \\ 1 & 0 & 3 & 0 \end{bmatrix} = A.$

(b) A is the coefficient matrix of part (a); $\mathbf{x} = \begin{bmatrix} x \\ y \\ z \\ w \end{bmatrix}$, $\mathbf{b} = \begin{bmatrix} 7 \\ -2 \\ 3 \\ 5 \end{bmatrix}$, $A\mathbf{x} = \mathbf{b}.$

$$\begin{bmatrix} 2 & 0 & 0 & 1 \\ 3 & 2 & 3 & 0 \\ 2 & 3 & -4 & 0 \\ 1 & 0 & 3 & 0 \end{bmatrix} \cdot \begin{bmatrix} x \\ y \\ z \\ w \end{bmatrix} = \begin{bmatrix} 7 \\ -2 \\ 3 \\ 5 \end{bmatrix}.$$

(c) $\begin{bmatrix} 2 & 0 & 0 & 1 & \vdots & 7 \\ 3 & 2 & 3 & 0 & \vdots & -2 \\ 2 & 3 & -4 & 0 & \vdots & 3 \\ 1 & 0 & 3 & 0 & \vdots & 5 \end{bmatrix}.$

21. $\begin{bmatrix} 2 & 0 & -4 \\ 0 & 1 & 2 \\ 1 & 3 & 4 \end{bmatrix} \begin{bmatrix} x \\ y \\ z \end{bmatrix} = \begin{bmatrix} 3 \\ 5 \\ -1 \end{bmatrix}$ is the matrix form of the linear system

$$\begin{aligned} 2x \quad\quad - 4z &= \quad 3 \\ y + 2z &= \quad 5 \\ x + 3y + 4z &= -1. \end{aligned}$$

23. $\begin{bmatrix} 1 & 2 & 3 \\ 2 & 3 & 6 \end{bmatrix} \begin{bmatrix} x \\ y \\ z \end{bmatrix} = \begin{bmatrix} -1 \\ 2 \end{bmatrix}$ is the matrix form of the linear system

$$\begin{aligned} x + 2y + 3z &= -1 \\ 2x + 3y + 6z &= \quad 2. \end{aligned}$$

$\begin{bmatrix} 1 & 2 & 3 \\ 2 & 3 & 6 \\ 0 & 0 & 0 \end{bmatrix} \begin{bmatrix} x \\ y \\ z \end{bmatrix} = \begin{bmatrix} -1 \\ 2 \\ 0 \end{bmatrix}$ is the matrix form of the linear system

$$\begin{aligned} x + 2y + 3z &= -1 \\ 2x + 3y + 6z &= \quad 2 \\ 0x + 0y + 0z &= \quad 0. \end{aligned}$$

The last equation here can be deleted so the linear systems are equivalent.

25. (a) $x + 2y = 3$
$ 2x - y = 5$

$$\begin{bmatrix} x \\ 2x \end{bmatrix} + \begin{bmatrix} 2y \\ -y \end{bmatrix} = \begin{bmatrix} 3 \\ 5 \end{bmatrix}$$

$$x \begin{bmatrix} 1 \\ 2 \end{bmatrix} + y \begin{bmatrix} 2 \\ -1 \end{bmatrix} = \begin{bmatrix} 3 \\ 5 \end{bmatrix}.$$

(b) $2x - 3y + 5z = -2$
$ x + 4y - z = 3$

$$\begin{bmatrix} 2x \\ x \end{bmatrix} + \begin{bmatrix} -3y \\ 4y \end{bmatrix} + \begin{bmatrix} 5z \\ -z \end{bmatrix} = \begin{bmatrix} -2 \\ 3 \end{bmatrix}$$

$$x \begin{bmatrix} 2 \\ 1 \end{bmatrix} + y \begin{bmatrix} -3 \\ 4 \end{bmatrix} + z \begin{bmatrix} 5 \\ -1 \end{bmatrix} = \begin{bmatrix} -2 \\ 3 \end{bmatrix}.$$

27. (a) $AB^T = \begin{bmatrix} r & 1 & -2 \end{bmatrix} \begin{bmatrix} 1 \\ 3 \\ -1 \end{bmatrix} = r(1) + 1(3) + (-2)(-1) = r + 3 + 2 = 0 \implies r = -5$

(b) $BA^T = \begin{bmatrix} 1 & 3 & -1 \end{bmatrix} \begin{bmatrix} r \\ 1 \\ -2 \end{bmatrix} = 1(r) + 3(1) + (-1)(-2) = r + 3 + 2 = 0$

Note: $AB^T = BA^T$.

29. In order to add partitioned matrices, each matrix must be partitioned in the same manner. We partition matrices A and B as follows:

$$A = \begin{bmatrix} 1 & 3 & -1 \\ 2 & 1 & 0 \\ \hline 2 & -3 & 1 \end{bmatrix} = \begin{bmatrix} A_{11} & A_{12} \\ A_{21} & A_{22} \end{bmatrix}, \qquad B = \begin{bmatrix} 3 & 2 & 1 \\ -2 & 3 & 1 \\ \hline 4 & 1 & 5 \end{bmatrix} = \begin{bmatrix} B_{11} & B_{12} \\ B_{21} & B_{22} \end{bmatrix}.$$

Then

$$A + B = \begin{bmatrix} A_{11} + B_{11} & A_{12} + B_{12} \\ A_{21} + B_{21} & A_{22} + B_{22} \end{bmatrix} = \begin{bmatrix} 4 & 5 & 0 \\ 0 & 4 & 1 \\ \hline 6 & -2 & 6 \end{bmatrix}.$$

Alternatively, we could partition A and B as follows:

$$A = \begin{bmatrix} 1 & 3 & -1 \\ 2 & 1 & 0 \\ 2 & -3 & 1 \end{bmatrix} = \begin{bmatrix} A_{11} & A_{12} \\ A_{21} & A_{22} \end{bmatrix}, \qquad B = \begin{bmatrix} 3 & 2 & 1 \\ -2 & 3 & 1 \\ 4 & 1 & 5 \end{bmatrix} = \begin{bmatrix} B_{11} & B_{12} \\ B_{21} & B_{22} \end{bmatrix}.$$

Then

$$A + B = \begin{bmatrix} A_{11} + B_{11} & A_{12} + B_{12} \\ A_{21} + B_{21} & A_{22} + B_{22} \end{bmatrix} = \begin{bmatrix} 4 & 5 & 0 \\ 0 & 4 & 1 \\ 6 & -2 & 6 \end{bmatrix}.$$

31. $AB = \begin{bmatrix} 2 & 2 \\ 3 & 4 \end{bmatrix} \begin{bmatrix} 9 & 10 \\ 10 & 12 \end{bmatrix} = \begin{bmatrix} 38 & 44 \\ 67 & 78 \end{bmatrix}.$

The entries of the matrix product AB tell the manufacturer the total cost of producing chairs and tables in each city.

$$B = \begin{bmatrix} \overset{\text{Salt Lake City}}{38} & \overset{\text{Chicago}}{44} \\ 67 & 78 \end{bmatrix} \begin{matrix} \text{Chair} \\ \text{Table.} \end{matrix}$$

33. (a) $\text{row}_1(A) \cdot \text{col}_1(B) = 80(20) + 120(10) = 2800$ grams of protein consumed daily by the males.

(b) $\text{row}_2(A) \cdot \text{col}_2(B) = 100(20) + 200(20) = 6000$ grams of fat consumed daily by the females.

35. (a) The matrix

$$P = \begin{bmatrix} \mathbf{s}_1 \\ \mathbf{s}_2 \end{bmatrix} = \begin{bmatrix} 18.95 & 14.75 & 8.98 \\ 17.80 & 13.50 & 10.79 \end{bmatrix}$$

represents the combined information about the prices of the items at the two stores.

(b) If each store announces a sale so that the price of each of the three items is reduced by 20%, then

$$0.80P = \begin{bmatrix} 0.80\mathbf{s}_1 \\ 0.80\mathbf{s}_2 \end{bmatrix} = \begin{bmatrix} 0.80(18.95) & 0.80(14.75) & 0.80(8.98) \\ 0.80(17.80) & 0.80(13.50) & 0.80(10.79) \end{bmatrix} = \begin{bmatrix} 15.16 & 11.80 & 7.18 \\ 14.24 & 10.80 & 8.63 \end{bmatrix}$$

represents the sale prices at the stores.

For bit vectors and bit matrices use Table 1.1 and Table 1.2 in Section 1.2.

37. (a) $\mathbf{a} \cdot \mathbf{b} = \begin{bmatrix} 1 & 1 & 0 \end{bmatrix} \cdot \begin{bmatrix} 1 \\ 0 \\ 1 \end{bmatrix} = 1 + 0 + 0 = 1.$

  (b) $\mathbf{a} \cdot \mathbf{b} = \begin{bmatrix} 0 & 1 & 1 & 0 \end{bmatrix} \cdot \begin{bmatrix} 1 \\ 1 \\ 1 \\ 0 \end{bmatrix} = 0 + 1 + 1 + 0 = 0.$

39. $AB = \begin{bmatrix} 1 & 1 & x \\ 0 & y & 1 \end{bmatrix} \begin{bmatrix} 1 \\ 1 \\ 1 \end{bmatrix} = \begin{bmatrix} 1 + 1 + x \\ 0 + y + 1 \end{bmatrix} = \begin{bmatrix} x \\ y + 1 \end{bmatrix} = \begin{bmatrix} 0 \\ 0 \end{bmatrix}.$

Equate corresponding elements and solve each equation:

$$x = 0$$
$$y + 1 = 0 \quad \Longrightarrow \quad y = 1$$

Thus $x = 0$ and $y = 1$.

41. Let $B = \begin{bmatrix} a & b \\ c & d \end{bmatrix}$. We have

$$AB = \begin{bmatrix} 1 & 1 \\ 0 & 1 \end{bmatrix} \begin{bmatrix} a & b \\ c & d \end{bmatrix} = \begin{bmatrix} a + c & b + d \\ c & d \end{bmatrix} = \begin{bmatrix} 1 & 0 \\ 0 & 1 \end{bmatrix}.$$

Equate corresponding elements:

$$a + c = 1$$
$$b + d = 0$$
$$c = 0$$
$$d = 1$$

Substituting $c = 0$, $d = 1$ into the first two equations, we find that $a = 1$, $b = 1$. Therefore

$$B = \begin{bmatrix} 1 & 1 \\ 0 & 1 \end{bmatrix}.$$

T.1. (a) No. If $\mathbf{x} = (x_1, x_2, \ldots, x_n)$, then $\mathbf{x} \cdot \mathbf{x} = x_1^2 + x_2^2 + \cdots + x_n^2 \geq 0.$
  (b) $\mathbf{x} = \mathbf{0}.$

T.3. Let $A = \begin{bmatrix} a_{ij} \end{bmatrix}$ be $m \times p$ and $B = \begin{bmatrix} b_{ij} \end{bmatrix}$ be $p \times n$.

  (a) Let the ith row of A consist entirely of zeros, so $a_{ik} = 0$ for $k = 1, 2, \ldots, p$. Then the (i, j) entry in AB is

$$\sum_{k=1}^{p} a_{ik} b_{kj} = 0 \qquad \text{for } j = 1, 2, \ldots, n.$$

(b) Let the jth column of B consist entirely of zeros, so $b_{kj} = 0$ for $k = 1, 2, \ldots, p$. Then again the (i, j) entry of AB is 0 for $i = 1, 2, \ldots, m$.

T.5. Let A and B be scalar matrices, so that $a_{ij} = a$ and $b_{ij} = b$ for all $i = j$. If $C = AB$, then

$$c_{ij} = \sum_{k=1}^{n} a_{ik} b_{kj} = \begin{cases} 0 & \text{whenever } i \neq j \\ a_{ii} b_{jj} & \text{for } i = j \end{cases}$$

since $a_{ii} b_{ii}$ is the only nonzero term in the expansion for c_{ij}. So C is a scalar matrix.

T.7. Yes. If $A = [a_{ij}]$ and $B = [b_{ij}]$ are diagonal matrices, then $C = [c_{ij}]$ is diagonal by Exercise T.4. Moreover, $c_{ii} = a_{ii} b_{ii}$. Similarly, if $D = BA$, then $d_{ii} = b_{ii} a_{ii}$. Thus, $C = D$.

T.9. (a) The jth column of AB is

$$\begin{bmatrix} \sum_{k} a_{1k} b_{kj} \\ \sum_{k} a_{2k} b_{kj} \\ \vdots \\ \sum_{k} a_{mk} b_{kj} \end{bmatrix} = A \cdot \mathbf{col}_j(B).$$

(b) The ith row of AB is

$$\begin{bmatrix} \sum_{k} a_{ik} b_{k1} & \sum_{k} a_{ik} b_{k2} & \cdots & \sum_{k} a_{ik} b_{kn} \end{bmatrix} = \mathbf{row}_i(A) \cdot B.$$

T.11. (a) $\displaystyle\sum_{i=1}^{n} (r_i + s_i) a_i = (r_1 + s_1) a_1 + (r_2 + s_2) a_2 + \cdots + (r_n + s_n) a_n$

$$= r_1 a_1 + s_1 a_1 + r_2 a_2 + s_2 a_2 + \cdots + r_n a_n + s_n a_n$$

$$= (r_1 a_1 + r_2 a_2 + \cdots + r_n a_n) + (s_1 a_1 + s_2 a_2 + \cdots + s_n a_n) = \sum_{i=1}^{n} r_i a_i + \sum_{i=1}^{n} s_i a_i$$

(b) $\displaystyle\sum_{i=1}^{n} c(r_i a_i) = cr_1 a_1 + cr_2 a_2 + \cdots + cr_n a_n = c(r_1 a_1 + r_2 a_2 + \cdots + r_n a_n) = c \sum_{i=1}^{n} r_i a_i.$

T.13. (a) True. $\displaystyle\sum_{i=1}^{n} (a_i + 1) = \sum_{i=1}^{n} a_i + \sum_{i=1}^{n} 1 = \sum_{i=1}^{n} a_i + n.$

(b) True. $\displaystyle\sum_{i=1}^{n} \sum_{j=1}^{m} 1 = \sum_{i=1}^{n} \left(\sum_{j=1}^{m} 1 \right) = \sum_{i=1}^{n} m = mn.$

(c) True. $\displaystyle\left[\sum_{i=1}^{n} a_i \right] \left[\sum_{j=1}^{m} b_j \right] = a_1 \sum_{j=1}^{m} b_j + a_2 \sum_{j=1}^{m} b_j + \cdots + a_n \sum_{j=1}^{m} b_j$

$$= (a_1 + a_2 + \cdots + a_n) \sum_{j=1}^{m} b_j$$

$$= \sum_{i=1}^{n} a_i \sum_{j=1}^{m} b_j = \sum_{j=1}^{m} \sum_{i=1}^{n} a_i b_j$$

ML.1   (a) **A ∗ C**

ans =

4.5000	2.2500	3.7500
1.5833	0.9167	1.5000
0.9667	0.5833	0.9500

(b) **A ∗ B**

```
??? Error using ===> *
Inner matrix dimensions must agree.
```

(c) **A + C′**

ans =

5.0000	1.5000
1.5833	2.2500
2.4500	3.1667

(d) **B ∗ A − C′ ∗ A**

```
??? Error using ===> *
Inner matrix dimensions must agree.
```

(e) **(2 ∗ C − 6 ∗ A′) ∗ B′**

```
??? Error using ===> *
Inner matrix dimensions must agree.
```

(f) **A ∗ C − C ∗ A**

```
??? Error using ===> −
Inner matrix dimensions must agree.
```

(g) **A ∗ A′ + C′ ∗ C**

ans =

18.2500	7.4583	12.2833
7.4583	5.7361	8.9208
12.2833	8.9208	14.1303

ML.3. aug =

4	−3	2	−1	−5
2	1	−3	0	7
−1	4	1	2	8

ML.5. (a) **diag([1 2 3 4])**

ans =

1	0	0	0
0	2	0	0
0	0	3	0
0	0	0	4

(b) **diag([0 1 1/2 1/3 1/4])**

ans =

0	0	0	0	0
0	1.0000	0	0	0
0	0	0.5000	0	0
0	0	0	0.3333	0
0	0	0	0	0.2500

(c) **diag([5 5 5 5 5 5])**

```
ans =
        5   0   0   0   0   0
        0   5   0   0   0   0
        0   0   5   0   0   0
        0   0   0   5   0   0
        0   0   0   0   5   0
        0   0   0   0   0   5
```

ML.7. Let $\mathbf{A} = [1 \; 1 \; 0; 0 \; 1 \; 0]$, $\mathbf{B} = [0 \; 1 \; 0; 1 \; 1 \; 0; 1 \; 0 \; 1]$

>>**binprod(A,B)**

```
ans  =
        1   0   0
        1   1   0
        1   0   1
```

>>**binprod(B,A)**

```
ans  =
        0   1   0
        1   0   0
        1   1   1
```

ML.9. (a) **B =bingen(0,7,3)**

```
B  =
        0   0   0   0   1   1   1   1
        0   0   1   1   0   0   1   1
        0   1   0   1   0   1   0   1
```

(b) **A =ones(3)**

```
A  =
        1   1   1
        1   1   1
        1   1   1
```

>>**binprod(A,B)**

```
ans  =
        0   1   1   0   1   0   0   1
        0   1   1   0   1   0   0   1
        0   1   1   0   1   0   0   1
```

(c) Under binary matrix multiplication, the rows of A compute the parity of the columns of B. So each column of AB represents the evenness or oddness of the corresponding column of B.

ML.11. (a) Let $n = 2$. Let **B=ones(2)**. Then

>> B * B

```
ans  =
        2   2
        2   2
```

(b) Let $n = 3$. Let **B=ones(3)**. Then

>> B * B

ans =

    3 3 3

    3 3 3

    3 3 3

(c) Let $n = 4$. Let **B=ones(4)**. Then

>> B * B

ans =

    4 4 4 4

    4 4 4 4

    4 4 4 4

    4 4 4 4

(d) Let $n = k$. Let **B=ones(k)**. Then

$$B * B = \begin{bmatrix} k & k & \cdots & k \\ k & k & \cdots & k \\ \vdots & \vdots & & \vdots \\ k & k & \cdots & k \end{bmatrix}$$

a $k \times k$ matrix each of whose entries is k. The ijth entry is

$$\sum_{l=1}^{k} b_{il}b_{lj} = \sum_{l=1}^{k} 1 \cdot 1 = k.$$

Section 1.4, p. 49

1. (a) $A + B = \begin{bmatrix} 1 & 2 & -2 \\ 3 & 4 & 5 \end{bmatrix} + \begin{bmatrix} 2 & 0 & 1 \\ 3 & -2 & 5 \end{bmatrix} = \begin{bmatrix} 3 & 2 & -1 \\ 6 & 2 & 10 \end{bmatrix}.$

$B + A = \begin{bmatrix} 2 & 0 & 1 \\ 3 & -2 & 5 \end{bmatrix} + \begin{bmatrix} 1 & 2 & -2 \\ 3 & 4 & 5 \end{bmatrix} = \begin{bmatrix} 3 & 2 & -1 \\ 6 & 2 & 10 \end{bmatrix}.$

(b) $\begin{aligned} A + (B + C) &= \begin{bmatrix} 1 & 2 & -2 \\ 3 & 4 & 5 \end{bmatrix} + \left(\begin{bmatrix} 2 & 0 & 1 \\ 3 & -2 & 5 \end{bmatrix} + \begin{bmatrix} -4 & -6 & 1 \\ 2 & 3 & 0 \end{bmatrix} \right) \\ &= \begin{bmatrix} 1 & 2 & -2 \\ 3 & 4 & 5 \end{bmatrix} + \begin{bmatrix} -2 & -6 & 2 \\ 5 & 1 & 5 \end{bmatrix} = \begin{bmatrix} -1 & -4 & 0 \\ 8 & 5 & 10 \end{bmatrix}. \end{aligned}$

$\begin{aligned} (A + B) + C &= \left(\begin{bmatrix} 1 & 2 & -2 \\ 3 & 4 & 5 \end{bmatrix} + \begin{bmatrix} 2 & 0 & 1 \\ 3 & -2 & 5 \end{bmatrix} \right) + \begin{bmatrix} -4 & -6 & 1 \\ 2 & 3 & 0 \end{bmatrix} \\ &= \begin{bmatrix} 3 & 2 & -1 \\ 6 & 2 & 10 \end{bmatrix} + \begin{bmatrix} -4 & -6 & 1 \\ 2 & 3 & 0 \end{bmatrix} = \begin{bmatrix} -1 & -4 & 0 \\ 8 & 5 & 10 \end{bmatrix}. \end{aligned}$

(c) $A + O = \begin{bmatrix} 1 & 2 & -2 \\ 3 & 4 & 5 \end{bmatrix} + \begin{bmatrix} 0 & 0 & 0 \\ 0 & 0 & 0 \end{bmatrix} = \begin{bmatrix} 1 & 2 & -2 \\ 3 & 4 & 5 \end{bmatrix}.$

(d) $A + (-A) = \begin{bmatrix} 1 & 2 & -2 \\ 3 & 4 & 5 \end{bmatrix} + \begin{bmatrix} -1 & -2 & 2 \\ -3 & -4 & -5 \end{bmatrix} = \begin{bmatrix} 0 & 0 & 0 \\ 0 & 0 & 0 \end{bmatrix}.$

3. $A(B+C) = \begin{bmatrix} 1 & -3 \\ -3 & 4 \end{bmatrix} \left(\begin{bmatrix} 2 & -3 & 2 \\ 3 & -1 & -2 \end{bmatrix} + \begin{bmatrix} 0 & 1 & 2 \\ 1 & 3 & -2 \end{bmatrix} \right)$

$\qquad = \begin{bmatrix} 1 & -3 \\ -3 & 4 \end{bmatrix} \begin{bmatrix} 2 & -2 & 4 \\ 4 & 2 & -4 \end{bmatrix} = \begin{bmatrix} -10 & -8 & 16 \\ 10 & 14 & -28 \end{bmatrix}.$

$\quad AB + AC = \begin{bmatrix} 1 & -3 \\ -3 & 4 \end{bmatrix} \begin{bmatrix} 2 & -3 & 2 \\ 3 & -1 & -2 \end{bmatrix} + \begin{bmatrix} 1 & -3 \\ -3 & 4 \end{bmatrix} \begin{bmatrix} 0 & 1 & 2 \\ 1 & 3 & -2 \end{bmatrix}$

$\qquad = \begin{bmatrix} -7 & 0 & 8 \\ 6 & 5 & -14 \end{bmatrix} + \begin{bmatrix} -3 & -8 & 8 \\ 4 & 9 & -14 \end{bmatrix} = \begin{bmatrix} -10 & -8 & 16 \\ 10 & 14 & -28 \end{bmatrix}.$

5. $A(rB) = \begin{bmatrix} 1 & 3 \\ 2 & -1 \end{bmatrix} \left(-3 \begin{bmatrix} -1 & 3 & 2 \\ 1 & -3 & 4 \end{bmatrix} \right) = \begin{bmatrix} 1 & 3 \\ 2 & -1 \end{bmatrix} \begin{bmatrix} 3 & -9 & -6 \\ -3 & 9 & -12 \end{bmatrix} = \begin{bmatrix} -6 & 18 & -42 \\ 9 & -27 & 0 \end{bmatrix}.$

$\quad r(AB) = -3 \left(\begin{bmatrix} 1 & 3 \\ 2 & -1 \end{bmatrix} \begin{bmatrix} -1 & 3 & 2 \\ 1 & -3 & 4 \end{bmatrix} \right) = -3 \begin{bmatrix} 2 & -6 & 14 \\ -3 & 9 & 0 \end{bmatrix} = \begin{bmatrix} -6 & 18 & -42 \\ 9 & -27 & 0 \end{bmatrix}.$

$\quad (rA)B = \left(-3 \begin{bmatrix} 1 & 3 \\ 2 & -1 \end{bmatrix} \right) \begin{bmatrix} -1 & 3 & 2 \\ 1 & -3 & 4 \end{bmatrix} = \begin{bmatrix} -3 & -9 \\ -6 & 3 \end{bmatrix} \begin{bmatrix} -1 & 3 & 2 \\ 1 & -3 & 4 \end{bmatrix} = \begin{bmatrix} -6 & 18 & -42 \\ 9 & -27 & 0 \end{bmatrix}.$

7. $(AB)^T = \left(\begin{bmatrix} 1 & 3 & 2 \\ 2 & 1 & -3 \end{bmatrix} \begin{bmatrix} 3 & -1 \\ 2 & 4 \\ 1 & 2 \end{bmatrix} \right)^T = \begin{bmatrix} 11 & 15 \\ 5 & -4 \end{bmatrix}^T = \begin{bmatrix} 11 & 5 \\ 15 & -4 \end{bmatrix}.$

$\quad B^T A^T = \begin{bmatrix} 3 & -1 \\ 2 & 4 \\ 1 & 2 \end{bmatrix}^T \begin{bmatrix} 1 & 3 & 2 \\ 2 & 1 & -3 \end{bmatrix}^T = \begin{bmatrix} 3 & 2 & 1 \\ -1 & 4 & 2 \end{bmatrix} \begin{bmatrix} 1 & 2 \\ 3 & 1 \\ 2 & -3 \end{bmatrix} = \begin{bmatrix} 11 & 5 \\ 15 & -4 \end{bmatrix}.$

9. (a) $(3C - 2E)^T B = \left(3 \begin{bmatrix} 2 & 1 & 3 \\ -1 & 2 & 4 \\ 3 & 1 & 0 \end{bmatrix} - 2 \begin{bmatrix} 1 & 1 & 2 \\ 2 & -1 & 3 \\ -3 & 2 & -1 \end{bmatrix} \right)^T \begin{bmatrix} 2 & -1 \\ 3 & 4 \\ 1 & -2 \end{bmatrix}$

$\qquad = \begin{bmatrix} 4 & 1 & 5 \\ -7 & 8 & 6 \\ 15 & -1 & 2 \end{bmatrix}^T \begin{bmatrix} 2 & -1 \\ 3 & 4 \\ 1 & -2 \end{bmatrix} = \begin{bmatrix} 4 & -7 & 15 \\ 1 & 8 & -1 \\ 5 & 6 & 2 \end{bmatrix} \begin{bmatrix} 2 & -1 \\ 3 & 4 \\ 1 & -2 \end{bmatrix} = \begin{bmatrix} 2 & -62 \\ 25 & 33 \\ 30 & 15 \end{bmatrix}.$

(b) $A^T(D + F) = \begin{bmatrix} 2 & 1 & -2 \\ 3 & 2 & 5 \end{bmatrix}^T \left(\begin{bmatrix} 2 & -1 \\ -3 & 2 \end{bmatrix} + \begin{bmatrix} 1 & 0 \\ 2 & -3 \end{bmatrix} \right)$

$\qquad = \begin{bmatrix} 2 & 3 \\ 1 & 2 \\ -2 & 5 \end{bmatrix} \begin{bmatrix} 3 & -1 \\ -1 & -1 \end{bmatrix} = \begin{bmatrix} 3 & -5 \\ 1 & -3 \\ -11 & -3 \end{bmatrix}.$

(c) $B^T C + A = \begin{bmatrix} 2 & -1 \\ 3 & 4 \\ 1 & -2 \end{bmatrix}^T \begin{bmatrix} 2 & 1 & 3 \\ -1 & 2 & 4 \\ 3 & 1 & 0 \end{bmatrix} + \begin{bmatrix} 2 & 1 & -2 \\ 3 & 2 & 5 \end{bmatrix}$

$\qquad = \begin{bmatrix} 2 & 3 & 1 \\ -1 & 4 & -2 \end{bmatrix} \begin{bmatrix} 2 & 1 & 3 \\ -1 & 2 & 4 \\ 3 & 1 & 0 \end{bmatrix} + \begin{bmatrix} 2 & 1 & -2 \\ 3 & 2 & 5 \end{bmatrix}$

$\qquad = \begin{bmatrix} 4 & 9 & 18 \\ -12 & 5 & 13 \end{bmatrix} + \begin{bmatrix} 2 & 1 & -2 \\ 3 & 2 & 5 \end{bmatrix} = \begin{bmatrix} 6 & 10 & 16 \\ -9 & 7 & 18 \end{bmatrix}.$

(d) $(2E)A^T = \left(2 \begin{bmatrix} 1 & 1 & 2 \\ 2 & -1 & 3 \\ -3 & 2 & -1 \end{bmatrix} \right) \begin{bmatrix} 2 & 1 & -2 \\ 3 & 2 & 5 \end{bmatrix}^T = \begin{bmatrix} 2 & 2 & 4 \\ 4 & -2 & 6 \\ -6 & 4 & -2 \end{bmatrix} \begin{bmatrix} 2 & 3 \\ 1 & 2 \\ -2 & 5 \end{bmatrix} = \begin{bmatrix} -2 & 30 \\ -6 & 38 \\ -4 & -20 \end{bmatrix}.$

(e) $(B^T + A)C = \left(\begin{bmatrix} 2 & -1 \\ 3 & 4 \\ 1 & -2 \end{bmatrix}^T + \begin{bmatrix} 2 & 1 & -2 \\ 3 & 2 & 5 \end{bmatrix} \right) \begin{bmatrix} 2 & 1 & 3 \\ -1 & 2 & 4 \\ 3 & 1 & 0 \end{bmatrix}$

$= \left(\begin{bmatrix} 2 & 3 & 1 \\ -1 & 4 & -2 \end{bmatrix} + \begin{bmatrix} 2 & 1 & -2 \\ 3 & 2 & 5 \end{bmatrix} \right) \begin{bmatrix} 2 & 1 & 3 \\ -1 & 2 & 4 \\ 3 & 1 & 0 \end{bmatrix}$

$= \begin{bmatrix} 4 & 4 & -1 \\ 2 & 6 & 3 \end{bmatrix} \begin{bmatrix} 2 & 1 & 3 \\ -1 & 2 & 4 \\ 3 & 1 & 0 \end{bmatrix} = \begin{bmatrix} 1 & 11 & 28 \\ 7 & 17 & 30 \end{bmatrix}.$

11. $AB = \begin{bmatrix} -2 & 3 \\ 2 & -3 \end{bmatrix} \begin{bmatrix} -1 & 3 \\ 2 & 0 \end{bmatrix} = \begin{bmatrix} 8 & -6 \\ -8 & 6 \end{bmatrix};\ AC = \begin{bmatrix} -2 & 3 \\ 2 & -3 \end{bmatrix} \begin{bmatrix} -4 & -3 \\ 0 & -4 \end{bmatrix} = \begin{bmatrix} 8 & -6 \\ -8 & 6 \end{bmatrix}.$

13. (a) $A^2 + 3A = AA + 3A = \begin{bmatrix} 4 & 2 \\ 1 & 3 \end{bmatrix} \begin{bmatrix} 4 & 2 \\ 1 & 3 \end{bmatrix} + 3 \begin{bmatrix} 4 & 2 \\ 1 & 3 \end{bmatrix} = \begin{bmatrix} 18 & 14 \\ 7 & 11 \end{bmatrix} + \begin{bmatrix} 12 & 6 \\ 3 & 9 \end{bmatrix} = \begin{bmatrix} 30 & 20 \\ 10 & 20 \end{bmatrix}.$

(b) $2A^3 + 3A^2 + 4A + 5I_2 = 2AAA + 3AA + 4A + 5I_2$

$= 2 \begin{bmatrix} 4 & 2 \\ 1 & 3 \end{bmatrix} \begin{bmatrix} 4 & 2 \\ 1 & 3 \end{bmatrix} \begin{bmatrix} 4 & 2 \\ 1 & 3 \end{bmatrix} + 3 \begin{bmatrix} 4 & 2 \\ 1 & 3 \end{bmatrix} \begin{bmatrix} 4 & 2 \\ 1 & 3 \end{bmatrix}$

$+ 4 \begin{bmatrix} 4 & 2 \\ 1 & 3 \end{bmatrix} + 5 \begin{bmatrix} 1 & 0 \\ 0 & 1 \end{bmatrix}$

$= 2 \begin{bmatrix} 18 & 14 \\ 7 & 11 \end{bmatrix} \begin{bmatrix} 4 & 2 \\ 1 & 3 \end{bmatrix} + 3 \begin{bmatrix} 18 & 14 \\ 7 & 11 \end{bmatrix} + \begin{bmatrix} 16 & 8 \\ 4 & 12 \end{bmatrix} + \begin{bmatrix} 5 & 0 \\ 0 & 5 \end{bmatrix}$

$= 2 \begin{bmatrix} 86 & 78 \\ 39 & 47 \end{bmatrix} + \begin{bmatrix} 54 & 42 \\ 21 & 33 \end{bmatrix} + \begin{bmatrix} 21 & 8 \\ 4 & 17 \end{bmatrix}$

$= \begin{bmatrix} 172 & 156 \\ 78 & 94 \end{bmatrix} + \begin{bmatrix} 75 & 50 \\ 25 & 50 \end{bmatrix} = \begin{bmatrix} 247 & 206 \\ 103 & 144 \end{bmatrix}.$

15. Compute $A\mathbf{x} = \begin{bmatrix} 2 & 1 \\ 1 & 2 \end{bmatrix} \begin{bmatrix} 1 \\ 1 \end{bmatrix} = \begin{bmatrix} 3 \\ 3 \end{bmatrix}$. Comparing this with $\mathbf{x}$, we see that $A\mathbf{x} = 3\mathbf{x}$. Thus $r = 3$.

17. $\mathbf{a}_1 = \text{col}_1(A) = \begin{bmatrix} -3 \\ 4 \end{bmatrix}$, $\mathbf{a}_1^T = \begin{bmatrix} -3 & 4 \end{bmatrix}.$

$\mathbf{a}_2 = \text{col}_2(A) = \begin{bmatrix} 2 \\ 5 \end{bmatrix}$, $\mathbf{a}_2^T = \begin{bmatrix} 2 & 5 \end{bmatrix}.$

$\mathbf{a}_3 = \text{col}_3(A) = \begin{bmatrix} 1 \\ 0 \end{bmatrix}$, $\mathbf{a}_3^T = \begin{bmatrix} 1 & 0 \end{bmatrix}.$

$A^T A = \begin{bmatrix} -3 & 4 \\ 2 & 5 \\ 1 & 0 \end{bmatrix} \begin{bmatrix} -3 & 2 & 1 \\ 4 & 5 & 0 \end{bmatrix} = \begin{bmatrix} \mathbf{a}_1^T \\ \mathbf{a}_2^T \\ \mathbf{a}_3^T \end{bmatrix} \begin{bmatrix} \mathbf{a}_1 & \mathbf{a}_2 & \mathbf{a}_3 \end{bmatrix}$

$= \begin{bmatrix} \mathbf{a}_1^T \cdot \mathbf{a}_1 & \mathbf{a}_1^T \cdot \mathbf{a}_2 & \mathbf{a}_1^T \cdot \mathbf{a}_3 \\ \mathbf{a}_2^T \cdot \mathbf{a}_1 & \mathbf{a}_2^T \cdot \mathbf{a}_2 & \mathbf{a}_2^T \cdot \mathbf{a}_3 \\ \mathbf{a}_3^T \cdot \mathbf{a}_1 & \mathbf{a}_3^T \cdot \mathbf{a}_2 & \mathbf{a}_3^T \cdot \mathbf{a}_3 \end{bmatrix} = \begin{bmatrix} 25 & 14 & -3 \\ 14 & 29 & 2 \\ -3 & 2 & 1 \end{bmatrix}.$

19. Matrix A displays the information:

$$A = \begin{bmatrix} \frac{1}{3} & \frac{1}{2} \\ \frac{2}{3} & \frac{1}{2} \end{bmatrix} \begin{matrix} M \\ N \end{matrix} \quad \begin{matrix} M \quad N \end{matrix}.$$

$\mathbf{x}_0 = \begin{bmatrix} \frac{1}{3} \\ \frac{2}{3} \end{bmatrix}$ is the initial distribution of the market.

(a) The distribution of the market after 1 year is

$$\mathbf{x}_1 = A\mathbf{x}_0 = \begin{bmatrix} \frac{1}{3} & \frac{1}{2} \\ \frac{2}{3} & \frac{1}{2} \end{bmatrix} \begin{bmatrix} \frac{1}{3} \\ \frac{2}{3} \end{bmatrix} = \begin{bmatrix} \frac{4}{9} \\ \frac{5}{9} \end{bmatrix}.$$

(b) The distribution of a market

$$\mathbf{x}_0 = \begin{bmatrix} a \\ b \end{bmatrix}$$

is said to be stable if $A\mathbf{x}_0 = \mathbf{x}_0$. Since the market is controlled by the two companies M and N, $a + b = 1$. The equation $A\mathbf{x}_0 = \mathbf{x}_0$ in matrix form is

$$\begin{bmatrix} \frac{1}{3} & \frac{1}{2} \\ \frac{2}{3} & \frac{1}{2} \end{bmatrix} \begin{bmatrix} a \\ b \end{bmatrix} = \begin{bmatrix} a \\ b \end{bmatrix} \implies \begin{matrix} \frac{1}{3}a + \frac{1}{2}b = a \\ \frac{2}{3}a + \frac{1}{2}b = b \end{matrix} \implies \begin{matrix} -\frac{2}{3}a + \frac{1}{2}b = 0 \\ \frac{2}{3}a - \frac{1}{2}b = 0. \end{matrix}$$

The two equations are the same so using one equation and the requirement that $a + b = 1$ we solve for a and b.

$$-\tfrac{2}{3}a + \tfrac{1}{2}b = 0$$
$$a + b = 1.$$

To eliminate a take $\left(\frac{2}{3}\right)$ times the second equation and add the two equations:

$$\left(\tfrac{1}{2} + \tfrac{2}{3}\right)b = \tfrac{2}{3} \implies \tfrac{7}{6}b = \tfrac{2}{3} \implies b = \tfrac{4}{7}.$$

Therefore $a = 1 - b = 1 - \frac{4}{7} = \frac{3}{7}$. Thus the stable distribution of the market is $\begin{bmatrix} \frac{3}{7} \\ \frac{4}{7} \end{bmatrix}$.

21. (a) The distribution of the market after 1 year is

$$\mathbf{x}_1 = A\mathbf{x}_0 = \begin{bmatrix} 0.4 & 0 & 0.4 \\ 0 & 0.5 & 0.4 \\ 0.6 & 0.5 & 0.2 \end{bmatrix} \begin{bmatrix} \frac{1}{3} \\ \frac{1}{3} \\ \frac{1}{3} \end{bmatrix} = \begin{bmatrix} \frac{2}{5} & 0 & \frac{2}{5} \\ 0 & \frac{1}{2} & \frac{2}{5} \\ \frac{3}{5} & \frac{1}{2} & \frac{1}{5} \end{bmatrix} \begin{bmatrix} \frac{1}{3} \\ \frac{1}{3} \\ \frac{1}{3} \end{bmatrix} = \begin{bmatrix} \frac{4}{15} \\ \frac{3}{10} \\ \frac{13}{30} \end{bmatrix} \approx \begin{bmatrix} 0.2666 \\ 0.3000 \\ 0.4333 \end{bmatrix}.$$

The distribution of the market after 2 years is

$$\mathbf{x}_2 = A\mathbf{x}_1 = = \begin{bmatrix} \frac{2}{5} & 0 & \frac{2}{5} \\ 0 & \frac{1}{2} & \frac{2}{5} \\ \frac{3}{5} & \frac{1}{2} & \frac{1}{5} \end{bmatrix} \begin{bmatrix} \frac{4}{15} \\ \frac{3}{10} \\ \frac{13}{30} \end{bmatrix} = \begin{bmatrix} \frac{7}{25} \\ \frac{97}{300} \\ \frac{119}{300} \end{bmatrix} \approx= \begin{bmatrix} 0.2800 \\ 0.3233 \\ 0.3967 \end{bmatrix}.$$

Note: $\mathbf{x}_2 = A^2\mathbf{x}_0$.

(b) The distribution of a market

$$\mathbf{x}_0 = \begin{bmatrix} a \\ b \\ c \end{bmatrix}$$

is said to be stable if $A\mathbf{x}_0 = \mathbf{x}_0$. Since the market is controlled by companies R, S, and T, $a + b + c = 1$. The equation $A\mathbf{x}_0 = \mathbf{x}_0$ in matrix form

$$\begin{bmatrix} \frac{2}{5} & 0 & \frac{2}{5} \\ 0 & \frac{1}{2} & \frac{2}{5} \\ \frac{3}{5} & \frac{1}{2} & \frac{1}{5} \end{bmatrix} \begin{bmatrix} a \\ b \\ c \end{bmatrix} = \begin{bmatrix} a \\ b \\ c \end{bmatrix}$$

leads to the linear system

$$\begin{aligned} \tfrac{2}{5}a \quad\quad + \tfrac{2}{5}c &= a \\ \tfrac{1}{2}b + \tfrac{2}{5}c &= b \\ \tfrac{3}{5}a + \tfrac{1}{2}b + \tfrac{1}{5}c &= c. \end{aligned}$$

In addition, using the requirement that $a + b + c = 1$, we have

$$\begin{aligned} -\tfrac{3}{5}a \quad\quad\quad + \tfrac{2}{5}c &= 0 \quad \text{(i)} \\ -\tfrac{1}{2}b + \tfrac{2}{5}c &= 0 \quad \text{(ii)} \\ \tfrac{3}{5}a + \tfrac{1}{2}b - \tfrac{4}{5}c &= 0 \quad \text{(iii)} \\ a + \quad b + \quad c &= 1. \quad \text{(iv)} \end{aligned}$$

To solve, multiply $\left(-\tfrac{1}{2}\right)$ times equation (iv) and add the result to equation (iii). We then have

$$\frac{1}{10}a - \frac{13}{10}c = -\frac{1}{2}.$$

Using equation (i), solve the linear system

$$\begin{aligned} \tfrac{1}{10}a - \tfrac{13}{10}c &= -\tfrac{1}{2} \\ -\tfrac{3}{5}a + \tfrac{2}{5}c &= \quad 0. \end{aligned}$$

To eliminate a, multiply the first equation by (6) and add the two equations. The resulting solution is $c = \frac{15}{37}$ and $a = \frac{10}{37}$. Using equation (ii), we find that $b = \frac{12}{37}$. Thus, the stable distribution of the market is

$$\begin{bmatrix} \frac{10}{37} \\ \frac{12}{37} \\ \frac{15}{37} \end{bmatrix}.$$

(c) T will gain the most market share over a long period of time. The percent of the market gained by the company is

$$\left(\frac{15}{37} - \frac{1}{3} \right) 100 \approx 7.21\%.$$

23. $A^2 = \begin{bmatrix} 1 & 1 \\ 0 & 1 \end{bmatrix} \begin{bmatrix} 1 & 1 \\ 0 & 1 \end{bmatrix} = \begin{bmatrix} 1 & 0 \\ 0 & 1 \end{bmatrix} = I_2.$

25. (a) $A^2 + A = \begin{bmatrix} 0 & 0 \\ 1 & 1 \end{bmatrix} \begin{bmatrix} 0 & 0 \\ 1 & 1 \end{bmatrix} + \begin{bmatrix} 0 & 0 \\ 1 & 1 \end{bmatrix} = \begin{bmatrix} 0 & 0 \\ 1 & 1 \end{bmatrix} + \begin{bmatrix} 0 & 0 \\ 1 & 1 \end{bmatrix} = \begin{bmatrix} 0 & 0 \\ 0 & 0 \end{bmatrix}.$

(b) From part (a) we have $A^2 = A$ and $A^2 + A = \begin{bmatrix} 0 & 0 \\ 0 & 0 \end{bmatrix}$.

$$A^4 + (A^3 + A^2) = (A^2)^2 + A(A^2 + A) = A^2 + A \begin{bmatrix} 0 & 0 \\ 0 & 0 \end{bmatrix}$$
$$= \begin{bmatrix} 0 & 0 \\ 1 & 1 \end{bmatrix} + \begin{bmatrix} 0 & 0 \\ 0 & 0 \end{bmatrix} = \begin{bmatrix} 0 & 0 \\ 1 & 1 \end{bmatrix} = A.$$

T.1. (b) The (i, j) entry of $A + (B + C)$ is $a_{ij} + (b_{ij} + c_{ij})$, that of $(A + B) + C$ is $(a_{ij} + b_{ij}) + c_{ij}$. These two entries are equal because of the associative law for addition of real numbers.

(d) For each (i, j) let $d_{ij} = -a_{ij}$, $D = [d_{ij}]$. Then $A + D = D + A = O$.

T.3. (b) $\displaystyle\sum_{k=1}^{p} a_{ik}(b_{kj} + c_{kj}) = \sum_{k=1}^{p}(a_{ik}b_{kj} + a_{ik}c_{kj}) = \sum_{k=1}^{p} a_{ik}b_{kj} + \sum_{k=1}^{p} a_{ik}c_{kj}$

(c) $\displaystyle\sum_{k=1}^{p}(a_{ik} + b_{ik})c_{kj} = \sum_{k=1}^{p}(a_{ik}c_{kj} + b_{ik}c_{kj}) = \sum_{k=1}^{p} a_{ik}c_{kj} + \sum_{k=1}^{p} b_{ik}c_{kj}.$

T.5. $A^p A^q = \underbrace{(A \cdot A \cdots A)}_{p \text{ factors}} \cdot \underbrace{(A \cdot A \cdots A)}_{q \text{ factors}} = A^{p+q}$, $\quad (A^p)^q = \underbrace{A^p \cdot A^p \cdot A^p \cdots A^p}_{q \text{ factors}} = A^{\overbrace{p + p + \cdots + p}^{q \text{ summands}}} = A^{pq}$.

T.7. From Exercise T.4 in Section 1.3 we know that the product of two diagonal matrices is a diagonal matrix. Let $A = [a_{ij}]$, $B = [b_{ij}]$, $AB = C = [c_{ij}]$ and $BA = D = [d_{ij}]$. Then

$$c_{ii} = \sum_{k=1}^{n} a_{ik}b_{ki} = a_{ii}b_{ii}; \qquad d_{ii} = \sum_{k=1}^{n} b_{ik}a_{ki} = b_{ii}a_{ii}$$

so $c_{ii} = d_{ii}$ for $i = 1, 2, \ldots, n$. Hence, $C = D$.

T.9. Set the entries of the product matrices equal to each other and solve for a, b, c, and d.

$$\begin{bmatrix} 1 & 2 \\ 0 & 1 \end{bmatrix} \begin{bmatrix} a & b \\ c & d \end{bmatrix} = \begin{bmatrix} a & b \\ c & d \end{bmatrix} \begin{bmatrix} 1 & 2 \\ 0 & 1 \end{bmatrix}$$

Possible answers: $B = \begin{bmatrix} a & b \\ 0 & a \end{bmatrix}$. There are infinitely many such matrices B.

T.11. For $p = 0$, $(cA)^0 = I_n = 1 \cdot I_n = c^0 \cdot A^0$. For $p = 1$, $(cA)^1 = cA$. Assume the result true for $p = k$: $(cA)^k = c^k A^k$. Then for $p = k + 1$, we have

$$(cA)^{k+1} = (cA)^k(cA) = c^k A^k \cdot cA = c^k(A^k c)A = c^k(cA^k)A = (c^k c)(A^k A) = c^{k+1} A^{k+1}.$$

Therefore the result is true for all positive integers p.

T.13. $(-1)a_{ij} = -a_{ij}$ (see Exercise T.1.).

T.15. Let $A = [a_{ij}]$ and $B = [b_{ij}]$. Then $A - B = [c_{ij}]$, where $c_{ij} = a_{ij} - b_{ij}$. Then $(A - B)^T = [c_{ij}]^T$, so

$$c_{ij}^T = c_{ji} = a_{ji} - b_{ji} = a_{ij}^T - b_{ij}^T = \text{the } i, j\text{th entry in } A^T - B^T.$$

T.17. If A is symmetric, then $A^T = A$. Thus $a_{ji} = a_{ij}$ for all i and j. Conversely, if $a_{ji} = a_{ij}$ for all i and j, then $A^T = A$ and A is symmetric.

T.19. If $A\mathbf{x} = \mathbf{0}$ for all $n \times 1$ matrices $\mathbf{x}$, then $A\mathbf{e}_j = \mathbf{0}$, $j = 1, 2, \ldots, n$, where $\mathbf{e}_j = $ column j of I_n. But then

$$A\mathbf{e}_j = \begin{bmatrix} a_{1j} \\ a_{2j} \\ \vdots \\ a_{nj} \end{bmatrix} = \mathbf{0}.$$

Hence column j of A is equal to $\mathbf{0}$ for each j and it follows that $A = O$.

T.21. Given that $AA^T = O$, we have that each entry of AA^T is zero. In particular then, each diagonal entry of AA^T is zero. Hence

$$0 = \text{row}_i(A) \cdot \text{col}_i(A^T) = \begin{bmatrix} a_{i1} & a_{i2} & \cdots & a_{in} \end{bmatrix} \begin{bmatrix} a_{i1} \\ a_{i2} \\ \vdots \\ a_{in} \end{bmatrix} = \sum_{j=1}^{n} (a_{ij})^2.$$

(Recall that $\text{col}_i(A^T)$ is $\text{row}_i(A)$ written in column form.) A sum of squares is zero only if each member of the sum is zero, hence $a_{i1} = a_{i2} = \cdots = a_{in} = 0$, which means that $\text{row}_i(A)$ consists of all zeros. The previous argument holds for each diagonal entry, hence each row of A contains all zeros. Thus it follows that $A = O$.

T.23.   (a) $(A + B)^T = A^T + B^T = A + B$, so $A + B$ is symmetric.

   (b) Suppose that AB is symmetric. Then

$$\begin{aligned} (AB)^T &= AB \\ B^T A^T &= AB \qquad \text{[Thm. 1.4(c)]} \\ BA &= AB \qquad (A \text{ and } B \text{ are each symmetric}) \end{aligned}$$

   Thus A and B commute. Conversely, if A and B commute, then $(AB)^T = B^T A^T = BA = AB$ so AB is symmetric.

T.25. If A is a scalar matrix, then $A = rI_n$. If A is skew symmetric, then $A^T = -A$, i.e., $(rI_n)^T = -rI_n$ so $rI_n = -rI_n$. Hence, $r = -r$, which implies that $r = 0$. That is, $A = O$.

T.27.   (a) A matrix P is symmetric if $P^T = P$. So consider $(A + A^T)^T = A^T + (A^T)^T = A^T + A = A + A^T$.

   (b) A matrix P is skew symmetric if $P^T = -P$. So consider $(A - A^T)^T = A^T - (A^T)^T = A^T - A = -(A - A^T)$.

T.29. If the diagonal entries of A are r, then since $r = r \cdot 1$, $A = rI_n$.

T.31. Suppose $r \neq 0$. The i, jth entry of rA is ra_{ij}. Since $r \neq 0$, $a_{ij} = 0$ for all i and j. Thus $A = O$.

T.33. Suppose $A = \begin{bmatrix} a & b \\ c & d \end{bmatrix}$ satisfies $AB = BA$ for any 2×2 matrix B. Choosing $B = \begin{bmatrix} 1 & 0 \\ 0 & 0 \end{bmatrix}$ we get

$$\begin{bmatrix} a & b \\ c & d \end{bmatrix} \begin{bmatrix} 1 & 0 \\ 0 & 0 \end{bmatrix} = \begin{bmatrix} 1 & 0 \\ 0 & 0 \end{bmatrix} \begin{bmatrix} a & b \\ c & d \end{bmatrix}$$

$$\begin{bmatrix} a & 0 \\ c & 0 \end{bmatrix} = \begin{bmatrix} a & b \\ 0 & 0 \end{bmatrix}$$

which implies $b = c = 0$. Thus $A = \begin{bmatrix} a & 0 \\ 0 & d \end{bmatrix}$ is diagonal. Next choosing $B = \begin{bmatrix} 0 & 1 \\ 0 & 0 \end{bmatrix}$ we get

$$\begin{bmatrix} 0 & a \\ 0 & 0 \end{bmatrix} = \begin{bmatrix} 0 & d \\ 0 & 0 \end{bmatrix},$$

or $a = d$. Thus $A = \begin{bmatrix} a & 0 \\ 0 & a \end{bmatrix}$ is a scalar matrix.

T.35. A symmetric matrix. To show this, let $A_1, \ldots, A_n$ be symmetric matrices and let $c_1, \ldots, c_n$ be scalars. Then $A_1^T = A_1, \ldots, A_n^T = A_n$. Therefore

$$(c_1 A_1 + \cdots + c_n A_n)^T = (c_1 A_1)^T + \cdots + (c_n A_n)^T$$
$$= c_1 A_1^T + \cdots + c_n A_n^T$$
$$= c_1 A_1 + \cdots + c_n A_n.$$

Hence the linear combination $c_1 A_1 + \cdots + c_n A_n$ is symmetric.

T.37. Let c_{ij} be the i, jth element of AB. Since $a_{ij} = 0$ whenever $i \neq j$,

$$c_{ij} = \sum_{k=1}^{n} a_{ik} b_{kj}$$

has a nonzero term only when $k = i$. Hence $c_{ij} = a_{ii} b_{ij} = r b_{ij}$. So $AB = rB$.

T.39. The easiest method is exhaustive search.

$$\begin{bmatrix} 1 & 1 \\ 1 & 1 \end{bmatrix}, \quad \begin{bmatrix} 0 & 0 \\ 0 & 0 \end{bmatrix}, \quad \begin{bmatrix} 0 & 1 \\ 0 & 0 \end{bmatrix}, \quad \begin{bmatrix} 0 & 0 \\ 1 & 0 \end{bmatrix}.$$

ML.1. (a) **A^2**
       ans =

       0 1 0
       0 0 1
       1 0 0

   A^3
       ans =

       1 0 0
       0 1 0
       0 0 1

   Thus $k = 3$.

   (b) **A^2**
       ans =

       −1   0   0   0
        0 −1   0   0
        0   0   1   0
        0   0   0   1

   A^3
       ans =

       0 −1 0 0
       1   0 0 0
       0   0 0 1
       0   0 1 0

   A^4
       ans =

       1 0 0 0
       0 1 0 0
       0 0 1 0
       0 0 0 1

   Thus $k = 4$.

ML.3. (a) Define the vector of coefficients
       v = [1 −1 1 0 2];
       then we have
       polyvalm(v,A)
       ans =

        0 −2   4
        4   0 −2
       −2   4   0

   (b) Define the vector of coefficients
       v = [1 −3 3 0];
       then we have
       polyvalm(v,A)
       ans =

       0 0 0
       0 0 0
       0 0 0

ML.5. The sequence seems to be converging to

    1.0000   0.7500

        0   0

since

$>>$ **A^5**

ans =

        1.000   0.7469

           0   0.0041

$>>$ **A^{10}**

ans =

        1.000   0.7500

           0   0.0000

$>>$ **A^{20}**

ans =

        1.000   0.7500

           0   0.0000

ML.7. (a) **A$'$ $*$ A**　　　　　　　　　　　**A $*$ A$'$**

    ans = 　　　　　　　　　　　　ans =

2	−3	−1
−3	9	2
−1	2	6

6	−1	−3
−1	6	4
−3	4	5

$A^T A$ and $A A^T$ are not equal.

(b) **B $=$ A $+$ A$'$**　　　　　　　　**C $=$ A $-$ A$'$**

    B = 　　　　　　　　　　　　C =

2	−3	1
−3	2	4
1	4	2

0	−1	1
1	0	0
−1	0	0

Just observe that $B = B^T$ and that $C^T = -C$.

(c) **B $+$ C**

    ans =

2	−4	2
−2	2	4
0	4	2

We see that $B + C = 2A$.

ML.9. $>>$ **B $=$ triu(ones(3))**

    B =

      1 1 1

      0 1 1

      0 0 1

$>>$ **B2 $=$ binprod(B, B)**

    B2 =

      1 0 1

      0 1 0

      0 0 1

>> **B3** = **binprod(B2, B)**

B3 =

$$
\begin{array}{ccc}
1 & 1 & 0 \\
0 & 1 & 1 \\
0 & 0 & 1
\end{array}
$$

>> **B4** = **binprod(B3, B)**

B4 =

$$
\begin{array}{ccc}
1 & 0 & 0 \\
0 & 1 & 0 \\
0 & 0 & 1
\end{array}
$$

the identity matrix. So $k = 4$.

ML.11. >> **B** = **triu(ones(5))**

B =

$$
\begin{array}{ccccc}
1 & 1 & 1 & 1 & 1 \\
0 & 1 & 1 & 1 & 1 \\
0 & 0 & 1 & 1 & 1 \\
0 & 0 & 0 & 1 & 1 \\
0 & 0 & 0 & 0 & 1
\end{array}
$$

>> **B2** = **binprod(B, B)**

B2 =

$$
\begin{array}{ccccc}
1 & 0 & 1 & 0 & 1 \\
0 & 1 & 0 & 1 & 0 \\
0 & 0 & 1 & 0 & 1 \\
0 & 0 & 0 & 1 & 0 \\
0 & 0 & 0 & 0 & 1
\end{array}
$$

>> **B3** = **binprod(B2, B)**

B3 =

$$
\begin{array}{ccccc}
1 & 1 & 0 & 0 & 1 \\
0 & 1 & 1 & 0 & 0 \\
0 & 0 & 1 & 1 & 0 \\
0 & 0 & 0 & 1 & 1 \\
0 & 0 & 0 & 0 & 1
\end{array}
$$

>> **B4** = **binprod(B3, B)**

B4 =

$$
\begin{array}{ccccc}
1 & 0 & 0 & 0 & 1 \\
0 & 1 & 0 & 0 & 0 \\
0 & 0 & 1 & 0 & 0 \\
0 & 0 & 0 & 1 & 0 \\
0 & 0 & 0 & 0 & 1
\end{array}
$$

>> **B5** = **binprod(B4, B)**

```
B5  =

        1   1   1   1   0
        0   1   1   1   1
        0   0   1   1   1
        0   0   0   1   1
        0   0   0   0   1
```

$>> \mathbf{B6} = \mathbf{binprod(B5, B)}$

```
B6  =

        1   0   1   0   0
        0   1   0   1   0
        0   0   1   0   1
        0   0   0   1   0
        0   0   0   0   1
```

$>> \mathbf{B7} = \mathbf{binprod(B6, B)}$

```
B7  =

        1   1   0   0   0
        0   1   1   0   0
        0   0   1   1   0
        0   0   0   1   1
        0   0   0   0   1
```

$>> \mathbf{B8} = \mathbf{binprod(B7, B)}$

```
B8  =

        1   0   0   0   0
        0   1   0   0   0
        0   0   1   0   0
        0   0   0   1   0
        0   0   0   0   1
```

So $k = 8$.

Section 1.5, p. 61

1. The matrix transformation $f\colon R^2 \to R^2$ defined by

$$f(\mathbf{u}) = \begin{bmatrix} 1 & 0 \\ 0 & -1 \end{bmatrix} \mathbf{u}$$

is a reflection with respect to the x-axis in R^2. Thus

$$f\left(\begin{bmatrix} 2 \\ 3 \end{bmatrix}\right) = \begin{bmatrix} 1 & 0 \\ 0 & -1 \end{bmatrix} \begin{bmatrix} 2 \\ 3 \end{bmatrix} = \begin{bmatrix} 2 \\ -3 \end{bmatrix}$$

as shown in the figure.

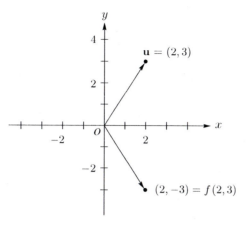

3. The matrix transformation $f\colon R^2 \to R^2$ defined by

$$f(\mathbf{u}) = \begin{bmatrix} \cos\frac{\pi}{6} & -\sin\frac{\pi}{6} \\ \sin\frac{\pi}{6} & \cos\frac{\pi}{6} \end{bmatrix} \mathbf{u}$$

is a counterclockwise rotation through an angle of $\frac{\pi}{6}$ radians. Thus

$$f\left(\begin{bmatrix} -1 \\ 3 \end{bmatrix}\right) = \begin{bmatrix} \frac{\sqrt{3}}{2} & -\frac{1}{2} \\ \frac{1}{2} & \frac{\sqrt{3}}{2} \end{bmatrix} \begin{bmatrix} -1 \\ 3 \end{bmatrix} = \begin{bmatrix} \frac{-3-\sqrt{3}}{2} \\ \frac{-1+3\sqrt{3}}{2} \end{bmatrix} \approx \begin{bmatrix} -2.366 \\ 2.098 \end{bmatrix}$$

as shown in the figure.

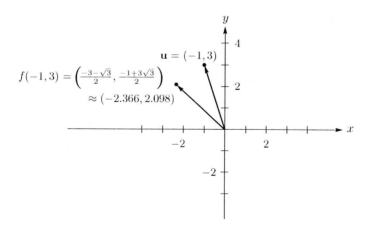

5. The matrix transformation $f\colon R^2 \to R^2$ defined by

$$f(\mathbf{u}) = \begin{bmatrix} -1 & 0 \\ 0 & -1 \end{bmatrix} \mathbf{u}$$

is a reflection about the origin. Thus

$$f\left(\begin{bmatrix} 3 \\ 2 \end{bmatrix}\right) = \begin{bmatrix} -1 & 0 \\ 0 & -1 \end{bmatrix} \begin{bmatrix} 3 \\ 2 \end{bmatrix} = \begin{bmatrix} -3 \\ -2 \end{bmatrix}$$

as shown in the figure.

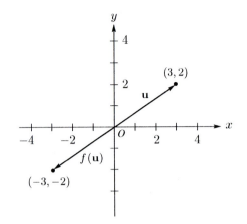

7. The matrix transformation $f: R^3 \to R^3$ is defined by

$$f(\mathbf{u}) = \begin{bmatrix} 1 & 0 & 0 \\ 1 & -1 & 0 \\ 0 & 0 & 0 \end{bmatrix} \mathbf{u}.$$

Thus

$$f\left(\begin{bmatrix} 2 \\ -1 \\ 3 \end{bmatrix}\right) = \begin{bmatrix} 1 & 0 & 0 \\ 1 & -1 & 0 \\ 0 & 0 & 0 \end{bmatrix} \begin{bmatrix} 2 \\ -1 \\ 3 \end{bmatrix} = \begin{bmatrix} 2 \\ 3 \\ 0 \end{bmatrix}$$

as shown in the figure.

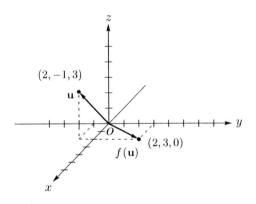

For problems 9 and 11 the matrix transformation $f: R^2 \to R^2$ is defined by

$$f(\mathbf{x}) = \begin{bmatrix} 1 & 3 \\ -1 & 2 \end{bmatrix} \mathbf{x}.$$

Vector $\mathbf{w}$ is in the range of f if we can find a vector $\mathbf{x} = \begin{bmatrix} a \\ b \end{bmatrix}$ such that

$$f(\mathbf{x}) = \begin{bmatrix} 1 & 3 \\ -1 & 2 \end{bmatrix} \mathbf{x} = \mathbf{w}.$$

9. Yes. $\begin{bmatrix} 1 & 3 \\ -1 & 2 \end{bmatrix}\begin{bmatrix} a \\ b \end{bmatrix} = \begin{bmatrix} a + 3b \\ -a + 2b \end{bmatrix} = \begin{bmatrix} 7 \\ 3 \end{bmatrix}.$

Solve the corresponding linear system

$$a + 3b = 7$$
$$-a + 2b = 3.$$

Add the equations to eliminate a:

$$5b = 10 \quad \Longrightarrow \quad b = 2.$$

Then $a = -5$. Thus $\mathbf{x} = \begin{bmatrix} -5 \\ 2 \end{bmatrix}$ and $\mathbf{w}$ is in the range of f.

11. Yes. $\begin{bmatrix} 1 & 3 \\ -1 & 2 \end{bmatrix}\begin{bmatrix} a \\ b \end{bmatrix} = \begin{bmatrix} a + 3b \\ -a + 2b \end{bmatrix} = \begin{bmatrix} -1 \\ -9 \end{bmatrix}.$

Solve the corresponding linear system

$$a + 3b = -1$$
$$-a + 2b = -9.$$

Add the equations to eliminate a:

$$5b = -10 \quad \Longrightarrow \quad b = -2.$$

Then $a = 5$. Thus $\mathbf{x} = \begin{bmatrix} 5 \\ -2 \end{bmatrix}$ and $\mathbf{w}$ is in the range of f.

13. No. The matrix transformation $f : R^2 \to R^3$ is defined by

$$f(\mathbf{x}) = \begin{bmatrix} 1 & 2 \\ 0 & 1 \\ 1 & 1 \end{bmatrix} \mathbf{x}.$$

Find a vector $\mathbf{x} = \begin{bmatrix} a \\ b \end{bmatrix}$ such that

$$f(\mathbf{x}) = \begin{bmatrix} 1 & 2 \\ 0 & 1 \\ 1 & 1 \end{bmatrix} \mathbf{x} = \mathbf{w}.$$

$$\begin{bmatrix} 1 & 2 \\ 0 & 1 \\ 1 & 1 \end{bmatrix}\begin{bmatrix} a \\ b \end{bmatrix} = \begin{bmatrix} a + 2b \\ b \\ a + b \end{bmatrix} = \begin{bmatrix} 1 \\ 1 \\ 1 \end{bmatrix}$$

The corresponding linear system

$$a + 2b = 1$$
$$b = 1$$
$$a + b = 1$$

is inconsistent so vector $\mathbf{w}$ is not in the range of f.

15. (a) $f(\mathbf{u}) = A\mathbf{u} = \begin{bmatrix} -1 & 0 \\ 0 & 1 \end{bmatrix} \mathbf{u}.$

If $\mathbf{u} = \begin{bmatrix} x \\ y \end{bmatrix}$, then $\begin{bmatrix} -1 & 0 \\ 0 & 1 \end{bmatrix} \begin{bmatrix} x \\ y \end{bmatrix} = \begin{bmatrix} -x \\ y \end{bmatrix}$. Therefore this matrix transformation is a reflection about the y-axis.

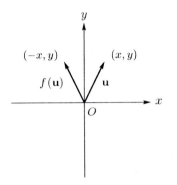

(b) If $\mathbf{u} = \begin{bmatrix} x \\ y \end{bmatrix}$, then $\begin{bmatrix} 0 & -1 \\ 1 & 0 \end{bmatrix} \begin{bmatrix} x \\ y \end{bmatrix} = \begin{bmatrix} -y \\ x \end{bmatrix}$. Using elementary geometry, it can be shown that this matrix transformation is a counterclockwise rotation through $\frac{\pi}{2}$ radians. (*Note:* The triangles are congruent.)

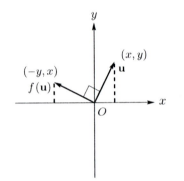

17. (a) $f(\mathbf{u}) = A\mathbf{u} = \begin{bmatrix} 1 & 0 \\ 0 & 0 \end{bmatrix} \mathbf{u}$.

If $\mathbf{u} = \begin{bmatrix} x \\ y \end{bmatrix}$, then $\begin{bmatrix} 1 & 0 \\ 0 & 0 \end{bmatrix} \begin{bmatrix} x \\ y \end{bmatrix} = \begin{bmatrix} x \\ 0 \end{bmatrix}$. This matrix transformation is a projection onto the x-axis.

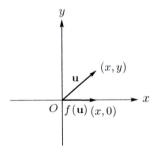

(b) $f(\mathbf{u}) = A\mathbf{u} = \begin{bmatrix} 0 & 0 \\ 0 & 1 \end{bmatrix} \mathbf{u}$.

If $\mathbf{u} = \begin{bmatrix} x \\ y \end{bmatrix}$, then $\begin{bmatrix} 0 & 0 \\ 0 & 1 \end{bmatrix} \begin{bmatrix} x \\ y \end{bmatrix} = \begin{bmatrix} 0 \\ y \end{bmatrix}$. This matrix transformation is a projection onto the y-axis.

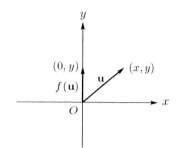

19. The matrix transformation $f \colon R^2 \to R^2$ is defined by

$$f(\mathbf{u}) = A\mathbf{u} = \begin{bmatrix} \cos \frac{\pi}{6} & -\sin \frac{\pi}{6} \\ \sin \frac{\pi}{6} & \cos \frac{\pi}{6} \end{bmatrix} \mathbf{u} = \begin{bmatrix} \frac{\sqrt{3}}{2} & -\frac{1}{2} \\ \frac{1}{2} & \frac{\sqrt{3}}{2} \end{bmatrix} \mathbf{u}.$$

(a) $T_1(\mathbf{u}) = A^2\mathbf{u} = \begin{bmatrix} \frac{\sqrt{3}}{2} & -\frac{1}{2} \\ \frac{1}{2} & \frac{\sqrt{3}}{2} \end{bmatrix} \begin{bmatrix} \frac{\sqrt{3}}{2} & -\frac{1}{2} \\ \frac{1}{2} & \frac{\sqrt{3}}{2} \end{bmatrix} \mathbf{u} = \begin{bmatrix} \frac{1}{2} & -\frac{\sqrt{3}}{2} \\ \frac{\sqrt{3}}{2} & \frac{1}{2} \end{bmatrix} \mathbf{u} = \begin{bmatrix} \cos \frac{\pi}{3} & -\sin \frac{\pi}{3} \\ \sin \frac{\pi}{3} & \cos \frac{\pi}{3} \end{bmatrix} \mathbf{u}$

Thus, the matrix transformation T_1 is a counterclockwise rotation by $\frac{\pi}{3}$ radians or $60°$.

(b) $T_2(\mathbf{u}) = A^{-1}(\mathbf{u})$.

To find A^{-1}, we want to find a matrix $\begin{bmatrix} a & b \\ c & d \end{bmatrix}$ such that

$$\begin{bmatrix} \frac{\sqrt{3}}{2} & -\frac{1}{2} \\ \frac{1}{2} & \frac{\sqrt{3}}{2} \end{bmatrix} \begin{bmatrix} a & b \\ c & d \end{bmatrix} = \begin{bmatrix} 1 & 0 \\ 0 & 1 \end{bmatrix} = I_2.$$

$$\begin{bmatrix} \frac{\sqrt{3}}{2}a - \frac{1}{2}c & \frac{\sqrt{3}}{2}b - \frac{1}{2}d \\ \frac{1}{2}a + \frac{\sqrt{3}}{2}c & \frac{1}{2}b + \frac{\sqrt{3}}{2}d \end{bmatrix} = \begin{bmatrix} 1 & 0 \\ 0 & 1 \end{bmatrix}.$$

The corresponding linear systems are

$$\begin{array}{ccc} \frac{\sqrt{3}}{2}a - \frac{1}{2}c = 1 & & \frac{\sqrt{3}}{2}b - \frac{1}{2}d = 0 \\ & \text{and} & \\ \frac{1}{2}a + \frac{\sqrt{3}}{2}c = 0 & & \frac{1}{2}b + \frac{\sqrt{3}}{2}d = 1. \end{array}$$

Solving each pair of equations by the method of elimination, we find that

$$a = \frac{\sqrt{3}}{2}, \quad c = -\frac{1}{2}, \quad b = \frac{1}{2} \quad \text{and} \quad d = \frac{\sqrt{3}}{2}.$$

Hence

$$A^{-1} = \begin{bmatrix} \frac{\sqrt{3}}{2} & \frac{1}{2} \\ -\frac{1}{2} & \frac{\sqrt{3}}{2} \end{bmatrix}$$

and this matrix transformation, T_2, is a clockwise rotation by $30°$.

(c) If $T(\mathbf{u}) = A^k\mathbf{u} = \mathbf{u}$, then $A^k = \begin{bmatrix} 1 & 0 \\ 0 & 1 \end{bmatrix}$. Hence this matrix transformation T describes a counterclockwise rotation of $360°$. Since $A\mathbf{u}$ describes a counterclockwise rotation of $30°$, if $k = 12$ you have a counterclockwise rotation of $360°$.

21. For any real numbers c and d, we have

$$f(c\mathbf{u} + d\mathbf{v}) = A(c\mathbf{u} + d\mathbf{v}) = A(c\mathbf{u}) + A(d\mathbf{v}) = c(A\mathbf{u}) + d(A\mathbf{v}) = cf(\mathbf{u}) + df(\mathbf{v}) = c\mathbf{0} + d\mathbf{0} = \mathbf{0} + \mathbf{0} = \mathbf{0}.$$

T.1.   (a) $f(\mathbf{u} + \mathbf{v}) = A(\mathbf{u} + \mathbf{v}) = A\mathbf{u} + A\mathbf{v}$ by Theorem 1.2(b), Sec. 1.4
$$= f(\mathbf{u}) + f(\mathbf{v}).$$

   (b) $f(c\mathbf{u}) = A(c\mathbf{u}) = c(A\mathbf{u})$ by Theorem 1.3(d), Sec. 1.4
$$= cf(\mathbf{u}).$$

   (c) $f(c\mathbf{u} + d\mathbf{v}) = f(c\mathbf{u}) + f(d\mathbf{v})$ by (a) above
$$= cf(\mathbf{u}) + df(\mathbf{v})$$ by (b) above

T.3.   (a)   Let $O = \left[b_{ij}\right]_{m \times n}$ where $b_{ij} = 0$ for all i, j. Let $O\mathbf{u} = \left[c_i\right]_{m \times 1}$. Then

$$c_i = \sum_{k=1}^{n} b_{ik} u_k = 0$$

for all i. Hence $O\mathbf{u} = \mathbf{0}$ for all $\mathbf{u}$ in R^n.

   (b)   Let $I_n = \left[a_{ij}\right]_{n \times n}$ where

$$a_{ij} = \begin{cases} 0 & \text{if } i \neq j \\ 1 & \text{if } i = j \end{cases}.$$

Let $I(\mathbf{u}) = I_n\mathbf{u} = \left[c_i\right]_{n \times 1}$. Then

$$c_i = \sum_{k=1}^{n} a_{ik} u_k = \begin{cases} a_{ii} u_i & \text{if } i = k \\ 0 & \text{if } i \neq k \end{cases}.$$

But $a_{ii} = 1$. So $c_i = u_i$, and $I_n\mathbf{u} = \mathbf{u}$.

Section 1.6, p. 85

1. This matrix is in reduced row echelon form.

3. This matrix is in reduced row echelon form.

5. This matrix is in row echelon form. It is not in reduced row echelon form since the second column contains a leading 1 in position (2,2) but the other entries of the column are not all zero.

7. Neither. The matrix is not in reduced row echelon form since its first row has all zeros but it does not appear at the bottom of the matrix. It is not in row echelon form for the same reason.

9. (a) $\begin{bmatrix} 4 & 2 & 2 \\ 5 & -1 & 5 \\ 1 & 0 & 3 \\ -3 & 1 & 4 \end{bmatrix}.$ (b) $\begin{bmatrix} 1 & 0 & 3 \\ -3 & 1 & 4 \\ 12 & 6 & 6 \\ 5 & -1 & 5 \end{bmatrix}.$ (c) $\begin{bmatrix} 1 & 0 & 3 \\ -3 & 1 & 4 \\ 4 & 2 & 2 \\ 2 & -1 & -4 \end{bmatrix}.$

11. Possible answers: $\begin{bmatrix} 2 & -1 & 3 & 4 \\ 5 & 2 & -3 & 4 \\ 0 & 1 & 2 & -1 \end{bmatrix}$, $\begin{bmatrix} 4 & -2 & 6 & 8 \\ 0 & 1 & 2 & -1 \\ 5 & 2 & -3 & 4 \end{bmatrix}$, $\begin{bmatrix} 2 & -1 & 3 & 4 \\ 0 & 1 & 2 & -1 \\ 7 & 1 & 0 & 8 \end{bmatrix}.$

13. Possible answer: $\begin{bmatrix} 1 & 3 & -1 & 2 \\ 0 & 1 & -2 & -3 \\ 0 & 0 & 1 & \frac{26}{7} \\ 0 & 0 & 0 & 1 \end{bmatrix}.$

15. Possible answer: $\begin{bmatrix} 1 & 2 & -3 & 1 \\ 0 & 1 & 2 & -1 \\ 0 & 0 & 1 & -\frac{7}{4} \\ 0 & 0 & 0 & 1 \end{bmatrix}$.

17. For Exercise 13: $\begin{bmatrix} 1 & 0 & 0 & 0 \\ 0 & 1 & 0 & 0 \\ 0 & 0 & 1 & 0 \\ 0 & 0 & 0 & 1 \end{bmatrix}$.

For Exercise 14: $\begin{bmatrix} 1 & 0 & 0 & 0 \\ 0 & 1 & 0 & 0 \\ 0 & 0 & 1 & 0 \\ 0 & 0 & 0 & 1 \\ 0 & 0 & 0 & 0 \end{bmatrix}$.

For Exercise 15: $\begin{bmatrix} 1 & 0 & 0 & 0 \\ 0 & 1 & 0 & 0 \\ 0 & 0 & 1 & 0 \\ 0 & 0 & 0 & 1 \end{bmatrix}$.

For Exercise 16: $\begin{bmatrix} 1 & 0 & -\frac{1}{3} & -\frac{2}{3} & \frac{11}{3} \\ 0 & 1 & -\frac{2}{3} & -\frac{7}{3} & \frac{10}{3} \\ 0 & 0 & 0 & 0 & 0 \\ 0 & 0 & 0 & 0 & 0 \end{bmatrix}$.

19. (a) Yes;

$$A\mathbf{x} = \begin{bmatrix} 1 & 2 & -1 & 3 \\ 1 & 3 & 0 & 2 \\ -1 & 2 & 1 & 3 \end{bmatrix} \begin{bmatrix} 5 \\ -3 \\ 5 \\ 2 \end{bmatrix} = \begin{bmatrix} 0 \\ 0 \\ 0 \end{bmatrix} = \mathbf{0}.$$

(b) No;

$$A\mathbf{x} = \begin{bmatrix} 1 & 2 & -1 & 3 \\ 1 & 3 & 0 & 2 \\ -1 & 2 & 1 & 3 \end{bmatrix} \begin{bmatrix} 1 \\ 2 \\ 3 \\ 4 \end{bmatrix} = \begin{bmatrix} 14 \\ 15 \\ 18 \end{bmatrix} \neq \mathbf{0}.$$

(c) Yes;

$$A\mathbf{x} = \begin{bmatrix} 1 & 2 & -1 & 3 \\ 1 & 3 & 0 & 2 \\ -1 & 2 & 1 & 3 \end{bmatrix} \begin{bmatrix} 1 \\ -\frac{3}{5} \\ 1 \\ \frac{2}{5} \end{bmatrix} = \begin{bmatrix} 0 \\ 0 \\ 0 \end{bmatrix} = \mathbf{0}.$$

(d) No;

$$A\mathbf{x} = \begin{bmatrix} 1 & 2 & -1 & 3 \\ 1 & 3 & 0 & 2 \\ -1 & 2 & 1 & 3 \end{bmatrix} \begin{bmatrix} 1 \\ 0 \\ 0 \\ -1 \end{bmatrix} = \begin{bmatrix} -2 \\ -1 \\ -4 \end{bmatrix} \neq \mathbf{0}.$$

21. (a) $x + y + 2z + 3w = 13$
           $x - 2y + z + w = 8$
           $3x + y + z - w = 1$

We form the augmented matrix and use row operations to form equivalent systems.

$$\begin{bmatrix} 1 & 1 & 2 & 3 & 13 \\ 1 & -2 & 1 & 1 & 8 \\ 3 & 1 & 1 & -1 & 1 \end{bmatrix}$$

Applying row operations

(-1) times row 1 added to row 2 and

(-3) times row 1 added to row 3,

we obtain

$$\begin{bmatrix} 1 & 1 & 2 & 3 & 13 \\ 0 & -3 & -1 & -2 & -5 \\ 0 & -2 & -5 & -10 & -38 \end{bmatrix}.$$

Next multiply row 2 by $\left(-\frac{1}{3}\right)$ to obtain

$$\begin{bmatrix} 1 & 1 & 2 & 3 & 13 \\ 0 & 1 & \frac{1}{3} & \frac{2}{3} & \frac{5}{3} \\ 0 & -2 & -5 & -10 & -38 \end{bmatrix}.$$

Applying row operations (2) times row 2 added to row 3, we obtain

$$\begin{bmatrix} 1 & 1 & 2 & 3 & 13 \\ 0 & 1 & \frac{1}{3} & \frac{2}{3} & \frac{5}{3} \\ 0 & 0 & -\frac{13}{3} & -\frac{26}{3} & -\frac{104}{3} \end{bmatrix}.$$

Multiply row 3 by $\left(-\frac{3}{13}\right)$ to put a 1 into the pivot position. The result is the matrix

$$\begin{bmatrix} 1 & 1 & 2 & 3 & 13 \\ 0 & 1 & \frac{1}{3} & \frac{2}{3} & \frac{5}{3} \\ 0 & 0 & 1 & 2 & 8 \end{bmatrix}.$$

Add (-1) times row 2 to row 1 to obtain

$$\begin{bmatrix} 1 & 0 & \frac{5}{3} & \frac{7}{3} & \frac{34}{3} \\ 0 & 1 & \frac{1}{3} & \frac{2}{3} & \frac{5}{3} \\ 0 & 0 & 1 & 2 & 8 \end{bmatrix}.$$

Finally, apply row operations

$\left(-\frac{5}{3}\right)$ times row 3 added to row 1

$\left(-\frac{1}{3}\right)$ times row 3 added to row 2

to obtain

$$\begin{bmatrix} 1 & 0 & 0 & -1 & -2 \\ 0 & 1 & 0 & 0 & -1 \\ 0 & 0 & 1 & 2 & 8 \end{bmatrix}.$$

This matrix is in reduced row echelon form and represents the equivalent linear system

$$\begin{aligned} x \quad\quad - w &= -2 \\ y \quad\quad\quad &= -1 \\ z + 2w &= \quad 8. \end{aligned}$$

We solve each equation for the unknown that corresponds to a leading 1. Hence we have

$$\begin{aligned} x &= -2 + w \\ y &= -1 \\ z &= \quad 8 - 2w. \end{aligned}$$

The unknown w can be chosen arbitrarily. Let $w = r$, where r is any real number. Then the solution of the linear system is given by

$$x = -2 + r, \quad y = -1, \quad z = 8 - 2r, \quad w = r.$$

Since r can be any real number, there are infinitely many solutions.

(b) $\begin{aligned} x + y + \quad z &= 1 \\ x + y - 2z &= 3 \\ 2x + y + \quad z &= 2 \end{aligned}$

We form the augmented matrix and use row operations to form equivalent systems.

$$\begin{bmatrix} 1 & 1 & 1 & 1 \\ 1 & 1 & -2 & 3 \\ 2 & 1 & 1 & 2 \end{bmatrix}$$

Applying row operations

$$(-1) \text{ times row 1 added to row 2 and}$$
$$(-2) \text{ times row 1 added to row 3,}$$

we obtain

$$\begin{bmatrix} 1 & 1 & 1 & 1 \\ 0 & 0 & -3 & 2 \\ 0 & -1 & -1 & 0 \end{bmatrix}.$$

The $(2, 2)$ entry is zero, hence we interchange rows 2 and 3 to obtain a nonzero pivot element. Thus the pivot element will be (-1). We next multiply row 2 by (-1). The resulting matrix is

$$\begin{bmatrix} 1 & 1 & 1 & 1 \\ 0 & 1 & 1 & 0 \\ 0 & 0 & -3 & 2 \end{bmatrix}.$$

Next, we create a 1 in the $(3, 3)$ entry by multiplying row 3 by $\left(-\frac{1}{3}\right)$ to obtain

$$\begin{bmatrix} 1 & 1 & 1 & 1 \\ 0 & 1 & 1 & 0 \\ 0 & 0 & 1 & -\frac{2}{3} \end{bmatrix}.$$

Now multiply row 2 by (-1) and add to row 1:

$$\begin{bmatrix} 1 & 0 & 0 & 1 \\ 0 & 1 & 1 & 0 \\ 0 & 0 & 1 & -\frac{2}{3} \end{bmatrix}.$$

Finally, multiply row 3 by (-1) and add to row 2 to obtain

$$\begin{bmatrix} 1 & 0 & 0 & 1 \\ 0 & 1 & 0 & \frac{2}{3} \\ 0 & 0 & 1 & -\frac{2}{3} \end{bmatrix}.$$

The unique solution of this system is

$$x = 1, \quad y = \tfrac{2}{3}, \quad \text{and} \quad z = -\tfrac{2}{3}.$$

(c)
$$\begin{aligned} 2x + y + z - 2w &= 1 \\ 3x - 2y + z - 6w &= -2 \\ x + y - z - w &= -1 \\ 6x \quad\;\; + z - 9w &= -2 \\ 5x - y + 2z - 8w &= 3 \end{aligned}$$

We form the augmented matrix and use row operations to form equivalent systems.

$$\begin{bmatrix} 2 & 1 & 1 & -2 & 1 \\ 3 & -2 & 1 & -6 & -2 \\ 1 & 1 & -1 & -1 & -1 \\ 6 & 0 & 1 & -9 & -2 \\ 5 & -1 & 2 & -8 & 3 \end{bmatrix}$$

In an effort to keep the arithmetic simple, we first interchange rows 1 and 3. This operation gives us a 1 in the $(1,1)$ entry to use as the first pivot. The resulting matrix is

$$\begin{bmatrix} 1 & 1 & -1 & -1 & -1 \\ 3 & -2 & 1 & -6 & -2 \\ 2 & 1 & 1 & -2 & 1 \\ 6 & 0 & 1 & -9 & -2 \\ 5 & -1 & 2 & -8 & 3 \end{bmatrix}.$$

Applying row operations

(-3) times row 1 added to row 2,
(-2) times row 1 added to row 3,
(-6) times row 1 added to row 4,
(-5) times row 1 added to row 5,

we obtain

$$\begin{bmatrix} 1 & 1 & -1 & -1 & -1 \\ 0 & -5 & 4 & -3 & 1 \\ 0 & -1 & 3 & 0 & 3 \\ 0 & -6 & 7 & -3 & 4 \\ 0 & -6 & 7 & -3 & 8 \end{bmatrix}.$$

To simplify the arithmetic and avoid fractions, we interchange rows 2 and 3, and then multiply row 2 by (-1). This puts a 1 into the $(2,2)$ entry to use as the next pivot. The resulting matrix is

$$\begin{bmatrix} 1 & 1 & -1 & -1 & -1 \\ 0 & 1 & -3 & 0 & -3 \\ 0 & -5 & 4 & -3 & 1 \\ 0 & -6 & 7 & -3 & 4 \\ 0 & -6 & 7 & -3 & 8 \end{bmatrix}.$$

Applying row operations

(5) times row 2 added to row 3,
(6) times row 2 added to row 4,
(6) times row 2 added to row 5,

we obtain

$$\begin{bmatrix} 1 & 1 & -1 & -1 & -1 \\ 0 & 1 & -3 & 0 & -3 \\ 0 & 0 & -11 & -3 & -14 \\ 0 & 0 & -11 & -3 & -14 \\ 0 & 0 & -11 & -3 & -10 \end{bmatrix}.$$

We observe that the last three equations have identical coefficients, but the entries in the augmented column are not all the same. Thus we use row operations

(-1) times row 3 added to row 4 and
(-1) times row 3 added to row 5

to obtain

$$\begin{bmatrix} 1 & 1 & -1 & -1 & -1 \\ 0 & 1 & -3 & 0 & -3 \\ 0 & 0 & -11 & -3 & -14 \\ 0 & 0 & 0 & 0 & 0 \\ 0 & 0 & 0 & 0 & 4 \end{bmatrix}.$$

The fifth row is equivalent to the equation

$$0x + 0y + 0z + 0w = 4$$

which has no solution. Thus the linear system is inconsistent.

(d) $\quad x + 2y + 3z - w = \quad 0$
$\quad 2x + \ y - \ z + w = \quad 3$
$\quad x - \ y \qquad + w = -2.$

We form the augmented matrix and use row operations to form equivalent systems.

$$\begin{bmatrix} 1 & 2 & 3 & -1 & 0 \\ 2 & 1 & -1 & 1 & 3 \\ 1 & -1 & 0 & 1 & -2 \end{bmatrix}$$

Applying row operations

(-2) times row 1 added to row 2,
(-1) times row 1 added to row 3,

we obtain

$$\begin{bmatrix} 1 & 2 & 3 & -1 & 0 \\ 0 & -3 & -7 & 3 & 3 \\ 0 & -3 & -3 & 2 & -2 \end{bmatrix}.$$

Add (-1) times row 2 to row 3 to obtain:

$$\begin{bmatrix} 1 & 2 & 3 & -1 & 0 \\ 0 & -3 & -7 & 3 & 3 \\ 0 & 0 & 4 & -1 & -5 \end{bmatrix}.$$

Add $\left(-\frac{1}{3}\right)$ times row 2 and $\left(\frac{1}{4}\right)$ times row 3 to obtain:

$$\begin{bmatrix} 1 & 2 & 3 & -1 & 0 \\ 0 & 1 & \frac{7}{3} & -1 & -1 \\ 0 & 0 & 1 & -\frac{1}{4} & -\frac{5}{4} \end{bmatrix}.$$

Add (-2) times row 2 to row 1 to obtain:

$$\begin{bmatrix} 1 & 0 & -\frac{5}{3} & 1 & 2 \\ 0 & 1 & \frac{7}{3} & -1 & -1 \\ 0 & 0 & 1 & -\frac{1}{4} & -\frac{5}{4} \end{bmatrix}.$$

Finally,

$$\left(\tfrac{5}{3}\right) \text{ times row 3 added to row 1,}$$
$$\text{and } \left(-\tfrac{7}{3}\right) \text{ times row 3 added to row 2,}$$

gives

$$\begin{bmatrix} 1 & 0 & 0 & \frac{7}{12} & -\frac{1}{12} \\ 0 & 1 & 0 & -\frac{5}{12} & \frac{23}{12} \\ 0 & 0 & 1 & -\frac{1}{4} & -\frac{5}{4} \end{bmatrix}.$$

This matrix is in reduced row echelon form and represents the equivalent linear system

$$x \qquad + \tfrac{7}{12}w = -\tfrac{1}{12}$$
$$y \quad - \tfrac{5}{12}w = \tfrac{23}{12}$$
$$z - \tfrac{1}{4}w = -\tfrac{5}{4}.$$

We solve each equation for the unknown that corresponds to a leading 1. Hence we have

$$x = -\tfrac{1}{12} - \tfrac{7}{12}w$$
$$y = \tfrac{23}{12} + \tfrac{5}{12}w$$
$$z = -\tfrac{5}{4} + \tfrac{1}{4}w.$$

The unknown w can be chosen arbitrarily. Let $w = r$, where r is any real number. Then the solution of the linear system is given by

$$x = -\tfrac{1}{12} - \tfrac{7}{12}r, \quad y = \tfrac{23}{12} + \tfrac{5}{12}r, \quad z = -\tfrac{5}{4} + \tfrac{1}{4}r, \quad w = r.$$

Since r can be any real number, there are infinitely many solutions.

In Exercises 23 and 25 we form the augmented matrix and then use row operations to reduce the system to row echelon form. We show how to choose values of a to obtain (a) no solution, (b) a unique solution, and (c) infinitely many solutions.

23. $\begin{bmatrix} 1 & 1 & -1 & \vdots & 2 \\ 1 & 2 & 1 & \vdots & 3 \\ 1 & 1 & a^2 - 5 & \vdots & a \end{bmatrix}$

Applying row operations (-1) times row 1 added to row 2 and (-1) times row 1 added to row 3, we obtain

$$\begin{bmatrix} 1 & 1 & -1 & \vdots & 2 \\ 0 & 1 & 2 & \vdots & 1 \\ 0 & 0 & a^2 - 4 & \vdots & a - 2 \end{bmatrix}.$$

(a) There will be no solution if a is chosen so that row 3 has the form

$$\begin{bmatrix} 0 & 0 & 0 & \vdots & * \end{bmatrix}$$

where $* \neq 0$. We see that $a^2 - 4 = 0$ if and only if $a = 2$ or $a = -2$. To have $a - 2 \neq 0$, a must be chosen as (-2).

(b) There will be a unique solution if a is chosen so that $a^2 - 4 \neq 0$. In this case the $(3, 3)$ entry can be used as pivot. Thus $a \neq 2$ and $a \neq -2$.

(c) There will be infinitely many solutions if the third row is a zero row. In that case variable z can be chosen arbitrarily. Thus $a = 2$.

25. $\begin{bmatrix} 1 & 1 & 1 & \vdots & 2 \\ 1 & 2 & 1 & \vdots & 3 \\ 1 & 1 & a^2 - 5 & \vdots & a \end{bmatrix}$

Applying row operations (-1) times row 1 added to row 2 and (-1) times row 1 added to row 3, we obtain

$$\begin{bmatrix} 1 & 1 & 1 & \vdots & 2 \\ 0 & 1 & 0 & \vdots & 1 \\ 0 & 0 & a^2 - 6 & \vdots & a - 2 \end{bmatrix}.$$

(a) There will be no solution if a is chosen so that row 3 has the form

$$\begin{bmatrix} 0 & 0 & 0 & \vdots & * \end{bmatrix}$$

where $* \neq 0$. We see that $a^2 - 6 = 0$ if and only if $a = \pm\sqrt{6}$. In either case $a - 2 \neq 0$.

(b) There will be a unique solution if a is chosen so that $a^2 - 6 \neq 0$. In this case the $(3, 3)$ entry can be used as pivot. Thus $a \neq \pm\sqrt{6}$.

(c) There will be infinitely many solutions if the third row is a zero row. In that case variable z can be chosen arbitrarily. There is no way to choose a so that both $a^2 - 6$ and $a - 2$ are zero.

In Exercises 27 and 29 we put the augmented matrix in reduced row echelon form and then determine its solution, if any exists.

27. (a) Applying row operation (-1) times row 1 added to row 2, we obtain

$$\begin{bmatrix} 1 & 1 & 1 & \vdots & 0 \\ 0 & 0 & -1 & \vdots & 3 \\ 0 & 1 & 1 & \vdots & 1 \end{bmatrix}.$$

Interchange rows 2 and 3 to get 1 in the $(2,2)$ position to use as a pivot. The resulting matrix is

$$\left[\begin{array}{ccc:c} 1 & 1 & 1 & 0 \\ 0 & 1 & 1 & 1 \\ 0 & 0 & -1 & 3 \end{array}\right].$$

Multiply row 3 by (-1) to obtain

$$\left[\begin{array}{ccc:c} 1 & 1 & 1 & 0 \\ 0 & 1 & 1 & 1 \\ 0 & 0 & 1 & -3 \end{array}\right].$$

Add (-1) times row 2 to row 1 to obtain:

$$\left[\begin{array}{ccc:c} 1 & 0 & 0 & -1 \\ 0 & 1 & 1 & 1 \\ 0 & 0 & 1 & -3 \end{array}\right].$$

Finally, (-1) times row 3 added to row 2 gives:

$$\left[\begin{array}{ccc:c} 1 & 0 & 0 & -1 \\ 0 & 1 & 0 & 4 \\ 0 & 0 & 1 & -3 \end{array}\right].$$

The linear system has a unique solution given by

$$x = -1, \qquad y = 4, \qquad z = -3.$$

(b) Applying row operations (-1) times row 1 added to row 2, (-1) times row 1 added to row 3, (-1) times row 1 added to row 4, and (-1) times row 1 added to row 5, we obtain

$$\left[\begin{array}{ccc:c} 1 & 2 & 3 & 0 \\ 0 & -1 & -2 & 0 \\ 0 & -1 & -1 & 0 \\ 0 & 1 & 0 & 0 \end{array}\right].$$

We multiply row 2 by (-1), then use it as a pivot row with row operations (1) times row 2 added to row 3, and (-1) times row 2 added to row 4, to obtain

$$\left[\begin{array}{ccc:c} 1 & 2 & 3 & 0 \\ 0 & 1 & 2 & 0 \\ 0 & 0 & 1 & 0 \\ 0 & 0 & -2 & 0 \end{array}\right].$$

Applying row operations (-2) times row 3 added to row 2, and (2) times row 3 added to row 4, we obtain

$$\left[\begin{array}{ccc:c} 1 & 2 & 3 & 0 \\ 0 & 1 & 0 & 0 \\ 0 & 0 & 1 & 0 \\ 0 & 0 & 0 & 0 \end{array}\right].$$

Add (-2) times row 2 to row 1 to obtain:

$$\left[\begin{array}{ccc:c} 1 & 0 & -1 & 0 \\ 0 & 1 & 2 & 0 \\ 0 & 0 & 1 & 0 \\ 0 & 0 & 0 & 0 \end{array}\right].$$

Finally, 1 times row 3 added to row 1, then (-2) times row 3 added to row 2 gives

$$\begin{bmatrix} 1 & 0 & 0 & \vdots & 0 \\ 0 & 1 & 0 & \vdots & 0 \\ 0 & 0 & 1 & \vdots & 0 \\ 0 & 0 & 0 & \vdots & 0 \end{bmatrix}.$$

Thus this homogeneous linear system has only the trivial solution: $x = 0$, $y = 0$, $z = 0$.

29. (a) Applying row operations (-1) times row 1 added to row 2 and (-1) times row 1 added to row 3, we obtain

$$\begin{bmatrix} 1 & 2 & 3 & 1 & \vdots & 8 \\ 0 & 1 & -3 & 0 & \vdots & -1 \\ 0 & -2 & -1 & 0 & \vdots & -5 \end{bmatrix}.$$

Applying row operation (2) times row 2 added to row 3, we obtain

$$\begin{bmatrix} 1 & 2 & 3 & 1 & \vdots & 8 \\ 0 & 1 & -3 & 0 & \vdots & -1 \\ 0 & 0 & -7 & 0 & \vdots & -7 \end{bmatrix}.$$

Multiply row 3 by $\left(-\frac{1}{7}\right)$. The resulting matrix is

$$\begin{bmatrix} 1 & 2 & 3 & 1 & \vdots & 8 \\ 0 & 1 & -3 & 0 & \vdots & -1 \\ 0 & 0 & 1 & 0 & \vdots & 1 \end{bmatrix}.$$

Next we use row operation (-2) times row 2 added to row 1 to obtain

$$\begin{bmatrix} 1 & 0 & 9 & 1 & \vdots & 10 \\ 0 & 1 & -3 & 0 & \vdots & -1 \\ 0 & 0 & 1 & 0 & \vdots & 1 \end{bmatrix}.$$

Finally, add (-9) times row 3 to row 1, and 3 times row 3 to row 2 gives

$$\begin{bmatrix} 1 & 0 & 0 & 1 & \vdots & 1 \\ 0 & 1 & 0 & 0 & \vdots & 2 \\ 0 & 0 & 1 & 0 & \vdots & 1 \end{bmatrix}.$$

This matrix represents the linear system

$$\begin{aligned} x \qquad\quad + w &= 1 \\ y \qquad\quad &= 2 \\ z \qquad\quad &= 1. \end{aligned}$$

Thus we have $x = 1 - w$, $y = 2$, and $z = 1$. Let $w = r$, where r is any real number. Then the infinitely many solutions of this linear system have the form

$$x = 1 - r, \qquad y = 2, \qquad z = 1, \qquad w = r.$$

(b) Applying row operations (-2) times row 1 added to row 2, (-3) times row 1 added to row 3, and (-3) times row 1 added to row 4, we obtain

$$\begin{bmatrix} 1 & -2 & 3 & \vdots & 4 \\ 0 & 3 & -9 & \vdots & -3 \\ 0 & 6 & -8 & \vdots & -10 \\ 0 & 3 & -9 & \vdots & -5 \end{bmatrix}.$$

Multiply row 2 by $\left(\frac{1}{3}\right)$. The resulting matrix is

$$\begin{bmatrix} 1 & -2 & 3 & \vdots & 4 \\ 0 & 1 & -3 & \vdots & -1 \\ 0 & 6 & -8 & \vdots & -10 \\ 0 & 3 & -9 & \vdots & -5 \end{bmatrix}.$$

Applying row operations (-6) times row 2 added to row 3 and (-3) times row 2 added to row 4, we obtain

$$\begin{bmatrix} 1 & -2 & 3 & \vdots & 4 \\ 0 & 1 & -3 & \vdots & -1 \\ 0 & 0 & 10 & \vdots & -4 \\ 0 & 0 & 0 & \vdots & -2 \end{bmatrix}.$$

We see that row 4 is in the form

$$\begin{bmatrix} 0 & 0 & 0 & \vdots & * \end{bmatrix}$$

where $*$ is not zero. This indicates that the corresponding equation has no solution. Thus this system has no solution.

31. The equation

$$f\left(\begin{bmatrix} x \\ y \\ z \end{bmatrix}\right) = \begin{bmatrix} 4 & 1 & 3 \\ 2 & -1 & 3 \\ 2 & 2 & 0 \end{bmatrix} \begin{bmatrix} x \\ y \\ z \end{bmatrix} = \begin{bmatrix} 4x + y + 3z \\ 2x - y + 3z \\ 2x + 2y \end{bmatrix} = \begin{bmatrix} 4 \\ 5 \\ -1 \end{bmatrix}$$

leads to the linear system

$$\begin{aligned} 4x + y + 3z &= 4 \\ 2x - y + 3z &= 5 \\ 2x + 2y &= -1 \end{aligned}$$

whose augmented matrix is

$$\begin{bmatrix} 4 & 1 & 3 & \vdots & 4 \\ 2 & -1 & 3 & \vdots & 5 \\ 2 & 2 & 0 & \vdots & -1 \end{bmatrix}.$$

To find the reduced row echelon matrix that is equivalent to the augmented matrix first add rows 1 and row 2 to obtain

$$\begin{bmatrix} 6 & 0 & 6 & \vdots & 9 \\ 2 & -1 & 3 & \vdots & 5 \\ 2 & 2 & 0 & \vdots & -1 \end{bmatrix}.$$

Next, multiply row 1 by $\left(\frac{1}{6}\right)$

$$\begin{bmatrix} 1 & 0 & 1 & \vdots & \frac{3}{2} \\ 2 & -1 & 3 & \vdots & 5 \\ 2 & 2 & 0 & \vdots & -1 \end{bmatrix}.$$

Then add (-2) times row 1 to row 2 and (-2) times row 1 to row 3. The resulting matrix is

$$\begin{bmatrix} 1 & 0 & 1 & \vdots & \frac{3}{2} \\ 0 & -1 & 1 & \vdots & 2 \\ 0 & 2 & -2 & \vdots & -4 \end{bmatrix}.$$

Finally, add (2) times row 2 to row 3 and then multiply (-1) times row 2 to obtain

$$\begin{bmatrix} 1 & 0 & 1 & \vdots & \frac{3}{2} \\ 0 & 1 & -1 & \vdots & -2 \\ 0 & 0 & 0 & \vdots & 0 \end{bmatrix}.$$

The linear system represented by this matrix is

$$x \quad + z = \quad \tfrac{3}{2}$$
$$y - z = -2.$$

Solving each equation for the unknown that corresponds to the leading entry in each row we have

$$x = \tfrac{3}{2} - z$$
$$y = -2 + z.$$

If $z = t =$ any real number, the solution is $x = \tfrac{3}{2} - t$, $y = -2 + t$, $z = t$.

33. The equation

$$f\left(\begin{bmatrix} x \\ y \\ z \end{bmatrix}\right) = \begin{bmatrix} 4 & 1 & 3 \\ 2 & -1 & 3 \\ 2 & 2 & 0 \end{bmatrix} \begin{bmatrix} x \\ y \\ z \end{bmatrix} = \begin{bmatrix} 4x + y + 3z \\ 2x - y + 3z \\ 2x + 2y \end{bmatrix} = \begin{bmatrix} a \\ b \\ c \end{bmatrix}$$

leads to the linear system

$$4x + \quad y + 3z = a$$
$$2x - \quad y + 3z = b.$$
$$2x + 2y \quad \quad = c$$

The augmented matrix of this linear system is

$$\begin{bmatrix} 4 & 1 & 3 & \vdots & a \\ 2 & -1 & 3 & \vdots & b \\ 2 & 2 & 0 & \vdots & c \end{bmatrix}.$$

Use the elementary row operations described in Exercise 31 to obtain the matrix that is row equivalent.

$$\begin{bmatrix} 1 & 0 & 1 & & \frac{a+b}{6} \\ 0 & 1 & -1 & \vdots & \frac{a-2b}{3} \\ 0 & 0 & 0 & \vdots & -a+b+c \end{bmatrix}.$$

This system will only have a solution if $-a + b + c = 0$.

35. Solve the first linear system separately.

$$A\mathbf{x} = \begin{bmatrix} 1 & -1 \\ 2 & 3 \end{bmatrix} \begin{bmatrix} x \\ y \end{bmatrix} = \begin{bmatrix} 1 \\ -8 \end{bmatrix}.$$

The augmented matrix of this linear system is

$$\begin{bmatrix} 1 & -1 & \vdots & 1 \\ 2 & 3 & \vdots & -8 \end{bmatrix}.$$

Using elementary row operations we find the augmented matrix is row equivalent to the matrix

$$\begin{bmatrix} 1 & 0 & \vdots & -1 \\ 0 & 1 & \vdots & -2 \end{bmatrix}.$$

Thus $x = -1$ and $y = -2$ is the solution.

Now solve the second linear system

$$A\mathbf{x} = \begin{bmatrix} 1 & -1 \\ 2 & 3 \end{bmatrix} \begin{bmatrix} x \\ y \end{bmatrix} = \begin{bmatrix} 5 \\ -5 \end{bmatrix}.$$

The augmented matrix of this linear system is

$$\begin{bmatrix} 1 & -1 & \vdots & 5 \\ 2 & 3 & \vdots & -5 \end{bmatrix}.$$

The reduced row echelon form of the matrix equivalent to the augmented matrix is

$$\begin{bmatrix} 1 & 0 & \vdots & 2 \\ 0 & 1 & \vdots & -3 \end{bmatrix}.$$

Thus $x = 2$ and $y = -3$.

Finally, we form the augmented matrix

$$[A \mathrel{\vdots} \mathbf{b}_1 \mathrel{\vdots} \mathbf{b}_2] = \begin{bmatrix} 1 & -1 & \vdots & 1 & \vdots & 5 \\ 2 & 3 & \vdots & -8 & \vdots & -5 \end{bmatrix}.$$

The reduced row echelon form of this matrix is

$$\begin{bmatrix} 1 & 0 & \vdots & -1 & \vdots & 2 \\ 0 & 1 & \vdots & -2 & \vdots & -3 \end{bmatrix}$$

which corresponds to the two linear systems

$$\begin{bmatrix} 1 & 0 \\ 0 & 1 \end{bmatrix} \begin{bmatrix} x \\ y \end{bmatrix} = \begin{bmatrix} -1 \\ -2 \end{bmatrix} \quad \text{and} \quad \begin{bmatrix} 1 & 0 \\ 0 & 1 \end{bmatrix} \begin{bmatrix} x \\ y \end{bmatrix} = \begin{bmatrix} 2 \\ -3 \end{bmatrix}$$

solved above and so have the same solutions as before.

37. The homogeneous system $(-4I_3 - A)\mathbf{x} = \mathbf{0}$ has augmented matrix

$$\begin{bmatrix} -5 & 0 & -5 & \vdots & 0 \\ -1 & -5 & -1 & \vdots & 0 \\ 0 & -1 & 0 & \vdots & 0 \end{bmatrix}.$$

Multiplying row 1 by $\left(-\frac{1}{5}\right)$ we obtain the matrix

$$\begin{bmatrix} 1 & 0 & 1 & \vdots & 0 \\ -1 & -5 & -1 & \vdots & 0 \\ 0 & -1 & 0 & \vdots & 0 \end{bmatrix}.$$

Applying row operation (1) times row 1 added to row 2, we obtain

$$\begin{bmatrix} 1 & 0 & 1 & \vdots & 0 \\ 0 & -5 & 0 & \vdots & 0 \\ 0 & -1 & 0 & \vdots & 0 \end{bmatrix}.$$

Add $\left(-\frac{1}{5}\right)$ times row 2 to row 3 then $\left(-\frac{1}{5}\right)$ times row 2 to obtain

$$\begin{bmatrix} 1 & 0 & 1 & \vdots & 0 \\ 0 & 1 & 0 & \vdots & 0 \\ 0 & 0 & 0 & \vdots & 0 \end{bmatrix}.$$

The corresponding linear system is

$$\begin{aligned} x \quad + z &= 0 \\ y \quad &= 0. \end{aligned}$$

Thus $x = -z$ and $y = 0$. Let $z = r$, where r is any real number. Then the infinitely many nontrivial solutions of this linear system have the form

$$x = -r, \qquad y = 0, \qquad z = r,$$

where $r \neq 0$.

39. The augmented matrix of the given linear system is

$$\begin{bmatrix} 1 & 2 & -3 & \vdots & a \\ 2 & 3 & 3 & \vdots & b \\ 5 & 9 & -6 & \vdots & c \end{bmatrix}.$$

To solve this system we find the reduced row echelon form of this matrix, obtaining:

$$\begin{bmatrix} 1 & 2 & -3 & \vdots & a \\ 0 & 1 & -9 & \vdots & 2a - b \\ 0 & 0 & 0 & \vdots & c - b - 3a \end{bmatrix}.$$

Therefore, to be consistent we must have $c - b - 3a = 0$.

41. Since $A\mathbf{x} = 4\mathbf{x}$, we have

$$4\mathbf{x} - A\mathbf{x} = 4I_2\mathbf{x} - A\mathbf{x} = (4I_2 - A)\mathbf{x} = \mathbf{0}.$$

Then

$$\left(4 \begin{bmatrix} 1 & 0 \\ 0 & 1 \end{bmatrix} - \begin{bmatrix} 4 & 1 \\ 0 & 2 \end{bmatrix} \right) \begin{bmatrix} x_1 \\ x_2 \end{bmatrix} = \begin{bmatrix} 0 & -1 \\ 0 & 2 \end{bmatrix} \begin{bmatrix} x_1 \\ x_2 \end{bmatrix} = \begin{bmatrix} 0 \\ 0 \end{bmatrix}.$$

Now form the augmented matrix and find its reduced row echelon form:

$$\begin{bmatrix} 0 & -1 & \vdots & 0 \\ 0 & 2 & \vdots & 0 \end{bmatrix} \quad \longrightarrow \quad \begin{bmatrix} 0 & 1 & \vdots & 0 \\ 0 & 2 & \vdots & 0 \end{bmatrix} \quad \longrightarrow \quad \begin{bmatrix} 0 & 1 & \vdots & 0 \\ 0 & 0 & \vdots & 0 \end{bmatrix}$$

$$(-1) \text{ times row 1} \qquad (-2) \text{ times row 1}$$
$$\text{added to row 2}$$

The equivalent linear system is $x_2 = 0$. Hence the solution is $x_1 = r$, $x_2 = 0$, where r is any real number. In matrix form,

$$\mathbf{x} = \begin{bmatrix} r \\ 0 \end{bmatrix},$$

where $r \neq 0$.

43. Rearrange the matrix equation $A\mathbf{x} = 3\mathbf{x}$ as follows:

$$3\mathbf{x} - A\mathbf{x} = 3I_3\mathbf{x} - A\mathbf{x} = (3I_3 - A)\mathbf{x} = \mathbf{0}.$$

Then

$$\left(3 \begin{bmatrix} 1 & 0 & 0 \\ 0 & 1 & 0 \\ 0 & 0 & 1 \end{bmatrix} - \begin{bmatrix} 1 & 2 & -1 \\ 1 & 0 & 1 \\ 4 & -4 & 5 \end{bmatrix} \right) \begin{bmatrix} x_1 \\ x_2 \\ x_3 \end{bmatrix} = \begin{bmatrix} 2 & -2 & 1 \\ -1 & 3 & -1 \\ -4 & 4 & -2 \end{bmatrix} \begin{bmatrix} x_1 \\ x_2 \\ x_3 \end{bmatrix} = \begin{bmatrix} 0 \\ 0 \\ 0 \end{bmatrix}.$$

Now form the augmented matrix and find its reduced row echelon form:

$$\begin{bmatrix} 2 & -2 & 1 & \vdots & 0 \\ -1 & 3 & -1 & \vdots & 0 \\ -4 & 4 & -2 & \vdots & 0 \end{bmatrix} \longrightarrow \begin{bmatrix} -1 & 3 & -1 & \vdots & 0 \\ 2 & -2 & 1 & \vdots & 0 \\ -4 & 4 & -2 & \vdots & 0 \end{bmatrix} \longrightarrow \begin{bmatrix} 1 & -3 & 1 & \vdots & 0 \\ 2 & -2 & 1 & \vdots & 0 \\ -4 & 4 & -2 & \vdots & 0 \end{bmatrix} \longrightarrow$$

Interchange rows 1 and 2 $\qquad$ (-1) times row 1 $\qquad$ (-2) times row 1 added to row 2
$$\qquad\qquad\qquad\qquad\qquad\qquad\qquad\qquad\qquad\qquad\qquad (4) \text{ times row 1 added to row 3}$$

$$\begin{bmatrix} 1 & -3 & 1 & \vdots & 0 \\ 0 & 4 & -1 & \vdots & 0 \\ 0 & -8 & 2 & \vdots & 0 \end{bmatrix} \longrightarrow \begin{bmatrix} 1 & -3 & 1 & \vdots & 0 \\ 0 & 1 & -\frac{1}{4} & \vdots & 0 \\ 0 & 1 & -\frac{1}{4} & \vdots & 0 \end{bmatrix} \longrightarrow \begin{bmatrix} 1 & 0 & \frac{1}{4} & \vdots & 0 \\ 0 & 1 & -\frac{1}{4} & \vdots & 0 \\ 0 & 0 & 0 & \vdots & 0 \end{bmatrix}$$

$\left(\frac{1}{4}\right)$ times row 2   (3) times row 2 added to row 1

$\left(-\frac{1}{8}\right)$ times row 3 (-1) times row 2 added to row 3

The equivalent linear system is

$$\begin{aligned} x_1 \quad + \tfrac{1}{4}x_3 &= 0 \\ x_2 - \tfrac{1}{4}x_3 &= 0. \end{aligned}$$

Hence the solution is $x_1 = -\frac{1}{4}r$, $x_2 = \frac{1}{4}r$, $x_3 = r$, where r is any real number. In matrix form,

$$\mathbf{x} = \begin{bmatrix} -\frac{1}{4}r \\ \frac{1}{4}r \\ r \end{bmatrix}, \quad r \neq 0.$$

45. The augmented matrix of this linear system is

$$\begin{bmatrix} 1 & 2 & -1 & -2 & \vdots & 2 \\ 2 & 1 & -2 & 3 & \vdots & 2 \\ 1 & 2 & 3 & 4 & \vdots & 5 \\ 4 & 5 & -4 & -1 & \vdots & 6 \end{bmatrix}.$$

The reduced row echelon form of this matrix is

$$\begin{bmatrix} 1 & 0 & 0 & \frac{25}{6} & \vdots & \frac{17}{12} \\ 0 & 1 & 0 & -\frac{7}{3} & \vdots & \frac{2}{3} \\ 0 & 0 & 1 & \frac{3}{2} & \vdots & \frac{3}{4} \\ 0 & 0 & 0 & 0 & \vdots & 0 \end{bmatrix}.$$

The equivalent linear system is

$$\begin{aligned} x \quad\quad + \tfrac{25}{6}w &= \tfrac{17}{12} \\ y \quad - \tfrac{7}{3}w &= \tfrac{2}{3} \\ z + \tfrac{3}{2}w &= \tfrac{3}{4}. \end{aligned}$$

solving each equation for the unknown that corresponds to the leading entry, we obtain

$$\begin{aligned} x &= \tfrac{17}{12} - \tfrac{25}{6}w \\ y &= \tfrac{2}{3} + \tfrac{7}{3}w \\ z &= \tfrac{3}{4} - \tfrac{3}{2}w. \end{aligned}$$

If we let $w = r$, where r is any real number, then the solution to the system is

$$\begin{aligned} x &= \tfrac{17}{12} - \tfrac{25}{6}r \\ y &= \tfrac{2}{3} + \tfrac{7}{3}r \\ z &= \tfrac{3}{4} - \tfrac{3}{2}r \\ w &= \ r. \end{aligned}$$

Let

$$\mathbf{x} = \begin{bmatrix} x \\ y \\ z \\ w \end{bmatrix}.$$

Then the solution can be expressed as

$$\mathbf{x} = \begin{bmatrix} \frac{17}{12} - \frac{25}{6}r \\ \frac{2}{3} + \frac{7}{3}r \\ \frac{3}{4} - \frac{3}{2}r \\ r \end{bmatrix} = \begin{bmatrix} \frac{17}{12} \\ \frac{2}{3} \\ \frac{3}{4} \\ 0 \end{bmatrix} + \begin{bmatrix} -\frac{25}{6}r \\ \frac{7}{3}r \\ -\frac{3}{2}r \\ r \end{bmatrix}.$$

Let

$$\mathbf{x}_p = \begin{bmatrix} \frac{17}{12} \\ \frac{2}{3} \\ \frac{3}{4} \\ 0 \end{bmatrix} \quad \text{and} \quad \mathbf{x}_h = \begin{bmatrix} -\frac{25}{6}r \\ \frac{7}{3}r \\ -\frac{3}{2}r \\ r \end{bmatrix}.$$

Then $\mathbf{x} = \mathbf{x}_p + \mathbf{x}_h$, where $\mathbf{x}_p$ is a particular solution to the given system and $\mathbf{x}_h$ is a solution to the associated homogeneous system.

47. Substitute the given points in the quadratic polynomial $y = a_2x^2 + a_1x + a_0$ to obtain the linear system

$$\begin{aligned} a_2 + \ a_1 + a_0 &= 2 \\ 9a_2 + 3a_1 + a_0 &= 3 \\ 25a_2 + 5a_1 + a_0 &= 8. \end{aligned}$$

Now form the augmented matrix and find its reduced row echelon form:

$$\begin{bmatrix} 1 & 1 & 1 & \vdots & 2 \\ 9 & 3 & 1 & \vdots & 3 \\ 25 & 5 & 1 & \vdots & 8 \end{bmatrix} \longrightarrow \begin{bmatrix} 1 & 1 & 1 & \vdots & 2 \\ 0 & -6 & -8 & \vdots & -15 \\ 0 & -20 & -24 & \vdots & -42 \end{bmatrix} \longrightarrow$$

(-9) times row 1 added to row 2 $\qquad$ $\left(-\frac{1}{6}\right)$ times row 2
(-25) times row 1 added to row 3 $\qquad$ $\left(-\frac{1}{20}\right)$ times row 3

$$\begin{bmatrix} 1 & 1 & 1 & \vdots & 2 \\ 0 & 1 & \frac{4}{3} & \vdots & \frac{5}{2} \\ 0 & 1 & \frac{6}{5} & \vdots & \frac{21}{10} \end{bmatrix} \longrightarrow \begin{bmatrix} 1 & 1 & 1 & \vdots & 2 \\ 0 & 1 & \frac{4}{3} & \vdots & \frac{5}{2} \\ 0 & 0 & -\frac{2}{15} & \vdots & -\frac{2}{5} \end{bmatrix} \longrightarrow$$

(-1) times row 2 added to row 3 $\qquad$ $\left(-\frac{15}{2}\right)$ times row 3
$\qquad\qquad\qquad\qquad\qquad\qquad$ (-1) times row 2 added to row 1

$$\begin{bmatrix} 1 & 0 & -\frac{1}{3} & \vdots & -\frac{1}{2} \\ 0 & 1 & \frac{4}{3} & \vdots & \frac{5}{2} \\ 0 & 0 & 1 & \vdots & 3 \end{bmatrix} \longrightarrow \begin{bmatrix} 1 & 0 & 0 & \vdots & \frac{1}{2} \\ 0 & 1 & 0 & \vdots & -\frac{3}{2} \\ 0 & 0 & 1 & \vdots & 3 \end{bmatrix}$$

$\left(\frac{1}{3}\right)$ times row 3 added to row 1

$\left(-\frac{4}{3}\right)$ times row 3 added to row 2

Hence the solution of the linear system is $a_2 = \frac{1}{2}$, $a_1 = -\frac{3}{2}$, $a_0 = 3$ and the quadratic interpolating polynomial is $y = \frac{1}{2}x^2 - \frac{3}{2}x + 3$.

49. Substitute the given points in the cubic polynomial $y = a_3x^3 + a_2x^2 + a_1x + a_0$ to obtain the linear system

$$\begin{aligned} -a_3 + a_2 - a_1 + a_0 &= -6 \\ a_3 + a_2 + a_1 + a_0 &= 0 \\ 8a_3 + 4a_2 + 2a_1 + a_0 &= 8 \\ 27a_3 + 9a_2 + 3a_1 + a_0 &= 34. \end{aligned}$$

Now form the augmented matrix and find its reduced row echelon form:

$$\begin{bmatrix} -1 & 1 & -1 & 1 & \vdots & -6 \\ 1 & 1 & 1 & 1 & \vdots & 0 \\ 8 & 4 & 2 & 1 & \vdots & 8 \\ 27 & 9 & 3 & 1 & \vdots & 34 \end{bmatrix} \longrightarrow \begin{bmatrix} 1 & 1 & 1 & 1 & \vdots & 0 \\ -1 & 1 & -1 & 1 & \vdots & -6 \\ 8 & 4 & 2 & 1 & \vdots & 8 \\ 27 & 9 & 3 & 1 & \vdots & 34 \end{bmatrix} \longrightarrow$$

interchange rows 1 and 2

row 1 is added to row 2

(-8) times row 1 added to row 3

(-27) times row 1 added to row 4

$$\begin{bmatrix} 1 & 1 & 1 & 1 & \vdots & 0 \\ 0 & 2 & 0 & 2 & \vdots & -6 \\ 0 & -4 & -6 & -7 & \vdots & 8 \\ 0 & -18 & -24 & -26 & \vdots & 34 \end{bmatrix} \longrightarrow \begin{bmatrix} 1 & 1 & 1 & 1 & \vdots & 0 \\ 0 & 1 & 0 & 1 & \vdots & -3 \\ 0 & -4 & -6 & -7 & \vdots & 8 \\ 0 & -18 & -24 & -26 & \vdots & 34 \end{bmatrix} \longrightarrow$$

$\left(\frac{1}{2}\right)$ times row 2

(4) times row 2 added to row 3

(18) times row 2 added to row 4

(-1) times row 2 added to row 1

$$\begin{bmatrix} 1 & 0 & 1 & 0 & \vdots & 3 \\ 0 & 1 & 0 & 1 & \vdots & -3 \\ 0 & 0 & -6 & -3 & \vdots & -4 \\ 0 & 0 & -24 & -8 & \vdots & -20 \end{bmatrix} \longrightarrow \begin{bmatrix} 1 & 0 & 1 & 0 & \vdots & 3 \\ 0 & 1 & 0 & 1 & \vdots & -3 \\ 0 & 0 & 1 & \frac{1}{2} & \vdots & \frac{2}{3} \\ 0 & 0 & -24 & -8 & \vdots & -20 \end{bmatrix} \longrightarrow$$

$\left(-\frac{1}{6}\right)$ times row 3

(-1) times row 3 added to row 1

(24) times row 3 added to row 4

$$\begin{bmatrix} 1 & 0 & 0 & -\frac{1}{2} & \vdots & \frac{7}{3} \\ 0 & 1 & 0 & 1 & \vdots & -3 \\ 0 & 0 & 1 & \frac{1}{2} & \vdots & \frac{2}{3} \\ 0 & 0 & 0 & 4 & \vdots & -4 \end{bmatrix} \longrightarrow \begin{bmatrix} 1 & 0 & 0 & -\frac{1}{2} & \vdots & \frac{7}{3} \\ 0 & 1 & 0 & 1 & \vdots & -3 \\ 0 & 0 & 1 & \frac{1}{2} & \vdots & \frac{2}{3} \\ 0 & 0 & 0 & 1 & \vdots & -1 \end{bmatrix} \longrightarrow$$

$\left(\frac{1}{4}\right)$ times row 4 $\left(\frac{1}{2}\right)$ times row 4 added to row 1

(-1) times row 4 added to row 2

$\left(-\frac{1}{2}\right)$ times row 4 added to row 3

$$\longrightarrow \begin{bmatrix} 1 & 0 & 0 & 0 & \vdots & \frac{11}{6} \\ 0 & 1 & 0 & 0 & \vdots & -2 \\ 0 & 0 & 1 & 0 & \vdots & \frac{7}{6} \\ 0 & 0 & 0 & 1 & \vdots & -1 \end{bmatrix}$$

Hence the solution of the linear system is $a_3 = \frac{11}{6}$, $a_2 = -2$, $a_1 = \frac{7}{6}$, $a_0 = -1$ and the cubic interpolating polynomial is $y = \frac{11}{6}x^3 - 2x^2 + \frac{7}{6}x - 1$.

51. Let

$$x_1 = \text{the number of chairs made per week}$$
$$x_2 = \text{the number of coffee tables made per week}$$
$$x_3 = \text{the number of dining room tables made per week.}$$

We convert all the times to minutes and develop an equation for use of the sanding bench, the staining bench, and the varnishing bench. The equations are, respectively:

$$10x_1 + 12x_2 + 15x_3 = 960$$
$$6x_1 + 8x_2 + 12x_3 = 660$$
$$12x_1 + 12x_2 + 18x_3 = 1080.$$

We now form the augmented matrix and use row operations to obtain the reduced row echelon form.

$$\begin{bmatrix} 10 & 12 & 15 & \vdots & 960 \\ 6 & 8 & 12 & \vdots & 660 \\ 12 & 12 & 18 & \vdots & 1080 \end{bmatrix}$$

Multiply the first row by $\left(\frac{1}{10}\right)$ to put a 1 in the $(1,1)$ position to use as a pivot. Writing the resulting matrix in decimal form, we have

$$\begin{bmatrix} 1 & 1.2 & 1.5 & \vdots & 96 \\ 6 & 8 & 12 & \vdots & 660 \\ 12 & 12 & 18 & \vdots & 1080 \end{bmatrix}.$$

Applying row operations (-6) times row 1 added to row 2 and (-12) times row 1 added to row 3, we obtain

$$\begin{bmatrix} 1 & 1.2 & 1.5 & \vdots & 96 \\ 0 & .8 & 3 & \vdots & 84 \\ 0 & -2.4 & 0 & \vdots & -72 \end{bmatrix}.$$

Multiply row 2 by $\left(\frac{1}{0.8}\right)$ to put a 1 in the $(2,2)$ position to use as a pivot. The resulting matrix is

$$\begin{bmatrix} 1 & 1.2 & 1.5 & \vdots & 96 \\ 0 & 1 & 3.75 & \vdots & 105 \\ 0 & -2.4 & 0 & \vdots & -72 \end{bmatrix}.$$

Applying row operation (2.4) times row 2 added to row 3, we obtain

$$\begin{bmatrix} 1 & 1.2 & 1.5 & \vdots & 96 \\ 0 & 1 & 3.75 & \vdots & 105 \\ 0 & 0 & 9 & \vdots & 180 \end{bmatrix}.$$

Multiply row 3 by $\left(\frac{1}{9}\right)$ to put a 1 in the $(3,3)$ position. The resulting matrix is

$$\begin{bmatrix} 1 & 1.2 & 1.5 & \vdots & 96 \\ 0 & 1 & 3.75 & \vdots & 105 \\ 0 & 0 & 1 & \vdots & 20 \end{bmatrix}.$$

Applying row operations (-1.5) times row 3 added to row 1 and (-3.75) times row 3 added to row 2, we obtain

$$\begin{bmatrix} 1 & 1.2 & 0 & \vdots & 66 \\ 0 & 1 & 0 & \vdots & 30 \\ 0 & 0 & 1 & \vdots & 20 \end{bmatrix}.$$

Finally, (-1.2) times row 2 added to row 1 gives

$$\begin{bmatrix} 1 & 0 & 0 & \vdots & 30 \\ 0 & 1 & 0 & \vdots & 30 \\ 0 & 0 & 1 & \vdots & 20 \end{bmatrix}.$$

Thus we have $x_1 = 30$, $x_2 = 30$, and $x_3 = 20$.

53. Given the quadratic polynomial
$$p(x) = ax^2 + bx + c,$$
First compute $p(0)$, $p'(0)$, and $p''(0)$:

$$\begin{aligned} p(x) = ax^2 + bx + c &\implies p(0) = c \\ p'(x) = 2ax + b &\implies p'(0) = b \\ p''(x) = 2a &\implies p''(0) = 2a \end{aligned}$$

For the function $f(x) = e^{2x}$, we have $f'(x) = 2e^{2x}$ and $f''(x) = 4e^{2x}$. Hence

$$\begin{aligned} p(0) = f(0) &\implies c = e^0 = 1 \\ p'(0) = f'(0) &\implies b = e^{2x}(2)\big|_{x=0} = 2 \\ p''(0) = f''(0) &\implies 2a = e^{2x}(4)\big|_{x=0} = 4 \end{aligned}$$

Therefore $a = 2$, $b = 2$, $c = 1$, and hence $p(x) = 2x^2 + 2x + 1$.

55. Using the averaging rule, we have:

$$\begin{aligned} T_1 = \frac{50 + 30 + T_2 + T_3}{4} &\implies 4T_1 - T_2 - T_3 = 80 \\ T_2 = \frac{30 + 50 + T_4 + T_1}{4} &\implies -T_1 + 4T_2 - T_4 = 80 \\ T_3 = \frac{0 + 50 + T_1 + T_4}{4} &\implies -T_1 + 4T_3 - T_4 = 50 \\ T_4 = \frac{0 + T_3 + T_2 + 50}{4} &\implies -T_2 - T_3 + 4T_4 = 50 \end{aligned}$$

The augmented matrix of this linear system is

$$\begin{bmatrix} 4 & -1 & -1 & 0 & \vdots & 80 \\ -1 & 4 & 0 & -1 & \vdots & 80 \\ -1 & 0 & 4 & -1 & \vdots & 50 \\ 0 & -1 & -1 & 4 & \vdots & 50 \end{bmatrix}$$

and the reduced echelon form is

$$\begin{bmatrix} 1 & 0 & 0 & 0 & \vdots & 36.25 \\ 0 & 1 & 0 & 0 & \vdots & 36.25 \\ 0 & 0 & 1 & 0 & \vdots & 28.75 \\ 0 & 0 & 0 & 1 & \vdots & 28.75 \end{bmatrix}.$$

Therefore $T_1 = 36.25°$, $T_2 = 36.25°$, $T_3 = 28.75°$, and $T_4 = 28.75°$.

57. (a) The augmented matrix is

$$\begin{bmatrix} 1 & 1 & 0 & 1 & \vdots & 0 \\ 1 & 0 & 1 & 1 & \vdots & 1 \\ 0 & 1 & 1 & 1 & \vdots & 1 \end{bmatrix}.$$

Add row 1 to row 2:

$$\begin{bmatrix} 1 & 1 & 0 & 1 & \vdots & 0 \\ 0 & 1 & 1 & 0 & \vdots & 1 \\ 0 & 1 & 1 & 1 & \vdots & 1 \end{bmatrix}.$$

Add row 2 to row 1 and row 3:

$$\begin{bmatrix} 1 & 0 & 1 & 1 & \vdots & 1 \\ 0 & 1 & 1 & 0 & \vdots & 1 \\ 0 & 0 & 0 & 1 & \vdots & 0 \end{bmatrix}.$$

Add row 3 to row 1:

$$\begin{bmatrix} 1 & 0 & 1 & 0 & \vdots & 1 \\ 0 & 1 & 1 & 0 & \vdots & 1 \\ 0 & 0 & 0 & 1 & \vdots & 0 \end{bmatrix}.$$

Therefore $x = 1 - r$, $y = 1 - r$, $z = r$, $w = 0$, where r is 0 or 1. There are two possible solutions:

$$r = 0: \begin{bmatrix} 1 \\ 1 \\ 0 \\ 0 \end{bmatrix} \quad \text{and} \quad r = 1: \begin{bmatrix} 0 \\ 0 \\ 1 \\ 0 \end{bmatrix}.$$

(b) The augmented matrix is

$$\begin{bmatrix} 1 & 1 & 0 & 0 & \vdots & 0 \\ 1 & 1 & 1 & 0 & \vdots & 1 \\ 1 & 1 & 1 & 1 & \vdots & 0 \end{bmatrix}.$$

Add row 1 to row 2 and row 3:

$$\begin{bmatrix} 1 & 1 & 0 & 0 & \vdots & 0 \\ 0 & 0 & 1 & 0 & \vdots & 1 \\ 0 & 0 & 1 & 1 & \vdots & 0 \end{bmatrix}.$$

Add row 2 to row 3:

$$\begin{bmatrix} 1 & 1 & 0 & 0 & \vdots & 0 \\ 0 & 0 & 1 & 0 & \vdots & 1 \\ 0 & 0 & 0 & 1 & \vdots & 1 \end{bmatrix}.$$

Therefore $x = -r$, $y = r$, $z = 1$, $w = 1$, where r is 0 or 1. There are two solutions:

$$r = 0: \begin{bmatrix} 0 \\ 0 \\ 1 \\ 1 \end{bmatrix} \quad \text{and} \quad r = 1: \begin{bmatrix} 1 \\ 1 \\ 1 \\ 1 \end{bmatrix}.$$

59. (a) The augmented matrix is

$$\begin{bmatrix} 1 & 1 & 0 & \vdots & 0 \\ 0 & 1 & 0 & \vdots & 1 \\ 1 & 1 & 1 & \vdots & 0 \end{bmatrix}$$

and the reduced row echelon form is

$$\begin{bmatrix} 1 & 0 & 0 & \vdots & 1 \\ 0 & 1 & 0 & \vdots & 1 \\ 0 & 0 & 1 & \vdots & 0 \end{bmatrix}.$$

The solution is 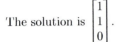 $\begin{bmatrix} 1 \\ 1 \\ 0 \end{bmatrix}$.

(b) The augmented matrix is

$$\begin{bmatrix} 1 & 1 & 0 & 1 & \vdots & 1 \\ 1 & 0 & 1 & 1 & \vdots & 0 \\ 0 & 0 & 1 & 1 & \vdots & 0 \\ 0 & 1 & 1 & 0 & \vdots & 0 \end{bmatrix}$$

and the reduced row echelon form is

$$\begin{bmatrix} 1 & 0 & 0 & 0 & \vdots & 0 \\ 0 & 1 & 0 & 1 & \vdots & 1 \\ 0 & 0 & 1 & 1 & \vdots & 0 \\ 0 & 0 & 0 & 0 & \vdots & 1 \end{bmatrix}.$$

The last row indicates that the linear system is inconsistent. There is no solution.

T.1. Suppose the leading entry of the ith row occurs in the jth column. Since leading ones of rows $i+1, i+2, \ldots$ are to the right of that of the ith row, and in any nonzero row, the leading one is the first nonzero element, all entries in the jth column below the ith row must be zero.

T.3. The sequence of elementary row operations which takes A to B, when applied to the augmented matrix $\begin{bmatrix} A & \vdots & \mathbf{0} \end{bmatrix}$, yields the augmented matrix $\begin{bmatrix} B & \vdots & \mathbf{0} \end{bmatrix}$. Thus both systems have the same solutions, by Theorem 1.7.

T.5. In order to show that if A is row equivalent to I_2 then $ad - bc \neq 0$, we argue the contrapositive, i.e., if $ad - bc = 0$, then A is *not* row equivalent to I_2.

Suppose $ad - bc = 0$, then two rows of

$$A = \begin{bmatrix} a & b \\ c & d \end{bmatrix}$$

are multiples of one another:

$$c \begin{bmatrix} a & b \end{bmatrix} = \begin{bmatrix} ac & bc \end{bmatrix} \quad \text{and} \quad a \begin{bmatrix} c & d \end{bmatrix} = \begin{bmatrix} ac & ad \end{bmatrix} \quad \text{and} \quad bc = ad.$$

Any elementary row operation applied to A will produce a matrix with rows that are multiples of each other. In particular, elementary row operations cannot produce I_2, and so I_2 is not row equivalent to A.

In order to argue the converse, we show that if $ad - bc \neq 0$, then A is row equivalent to I_2.

Suppose $ad - bc \neq 0$, then a and c are not both 0. Suppose $a \neq 0$.

$$
\begin{array}{cc|cc}
a & b & 1 & 0 \\[2mm]
c & d & 0 & 1
\end{array}
$$

Multiply the first row by $\frac{1}{a}$, and add $(-c)$ times the first row to the second row.

$$
\begin{array}{cc|cc}
1 & \dfrac{b}{a} & \dfrac{1}{a} & 0 \\[4mm]
0 & d - \dfrac{bc}{a} & -\dfrac{c}{a} & 1
\end{array}
$$

Multiply the second row by $\frac{a}{ad-bc}$.

$$
\begin{array}{cc|cc}
1 & \dfrac{b}{a} & \dfrac{1}{a} & 0 \\[4mm]
0 & 1 & \dfrac{-c}{ad-bc} & \dfrac{1}{ad-bc}
\end{array}
$$

Add $\left(-\frac{b}{a}\right)$ times the second row to the first row.

$$
\begin{array}{cc|cc}
1 & 0 & \dfrac{d}{ad-bc} & \dfrac{-b}{ad-bc} \\[4mm]
0 & 1 & \dfrac{-c}{ad-bc} & \dfrac{a}{ad-bc}
\end{array}
$$

T.7. For any angle θ, $\cos\theta$ and $\sin\theta$ are not both zero. Assume that $\cos\theta \neq 0$ and proceed as follows. The row operation $\frac{1}{\cos\theta}$ times row 1 gives

$$
\begin{bmatrix}
1 & \dfrac{\sin\theta}{\cos\theta} \\[3mm]
-\sin\theta & \cos\theta
\end{bmatrix}.
$$

Applying row operation $\sin\theta$ times row 1 added to row 2 we obtain

$$
\begin{bmatrix}
1 & \dfrac{\sin\theta}{\cos\theta} \\[3mm]
0 & \cos\theta + \dfrac{\sin^2\theta}{\cos\theta}
\end{bmatrix}.
$$

Simplifying the $(2,2)$-entry we have

$$
\cos\theta + \frac{\sin^2\theta}{\cos\theta} = \frac{\cos^2\theta + \sin^2\theta}{\cos\theta} = \frac{1}{\cos\theta}
$$

and hence our matrix is

$$
\begin{bmatrix}
1 & \dfrac{\sin\theta}{\cos\theta} \\[3mm]
0 & \dfrac{1}{\cos\theta}
\end{bmatrix}.
$$

Applying row operations $\cos\theta$ times row 2 followed by $\left(-\frac{\sin\theta}{\cos\theta}\right)$ times row 2 added to row 1 gives us I_2. Hence the reduced row echelon form is the 2×2 identity matrix. (If $\cos\theta = 0$, then interchange rows and proceed in a similar manner.)

T.9. Let A be in reduced row echelon form and assume $A \neq I_n$. Thus there is at least one row of A without a leading 1. From the definition of reduced row echelon form, this row must be a zero row.

T.11. (a) $A(\mathbf{u} + \mathbf{v}) = A\mathbf{u} + A\mathbf{v} = \mathbf{0} + \mathbf{0} = \mathbf{0}$.

(b) $A(\mathbf{u} - \mathbf{v}) = A\mathbf{u} - A\mathbf{v} = \mathbf{0} - \mathbf{0} = \mathbf{0}$.

(c) $A(r\mathbf{u}) = r(A\mathbf{u}) = r\mathbf{0} = \mathbf{0}$.

(d) $A(r\mathbf{u} + s\mathbf{v}) = r(A\mathbf{u}) + s(A\mathbf{v}) = r\mathbf{0} + s\mathbf{0} = \mathbf{0}$.

T.13. Let $\mathbf{s}$ and $\mathbf{x}_p$ both be solutions to $A\mathbf{x} = \mathbf{b}$, and let $\mathbf{s} - \mathbf{x}_p$ be the candidate for $\mathbf{x}_h$. We must argue that $\mathbf{s} - \mathbf{x}_p$ satisfies the associated homogeneous system $A\mathbf{x} = \mathbf{0}$.

$$A(\mathbf{s} - \mathbf{x}_p) = A\mathbf{s} - A\mathbf{x}_p = \mathbf{b} - \mathbf{b} = \mathbf{0}.$$

So $\mathbf{s} = \mathbf{x}_p + (\mathbf{s} - \mathbf{x}_p) = \mathbf{x}_p + \mathbf{x}_h$, a solution of the nonhomogeneous system plus a solution of the homogeneous.

ML.1. Enter A into MATLAB and use the following MATLAB commands.

(a) **A(1,:) = (1/4) * A(1,:)**
```
A =
        1.0000    0.5000    0.5000
       -3.0000    1.0000    4.0000
        1.0000         0    3.0000
        5.0000   -1.0000    5.0000
```

(b) **A(2,:) = 3 * A(1,:) + A(2,:)**
```
A =
        1.0000    0.5000    0.5000
             0    2.5000    5.5000
        1.0000         0    3.0000
        5.0000   -1.0000    5.0000
```

(c) **A(3,:) = −1 * A(1,:) + A(3,:)**
```
A =
        1.0000    0.5000    0.5000
             0    2.5000    5.5000
             0   -0.5000    2.5000
        5.0000   -1.0000    5.0000
```

(d) **A(4,:) = −5 * A(1,:) + A(4,:)**
```
A =
        1.0000    0.5000    0.5000
             0    2.5000    5.5000
             0   -0.5000    2.5000
             0   -3.5000    2.5000
```

(e) **temp = A(2,:)**
```
temp =
             0    2.5000    5.5000
```
A(2,:) = A(4,:)
```
A =
        1.0000    0.5000    0.5000
             0   -3.5000    2.5000
             0   -0.5000    2.5000
             0   -3.5000    2.5000
```

$$\mathbf{A(4,:)} = \mathbf{temp}$$
$$\mathbf{A} =$$

1.0000	0.5000	0.5000
0	−3.5000	2.5000
0	−0.5000	2.5000
0	2.5000	5.5000

ML.3. Enter A into MATLAB, then type **reduce(A)**. Use the menu to select row operations. There are many different sequences of row operations that can be used to obtain the reduced row echelon form. However, the reduced row echelon form is unique and is

ans =

1	0	0
0	1	0
0	0	1
0	0	0

ML.5. Enter the augmented matrix **aug** into MATLAB. Then use command **reduce(aug)** to construct row operations to obtain the reduced row echelon form. We obtain

ans =

1	0	0	−1	−2
0	1	0	0	−1
0	0	1	2	8

We write the equations equivalent to rows of the reduced row echelon form and use back substitution to determine the solution. The last row corresponds to the equation $z - 2w = 8$. Hence we can choose w arbitrarily, $w = r$, r any real number. Then $z = 8 + 2r$. The second row corresponds to the equation $y = -1$. The first row corresponds to the equation $x - w = -2$ hence $x = -2 + w = -2 + r$. Thus the solution is given by

$$x = -2 + r$$
$$y = -1$$
$$z = 8 - 2r$$
$$w = r.$$

ML.7. Enter the augmented matrix **aug** into MATLAB. Then use command **reduce(aug)** to construct row operations to obtain the reduced row echelon form. We obtain

ans =

1	0	0	0
0	1	0	0
0	0	1	0
0	0	0	0

It follows that this system has only the trivial solution.

ML.9. After entering A into MATLAB, use command **reduce(5 ∗ eye(size(A)) − A)**. Selecting row operations, we can show that the reduced row echelon form of $5I_2 - A$ is

$$\begin{bmatrix} 1 & -\frac{1}{2} \\ 0 & 0 \end{bmatrix}.$$

Thus the solution to the homogeneous system is

$$\mathbf{x} = \begin{bmatrix} .5r \\ r \end{bmatrix}.$$

Hence for any real number r, not zero, we obtain a nontrivial solution.

ML.11. For a linear system enter the augmented matrix **aug** and use the command **rref**. Then write out the solution

For 27(a):
rref(aug)

ans =

$$\begin{array}{cccc} 1 & 0 & 0 & -1 \\ 0 & 1 & 0 & 4 \\ 0 & 0 & 1 & -3 \end{array}$$

It follows that there is a unique solution $x = -1$, $y = 4$, $z = -3$.

For 27(b):
rref(aug)

ans =

$$\begin{array}{cccc} 1 & 0 & 0 & 0 \\ 0 & 1 & 0 & 0 \\ 0 & 0 & 1 & 0 \\ 0 & 0 & 0 & 0 \end{array}$$

It follows that the only solution is the trivial solution.

For 28(a):
rref(aug)

ans =

$$\begin{array}{cccc} 1 & 0 & -1 & 0 \\ 0 & 1 & 2 & 0 \\ 0 & 0 & 0 & 0 \end{array}$$

It follows that $x = r$, $y = -2r$, $z = r$, where r is any real number.

For 28(b):
rref(aug)

ans =

$$\begin{array}{cccc} 1 & 0 & 0 & 1 \\ 0 & 1 & 0 & 2 \\ 0 & 0 & 1 & 2 \\ 0 & 0 & 0 & 0 \\ 0 & 0 & 0 & 0 \end{array}$$

It follows that there is a unique solution $x = 1$, $y = 2$, $z = 2$.

ML.13. **A = [1 2 3;4 5 6;7 8 9];**
b = [1 0 0]′;
aug = [A b]
rref(aug) =

$$\begin{array}{cccc} 1 & 0 & -1 & 0 \\ 0 & 1 & 2 & 0 \\ 0 & 0 & 0 & 1 \end{array}$$

This augmented matrix implies that the system is inconsistent. We can also infer that the coefficient matrix is singular.

x = A\b
Warning: Matrix is close to singular or badly scaled. Results may be inaccurate.

```
RCOND=2.937385e-018.
X =
    1.0e + 015*
        3.1522
       -6.3044
        3.1522
```

Each element of the solution displayed using \ is huge. This, together with the warning, suggests that errors due to using computer arithmetic were magnified in the solution process. MATLAB uses an LU-factorization procedure when \ is used to solve linear systems (see Section 1.7), while **rref** actually rounds values before displaying them.

ML.15. Exercise 56(a) $\mathbf{A} = [\mathbf{1}\ \mathbf{1}\ \mathbf{1}\ \mathbf{0}; \mathbf{0}\ \mathbf{1}\ \mathbf{1}\ \mathbf{1}; \mathbf{1}\ \mathbf{1}\ \mathbf{0}\ \mathbf{1}]$. There are many ways to row reduce the matrix A. The final form is
$>>$ **binreduce(A)**...

```
***** ==> REDUCE is over.  Your final matrix is:
A  =
     1  0  0  0
     0  1  0  1
     0  0  1  0
```

The solution is $x = z = 0$ and $y = 1$.

Exercise 56(b) $\mathbf{A} = [\mathbf{1}\ \mathbf{1}\ \mathbf{1}\ \mathbf{1}; \mathbf{1}\ \mathbf{0}\ \mathbf{1}\ \mathbf{0}; \mathbf{0}\ \mathbf{1}\ \mathbf{1}\ \mathbf{1}]$. There are many ways to row reduce the matrix A. The final form is
$>>$ **binreduce(A)**...

```
***** ==> REDUCE is over.  Your final matrix is:
A  =
     1  0  0  1
     0  1  0  0
     0  0  1  1
```

The solution is $x = z = 1$ and $y = 0$.

Exercise 57(a) $\mathbf{A} = [\mathbf{1}\ \mathbf{1}\ \mathbf{0}\ \mathbf{1}\ \mathbf{0}; \mathbf{1}\ \mathbf{0}\ \mathbf{1}\ \mathbf{1}\ \mathbf{1}; \mathbf{0}\ \mathbf{1}\ \mathbf{1}\ \mathbf{1}\ \mathbf{1}]$. The final form of A is
$>>$ **binreduce(A)**...

```
***** ==> REDUCE is over.  Your final matrix is:
A  =
     1  0  1  0  1
     0  1  1  0  1
     0  0  0  1  0
```

The solution is $w = 0$, $z = r$, $y = 1 - r$, and $x = 1 - r$.

Exercise 57(b) $\mathbf{A} = [\mathbf{1}\ \mathbf{1}\ \mathbf{0}\ \mathbf{0}\ \mathbf{0}; \mathbf{1}\ \mathbf{1}\ \mathbf{1}\ \mathbf{0}\ \mathbf{1}; \mathbf{1}\ \mathbf{1}\ \mathbf{1}\ \mathbf{1}\ \mathbf{0}]$. The final form of A is
$>>$ **binreduce(A)**...

```
***** ==> REDUCE is over.  Your final matrix is:
A  =
     1  1  0  0  0
     0  0  1  0  1
     0  0  0  1  1
```

The solution is $w = 1$, $z = 1$, $y = r$, and $x = -r$.

Exercise 58(a) $\mathbf{A} = [\mathbf{1}\ \ \mathbf{1}\ \ \mathbf{1}\ \ \mathbf{0}\ \ \mathbf{1}; \mathbf{0}\ \ \mathbf{1}\ \ \mathbf{1}\ \ \mathbf{1}\ \ \mathbf{1}; \mathbf{1}\ \ \mathbf{0}\ \ \mathbf{0}\ \ \mathbf{1}\ \ \mathbf{1}]$. The final form of A is
>> **binreduce(A)**...

***** ==> REDUCE is over. Your final matrix is:

```
A   =
        1  0  0  1  0
        0  1  1  1  1
        0  0  0  0  1
```

The system is inconsistent.

Exercise 58(b) $\mathbf{A} = [\mathbf{1}\ \ \mathbf{1}\ \ \mathbf{1}\ \ \mathbf{0}\ \ \mathbf{0}; \mathbf{0}\ \ \mathbf{1}\ \ \mathbf{1}\ \ \mathbf{1}\ \ \mathbf{0}; \mathbf{1}\ \ \mathbf{0}\ \ \mathbf{0}\ \ \mathbf{1}\ \ \mathbf{0}]$. The final form of A is
>> **binreduce(A)**...

***** ==> REDUCE is over. Your final matrix is:

```
A   =
        1  0  0  1  0
        0  1  1  1  0
        0  0  0  0  0
```

The solution is $w = r$, $z = s$, $y = -r - s$, and $x = -r$.

Exercise 59(a) $\mathbf{A} = [\mathbf{1}\ \ \mathbf{1}\ \ \mathbf{0}\ \ \mathbf{0}; \mathbf{0}\ \ \mathbf{1}\ \ \mathbf{0}\ \ \mathbf{1}; \mathbf{1}\ \ \mathbf{1}\ \ \mathbf{1}\ \ \mathbf{0}]$. The final form of A is
>> **binreduce(A)**...

***** ==> REDUCE is over. Your final matrix is:

```
A   =
        1  0  0  1
        0  1  0  1
        0  0  1  0
```

The solution is $z = 0$, $y = 1$, and $x = 1$.

Section 1.7, p. 105

Exercises 1 and 3 use the method of Examples 2 and 3. Represent A^{-1} as a matrix of unknowns which are to be determined. From the equation $AA^{-1} = I$ construct a system of equations to determine the entries of A^{-1} by equating corresponding entries.

1. $A = \begin{bmatrix} 2 & 1 \\ -2 & 3 \end{bmatrix}$. Let $A^{-1} = \begin{bmatrix} a & b \\ c & d \end{bmatrix}$. Forming the product AA^{-1} and setting it equal to I_2, we obtain

$$\begin{bmatrix} 2a + c & 2b + d \\ -2a + 3c & -2b + 3d \end{bmatrix} = \begin{bmatrix} 1 & 0 \\ 0 & 1 \end{bmatrix}.$$

Equating corresponding entries, we obtain the two linear systems

$$\begin{array}{ll} 2a + \ c = 1 \\ -2a + 3c = 0 \end{array} \quad \text{and} \quad \begin{array}{ll} 2b + \ d = 0 \\ -2b + 3d = 1. \end{array}$$

Adding the equations in the first system gives $4c = 1$, which implies $c = \frac{1}{4}$. Substituting $c = \frac{1}{4}$ into either equation in the first system gives $a = \frac{3}{8}$. Adding the equations in the second system gives

$4d = 1$, which implies $d = \frac{1}{4}$. Then, substituting $d = \frac{1}{4}$ into either equation in the second system gives $b = -\frac{1}{8}$. Thus we have

$$\begin{bmatrix} a & b \\ c & d \end{bmatrix} = \begin{bmatrix} \frac{3}{8} & -\frac{1}{8} \\ \frac{1}{4} & \frac{1}{4} \end{bmatrix}.$$

In addition,

$$\begin{bmatrix} \frac{3}{8} & -\frac{1}{8} \\ \frac{1}{4} & \frac{1}{4} \end{bmatrix} \begin{bmatrix} 2 & 1 \\ -2 & 3 \end{bmatrix} = I_2,$$

hence we conclude that A is nonsingular and that

$$A^{-1} = \begin{bmatrix} \frac{3}{8} & -\frac{1}{8} \\ \frac{1}{4} & \frac{1}{4} \end{bmatrix}.$$

3. $A = \begin{bmatrix} 1 & 1 \\ 3 & 4 \end{bmatrix}$. Let $A^{-1} = \begin{bmatrix} a & b \\ c & d \end{bmatrix}$. Forming the product AA^{-1} and setting it equal to I_2, we obtain

$$\begin{bmatrix} a+c & b+d \\ 3a+4c & 3b+4d \end{bmatrix} = \begin{bmatrix} 1 & 0 \\ 0 & 1 \end{bmatrix}.$$

Equating corresponding entries, we obtain the two linear systems

$$\begin{array}{cc} a + c = 1 \\ 3a + 4c = 0 \end{array} \quad \text{and} \quad \begin{array}{cc} b + d = 0 \\ 3b + 4d = 1. \end{array}$$

To eliminate a in the first system, we take (-3) times the first equation and add it to the second. The result is $c = -3$. Substitute $c = -3$ into either equation in the first system to obtain $a = 4$. To eliminate b in the second system, we take (-3) times the first equation and add it to the second. The result is $d = 1$. Substituting $d = 1$ into either equation in the second system gives $b = -1$. Thus we have

$$\begin{bmatrix} a & b \\ c & d \end{bmatrix} = \begin{bmatrix} 4 & -1 \\ -3 & 1 \end{bmatrix}.$$

In addition,

$$\begin{bmatrix} 4 & -1 \\ -3 & 1 \end{bmatrix} \begin{bmatrix} 1 & 1 \\ 3 & 4 \end{bmatrix} = I_2$$

hence we conclude that A is nonsingular and that

$$A^{-1} = \begin{bmatrix} 4 & -1 \\ -3 & 1 \end{bmatrix}.$$

Exercises 5, 7, and 9 use the method of Examples 5 and 6.

5. (a) $A = \begin{bmatrix} 1 & 3 \\ -2 & 6 \end{bmatrix}$. Form the matrix

$$\begin{bmatrix} A & \vdots & I_2 \end{bmatrix} = \begin{bmatrix} 1 & 3 & \vdots & 1 & 0 \\ -2 & 6 & \vdots & 0 & 1 \end{bmatrix}.$$

Applying row operation (2) times row 1 added to row 2, we obtain

$$\begin{bmatrix} 1 & 3 & \vdots & 1 & 0 \\ 0 & 12 & \vdots & 2 & 1 \end{bmatrix}.$$

Multiply row 2 by $\left(\frac{1}{12}\right)$, then take (-3) times row 2 added to row 1. The result is the row equivalent matrices

$$\begin{bmatrix} 1 & 3 & \vdots & 1 & 0 \\ 0 & 1 & \vdots & \frac{1}{6} & \frac{1}{12} \end{bmatrix} \quad \text{and} \quad \begin{bmatrix} 1 & 0 & \vdots & \frac{1}{2} & -\frac{1}{4} \\ 0 & 1 & \vdots & \frac{1}{6} & \frac{1}{12} \end{bmatrix}.$$

Thus $A^{-1} = \begin{bmatrix} \frac{1}{2} & -\frac{1}{4} \\ \frac{1}{6} & \frac{1}{12} \end{bmatrix}$.

(b) $A = \begin{bmatrix} 1 & 2 & 3 \\ 1 & 1 & 2 \\ 0 & 1 & 2 \end{bmatrix}$. Form the matrix

$$[A \vdots I_3] = \begin{bmatrix} 1 & 2 & 3 & \vdots & 1 & 0 & 0 \\ 1 & 1 & 2 & \vdots & 0 & 1 & 0 \\ 0 & 1 & 2 & \vdots & 0 & 0 & 1 \end{bmatrix}.$$

Applying row operation (-1) times row 1 added to row 2, we obtain

$$\begin{bmatrix} 1 & 2 & 3 & \vdots & 1 & 0 & 0 \\ 0 & -1 & -1 & \vdots & -1 & 1 & 0 \\ 0 & 1 & 2 & \vdots & 0 & 0 & 1 \end{bmatrix}.$$

Multiply row 2 by (-1), then use row operations (-1) times row 2 added to row 3. The result is the row equivalent matrices

$$\begin{bmatrix} 1 & 2 & 3 & \vdots & 1 & 0 & 0 \\ 0 & 1 & 1 & \vdots & 1 & -1 & 0 \\ 0 & 1 & 2 & \vdots & 0 & 0 & 1 \end{bmatrix} \quad \text{and} \quad \begin{bmatrix} 1 & 2 & 3 & \vdots & 1 & 0 & 0 \\ 0 & 1 & 1 & \vdots & 1 & -1 & 0 \\ 0 & 0 & 1 & \vdots & -1 & 1 & 1 \end{bmatrix}.$$

Applying row operations (-3) times row 3 added to row 1 and (-1) times row 3 added to row 2, we obtain

$$\begin{bmatrix} 1 & 2 & 0 & \vdots & 4 & -3 & -3 \\ 0 & 1 & 0 & \vdots & 2 & -2 & -1 \\ 0 & 0 & 1 & \vdots & -1 & 1 & 1 \end{bmatrix}.$$

Finally, (-2) times row 2 added to row 1 gives

$$\begin{bmatrix} 1 & 0 & 0 & \vdots & 0 & 1 & -1 \\ 0 & 1 & 0 & \vdots & 2 & -2 & -1 \\ 0 & 0 & 1 & \vdots & -1 & 1 & 1 \end{bmatrix}.$$

Thus $A^{-1} = \begin{bmatrix} 0 & 1 & -1 \\ 2 & -2 & -1 \\ -1 & 1 & 1 \end{bmatrix}$.

(c) $A = \begin{bmatrix} 1 & 1 & 1 & 1 \\ 1 & 2 & -1 & 2 \\ 1 & -1 & 2 & 1 \\ 1 & 3 & 3 & 2 \end{bmatrix}$. Form the matrix

$$[A \vdots I_4] = \begin{bmatrix} 1 & 1 & 1 & 1 & \vdots & 1 & 0 & 0 & 0 \\ 1 & 2 & -1 & 2 & \vdots & 0 & 1 & 0 & 0 \\ 1 & -1 & 2 & 1 & \vdots & 0 & 0 & 1 & 0 \\ 1 & 3 & 3 & 2 & \vdots & 0 & 0 & 0 & 1 \end{bmatrix}.$$

Applying row operations (-1) times row 1 added to row j, for $j = 2, 3, 4$, we obtain

$$\left[\begin{array}{cccc:cccc} 1 & 1 & 1 & 1 & 1 & 0 & 0 & 0 \\ 0 & 1 & -2 & 1 & -1 & 1 & 0 & 0 \\ 0 & -2 & 1 & 0 & -1 & 0 & 1 & 0 \\ 0 & 2 & 2 & 1 & -1 & 0 & 0 & 1 \end{array}\right].$$

Applying row operations 2 times row 2 added to row 3 and (-2) times row 2 added to row 4, we obtain

$$\left[\begin{array}{cccc:cccc} 1 & 1 & 1 & 1 & 1 & 0 & 0 & 0 \\ 0 & 1 & -2 & 1 & -1 & 1 & 0 & 0 \\ 0 & 0 & -3 & 2 & -3 & 2 & 1 & 0 \\ 0 & 0 & 6 & -1 & 1 & -2 & 0 & 1 \end{array}\right].$$

Multiply row 3 by $\left(-\frac{1}{3}\right)$, then apply row operation (-6) times row 3 added to row 4, we obtain

$$\left[\begin{array}{cccc:cccc} 1 & 1 & 1 & 1 & 1 & 0 & 0 & 0 \\ 0 & 1 & -2 & 1 & 1 & -1 & 1 & 0 \\ 0 & 0 & 1 & -\frac{2}{3} & 1 & -\frac{2}{3} & -\frac{1}{3} & 0 \\ 0 & 0 & 0 & 3 & -5 & 2 & 2 & 1 \end{array}\right].$$

Multiply row 4 by $\left(\frac{1}{3}\right)$, and then apply row operations (-1) times row 4 added to row 1 and row 2, and $\left(\frac{2}{3}\right)$ times row 4 added to row 3. The result is

$$\left[\begin{array}{cccc:cccc} 1 & 1 & 1 & 0 & \frac{8}{3} & -\frac{2}{3} & -\frac{2}{3} & -\frac{1}{3} \\ 0 & 1 & -2 & 0 & \frac{2}{3} & \frac{1}{3} & -\frac{2}{3} & -\frac{1}{3} \\ 0 & 0 & 1 & 0 & -\frac{1}{9} & -\frac{2}{9} & \frac{1}{9} & \frac{2}{9} \\ 0 & 0 & 0 & 1 & -\frac{5}{3} & \frac{2}{3} & \frac{2}{3} & \frac{1}{3} \end{array}\right].$$

Now, (-1) times row 3 added to row 1, and then 2 times row 3 added to row 2 gives

$$\left[\begin{array}{cccc:cccc} 1 & 1 & 0 & 0 & \frac{25}{9} & -\frac{4}{9} & -\frac{5}{9} & -\frac{5}{9} \\ 0 & 1 & 0 & 0 & \frac{4}{9} & -\frac{1}{9} & -\frac{4}{9} & \frac{1}{9} \\ 0 & 0 & 1 & 0 & -\frac{1}{9} & -\frac{2}{9} & \frac{1}{9} & \frac{2}{9} \\ 0 & 0 & 0 & 1 & -\frac{5}{3} & \frac{2}{3} & \frac{2}{3} & \frac{1}{3} \end{array}\right].$$

Finally, (-1) times row 2 added to row 1 gives:

$$\left[\begin{array}{cccc:cccc} 1 & 0 & 0 & 0 & \frac{7}{3} & -\frac{1}{3} & -\frac{1}{3} & -\frac{2}{3} \\ 0 & 1 & 0 & 0 & \frac{4}{9} & -\frac{1}{9} & -\frac{4}{9} & \frac{1}{9} \\ 0 & 0 & 1 & 0 & -\frac{1}{9} & -\frac{2}{9} & \frac{1}{9} & \frac{2}{9} \\ 0 & 0 & 0 & 1 & -\frac{5}{3} & \frac{2}{3} & \frac{2}{3} & \frac{1}{3} \end{array}\right].$$

Thus $A^{-1} = \left[\begin{array}{cccc} \frac{7}{3} & -\frac{1}{3} & -\frac{1}{3} & -\frac{2}{3} \\ \frac{4}{9} & -\frac{1}{9} & -\frac{4}{9} & \frac{1}{9} \\ -\frac{1}{9} & -\frac{2}{9} & \frac{1}{9} & \frac{2}{9} \\ -\frac{5}{3} & \frac{2}{3} & \frac{2}{3} & \frac{1}{3} \end{array}\right]$

7. (a) $A = \begin{bmatrix} 1 & 3 \\ 2 & 4 \end{bmatrix}$. Form the matrix

$$[A \mid I_2] = \begin{bmatrix} 1 & 3 & \vdots & 1 & 0 \\ 2 & 4 & \vdots & 0 & 1 \end{bmatrix}.$$

Applying row operation (-2) times row 1 added to row 2, we obtain

$$\begin{bmatrix} 1 & 3 & \vdots & 1 & 0 \\ 0 & -2 & \vdots & -2 & 1 \end{bmatrix}.$$

Multiply row 2 by $\left(-\frac{1}{2}\right)$, then apply row operation (-3) times row 2 added to row 1. The result is the row equivalent matrices

$$\begin{bmatrix} 1 & 3 & \vdots & 1 & 0 \\ 0 & 1 & \vdots & 1 & -\frac{1}{2} \end{bmatrix} \quad \text{and} \quad \begin{bmatrix} 1 & 0 & \vdots & -2 & \frac{3}{2} \\ 0 & 1 & \vdots & 1 & -\frac{1}{2} \end{bmatrix}.$$

Thus $A^{-1} = \begin{bmatrix} -2 & \frac{3}{2} \\ 1 & -\frac{1}{2} \end{bmatrix}$.

(b) $A = \begin{bmatrix} 1 & 1 & 1 & 1 \\ 1 & 3 & 1 & 2 \\ 1 & 2 & -1 & 1 \\ 5 & 9 & 1 & 6 \end{bmatrix}$. Form the matrix

$$[A \mid I_4] = \begin{bmatrix} 1 & 1 & 1 & 1 & \vdots & 1 & 0 & 0 & 0 \\ 1 & 3 & 1 & 2 & \vdots & 0 & 1 & 0 & 0 \\ 1 & 2 & -1 & 1 & \vdots & 0 & 0 & 1 & 0 \\ 5 & 9 & 1 & 6 & \vdots & 0 & 0 & 0 & 1 \end{bmatrix}.$$

Applying the row operations (-1) times row 1 added to row 2, (-1) times row 1 added to row 3, and (-5) times row 1 added to row 4, we obtain

$$\begin{bmatrix} 1 & 1 & 1 & 1 & \vdots & 1 & 0 & 0 & 0 \\ 0 & 2 & 0 & 1 & \vdots & -1 & 1 & 0 & 0 \\ 0 & 1 & -2 & 0 & \vdots & -1 & 0 & 1 & 0 \\ 0 & 4 & -4 & 1 & \vdots & -5 & 0 & 0 & 1 \end{bmatrix}.$$

Interchange rows 2 and 3 to get a 1 in the $(2, 2)$ position to use as a pivot (and avoid fractions). Then apply row operations (-2) times row 2 added to row 3 and (-4) times row 2 added to row 4 to obtain

$$\begin{bmatrix} 1 & 1 & 1 & 1 & \vdots & 1 & 0 & 0 & 0 \\ 0 & 1 & -2 & 0 & \vdots & -1 & 0 & 1 & 0 \\ 0 & 0 & 4 & 1 & \vdots & 1 & 1 & -2 & 0 \\ 0 & 0 & 4 & 1 & \vdots & -1 & 0 & -4 & 1 \end{bmatrix}.$$

Applying the row operation (-1) times row 3 added to row 4 gives us a matrix with a zero row that is row equivalent to A. Thus A has no inverse. Matrix A is singular.

(c) $A = \begin{bmatrix} 1 & 2 & 1 \\ 1 & 3 & 2 \\ 1 & 0 & 1 \end{bmatrix}$. Form the matrix

$$[A \mid I_3] = \begin{bmatrix} 1 & 2 & 1 & \vdots & 1 & 0 & 0 \\ 1 & 3 & 2 & \vdots & 0 & 1 & 0 \\ 1 & 0 & 1 & \vdots & 0 & 0 & 1 \end{bmatrix}.$$

Applying row operations (-1) times row 1 added to rows 2 and 3, we obtain

$$\begin{bmatrix} 1 & 2 & 1 & \vdots & 1 & 0 & 0 \\ 0 & 1 & 1 & \vdots & -1 & 1 & 0 \\ 0 & -2 & 0 & \vdots & -1 & 0 & 1 \end{bmatrix}.$$

Applying row operation 2 times row 2 added to row 3, we obtain

$$\begin{bmatrix} 1 & 2 & 1 & \vdots & 1 & 0 & 0 \\ 0 & 1 & 1 & \vdots & -1 & 1 & 0 \\ 0 & 0 & 2 & \vdots & -3 & 2 & 1 \end{bmatrix}.$$

Multiply row 3 by $\left(\frac{1}{2}\right)$ to obtain

$$\begin{bmatrix} 1 & 2 & 1 & \vdots & 1 & 0 & 0 \\ 0 & 1 & 1 & \vdots & -1 & 1 & 0 \\ 0 & 0 & 1 & \vdots & -\frac{3}{2} & 1 & \frac{1}{2} \end{bmatrix}.$$

Now, (-1) times row 3 added to row 1 and (-1) times row 3 added to row 2 gives

$$\begin{bmatrix} 1 & 2 & 0 & \vdots & \frac{5}{2} & -1 & -\frac{1}{2} \\ 0 & 1 & 0 & \vdots & \frac{1}{2} & 0 & -\frac{1}{2} \\ 0 & 0 & 1 & \vdots & -\frac{3}{2} & 1 & \frac{1}{2} \end{bmatrix}.$$

Finally, (-2) times row 2 added to row 1 gives:

$$\begin{bmatrix} 1 & 0 & 0 & \vdots & \frac{3}{2} & -1 & \frac{1}{2} \\ 0 & 1 & 0 & \vdots & \frac{1}{2} & 0 & -\frac{1}{2} \\ 0 & 0 & 1 & \vdots & -\frac{3}{2} & 1 & \frac{1}{2} \end{bmatrix}.$$

Thus $A^{-1} = \begin{bmatrix} \frac{3}{2} & -1 & \frac{1}{2} \\ \frac{1}{2} & 0 & -\frac{1}{2} \\ -\frac{3}{2} & 1 & \frac{1}{2} \end{bmatrix}.$

9. (a) $A = \begin{bmatrix} 1 & 2 & -3 & 1 \\ -1 & 3 & -3 & -2 \\ 2 & 0 & 1 & 5 \\ 3 & 1 & -2 & 5 \end{bmatrix}$. Form the matrix

$$[A \vdots I_4] = \begin{bmatrix} 1 & 2 & -3 & 1 & \vdots & 1 & 0 & 0 & 0 \\ -1 & 3 & -3 & -2 & \vdots & 0 & 1 & 0 & 0 \\ 2 & 0 & 1 & 5 & \vdots & 0 & 0 & 1 & 0 \\ 3 & 1 & -2 & 5 & \vdots & 0 & 0 & 0 & 1 \end{bmatrix}.$$

Applying row operations 1 times row 1 added to row 2, (-2) times row 1 added to row 3, and (-3) times row 1 added to row 4, we obtain

$$\begin{bmatrix} 1 & 2 & -3 & 1 & \vdots & 1 & 0 & 0 & 0 \\ 0 & 5 & -6 & -1 & \vdots & 1 & 1 & 0 & 0 \\ 0 & -4 & 7 & 3 & \vdots & -2 & 0 & 1 & 0 \\ 0 & -5 & 7 & 2 & \vdots & -3 & 0 & 0 & 1 \end{bmatrix}.$$

In order to continue the reduction, we want to obtain a 1 to use as a pivot in the $(2,2)$ entry. Obviously, we can multiply row 2 by $\left(\frac{1}{5}\right)$ to achieve this goal. However, to avoid fractions, we observe that if we add row 3 to row 2 the $(2,2)$ entry will then contain a 1. The result is

$$\begin{bmatrix} 1 & 2 & -3 & 1 & \vdots & 1 & 0 & 0 & 0 \\ 0 & 1 & 1 & 2 & \vdots & -1 & 1 & 1 & 0 \\ 0 & -4 & 7 & 3 & \vdots & -2 & 0 & 1 & 0 \\ 0 & -5 & 7 & 2 & \vdots & -3 & 0 & 0 & 1 \end{bmatrix}.$$

Performing row operations 4 times row 2 added to row 3, and 5 times row 2 added to row 4, we obtain

$$\begin{bmatrix} 1 & 2 & -3 & 1 & \vdots & 1 & 0 & 0 & 0 \\ 0 & 1 & 1 & 2 & \vdots & -1 & 1 & 1 & 0 \\ 0 & 0 & 11 & 11 & \vdots & -6 & 4 & 5 & 0 \\ 0 & 0 & 12 & 12 & \vdots & -8 & 5 & 5 & 1 \end{bmatrix}.$$

Next we see that $\left(-\frac{12}{11}\right)$ times row 3 added to row 4 will produce zeros in the first four entries of row 4. Thus A will be row equivalent to a matrix with a zero row. Hence, A is singular.

(b) $A = \begin{bmatrix} 3 & 1 & 2 \\ 2 & 1 & 2 \\ 1 & 2 & 2 \end{bmatrix}$. Form the matrix

$$\left[A \,\vdots\, I_3\right] = \begin{bmatrix} 3 & 1 & 2 & \vdots & 1 & 0 & 0 \\ 2 & 1 & 2 & \vdots & 0 & 1 & 0 \\ 1 & 2 & 2 & \vdots & 0 & 0 & 1 \end{bmatrix}.$$

To avoid fractions and simplify the arithmetic, interchange rows 1 and 3. This puts a 1 in the $(1,1)$ entry to use as a pivot. Then apply row operations (-2) times row 1 added to row 2 and -3 times row 1 added to row 3 to obtain

$$\begin{bmatrix} 1 & 2 & 2 & \vdots & 0 & 0 & 1 \\ 2 & 1 & 2 & \vdots & 0 & 1 & 0 \\ 3 & 1 & 2 & \vdots & 1 & 0 & 0 \end{bmatrix} \quad \text{and} \quad \begin{bmatrix} 1 & 2 & 2 & \vdots & 0 & 0 & 1 \\ 0 & -3 & -2 & \vdots & 0 & 1 & -2 \\ 0 & -5 & -4 & \vdots & 1 & 0 & -3 \end{bmatrix}.$$

Multiply row 2 by $\left(-\frac{1}{3}\right)$ to get a 1 in the $(2,2)$ position:

$$\begin{bmatrix} 1 & 2 & 2 & \vdots & 0 & 0 & 1 \\ 0 & 1 & \frac{2}{3} & \vdots & 0 & -\frac{1}{3} & \frac{2}{3} \\ 0 & -5 & -4 & \vdots & 1 & 0 & -3 \end{bmatrix}.$$

Applying row operation 5 times row 2 added to row 3, we obtain

$$\begin{bmatrix} 1 & 2 & 2 & \vdots & 0 & 0 & 1 \\ 0 & 1 & \frac{2}{3} & \vdots & 0 & -\frac{1}{3} & \frac{2}{3} \\ 0 & 0 & -\frac{2}{3} & \vdots & 1 & -\frac{5}{3} & \frac{1}{3} \end{bmatrix}.$$

Add row 3 to row 2, then multiply row 3 by $\left(-\frac{3}{2}\right)$. The result is

$$\begin{bmatrix} 1 & 2 & 2 & \vdots & 0 & 0 & 1 \\ 0 & 1 & 0 & \vdots & 1 & -2 & 1 \\ 0 & 0 & 1 & \vdots & -\frac{3}{2} & \frac{5}{2} & -\frac{1}{2} \end{bmatrix}.$$

Finally, (-2) times row 2 added to row 1 gives:

$$\begin{bmatrix} 1 & 0 & 0 & \vdots & 1 & -1 & 0 \\ 0 & 1 & 0 & \vdots & 1 & -2 & 1 \\ 0 & 0 & 1 & \vdots & -\frac{3}{2} & \frac{5}{2} & -\frac{1}{2} \end{bmatrix}.$$

Thus $A^{-1} = \begin{bmatrix} 1 & -1 & 0 \\ 1 & -2 & 1 \\ -\frac{3}{2} & \frac{5}{2} & -\frac{1}{2} \end{bmatrix}.$

(c) $A = \begin{bmatrix} 1 & 2 & 3 \\ 1 & 1 & 2 \\ 1 & 1 & 0 \end{bmatrix}.$ Form the matrix

$$[A \vdots I_3] = \begin{bmatrix} 1 & 2 & 3 & \vdots & 1 & 0 & 0 \\ 1 & 1 & 2 & \vdots & 0 & 1 & 0 \\ 1 & 1 & 0 & \vdots & 0 & 0 & 1 \end{bmatrix}.$$

Applying row operations (-1) times row 1 added to rows 2 and 3, we obtain

$$\begin{bmatrix} 1 & 2 & 3 & \vdots & 1 & 0 & 0 \\ 0 & -1 & -1 & \vdots & -1 & 1 & 0 \\ 0 & -1 & -3 & \vdots & -1 & 0 & 1 \end{bmatrix}.$$

Multiply row 2 by (-1), then apply row operation 1 times row 2 added to row 3:

$$\begin{bmatrix} 1 & 2 & 3 & \vdots & 1 & 0 & 0 \\ 0 & 1 & 1 & \vdots & 1 & -1 & 0 \\ 0 & 0 & -2 & \vdots & 0 & -1 & 1 \end{bmatrix}.$$

Multiply row 3 by $\left(-\frac{1}{2}\right)$, then apply row operations (-3) times row 3 added to row 1 and (-1) times row 3 added to row 2:

$$\begin{bmatrix} 1 & 2 & 0 & \vdots & 1 & -\frac{3}{2} & \frac{3}{2} \\ 0 & 1 & 0 & \vdots & 1 & -\frac{3}{2} & \frac{1}{2} \\ 0 & 0 & 1 & \vdots & 0 & \frac{1}{2} & -\frac{1}{2} \end{bmatrix}.$$

Finally, (-2) times row 2 added to row 1 gives

$$\begin{bmatrix} 1 & 0 & 0 & \vdots & -1 & \frac{3}{2} & \frac{1}{2} \\ 0 & 1 & 0 & \vdots & 1 & -\frac{3}{2} & \frac{1}{2} \\ 0 & 0 & 1 & \vdots & 0 & \frac{1}{2} & -\frac{1}{2} \end{bmatrix}.$$

Thus $A^{-1} = \begin{bmatrix} -1 & \frac{3}{2} & \frac{1}{2} \\ 1 & -\frac{3}{2} & \frac{1}{2} \\ 0 & \frac{1}{2} & -\frac{1}{2} \end{bmatrix}.$

For Exercise 11 we use Theorem 1.13. Row reduce the coefficient matrix and determine whether it is row equivalent to I_n or not. If it is not row equivalent to I_n, then the system has a nontrivial solution.

11. (a) The linear system in matrix form is

$$Ax = \begin{bmatrix} 1 & 2 & 3 \\ 0 & 2 & 2 \\ 1 & 2 & 3 \end{bmatrix} \begin{bmatrix} x \\ y \\ z \end{bmatrix} = \mathbf{0}.$$

Applying row operation (-1) times row 1 added to row 3 of matrix A, we obtain the row equivalent matrix

$$\begin{bmatrix} 1 & 2 & 3 \\ 0 & 2 & 2 \\ 0 & 0 & 0 \end{bmatrix}.$$

Thus A is row equivalent to a matrix with a zero row. Hence A is singular and the linear system $Ax = \mathbf{0}$ has a nontrivial solution.

(b) The linear system in matrix form is

$$Ax = \begin{bmatrix} 2 & 1 & -1 \\ 1 & -2 & -3 \\ -3 & -1 & 2 \end{bmatrix} \begin{bmatrix} x \\ y \\ z \end{bmatrix} = \mathbf{0}.$$

Interchange rows 1 and 2 of A. Then apply row operations (-2) times row 1 added to row 2 and 3 times row 1 added to row 3. The result is the row equivalent matrices

$$\begin{bmatrix} 1 & -2 & -3 \\ 2 & 1 & -1 \\ -3 & -1 & 2 \end{bmatrix} \quad \text{and} \quad \begin{bmatrix} 1 & -2 & -3 \\ 0 & 5 & 5 \\ 0 & -7 & -7 \end{bmatrix}.$$

Multiply row 2 by $\left(\frac{1}{5}\right)$, then apply row operation 7 times row 2 added to row 3. The result is the matrix

$$\begin{bmatrix} 1 & -2 & -3 \\ 0 & 1 & 1 \\ 0 & 0 & 0 \end{bmatrix}.$$

Thus A is row equivalent to a matrix with a zero row. Hence A is singular. It follows that the linear system $Ax = \mathbf{0}$ has a nontrivial solution.

13. To find A, given A^{-1}, we find the inverse of A^{-1}. This follows from Theorem 1.10(a): $(A^{-1})^{-1} = A$. Using the technique from Example 5, we proceed as follows. Form the matrix

$$\begin{bmatrix} A^{-1} & \vdots & I_2 \end{bmatrix} = \begin{bmatrix} 2 & 3 & \vdots & 1 & 0 \\ 1 & 4 & \vdots & 0 & 1 \end{bmatrix}.$$

Use row operations to obtain the reduced row echelon form. Interchange rows 1 and 2, then apply the row operation (-2) times row 1 added to row 2. The result is the matrix

$$\begin{bmatrix} 1 & 4 & \vdots & 0 & 1 \\ 0 & -5 & \vdots & 1 & -2 \end{bmatrix}.$$

Multiply row 2 by $\left(-\frac{1}{5}\right)$, then apply row operation (-4) times row 2 added to row 1. The result is

$$\begin{bmatrix} 1 & 0 & \vdots & \frac{4}{5} & -\frac{3}{5} \\ 0 & 1 & \vdots & -\frac{1}{5} & \frac{2}{5} \end{bmatrix}.$$

Thus $A = \begin{bmatrix} \frac{4}{5} & -\frac{3}{5} \\ -\frac{1}{5} & \frac{2}{5} \end{bmatrix}.$

15. Let A be a matrix with a row of zeros. Then for any matrix B such that BA is defined, the product BA also has a row of zeros. (See Exercise T.3(a) in Section 1.3.) Hence there does not exist a matrix B such that $BA = I_n$, since I_n has no row of zeros. It follows that A has no inverse matrix, hence A is singular. A similar argument holds for a column of zeros using the product AB.

17. Following Example 7, the input matrix $\mathbf{x} = A^{-1}B$. We compute A^{-1} and form the product for each output matrix B. Form the matrix

$$[A \mid I_3] = \begin{bmatrix} 2 & 1 & 3 & \vdots & 1 & 0 & 0 \\ 3 & 2 & -1 & \vdots & 0 & 1 & 0 \\ 2 & 1 & 1 & \vdots & 0 & 0 & 1 \end{bmatrix}.$$

Multiply row 1 by $\left(\frac{1}{2}\right)$, then apply row operations (-3) times row 1 added to row 2 and (-2) times row 1 added to row 3. The result is the matrix

$$\begin{bmatrix} 1 & \frac{1}{2} & \frac{3}{2} & \vdots & \frac{1}{2} & 0 & 0 \\ 0 & \frac{1}{2} & -\frac{11}{2} & \vdots & -\frac{3}{2} & 1 & 0 \\ 0 & 0 & -2 & \vdots & -1 & 0 & 1 \end{bmatrix}.$$

Now multiply row 2 by 2 and row 3 by $\left(-\frac{1}{2}\right)$ to obtain

$$\begin{bmatrix} 1 & \frac{1}{2} & \frac{3}{2} & \vdots & \frac{1}{2} & 0 & 0 \\ 0 & 1 & -11 & \vdots & -3 & 2 & 0 \\ 0 & 0 & 1 & \vdots & \frac{1}{2} & 0 & -\frac{1}{2} \end{bmatrix}.$$

Add $\left(-\frac{3}{2}\right)$ times row 3 to row 1, then 11 times row 3 added to row 2 to obtain

$$\begin{bmatrix} 1 & \frac{1}{2} & 0 & \vdots & -\frac{1}{4} & 0 & \frac{3}{4} \\ 0 & 1 & 0 & \vdots & \frac{5}{2} & 2 & -\frac{11}{2} \\ 0 & 0 & 1 & \vdots & \frac{1}{2} & 0 & -\frac{1}{2} \end{bmatrix}.$$

Finally, $\left(-\frac{1}{2}\right)$ times row 2 added to row 1 gives:

$$\begin{bmatrix} 1 & 0 & 0 & \vdots & -\frac{3}{2} & -1 & \frac{7}{2} \\ 0 & 1 & 0 & \vdots & \frac{5}{2} & 2 & -\frac{11}{2} \\ 0 & 0 & 1 & \vdots & \frac{1}{2} & 0 & -\frac{1}{2} \end{bmatrix}.$$

Thus $A^{-1} = \begin{bmatrix} -\frac{3}{2} & -1 & \frac{7}{2} \\ \frac{5}{2} & 2 & -\frac{11}{2} \\ \frac{1}{2} & 0 & -\frac{1}{2} \end{bmatrix}$

(a) For output matrix $B = \begin{bmatrix} 30 \\ 20 \\ 10 \end{bmatrix}$, the input matrix is

$$\mathbf{x} = A^{-1}B = \begin{bmatrix} -30 \\ 60 \\ 10 \end{bmatrix}.$$

(b) For output matrix $B = \begin{bmatrix} 12 \\ 8 \\ 14 \end{bmatrix}$, the input matrix is

$$\mathbf{x} = A^{-1}B = \begin{bmatrix} 23 \\ -31 \\ -1 \end{bmatrix}.$$

19. Yes. $(A^{-1})^T A = (A^{-1})^T A^T = (AA^{-1})^T = I_n^T = I_n$. By Theorem 1.9, $(A^{-1})^T = A^{-1}$. That is, A^{-1} is symmetric.

21. The homogeneous system will have a nontrivial solution provided the coefficient matrix

$$A = \begin{bmatrix} \lambda - 1 & 2 \\ 2 & \lambda - 1 \end{bmatrix}$$

is singular. We row reduce A and determine values of λ so that A has a zero row. Interchange rows 1 and 2 and multiply the new row 1 by $\left(\frac{1}{2}\right)$ to obtain

$$\begin{bmatrix} 1 & \frac{\lambda-1}{2} \\ \lambda - 1 & 1 \end{bmatrix}.$$

Applying row operation $[-(\lambda - 1)]$ times row 1 added to row 2, we obtain

$$\begin{bmatrix} 1 & \frac{\lambda-1}{2} \\ 0 & \frac{-(\lambda-1)(\lambda-1)}{2} + 2 \end{bmatrix} = \begin{bmatrix} 1 & \frac{\lambda-1}{2} \\ 0 & \frac{-\lambda^2+2\lambda+3}{2} \end{bmatrix}.$$

The second row will be a zero row provided

$$-(\lambda^2 - 2\lambda - 3) = -(\lambda - 3)(\lambda + 1) = 0.$$

That is, if $\lambda = 3$ or $\lambda = -1$.

23. Form the matrix $\begin{bmatrix} D & \vdots & I_3 \end{bmatrix}$. Applying row operations $\left(\frac{1}{4}\right)$ times row 1, $\left(-\frac{1}{2}\right)$ times row 2, and $\left(\frac{1}{3}\right)$ times row 3, we obtain

$$\begin{bmatrix} 1 & 0 & 0 & \vdots & \frac{1}{4} & 0 & 0 \\ 0 & 1 & 0 & \vdots & 0 & -\frac{1}{2} & 0 \\ 0 & 0 & 1 & \vdots & 0 & 0 & \frac{1}{3} \end{bmatrix}.$$

Thus $D^{-1} = \begin{bmatrix} \frac{1}{4} & 0 & 0 \\ 0 & -\frac{1}{2} & 0 \\ 0 & 0 & \frac{1}{3} \end{bmatrix}.$

25. $\mathbf{x} = A^{-1}\mathbf{b} = \begin{bmatrix} 2 & 3 \\ 4 & 1 \end{bmatrix} \begin{bmatrix} 5 \\ 3 \end{bmatrix} = \begin{bmatrix} 19 \\ 23 \end{bmatrix}.$

27. Form the matrix

$$\begin{bmatrix} 5 & 2 & 0 & \vdots & 1 & 0 & 0 \\ 3 & 1 & 0 & \vdots & 0 & 1 & 0 \\ 0 & 0 & -4 & \vdots & 0 & 0 & 1 \end{bmatrix}.$$

Row reducing this matrix, we obtain

$$\left[\begin{array}{ccc:ccc} 1 & 0 & 0 & -1 & 0 & 0 \\ 0 & 1 & 0 & 3 & -5 & 0 \\ 0 & 0 & 1 & 0 & 0 & -\frac{1}{4} \end{array}\right].$$

Therefore

$$\left[\begin{array}{cc:c} 5 & 2 & 0 \\ 3 & 1 & 0 \\ \hdashline 0 & 0 & -4 \end{array}\right]^{-1} = \left[\begin{array}{ccc} -1 & 0 & 0 \\ 3 & -5 & 0 \\ 0 & 0 & -\frac{1}{4} \end{array}\right] = \left[\begin{array}{cc:c} -1 & 0 & 0 \\ 3 & -5 & 0 \\ \hdashline 0 & 0 & -\frac{1}{4} \end{array}\right].$$

Observe that (verify)

$$\left[\begin{array}{cc} 5 & 2 \\ 3 & 1 \end{array}\right]^{-1} = \left[\begin{array}{cc} -1 & 0 \\ 3 & -5 \end{array}\right] \quad \text{and} \quad \left[-4\right]^{-1} = \left[-\frac{1}{4}\right].$$

29. (a) Form the matrix

$$\left[A \quad I_3\right] = \left[\begin{array}{ccc:ccc} 1 & 1 & 1 & 1 & 0 & 0 \\ 1 & 1 & 0 & 0 & 1 & 0 \\ 1 & 0 & 0 & 0 & 0 & 1 \end{array}\right].$$

Add row 1 to row 2 and to row 3:

$$\left[\begin{array}{ccc:ccc} 1 & 1 & 1 & 1 & 0 & 0 \\ 0 & 0 & 1 & 1 & 1 & 0 \\ 0 & 1 & 1 & 1 & 0 & 1 \end{array}\right].$$

Interchange rows 2 and 3:

$$\left[\begin{array}{ccc:ccc} 1 & 1 & 1 & 1 & 0 & 0 \\ 0 & 1 & 1 & 1 & 0 & 1 \\ 0 & 0 & 1 & 1 & 1 & 0 \end{array}\right].$$

Add row 3 to row 2:

$$\left[\begin{array}{ccc:ccc} 1 & 1 & 0 & 0 & 1 & 0 \\ 0 & 1 & 0 & 0 & 1 & 1 \\ 0 & 0 & 1 & 1 & 1 & 0 \end{array}\right].$$

Add row 2 to row 1:

$$\left[\begin{array}{ccc:ccc} 1 & 0 & 0 & 0 & 0 & 1 \\ 0 & 1 & 0 & 0 & 1 & 1 \\ 0 & 0 & 1 & 1 & 1 & 0 \end{array}\right].$$

Therefore

$$\left[\begin{array}{ccc} 1 & 1 & 1 \\ 1 & 1 & 0 \\ 1 & 0 & 0 \end{array}\right]^{-1} = \left[\begin{array}{ccc} 0 & 0 & 1 \\ 0 & 1 & 1 \\ 1 & 1 & 0 \end{array}\right].$$

(b) Form the matrix

$$\left[\begin{array}{ccc:ccc} 1 & 0 & 1 & 1 & 0 & 0 \\ 1 & 1 & 1 & 0 & 1 & 0 \\ 0 & 1 & 0 & 0 & 0 & 1 \end{array}\right].$$

Add row 1 to row 2:

$$\left[\begin{array}{ccc:ccc} 1 & 0 & 1 & 1 & 0 & 0 \\ 0 & 1 & 0 & 1 & 1 & 0 \\ 0 & 1 & 0 & 0 & 0 & 1 \end{array}\right].$$

Add row 2 to row 3:

$$\begin{bmatrix} 1 & 0 & 1 & \vdots & 1 & 0 & 0 \\ 0 & 1 & 0 & \vdots & 1 & 1 & 0 \\ 0 & 0 & 0 & \vdots & 1 & 1 & 1 \end{bmatrix}.$$

The row of zeros indicates that the given matrix does not have an inverse. The matrix is singular.

(c) Form the matrix

$$\begin{bmatrix} 1 & 1 & 0 & 0 & \vdots & 1 & 0 & 0 & 0 \\ 1 & 1 & 1 & 0 & \vdots & 0 & 1 & 0 & 0 \\ 0 & 1 & 1 & 1 & \vdots & 0 & 0 & 1 & 0 \\ 0 & 0 & 1 & 1 & \vdots & 0 & 0 & 0 & 1 \end{bmatrix}.$$

Add row 1 to row 2:

$$\begin{bmatrix} 1 & 1 & 0 & 0 & \vdots & 1 & 0 & 0 & 0 \\ 0 & 0 & 1 & 0 & \vdots & 1 & 1 & 0 & 0 \\ 0 & 1 & 1 & 1 & \vdots & 0 & 0 & 1 & 0 \\ 0 & 0 & 1 & 1 & \vdots & 0 & 0 & 0 & 1 \end{bmatrix}.$$

Add row 3 to row 1:

$$\begin{bmatrix} 1 & 0 & 1 & 1 & \vdots & 1 & 0 & 1 & 0 \\ 0 & 0 & 1 & 0 & \vdots & 1 & 1 & 0 & 0 \\ 0 & 1 & 1 & 1 & \vdots & 0 & 0 & 1 & 0 \\ 0 & 0 & 1 & 1 & \vdots & 0 & 0 & 0 & 1 \end{bmatrix}.$$

Add row 4 to rows 1, 2, and 3:

$$\begin{bmatrix} 1 & 0 & 0 & 0 & \vdots & 1 & 0 & 1 & 1 \\ 0 & 0 & 0 & 1 & \vdots & 1 & 1 & 0 & 1 \\ 0 & 1 & 0 & 0 & \vdots & 0 & 0 & 1 & 1 \\ 0 & 0 & 1 & 1 & \vdots & 0 & 0 & 0 & 1 \end{bmatrix}.$$

Interchange rows 2 and 3, then rows 3 and 4:

$$\begin{bmatrix} 1 & 0 & 0 & 0 & \vdots & 1 & 0 & 1 & 1 \\ 0 & 1 & 0 & 0 & \vdots & 0 & 0 & 1 & 1 \\ 0 & 0 & 1 & 1 & \vdots & 0 & 0 & 0 & 1 \\ 0 & 0 & 0 & 1 & \vdots & 1 & 1 & 0 & 1 \end{bmatrix}.$$

Add row 4 to row 3:

$$\begin{bmatrix} 1 & 0 & 0 & 0 & \vdots & 1 & 0 & 1 & 1 \\ 0 & 1 & 0 & 0 & \vdots & 0 & 0 & 1 & 1 \\ 0 & 0 & 1 & 0 & \vdots & 1 & 1 & 0 & 0 \\ 0 & 0 & 0 & 1 & \vdots & 1 & 1 & 0 & 1 \end{bmatrix}.$$

Therefore

$$\begin{bmatrix} 1 & 1 & 0 & 0 \\ 1 & 1 & 1 & 0 \\ 0 & 1 & 1 & 1 \\ 0 & 0 & 1 & 1 \end{bmatrix}^{-1} = \begin{bmatrix} 1 & 0 & 1 & 1 \\ 0 & 0 & 1 & 1 \\ 1 & 1 & 0 & 0 \\ 1 & 1 & 0 & 1 \end{bmatrix}.$$

31. (a) The coefficient matrix is

$$A = \begin{bmatrix} 1 & 1 & 1 \\ 1 & 0 & 1 \\ 0 & 1 & 0 \end{bmatrix}$$

whose reduced row echelon form is

$$\begin{bmatrix} 1 & 0 & 1 \\ 0 & 1 & 0 \\ 0 & 0 & 0 \end{bmatrix}.$$

The row of zeros indicates that the matrix A is singular and it follows that the bit linear system $A\mathbf{x} = \mathbf{0}$ has a nontrivial solution.

(b) The coefficient matrix is

$$A = \begin{bmatrix} 1 & 0 & 0 \\ 1 & 1 & 1 \\ 1 & 0 & 1 \end{bmatrix}$$

whose reduced row echelon form is

$$\begin{bmatrix} 1 & 0 & 0 \\ 0 & 1 & 0 \\ 0 & 0 & 1 \end{bmatrix}.$$

Therefore A is nonsingular and the only solution is $\mathbf{x} = \mathbf{0}$, i.e., the trivial solution.

T.1. B is nonsingular, so B^{-1} exists, and

$$A = AI_n = A(BB^{-1}) = (AB)B^{-1} = OB^{-1} = O.$$

T.3. A is row equivalent to a matrix B in reduced row echelon form which, by Theorem 1.12 is not I_n. Thus B has fewer than n nonzero rows, and fewer than n unknowns corresponding to pivotal columns of B. Choose one of the free unknowns — unknowns not corresponding to pivotal columns of B. Assign any nonzero value to that unknown. This leads to a nontrivial solution to the homogeneous system $A\mathbf{x} = \mathbf{0}$.

T.5. For any angle θ, $\cos\theta$ and $\sin\theta$ are never simultaneously zero. Thus at least one element in column 1 is not zero. Assume $\cos\theta \neq 0$. (If $\cos\theta = 0$, then interchange rows 1 and 2 and proceed in a similar manner to that described below.) To show that the matrix is nonsingular and determine its inverse, we put

$$\left[\begin{array}{cc:cc} \cos\theta & \sin\theta & 1 & 0 \\ -\sin\theta & \cos\theta & 0 & 1 \end{array} \right]$$

into reduced row echelon form. Apply row operations $\frac{1}{\cos\theta}$ times row 1 and $\sin\theta$ times row 1 added to row 2 to obtain

$$\left[\begin{array}{cc:cc} 1 & \dfrac{\sin\theta}{\cos\theta} & \dfrac{1}{\cos\theta} & 0 \\ 0 & \dfrac{\sin^2\theta}{\cos\theta} + \cos\theta & \dfrac{\sin\theta}{\cos\theta} & 1 \end{array} \right].$$

Since

$$\frac{\sin^2\theta}{\cos\theta} + \cos\theta = \frac{\sin^2\theta + \cos^2\theta}{\cos\theta} = \frac{1}{\cos\theta},$$

the $(2,2)$-element is not zero. Applying row operations $\cos\theta$ times row 2 and $\left(-\frac{\sin\theta}{\cos\theta}\right)$ times row 2 added to row 1 we obtain

$$\left[\begin{array}{cc:cc} 1 & 0 & \cos\theta & -\sin\theta \\ 0 & 1 & \sin\theta & \cos\theta \end{array} \right].$$

It follows that the matrix is nonsingular and its inverse is

$$\begin{bmatrix} \cos\theta & -\sin\theta \\ \sin\theta & \cos\theta \end{bmatrix}.$$

T.7. Let **u** be one solution to $A\mathbf{x} = \mathbf{b}$. Since A is singular, the homogeneous system $A\mathbf{x} = \mathbf{0}$ has a nontrivial solution $\mathbf{u}_0$. Then for any real number r, $\mathbf{v} = r\mathbf{u}_0$ is also a solution to the homogeneous system. Finally, by Exercise T.13(a), Sec. 1.6, for each of the infinitely many matrices $\mathbf{v}$, the matrix $\mathbf{w} = \mathbf{u}_0 + \mathbf{v}$ is a solution to the nonhomogeneous system $A\mathbf{x} = \mathbf{b}$.

T.9. Let $A = \begin{bmatrix} a_{ij} \end{bmatrix}$ be a diagonal matrix with nonzero diagonal entries $a_{11}, a_{22}, \ldots, a_{nn}$. That is, $a_{ij} \neq 0$ if $i = j$ and 0 otherwise. We seek an $n \times n$ matrix $B = \begin{bmatrix} b_{ij} \end{bmatrix}$ such that $AB = I_n$. The (i, j) entry in AB is $\sum_{k=1}^{n} a_{ik} b_{kj}$, so $\sum_{k=1}^{n} a_{ik} b_{kj} = 1$ if $i = j$ and 0 otherwise. If $i \neq j$ then the equation

$$a_{i1} b_{1j} + a_{i2} b_{2j} + a_{i3} b_{3j} + \cdots + a_{in} b_{nj} = 0$$

is true only when $b_{ij} = 0$ since the conditions on the a_{ij} leave only one nonzero term, $a_{ii} b_{ij}$. This implies $b_{ij} = 0$ whenever $i \neq j$ and a similar srgument implies $b_{ii} = \frac{1}{a_{ii}}$ when $i = j$. Hence, A is nonsingular and $A^{-1} = B$.

T.11. (See Exercise T.13 in Sec. 1.2 for the list)

Nonsingular matrices are

$$\begin{bmatrix} 0 & 1 \\ 1 & 0 \end{bmatrix}, \quad \begin{bmatrix} 0 & 1 \\ 1 & 1 \end{bmatrix}, \quad \begin{bmatrix} 1 & 0 \\ 0 & 1 \end{bmatrix}, \quad \begin{bmatrix} 1 & 0 \\ 1 & 1 \end{bmatrix}, \quad \begin{bmatrix} 1 & 1 \\ 0 & 1 \end{bmatrix}, \quad \begin{bmatrix} 1 & 1 \\ 1 & 0 \end{bmatrix}$$

Their corresponding inverses are

$$\begin{bmatrix} 0 & 1 \\ 1 & 0 \end{bmatrix}, \quad \begin{bmatrix} 1 & 1 \\ 1 & 0 \end{bmatrix}, \quad \begin{bmatrix} 1 & 0 \\ 0 & 1 \end{bmatrix}, \quad \begin{bmatrix} 1 & 0 \\ 1 & 1 \end{bmatrix}, \quad \begin{bmatrix} 1 & 1 \\ 0 & 1 \end{bmatrix}, \quad \begin{bmatrix} 0 & 1 \\ 1 & 1 \end{bmatrix}.$$

T.13. (See Exercise T.13 in Sec. 1.2 for the list) Exhaustive search.

$$\begin{bmatrix} 0 & 0 \\ 0 & 0 \end{bmatrix}, \quad \begin{bmatrix} 0 & 0 \\ 1 & 0 \end{bmatrix}, \quad \begin{bmatrix} 0 & 1 \\ 0 & 0 \end{bmatrix}, \quad \begin{bmatrix} 1 & 1 \\ 1 & 1 \end{bmatrix}$$

ML.1. We use the fact that A is nonsingular if **rref(A)** is the identity matrix.

(a) **A = [1 2; − 2 1];**
rref(A)
ans =
     1   0
     0   1
Thus A is nonsingular.

(b) **A = [1 2 3;4 5 6;7 8 9];**
rref(A)
ans =
       1   0 −1
       0   1   2
       0   0   0
Thus A is singular.

(c) **A = [1 2 3;4 5 6;7 8 0];**
rref(A)
ans =
       1   0   0
       0   1   0
       0   0   1
Thus A is nonsingular.

ML.3. (a) **A = [1 3;1 2];**
rref([A eye(size(A))])   (We augment A with the appropriate size identity matrix)
ans =

$$\begin{array}{cccc} 1 & 0 & -2 & 3 \\ 0 & 1 & 1 & -1 \end{array}$$

so $A^{-1} = \begin{bmatrix} -2 & 3 \\ 1 & -1 \end{bmatrix}$.

(b) **A = [1 1 2;2 1 1;1 2 1];**
rref(A eye(size(A)))
ans =

$$\begin{array}{cccccc} 1.0000 & 0 & 0 & -0.2500 & 0.7500 & -0.2500 \\ 0 & 1.0000 & 0 & -0.2500 & -0.2500 & 0.7500 \\ 0 & 0 & 1.0000 & 0.7500 & -0.2500 & -0.2500 \end{array}$$

format rat, ans   i.e., change the format from a decimal representation to a rational fraction
ans =

$$\begin{array}{cccccc} 1 & 0 & 0 & -1/4 & 3/4 & -1/4 \\ 0 & 1 & 0 & -1/4 & -1/4 & 3/4 \\ 0 & 0 & 1 & 3/4 & -1/4 & -1/4 \end{array}$$

format
so A^{-1} with rational entries is

$$\begin{bmatrix} -1/4 & 3/4 & -1/4 \\ -1/4 & -1/4 & 3/4 \\ -3/4 & -1/4 & -1/4 \end{bmatrix}.$$

ML.5. We experiment choosing successive values of t then computing the **rref** of

$$(t * \mathbf{eye(size(A))}) - \mathbf{A}.$$

(a) **A = [1 3;3 1];**
t = 1; rref(t * eye(size(A)) − A)
(Use the up arrow key to recall and then revise it for use below.)
ans =

$$\begin{array}{cc} 1 & 0 \\ 0 & 1 \end{array}$$

t = 2;rref(t * eye(size(A)) − A)
ans =

$$\begin{array}{cc} 1 & 0 \\ 0 & 1 \end{array}$$

t = 3;rref(t * eye(size(A)) − A)
ans =

$$\begin{array}{cc} 1 & 0 \\ 0 & 1 \end{array}$$

t = 4;rref(t * eye(size(A)) − A)
ans =

$$\begin{array}{cc} 1 & -1 \\ 0 & 0 \end{array}$$

Thus $t = 4$.

(b) **A = [4 1 2;1 4 1;0 0 − 4];**
t = 1;rref(t * eye(size(A)) − A)
ans =

$$\begin{array}{ccc} 1 & 0 & 0 \\ 0 & 1 & 0 \\ 0 & 0 & 1 \end{array}$$

t = 2;rref(t * eye(size(A)) − A)

```
ans =
        1   0   0
        0   1   0
        0   0   1
```
t = 3;rref(t ∗ eye(size(A)) − A)
```
ans =
        1   1   0
        0   0   1
        0   0   0
```
Thus $t = 3$.

ML.7. Exercise 31(a) <<**A** = [1 1 1 0; 1 0 1 0; 0 1 0 0] with final form

***** ==> REDUCE is over. Your final matrix is:

```
A   =
        1   0   1   0
        0   1   0   0
        0   0   0   0
```

which has the nontrivial solution $z = r$, $y = 0$, and $x = -r$.

Exercise 31(b) <<**A** = [1 0 0 0; 1 1 1 0; 1 0 1 0] with final form

***** ==> REDUCE is over. Your final matrix is:

```
A   =
        1   0   0   0
        0   1   0   0
        0   0   1   0
```

which has the trivial solution.

Exercise 32(a) <<**A** = [1 1 0 0; 1 1 1 0; 0 1 1 0] with final form

***** ==> REDUCE is over. Your final matrix is:

```
A   =
        1   0   0   0
        0   1   0   0
        0   0   1   0
```

which has the trivial solution.

Exercise 32(b) <<**A** = [0 1 1 0; 1 0 1 0; 1 1 0 0] with final form

***** ==> REDUCE is over. Your final matrix is:

```
A   =
        1   0   1   0
        0   1   1   0
        0   0   0   0
```

which has the nontrivial solution $z = r$, $y = -r$, and $x = -r$.

ML.9. To find the inverse to matrix A in binary arithmetic, construct the augmented matrix $\begin{bmatrix} A & I_3 \end{bmatrix}$ and **binreduce**.

>> **bingen(1, 7, 3)**

```
ans    =
        0   0   0   1   1   1   1
        0   1   1   0   0   1   1
        1   0   1   0   1   0   1
```

(a) Choose the 2nd, 3rd, and 4th columns, and input the augmented matrix
The current matrix is:

```
A  =
       0  0  1  1  0  0
       1  1  0  0  1  0
       0  1  0  0  0  1
***** ==> REDUCE is over.  Your final matrix is:
A  =
       1  0  0  0  1  1
       0  1  0  0  0  1
       0  0  1  1  0  0
```

so the inverse matrix sits on the right:

$$\begin{bmatrix} 0 & 1 & 1 \\ 0 & 0 & 1 \\ 1 & 0 & 0 \end{bmatrix}.$$

(b) Construct the matrix B composed of the 5th, 6th, and 7th columns of **bingen(1,7,3)**.

```
>> B = [1  1  1;0  1  1;1  0  1]
B  =
       1  1  1
       0  1  1
       1  0  1
>> C = [B eye(3)]
C  =
       1  1  1  1  0  0
       0  1  1  0  1  0
       1  0  1  0  0  1
```

the row reduced form of C is

```
***** ==> REDUCE is over.  Your final matrix is:
C  =
       1  0  0  1  1  0
       0  1  0  1  0  1
       0  0  1  1  1  1
```

so the inverse is

$$\begin{bmatrix} 1 & 1 & 0 \\ 1 & 0 & 1 \\ 1 & 1 & 1 \end{bmatrix}.$$

(c) Construct the matrix B composed of the 1st, 2nd, and 3rd columns of **bingen(1,7,3)**.

```
>> B = [0  0  0;0  1  1;1  0  1]
B  =
       0  0  0
       0  1  1
       1  0  1
>> C = [B eye(3)]
```

```
C  =
        0  0  0  1  0  0
        0  1  1  0  1  0
        1  0  1  0  0  1
```

$>>$ **binreduce(C)**. The final form is

```
***** ==> REDUCE is over.  Your final matrix is:
C  =
        1  0  1  0  0  1
        0  1  1  0  1  0
        0  0  0  1  0  0
```

and since the left-hand side did not reduce to I_3, this matrix is singular.

(d) Construct the matrix B composed of the 1st, 4th, and 5th columns of **bingen(1,7,3)**.

$>>$ **B** $= [0\ 1\ 1; 0\ 0\ 0; 1\ 0\ 1]$

```
B  =
        0  1  1
        0  0  0
        1  0  1
```

$>>$ **C** $= [$**B eye(3)**$]$

```
C  =
        0  1  1  1  0  0
        0  0  0  0  1  0
        1  0  1  0  0  1
```

$>>$ **binreduce(C)**. The final form is

```
***** ==> REDUCE is over.  Your final matrix is:
C  =
        1  0  1  0  0  1
        0  1  1  1  0  0
        0  0  0  0  1  0
```

Since the left-hand side did not reduce to I_3, this matrix is singular.

Section 1.8, p. 113

1. Solve $L\mathbf{z} = \begin{bmatrix} 2 & 0 & 0 \\ 2 & -3 & 0 \\ 1 & -1 & 4 \end{bmatrix} \begin{bmatrix} z_1 \\ z_2 \\ z_3 \end{bmatrix} = \mathbf{b} = \begin{bmatrix} 18 \\ 3 \\ 12 \end{bmatrix}$ by forward substitution:

$$z_1 = \frac{18}{2} = 9$$

$$z_2 = \frac{3 - 2z_1}{-3} = \frac{-15}{-3} = 5$$

$$z_3 = \frac{12 + z_2 + z_1}{4} = \frac{8}{4} = 2.$$

Solve $U\mathbf{x} = \begin{bmatrix} 1 & 4 & 0 \\ 0 & 2 & 1 \\ 0 & 0 & 2 \end{bmatrix} \begin{bmatrix} x_1 \\ x_2 \\ x_3 \end{bmatrix} = \mathbf{z} = \begin{bmatrix} 9 \\ 5 \\ 2 \end{bmatrix}$ by back substitution:

$$x_3 = \frac{2}{2} = 1$$

$$x_2 = \frac{5 - x_3}{2} = \frac{4}{2} = 2$$

$$x_1 = \frac{9 - 0x_3 - 4x_2}{1} = \frac{1}{1} = 1.$$

Thus the solution is $\mathbf{x} = \begin{bmatrix} 1 \\ 2 \\ 1 \end{bmatrix}$.

3. Solve $L\mathbf{z} = \begin{bmatrix} 1 & 0 & 0 & 0 \\ 2 & 1 & 0 & 0 \\ -1 & 3 & 1 & 0 \\ 4 & 3 & 2 & 1 \end{bmatrix} \begin{bmatrix} z_1 \\ z_2 \\ z_3 \\ z_4 \end{bmatrix} = \mathbf{b} = \begin{bmatrix} -2 \\ -2 \\ -16 \\ -66 \end{bmatrix}$ by forward substitution:

$$z_1 = \frac{-2}{1} = -2$$

$$z_2 = \frac{-2 - 2z_1}{1} = \frac{2}{1} = 2$$

$$z_3 = \frac{-16 - 3z_2 + z_1}{1} = \frac{-24}{1} = -24$$

$$z_4 = \frac{-66 - 2z_3 - 3z_2 - 4z_1}{1} = \frac{-16}{1} = -16.$$

Solve $U\mathbf{x} = \begin{bmatrix} 2 & 3 & 0 & 1 \\ 0 & -1 & 3 & 1 \\ 0 & 0 & -2 & 5 \\ 0 & 0 & 0 & 4 \end{bmatrix} \begin{bmatrix} x_1 \\ x_2 \\ x_3 \\ x_4 \end{bmatrix} = \mathbf{z} = \begin{bmatrix} -2 \\ 2 \\ -24 \\ -16 \end{bmatrix}$ by back substitution:

$$x_4 = \frac{-16}{4} = -4$$

$$x_3 = \frac{-24 - 5x_4}{-2} = \frac{-4}{-2} = 2$$

$$x_2 = \frac{2 - x_4 - 3x_3}{-1} = \frac{0}{-1} = 0$$

$$x_1 = \frac{-2 - x_4 + 0x_3 - 3x_2}{2} = \frac{2}{2} = 1.$$

Thus the solution is $\mathbf{x} = \begin{bmatrix} 1 \\ 0 \\ 2 \\ -4 \end{bmatrix}$.

5. To find an LU-factorization of

$$A = \begin{bmatrix} 2 & 3 & 4 \\ 4 & 5 & 10 \\ 4 & 8 & 2 \end{bmatrix}$$

we follow the procedure used in Example 3.

Step 1. "Zero out" below the first diagonal entry of A. Add (-2) times the first row of A to the second row of A. Add (-2) times the first row of A to the third row of A. Call the new matrix U_1:

$$U_1 = \begin{bmatrix} 2 & 3 & 4 \\ 0 & -1 & 2 \\ 0 & 2 & -6 \end{bmatrix}$$

We begin building a lower triangular matrix, with 1's on the main diagonal, to record the row operations. Enter the negatives of the multipliers used in the row operations in the first column of L_1, below the first diagonal entry of L_1:

$$L_1 = \begin{bmatrix} 1 & 0 & 0 \\ 2 & 1 & 0 \\ 2 & * & 1 \end{bmatrix}$$

Step 2. "Zero out" below the second diagonal entry of U_1. Add 2 times the second row of U_1 to the third row of U_1. Call the new matrix U_2:

$$U_2 = \begin{bmatrix} 2 & 3 & 4 \\ 0 & -1 & 2 \\ 0 & 0 & -2 \end{bmatrix}$$

Enter the negatives of the multipliers from the row operations below the second diagonal entry of L_1. Call the new matrix L_2:

$$L_2 = \begin{bmatrix} 1 & 0 & 0 \\ 2 & 1 & 0 \\ 2 & -2 & 1 \end{bmatrix}$$

Let $L = L_2$ and $U = U_2$. Solve

$$L\mathbf{z} = \begin{bmatrix} 1 & 0 & 0 \\ 2 & 1 & 0 \\ 2 & -2 & 1 \end{bmatrix} \begin{bmatrix} z_1 \\ z_2 \\ z_3 \end{bmatrix} = \mathbf{b} = \begin{bmatrix} 6 \\ 16 \\ 2 \end{bmatrix}$$

by forward substitution:

$$z_1 = 6$$
$$z_2 = 16 - 2z_1 = 4$$
$$z_3 = 2 + 2z_2 - 2z_1 = -2.$$

Solve

$$U\mathbf{x} = \begin{bmatrix} 2 & 3 & 4 \\ 0 & -1 & 2 \\ 0 & 0 & -2 \end{bmatrix} \begin{bmatrix} x_1 \\ x_2 \\ x_3 \end{bmatrix} = \mathbf{z} = \begin{bmatrix} 6 \\ 4 \\ -2 \end{bmatrix}$$

by back substitution:

$$x_3 = \frac{-2}{-2} = 1$$

$$x_2 = \frac{4 - 2x_3}{-1} = -2$$

$$x_1 = \frac{6 - 4x_3 - 3x_2}{2} = 4.$$

Thus the solution is $\mathbf{x} = \begin{bmatrix} 4 \\ -2 \\ 1 \end{bmatrix}$.

7. To find an *LU*-factorization of

$$A = \begin{bmatrix} 4 & 2 & 3 \\ 2 & 0 & 5 \\ 1 & 2 & 1 \end{bmatrix}$$

we follow the procedure used in Example 3.

Step 1. "Zero out" below the first diagonal entry of A. Add $\left(-\frac{1}{2}\right)$ times the first row of A to the second row of A. Add $\left(-\frac{1}{4}\right)$ times the first row of A to the third row of A. Call the new matrix U_1:

$$U_1 = \begin{bmatrix} 4 & 2 & 3 \\ 0 & -1 & \frac{7}{2} \\ 0 & \frac{3}{2} & \frac{1}{4} \end{bmatrix}$$

We begin building a lower triangular matrix, with 1's on the main diagonal, to record the row operations. Enter the negatives of the multipliers used in the row operations in the first column of L_1, below the first diagonal entry of L_1:

$$L_1 = \begin{bmatrix} 1 & 0 & 0 \\ \frac{1}{2} & 1 & 0 \\ \frac{1}{4} & * & 1 \end{bmatrix}$$

Step 2. "Zero out" below the second diagonal entry of U_1. Add $\frac{3}{2}$ times the second row of U_1 to the third row of U_1. Call the new matrix U_2:

$$U_2 = \begin{bmatrix} 4 & 2 & 3 \\ 0 & -1 & \frac{7}{2} \\ 0 & 0 & \frac{11}{2} \end{bmatrix}$$

Enter the negatives of the multipliers from the row operations below the second diagonal entry of L_1. Call the new matrix L_2:

$$L_2 = \begin{bmatrix} 1 & 0 & 0 \\ \frac{1}{2} & 1 & 0 \\ \frac{1}{4} & -\frac{3}{2} & 1 \end{bmatrix}$$

Let $L = L_2$ and $U = U_2$. Solve

$$L\mathbf{z} = \begin{bmatrix} 1 & 0 & 0 \\ \frac{1}{2} & 1 & 0 \\ \frac{1}{4} & -\frac{3}{2} & 1 \end{bmatrix} \begin{bmatrix} z_1 \\ z_2 \\ z_3 \end{bmatrix} = \mathbf{b} = \begin{bmatrix} 1 \\ -1 \\ -3 \end{bmatrix}$$

by using forward substitution :

$$z_1 = 1$$

$$z_2 = -1 - \tfrac{1}{2}z_1 = -\tfrac{3}{2}$$

$$z_3 = -3 + \tfrac{3}{2}z_2 - \tfrac{1}{4}z_1 = -\frac{11}{2}.$$

Solve

$$U\mathbf{x} = \begin{bmatrix} 4 & 2 & 3 \\ 0 & -1 & \tfrac{7}{2} \\ 0 & 0 & \tfrac{11}{2} \end{bmatrix} \begin{bmatrix} x_1 \\ x_2 \\ x_3 \end{bmatrix} = \mathbf{z} = \begin{bmatrix} 1 \\ -\tfrac{3}{2} \\ -\tfrac{11}{2} \end{bmatrix}$$

by back substitution:

$$x_3 = -1$$

$$x_2 = \frac{-\tfrac{3}{2} - \tfrac{7}{2}x_3}{-1} = -2$$

$$x_1 = \frac{1 - 3x_3 - 2x_2}{4} = 2.$$

Thus the solution is $\mathbf{x} = \begin{bmatrix} 2 \\ -2 \\ -1 \end{bmatrix}$.

9. To find an *LU*-factorization of

$$A = \begin{bmatrix} 2 & 1 & 0 & -4 \\ 1 & 0 & .25 & -1 \\ -2 & -1.1 & .25 & 6.2 \\ 4 & 2.2 & .30 & -2.4 \end{bmatrix}$$

we follow the procedure in Example 3.

Step 1. "Zero out" below the first diagonal entry of A. Add $(-.5)$ times the first row of A to the second row of A. Add 1 times the first row of A to the third row of A. Add (-2) times the first row of A to the fourth row of A. Call the new matrix U_1:

$$U_1 = \begin{bmatrix} 2 & 1 & 0 & -4 \\ 0 & -.5 & .25 & 1 \\ 0 & -.1 & .25 & 2.2 \\ 0 & .2 & .30 & 5.6 \end{bmatrix}$$

We begin building a lower triangular matrix, with 1's on the main diagonal, to record the row operations. Enter the negatives of the multipliers used in the row operations in the first column of L_1, below the first diagonal entry of L_1:

$$L_1 = \begin{bmatrix} 1 & 0 & 0 & 0 \\ .5 & 1 & 0 & 0 \\ -1 & * & 1 & 0 \\ 2 & * & * & 1 \end{bmatrix}$$

Step 2. "Zero out" below the second diagonal entry of U_1. Add $(-.2)$ times the second row of U_1 to the third row of U_1. Add .4 times the second row of U_1 to the fourth row of U_1. Call the new matrix U_2:

$$U_2 = \begin{bmatrix} 2 & 1 & 0 & -4 \\ 0 & -.5 & .25 & 1 \\ 0 & 0 & .20 & 2 \\ 0 & 0 & .40 & 6 \end{bmatrix}$$

Enter the negatives of the multipliers from the row operations below the second diagonal entry of L_1. Call the new matrix L_2.

$$L_2 = \begin{bmatrix} 1 & 0 & 0 & 0 \\ .5 & 1 & 0 & 0 \\ -1 & .2 & 1 & 0 \\ 2 & -.4 & * & 1 \end{bmatrix}$$

Step 3. "Zero out" below the third diagonal entry of U_2. Add (-2) times the third row of U_2 to the fourth row of U_2. Call the new matrix U_3:

$$U_3 = \begin{bmatrix} 2 & 1 & 0 & -4 \\ 0 & -.5 & .25 & 1 \\ 0 & 0 & .20 & 2 \\ 0 & 0 & 0 & 2 \end{bmatrix}$$

Enter the negatives of the multipliers from the row operations below the third diagonal entry of L_2. Call the new matrix L_3:

$$L_3 = \begin{bmatrix} 1 & 0 & 0 & 0 \\ .5 & 1 & 0 & 0 \\ -1 & .2 & 1 & 0 \\ 2 & -.4 & 2 & 1 \end{bmatrix}$$

Let $L = L_3$ and $U = U_3$. Solve

$$L\mathbf{z} = \begin{bmatrix} 1 & 0 & 0 & 0 \\ .5 & 1 & 0 & 0 \\ -1 & .2 & 1 & 0 \\ 2 & -.4 & 2 & 1 \end{bmatrix} \begin{bmatrix} z_1 \\ z_2 \\ z_3 \end{bmatrix} = \mathbf{b} = \begin{bmatrix} -3 \\ -1.5 \\ 5.6 \\ 2.2 \end{bmatrix}$$

by using forward substitution :

$$z_1 = -3$$
$$z_2 = -1.5 - .5z_1 = 0$$
$$z_3 = 5.6 - .2z_2 + z_1 = 2.6$$
$$z_4 = 2.2 - 2z_3 + .4z_2 - 2z_1 = 3.$$

Solve

$$U\mathbf{x} = \begin{bmatrix} 2 & 1 & 0 & -4 \\ 0 & -.5 & .25 & 1 \\ 0 & 0 & .20 & 2 \\ 0 & 0 & 0 & 2 \end{bmatrix} \begin{bmatrix} x_1 \\ x_2 \\ x_3 \\ x_4 \end{bmatrix} = \mathbf{z} = \begin{bmatrix} -3 \\ 0 \\ 2.6 \\ 3 \end{bmatrix}$$

by back substitution:

$$x_4 = 1.5$$

$$x_3 = \frac{2.6 - 2x_4}{.2} = -2$$

$$x_2 = \frac{0 - x_4 - .25x_3}{-.5} = 2$$

$$x_1 = \frac{-3 + 4x_4 - 0x_3 - x_2}{2} = 0.5.$$

Thus the solution is $\mathbf{x} = \begin{bmatrix} 0.5 \\ 2 \\ -2 \\ 1.5 \end{bmatrix}$.

ML.1. We show the first few steps of the LU-factorization using routine **lupr** and then display the matrices L and U.

[L,U] = lupr(A)

```
+++++++++++++++++++++++++++++++++++++++++++++++++++++++++++++++++++++++
                * * * * * Find an LU-FACTORIZATION by Row Reduction * * * * *
L =                 U =
   1  0  0            2    8    0
   0  1  0            2    2   -3
   0  0  1            1    2    7
                            OPTIONS
<1> Insert element into L.    <-1> Undo previous operation.    <0> Quit.
ENTER your choice ===> 1

Enter multiplier.  -1
Enter first row number.   1
Enter number of row that changes.   2


+++++++++++++++++++++++++++++++++++++++++++++++++++++++++++++++++++++++
               Replacement by Linear Combination Complete
L =                 U =
   1  0  0            2    8    0
   0  1  0            0   -6   -3
   0  0  1            1    2    7
```

You just performed operation $-1 * \text{Row}(1) + \text{Row}(2)$.

```
                            OPTIONS
<1> Insert element into L.    <-1> Undo previous operation.    <0> Quit.
ENTER your choice ===> 1


+++++++++++++++++++++++++++++++++++++++++++++++++++++++++++++++++++++++
```

Insert a value in L in the position you just eliminated in U. Let the multiplier you just used be called num. It has the value -1.

```
Enter row number of L to change.  2
Enter column number of L to change.  1
Value of L(2,1) = -num
Correct:  L(2,1)=1
```

$+++$

Continuing the factorization gives

```
L =                     U =
    1       0   0           2    8    0
    1       1   0           0   -6   -3
  0.5  0.3333   1           0    0    8
```

ML.3. We first use **lupr** to find an LU-factorization of A. The matrices L and U that we find are different from those stated in Example 2. There can be many LU-factorizations for a matrix. We omit the details from **lupr**. It is assumed that A and $\mathbf{b}$ have been entered.

```
L =                                          U =
   1.0000        0        0        0          6   -2   -4    4
   0.5000   1.0000        0        0          0   -2   -4   -1
  -2.0000  -2.0000   1.0000        0          0    0    5   -2
  -1.0000   1.0000  -2.0000   1.0000          0    0    0    8
```

$\mathbf{z} = \textbf{forsub(L,b)}$

```
z =
      2
     -5
      2
    -32
```

$\mathbf{x} = \textbf{bksub(U,z)}$

```
x =
    4.5000
    6.9000
   -1.2000
   -4.0000
```

Supplementary Exercises, p. 114

1. $2A + BC = 2 \begin{bmatrix} 1 & 2 \\ 3 & -2 \end{bmatrix} + \begin{bmatrix} 3 & -5 \\ 2 & 4 \end{bmatrix} \begin{bmatrix} 4 & 1 \\ 3 & 2 \end{bmatrix} = \begin{bmatrix} 2 & 4 \\ 6 & -4 \end{bmatrix} + \begin{bmatrix} -3 & -7 \\ 20 & 10 \end{bmatrix} = \begin{bmatrix} -1 & -3 \\ 26 & 6 \end{bmatrix}.$

3. $A^T + B^T C = \begin{bmatrix} 1 & 3 \\ 2 & -2 \end{bmatrix} + \begin{bmatrix} 3 & 2 \\ -5 & 4 \end{bmatrix} \begin{bmatrix} 4 & 1 \\ 3 & 2 \end{bmatrix} = \begin{bmatrix} 1 & 3 \\ 2 & -2 \end{bmatrix} + \begin{bmatrix} 18 & 7 \\ -8 & 3 \end{bmatrix} = \begin{bmatrix} 19 & 10 \\ -6 & 1 \end{bmatrix}.$

5. (a) The augmented matrix is $\begin{bmatrix} 1 & 2 & -1 & 1 & \vdots & 7 \\ 2 & -1 & 0 & 2 & \vdots & -8 \end{bmatrix}.$

(b) The corresponding linear system is

$$\begin{aligned}
3x + 2y &= -4 \\
5x + y &= 2 \\
3x + 2y &= 6.
\end{aligned}$$

7. For **w** to be in the range of f we must find all values of k and t so that

$$\begin{bmatrix} 0 & 2 & 1 & 0 \\ 1 & 0 & 2 & 1 \\ 1 & 1 & k & t \end{bmatrix} \begin{bmatrix} x \\ y \\ z \\ w \end{bmatrix} = \begin{bmatrix} 4 \\ 2 \\ 6 \end{bmatrix}.$$

The augmented matrix of the linear system is

$$\left[\begin{array}{cccc:c} 0 & 2 & 1 & 0 & 4 \\ 1 & 0 & 2 & 1 & 2 \\ 1 & 1 & k & t & 6 \end{array}\right].$$

To find the reduced row echelon matrix that is equivalent to the augmented matrix first interchange rows 1 and 2. The resulting matrix is

$$\left[\begin{array}{cccc:c} 1 & 0 & 2 & 1 & 2 \\ 0 & 2 & 1 & 0 & 4 \\ 1 & 1 & k & t & 6 \end{array}\right].$$

Now add (-1) times row 1 to row 3 and multiply row 2 by $\left(\frac{1}{2}\right)$ to obtain

$$\left[\begin{array}{cccc:c} 1 & 0 & 2 & 1 & 2 \\ 0 & 1 & \frac{1}{2} & 0 & 2 \\ 0 & 0 & k-2 & t-1 & 4 \end{array}\right].$$

Finally, add (-1) times row 2 to row 3 to obtain

$$\left[\begin{array}{cccc:c} 1 & 0 & 2 & 1 & 2 \\ 0 & 1 & \frac{1}{2} & 0 & 2 \\ 0 & 0 & k-\frac{5}{2} & t-1 & 4 \end{array}\right].$$

There will be no solution if row 3 has the form

$$\left[\begin{array}{cccc:c} 0 & 0 & 0 & 0 & * \end{array}\right],$$

where $* \neq 0$. Since $k - \frac{5}{2} = 0$ when $k = \frac{5}{2}$ and $t - 1 = 0$ when $t = 1$, we conclude that $k \neq \frac{5}{2}$ and $t \neq 1$.

9. We form the augmented matrix for the linear system.

$$\left[\begin{array}{ccc:c} 1 & 1 & -1 & 5 \\ 2 & 1 & 1 & 2 \\ 1 & -1 & -2 & 3 \end{array}\right].$$

Applying row operations (-2) times row 1 added to row 2 and (-1) times row 1 added to row 3, we obtain

$$\left[\begin{array}{ccc:c} 1 & 1 & -1 & 5 \\ 0 & -1 & 3 & -8 \\ 0 & -2 & -1 & -2 \end{array}\right].$$

Applying row operations (-1) times row 2, and then 2 times row 2 added to row 3, we obtain

$$\begin{bmatrix} 1 & 1 & -1 & \vdots & 5 \\ 0 & 1 & -3 & \vdots & 8 \\ 0 & 0 & -7 & \vdots & 14 \end{bmatrix}.$$

Now multiply row 3 by $\left(-\frac{1}{7}\right)$:

$$\begin{bmatrix} 1 & 1 & -1 & \vdots & 5 \\ 0 & 1 & -3 & \vdots & 8 \\ 0 & 0 & 1 & \vdots & -2 \end{bmatrix}.$$

Add 1 times row 3 to row 1, then 3 times row 3 added to row 2 to obtain:

$$\begin{bmatrix} 1 & 1 & 0 & \vdots & 3 \\ 0 & 1 & 0 & \vdots & 2 \\ 0 & 0 & 1 & \vdots & -2 \end{bmatrix}.$$

Finally, (-1) times row 2 added to row 1 gives:

$$\begin{bmatrix} 1 & 0 & 0 & \vdots & 1 \\ 0 & 1 & 0 & \vdots & 2 \\ 0 & 0 & 1 & \vdots & -2 \end{bmatrix}.$$

Thus we see that the solution of the linear system is

$$x = 1, \quad y = 2, \quad \text{and} \quad z = -2.$$

11. We form the augmented matrix and apply row operations to obtain a simpler equivalent system. We have

$$\begin{bmatrix} 1 & 1 & -1 & \vdots & 3 \\ 1 & -1 & 3 & \vdots & 4 \\ 1 & 1 & a^2 - 10 & \vdots & a \end{bmatrix}.$$

Applying row operations (-1) times row 1 added to rows 2 and 3, we obtain

$$\begin{bmatrix} 1 & 1 & -1 & \vdots & 3 \\ 0 & -2 & 4 & \vdots & 1 \\ 0 & 0 & a^2 - 9 & \vdots & a - 3 \end{bmatrix}.$$

(a) There will be no solutions if a is chosen so that row 3 has the form

$$\begin{bmatrix} 0 & 0 & 0 & * \end{bmatrix}$$

where $* \neq 0$. We see that $a^2 - 9 = 0$ if and only if $a = \pm 3$. If $a = -3$, then $a - 3 \neq 0$. Thus there will be no solutions if $a = -3$.

(b) There will be a unique solution in the case that a is chosen so that $a^2 - 9$ is not zero. In that case the $(3,3)$ entry can be used as a pivot. Thus there will be a unique solution if $a \neq 3$ or -3.

(c) There will be infinitely many solutions if the third row is a zero row. This is the case if $a = 3$.

13. To determine all the solutions of $(\lambda I_3 - A)\mathbf{x} = \mathbf{0}$, we row reduce the augmented matrix

$$\begin{bmatrix} \lambda I_3 - A & \vdots & \mathbf{0} \end{bmatrix} = \begin{bmatrix} 4I_3 - A & \vdots & \mathbf{0} \end{bmatrix} = \begin{bmatrix} -1 & -3 & -1 & \vdots & 0 \\ 0 & 0 & -2 & \vdots & 0 \\ 0 & 0 & 0 & \vdots & 0 \end{bmatrix}.$$

Applying row operations (-1) times row 1 and $\left(-\frac{1}{2}\right)$ times row 2, we obtain

$$\left[\begin{array}{ccc:c} 1 & 3 & 1 & 0 \\ 0 & 0 & 1 & 0 \\ 0 & 0 & 0 & 0 \end{array}\right].$$

Finally, applying (-1) times row 2 added to row 1 gives

$$\left[\begin{array}{ccc:c} 1 & 3 & 0 & 0 \\ 0 & 0 & 1 & 0 \\ 0 & 0 & 0 & 0 \end{array}\right].$$

The corresponding linear system is

$$\begin{aligned} x + 3y &= 0 \\ z &= 0. \end{aligned}$$

Let $y = r$, where r is any real number. Then the solution is given by

$$x = -3r, \quad y = r, \quad \text{and} \quad z = 0.$$

15. We form the matrix

$$\left[\begin{array}{ccc:ccc} 1 & 2 & 3 & 1 & 0 & 0 \\ 2 & 5 & 3 & 0 & 1 & 0 \\ 1 & 0 & 8 & 0 & 0 & 1 \end{array}\right].$$

We apply row operations to obtain the reduced row echelon form. Applying row operations (-2) times row 1 added to row 2 and (-1) times row 1 added to row 3, we obtain

$$\left[\begin{array}{ccc:ccc} 1 & 2 & 3 & 1 & 0 & 0 \\ 0 & 1 & -3 & -2 & 1 & 0 \\ 0 & -2 & 5 & -1 & 0 & 1 \end{array}\right].$$

Applying row operation (2) times row 2 added to row 3 gives

$$\left[\begin{array}{ccc:ccc} 1 & 2 & 3 & 1 & 0 & 0 \\ 0 & 1 & -3 & -2 & 1 & 0 \\ 0 & 0 & -1 & -5 & 2 & 1 \end{array}\right].$$

Applying row operations (-1) times row 3, then (3) times row 3 added to row 2, and (-3) times row 3 added to row 1, we obtain

$$\left[\begin{array}{ccc:ccc} 1 & 2 & 0 & -14 & 6 & 3 \\ 0 & 1 & 0 & 13 & -5 & -3 \\ 0 & 0 & 1 & 5 & -2 & -1 \end{array}\right].$$

Finally, (-2) times row 2 added to row 1 gives:

$$\left[\begin{array}{ccc:ccc} 1 & 0 & 0 & -40 & 16 & 9 \\ 0 & 1 & 0 & 13 & -5 & -3 \\ 0 & 0 & 1 & 5 & -2 & -1 \end{array}\right].$$

Thus the matrix has an inverse given by $\left[\begin{array}{ccc} -40 & 16 & 9 \\ 13 & -5 & -3 \\ 5 & -2 & -1 \end{array}\right].$

17. We row reduce the corresponding augmented matrix

$$\left[\begin{array}{rrr:r} 1 & -1 & 3 & 0 \\ 1 & 2 & -3 & 0 \\ 2 & 1 & 0 & 0 \end{array}\right].$$

Applying row operations (-1) times row 1 added to row 3 and (-2) times row 1 added to row 3, we obtain

$$\left[\begin{array}{rrr:r} 1 & -1 & 3 & 0 \\ 0 & 3 & -6 & 0 \\ 0 & 3 & -6 & 0 \end{array}\right].$$

Then adding (-1) times row 2 to row 3 gives

$$\left[\begin{array}{rrr:r} 1 & -1 & 3 & 0 \\ 0 & 3 & -6 & 0 \\ 0 & 0 & 0 & 0 \end{array}\right].$$

We see that the linear system equivalent to this last matrix will have 2 equations in 3 unknowns. Hence at least one of the variables can be chosen arbitrarily. Thus the system has a nontrivial solution.

19. $\mathbf{x} = A^{-1}\mathbf{b} = \begin{bmatrix} 1 & 2 & 0 \\ 0 & 1 & 0 \\ 3 & 1 & -1 \end{bmatrix} \begin{bmatrix} 2 \\ 1 \\ 3 \end{bmatrix} = \begin{bmatrix} 4 \\ 1 \\ 4 \end{bmatrix}.$

21. Assuming that $A^4 = O$, we verify that

$$(I_4 - A)^{-1} = I_4 + A + A^2 + A^3$$

by showing

$$(I_4 - A)(I_4 + A + A^2 + A^3) = (I_4 + A + A^2 + A^3)(I_4 - A) = I_4.$$

We have

$$(I_4 - A)(I_4 + A + A^2 + A^3) = I_4 + A + A^2 + A^3 - A - A^2 - A^3 - A^4 = I_4.$$

Similarly,

$$(I_4 + A + A^2 + A^3)(I_4 - A) = I_4 - A + A - A^2 + A^2 - A^3 + A^3 - A^4 = I_4.$$

23. Form the augmented matrix

$$\left[\begin{array}{rr:r} 1 & 1 & 3 \\ 5 & 5 & a \end{array}\right].$$

Applying row operation (-5) times row 1 added to row 2 gives the equivalent linear system

$$\left[\begin{array}{rr:r} 1 & 1 & 3 \\ 0 & 0 & a-15 \end{array}\right].$$

(a) For $a \neq 15$, the linear system is inconsistent, hence has no solution.

(b) There are no values of a such that the system has exactly one solution.

(c) There are infinitely many solutions for $a = 15$.

25. The coefficient matrix is

$$\begin{bmatrix} 1-a & 0 & 1 \\ 0 & -a & 1 \\ 0 & 1 & -a \end{bmatrix}.$$

Applying row operation (a) times row 3 added to row 2 gives the matrix

$$\begin{bmatrix} 1-a & 0 & 0 \\ 0 & 0 & 1-a^2 \\ 0 & 1 & -a \end{bmatrix}.$$

The corresponding homogeneous system will have a nontrivial solution provided $1 - a^2 = 0$; that is, for $a = 1$ or $a = -1$.

27. (a) For $k = 1$, $\mathbf{b} = \begin{bmatrix} b_1 \\ 0 \end{bmatrix}$. ($b_1$ arbitrary.)

For $k = 2$, $\mathbf{b} = \begin{bmatrix} b_{11} & b_{12} \\ 0 & 0 \end{bmatrix}$. ($b_{11}$ and b_{12} arbitrary.)

For $k = 3$, $\mathbf{b} = \begin{bmatrix} b_{11} & b_{12} & b_{13} \\ 0 & 0 & 0 \end{bmatrix}$. ($b_{11}$, b_{12}, and b_{13} arbitrary.)

For $k = 4$, $\mathbf{b} = \begin{bmatrix} b_{11} & b_{12} & b_{13} & b_{14} \\ 0 & 0 & 0 & 0 \end{bmatrix}$. ($b_{11}$, b_{12}, b_{13}, and b_{14} arbitrary.)

(b) The answers are not unique. The only requirement is that row 2 of $\mathbf{b}$ have all zero entries.

29. (a) Let $A = \begin{bmatrix} a & b \\ c & d \end{bmatrix}$. Then

$$A^2 = \begin{bmatrix} a^2 + bc & ab + bd \\ ac + dc & bc + d^2 \end{bmatrix} = \begin{bmatrix} 1 & 1 \\ 0 & 1 \end{bmatrix} = B.$$

Equating corresponding entries we have

$$a^2 + bc = 1 \qquad b(a + d) = 1$$
$$c(a + d) = 0 \qquad bc + d^2 = 1.$$

It follows that $a + d \neq 0$ and $c = 0$. Hence we have

$$a^2 = 1, \qquad b = \frac{1}{a + d}, \qquad d^2 = 1.$$

One solution is

$$A = \begin{bmatrix} 1 & \frac{1}{2} \\ 0 & 1 \end{bmatrix} \quad \text{and another is} \quad A = \begin{bmatrix} -1 & -\frac{1}{2} \\ 0 & -1 \end{bmatrix}.$$

(b) One solution is $A = B$.

(c) One solution is $A = I_4$.

(d) Let $\begin{bmatrix} a & b \\ c & d \end{bmatrix}$. Then

$$A^2 = \begin{bmatrix} a^2 + bc & ab + bd \\ ac + dc & bc + d^2 \end{bmatrix} = \begin{bmatrix} 1 & 1 \\ 0 & 1 \end{bmatrix} = B.$$

Equating corresponding entries we have

$$a^2 + bc = 1 \qquad b(a + d) = 1$$
$$c(a + d) = 0 \qquad bc + d^2 = 1.$$

It follows that $a + d \neq 0$ and $c = 0$. Thus

$$A^2 = \begin{bmatrix} a^2 & b(a+d) \\ 0 & d^2 \end{bmatrix} = \begin{bmatrix} 0 & 1 \\ 0 & 0 \end{bmatrix}.$$

Again equating corresponding entries, we have

$$a = 0, \qquad b = \frac{1}{a+d}, \qquad d = 0.$$

Hence, $a = d = 0$, which is a contradiction. B has no square root.

31. Inspect the sequence A^2, A^3, A^4, A^5, and higher powers if need be, to discover a pattern. We have

$$A^2 = \begin{bmatrix} 1 & \frac{3}{4} \\ 0 & \frac{1}{4} \end{bmatrix}, \qquad A^3 = \begin{bmatrix} 1 & \frac{7}{8} \\ 0 & \frac{1}{8} \end{bmatrix}, \qquad A^4 = \begin{bmatrix} 1 & \frac{15}{16} \\ 0 & \frac{1}{16} \end{bmatrix}, \qquad A^5 = \begin{bmatrix} 1 & \frac{31}{32} \\ 0 & \frac{1}{32} \end{bmatrix}.$$

It appears that

$$A^n = \begin{bmatrix} 1 & \frac{(2^n - 1)}{2^n} \\ 0 & \frac{1}{2^n} \end{bmatrix}.$$

33. (a) $\mathbf{w} = A^{-1}(C + F)\mathbf{v} \implies A\mathbf{w} = (C + F)\mathbf{v}.$
We first find

$$\mathbf{b} = (C + F)\mathbf{v} = \left(\begin{bmatrix} 1 & 1 & 1 \\ 2 & 3 & 1 \\ 1 & 2 & 1 \end{bmatrix} + \begin{bmatrix} 2 & 1 & 0 \\ -3 & 0 & 2 \\ -1 & 1 & 2 \end{bmatrix} \right) \begin{bmatrix} 6 \\ 7 \\ -3 \end{bmatrix} = \begin{bmatrix} 3 & 2 & 1 \\ -1 & 3 & 3 \\ 0 & 3 & 3 \end{bmatrix} \begin{bmatrix} 6 \\ 7 \\ -3 \end{bmatrix} = \begin{bmatrix} 29 \\ 6 \\ 12 \end{bmatrix}.$$

Now solve the linear system $A\mathbf{w} = \mathbf{b}$ with augmented matrix

$$\begin{bmatrix} 1 & 0 & -2 & \vdots & 29 \\ 1 & 1 & 0 & \vdots & 6 \\ 0 & 1 & 1 & \vdots & 12 \end{bmatrix}.$$

Add (-1) times row 1 to row 2 to obtain

$$\begin{bmatrix} 1 & 0 & -2 & \vdots & 29 \\ 0 & 1 & 2 & \vdots & -23 \\ 0 & 1 & 1 & \vdots & 12 \end{bmatrix}.$$

Add (-1) times row 2 to row 3 to obtain

$$\begin{bmatrix} 1 & 0 & -2 & \vdots & 29 \\ 0 & 1 & 2 & \vdots & -23 \\ 0 & 0 & -1 & \vdots & 35 \end{bmatrix}.$$

Add (2) times row 3 to row 2 and (-2) times row 3 to row 1 to obtain

$$\begin{bmatrix} 1 & 0 & 0 & \vdots & -41 \\ 0 & 1 & 0 & \vdots & 47 \\ 0 & 0 & -1 & \vdots & 35 \end{bmatrix}.$$

Finally multiply (-1) times row 3 to obtain

$$\begin{bmatrix} 1 & 0 & 0 & \vdots & -41 \\ 0 & 1 & 0 & \vdots & 47 \\ 0 & 0 & 1 & \vdots & -35 \end{bmatrix}.$$

Therefore $\mathbf{w} = \begin{bmatrix} -41 \\ 47 \\ -35 \end{bmatrix}$.

(b) $\mathbf{w} = (F + 2A)C^{-1}\mathbf{v}$

First let $\mathbf{u} = C^{-1}\mathbf{v}$. Then $C\mathbf{u} = \mathbf{v}$. Solve this linear system with augmented matrix

$$\begin{bmatrix} 1 & 1 & 1 & \vdots & 6 \\ 2 & 3 & 1 & \vdots & 7 \\ 1 & 2 & 1 & \vdots & -3 \end{bmatrix}.$$

Add (-2) times row 1 to row 2 and (-1) times row 1 to row 3 to obtain

$$\begin{bmatrix} 1 & 1 & 1 & \vdots & 6 \\ 0 & 1 & -1 & \vdots & -5 \\ 0 & 1 & 0 & \vdots & -9 \end{bmatrix}.$$

Add (-1) times row 3 to rows 1 and 2.

$$\begin{bmatrix} 1 & 0 & 1 & \vdots & 15 \\ 0 & 0 & -1 & \vdots & 4 \\ 0 & 1 & 0 & \vdots & -9 \end{bmatrix}.$$

Interchange rows 2 and 3. Then add row 3 to row 1.

$$\begin{bmatrix} 1 & 0 & 0 & \vdots & 19 \\ 0 & 1 & 0 & \vdots & -9 \\ 0 & 0 & -1 & \vdots & 4 \end{bmatrix}.$$

Finally multiply (-1) times row 3 to obtain

$$\begin{bmatrix} 1 & 0 & 0 & \vdots & 19 \\ 0 & 1 & 0 & \vdots & -9 \\ 0 & 0 & 1 & \vdots & -4 \end{bmatrix}.$$

Thus $\mathbf{v} = \begin{bmatrix} 19 \\ -9 \\ -4 \end{bmatrix}$. Use this vector $\mathbf{v}$ to find

$$\mathbf{w} = (F+2A)\mathbf{v} = \left(\begin{bmatrix} 2 & 1 & 0 \\ -3 & 0 & 2 \\ -1 & 1 & 2 \end{bmatrix} + 2 \begin{bmatrix} 1 & 0 & -2 \\ 1 & 1 & 0 \\ 0 & 1 & 1 \end{bmatrix} \right) \begin{bmatrix} 19 \\ -9 \\ -4 \end{bmatrix} = \begin{bmatrix} 4 & 1 & -4 \\ -1 & 2 & 2 \\ -1 & 3 & 4 \end{bmatrix} \begin{bmatrix} 19 \\ -9 \\ -4 \end{bmatrix} = \begin{bmatrix} 83 \\ -45 \\ -62 \end{bmatrix}.$$

Therefore $\mathbf{w} = \begin{bmatrix} 83 \\ -45 \\ -62 \end{bmatrix}$.

35. To find an LU-factorization of

$$A = \begin{bmatrix} -2 & 1 & -2 \\ 6 & 1 & 9 \\ -4 & 18 & 5 \end{bmatrix}$$

we follow the procedure used in Example 3 in Section 1.7.

Step 1. "Zero-out" below the first diagonal entry of A. Add 3 times the first row of A to the second row of A. That is, $3\mathbf{r}_1 + \mathbf{r}_2$. Add (-2) times the first row of A to the third row of A. That is, $-2\mathbf{r}_1 + \mathbf{r}_3$. Call the new matrix U_1.

$$U_1 = \begin{bmatrix} -2 & 1 & -2 \\ 0 & 4 & 3 \\ 0 & 16 & 9 \end{bmatrix}.$$

We begin building a lower triangular matrix, with 1's on the main diagonal, to record the row operations. Enter the negatives of the multipliers used in the row operations in the first column of L_1, below the first diagonal entry of L_1.

$$L_1 = \begin{bmatrix} 1 & 0 & 0 \\ -3 & 1 & 0 \\ 2 & * & 1 \end{bmatrix}.$$

Step 2. "Zero out" below the second diagonal entry of U_1. Add (-4) times the second row of U_1 to the third row of U_1. That is, $-4\mathbf{r}_2 + \mathbf{r}_3$. Call the new matrix U_2.

$$U_2 = \begin{bmatrix} -2 & 1 & -2 \\ 0 & 4 & 3 \\ 0 & 0 & -3 \end{bmatrix}.$$

Enter the negatives of the multipliers from the row operations below the second diagonal entry of L_1. Call the new matrix L_2.

$$L_2 = \begin{bmatrix} 1 & 0 & 0 \\ -3 & 1 & 0 \\ 2 & 4 & 1 \end{bmatrix}.$$

Let $L = L_2$ and $U = U_2$. Solve

$$L\mathbf{z} = \begin{bmatrix} 1 & 0 & 0 \\ -3 & 1 & 0 \\ 2 & 4 & 1 \end{bmatrix} \begin{bmatrix} z_1 \\ z_2 \\ z_3 \end{bmatrix} = \mathbf{b} = \begin{bmatrix} -6 \\ 19 \\ -17 \end{bmatrix}$$

by forward substitution:

$$z_1 = -6$$
$$z_2 = 19 + 3z_1 = 1$$
$$z_3 = -17 - 4z_2 - 2z_1 = -9.$$

Solve

$$U\mathbf{x} = \begin{bmatrix} -2 & 1 & -2 \\ 0 & 4 & 3 \\ 0 & 0 & -3 \end{bmatrix} \begin{bmatrix} x_1 \\ x_2 \\ x_3 \end{bmatrix} = \mathbf{z} = \begin{bmatrix} -6 \\ 1 \\ -9 \end{bmatrix}$$

by back substitution:

$$x_3 = \frac{-9}{-3} = 3$$

$$x_2 = \frac{1 - 3x_3}{4} = -2$$

$$x_1 = \frac{-6 + 2x_3 - x_2}{-2} = -1.$$

Thus the solution is $\mathbf{x} = \begin{bmatrix} -1 \\ -2 \\ 3 \end{bmatrix}.$

T.1. (a) $\text{Tr}(cA) = \sum_{i=1}^{n} ca_{ii} = c\sum_{i=1}^{n} a_{ii} = c\,\text{Tr}(A)$.

(b) $\text{Tr}(A + B) = \sum_{i=1}^{n}(a_{ii} + b_{ii}) = \sum_{i=1}^{n} a_{ii} + \sum_{i=1}^{n} b_{ii} = \text{Tr}(A) + \text{Tr}(B)$.

(c) Let $AB = C = \begin{bmatrix} c_{ij} \end{bmatrix}$. Then

$$\text{Tr}(AB) = \text{Tr}(C) = \sum_{i=1}^{n} c_{ii} = \sum_{i=1}^{n}\sum_{k=1}^{n} a_{ik}b_{ki} = \sum_{k=1}^{n}\sum_{i=1}^{n} b_{ki}a_{ik} = \text{Tr}(BA).$$

(d) $\text{Tr}(A^T) = \sum_{i=1}^{n} a_{ii} = \text{Tr}(A)$.

(e) $\text{Tr}(A^T A)$ is the sum of the diagonal entries of $A^T A$. The ith diagonal entry of $A^T A$ is $\sum_{j=1}^{n} a_{ji}^2$, so

$$\text{Tr}(A^T A) = \sum_{i=1}^{n}\left[\sum_{j=1}^{n} a_{ji}^2\right] = \text{sum of the squares of all entries of } A.$$

Hence, $\text{Tr}(A^T A) \geq 0$.

T.3. If $A\mathbf{x} = B\mathbf{x}$ for all $n \times 1$ matrices $\mathbf{x}$, then $A\mathbf{e}_j = B\mathbf{e}_j$, $j = 1, 2, \ldots, n$, where $\mathbf{e}_j = $ column j of I_n. But then

$$A\mathbf{e}_j = \begin{bmatrix} a_{1j} \\ a_{2j} \\ \vdots \\ a_{nj} \end{bmatrix} = B\mathbf{e}_j = \begin{bmatrix} b_{1j} \\ b_{2j} \\ \vdots \\ b_{nj} \end{bmatrix}.$$

Hence column j of $A = $ column j of B for each j and it follows that $A = B$.

T.5. Suppose that A is skew symmetric, so $A^T = -A$. Then $(A^k)^T = (A^T)^k = (-A)^k = -A^k$ if k is a positive odd integer, so A^k is skew symmetric.

T.7. If A is symmetric and upper (lower) triangular, then $a_{ij} = a_{ji}$ and $a_{ij} = 0$ for $j > i$ ($j < i$). Thus, $a_{ij} = 0$, $i \neq j$, so A is diagonal.

T.9. Suppose that $A \neq O$ but row equivalent to O. Then in the reduction process some row operation must have transformed a nonzero matrix into the zero matrix. However, considering the types of row operations this is impossible. Thus $A = O$. The converse follows immediately.

T.11 Proof by contradiction: Assume that B is singular. Then by Theorem 1.13, Sec. 1.7, there exists $\mathbf{x} \neq \mathbf{0}$ such that $B\mathbf{x} = \mathbf{0}$. Then $(AB)\mathbf{x} = A\mathbf{0} = \mathbf{0}$, which means that the homogeneous system $(AB)\mathbf{x} = \mathbf{0}$ has a nontrivial solution. Theorem 1.13 implies that AB is singular, but this is a contradiction. Suppose now that A is singular and B is nonsingular. Then there exists a $\mathbf{y} \neq \mathbf{0}$ such that $A\mathbf{y} = \mathbf{0}$. Since B is nonsingular we can find $\mathbf{x} \neq \mathbf{0}$ such that $\mathbf{y} = B\mathbf{x}$ ($\mathbf{x} = B^{-1}\mathbf{y}$). Then $\mathbf{0} = A\mathbf{y} = (AB)\mathbf{x}$, which again implies that AB is singular, a contradiction.

T.13. We want to show that $(A^{-1})^T = -A^{-1}$, i.e., A^{-1} is skew symmetric. We are given that $A^T = -A$ and A^{-1} exists, so

$$(A^{-1})^T = (A^T)^{-1} = (-A)^{-1} = -A^{-1}.$$

Hence A^{-1} is skew symmetric.

T.15. (a) Let A be nilpotent. If A were nonsingular, then products of A with itself are also nonsingular. (See T.11 above.) But $A^k = O$, hence A^k is singular. Thus A must be singular.

(b) $A^2 = \begin{bmatrix} 0 & 0 & 1 \\ 0 & 0 & 0 \\ 0 & 0 & 0 \end{bmatrix}$, $A^3 = O$.

(c) $k = 1$, $A = O$; $I_n - A = I_n$; $(I_n - A)^{-1} = I_n$.
$k = 2$, $A^2 = O$; $(I_n - A)(I_n + A) = I_n - A^2 = I_n$; $(I_n - A)^{-1} = I_n + A$.
$k = 3$, $A^3 = O$; $(I_n - A)(I_n + A + A^2) = I_n - A^3 = I_n$; $(I_n - A)^{-1} = I_n + A + A^2$, etc.

T.17. Let

$$A = \begin{bmatrix} 0 & a \\ -a & 0 \end{bmatrix} \quad \text{and} \quad B = \begin{bmatrix} 0 & b \\ -b & 0 \end{bmatrix}.$$

Then A and B are skew symmetric and

$$AB = \begin{bmatrix} -ab & 0 \\ 0 & -ab \end{bmatrix}$$

which is diagonal. The result is not true for $n > 2$. For example, let

$$A = \begin{bmatrix} 0 & 1 & 2 \\ -1 & 0 & 3 \\ -2 & -3 & 0 \end{bmatrix}.$$

Then

$$A^2 = \begin{bmatrix} 5 & 6 & -3 \\ 6 & 10 & 2 \\ -3 & 2 & 13 \end{bmatrix}.$$

T.19. (a) $\mathbf{x}\mathbf{y}^T = \begin{bmatrix} 4 & 5 & 6 \\ 8 & 10 & 12 \\ 12 & 15 & 18 \end{bmatrix}.$ (b) $\mathbf{x}\mathbf{y}^T = \begin{bmatrix} -1 & 0 & 3 & 5 \\ -2 & 0 & 6 & 10 \\ -1 & 0 & 3 & 5 \\ -2 & 0 & 6 & 10 \end{bmatrix}.$

T.21. $\text{Tr}(\mathbf{x}\mathbf{y}^T) = x_1 y_1 + x_2 y_2 + \cdots + x_n y_n = \mathbf{x}^T \mathbf{y}$. (See discussion preceding Exercises T.19–T.22.)

T.23. (a) $H^T = (I_n - 2\mathbf{w}\mathbf{w}^T)^T = I_n^T - 2(\mathbf{w}\mathbf{w}^T)^T = I_n - 2(\mathbf{w}^T)^T \mathbf{w}^T = I_n - 2\mathbf{w}\mathbf{w}^T = H$.

(b) $HH^T = HH = (I_n - 2\mathbf{w}\mathbf{w}^T)(I_n - 2\mathbf{w}\mathbf{w}^T)$
$= I_n - 4\mathbf{w}\mathbf{w}^T + 4\mathbf{w}\mathbf{w}^T\mathbf{w}\mathbf{w}^T$
$= I_n - 4\mathbf{w}\mathbf{w}^T + 4\mathbf{w}(\mathbf{w}^T\mathbf{w})\mathbf{w}^T$
$= I_n - 4\mathbf{w}\mathbf{w}^T + 4\mathbf{w}(I_n)\mathbf{w}^T = I_n$

Thus, $H^T = H^{-1}$.

Chapter 2

Applications of Linear Equations and Matrices (Optional)

Section 2.1, p. 123

1. Let e be the function from B^3 to B^4 given by

$$e(b_1 b_2 b_3) = b_1 b_2 b_3 b_4, \quad \text{where } b_4 = b_1 + b_3.$$

   (a) We have

$$e(000) = 0000$$
$$e(100) = 1001$$
$$e(010) = 0100$$
$$e(001) = 0011$$
$$e(110) = 1101$$
$$e(101) = 1010$$
$$e(011) = 0111$$
$$e(111) = 1110$$

   and it follows that e is one-to-one.

   (b) Let $A = \begin{bmatrix} 1 & 0 & 0 \\ 0 & 1 & 0 \\ 0 & 0 & 1 \\ 1 & 0 & 1 \end{bmatrix}$. Then

$$e(b_1 b_2 b_3) = \begin{bmatrix} 1 & 0 & 0 \\ 0 & 1 & 0 \\ 0 & 0 & 1 \\ 1 & 0 & 1 \end{bmatrix} \begin{bmatrix} b_1 \\ b_2 \\ b_3 \end{bmatrix} = \begin{bmatrix} b_1 \\ b_2 \\ b_3 \\ b_1 + b_3 \end{bmatrix} = \begin{bmatrix} b_1 \\ b_2 \\ b_3 \\ b_4 \end{bmatrix}.$$

3. Let e be the function from B^3 to B^2 given by

$$e(b_1 b_2 b_3) = b_1 b_2.$$

   (a) We have, for example,

$$e(000) = 00 \quad \text{and} \quad e(001) = 00$$
$$e(100) = 10 \quad \text{and} \quad e(101) = 10,$$

so e is clearly not one-to-one. In fact,

$$e(b_1 b_2 0) = e(b_1 b_2 1) = b_1 b_2$$

but $b_1 b_2 0 \neq b_1 b_2 1$.

(b) Let $A = \begin{bmatrix} 1 & 0 & 0 \\ 0 & 1 & 0 \end{bmatrix}$. Then

$$e(b_1 b_2 b_3) = \begin{bmatrix} 1 & 0 & 0 \\ 0 & 1 & 0 \end{bmatrix} \begin{bmatrix} b_1 \\ b_2 \\ b_3 \end{bmatrix} = \begin{bmatrix} b_1 \\ b_2 \end{bmatrix}.$$

5. The weight of $\mathbf{x}$, denoted $|\mathbf{x}|$, is the number of 1s in $\mathbf{x}$.

   (a) $|01110| = 3$
   (b) $|10101| = 3$
   (c) $|11000| = 2$
   (d) $|00010| = 1$

7. (a) $|1101| = 3$ so 1101 has odd parity
   (b) $|0011| = 2$ so 0011 has even parity
   (c) $|0100| = 1$ so 0100 has odd parity
   (d) $|0000| = 0$ so 0000 has even parity

9. The parity $(4, 5)$ check code detects an odd number of errors.

   (a) $\mathbf{x}_t = 10100$, $|\mathbf{x}_t| = 2$. Since $|1010| = 2$ and $\mathbf{x}_t$ has even parity an error will not be detected.
   (b) $\mathbf{x}_t = 01101$, $|\mathbf{x}_t| = 3$. Since $|0110| = 2$ and $\mathbf{x}_t$ has odd parity an error will be detected.
   (c) $\mathbf{x}_t = 11110$, $|\mathbf{x}_t| = 4$. Since $|1111| = 4$ and $\mathbf{x}_t$ has even parity an error will not be detected.
   (d) $\mathbf{x}_t = 10000$, $|\mathbf{x}_t| = 1$. Since $|1000| = 1$ and $\mathbf{x}_t$ has odd parity an error will be detected.

11. (a) The parity $(2, 3)$ check code produces the code words

    $$e(00) = 000, \quad e(01) = 011, \quad e(10) = 101, \quad \text{and} \quad e(11) = 110.$$

    (b) (i) $\mathbf{x}_t = 011$, $|\mathbf{x}_t| = 2$. Since $|01| = 1$ and $\mathbf{x}_t$ has even parity an error will not be detected.
    (ii) $\mathbf{x}_t = 111$, $|\mathbf{x}_t| = 3$. Since $|11| = 2$ and $\mathbf{x}_t$ has odd parity an error will be detected.
    (iii) $\mathbf{x}_t = 010$, $|\mathbf{x}_t| = 1$. Since $|01| = 1$ and $\mathbf{x}_t$ has odd parity an error will be detected.
    (iv) $\mathbf{x}_t = 001$, $|\mathbf{x}_t| = 1$. Since $|00| = 0$ and $\mathbf{x}_t$ has odd parity an error will be detected.

T.1. # of words of weight 0 in $B^2 = 1$: $\{00\}$
# of words of weight 1 in $B^2 = 2$: $\{10, 01\}$
# of words of weight 2 in $B^2 = 1$: $\{11\}$

T.3 # of words of weight 1 in $B^n = n$: all 0-1 strings of length n with exactly $n - 1$ zeros and 1 one.

# of words of weight 2 in $B^n = \binom{n}{2} = \frac{n(n-1)}{2}$: all 0-1 strings of length n with exactly $n - 2$ zeros and 2 ones.

T.5. (a) $e(00) = 0000$, $e(01) = 0101$, $e(10) = 1010$, $e(11) = 1111$.

(b) Yes.

(c) A parity check would find only an odd number of errors. It would not be able to see 1100, a word with 2 errors, as a problem.

ML.1. (a) >> **M** = **bingen(0, 15, 4)**

```
M   =
      0 0 0 0 0 0 0 0 1 1 1 1 1 1 1 1
      0 0 0 0 1 1 1 1 0 0 0 0 1 1 1 1
      0 0 1 1 0 0 1 1 0 0 1 1 0 0 1 1
      0 1 0 1 0 1 0 1 0 1 0 1 0 1 0 1
```

(b) >> **s** = **sum(M)**

```
s   =
      0  1  1  2  1  2  2  3  1  2  2  3  2  3  3  4
```

(c) >> **w** = **[s]**

```
w   =
      0  1  1  2  1  2  2  3  1  2  2  3  2  3  3  4
```

(d) >> **C** = **[M′ w′]′**

```
C   =
      0 0 0 0 0 0 0 0 1 1 1 1 1 1 1 1
      0 0 0 0 1 1 1 1 0 0 0 0 1 1 1 1
      0 0 1 1 0 0 1 1 0 0 1 1 0 0 1 1
      0 1 0 1 0 1 0 1 0 1 0 1 0 1 0 1
      0 1 1 2 1 2 2 3 1 2 2 3 2 3 3 4
```

Section 2.2, p. 134

1. (a)

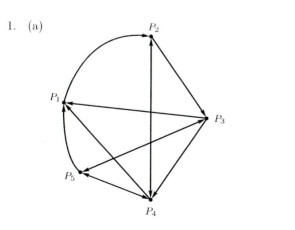

(b)

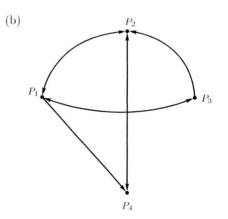

3.

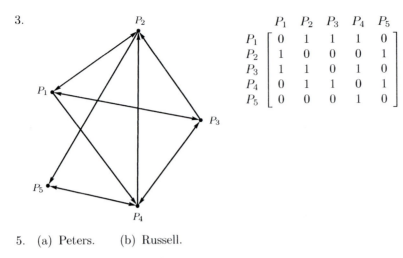

$$\begin{array}{c c c c c c} & P_1 & P_2 & P_3 & P_4 & P_5 \\ \begin{array}{c} P_1 \\ P_2 \\ P_3 \\ P_4 \\ P_5 \end{array} & \left[\begin{array}{c c c c c} 0 & 1 & 1 & 1 & 0 \\ 1 & 0 & 0 & 0 & 1 \\ 1 & 1 & 0 & 1 & 0 \\ 0 & 1 & 1 & 0 & 1 \\ 0 & 0 & 0 & 1 & 0 \end{array}\right] \end{array}$$

5. (a) Peters.   (b) Russell.

7. (a) No.   (b) 3.   (c) 5.

9. P_1, P_4, P_5, and P_6.

11. (a) No.   (b) 3.   (c) 4.

13. (a) Strongly connected.   (b) Not strongly connected.

T.1. In a dominance digraph, for each i and j, it is not the case that both P_i dominates P_j and P_j dominates P_i.

T.3. The implication in one direction is proved in the discussion following the theorem. Next suppose P_i belongs to the clique $\{P_i, P_j, P_k, \ldots, P_m\}$. According to the definition of clique, it contains at least three vertices so we may assume P_i, P_j and P_k all exist in the clique. Then $s_{ij} = s_{ji} = s_{jk} = s_{kj} = s_{ik} = s_{ki} = 1$ and $s_{ii}^{(3)}$ is a sum of nonnegative integer terms including the positive term which represents three stage access from P_i to P_j to P_k to P_i. Thus $s_{ii}^{(3)}$ is positive.

ML.1. **A = [0 0 0 0 0;1 0 1 1 1;0 1 0 1 0;1 1 1 0 0;0 0 1 1 0];**

S = zeros(size(A)); for i = 1:5, for j = 1:5,

if A(i,j) == 1&A(j,i) == 1&j ~= i,S(i,j) = 1;S(j,i) = 1;end,end,end,S

S =

```
    0   0   0   0   0
    0   0   1   1   0
    0   1   0   1   0
    0   1   1   0   0
    0   0   0   0   0
```

Next we compute S3 as follows.

S^3

ans =

```
    0   0   0   0   0
    0   2   3   3   0
    0   3   2   3   0
    0   3   3   2   0
    0   0   0   0   0
```

It follows that P2, P3, and P4 form a clique.

ML.3. We use Theorem 2.3.

(a) $\mathbf{A} = [0 \ \ 0 \ \ 1 \ \ 1 \ \ 1; 1 \ \ 0 \ \ 1 \ \ 1 \ \ 0; 0 \ \ 1 \ \ 0 \ \ 0 \ \ 0; 0 \ \ 1 \ \ 0 \ \ 0 \ \ 1; 1 \ \ 1 \ \ 0 \ \ 0 \ \ 0];$

Here $n = 5$, so we form

$\mathbf{E} = \mathbf{A} + \mathbf{A}^\wedge 2 + \mathbf{A}^\wedge 3 + \mathbf{A}^\wedge 4$

$\mathbf{E} =$

7	13	10	10	11
10	11	12	12	9
5	7	4	4	3
8	13	7	7	8
11	11	9	9	6

Since E has no zero entries the digraph represented by A is strongly connected.

(b) $\mathbf{A} = [0 \ \ 0 \ \ 0 \ \ 0 \ \ 1; 0 \ \ 0 \ \ 1 \ \ 1 \ \ 0; 0 \ \ 1 \ \ 0 \ \ 0 \ \ 1; 1 \ \ 0 \ \ 0 \ \ 0 \ \ 0; 0 \ \ 0 \ \ 0 \ \ 1 \ \ 0];$

Here $n = 5$, so we form

$\mathbf{E} = \mathbf{A} + \mathbf{A}^\wedge 2 + \mathbf{A}^\wedge 3 + \mathbf{A}^\wedge 4$

$\mathbf{E} =$

1	0	0	1	2
3	2	2	4	3
2	2	2	4	4
2	0	0	1	1
1	0	0	2	1

Since E has zero entries the digraph represented by A is not strongly connected.

Section 2.3, p. 141

1. Let $A = \begin{bmatrix} -1 & 0 \\ 0 & 1 \end{bmatrix}$. Then $f(\mathbf{v}) = A\mathbf{v}$ is a reflection with respect to the y-axis. Let R be the rectangle with vertices $(1,1)$, $(2,1)$, $(1,3)$, $(2,3)$. Then

$$f\left(\begin{bmatrix} 1 \\ 1 \end{bmatrix}\right) = \begin{bmatrix} -1 & 0 \\ 0 & 1 \end{bmatrix} \begin{bmatrix} 1 \\ 1 \end{bmatrix} = \begin{bmatrix} -1 \\ 1 \end{bmatrix}$$

$$f\left(\begin{bmatrix} 2 \\ 1 \end{bmatrix}\right) = \begin{bmatrix} -1 & 0 \\ 0 & 1 \end{bmatrix} \begin{bmatrix} 2 \\ 1 \end{bmatrix} = \begin{bmatrix} -2 \\ 1 \end{bmatrix}$$

$$f\left(\begin{bmatrix} 1 \\ 3 \end{bmatrix}\right) = \begin{bmatrix} -1 & 0 \\ 0 & 1 \end{bmatrix} \begin{bmatrix} 1 \\ 3 \end{bmatrix} = \begin{bmatrix} -1 \\ 3 \end{bmatrix}$$

$$f\left(\begin{bmatrix} 2 \\ 3 \end{bmatrix}\right) = \begin{bmatrix} -1 & 0 \\ 0 & 1 \end{bmatrix} \begin{bmatrix} 2 \\ 3 \end{bmatrix} = \begin{bmatrix} -2 \\ 3 \end{bmatrix}$$

The image of R is a rectangle with vertices $(-1,1)$, $(-2,1)$, $(-1,3)$, and $(-2,3)$, as shown in the figure.

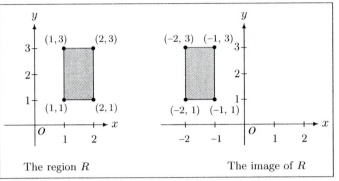

3. Let $A = \begin{bmatrix} 1 & 0 \\ -2 & 1 \end{bmatrix}$. Then $f(\mathbf{v}) = A\mathbf{v}$ is a shear in the y-direction. Let R be the rectangle with vertices $(1,1)$, $(1,4)$, $(3,1)$, $(3,4)$. Then

$$f\left(\begin{bmatrix} 1 \\ 1 \end{bmatrix}\right) = \begin{bmatrix} 1 & 0 \\ -2 & 1 \end{bmatrix}\begin{bmatrix} 1 \\ 1 \end{bmatrix} = \begin{bmatrix} 1 \\ -1 \end{bmatrix} \qquad f\left(\begin{bmatrix} 1 \\ 4 \end{bmatrix}\right) = \begin{bmatrix} 1 & 0 \\ -2 & 1 \end{bmatrix}\begin{bmatrix} 1 \\ 4 \end{bmatrix} = \begin{bmatrix} 1 \\ 2 \end{bmatrix}$$

$$f\left(\begin{bmatrix} 3 \\ 1 \end{bmatrix}\right) = \begin{bmatrix} 1 & 0 \\ -2 & 1 \end{bmatrix}\begin{bmatrix} 3 \\ 1 \end{bmatrix} = \begin{bmatrix} 3 \\ -5 \end{bmatrix} \qquad f\left(\begin{bmatrix} 3 \\ 4 \end{bmatrix}\right) = \begin{bmatrix} 1 & 0 \\ -2 & 1 \end{bmatrix}\begin{bmatrix} 3 \\ 4 \end{bmatrix} = \begin{bmatrix} 3 \\ -2 \end{bmatrix}$$

5. Let $A = \begin{bmatrix} 2 & 0 \\ 0 & 1 \end{bmatrix}$. Then $f(\mathbf{v}) = A\mathbf{v}$ is a dilation in the x-direction. Let R be the unit square with vertices $(0,0)$, $(0,1)$, $(1,0)$, $(1,1)$. Then

$$f\left(\begin{bmatrix} 0 \\ 0 \end{bmatrix}\right) = \begin{bmatrix} 2 & 0 \\ 0 & 1 \end{bmatrix}\begin{bmatrix} 0 \\ 0 \end{bmatrix} = \begin{bmatrix} 0 \\ 0 \end{bmatrix} \qquad f\left(\begin{bmatrix} 1 \\ 0 \end{bmatrix}\right) = \begin{bmatrix} 2 & 0 \\ 0 & 1 \end{bmatrix}\begin{bmatrix} 1 \\ 0 \end{bmatrix} = \begin{bmatrix} 2 \\ 0 \end{bmatrix}$$

$$f\left(\begin{bmatrix} 0 \\ 1 \end{bmatrix}\right) = \begin{bmatrix} 2 & 0 \\ 0 & 1 \end{bmatrix}\begin{bmatrix} 0 \\ 1 \end{bmatrix} = \begin{bmatrix} 0 \\ 1 \end{bmatrix} \qquad f\left(\begin{bmatrix} 1 \\ 1 \end{bmatrix}\right) = \begin{bmatrix} 2 & 0 \\ 0 & 1 \end{bmatrix}\begin{bmatrix} 1 \\ 1 \end{bmatrix} = \begin{bmatrix} 2 \\ 1 \end{bmatrix}$$

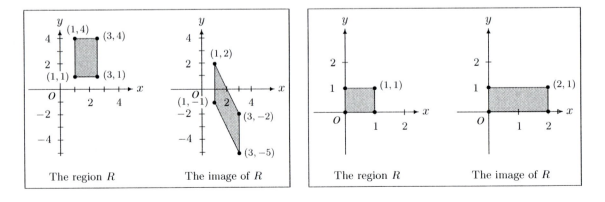

Exercise 3   Exercise 5

7. Let T be the triangle with vertices $(5,0)$, $(0,3)$, $(2,-1)$, and $A = \begin{bmatrix} -2 & 1 \\ 3 & 4 \end{bmatrix}$. Then $f(\mathbf{v}) = A\mathbf{v}$. We have

$$f\left(\begin{bmatrix} 5 \\ 0 \end{bmatrix}\right) = \begin{bmatrix} -2 & 1 \\ 3 & 4 \end{bmatrix} \begin{bmatrix} 5 \\ 0 \end{bmatrix} = \begin{bmatrix} -10 \\ 15 \end{bmatrix}$$

$$f\left(\begin{bmatrix} 0 \\ 3 \end{bmatrix}\right) = \begin{bmatrix} -2 & 1 \\ 3 & 4 \end{bmatrix} \begin{bmatrix} 0 \\ 3 \end{bmatrix} = \begin{bmatrix} 3 \\ 12 \end{bmatrix}$$

$$f\left(\begin{bmatrix} 2 \\ -1 \end{bmatrix}\right) = \begin{bmatrix} -2 & 1 \\ 3 & 4 \end{bmatrix} \begin{bmatrix} 2 \\ -1 \end{bmatrix} = \begin{bmatrix} -5 \\ 2 \end{bmatrix}.$$

The coordinates of the vertices of the image of T are therefore $(-10, 15)$, $(3, 12)$, and $(-5, 2)$

9. Let

$$A = \begin{bmatrix} \cos 60° & -\sin 60° \\ \sin 60° & \cos 60° \end{bmatrix} = \begin{bmatrix} \frac{1}{2} & -\frac{\sqrt{3}}{2} \\ \frac{\sqrt{3}}{2} & \frac{1}{2} \end{bmatrix}.$$

Then $f(\mathbf{v}) = A\mathbf{v}$ is a counterclockwise rotation through $60°$. (See Example 9 in Section 1.5). Let T be the triangle with vertices $(1,1)$, $(-3,-3)$, $(2,-1)$. Then

$$f\left(\begin{bmatrix} 1 \\ 1 \end{bmatrix}\right) = \begin{bmatrix} \frac{1}{2} & -\frac{1}{2}\sqrt{3} \\ \frac{1}{2}\sqrt{3} & \frac{1}{2} \end{bmatrix} \begin{bmatrix} 1 \\ 1 \end{bmatrix} = \begin{bmatrix} \frac{1}{2} - \frac{1}{2}\sqrt{3} \\ \frac{1}{2}\sqrt{3} + \frac{1}{2} \end{bmatrix} \approx \begin{bmatrix} -0.366 \\ 1.366 \end{bmatrix}$$

$$f\left(\begin{bmatrix} -3 \\ -3 \end{bmatrix}\right) = \begin{bmatrix} \frac{1}{2} & -\frac{1}{2}\sqrt{3} \\ \frac{1}{2}\sqrt{3} & \frac{1}{2} \end{bmatrix} \begin{bmatrix} -3 \\ -3 \end{bmatrix} = \begin{bmatrix} -\frac{3}{2} + \frac{3}{2}\sqrt{3} \\ -\frac{3}{2}\sqrt{3} - \frac{3}{2} \end{bmatrix} \approx \begin{bmatrix} 1.098 \\ -4.098 \end{bmatrix}$$

$$f\left(\begin{bmatrix} 2 \\ -1 \end{bmatrix}\right) = \begin{bmatrix} \frac{1}{2} & -\frac{1}{2}\sqrt{3} \\ \frac{1}{2}\sqrt{3} & \frac{1}{2} \end{bmatrix} \begin{bmatrix} 2 \\ -1 \end{bmatrix} = \begin{bmatrix} 1 + \frac{1}{2}\sqrt{3} \\ \sqrt{3} - \frac{1}{2} \end{bmatrix} \approx \begin{bmatrix} 1.866 \\ 1.232 \end{bmatrix}$$

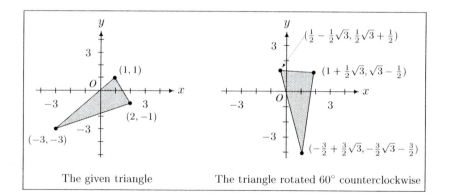

The given triangle   The triangle rotated $60°$ counterclockwise

Exercise 9

11. Let $A = \begin{bmatrix} 1 & 2 \\ 2 & 4 \end{bmatrix}$. Then $f(\mathbf{v}) = A\mathbf{v}$. Let T be the triangle with vertices $(1,1)$, $(-3,-3)$, $(2,-1)$.

Then

$$f\left(\begin{bmatrix}1\\1\end{bmatrix}\right) = \begin{bmatrix}1&2\\2&4\end{bmatrix}\begin{bmatrix}1\\1\end{bmatrix} = \begin{bmatrix}3\\6\end{bmatrix}$$

$$f\left(\begin{bmatrix}-3\\-3\end{bmatrix}\right) = \begin{bmatrix}1&2\\2&4\end{bmatrix}\begin{bmatrix}-3\\-3\end{bmatrix} = \begin{bmatrix}-9\\-18\end{bmatrix}$$

$$f\left(\begin{bmatrix}2\\-1\end{bmatrix}\right) = \begin{bmatrix}1&2\\2&4\end{bmatrix}\begin{bmatrix}2\\-1\end{bmatrix} = \begin{bmatrix}0\\0\end{bmatrix}.$$

The images of the vertices of T are the points $(3,6)$, $(-9,-18)$, and $(0,0)$ which are colinear. Thus the image of T is a line segment.

13. Let $A = \begin{bmatrix}1 & -1\\2 & 3\end{bmatrix}$. Then $f(\mathbf{v}) = A\mathbf{v}$. Let R be the unit square with vertices $(0,0)$, $(0,1)$, $(1,0)$, $(1,1)$. Then

We have

$$f\left(\begin{bmatrix}0\\0\end{bmatrix}\right) = \begin{bmatrix}1&-1\\2&3\end{bmatrix}\begin{bmatrix}0\\0\end{bmatrix} = \begin{bmatrix}0\\0\end{bmatrix} \qquad f\left(\begin{bmatrix}1\\0\end{bmatrix}\right) = \begin{bmatrix}1&-1\\2&3\end{bmatrix}\begin{bmatrix}1\\0\end{bmatrix} = \begin{bmatrix}1\\2\end{bmatrix}$$

$$f\left(\begin{bmatrix}0\\1\end{bmatrix}\right) = \begin{bmatrix}1&-1\\2&3\end{bmatrix}\begin{bmatrix}0\\1\end{bmatrix} = \begin{bmatrix}-1\\3\end{bmatrix} \qquad f\left(\begin{bmatrix}1\\1\end{bmatrix}\right) = \begin{bmatrix}1&-1\\2&3\end{bmatrix}\begin{bmatrix}1\\1\end{bmatrix} = \begin{bmatrix}0\\5\end{bmatrix}$$

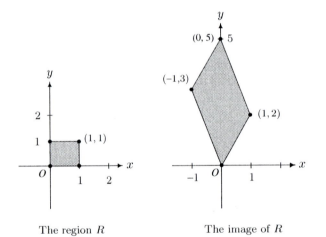

The region R The image of R

15. (a) Method 1. First apply f_1 ($\phi = 90°$):

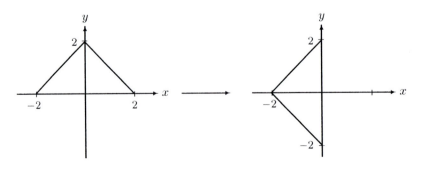

Then apply f_3:

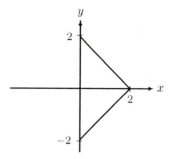

Method 2. First apply f_2:

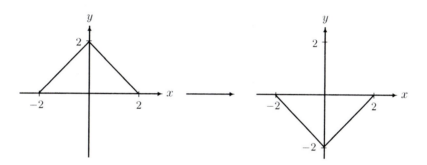

Then apply f_1 ($\phi = 90°$):

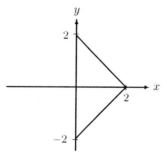

(b) Method 1. Apply f_1 ($\phi = -135°$):

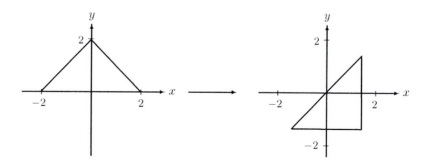

Method 2. First apply f_1 ($\phi = 45°$):

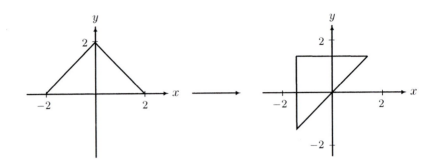

Then apply f_4:

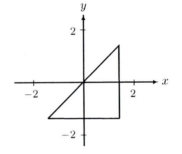

ML.1. (c) The circle was compressed in the x-coordinates so that it looked flattened along the vertical axis. The y-coordinates were unchanged. The figure from (b) was more compressed than (a).

(d) The new image would lie inside the previous one, touching the other images only at the points $(0, 1)$ and $(0, -1)$.

ML.3. (a) The area of the house is 5. The area of the sheared house with $k = 1$ is 5.

(b) When $k = .5$, the area is 5.

(c) When $k = 2$, the area is 5. To see this consider the base of the house. Without shearing, the shape is a rectangle with base 2 and height 2 and hence area 4. After a shear with $k = 1$, the rectangle changes to a parallelogram with the same base and height and hence the same area. A similar argument works for the other k-values. One can think of the triangle that forms the top of the house as half of another parallelogram.

ML.7. (a) $\text{proj}_{\mathbf{w}}\mathbf{u}$ is longer than $\mathbf{w}$ and in the same direction.

(b) $\text{proj}_{\mathbf{w}}\mathbf{u}$ is shorter than $\mathbf{w}$ and in the opposite direction.

(c) $\text{proj}_{\mathbf{w}}\mathbf{u}$ is shorter than $\mathbf{w}$ and in the same direction.

(d) $\text{proj}_{\mathbf{w}}\mathbf{u}$ is shorter than $\mathbf{w}$ and in the same direction.

Section 2.4, p. 148

1. We first assign directions to the currents as shown in Figure A.

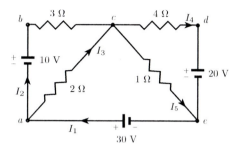

Figure A

We now obtain the following linear equations:

$$
\begin{aligned}
I_1 - I_2 - I_3 &= 0 & \text{(node } a\text{)} \\
I_1 - I_4 - I_5 &= 0 & \text{(node } e\text{)} \\
30 - 2I_3 - I_5 &= 0 & \text{(loop } acea\text{)} \\
10 - 3I_2 + 2I_3 &= 0 & \text{(loop } abca\text{)} \\
20 + 4I_4 - I_5 &= 0 & \text{(loop } edce\text{)},
\end{aligned}
$$

a linear system that can be written in matrix form as

$$
\begin{bmatrix}
1 & -1 & -1 & 0 & 0 \\
1 & 0 & 0 & -1 & -1 \\
0 & 0 & 2 & 0 & 1 \\
0 & 3 & -2 & 0 & 0 \\
0 & 0 & 0 & -4 & 1
\end{bmatrix}
\begin{bmatrix}
I_1 \\ I_2 \\ I_3 \\ I_4 \\ I_5
\end{bmatrix}
=
\begin{bmatrix}
0 \\ 0 \\ 30 \\ 10 \\ 20
\end{bmatrix}.
$$

Solving this linear system by Gauss–Jordan reduction, we have

$$
\begin{aligned}
&I_1 = 15 \text{ A from } e \text{ to } a, &&I_2 = 8 \text{ A from } a \text{ to } b, \\
&I_3 = 7 \text{ A from } a \text{ to } c, &&I_4 = -1 \text{ A from } c \text{ to } d \text{ (or 1 A from } d \text{ to } c\text{)}, \\
&I_5 = 16 \text{ A from } c \text{ to } e.
\end{aligned}
$$

3. We first assign directions to the currents as shown in Figure B.

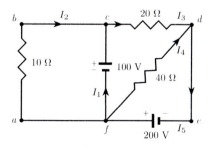

Figure B

We now obtain the following linear system:

$$I_1 + I_2 - I_3 = 0 \qquad \text{(node } c)$$
$$I_3 + I_4 - I_5 = 0 \qquad \text{(node } d)$$
$$100 + 10I_2 = 0 \qquad \text{(loop } fcbaf)$$
$$200 - 40I_4 = 0 \qquad \text{(loop } fdef)$$
$$-40I_4 + 20I_3 - 100 = 0 \qquad \text{(loop } efabcde),$$

which can be written in matrix form as

$$\begin{bmatrix} 1 & 1 & -1 & 0 & 0 \\ 0 & 0 & 1 & 1 & -1 \\ 0 & 10 & 0 & 0 & 0 \\ 0 & 0 & 20 & -40 & 0 \\ 0 & 0 & 0 & 40 & 0 \end{bmatrix} \begin{bmatrix} I_1 \\ I_2 \\ I_3 \\ I_4 \\ I_5 \end{bmatrix} = \begin{bmatrix} 0 \\ 0 \\ -100 \\ 100 \\ 200 \end{bmatrix}.$$

Solving this linear system by Gauss–Jordan reduction, we have

$$I_1 = 25 \text{ A from } f \text{ to } c, \qquad I_2 = -10 \text{ A from } b \text{ to } c \text{ (or 10 A from } c \text{ to } b),$$
$$I_3 = 15 \text{ A from } c \text{ to } d, \qquad I_4 = 5 \text{ A from } f \text{ to } d,$$
$$I_5 = 20 \text{ A from } e \text{ to } f.$$

5. We first assign directions to the currents as shown in Figure C.

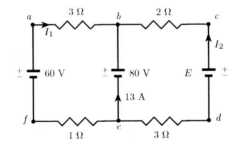

Figure C

We now obtain the following linear equations

$$I_1 + 13 + I_2 = 0 \qquad \text{(node } b)$$
$$60 - 3I_1 - 80 - I_1 = 0 \qquad \text{(loop } abefa)$$
$$80 + 2I_2 - E + 3I_2 = 0 \qquad \text{(loop } bcdeb),$$

which can be written in matrix form as

$$\begin{bmatrix} 1 & 1 & 0 \\ 4 & 0 & 0 \\ 0 & 5 & -1 \end{bmatrix} \begin{bmatrix} I_1 \\ I_2 \\ E \end{bmatrix} = \begin{bmatrix} -13 \\ -20 \\ -80 \end{bmatrix}.$$

Solving this linear system by Gauss-Jordan reduction we have

$$I_1 = -5 \text{ A from } a \text{ to } b \text{ (or 5 A from } b \text{ to } a),$$
$$I_2 = -8 \text{ A from } d \text{ to } c \text{ (or 8 A from } c \text{ to } d),$$
$$E = 40 \text{ V}.$$

7. We first assign directions to the currents as shown in Figure D.

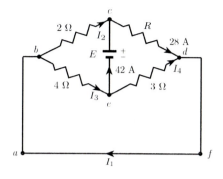

Figure D

We now obtain the following linear equations

$$I_1 - I_2 - I_3 = 0 \quad \text{(node } b)$$
$$I_2 + 42 - 28 = 0 \quad \text{(node } c)$$
$$-I_1 + I_4 + 28 = 0 \quad \text{(node } d)$$
$$E + 2I_2 - 4I_3 = 0 \quad \text{(loop } cbec)$$
$$-4I_3 - 3I_4 = 0 \quad \text{(loop } abedfa)$$
$$-2I_2 - 28R = 0 \quad \text{(loop } abcdfa),$$

which can be written in matrix form as

$$\begin{bmatrix} 1 & -1 & -1 & 0 & 0 & 0 \\ 0 & 1 & 0 & 0 & 0 & 0 \\ 1 & 0 & 0 & -1 & 0 & 0 \\ 0 & 2 & -4 & 0 & 0 & 1 \\ 0 & 0 & 4 & 3 & 0 & 0 \\ 0 & 2 & 0 & 0 & 28 & 0 \end{bmatrix} \begin{bmatrix} I_1 \\ I_2 \\ I_3 \\ I_4 \\ R \\ E \end{bmatrix} = \begin{bmatrix} 0 \\ -14 \\ 28 \\ 0 \\ 0 \\ 0 \end{bmatrix}.$$

Solving this linear system by Gauss–Jordan reduction, we have

$$I_1 = 4 \text{ A from } f \text{ to } a, \qquad I_2 = -14 \text{ A from } b \text{ to } c \text{ (or 14 A from } c \text{ to } b),$$
$$I_3 = 18 \text{ A from } b \text{ to } e, \qquad I_4 = 24 \text{ A from } d \text{ to } e,$$
$$R = 1\Omega \qquad E = 100 \text{ V}.$$

T.1. We choose the following directions for the currents:

$$I : a \text{ to } b$$
$$I_1 : b \text{ to } e$$
$$I_2 : b \text{ to } c.$$

Then we have the following linear equations

$$I - \quad I_1 - \quad I_2 = 0 \quad \text{(node } b)$$
$$- R_1 I_1 + R_2 I_2 = 0 \quad \text{(loop } bcdeb)$$

which leads to the linear system

$$\begin{bmatrix} 1 & 1 \\ R_1 & -R_2 \end{bmatrix} \begin{bmatrix} I_1 \\ I_2 \end{bmatrix} = \begin{bmatrix} I \\ 0 \end{bmatrix}$$

whose solution is

$$I_1 = \left(\frac{R_2}{R_1 + R_2}\right) I = \frac{R}{R_1} I \quad \text{and} \quad I_2 = \left(\frac{R_1}{R_1 + R_2}\right) I = \frac{R}{R_2} I,$$

where

$$R = \frac{R_1 R_2}{R_1 + R_2},$$

so

$$\frac{1}{R} = \frac{1}{R_1} + \frac{1}{R_2}.$$

Section 2.5, p. 157

1. T is a transition matrix if and only if $0 \le t_{ij} \le 1$ for $1 \le i, j \le n$ and the sum of the entries in each column is 1.

   (a) Let $T = \begin{bmatrix} .3 & .7 \\ .4 & .6 \end{bmatrix}$. T is not a transition matrix because the sum of the entries in each column is not 1.

   (b) Let $T = \begin{bmatrix} .2 & .3 & .1 \\ .8 & .5 & .7 \\ 0 & .2 & .2 \end{bmatrix}$. T is a transition matrix.

   (c) Let $T = \begin{bmatrix} .55 & .33 \\ .45 & .67 \end{bmatrix}$. T is a transition matrix.

   (d) Let $T = \begin{bmatrix} .3 & .4 & .2 \\ .2 & 0 & .8 \\ .1 & .3 & .6 \end{bmatrix}$. T is not a transition matrix because the sum of the entries in each column is not 1.

3. The sum of the entries in column 2 must equal 1. Therefore the missing entry is equal to 0.4. Similarly, the missing entry in column 3 is equal to 0.2. In column 1, the two missing entries must be nonnegative and add up to 0.7. There are infinitely many possibilities; for example,

$$\begin{bmatrix} 0.1 & 0.4 & 0.3 \\ 0.3 & 0.4 & 0.5 \\ 0.6 & 0.2 & 0.2 \end{bmatrix}, \quad \begin{bmatrix} 0.2 & 0.4 & 0.3 \\ 0.3 & 0.4 & 0.5 \\ 0.5 & 0.2 & 0.2 \end{bmatrix}, \quad \text{and} \quad \begin{bmatrix} 0.5 & 0.4 & 0.3 \\ 0.3 & 0.4 & 0.5 \\ 0.2 & 0.2 & 0.2 \end{bmatrix}.$$

5. Let $T = \begin{bmatrix} .7 & .4 \\ .3 & .6 \end{bmatrix}$.

   (a) For $\mathbf{x}^{(0)} = \begin{bmatrix} 1 \\ 0 \end{bmatrix}$,

$$\mathbf{x}^{(1)} = T\mathbf{x}^{(0)} = \begin{bmatrix} .700 \\ .300 \end{bmatrix}, \quad \mathbf{x}^{(2)} = T\mathbf{x}^{(1)} = \begin{bmatrix} .610 \\ .390 \end{bmatrix}, \quad \mathbf{x}^{(3)} = T\mathbf{x}^{(2)} = \begin{bmatrix} .583 \\ .417 \end{bmatrix}.$$

   (b) All the entries of T are positive thus T is regular. Following the method in Example 8, we solve for a vector $\mathbf{u}$ such that $T\mathbf{u} = \mathbf{u}$. Hence solve the homogeneous system

$$(I_3 - T)\mathbf{u} = \begin{bmatrix} .3 & -.4 \\ -.3 & .4 \end{bmatrix} \begin{bmatrix} u_1 \\ u_2 \end{bmatrix} = \mathbf{0}.$$

Row reduce the coefficient matrix to obtain

$$\begin{bmatrix} 1 & -\frac{4}{3} \\ 0 & 0 \end{bmatrix}.$$

The solution is $u_1 = \frac{4}{3}r$, $u_2 = r$, where r is any real number. Since $\mathbf{u}$ is to be a probability vector, we require that

$$u_1 + u_2 = \frac{4}{3}r + r = \frac{7}{3}r = 1.$$

Set $r = \frac{3}{7}$. Then the steady-state vector is

$$\mathbf{u} = \begin{bmatrix} \frac{4}{7} \\ \frac{3}{7} \end{bmatrix}.$$

7. (a) $T = \begin{bmatrix} 0 & \frac{1}{2} \\ 1 & \frac{1}{2} \end{bmatrix}$ is regular since $T^2 = \begin{bmatrix} \frac{1}{2} & \frac{1}{4} \\ \frac{1}{2} & \frac{3}{4} \end{bmatrix} > \mathbf{0}$.

(b) $T = \begin{bmatrix} \frac{1}{2} & 0 & 0 \\ 0 & 1 & \frac{1}{2} \\ \frac{1}{2} & 0 & \frac{1}{2} \end{bmatrix}$ is not regular since T^k will always have $\begin{bmatrix} 0 \\ 1 \\ 0 \end{bmatrix}$ as its second column.

(c) $T = \begin{bmatrix} 1 & \frac{1}{3} & 0 \\ 0 & \frac{1}{3} & 1 \\ 0 & \frac{1}{3} & 0 \end{bmatrix}$ is not regular since T^k will always have $\begin{bmatrix} 1 \\ 0 \\ 0 \end{bmatrix}$ as its first column.

(d) $T = \begin{bmatrix} \frac{1}{4} & \frac{3}{5} & \frac{1}{2} \\ \frac{1}{2} & 0 & 0 \\ \frac{1}{4} & \frac{2}{5} & \frac{1}{2} \end{bmatrix}$ is regular since $T^2 = \begin{bmatrix} .4875 & .35 & .375 \\ .1250 & .30 & .250 \\ .3875 & .35 & .375 \end{bmatrix} > \mathbf{0}$.

9. Let $T = \begin{bmatrix} \frac{1}{2} & 0 \\ \frac{1}{2} & 1 \end{bmatrix}$.

(a) T^k will have as its second column $\begin{bmatrix} 0 \\ 1 \end{bmatrix}$ for all k. Thus T is not regular.

(b) We have

$$T^2 = \begin{bmatrix} \frac{1}{2^2} & 0 \\ \frac{1}{2^2} + \frac{1}{2} & 1 \end{bmatrix},$$

$$T^3 = \begin{bmatrix} \frac{1}{2^3} & 0 \\ \frac{1}{2^3} + \frac{1}{2^2} + \frac{1}{2} & 1 \end{bmatrix},$$

$$\vdots$$

$$T^n = \begin{bmatrix} \frac{1}{2^n} & 0 \\ \frac{1}{2^n} + \frac{1}{2^{n-1}} + \cdots + \frac{1}{2} & 1 \end{bmatrix} = \begin{bmatrix} \frac{1}{2^n} & 0 \\ \frac{\frac{1}{2}\left(1 - \frac{1}{2^n}\right)}{1 - \frac{1}{2}} & 1 \end{bmatrix}.$$

This follows since

$$\frac{1}{2^n} + \frac{1}{2^{n-1}} + \cdots + \frac{1}{2}$$

is a geometric progression. Thus for

$$\mathbf{x} = \begin{bmatrix} x_1 \\ x_2 \end{bmatrix}$$

with $x_1 + x_2 = 1$, we have

$$T^n \mathbf{x} = \begin{bmatrix} \frac{1}{2^n} x_1 \\ \frac{\frac{1}{2}\left(1 - \frac{1}{2^n}\right)}{1 - \frac{1}{2}} x_1 + x_2 \end{bmatrix} \rightarrow \begin{bmatrix} 0 \\ x_1 + x_2 \end{bmatrix} = \begin{bmatrix} 0 \\ 1 \end{bmatrix}$$

as $n \to \infty$. Hence $\begin{bmatrix} 0 \\ 1 \end{bmatrix}$ is a unique steady-state vector.

11. (a) $\begin{array}{cc} A & B \end{array}$
$$T = \begin{bmatrix} .3 & .4 \\ .7 & .6 \end{bmatrix} \begin{array}{c} A \\ B \end{array}$$

(b) Compute $T\mathbf{x}^{(2)}$, where $\mathbf{x}^{(0)} = \begin{bmatrix} \frac{1}{2} \\ \frac{1}{2} \end{bmatrix}$:

$$T\mathbf{x}^{(0)} = \mathbf{x}^{(1)} = \begin{bmatrix} .35 \\ .65 \end{bmatrix}, \qquad T\mathbf{x}^{(1)} = \mathbf{x}^{(2)} = \begin{bmatrix} .365 \\ .635 \end{bmatrix}, \qquad T\mathbf{x}^{(2)} = \mathbf{x}^{(3)} = \begin{bmatrix} .364 \\ .636 \end{bmatrix}.$$

The probability of the rat going through door A on the third day is $p_1^{(3)} = .364$.

(c) Solve

$$(I_2 - T)\mathbf{u} = \begin{bmatrix} .7 & -.4 \\ -.7 & .4 \end{bmatrix} \begin{bmatrix} u_1 \\ u_2 \end{bmatrix} = \mathbf{0}.$$

Row reduce the coefficient matrix to obtain

$$\begin{bmatrix} 1 & -\frac{4}{7} \\ 0 & 0 \end{bmatrix}.$$

Thus the solution is $u_1 = \frac{4}{7}r$, $u_2 = r$, where r is any real number. Since **u** is a probability vector, $u_1 + u_2 = \frac{4}{7}r + r = 1$. Hence $r = \frac{7}{11}$ and

$$\mathbf{u} = \begin{bmatrix} \frac{4}{11} \\ \frac{7}{11} \end{bmatrix}$$

which is approximately $\begin{bmatrix} .364 \\ .636 \end{bmatrix}$.

13. Let $T = \begin{bmatrix} .8 & .3 & .2 \\ .1 & .5 & .2 \\ .1 & .2 & .6 \end{bmatrix}$.

   (a) We have that an individual is a professional. Thus the current state is

$$\mathbf{x}^{(0)} = \begin{bmatrix} 1 \\ 0 \\ 0 \end{bmatrix}.$$

   We are asked to determine the probability that a grandson will be a professional. This represents a transition through two stages:

$$\mathbf{x}^{(2)} = T\mathbf{x}^{(1)} = T(T\mathbf{x}^{(0)}) = T\begin{bmatrix} .8 \\ .1 \\ .1 \end{bmatrix} = \begin{bmatrix} .69 \\ .15 \\ .16 \end{bmatrix}.$$

   Thus the probability that a grandson of a professional will be a professional is .69.

   (b) The "long-run" distribution is the steady-state vector. Thus solve

$$(I_3 - T)\mathbf{u} = \begin{bmatrix} .2 & -.3 & -.2 \\ -.1 & .5 & -.2 \\ -.1 & -.2 & .4 \end{bmatrix} \begin{bmatrix} u_1 \\ u_2 \\ u_3 \end{bmatrix} = \mathbf{0}.$$

   Row reduce the coefficient matrix to obtain

$$\begin{bmatrix} 1 & 0 & -\frac{16}{7} \\ 0 & 1 & -\frac{6}{7} \\ 0 & 0 & 0 \end{bmatrix}.$$

   The solution is $u_1 = \frac{16}{7}r$, $u_2 = \frac{6}{7}r$, $u_3 = r$, where r is any real number. Since the steady-state vector is a probability vector we have

$$u_1 + u_2 + u_3 = \frac{16}{7}r + \frac{6}{7}r + r = 1.$$

   Thus $r = \frac{7}{29}$ and

$$\mathbf{u} = \begin{bmatrix} \frac{16}{29} \\ \frac{6}{29} \\ \frac{7}{29} \end{bmatrix} \approx \begin{bmatrix} .552 \\ .207 \\ .241 \end{bmatrix}.$$

   The proportion of the population that will be farmers is .207 or 20.7%.

15. Let

$$T = \begin{bmatrix} .7 & .2 \\ .3 & .8 \end{bmatrix}.$$

The initial state is that 30% of commuters use mass transit and 70% use their automobiles. Hence the initial state is

$$\mathbf{x}^{(0)} = \begin{bmatrix} .3 \\ .7 \end{bmatrix}.$$

(a) The state 1 year from now is calculated as

$$\mathbf{x}^{(1)} = T\mathbf{x}^{(0)} = \begin{bmatrix} .35 \\ .65 \end{bmatrix}.$$

The state 2 years from now is

$$\mathbf{x}^{(2)} = T\mathbf{x}^{(1)} = \begin{bmatrix} .375 \\ .625 \end{bmatrix}.$$

Hence after one year 35% of the commuters will be using mass transit and at the end of two years 37.5% will be using mass transit.

(b) Find the steady-state vector. Solve the homogeneous linear system

$$(I_2 - T)\mathbf{u} = \begin{bmatrix} .3 & -.2 \\ -.3 & .2 \end{bmatrix} \begin{bmatrix} u_1 \\ u_2 \end{bmatrix} = \mathbf{0}.$$

Row reduce the coefficient matrix to obtain

$$\begin{bmatrix} 1 & -\frac{2}{3} \\ 0 & 0 \end{bmatrix}.$$

The solution is $u_1 = \frac{2}{3}r$, $u_2 = r$, where r is any real number. Since the steady-state vector is a probability vector, $u_1 + u_2 = \frac{5}{3}r = 1$. Thus $r = \frac{3}{5}$ and

$$\mathbf{u} = \begin{bmatrix} \frac{2}{5} \\ \frac{3}{5} \end{bmatrix}.$$

In the long run 40% of commuters will be using mass transit.

T.1. No. If the sum of the entries of each column is 1, it does not follow that the sum of the entries in each column of A^T will also be 1. For example, the transpose of the transition matrix T in Exercise 6 is not the transition matrix of a Markov chain because none of the columns add up to 1:

$$T = \begin{bmatrix} 0 & 0.2 & 0.0 \\ 0 & 0.3 & 0.3 \\ 1 & 0.5 & 0.7 \end{bmatrix} \implies T^T = \begin{bmatrix} 0 & 0 & 1 \\ 0.2 & 0.3 & 0.5 \\ 0.0 & 0.3 & 0.7 \end{bmatrix}.$$

ML.2. Enter the matrix T and initial state vector $\mathbf{x}^{(0)}$ into MATLAB.

T = [.5 .6 .4;.25 .3 .3;.25 .1 .3];

x0 = [.1 .3 .6];

State vector $\mathbf{x}^{(5)}$ is given by

$$x5 = T\verb|^|5 * x0$$

$$x5 =$$

$$0.5055$$
$$0.2747$$
$$0.2198$$

ML.3. The command **sum** operating on a matrix computes the sum of the entries in each column and displays these totals as a row vector. If the output from the **sum** command is a row of ones, then the matrix is a Markov matrix.

(a) **A = [2/3 1/3 1/2;1/3 1/3 1/4;0 1/3 1/4];sum(A)**

ans =

$$1 \quad 1 \quad 1$$

Hence A is a Markov matrix.

(b) **A = [.5 .6 .7;.3 .2 .3;.1 .2 0];sum(A)**

ans =

$$0.9000 \quad 1.0000 \quad 1.0000$$

A is not a Markov matrix.

(c) **A = [.66 .25 .125;.33 .25 .625;0 .5 .25];sum(A)**

ans =

$$0.9900 \quad 1.0000 \quad 1.0000$$

A is not a Markov matrix.

Section 2.6, p. 165

1. Apply the definition of an exchange matrix.

    (a) This is not an exchange matrix since it has a negative entry: $a_{32} = -\frac{1}{2}$.

    (b) This is an exchange matrix: all entries are nonnegative and the sum of the entries in each column is 1.

    (c) This is not an exchange matrix since it has a negative entry: $a_{12} = -\frac{2}{3}$.

    (d) This is an exchange matrix: all entries are nonnegative and the sum of the entries in each column is 1.

3. Following Example 1, we find a solution $\mathbf{p}$ of the homogeneous system $(I_3 - A)\mathbf{p} = \mathbf{0}$ such that $\mathbf{p} \geq \mathbf{0}$ with at least one positive component. Let

$$A = \begin{bmatrix} \frac{1}{2} & 1 & \frac{2}{3} \\ 0 & 0 & 0 \\ \frac{1}{2} & 0 & \frac{1}{3} \end{bmatrix}.$$

Then

$$(I_3 - A)\mathbf{p} = \begin{bmatrix} \frac{1}{2} & -1 & -\frac{2}{3} \\ 0 & 1 & 0 \\ -\frac{1}{2} & 0 & \frac{2}{3} \end{bmatrix} \begin{bmatrix} p_1 \\ p_2 \\ p_3 \end{bmatrix} = \begin{bmatrix} 0 \\ 0 \\ 0 \end{bmatrix}.$$

Row reduce the coefficient matrix by using row operations $2\mathbf{r}_1$, $\frac{1}{2}\mathbf{r}_1 + \mathbf{r}_3$, $2\mathbf{r}_2 + \mathbf{r}_1$, $\mathbf{r}_2 + \mathbf{r}_3$, to obtain

$$\begin{bmatrix} 1 & 0 & -\frac{4}{3} \\ 0 & 1 & 0 \\ 0 & 0 & 0 \end{bmatrix}.$$

The solution of the homogeneous system is $p_1 = \frac{4}{3}r$, $p_2 = 0$, $p_3 = r$, where r is any real number. Set $r = 3$, then one vector satisfying the requirements is

$$\mathbf{p} = \begin{bmatrix} 4 \\ 0 \\ 3 \end{bmatrix}.$$

5. Follow the method of Example 1. Let p_1 be the price per unit of food, p_2 the price per unit of housing, and p_3 the price per unit of clothes. We have the following relations:

$$\text{farmer:} \quad \tfrac{2}{5}p_1 + \tfrac{1}{3}p_2 + \tfrac{1}{2}p_3 = p_1$$
$$\text{carpenter:} \quad \tfrac{2}{5}p_1 + \tfrac{1}{3}p_2 + \tfrac{1}{2}p_3 = p_2$$
$$\text{tailor:} \quad \tfrac{1}{5}p_1 + \tfrac{1}{3}p_2 + 0p_3 = p_3$$

Then the exchange matrix is

$$A = \begin{bmatrix} \frac{2}{5} & \frac{1}{3} & \frac{1}{2} \\ \frac{2}{5} & \frac{1}{3} & \frac{1}{2} \\ \frac{1}{5} & \frac{1}{3} & 0 \end{bmatrix}.$$

We solve

$$(I_3 - A)\mathbf{p} = \begin{bmatrix} \frac{3}{5} & -\frac{1}{3} & -\frac{1}{2} \\ -\frac{2}{5} & \frac{2}{3} & -\frac{1}{2} \\ -\frac{1}{5} & -\frac{1}{3} & 1 \end{bmatrix} \begin{bmatrix} p_1 \\ p_2 \\ p_3 \end{bmatrix} = \begin{bmatrix} 0 \\ 0 \\ 0 \end{bmatrix}.$$

Row reduce the coefficient matrix by applying row operations $\frac{5}{3}\mathbf{r}_1$, $\frac{2}{5}\mathbf{r}_1 + \mathbf{r}_2$, $\frac{1}{5}\mathbf{r}_1 + \mathbf{r}_3$, $\frac{3}{4}\mathbf{r}_2$, $\frac{4}{3}\mathbf{r}_2 + \mathbf{r}_3$ $\left(-\frac{5}{3}\right)\mathbf{r}_2 + \mathbf{r}_1$, to obtain

$$\begin{bmatrix} 1 & 0 & -\frac{15}{8} \\ 0 & 1 & -\frac{15}{8} \\ 0 & 0 & 0 \end{bmatrix}.$$

The general solution of the homogeneous system is $p_1 = \frac{15}{8}r$, $p_2 = \frac{15}{8}r$, $p_3 = r$, where r is any real number. Set $r = 40$, then one vector satisfying the requirements is

$$\mathbf{p} = \begin{bmatrix} 75 \\ 75 \\ 40 \end{bmatrix}.$$

In Exercises 7 and 9, we determine if $(I_n - C)^{-1}$ exists, and if it does whether all the entries are nonnegative. If both conditions are satisfied C is called productive.

7. Let $C = \begin{bmatrix} \frac{1}{2} & \frac{1}{3} & 0 \\ 0 & \frac{2}{3} & 0 \\ 1 & 0 & 2 \end{bmatrix}$. Then

$$I_3 - C = \begin{bmatrix} \frac{1}{2} & -\frac{1}{3} & 0 \\ 0 & \frac{1}{3} & 0 \\ -1 & 0 & -1 \end{bmatrix}.$$

Compute the determinant: $|I_3 - C| = -\frac{1}{6} \neq 0$. Thus $I_3 - C$ is nonsingular. To compute its inverse we proceed as follows. Form the matrix $\begin{bmatrix} I_3 - C \vdots I_3 \end{bmatrix}$ and row reduce it. The result is

$$\begin{bmatrix} 1 & 0 & 0 & \vdots & 2 & 2 & 0 \\ 0 & 1 & 0 & \vdots & 0 & 3 & 0 \\ 0 & 0 & 1 & \vdots & -2 & -2 & -1 \end{bmatrix}.$$

Thus

$$(I_3 - C)^{-1} = \begin{bmatrix} 2 & 2 & 0 \\ 0 & 3 & 0 \\ -2 & -2 & -1 \end{bmatrix}.$$

Since there are negative entries C is not productive.

9. Let $C = \begin{bmatrix} 0 & \frac{1}{3} & \frac{1}{2} \\ \frac{1}{2} & 0 & \frac{1}{4} \\ \frac{1}{4} & \frac{1}{3} & 0 \end{bmatrix}$. Then

$$I_3 - C = \begin{bmatrix} 1 & -\frac{1}{3} & -\frac{1}{2} \\ -\frac{1}{2} & 1 & -\frac{1}{4} \\ -\frac{1}{4} & -\frac{1}{3} & 1 \end{bmatrix}.$$

Compute the determinant: $|I_3 - C| = \frac{25}{48} \neq 0$. Thus $I_3 - C$ is nonsingular. To compute its inverse we proceed as follows. Form the matrix $\begin{bmatrix} I_3 - C \vdots I_3 \end{bmatrix}$ and row reduce it. The result is

$$\begin{bmatrix} 1 & 0 & 0 & \vdots & \frac{44}{25} & \frac{24}{25} & \frac{28}{25} \\ 0 & 1 & 0 & \vdots & \frac{27}{25} & \frac{42}{25} & \frac{24}{25} \\ 0 & 0 & 1 & \vdots & \frac{4}{5} & \frac{4}{5} & \frac{8}{5} \end{bmatrix}.$$

Thus

$$(I_3 - C)^{-1} = \begin{bmatrix} \frac{44}{25} & \frac{24}{25} & \frac{28}{25} \\ \frac{27}{25} & \frac{42}{25} & \frac{24}{25} \\ \frac{4}{5} & \frac{4}{5} & \frac{8}{5} \end{bmatrix}.$$

Since all the entries are nonnegative, C is productive.

11. Let

$$C = \begin{bmatrix} \frac{1}{2} & \frac{1}{2} \\ \frac{1}{2} & \frac{1}{4} \end{bmatrix}.$$

Following Example 7, we find $(I_2 - C)^{-1}$. Form the matrix $\begin{bmatrix} I_2 - C & \vdots & I_2 \end{bmatrix}$ and row reduce it. We obtain

$$\begin{bmatrix} 1 & 0 & \vdots & 6 & 4 \\ 0 & 1 & \vdots & 4 & 4 \end{bmatrix}.$$

Thus $(I_2 - C)^{-1} = \begin{bmatrix} 6 & 4 \\ 4 & 4 \end{bmatrix}$.

(a) Let $\mathbf{d} = \begin{bmatrix} 1 \\ 3 \end{bmatrix}$. Then the production vector is

$$\mathbf{x} = (I_3 - C)^{-1}\mathbf{d} = \begin{bmatrix} 18 \\ 16 \end{bmatrix}.$$

(b) Let $\mathbf{d} = \begin{bmatrix} 2 \\ 0 \end{bmatrix}$. Then the production vector is

$$\mathbf{x} = (I_3 - C)^{-1}\mathbf{d} = \begin{bmatrix} 12 \\ 8 \end{bmatrix}.$$

T.1. We must show that for an exchange matrix A and vector $\mathbf{p}$, $A\mathbf{p} \le \mathbf{p}$ implies $A\mathbf{p} = \mathbf{p}$. Let $A = \begin{bmatrix} a_{ij} \end{bmatrix}$, $\mathbf{p} = \begin{bmatrix} p_j \end{bmatrix}$, and $A\mathbf{p} = \mathbf{y} = \begin{bmatrix} y_j \end{bmatrix}$. Then

$$\sum_{j=1}^{n} y_j = \sum_{j=1}^{n}\sum_{k=1}^{n} a_{jk}p_k = \sum_{k=1}^{n}\left(\sum_{j=1}^{n} a_{jk}\right)p_k = \sum_{k=1}^{n} p_k$$

since the sum of the entries in the kth column of A is 1.

Since $y_j \le p_j$ for $j = 1, \ldots, n$ and $\sum y_j = \sum p_j$, the respective entries must be equal: $y_j = p_j$ for $j = 1, \ldots, n$. Thus $A\mathbf{p} = \mathbf{p}$.

Section 2.7, p. 178

1. $\mathbf{v} = \begin{bmatrix} 87 & 81 & 62 & 64 \end{bmatrix}^T$

$$A_1\mathbf{v} = \begin{bmatrix} \frac{1}{2} & \frac{1}{2} & 0 & 0 \\ 0 & 0 & \frac{1}{2} & \frac{1}{2} \\ \frac{1}{2} & -\frac{1}{2} & 0 & 0 \\ 0 & 0 & \frac{1}{2} & -\frac{1}{2} \end{bmatrix}\begin{bmatrix} 87 \\ 81 \\ 62 \\ 64 \end{bmatrix} = \begin{bmatrix} 84 \\ 63 \\ 3 \\ -1 \end{bmatrix}$$

$$A_2 A_1\mathbf{v} = \begin{bmatrix} \frac{1}{2} & \frac{1}{2} & 0 & 0 \\ \frac{1}{2} & -\frac{1}{2} & 0 & 0 \\ 0 & 0 & 1 & 0 \\ 0 & 0 & 0 & 1 \end{bmatrix}\begin{bmatrix} 84 \\ 63 \\ 3 \\ -1 \end{bmatrix} = \begin{bmatrix} 73.5 \\ 10.5 \\ 3 \\ -1 \end{bmatrix}.$$

3. $\mathbf{v} = \begin{bmatrix} 87 & 81 & 62 & 64 & 76 & 78 & 68 & 54 \end{bmatrix}^T$.
   Here we have

$$A_1 = \begin{bmatrix} P_1 & P_2 & Z & Z \\ Z & Z & P_1 & P_2 \\ S_1 & S_2 & Z & Z \\ Z & Z & S_1 & S_2 \end{bmatrix} = \begin{bmatrix} \frac{1}{2} & \frac{1}{2} & 0 & 0 & 0 & 0 & 0 & 0 \\ 0 & 0 & \frac{1}{2} & \frac{1}{2} & 0 & 0 & 0 & 0 \\ 0 & 0 & 0 & 0 & \frac{1}{2} & \frac{1}{2} & 0 & 0 \\ 0 & 0 & 0 & 0 & 0 & 0 & 0 & 0 \\ \frac{1}{2} & -\frac{1}{2} & 0 & 0 & 0 & 0 & 0 & 0 \\ 0 & 0 & \frac{1}{2} & -\frac{1}{2} & 0 & 0 & 0 & 0 \\ 0 & 0 & 0 & 0 & \frac{1}{2} & -\frac{1}{2} & 0 & 0 \\ 0 & 0 & 0 & 0 & 0 & 0 & \frac{1}{2} & -\frac{1}{2} \end{bmatrix}$$

$$A_2 = \begin{bmatrix} P_1 & P_2 & Z & Z \\ S_1 & S_2 & Z & Z \\ Z & Z & I & Z \\ Z & Z & Z & I \end{bmatrix} = \begin{bmatrix} \frac{1}{2} & \frac{1}{2} & 0 & 0 & 0 & 0 & 0 & 0 \\ 0 & 0 & \frac{1}{2} & \frac{1}{2} & 0 & 0 & 0 & 0 \\ \frac{1}{2} & -\frac{1}{2} & 0 & 0 & 0 & 0 & 0 & 0 \\ 0 & 0 & \frac{1}{2} & -\frac{1}{2} & 0 & 0 & 0 & 0 \\ 0 & 0 & 0 & 0 & 1 & 0 & 0 & 0 \\ 0 & 0 & 0 & 0 & 0 & 1 & 0 & 0 \\ 0 & 0 & 0 & 0 & 0 & 0 & 1 & 0 \\ 0 & 0 & 0 & 0 & 0 & 0 & 0 & 1 \end{bmatrix}$$

$$A_3 = \begin{bmatrix} Q & Z & Z & Z \\ Z & I & Z & Z \\ Z & Z & I & Z \\ Z & Z & Z & I \end{bmatrix} = \begin{bmatrix} \frac{1}{2} & \frac{1}{2} & 0 & 0 & 0 & 0 & 0 & 0 \\ \frac{1}{2} & -\frac{1}{2} & 0 & 0 & 0 & 0 & 0 & 0 \\ 0 & 0 & 1 & 0 & 0 & 0 & 0 & 0 \\ 0 & 0 & 0 & 1 & 0 & 0 & 0 & 0 \\ 0 & 0 & 0 & 0 & 1 & 0 & 0 & 0 \\ 0 & 0 & 0 & 0 & 0 & 1 & 0 & 0 \\ 0 & 0 & 0 & 0 & 0 & 0 & 1 & 0 \\ 0 & 0 & 0 & 0 & 0 & 0 & 0 & 1 \end{bmatrix}.$$

Therefore,

$$A_1\mathbf{v} = \begin{bmatrix} \frac{1}{2} & \frac{1}{2} & 0 & 0 & 0 & 0 & 0 & 0 \\ 0 & 0 & \frac{1}{2} & \frac{1}{2} & 0 & 0 & 0 & 0 \\ 0 & 0 & 0 & 0 & \frac{1}{2} & \frac{1}{2} & 0 & 0 \\ 0 & 0 & 0 & 0 & 0 & 0 & 0 & 0 \\ \frac{1}{2} & -\frac{1}{2} & 0 & 0 & 0 & 0 & 0 & 0 \\ 0 & 0 & \frac{1}{2} & -\frac{1}{2} & 0 & 0 & 0 & 0 \\ 0 & 0 & 0 & 0 & \frac{1}{2} & -\frac{1}{2} & 0 & 0 \\ 0 & 0 & 0 & 0 & 0 & 0 & \frac{1}{2} & -\frac{1}{2} \end{bmatrix} \begin{bmatrix} 87 \\ 81 \\ 62 \\ 64 \\ 76 \\ 78 \\ 68 \\ 54 \end{bmatrix} = \begin{bmatrix} 84 \\ 63 \\ 77 \\ 61 \\ 3 \\ -1 \\ -1 \\ 7 \end{bmatrix}$$

$$A_2 A_1 \mathbf{v} = \begin{bmatrix} \frac{1}{2} & \frac{1}{2} & 0 & 0 & 0 & 0 & 0 & 0 \\ 0 & 0 & \frac{1}{2} & \frac{1}{2} & 0 & 0 & 0 & 0 \\ \frac{1}{2} & -\frac{1}{2} & 0 & 0 & 0 & 0 & 0 & 0 \\ 0 & 0 & \frac{1}{2} & -\frac{1}{2} & 0 & 0 & 0 & 0 \\ 0 & 0 & 0 & 0 & 1 & 0 & 0 & 0 \\ 0 & 0 & 0 & 0 & 0 & 1 & 0 & 0 \\ 0 & 0 & 0 & 0 & 0 & 0 & 1 & 0 \\ 0 & 0 & 0 & 0 & 0 & 0 & 0 & 1 \end{bmatrix} \begin{bmatrix} 84 \\ 63 \\ 77 \\ 61 \\ 3 \\ -1 \\ -1 \\ 7 \end{bmatrix} = \begin{bmatrix} 73.5 \\ 69 \\ 10.5 \\ 8 \\ 3 \\ -1 \\ -1 \\ 7 \end{bmatrix}$$

$$A_3 A_2 A_1 \mathbf{v} = \begin{bmatrix} \frac{1}{2} & \frac{1}{2} & 0 & 0 & 0 & 0 & 0 & 0 \\ \frac{1}{2} & -\frac{1}{2} & 0 & 0 & 0 & 0 & 0 & 0 \\ 0 & 0 & 1 & 0 & 0 & 0 & 0 & 0 \\ 0 & 0 & 0 & 1 & 0 & 0 & 0 & 0 \\ 0 & 0 & 0 & 0 & 1 & 0 & 0 & 0 \\ 0 & 0 & 0 & 0 & 0 & 1 & 0 & 0 \\ 0 & 0 & 0 & 0 & 0 & 0 & 1 & 0 \\ 0 & 0 & 0 & 0 & 0 & 0 & 0 & 1 \end{bmatrix} \begin{bmatrix} 73.5 \\ 69 \\ 10.5 \\ 8 \\ 3 \\ -1 \\ -1 \\ 7 \end{bmatrix} = \begin{bmatrix} 71.25 \\ 2.25 \\ 10.5 \\ 8 \\ 3 \\ -1 \\ -1 \\ 7 \end{bmatrix}.$$

5. The averages are computed using a pair of data. At the second stage we will have only three averages and hence cannot use the procedure as discussed. One remedy is to adjoin a pair of zeros after the original six items to give eight items. Then proceed as in the discussion in the text.

7. $A_3^{-1} = \begin{bmatrix} Q^{-1} & Z & Z & Z \\ Z & I & Z & Z \\ Z & Z & I & Z \\ Z & Z & Z & I \end{bmatrix}$, where $Q^{-1} = \begin{bmatrix} 1 & 1 \\ 1 & -1 \end{bmatrix}$.

Supplementary Exercises, p. 179

1. The parity $(4,5)$ check code produces the code words.

   (a) There will be $2^4 = 16$ code words.

   (b) The code words that contain two 1 bits are

$$10001, 01001, 00101, 00011, 11000, 01100, 00110, 10010, 10100, 01010.$$

   (c) There are 6 code words from part (b) that have even parity. They are

$$11000, 01100, 00110, 10010, 10100, 01010.$$

3. We want to find the number of ways that P_1 has 3-stage access to P_3. By Theorem 2.1, this number is the $(1,3)$ entry in $[A(G)]^3$. We have

$$[A(G)]^3 = \begin{bmatrix} 2 & 0 & 2 & 1 & 2 & 0 \\ 1 & 1 & 1 & 1 & 1 & 2 \\ 0 & 1 & 0 & 1 & 0 & 2 \\ 1 & 0 & 2 & 3 & 1 & 0 \\ 1 & 0 & 1 & 0 & 1 & 0 \\ 0 & 1 & 0 & 1 & 0 & 2 \end{bmatrix}.$$

Since the $(1,3)$ entry is 2, we conclude that P_1 has access to P_3 through two individuals in two ways.

5. (a) We have

$$T = \begin{array}{cc} \text{plays} & \text{does not} \\ & \text{play} \end{array} \\ T = \begin{bmatrix} 0.2 & 0.6 \\ 0.8 & 0.4 \end{bmatrix} \begin{array}{l} \text{plays} \\ \text{does not play} \end{array}$$

(b) If the average student plays on Monday, then the initial state vector is

$$\mathbf{x}^{(0)} = \begin{bmatrix} 1 \\ 0 \end{bmatrix}.$$

We are interested in the behavior on Friday, $k = 4$, and we need to find $p_1^{(4)}$. We have

$$\mathbf{x}^{(4)} = T^4 \mathbf{x}^{(0)} = \begin{bmatrix} 0.4432 \\ 0 \end{bmatrix}.$$

We conclude that the probability that the student will play a video game on Friday is 0.4432.

(c) As n increases, $\mathbf{x}^{(n)} = T^n \mathbf{x}^{(0)}$ approaches the vector

$$\begin{bmatrix} 0.4286 \\ 0 \end{bmatrix},$$

so the probability the average student will play a video game is $0.4286 \approx \frac{3}{7}$.

7. $\mathbf{v} = \begin{bmatrix} 2 & 4 & 5 & 1 \end{bmatrix}^T$.

(a) $A_2 A_1 \mathbf{v} = \begin{bmatrix} \frac{1}{2} & \frac{1}{2} & 0 & 0 \\ \frac{1}{2} & -\frac{1}{2} & 0 & 0 \\ 0 & 0 & 1 & 0 \\ 0 & 0 & 0 & 1 \end{bmatrix} \begin{bmatrix} \frac{1}{2} & \frac{1}{2} & 0 & 0 \\ 0 & 0 & \frac{1}{2} & \frac{1}{2} \\ \frac{1}{2} & -\frac{1}{2} & 0 & 0 \\ 0 & 0 & \frac{1}{2} & -\frac{1}{2} \end{bmatrix} \begin{bmatrix} 2 \\ 4 \\ 5 \\ 1 \end{bmatrix} = \begin{bmatrix} 3 \\ 0 \\ -1 \\ 2 \end{bmatrix}.$

Final average: 3
Detail coefficients: 0, -1, 2

(b) Using $\varepsilon = 1$, compressed data: 3, 0, 0,2. The corresponding wavelet is: $\widetilde{\mathbf{w}} = \begin{bmatrix} 3 & 0 & 0 & 2 \end{bmatrix}^T$.

(c) The wavelet approximation is

$$\widetilde{\mathbf{y}} = A_1^{-1} A_2^{-1} \widetilde{\mathbf{w}} = \begin{bmatrix} 1 & 0 & 1 & 0 \\ 1 & 0 & -1 & 0 \\ 0 & 1 & 0 & 1 \\ 0 & 1 & 0 & -1 \end{bmatrix} \begin{bmatrix} 1 & 1 & 0 & 0 \\ 1 & -1 & 0 & 0 \\ 0 & 0 & 1 & 0 \\ 0 & 0 & 0 & 1 \end{bmatrix} \begin{bmatrix} 3 \\ 0 \\ 0 \\ 2 \end{bmatrix} = \begin{bmatrix} 3 \\ 3 \\ 5 \\ 1 \end{bmatrix}.$$

o original data
+ wavelet approximation

T.1. Let $f_1(\mathbf{u}) = A\mathbf{u}$ and $f_2(\mathbf{u}) = B\mathbf{u}$, where A and B are 2×2 matrices. Then

$$(f_1 \circ f_2)(\mathbf{u}) = f_1(f_2(\mathbf{u})) = f_1(B\mathbf{u}) = A(B\mathbf{u}) = (AB)\mathbf{u}$$

which is matrix multiplication, a matrix transformation.

Chapter 3

Determinants

Section 3.1, p. 192

1. $S = \{1, 2, 3, 4, 5\}$.

   (a) Permutation 52134 has 5 inversions: 52, 51, 53, 54, 21.

   (b) Permutation 45213 has 7 inversions: 42, 41, 43, 52, 51, 53, 21.

   (c) Permutation 42135 has 4 inversions: 42, 41, 43, 21.

   (d) Permutation 13542 has 4 inversions: 32, 54, 52, 42.

   (e) Permutation 35241 has 7 inversions: 32, 31, 52, 54, 51, 21, 41.

   (f) Permutation 12345 has no inversions.

3. $S = \{1, 2, 3, 4, 5\}$

   (a) Since 25431 is an odd permutation (seven inversions: 21, 54, 53, 51, 43, 41, 31) a $-$ sign is associated with it.

   (b) Since 31245 is an even permutation (two inversions: 31, 32) a $+$ sign is associated with it.

   (c) Since 21345 is an odd permutation (one inversion: 21) a $-$ sign is associated with it.

   (d) Since 52341 is an odd permutation (seven inversions: 52, 53, 54, 51, 21, 31, 41) a $-$ sign is associated with it.

   (e) Since 34125 is an even permutation (four inversions: 31, 32, 41, 42) a $+$ sign is associated with it.

   (f) Since 41253 is an even permutation (four inversions: 41, 42, 43, 53) a $+$ sign is associated with it.

5. (a) $\begin{vmatrix} 2 & -1 \\ 3 & 2 \end{vmatrix} = (2)(2) - (-1)(3) = 7$ as in Example 5.

   (b) $\begin{vmatrix} 0 & 3 & 0 \\ 2 & 0 & 0 \\ 0 & 0 & -5 \end{vmatrix} = (0)(0)(-5) + (3)(0)(0) + (0)(2)(0) - (0)(0)(0) - (3)(2)(-5) - (0)(0)(0) = 30$ as in Example 6. (See Equation (3).)

   (c) $\begin{vmatrix} 4 & 2 & 0 \\ 0 & -2 & 5 \\ 0 & 0 & 3 \end{vmatrix} = (4)(-2)(3) + (2)(5)(0) + (0)(0)(0) - (4)(5)(0) - (2)(0)(3) - (0)(-2)(0) = -24$ as in Example 6. (See Equation (3).)

(d) $\begin{vmatrix} 4 & 2 & 2 & 0 \\ 2 & 0 & 0 & 0 \\ 3 & 0 & 0 & 1 \\ 0 & 0 & 1 & 0 \end{vmatrix}$ $\begin{aligned} = \ & (4)(0)(0)(0) + (4)(0)(0)(1) + (4)(0)(1)(0) \\ & + (2)(2)(1)(1) + (2)(0)(3)(0) + (2)(0)(0)(0) + (2)(2)(0)(0) \\ & + (2)(0)(1)(0) + (2)(0)(3)(0) + (0)(2)(0)(0) + (0)(0)(0)(1) \\ & + (0)(0)(0)(0) - (4)(0)(1)(1) - (4)(0)(0)(0) - (4)(0)(0)(0) \\ & - (2)(2)(0)(0) - (2)(0)(1)(0) - (2)(0)(3)(1) - (2)(2)(1)(0) \\ & - (2)(0)(3)(0) - (2)(0)(0)(0) - (0)(0)(0)(1) - (0)(0)(0)(0) \\ & - (0)(0)(3)(0) = 4. \end{aligned}$

7. Let $A = \begin{bmatrix} a_{ij} \end{bmatrix}$ be a 4×4 matrix. Using Equation (2) we have

$$\det(A) = \begin{vmatrix} a_{11} & a_{12} & a_{13} & a_{14} \\ a_{21} & a_{22} & a_{23} & a_{24} \\ a_{31} & a_{32} & a_{33} & a_{34} \\ a_{41} & a_{42} & a_{43} & a_{44} \end{vmatrix}$$

$$\begin{aligned} = \ & a_{11}a_{22}a_{33}a_{44} + a_{11}a_{24}a_{32}a_{43} + a_{11}a_{23}a_{34}a_{42} + a_{12}a_{21}a_{34}a_{43} \\ & + a_{12}a_{23}a_{31}a_{44} + a_{12}a_{24}a_{33}a_{41} + a_{13}a_{21}a_{32}a_{44} + a_{13}a_{22}a_{34}a_{41} \\ & + a_{13}a_{24}a_{31}a_{42} + a_{14}a_{21}a_{33}a_{42} + a_{14}a_{22}a_{31}a_{43} + a_{14}a_{23}a_{32}a_{41} \\ & - a_{11}a_{22}a_{34}a_{43} - a_{11}a_{23}a_{32}a_{44} - a_{11}a_{24}a_{33}a_{42} - a_{12}a_{21}a_{33}a_{44} \\ & - a_{12}a_{23}a_{34}a_{41} - a_{12}a_{24}a_{31}a_{43} - a_{13}a_{21}a_{34}a_{42} - a_{13}a_{22}a_{31}a_{44} \\ & - a_{13}a_{24}a_{32}a_{41} - a_{14}a_{21}a_{32}a_{43} - a_{14}a_{22}a_{33}a_{41} - a_{14}a_{23}a_{31}a_{42} \end{aligned}$$

To construct the preceding terms, take the set $S = \{1, 2, 3, 4\}$, write out the 24 permutations of S, and determine which are even and odd. Then use these permutations to fill in the blanks in 24 products of the form

$$a_{1\_}\, a_{2\_}\, a_{3\_}\, a_{4\_}.$$

The sign of each term is determined by whether the permutation is even or odd.

9. Given that

$$|A| = \begin{vmatrix} a_1 & a_2 & a_3 \\ b_1 & b_2 & b_3 \\ c_1 & c_2 & c_3 \end{vmatrix} = 3.$$

Matrix B is obtained from A by performing the row operations $2\mathbf{r}_2 + \mathbf{r}_1 \to \mathbf{r}_1$ and $-3\mathbf{r}_3 + \mathbf{r}_1 \to \mathbf{r}_1$. Thus by Theorem 3.6, $|B| = |A| = 3$. Matrix C is obtained from A by performing the operation $3\mathbf{c}_2 \to \mathbf{c}_2$. Thus by Theorem 3.5, $|C| = 3|A| = 9$. Matrix D is obtained from A by performing the operation $\mathbf{r}_2 \leftrightarrow \mathbf{r}_3$. Thus by Theorem 3.2, $|D| = -|A| = -3$.

11. (a) $\det\left(\begin{bmatrix} \lambda - 1 & 2 \\ 3 & \lambda - 2 \end{bmatrix}\right) = (\lambda - 1)(\lambda - 2) - (2)(3) = \lambda^2 - 3\lambda - 4.$

(b) $\det\left([\lambda I_2 - A]\right) = \begin{vmatrix} \lambda - 4 & -2 \\ 1 & \lambda - 1 \end{vmatrix} = (\lambda - 4)(\lambda - 1) - (-2)(1) = \lambda^2 - 5\lambda + 6.$

13. (a) $\det\left(\begin{bmatrix} \lambda - 1 & 2 \\ 3 & \lambda - 2 \end{bmatrix}\right) = \lambda^2 - 3\lambda - 4 = 0.$ Factoring, we have

$$\lambda^2 - 3\lambda - 4 = (\lambda - 4)(\lambda + 1) = 0.$$

Thus, $\lambda = 4$ or $\lambda = -1$.

(b) $\det\left(\left[\lambda I_2 - A\right]\right) = \begin{vmatrix} \lambda - 4 & -2 \\ 1 & \lambda - 1 \end{vmatrix} = \lambda^2 - 5\lambda + 6 = 0.$ Factoring, we have

$$\lambda^2 - 5\lambda + 6 = (\lambda - 3)(\lambda - 2) = 0.$$

Thus, $\lambda = 3$ or $\lambda = 2$.

15. Each of these matrices is upper triangular and hence the determinant is the product of the diagonal entries.

(a) $\begin{vmatrix} 0 & 2 & -5 \\ 0 & 4 & 6 \\ 0 & 0 & -1 \end{vmatrix} = (0)(4)(-1) = 0.$

(b) $\begin{vmatrix} 6 & 6 & 3 & -2 \\ 0 & 4 & 7 & 5 \\ 0 & 0 & -3 & 2 \\ 0 & 0 & 0 & 2 \end{vmatrix} = (6)(4)(-3)(2) = -144.$

(c) $\begin{vmatrix} 2 & 0 & 0 \\ 0 & 4 & 0 \\ 0 & 0 & 9 \end{vmatrix} = (2)(4)(9) = 72.$

In Exercises 17 and 19 we compute the determinant via reduction to triangular form.

17. (a) $\begin{vmatrix} 4 & -3 & 5 \\ 5 & 2 & 0 \\ 2 & 0 & 4 \end{vmatrix}_{-\frac{5}{4}\mathbf{r}_1 + \mathbf{r}_2 \to \mathbf{r}_2} = \begin{bmatrix} 4 & -3 & 5 \\ 0 & \frac{23}{4} & -\frac{25}{4} \\ 2 & 0 & 4 \end{bmatrix}_{-\frac{1}{2}\mathbf{r}_1 + \mathbf{r}_3 \to \mathbf{r}_3} = \begin{bmatrix} 4 & -3 & 5 \\ 0 & \frac{23}{4} & -\frac{25}{4} \\ 0 & \frac{3}{2} & \frac{3}{2} \end{bmatrix}_{-\frac{6}{23}\mathbf{r}_2 + \mathbf{r}_3 \to \mathbf{r}_3}$

$= \begin{vmatrix} 4 & -3 & 5 \\ 0 & \frac{23}{4} & -\frac{25}{4} \\ 0 & 0 & \frac{72}{23} \end{vmatrix} = (4)\left(\frac{23}{4}\right)\left(\frac{72}{23}\right) = 72.$

(b) $\begin{vmatrix} 2 & 0 & 1 & 4 \\ 3 & 2 & -4 & -2 \\ 2 & 3 & -1 & 0 \\ 11 & 8 & -4 & 6 \end{vmatrix}_{\mathbf{c}_1 \leftrightarrow \mathbf{c}_4} = (-1)\begin{vmatrix} 4 & 0 & 1 & 2 \\ -2 & 2 & -4 & 3 \\ 0 & 3 & -1 & 2 \\ 6 & 8 & -4 & 11 \end{vmatrix}_{3\mathbf{r}_2 + \mathbf{r}_4 \to \mathbf{r}_4}$

$= (-1)\begin{vmatrix} 4 & 0 & 1 & 2 \\ -2 & 2 & -4 & 3 \\ 0 & 3 & -1 & 2 \\ 0 & 14 & -16 & 20 \end{vmatrix}_{\mathbf{r}_1 \leftrightarrow \mathbf{r}_2} = \begin{vmatrix} -2 & 2 & -4 & 3 \\ 4 & 0 & 1 & 2 \\ 0 & 3 & -1 & 2 \\ 0 & 14 & -16 & 20 \end{vmatrix}_{2\mathbf{r}_1 + \mathbf{r}_2 \to \mathbf{r}_2}$

$= \begin{vmatrix} -2 & 2 & -4 & 3 \\ 0 & 4 & -7 & 8 \\ 0 & 3 & -1 & 2 \\ 0 & 14 & -16 & 20 \end{vmatrix}_{\frac{1}{4}\mathbf{r}_2 \to \mathbf{r}_2} = (4)\begin{vmatrix} -2 & 2 & -4 & 3 \\ 0 & 1 & -\frac{7}{4} & 2 \\ 0 & 3 & -1 & 2 \\ 0 & 14 & -16 & 20 \end{vmatrix}_{-3\mathbf{r}_2 + \mathbf{r}_3 \to \mathbf{r}_3}$

$$= (4) \begin{vmatrix} -2 & 2 & -4 & 3 \\ 0 & 1 & -\frac{7}{4} & 2 \\ 0 & 0 & \frac{17}{4} & -4 \\ 0 & 14 & -16 & 20 \end{vmatrix}_{-14\mathbf{r}_2+\mathbf{r}_4\to\mathbf{r}_4} \qquad = (4) \begin{vmatrix} -2 & 2 & -4 & 3 \\ 0 & 1 & -\frac{7}{4} & 2 \\ 0 & 0 & \frac{17}{4} & -4 \\ 0 & 0 & \frac{17}{2} & -8 \end{vmatrix}_{\frac{1}{2}\mathbf{r}_4\to\mathbf{r}_4}$$

$$= (2)(4) \begin{vmatrix} -2 & 2 & -4 & 3 \\ 0 & 1 & -\frac{7}{4} & 2 \\ 0 & 0 & \frac{17}{4} & -4 \\ 0 & 0 & \frac{17}{4} & -4 \end{vmatrix} = (4)(2)(0) \text{ since two rows are equal.}$$

(c) $\begin{vmatrix} 4 & 1 & 2 \\ 0 & 2 & 3 \\ 0 & 0 & -3 \end{vmatrix} = (4)(2)(-3) = -24$

19. (a) $\begin{vmatrix} 4 & 2 & 3 & -4 \\ 3 & -2 & 1 & 5 \\ -2 & 0 & 1 & -3 \\ 8 & -2 & 6 & 4 \end{vmatrix}_{\frac{1}{4}\mathbf{r}_1\to\mathbf{r}_1} = (4) \begin{vmatrix} 1 & \frac{1}{2} & \frac{3}{4} & -1 \\ 3 & -2 & 1 & 5 \\ -2 & 0 & 1 & -3 \\ 8 & -2 & 6 & 4 \end{vmatrix}_{\substack{-3\mathbf{r}_1 + \mathbf{r}_2 \to \mathbf{r}_2 \\ 2\mathbf{r}_1 + \mathbf{r}_3 \to \mathbf{r}_3 \\ -8\mathbf{r}_1 + \mathbf{r}_4 \to \mathbf{r}_4}}$

$$= (4) \begin{vmatrix} 1 & \frac{1}{2} & \frac{3}{4} & -1 \\ 0 & -\frac{7}{2} & -\frac{5}{4} & 8 \\ 0 & 1 & \frac{5}{2} & -5 \\ 0 & -6 & 0 & 12 \end{vmatrix}_{-\frac{2}{7}\mathbf{r}_2\to\mathbf{r}_2} \qquad = (4)\left(-\frac{7}{2}\right) \begin{vmatrix} 1 & \frac{1}{2} & \frac{3}{4} & -1 \\ 0 & 1 & \frac{5}{14} & -\frac{16}{7} \\ 0 & 1 & \frac{5}{2} & -5 \\ 0 & -6 & 0 & 12 \end{vmatrix}_{\substack{-1\mathbf{r}_2 + \mathbf{r}_3 \to \mathbf{r}_3 \\ 6\mathbf{r}_2 + \mathbf{r}_4 \to \mathbf{r}_4}}$$

$$= (4)\left(-\frac{7}{2}\right) \begin{vmatrix} 1 & \frac{1}{2} & \frac{3}{4} & -1 \\ 0 & 1 & \frac{5}{14} & -\frac{16}{7} \\ 0 & 0 & \frac{15}{7} & -\frac{19}{7} \\ 0 & 0 & \frac{15}{7} & -\frac{12}{7} \end{vmatrix}_{-1\mathbf{r}_3+\mathbf{r}_4\to\mathbf{r}_4} \qquad = (4)\left(-\frac{7}{2}\right) \begin{vmatrix} 1 & \frac{1}{2} & \frac{3}{4} & -1 \\ 0 & 1 & \frac{5}{14} & -\frac{16}{7} \\ 0 & 0 & \frac{15}{7} & \frac{19}{7} \\ 0 & 0 & 0 & 1 \end{vmatrix}$$

$$= (4)\left(-\frac{7}{2}\right)(1)(1)\left(\frac{15}{7}\right)(1) = -30.$$

(b) $\begin{vmatrix} 1 & 3 & -4 \\ -2 & 1 & 2 \\ -9 & 15 & 0 \end{vmatrix}_{\substack{2\mathbf{r}_1 + \mathbf{r}_2 \to \mathbf{r}_2 \\ 9\mathbf{r}_1 + \mathbf{r}_3 \to \mathbf{r}_3}} = \begin{vmatrix} 1 & 3 & -4 \\ 0 & 7 & -6 \\ 0 & 42 & -36 \end{vmatrix}_{-6\mathbf{r}_2+\mathbf{r}_3\to\mathbf{r}_3} = \begin{vmatrix} 1 & 3 & -4 \\ 0 & 7 & -6 \\ 0 & 0 & 0 \end{vmatrix} = (1)(7)(0) = 0.$

(c) $\begin{vmatrix} 1 & 1 & 2 \\ 0 & 2 & -2 \\ 0 & 0 & 3 \end{vmatrix} = (1)(2)(3) = 6$ by Theorem 3.7.

21. (a) $\det(A) = \begin{vmatrix} 1 & -2 & 3 \\ -2 & 3 & 1 \\ 0 & 1 & 0 \end{vmatrix}_{2\mathbf{r}_1+\mathbf{r}_2\to\mathbf{r}_2} = \begin{vmatrix} 1 & -2 & 3 \\ 0 & -1 & 7 \\ 0 & 1 & 0 \end{vmatrix}_{\mathbf{r}_2+\mathbf{r}_3\to\mathbf{r}_3} = \begin{vmatrix} 1 & -2 & 3 \\ 0 & -1 & 7 \\ 0 & 0 & 7 \end{vmatrix}$

$= (1)(-1)(7) = -7.$

$\det(B) = \begin{vmatrix} 1 & 0 & 2 \\ 3 & -2 & 5 \\ 2 & 1 & 3 \end{vmatrix}_{\substack{-3\mathbf{r}_1+\mathbf{r}_2\to\mathbf{r}_2 \\ -2\mathbf{r}_1+\mathbf{r}_3\to\mathbf{r}_3}} = \begin{vmatrix} 1 & 0 & 2 \\ 0 & -2 & -1 \\ 0 & 1 & -1 \end{vmatrix}_{\frac{1}{2}\mathbf{r}_2+\mathbf{r}_3\to\mathbf{r}_3} = \begin{vmatrix} 1 & 0 & 2 \\ 0 & -2 & -1 \\ 0 & 0 & -\frac{3}{2} \end{vmatrix}$

$= (1)(-2)\left(-\frac{3}{2}\right) = 3.$

$\det(AB) = \begin{vmatrix} 1 & 7 & 1 \\ 9 & -5 & 14 \\ 3 & -2 & 5 \end{vmatrix}_{\substack{-9\mathbf{r}_1+\mathbf{r}_2\to\mathbf{r}_2 \\ -3\mathbf{r}_1+\mathbf{r}_3\to\mathbf{r}_3}} = \begin{vmatrix} 1 & 7 & 1 \\ 0 & -68 & 5 \\ 0 & -23 & 2 \end{vmatrix}_{-\frac{23}{68}\mathbf{r}_2+\mathbf{r}_3\to\mathbf{r}_3} = \begin{vmatrix} 1 & 7 & 1 \\ 0 & -68 & 5 \\ 0 & 0 & \frac{21}{68} \end{vmatrix}$

$= (1)(-68)\left(\frac{21}{68}\right) = -21.$

Thus we see that $\det(A)\det(B) = (-7)(3) = -21 = \det(AB)$.

(b) $\det(A) = \begin{vmatrix} 2 & 3 & 6 \\ 0 & 3 & 2 \\ 0 & 0 & -4 \end{vmatrix} = (2)(3)(-4) = -24.$

$\det(B) = \begin{vmatrix} 3 & 0 & 0 \\ 4 & 5 & 0 \\ 2 & 1 & -2 \end{vmatrix} = (3)(5)(-2) = -30.$

$\det(AB) = \begin{vmatrix} 30 & 21 & -12 \\ 16 & 17 & -4 \\ -8 & -4 & 8 \end{vmatrix}_{-\frac{1}{8}\mathbf{r}_3\to\mathbf{r}_3} = (-8)\begin{vmatrix} 30 & 21 & -12 \\ 16 & 17 & -4 \\ 1 & \frac{1}{2} & -1 \end{vmatrix}_{\mathbf{r}_1\leftrightarrow\mathbf{r}_3}$

$= (-8)(-1)\begin{vmatrix} 1 & \frac{1}{2} & -1 \\ 16 & 17 & -4 \\ 30 & 21 & -12 \end{vmatrix}_{\substack{-16\mathbf{r}_1+\mathbf{r}_2\to\mathbf{r}_2 \\ -30\mathbf{r}_1+\mathbf{r}_3\to\mathbf{r}_3}} = (8)\begin{vmatrix} 1 & \frac{1}{2} & -1 \\ 0 & 9 & 12 \\ 0 & 6 & 18 \end{vmatrix}_{-\frac{6}{9}\mathbf{r}_2+\mathbf{r}_3\to\mathbf{r}_3}$

$= (8)\begin{vmatrix} 1 & \frac{1}{2} & -1 \\ 0 & 9 & 12 \\ 0 & 0 & 10 \end{vmatrix} = (8)(1)(9)(10) = 720.$

Thus we have $\det(A)\det(B) = (-24)(-30) = 720 = \det(AB)$.

23. Given $|A| = 2$ and $|B| = -3$, we find that

$$\left|A^{-1}B^T\right| = \left|A^{-1}\right|\left|B^T\right| = \frac{1}{|A|}|B| = \frac{-3}{2}.$$

25. (a) $\begin{vmatrix} 1 & 1 \\ 1 & 0 \end{vmatrix} = 1(0) - (1)(1) = 0 - 1 = 1$ (*Note*: The additive inverse of 1 is 1.)

(b) To obtain $\det(A)$, we will use the method given in Example 6. We find that

$\det(A) = (0)(1)(1) + (1)(1)(0) + (1)(1)(0) - (1)(1)(1) - (0)(1)(0) - (0)(1)(1) = -1 = 1.$

(c) Using the method of Example 6, we find that

$\det(A) = (1)(1)(0) + (0)(1)(1) + (0)(1)(1) - (0)(1)(0) - (1)(1)(1) - (0)(1)(1) = -1 = 1.$

27. (a) $\begin{vmatrix} 1 & 0 & 1 \\ 0 & 1 & 1 \\ 1 & 1 & 0 \end{vmatrix}_{\mathbf{r_1}+\mathbf{r_3}\to\mathbf{r_3}} = \begin{vmatrix} 1 & 0 & 1 \\ 0 & 1 & 1 \\ 0 & 1 & 1 \end{vmatrix}_{\mathbf{r_2}+\mathbf{r_3}\to\mathbf{r_3}} = \begin{vmatrix} 1 & 0 & 1 \\ 0 & 1 & 1 \\ 0 & 0 & 0 \end{vmatrix} = (1)(1)(0) = 0$

(b) $\begin{vmatrix} 1 & 1 & 1 & 0 \\ 1 & 1 & 0 & 1 \\ 1 & 0 & 1 & 1 \\ 0 & 0 & 0 & 1 \end{vmatrix}_{\substack{\mathbf{r_1}+\mathbf{r_2}\to\mathbf{r_2} \\ \mathbf{r_1}+\mathbf{r_3}\to\mathbf{r_3}}} = \begin{vmatrix} 1 & 1 & 1 & 0 \\ 0 & 0 & 1 & 1 \\ 0 & 1 & 0 & 1 \\ 0 & 0 & 0 & 1 \end{vmatrix}_{\mathbf{r_2}\leftrightarrow\mathbf{r_3}} = (-1)\begin{vmatrix} 1 & 1 & 1 & 0 \\ 0 & 1 & 0 & 1 \\ 0 & 0 & 1 & 1 \\ 0 & 0 & 0 & 1 \end{vmatrix} = (-1)(1)(1)(1)(1) = 1.$

T.1. If j_i and j_{i+1} are interchanged, all inversions between numbers distinct from j_i and j_{i+1} remain unchanged, and all inversions between one of j_i, j_{i+1} and some other number also remain unchanged. If originally $j_i < j_{i+1}$, then after interchange there is one additional inversion due to $j_{i+1}j_i$. If originally $j_i > j_{i+1}$, then after interchange there is one fewer inversion.

Suppose j_p and j_q are separated by k intervening numbers. Then k interchanges of adjacent numbers will move j_p next to j_q. One interchange switches j_p and j_q. Finally, k interchanges of adjacent numbers takes j_q back to j_p's original position. The total number of interchanges is the odd number $2k + 1$.

T.3. $cA = [ca_{ij}]$. By n applications of Theorem 3.5, the result follows.

T.5. By Theorem 3.8, $\det(AB) = \det(A)\det(B)$. Thus if $\det(AB) = 0$, then $\det(A)\det(B) = 0$ and either $\det(A) = 0$ or $\det(B) = 0$.

T.7. In the summation

$$\det(A) = \sum(\pm)a_{1j_1}a_{2j_2}\cdots a_{nj_n}$$

for the definition of $\det(A)$ there is exactly one nonzero term. Thus $\det(A) \neq 0$.

T.9. (a) $[\det(A)]^2 = \det(A)\det(A) = \det(A)\det(A^{-1}) = \det(AA^{-1}) = 1$.

(b) $[\det(A)]^2 = \det(A)\det(A) = \det(A)\det(A^T) = \det(A)\det(A^{-1}) = \det(AA^{-1}) = 1$.

T.11. $\det(A^T B^T) = \det(A^T)\det(B^T) = \det(A)\det(B^T) = \det(A^T)\det(B)$.

T.13. If A is nonsingular, by Corollary 3.2, $\det(A) \neq 0$ and $a_{ii} \neq 0$ for $i = 1, 2, \ldots, n$ since the determinant of an upper triangular matrix is the product of its diagonals. Conversely, if $a_{ii} \neq 0$ for $i = 1, \ldots, n$, then clearly A is row equivalent to I_n, and thus is nonsingular.

T.15. If $\det(A) \neq 0$, then since

$$0 = \det(O) = \det(A^n) = \det(A)\det(A^{n-1}),$$

by Exercise T.5 above, $\det(A^{n-1}) = 0$. Working downward, $\det(A^{n-2}) = 0$, $\ldots$, $\det(A^2) = 0$, $\det(A) = 0$, which is a contradiction.

T.17. Let

$$A = \begin{bmatrix} d_{11} & 0 & \cdots & 0 \\ 0 & d_{22} & \cdots & 0 \\ \vdots & \vdots & \ddots & \vdots \\ 0 & 0 & \cdots & d_{nn} \end{bmatrix}$$

be a diagonal matrix. Then A is a triangular matrix. Hence by Theorem 3.7, the determinant of A is the product of the diagonal entries; that is,

$$\det(A) = d_{11}d_{22}\cdots d_{nn}.$$

T.19. There are six 2×2 bit matrices having determinant 1. They are

$$\begin{bmatrix} 0 & 1 \\ 1 & 0 \end{bmatrix}, \quad \begin{bmatrix} 0 & 1 \\ 1 & 1 \end{bmatrix}, \quad \begin{bmatrix} 1 & 0 \\ 0 & 1 \end{bmatrix}, \quad \begin{bmatrix} 1 & 0 \\ 1 & 1 \end{bmatrix}, \quad \begin{bmatrix} 1 & 1 \\ 0 & 1 \end{bmatrix}, \quad \begin{bmatrix} 1 & 1 \\ 1 & 0 \end{bmatrix}.$$

The rest have determinant 0.

ML.1. There are many sequences of row operations that can be used. Here we record the value of the determinant so you may check your result.

(a) $\det(A) = -18$.　　(b) $\det(A) = 5$.

ML.3.　(a) **A = [1 − 1　1;1　1 − 1; − 1　1　1];**
　　　　　det(A)
　　　　　ans =
　　　　　　　4

　　(b) **A = [1　2　3　4;2　3　4　5;3　4　5　6;4　5　6　7];**
　　　　　det(A)
　　　　　ans =
　　　　　　　0

ML.5. **A = [5　2; − 1　2];**
　　　t = 1;
　　　det(t ∗ eye(size(A)) − A)
　　　ans =
　　　　　6
　　　t = 2;
　　　det(t ∗ eye(size(A)) − A)
　　　ans =
　　　　　2
　　　t = 3;
　　　det(t ∗ eye(size(A)) − A)
　　　ans =
　　　　　0

Section 3.2, p. 207

1. $A = \begin{bmatrix} 1 & 0 & -2 \\ 3 & 1 & 4 \\ 5 & 2 & -3 \end{bmatrix}$. The cofactors of A are:

$$A_{11} = (-1)^2 \begin{vmatrix} 1 & 4 \\ 2 & -3 \end{vmatrix} = -11, \qquad A_{12} = (-1)^3 \begin{vmatrix} 3 & 4 \\ 5 & -3 \end{vmatrix} = 29,$$

$$A_{13} = (-1)^4 \begin{vmatrix} 3 & 1 \\ 5 & 2 \end{vmatrix} = 1, \qquad A_{21} = (-1)^3 \begin{vmatrix} 0 & -2 \\ 2 & -3 \end{vmatrix} = -4,$$

$$A_{22} = (-1)^4 \begin{vmatrix} 1 & -2 \\ 5 & -3 \end{vmatrix} = 7, \qquad A_{23} = (-1)^5 \begin{vmatrix} 1 & 0 \\ 5 & 2 \end{vmatrix} = -2,$$

$$A_{31} = (-1)^4 \begin{vmatrix} 0 & -2 \\ 1 & 4 \end{vmatrix} = 2, \qquad A_{32} = (-1)^5 \begin{vmatrix} 1 & -2 \\ 3 & 4 \end{vmatrix} = -10,$$

$$A_{33} = (-1)^6 \begin{vmatrix} 1 & 0 \\ 3 & 1 \end{vmatrix} = 1.$$

To evaluate the determinants in Exercises 3 and 5 we use expansion by cofactors as in Theorem 3.9. To reduce the work involved, we choose a row or column to expand along that has one or more zeros, if possible. We present one approach for the computations. Many others are possible.

3. (a) $A = \begin{bmatrix} 1 & 2 & 3 \\ -1 & 5 & 2 \\ 3 & 2 & 0 \end{bmatrix}$.

Since row 3 contains a zero, we use expansion by cofactors along row 3. Thus, we need only cofactors A_{31} and A_{32}.

$$A_{31} = (-1)^4 \begin{vmatrix} 2 & 3 \\ 5 & 2 \end{vmatrix} = -11, \qquad A_{32} = (-1)^5 \begin{vmatrix} 1 & 3 \\ -1 & 2 \end{vmatrix} = -5.$$

Then

$$\det(A) = a_{31}A_{31} + a_{32}A_{32} + a_{33}A_{33} = 3(-11) + 2(-5) + 0(A_{33}) = -43.$$

(b) $A = \begin{bmatrix} 4 & -4 & 2 & 1 \\ 1 & 2 & 0 & 3 \\ 2 & 0 & 3 & 4 \\ 0 & -3 & 2 & 1 \end{bmatrix}$.

Since column 1 contains a zero, we use expansion by cofactors along column 1. Thus we need only cofactors A_{11}, A_{21}, and A_{31}.

$$A_{11} = (-1)^2 \begin{vmatrix} 2 & 0 & 3 \\ 0 & 3 & 4 \\ -3 & 2 & 1 \end{vmatrix}.$$

We evaluate this determinant using expansion by cofactors along the first row:

$$A_{11} = (-1)^2 \left[(-1)^2(2) \begin{vmatrix} 3 & 4 \\ 2 & 1 \end{vmatrix} + (-1)^4(3) \begin{vmatrix} 0 & 3 \\ -3 & 2 \end{vmatrix} \right] = (2)(-5) + (3)(9) = 17.$$

$$A_{21} = (-1)^3 \begin{vmatrix} -4 & 2 & 1 \\ 0 & 3 & 4 \\ -3 & 2 & 1 \end{vmatrix}.$$

We evaluate this determinant using expansion by cofactors along the first column:

$$A_{21} = (-1)^3 \left[(-1)^2(-4) \begin{vmatrix} 3 & 4 \\ 2 & 1 \end{vmatrix} + (-1)^4(-3) \begin{vmatrix} 2 & 1 \\ 3 & 4 \end{vmatrix} \right] = (-1)\left[(-4)(-5) + (-3)(5)\right] = -5.$$

$$A_{31} = (-1)^4 \begin{vmatrix} -4 & 2 & 1 \\ 2 & 0 & 3 \\ -3 & 2 & 1 \end{vmatrix}.$$

We evaluate this determinant using expansion by cofactors along the second row:

$$A_{31} = (-1)^4 \left[(-1)^3(2) \begin{vmatrix} 2 & 1 \\ 2 & 1 \end{vmatrix} + (-1)^5(3) \begin{vmatrix} -4 & 2 \\ -3 & 2 \end{vmatrix} \right]$$

$$= (-1)^4 \left[(-1)^3(2)(0) + (-1)^5(3)(-2) \right] = 6.$$

Then

$$\det(A) = a_{11}A_{11} + a_{21}A_{21} + a_{31}A_{31} + a_{41}A_{41} = (4)(17) + (1)(-5) + (2)(6) + (0)A_{41} = 75.$$

(c) $A = \begin{bmatrix} 4 & -2 & 0 \\ 0 & 2 & 4 \\ -1 & -1 & -3 \end{bmatrix}.$

Since row 1 contains a zero we use expansion by cofactors along row 1. Thus we need only cofactors A_{11} and A_{12}.

$$A_{11} = (-1)^2 \begin{vmatrix} 2 & 4 \\ -1 & -3 \end{vmatrix} = -2, \qquad A_{12} = (-1)^3 \begin{vmatrix} 0 & 4 \\ -1 & -3 \end{vmatrix} = -4.$$

Then

$$\det(A) = a_{11}A_{11} + a_{12}A_{12} + a_{13}A_{13} = (4)(-2) + (-2)(-4) + (0)A_{13} = 0.$$

5. (a) $A = \begin{bmatrix} 3 & 1 & 2 & -1 \\ 2 & 0 & 3 & -7 \\ 1 & 3 & 4 & -5 \\ 0 & -1 & 1 & -5 \end{bmatrix}.$

Expand by cofactors along column 1. We need only cofactors A_{11}, A_{21}, and A_{31}.

$$A_{11} = (-1)^2 \begin{vmatrix} 0 & 3 & -7 \\ 3 & 4 & -5 \\ -1 & 1 & -5 \end{vmatrix}.$$

Expand by cofactors along row 1:

$$A_{11} = (-1)^2 \left[(-1)^3(3) \begin{vmatrix} 3 & -5 \\ -1 & -5 \end{vmatrix} + (-1)^4(-7) \begin{vmatrix} 3 & 4 \\ -1 & 1 \end{vmatrix} \right] = (-1)(3)(-20) + (-7)(7) = 11.$$

$$A_{21} = (-1)^3 \begin{vmatrix} 1 & 2 & -1 \\ 3 & 4 & -5 \\ -1 & 1 & -5 \end{vmatrix}.$$

Expand by cofactors along row 1:

$$A_{21} = (-1)^3 \left[(-1)^2(1) \begin{vmatrix} 4 & -5 \\ 1 & -5 \end{vmatrix} + (-1)^3(2) \begin{vmatrix} 3 & -5 \\ -1 & -5 \end{vmatrix} + (-1)^4(-1) \begin{vmatrix} 3 & 4 \\ -1 & 1 \end{vmatrix} \right]$$

$$= (-1)\left[(-15) + (-2)(-20) + (-7)\right] = -18.$$

$$A_{31} = (-1)^4 \begin{vmatrix} 1 & 2 & -1 \\ 0 & 3 & -7 \\ -1 & 1 & -5 \end{vmatrix}.$$

Expand by cofactors along column 1:

$$A_{31} = (-1)^4 \left[(-1)^2(1) \begin{vmatrix} 3 & -7 \\ 1 & -5 \end{vmatrix} + (-1)^4(-1) \begin{vmatrix} 2 & -1 \\ 3 & -7 \end{vmatrix} \right] = -8 + 11 = 3.$$

Then

$$\det(A) = a_{11}A_{11} + a_{21}A_{21} + a_{31}A_{31} + a_{41}A_{41} = (3)(11) + (2)(-18) + (1)(3) + (0)A_{41} = 0.$$

(b) $A = \begin{bmatrix} 3 & 1 & 0 \\ 3 & 2 & 1 \\ 0 & 1 & -1 \end{bmatrix}.$

Expand by cofactors along row 1. Thus we need only cofactors A_{11} and A_{12}.

$$A_{11} = (-1)^2 \begin{vmatrix} 2 & 1 \\ 1 & -1 \end{vmatrix} = -3, \qquad A_{12} = (-1)^3 \begin{vmatrix} 3 & 1 \\ 0 & -1 \end{vmatrix} = 3.$$

Then

$$\det(A) = a_{11}A_{11} + a_{12}A_{12} + a_{13}A_{13} = (3)(-3) + (1)(3) + (0)A_{13} = -6.$$

(c) $A = \begin{bmatrix} 3 & -3 & 0 \\ 2 & 0 & 2 \\ 2 & 1 & -3 \end{bmatrix}.$

Expand by cofactors along row 1. Thus we need only cofactors A_{11} and A_{12}.

$$A_{11} = (-1)^2 \begin{vmatrix} 0 & 2 \\ 1 & -3 \end{vmatrix} = -2, \qquad A_{12} = (-1)^3 \begin{vmatrix} 2 & 2 \\ 2 & -3 \end{vmatrix} = 10.$$

Then

$$\det(A) = a_{11}A_{11} + a_{12}A_{12} + a_{13}A_{13} = (3)(-2) + (-3)(10) + (0)A_{13} = -36.$$

7. $A = \begin{bmatrix} -2 & 3 & 0 \\ 4 & 1 & -3 \\ 2 & 0 & 1 \end{bmatrix}.$

We have

$$A_{12} = (-1)^3 \begin{vmatrix} 4 & -3 \\ 2 & 1 \end{vmatrix} = -10, \quad A_{22} = (-1)^4 \begin{vmatrix} -2 & 0 \\ 2 & 1 \end{vmatrix} = -2, \quad A_{32} = (-1)^5 \begin{vmatrix} -2 & 0 \\ 4 & -3 \end{vmatrix} = -6.$$

Therefore

$$\det(A) = a_{11}A_{12} + a_{21}A_{22} + a_{31}A_{32} = (-2)(-10) + (4)(-2) + (2)(-6) = 0.$$

9. $A = \begin{bmatrix} 6 & 2 & 8 \\ -3 & 4 & 1 \\ 4 & -4 & 5 \end{bmatrix}.$

(a) adj $A = \begin{bmatrix} A_{11} & A_{21} & A_{31} \\ A_{12} & A_{22} & A_{32} \\ A_{13} & A_{23} & A_{33} \end{bmatrix}$, where $A_{11} = (-1)^2 \begin{vmatrix} 4 & 1 \\ -4 & 5 \end{vmatrix} = 24,$

$$A_{12} = (-1)^3 \begin{vmatrix} -3 & 1 \\ 4 & 5 \end{vmatrix} = 19, \qquad A_{13} = (-1)^4 \begin{vmatrix} -3 & 4 \\ 4 & -4 \end{vmatrix} = -4,$$

$$A_{21} = (-1)^3 \begin{vmatrix} 2 & 8 \\ -4 & 5 \end{vmatrix} = -42, \qquad A_{22} = (-1)^4 \begin{vmatrix} 6 & 8 \\ 4 & 5 \end{vmatrix} = -2,$$

$$A_{23} = (-1)^5 \begin{vmatrix} 6 & 2 \\ 4 & -4 \end{vmatrix} = 32, \qquad A_{31} = (-1)^4 \begin{vmatrix} 2 & 8 \\ 4 & 1 \end{vmatrix} = -30,$$

$$A_{32} = (-1)^5 \begin{vmatrix} 6 & 8 \\ -3 & 1 \end{vmatrix} = -30, \qquad A_{33} = (-1)^6 \begin{vmatrix} 6 & 2 \\ -3 & 4 \end{vmatrix} = 30.$$

Thus adj $= \begin{bmatrix} 24 & -42 & -30 \\ 19 & -2 & -30 \\ -4 & 32 & 30 \end{bmatrix}.$

(b) $\det(A) = \begin{vmatrix} 6 & 2 & 8 \\ -3 & 4 & 1 \\ 4 & -4 & 5 \end{vmatrix}.$

Expand by cofactors along row 1:

$$\det(A) = (-1)^2(6) \begin{vmatrix} 4 & 1 \\ -4 & 5 \end{vmatrix} + (-1)^3(2) \begin{vmatrix} -3 & 1 \\ 4 & 5 \end{vmatrix} + (-1)^4(8) \begin{vmatrix} -3 & 4 \\ 4 & -4 \end{vmatrix}$$
$$= (6)(24) + (-2)(-19) + (8)(-4) = 150.$$

(c) $A(\text{adj } A) = \begin{bmatrix} 6 & 2 & 8 \\ -3 & 4 & 1 \\ 4 & -4 & 5 \end{bmatrix} \begin{bmatrix} 24 & -42 & -30 \\ 19 & -2 & -30 \\ -4 & 32 & 30 \end{bmatrix} = \begin{bmatrix} 150 & 0 & 0 \\ 0 & 150 & 0 \\ 0 & 0 & 150 \end{bmatrix}.$

$(\text{adj } A)A = \begin{bmatrix} 24 & -42 & -30 \\ 19 & -2 & -30 \\ -4 & 32 & 30 \end{bmatrix} \begin{bmatrix} 6 & 2 & 8 \\ -3 & 4 & 1 \\ 4 & -4 & 5 \end{bmatrix} = \begin{bmatrix} 150 & 0 & 0 \\ 0 & 150 & 0 \\ 0 & 0 & 150 \end{bmatrix}.$

Thus $A(\text{adj } A) = (\text{adj } A)A = 150I_3 = \det(A)I_3.$

In Exercises 11 and 13 we first compute $|A|$. If $|A| \neq 0$, then we compute

$$A^{-1} = \frac{1}{\det(A)} \text{adj}(A).$$

11. (a) $A = \begin{bmatrix} 1 & 2 & -3 \\ -4 & -5 & 2 \\ -1 & 1 & -7 \end{bmatrix}.$

To find $\det(A)$ expand by cofactors along the first row:

$$\det(A) = \begin{vmatrix} 1 & 2 & -3 \\ -4 & -5 & 2 \\ -1 & 1 & -7 \end{vmatrix}$$
$$= (-1)^2(1) \begin{vmatrix} -5 & 2 \\ 1 & -7 \end{vmatrix} + (-1)^3(2) \begin{vmatrix} -4 & 2 \\ -1 & -7 \end{vmatrix} + (-1)^4(-3) \begin{vmatrix} -4 & -5 \\ -1 & 1 \end{vmatrix}$$
$$= 33 - 60 + 27 = 0.$$

Thus A is singular and has no inverse.

(b) $A = \begin{bmatrix} 2 & 3 \\ -1 & 2 \end{bmatrix}$.

Then $\det(A) = 7$ and the cofactors are

$$A_{11} = (-1)^2(2) = 2, \quad A_{12} = (-1)^3(-1) = 1, \quad A_{21} = (-1)^3(3) = -3, \quad A_{22} = (-1)^4(2) = 2.$$

Hence

$$\operatorname{adj} A = \begin{bmatrix} 2 & -3 \\ 1 & 2 \end{bmatrix} \quad \text{and} \quad A^{-1} = \frac{1}{\det(A)}(\operatorname{adj} A) = \begin{bmatrix} \frac{2}{7} & -\frac{3}{7} \\ \frac{1}{7} & \frac{2}{7} \end{bmatrix}.$$

(c) $A = \begin{bmatrix} 4 & 0 & 2 \\ 0 & 3 & 4 \\ 0 & 1 & -2 \end{bmatrix}$.

To find $\det(A)$, expand by cofactors along column 1:

$$\det(A) = (-1)^2(4) \begin{vmatrix} 3 & 4 \\ 1 & -2 \end{vmatrix} = (4)(-10) = -40.$$

The cofactors are $A_{11} = (-1)^2 \begin{vmatrix} 3 & 4 \\ 1 & -2 \end{vmatrix} = -10,$

$$A_{12} = (-1)^3 \begin{vmatrix} 0 & 4 \\ 0 & -2 \end{vmatrix} = 0, \qquad A_{13} = (-1)^4 \begin{vmatrix} 0 & 3 \\ 0 & 1 \end{vmatrix} = 0,$$

$$A_{21} = (-1)^3 \begin{vmatrix} 0 & 2 \\ 1 & -2 \end{vmatrix} = 2, \qquad A_{22} = (-1)^4 \begin{vmatrix} 4 & 2 \\ 0 & -2 \end{vmatrix} = -8,$$

$$A_{23} = (-1)^5 \begin{vmatrix} 4 & 0 \\ 0 & 1 \end{vmatrix} = -4, \qquad A_{31} = (-1)^4 \begin{vmatrix} 0 & 2 \\ 3 & 4 \end{vmatrix} = -6,$$

$$A_{32} = (-1)^5 \begin{vmatrix} 4 & 2 \\ 0 & 4 \end{vmatrix} = -16, \qquad A_{33} = (-1)^6 \begin{vmatrix} 4 & 0 \\ 0 & 3 \end{vmatrix} = 12.$$

Thus it follows that

$$\operatorname{adj} A = \begin{bmatrix} -10 & 2 & -6 \\ 0 & -8 & -16 \\ 0 & -4 & 12 \end{bmatrix} \quad \text{and} \quad A^{-1} = \frac{1}{\det(A)}(\operatorname{adj} A) = \begin{bmatrix} \frac{1}{4} & -\frac{1}{20} & \frac{3}{20} \\ 0 & \frac{1}{5} & \frac{2}{5} \\ 0 & \frac{1}{10} & -\frac{3}{10} \end{bmatrix}.$$

13. (a) $A = \begin{bmatrix} -3 & 1 \\ 2 & 0 \end{bmatrix}$. Then $\det(A) = -2$ and the cofactors are

$$A_{11} = (-1)^2(0) = 0, \qquad A_{12} = (-1)^3(2) = -2,$$
$$A_{21} = (-1)^3(1) = -1, \qquad A_{22} = (-1)^4(-3) = -3.$$

Thus

$$\operatorname{adj} A = \begin{bmatrix} 0 & -1 \\ -2 & -3 \end{bmatrix} \quad \text{and} \quad A^{-1} = \frac{1}{\det(A)}(\operatorname{adj} A) = \begin{bmatrix} 0 & \frac{1}{2} \\ 1 & \frac{3}{2} \end{bmatrix}.$$

(b) $A = \begin{bmatrix} 4 & 0 & 0 \\ 0 & -3 & 0 \\ 0 & 0 & 2 \end{bmatrix}$. Then $\det(A) = (4)(-3)(2) = -24$ and the cofactors are

$$A_{11} = (-1)^2 \begin{vmatrix} -3 & 0 \\ 0 & 2 \end{vmatrix} = -6, \qquad A_{12} = (-1)^3 \begin{vmatrix} 0 & 0 \\ 0 & 2 \end{vmatrix} = 0,$$

$$A_{13} = (-1)^4 \begin{vmatrix} 0 & -3 \\ 0 & 0 \end{vmatrix} = 0, \qquad A_{21} = (-1)^3 \begin{vmatrix} 0 & 0 \\ 0 & 2 \end{vmatrix} = 0,$$

$$A_{22} = (-1)^4 \begin{vmatrix} 4 & 0 \\ 0 & 2 \end{vmatrix} = 8, \qquad A_{23} = (-1)^5 \begin{vmatrix} 4 & 0 \\ 0 & 0 \end{vmatrix} = 0,$$

$$A_{31} = (-1)^4 \begin{vmatrix} 0 & 0 \\ -3 & 0 \end{vmatrix} = 0, \qquad A_{32} = (-1)^5 \begin{vmatrix} 4 & 0 \\ 0 & 0 \end{vmatrix} = 0,$$

$$A_{33} = (-1)^6 \begin{vmatrix} 4 & 0 \\ 0 & -3 \end{vmatrix} = -12.$$

Thus

$$\text{adj } A = \begin{bmatrix} -6 & 0 & 0 \\ 0 & 8 & 0 \\ 0 & 0 & -12 \end{bmatrix}$$

and

$$A^{-1} = \frac{1}{\det(A)}(\text{adj } A) = \begin{bmatrix} \frac{1}{4} & 0 & 0 \\ 0 & -\frac{1}{3} & 0 \\ 0 & 0 & \frac{1}{2} \end{bmatrix}.$$

(c) $A = \begin{bmatrix} 0 & 2 & 1 & 3 \\ 2 & -1 & 3 & 4 \\ -2 & 1 & 5 & 2 \\ 0 & 1 & 0 & 2 \end{bmatrix}$. To find $\det(A)$ expand by cofactors along column 1:

$$\det(A) = (-1)^3(2) \begin{vmatrix} 2 & 1 & 3 \\ 1 & 5 & 2 \\ 1 & 0 & 2 \end{vmatrix} + (-1)^4(-2) \begin{vmatrix} 2 & 1 & 3 \\ -1 & 3 & 4 \\ 1 & 0 & 2 \end{vmatrix}$$

$$= (-2)(5) + (-2)(9) = -28.$$

The cofactors are:

$$A_{11} = (-1)^2 \begin{vmatrix} -1 & 3 & 4 \\ 1 & 5 & 2 \\ 1 & 0 & 2 \end{vmatrix} = -30, \qquad A_{12} = (-1)^3 \begin{vmatrix} 2 & 3 & 4 \\ -2 & 5 & 2 \\ 0 & 0 & 2 \end{vmatrix} = -32,$$

$$A_{13} = (-1)^4 \begin{vmatrix} 2 & -1 & 4 \\ -2 & 1 & 2 \\ 0 & 1 & 2 \end{vmatrix} = -12, \qquad A_{14} = (-1)^5 \begin{vmatrix} 2 & -1 & 3 \\ -2 & 1 & 5 \\ 0 & 1 & 0 \end{vmatrix} = 16,$$

$$A_{21} = (-1)^3 \begin{vmatrix} 2 & 1 & 3 \\ 1 & 5 & 2 \\ 1 & 0 & 2 \end{vmatrix} = -5, \qquad A_{22} = (-1)^4 \begin{vmatrix} 0 & 1 & 3 \\ -2 & 5 & 2 \\ 0 & 0 & 2 \end{vmatrix} = 4,$$

$$A_{23} = (-1)^5 \begin{vmatrix} 0 & 2 & 3 \\ -2 & 1 & 2 \\ 0 & 1 & 2 \end{vmatrix} = -2, \qquad A_{24} = (-1)^6 \begin{vmatrix} 0 & 2 & 1 \\ -2 & 1 & 5 \\ 0 & 1 & 0 \end{vmatrix} = -2,$$

$$A_{31} = (-1)^4 \begin{vmatrix} 2 & 1 & 3 \\ -1 & 3 & 4 \\ 1 & 0 & 2 \end{vmatrix} = 9, \qquad A_{32} = (-1)^5 \begin{vmatrix} 0 & 1 & 3 \\ 2 & 3 & 4 \\ 0 & 0 & 2 \end{vmatrix} = 4,$$

$$A_{33} = (-1)^6 \begin{vmatrix} 0 & 2 & 3 \\ 2 & -1 & 4 \\ 0 & 1 & 2 \end{vmatrix} = -2, \qquad A_{34} = (-1)^7 \begin{vmatrix} 0 & 2 & 1 \\ 2 & -1 & 3 \\ 0 & 1 & 0 \end{vmatrix} = -2,$$

$$A_{41} = (-1)^5 \begin{vmatrix} 2 & 1 & 3 \\ -1 & 3 & 4 \\ 1 & 5 & 2 \end{vmatrix} = 46, \qquad A_{42} = (-1)^6 \begin{vmatrix} 0 & 1 & 3 \\ 2 & 3 & 4 \\ -2 & 5 & 2 \end{vmatrix} = 36,$$

$$A_{43} = (-1)^7 \begin{vmatrix} 0 & 2 & 3 \\ 2 & -1 & 4 \\ -2 & 1 & 2 \end{vmatrix} = 24, \qquad A_{44} = (-1)^8 \begin{vmatrix} 0 & 2 & 1 \\ 2 & -1 & 3 \\ -2 & 1 & 5 \end{vmatrix} = -32.$$

Thus

$$\text{adj } A = \begin{bmatrix} -30 & -5 & 9 & 46 \\ -32 & 4 & 4 & 36 \\ -12 & -2 & -2 & 24 \\ 16 & -2 & -2 & -32 \end{bmatrix} \quad \text{and} \quad A^{-1} = \frac{1}{\det(A)}(\text{adj } A) = \begin{bmatrix} \frac{15}{14} & \frac{5}{28} & -\frac{9}{28} & -\frac{23}{14} \\ \frac{8}{7} & -\frac{1}{7} & -\frac{1}{7} & -\frac{9}{7} \\ \frac{3}{7} & \frac{1}{14} & \frac{1}{14} & -\frac{6}{7} \\ -\frac{4}{7} & \frac{1}{14} & \frac{1}{14} & \frac{8}{7} \end{bmatrix}.$$

In Exercise 15 the determinants involved can be computed using either row or column operations as in Section 3.1 or expansion by cofactors. When possible we use the special forms for the determinants of 2×2 and 3×3 matrices as discussed in Examples 5 and 6 of Section 3.1 respectively.

15. (a) $A = \begin{bmatrix} 4 & 3 & -5 \\ -2 & -1 & 3 \\ 4 & 6 & -2 \end{bmatrix}$.

Using Example 6 of Section 3.1, $\det(A) = 0$. Thus A is singular.

(b) $A = \begin{bmatrix} 1 & 3 & -1 & 2 \\ 2 & -6 & 4 & 1 \\ 3 & 5 & -1 & 3 \\ 4 & -6 & 5 & 2 \end{bmatrix}$.

Using row operations we have

$$\det(A) = \begin{vmatrix} 1 & 3 & -1 & 2 \\ 2 & -6 & 4 & 1 \\ 3 & 5 & -1 & 3 \\ 4 & -6 & 5 & 2 \end{vmatrix} \underset{\substack{-2\mathbf{r}_1 + \mathbf{r}_2 \to \mathbf{r}_2 \\ -3\mathbf{r}_1 + \mathbf{r}_3 \to \mathbf{r}_3 \\ -4\mathbf{r}_1 + \mathbf{r}_4 \to \mathbf{r}_4}}{=} \begin{vmatrix} 1 & 3 & -1 & 2 \\ 0 & -12 & 6 & -3 \\ 0 & -4 & 2 & -3 \\ 0 & -18 & 9 & -6 \end{vmatrix}.$$

Then expanding by cofactors along the first column, we find

$$\det(A) = (-1)^2 (1) \begin{vmatrix} -12 & 6 & -3 \\ -4 & 2 & -3 \\ -18 & 9 & -6 \end{vmatrix}.$$

Using Example 6 from Section 3.1 to compute the previous determinant, we obtain $\det(A) = 0$ and hence A is singular.

(c) $A = \begin{bmatrix} 2 & 2 & -4 \\ 1 & 5 & 2 \\ 3 & 7 & -2 \end{bmatrix}$.

Using Example 6 from Section 3.1 we obtain $\det(A) = 0$. Thus A is singular.

(d) $A = \begin{bmatrix} 0 & 1 & 2 \\ 1 & 2 & 0 \\ 1 & 3 & 4 \end{bmatrix}$.

Using Example 6 of Section 3.1, $\det(A) = -2$. Thus A is nonsingular.

17. (a) To find all the values of λ for which

$$\det\left(\begin{bmatrix} \lambda - 1 & -4 \\ 0 & \lambda - 4 \end{bmatrix} \right) = 0,$$

compute the determinant and solve the resulting polynomial equation.

$$\begin{vmatrix} \lambda - 1 & -4 \\ 0 & \lambda - 4 \end{vmatrix} = (\lambda - 1)(\lambda - 4) = 0.$$

Thus $\lambda = 1$ or $\lambda = 4$.

(b) To find all the values of λ for which $\det\left(\begin{bmatrix} \lambda I_3 - A \end{bmatrix} \right) = 0$, where

$$A = \begin{bmatrix} -3 & -1 & -3 \\ 0 & 3 & 0 \\ -2 & -1 & -2 \end{bmatrix},$$

compute the determinant and solve the resulting polynomial equation.

$$\begin{aligned} \det(\lambda I_3 - A) &= \begin{vmatrix} \lambda + 3 & 1 & 3 \\ 0 & \lambda - 3 & 0 \\ 2 & 1 & \lambda + 2 \end{vmatrix} \\ &= (\lambda + 3)(\lambda - 3)(\lambda + 2) - 6(\lambda - 3) = \lambda^3 + 2\lambda^2 - 15\lambda \\ &= \lambda(\lambda^2 + 2\lambda - 15) \\ &= \lambda(\lambda + 5)(\lambda - 3) = 0. \end{aligned}$$

Thus $\lambda = 0$, $\lambda = -5$, or $\lambda = 3$.

In Exercise 19, write the linear system in matrix form $A\mathbf{x} = \mathbf{0}$ then compute $\det(A)$. If $\det(A) = 0$, then the system has a nontrivial solution.

19. (a) $A\mathbf{x} = \begin{bmatrix} 1 & 2 & -1 \\ 2 & 1 & 2 \\ 3 & -1 & 1 \end{bmatrix} \begin{bmatrix} x \\ y \\ z \end{bmatrix} = \mathbf{0}.$

Using Example 6 in Section 3.1, we find that $\det(A) = 16$. Thus, the linear system has only the trivial solution.

(b) $A\mathbf{x} = \begin{bmatrix} 1 & 1 & 2 & 1 \\ 2 & -1 & 1 & -1 \\ 3 & 1 & 2 & 3 \\ 2 & -1 & -1 & 1 \end{bmatrix} \begin{bmatrix} x \\ y \\ z \\ w \end{bmatrix} = \mathbf{0}.$

Then

$$\det(A) = \begin{vmatrix} 1 & 1 & 2 & 1 \\ 2 & -1 & 1 & -1 \\ 3 & 1 & 2 & 3 \\ 2 & -1 & -1 & 1 \end{vmatrix} \begin{array}{l} {} \\ {} \\ {} \\ {}_{\substack{-2\mathbf{r}_1 + \mathbf{r}_2 \to \mathbf{r}_2 \\ -3\mathbf{r}_1 + \mathbf{r}_3 \to \mathbf{r}_3 \\ -2\mathbf{r}_1 + \mathbf{r}_4 \to \mathbf{r}_4}} \end{array} = \begin{vmatrix} 1 & 1 & 2 & 1 \\ 0 & -3 & -3 & -3 \\ 0 & -2 & -4 & 0 \\ 0 & -3 & -5 & -1 \end{vmatrix}.$$

Using expansion by cofactors along the first column we have

$$\det(A) = (-1)^2(1) \begin{vmatrix} -3 & -3 & -3 \\ -2 & -4 & 0 \\ -3 & -5 & -1 \end{vmatrix} = 0.$$

Thus the linear system has nontrivial solutions.

In Exercises 21 and 23 first write the linear system in matrix form $A\mathbf{x} = \mathbf{b}$. Then use Cramer's Rule to solve the linear system, provided $\det(A) \neq 0$. The determinants of 3×3 matrices are found using Example 6 in Section 3.1.

21. $A\mathbf{x} = \begin{bmatrix} 1 & 1 & 1 & -2 \\ 0 & 2 & 1 & 3 \\ 2 & 1 & -1 & 2 \\ 1 & -1 & 0 & 1 \end{bmatrix} \begin{bmatrix} x \\ y \\ z \\ w \end{bmatrix} = \begin{bmatrix} -4 \\ 4 \\ 5 \\ 4 \end{bmatrix}.$

Using row operations we have

$$\det(A) = \begin{vmatrix} 1 & 1 & 1 & -2 \\ 0 & 2 & 1 & 3 \\ 2 & 1 & -1 & 2 \\ 1 & -1 & 0 & 1 \end{vmatrix} \begin{array}{l} {} \\ {} \\ {}_{\substack{-2\mathbf{r}_1 + \mathbf{r}_3 \to \mathbf{r}_3 \\ -1\mathbf{r}_1 + \mathbf{r}_4 \to \mathbf{r}_4}} \end{array} = \begin{vmatrix} 1 & 1 & 1 & -2 \\ 0 & 2 & 1 & 3 \\ 0 & -1 & -3 & 6 \\ 0 & -2 & -1 & 3 \end{vmatrix}.$$

Expanding by cofactors along the first column, we have

$$\det(A) = (-1)^2(1) \begin{vmatrix} 2 & 1 & 3 \\ -1 & -3 & 6 \\ -2 & -1 & 3 \end{vmatrix} = -30.$$

Then

$$\det(A_1) = \begin{vmatrix} -4 & 1 & 1 & -2 \\ 4 & 2 & 1 & 3 \\ 5 & 1 & -1 & 2 \\ 4 & -1 & 0 & 1 \end{vmatrix}_{\mathbf{c}_1 \leftrightarrow \mathbf{c}_3} = (-1) \begin{vmatrix} 1 & 1 & -4 & -2 \\ 1 & 2 & 4 & 3 \\ -1 & 1 & 5 & 2 \\ 0 & -1 & 4 & 1 \end{vmatrix}_{\substack{-1\mathbf{r}_1 + \mathbf{r}_2 \to \mathbf{r}_2 \\ 1\mathbf{r}_1 + \mathbf{r}_3 \to \mathbf{r}_3}}$$

$$= (-1) \begin{vmatrix} 1 & 1 & -4 & -2 \\ 0 & 1 & 8 & 5 \\ 0 & 2 & 1 & 0 \\ 0 & -1 & 4 & 1 \end{vmatrix} = (-1) \begin{vmatrix} 1 & 8 & 5 \\ 2 & 1 & 0 \\ -1 & 4 & 1 \end{vmatrix} = -30,$$

$$\det(A_2) = \begin{vmatrix} 1 & -4 & 1 & -2 \\ 0 & 4 & 1 & 3 \\ 2 & 5 & -1 & 2 \\ 1 & 4 & 0 & 1 \end{vmatrix}_{\substack{-2\mathbf{r}_1 + \mathbf{r}_3 \to \mathbf{r}_3 \\ -1\mathbf{r}_1 + \mathbf{r}_4 \to \mathbf{r}_4}} = \begin{vmatrix} 1 & -4 & 1 & -2 \\ 0 & 4 & 1 & 3 \\ 0 & 13 & -3 & 6 \\ 0 & 8 & -1 & 3 \end{vmatrix} = \begin{vmatrix} 4 & 1 & 3 \\ 13 & -3 & 6 \\ 8 & -1 & 3 \end{vmatrix} = 30,$$

$$\det(A_3) = \begin{vmatrix} 1 & 1 & -4 & -2 \\ 0 & 2 & 4 & 3 \\ 2 & 1 & 5 & 2 \\ 1 & -1 & 4 & 1 \end{vmatrix}_{\substack{-2\mathbf{r}_1 + \mathbf{r}_3 \to \mathbf{r}_3 \\ -1\mathbf{r}_1 + \mathbf{r}_4 \to \mathbf{r}_4}} = \begin{vmatrix} 1 & 1 & -4 & -2 \\ 0 & 2 & 4 & 3 \\ 0 & -1 & 13 & 6 \\ 0 & -2 & 8 & 3 \end{vmatrix} = \begin{vmatrix} 2 & 4 & 3 \\ -1 & 13 & 6 \\ -2 & 8 & 3 \end{vmatrix} = 0,$$

$$\det(A_4) = \begin{vmatrix} 1 & 1 & 1 & -4 \\ 0 & 2 & 1 & 4 \\ 2 & 1 & -1 & 5 \\ 1 & -1 & 0 & 4 \end{vmatrix}_{\substack{-2\mathbf{r}_1 + \mathbf{r}_3 \to \mathbf{r}_3 \\ -1\mathbf{r}_1 + \mathbf{r}_4 \to \mathbf{r}_4}} = \begin{vmatrix} 1 & 1 & 1 & -4 \\ 0 & 2 & 1 & 4 \\ 0 & -1 & -3 & 13 \\ 0 & -2 & -1 & 8 \end{vmatrix} = \begin{vmatrix} 2 & 1 & 4 \\ -1 & -3 & 13 \\ -2 & -1 & 8 \end{vmatrix} = -60.$$

Thus

$$x = \frac{\det(A_1)}{\det(A)} = \frac{-30}{-30} = 1, \qquad y = \frac{\det(A_2)}{\det(A)} = \frac{30}{-30} = -1,$$

$$z = \frac{\det(A_3)}{\det(A)} = \frac{0}{-30} = 0, \qquad w = \frac{\det(A_4)}{\det(A)} = \frac{-60}{-30} = 2.$$

23. $A\mathbf{x} = \begin{bmatrix} 2 & 3 & 7 \\ -2 & 0 & -4 \\ 1 & 2 & 4 \end{bmatrix} \begin{bmatrix} x \\ y \\ z \end{bmatrix} = \begin{bmatrix} 2 \\ 0 \\ 0 \end{bmatrix}.$

Then

$$\det(A) = \begin{vmatrix} 2 & 3 & 7 \\ -2 & 0 & -4 \\ 1 & 2 & 4 \end{vmatrix} = 0.$$

Thus the linear system has no solution.

25. We evaluate the determinants in this exercise by using the methods discussed in Examples 5 and 6 of Section 3.1.

(a) Let $A = \begin{bmatrix} 1 & 0 & 1 & 1 \\ 1 & 1 & 0 & 1 \\ 1 & 1 & 1 & 0 \\ 0 & 1 & 1 & 1 \end{bmatrix}$. Show that $\det(A) \neq 0$ by reducing A to triangular form.

$$\begin{vmatrix} 1 & 0 & 1 & 1 \\ 1 & 1 & 0 & 1 \\ 1 & 1 & 1 & 0 \\ 0 & 1 & 1 & 1 \end{vmatrix}_{\substack{\mathbf{r}_1 + \mathbf{r}_2 \to \mathbf{r}_2 \\ \mathbf{r}_1 + \mathbf{r}_3 \to \mathbf{r}_3}} = \begin{vmatrix} 1 & 0 & 1 & 1 \\ 0 & 1 & 1 & 0 \\ 0 & 1 & 0 & 1 \\ 0 & 1 & 1 & 1 \end{vmatrix}_{\substack{\mathbf{r}_2 + \mathbf{r}_3 \to \mathbf{r}_3 \\ \mathbf{r}_2 + \mathbf{r}_4 \to \mathbf{r}_4}} = \begin{vmatrix} 1 & 0 & 1 & 1 \\ 0 & 1 & 1 & 0 \\ 0 & 0 & 1 & 1 \\ 0 & 0 & 0 & 1 \end{vmatrix} = (1)(1)(1)(1) = 1 \neq 0.$$

Therefore A is nonsingular.

(b) Let $A = \begin{bmatrix} 0 & 1 & 1 & 1 & 1 \\ 1 & 0 & 1 & 1 & 1 \\ 1 & 1 & 0 & 1 & 1 \\ 1 & 1 & 1 & 0 & 1 \\ 1 & 1 & 1 & 1 & 0 \end{bmatrix}$. Show that $\det(A) = 0$ by reducing A to triangular form.

$$\begin{vmatrix} 0 & 1 & 1 & 1 & 1 \\ 1 & 0 & 1 & 1 & 1 \\ 1 & 1 & 0 & 1 & 1 \\ 1 & 1 & 1 & 0 & 1 \\ 1 & 1 & 1 & 1 & 0 \end{vmatrix}_{\mathbf{r}_1 \leftrightarrow \mathbf{r}_5} = (-1) \begin{vmatrix} 1 & 1 & 1 & 1 & 0 \\ 1 & 0 & 1 & 1 & 1 \\ 1 & 1 & 0 & 1 & 1 \\ 1 & 1 & 1 & 0 & 1 \\ 0 & 1 & 1 & 1 & 1 \end{vmatrix}_{\substack{\mathbf{r}_1 + \mathbf{r}_2 \to \mathbf{r}_2 \\ \mathbf{r}_1 + \mathbf{r}_3 \to \mathbf{r}_3 \\ \mathbf{r}_1 + \mathbf{r}_4 \to \mathbf{r}_4}} = (-1) \begin{vmatrix} 1 & 1 & 1 & 1 & 0 \\ 0 & 1 & 0 & 0 & 1 \\ 0 & 0 & 1 & 0 & 1 \\ 0 & 0 & 0 & 1 & 1 \\ 0 & 1 & 1 & 1 & 1 \end{vmatrix}_{\mathbf{r}_2 + \mathbf{r}_5 \to \mathbf{r}_5}$$

$$= (-1) \begin{vmatrix} 1 & 1 & 1 & 1 & 0 \\ 0 & 1 & 0 & 0 & 1 \\ 0 & 0 & 1 & 0 & 1 \\ 0 & 0 & 0 & 1 & 1 \\ 0 & 0 & 1 & 1 & 0 \end{vmatrix}_{\mathbf{r}_3 + \mathbf{r}_5 \to \mathbf{r}_5} = (-1) \begin{vmatrix} 1 & 1 & 1 & 1 & 0 \\ 0 & 1 & 0 & 0 & 1 \\ 0 & 0 & 1 & 0 & 1 \\ 0 & 0 & 0 & 1 & 1 \\ 0 & 0 & 0 & 1 & 1 \end{vmatrix} = 0.$$

The last determinant is equal to since since the last two rows are identical. Therefore A is singular.

T.1. Let A be upper triangular. Then

$$\det(A) = \begin{vmatrix} a_{11} & a_{12} & \cdots & a_{1n} \\ 0 & a_{22} & \cdots & a_{2n} \\ \vdots & \vdots & & \vdots \\ 0 & 0 & \cdots & a_{nn} \end{vmatrix} = a_{11}A_{11} = a_{11} \begin{vmatrix} a_{22} & \cdots & a_{2n} \\ 0 & \cdots & \\ 0 & \cdots & a_{nn} \end{vmatrix}$$

$$= a_{11}a_{22} \begin{vmatrix} a_{33} & \cdots & a_{3n} \\ & \ddots & \\ 0 & \cdots & a_{nn} \end{vmatrix} = \cdots = a_{11}a_{22} \cdots a_{nn}.$$

T.3. The i, j entry of adj A is $A_{ji} = (-1)^{j+i} \det(M_{ji})$, where M_{ji} is the submatrix of A obtained by deleting from A the jth row and ith column. Since A is symmetric, that submatrix is the transpose of M_{ij}. Thus

$$A_{ji} = (-1)^{j+i} \det(M_{ji}) = (-1)^{i+j} \det(M_{ij}) = j, i \text{ entry of adj } A.$$

Thus adj A is symmetric.

T.5. If $\det(A) = ad - bc \neq 0$, then by Corollary 3.3,

$$A^{-1} = \frac{1}{\det(A)}(\text{adj } A) = \frac{1}{ad - bc} \begin{vmatrix} d & -b \\ -c & a \end{vmatrix}.$$

T.7. If $A = O$, then adj $A = O$ and is singular. Suppose $A \neq O$ but is singular. By Theorem 3.11, $A(\text{adj } A) = \det(A)I_n = O$. Were adj A nonsingular, it should have an inverse, and

$$A = A(\text{adj } A)(\text{adj } A)^{-1} = O(\text{adj } A)^{-1} = O.$$

Contradiction.

T.9. $\begin{vmatrix} a - \lambda & b \\ c & d - \lambda \end{vmatrix} = (a - \lambda)(d - \lambda) - bc$. In other words, the determinant $= 0$, so the matrix is singular and thus has nontrivial solutions.

T.11. Since the entries of A are integers, the cofactors of entries of A are integers and adj A is a matrix of integer entries. Since $\det(A) = \pm 1$, A^{-1} is also a matrix of integers.

T.13. There are 6 of them.

$$\begin{bmatrix} 0 & 1 \\ 1 & 0 \end{bmatrix}, \quad \begin{bmatrix} 0 & 1 \\ 1 & 1 \end{bmatrix}, \quad \begin{bmatrix} 1 & 0 \\ 0 & 1 \end{bmatrix}, \quad \begin{bmatrix} 1 & 0 \\ 1 & 1 \end{bmatrix}, \quad \begin{bmatrix} 1 & 1 \\ 0 & 1 \end{bmatrix}, \quad \begin{bmatrix} 1 & 1 \\ 1 & 0 \end{bmatrix}.$$

ML.1. We present a sample of the cofactors.

A = [1 0 − 2;3 1 4;5 2 − 3];

cofactor(1,1,A)   **cofactor(2,3,A)**   **cofactor(3,1,A)**
```
ans =                     ans =                     ans =
     -11                       -2                        2
```

ML.3. **A = [4 0 − 1; − 2 2 − 1;0 4 − 3];**
detA = 4 ∗ cofactor(1,1,A) + (− 1) ∗ cofactor(1,3,A)
```
detA =
     0
```
We can check this using the **det** command.

ML.5. Before using the expression for the inverse in Corollary 3.3, check the value of the determinant to avoid division by zero.

(a) **A = [1 2 − 3; − 4 − 5 2; − 1 1 − 7];**
det(A)
```
ans =
     0
```
The matrix is singular.

(b) **A = [2 3; − 1 2];**
det(A)
```
ans =
     7
```
invA = (1/det(A)) ∗ adjoint(A)
```
invA =
     0.2857   -0.4286
     0.1429    0.2857
```
To see the inverse with rational entries proceed as follows.
format rat, ans
```
ans =
     2/7   -3/7
     1/7    2/7
```
format

(c) **A = [4 0 2;0 3 4;0 1 − 2];**
det(A)
```
ans =
     -40
```
invA = (1/det(A)) ∗ adjoint(A)
```
invA =
     0.2500   -0.0500    0.1500
          0    0.2000    0.4000
          0    0.1000   -0.3000
```

format rat, ans
ans =
$$\begin{array}{rrr} 1/4 & -1/20 & 3/20 \\ 0 & 1/5 & 2/5 \\ 0 & 1/10 & -3/10 \end{array}$$

Supplementary Exercises, p. 212

1. Use Equation (2).

   (a) $\begin{vmatrix} 0 & 2 & 0 \\ 0 & 0 & -3 \\ 4 & 0 & 0 \end{vmatrix} = (0)(0)(0) - (0)(-3)(0) - (2)(0)(0) + (2)(-3)(4) + (0)(0)(0) - (0)(0)(4) = -24.$

   (b) $\begin{vmatrix} 3 & 0 & 0 & 0 \\ 0 & -2 & 0 & 0 \\ 0 & 4 & 1 & 0 \\ 3 & 2 & -1 & -4 \end{vmatrix} = (3)(-2)(1)(-4) + (3)(0)(4)(-1)$

   $$+ (3)(0)(0)(2) + (0)(0)(0)(-1) + (0)(0)(0)(-4) + (0)(0)(1)(3)$$
   $$+ (0)(0)(4)(-4) + (0)(-2)(0)(3) + (0)(0)(0)(2) + (0)(0)(1)(2)$$
   $$+ (0)(-2)(0)(-1) + (0)(0)(4)(3) - (3)(-2)(0)(-1) - (3)(0)(4)(-4)$$
   $$- (3)(0)(1)(2) - (0)(0)(1)(-4) - (0)(0)(0)(3) - (0)(0)(0)(-1)$$
   $$- (0)(0)(0)(2) - (0)(-2)(0)(-4) - (0)(0)(4)(3) - (0)(0)(4)(-1)$$
   $$- (0)(-2)(1)(3) - (0)(0)(0)(2) = 24.$$

   (See also Exercise 7 in Section 3.1.)

3. A is 4×4 and $\det(A) = 5$.

   (a) Since $\det(A) \neq 0$, Theorem 3.12 implies that A is nonsingular. Then Corollary 3.2 implies that

   $$\det(A^{-1}) = \frac{1}{\det(A)} = \frac{1}{5}.$$

   (b) By Theorem 3.5, $\det(2A) = 2^4 \det(A) = (16)(5) = 80.$

   (c) By Theorem 3.5 and part (a), we have that

   $$\det(2A^{-1}) = 2^4 \det(A^{-1}) = (16)\frac{1}{5} = \frac{16}{5}.$$

   (d) By Corollary 3.2 and part (b),

   $$\det\left((2A)^{-1}\right) = \frac{1}{\det(2A)} = \frac{1}{80}.$$

5. Expanding by cofactors along row 3, we have

$$\det\left(\begin{bmatrix} \lambda + 2 & -1 & 3 \\ 2 & \lambda - 1 & 2 \\ 0 & 0 & \lambda + 4 \end{bmatrix}\right) = (\lambda + 4)(-1)^6 \begin{vmatrix} \lambda + 2 & -1 \\ 2 & \lambda - 1 \end{vmatrix}$$
$$= (\lambda + 4)[(\lambda + 2)(\lambda - 1) + 2]$$
$$= (\lambda + 4)(\lambda^2 + \lambda)$$
$$= (\lambda + 4)\lambda(\lambda + 1) = 0.$$

   Thus $\lambda = -4$, $\lambda = 0$, or $\lambda = -1$.

7. $\begin{vmatrix} 3 & 2 & -1 & 1 \\ 4 & 1 & 1 & 0 \\ -1 & 2 & 3 & 4 \\ -2 & 3 & 5 & 1 \end{vmatrix}_{\mathbf{c}_1 \leftrightarrow \mathbf{c}_4} = (-1) \begin{vmatrix} 1 & 2 & -1 & 3 \\ 0 & 1 & 1 & 4 \\ 4 & 2 & 3 & -1 \\ 1 & 3 & 5 & -2 \end{vmatrix}_{\substack{-4\mathbf{r}_1 + \mathbf{r}_3 \to \mathbf{r}_3 \\ -1\mathbf{r}_1 + \mathbf{r}_4 \to \mathbf{r}_4}}$

$= (-1) \begin{vmatrix} 1 & 2 & -1 & 3 \\ 0 & 1 & 1 & 4 \\ 0 & -6 & 7 & -13 \\ 0 & 1 & 6 & -5 \end{vmatrix}_{\substack{6\mathbf{r}_2 + \mathbf{r}_3 \to \mathbf{r}_3 \\ -1\mathbf{r}_2 + \mathbf{r}_4 \to \mathbf{r}_4}}$

$= (-1) \begin{vmatrix} 1 & 2 & -1 & 3 \\ 0 & 1 & 1 & 4 \\ 0 & 0 & 13 & 11 \\ 0 & 0 & 5 & -9 \end{vmatrix}_{-\frac{5}{13}\mathbf{r}_3 + \mathbf{r}_4 \to \mathbf{r}_4}$

$= (-1) \begin{vmatrix} 1 & 2 & -1 & 3 \\ 0 & 1 & 1 & 4 \\ 0 & 0 & 13 & 11 \\ 0 & 0 & 0 & -\frac{172}{13} \end{vmatrix} = 172.$

9. Since row 1 contains a zero, expand by cofactors along row 1. First compute the cofactors using Equation (3) of Section 3.1.

$$A_{11} = (-1)^2 \begin{vmatrix} 0 & 3 & 2 \\ 1 & 5 & -2 \\ 3 & 2 & -3 \end{vmatrix} = -35, \qquad A_{12} = (-1)^3 \begin{vmatrix} -1 & 3 & 2 \\ 4 & 5 & -2 \\ 1 & 2 & -3 \end{vmatrix} = -47,$$

$$A_{13} = (-1)^4 \begin{vmatrix} -1 & 0 & 2 \\ 4 & 1 & -2 \\ 1 & 3 & -3 \end{vmatrix} = 19.$$

Then

$$\det(A) = a_{11}A_{11} + a_{12}A_{12} + a_{13}A_{13} = (3)(-35) + (2)(-47) + (-1)(19) = -218.$$

11. We first use Equation (3) of Section 3.1 to compute $\det(A) = 5$. Thus, A is nonsingular and we proceed by finding adj A. The cofactors are:

$$A_{11} = (-1)^2 \begin{vmatrix} 1 & 2 \\ 1 & 2 \end{vmatrix} = 0, \qquad A_{12} = (-1)^3 \begin{vmatrix} 0 & 2 \\ -1 & 2 \end{vmatrix} = -2,$$

$$A_{13} = (-1)^4 \begin{vmatrix} 0 & 1 \\ -1 & 1 \end{vmatrix} = 1, \qquad A_{21} = (-1)^3 \begin{vmatrix} -1 & 3 \\ 1 & 2 \end{vmatrix} = 5,$$

$$A_{22} = (-1)^4 \begin{vmatrix} 2 & 3 \\ -1 & 2 \end{vmatrix} = 7, \qquad A_{23} = (-1)^5 \begin{vmatrix} 2 & -1 \\ -1 & 1 \end{vmatrix} = -1,$$

$$A_{31} = (-1)^4 \begin{vmatrix} -1 & 3 \\ 1 & 2 \end{vmatrix} = -5, \qquad A_{32} = (-1)^5 \begin{vmatrix} 2 & 3 \\ 0 & 2 \end{vmatrix} = -4,$$

$$A_{33} = (-1)^6 \begin{vmatrix} 2 & -1 \\ 0 & 1 \end{vmatrix} = 2.$$

Thus,

$$A^{-1} = \frac{1}{\det(A)}(\text{adj } A) = \frac{1}{5}\begin{bmatrix} 0 & 5 & -5 \\ -2 & 7 & -4 \\ 1 & -1 & 2 \end{bmatrix}.$$

13. By Corollary 3.4, the homogeneous system $A\mathbf{x} = \mathbf{0}$ has only the trivial solution when $\det(A) \neq 0$. We compute $\det(A)$, set it equal to zero, and determine the values of λ for which there are nontrivial solutions. Then for all other values of λ there is only the trivial solution.

$$\begin{vmatrix} \lambda & 0 & 1 \\ 1 & \lambda - 1 & 0 \\ 0 & 0 & \lambda + 1 \end{vmatrix} = \lambda(\lambda - 1)(\lambda + 1) = 0$$

when $\lambda = 0$, $\lambda = 1$, or $\lambda = -1$. Thus, there is only the trivial solution to $A\mathbf{x} = \mathbf{0}$ for $\lambda \neq 0, 1, -1$.

15. (a) $$\begin{vmatrix} a & 1 & b \\ b & 1 & c \\ c & 1 & a \end{vmatrix}_{-1\mathbf{c}_3+\mathbf{c}_1 \to \mathbf{c}_1} = \begin{vmatrix} a-b & 1 & b \\ b-c & 1 & c \\ c-a & 1 & a \end{vmatrix}_{\mathbf{c}_1+\mathbf{c}_3 \to \mathbf{c}_3} = \begin{vmatrix} a-b & 1 & a \\ b-c & 1 & b \\ c-a & 1 & c \end{vmatrix}$$

(b) $$\begin{vmatrix} 1 & a & a^2 \\ 1 & b & b^2 \\ 1 & c & c^2 \end{vmatrix}_{\substack{-\mathbf{r}_1+\mathbf{r}_2 \to \mathbf{r}_2 \\ -\mathbf{r}_1+\mathbf{r}_3 \to \mathbf{r}_3}} = \begin{vmatrix} 1 & a & a^2 \\ 0 & b-a & (b+a)(b-a) \\ 0 & c-a & (c+a)(c-a) \end{vmatrix}_{\substack{-a\mathbf{c}_1+\mathbf{c}_2 \to \mathbf{c}_2 \\ -a^2\mathbf{c}_1+\mathbf{c}_3 \to \mathbf{c}_3}}$$

$$= \begin{vmatrix} 1 & 0 & 0 \\ 0 & b-a & (b+a)(b-a) \\ 0 & c-a & (c+a)(c-a) \end{vmatrix}_{-(a+b+c)\mathbf{c}_2+\mathbf{c}_3 \to \mathbf{c}_3}$$

$$= \begin{vmatrix} 1 & 0 & 0 \\ 0 & b-a & -c(b-a) \\ 0 & c-a & -b(c-a) \end{vmatrix}_{\substack{ac\mathbf{c}_1+\mathbf{c}_2 \to \mathbf{c}_2 \\ bc\mathbf{c}_1+\mathbf{c}_3 \to \mathbf{c}_3}}$$

$$= \begin{vmatrix} 1 & a & bc \\ 0 & b-a & -c(b-a) \\ 0 & c-a & -b(c-a) \end{vmatrix}_{\substack{\mathbf{r}_1+\mathbf{r}_2 \to \mathbf{r}_2 \\ \mathbf{r}_1+\mathbf{r}_3 \to \mathbf{r}_3}} = \begin{vmatrix} 1 & a & bc \\ 1 & b & ca \\ 1 & c & ab \end{vmatrix}.$$

17. Let $A = \begin{bmatrix} a-2 & 2 \\ a-2 & a+2 \end{bmatrix}$. A is nonsingular if and only if

$$\det(A) = \begin{vmatrix} a-2 & 2 \\ a-2 & a+2 \end{vmatrix} = (a-2)a \neq 0.$$

Thus A is nonsingular for $a \neq 0$ and $a \neq 2$.

T.1. If rows i and j are proportional with $ta_{ik} = a_{jk}$, $k = 1, 2, \ldots, n$, then

$$\det(A) = \det(A)_{-t\mathbf{r}_i+\mathbf{r}_j \to \mathbf{r}_j} = 0$$

since this row operation makes row j all zeros.

T.3. $$\det(Q - nI_n) = \begin{vmatrix} 1-n & 1 & \cdots & 1 \\ 1 & 1-n & \cdots & 1 \\ \vdots & \vdots & & \vdots \\ 1 & 1 & \cdots & 1-n \end{vmatrix}_{\substack{\mathbf{r}_i+\mathbf{r}_1 \to \mathbf{r}_1 \\ i=2,3,\ldots,n}} = \begin{vmatrix} 0 & 0 & \cdots & 0 \\ 1 & 1-n & \cdots & 1 \\ \vdots & \vdots & & \vdots \\ 1 & 1 & \cdots & 1-n \end{vmatrix} = 0.$$

T.5. From Theorem 3.11, $A(\operatorname{adj} A) = \det(A)I_n$. Since A is singular, $\det(A) = 0$. Therefore $A(\operatorname{adj} A) = O$.

T.7. Compute

$$\begin{vmatrix} A & O \\ O & B \end{vmatrix}$$

by expanding along the first column and expand the resulting $(n-1) \times (n-1)$ determinants along the first column, etc.

T.9. Since $\det(A) \neq 0$, A is nonsingular. Hence the solution to $A\mathbf{x} = \mathbf{b}$ is $\mathbf{x} = A^{-1}\mathbf{b}$. By Exercise T.11 in Section 3.2 matrix A^{-1} has only integer entries. It follows that the product $A^{-1}\mathbf{b}$ has only integer entries.

Chapter 4

Vectors in R^n

Section 4.1, p. 227

1.

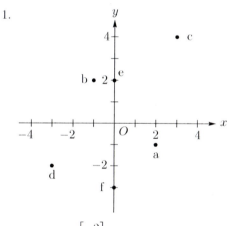

3. The vector $\begin{bmatrix} -2 \\ 5 \end{bmatrix}$ is to have its tail at $(x_1, y_1) = (3, 2)$. Let (x_2, y_2) represent the head of the vector. Then we must have

$$\begin{bmatrix} x_2 - x_1 \\ y_2 - y_1 \end{bmatrix} = \begin{bmatrix} -2 \\ 5 \end{bmatrix}.$$

Setting $x_1 = 3$ and $y_1 = 2$ and solving for x_2 and y_2, we have $x_2 = 1$ and $y_2 = 7$. The head of the vector is at $(1, 7)$.

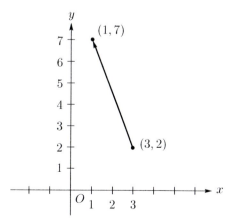

In Exercises 5 and 7 we use the following vector operations: for $\mathbf{u} = (x_1, y_1)$, $\mathbf{v} = (x_2, y_2)$, and scalar c,

$$\mathbf{u} + \mathbf{v} = (x_1 + x_2, y_1 + y_2), \quad \mathbf{u} - \mathbf{v} = (x_1 - x_2, y_1 - y_2), \quad \text{and} \quad c\mathbf{u} = (cx_1, cy_1).$$

5. (a) $\mathbf{u} = (2, 3)$, $\mathbf{v} = (-2, 5)$.
   $\mathbf{u} + \mathbf{v} = (2 + (-2), 3 + 5) = (0, 8)$.
   $\mathbf{u} - \mathbf{v} = (2 - (-2), 3 - 5) = (4, -2)$.
   $2\mathbf{u} = (2(2), 2(3)) = (4, 6)$.
   $3\mathbf{u} - 2\mathbf{v} = (3(2), 3(3)) - (2(-2), 2(5)) = (6, 9) - (-4, 10) = (6 - (-4), 9 - 10) = (10, -1)$.

   (b) $\mathbf{u} = (0, 3)$, $\mathbf{v} = (3, 2)$.
   $\mathbf{u} + \mathbf{v} = (0 + 3, 3 + 2) = (3, 5)$.
   $\mathbf{u} - \mathbf{v} = (0 - 3, 3 - 2) = (-3, 1)$.
   $2\mathbf{u} = (2(0), 2(3)) = (0, 6)$.
   $3\mathbf{u} - 2\mathbf{v} = (3(0), 3(3)) - (2(3), 2(2)) = (0, 9) - (6, 4) = (0 - 6, 9 - 4) = (-6, 5)$.

   (c) $\mathbf{u} = (2, 6)$, $\mathbf{v} = (3, 2)$.
   $\mathbf{u} + \mathbf{v} = (2 + 3, 6 + 2) = (5, 8)$.
   $\mathbf{u} - \mathbf{v} = (2 - 3, 6 - 2) = (-1, 4)$.
   $2\mathbf{u} = (2(2), 2(6)) = (4, 12)$.
   $3\mathbf{u} - 2\mathbf{v} = (3(2), 3(6)) - (2(3), 2(2)) = (6, 18) - (6, 4) = (6 - 6, 18 - 4) = (0, 14)$.

7. Let $\mathbf{u} = (1, 2)$, $\mathbf{v} = (-3, 4)$, $\mathbf{w} = (w_1, 4)$, and $\mathbf{x} = (-2, x_2)$. Determine w_1 or x_2 in each of the following cases.

   (a) Require $\mathbf{w} = 2\mathbf{u}$. Then $(w_1, 4) = (2(1), 2(2)) = (2, 4)$. Hence $w_1 = 2$.

   (b) Require $\frac{3}{2}\mathbf{x} = \mathbf{v}$. Then $\left(\frac{3}{2}(-2), \frac{3}{2}x_2\right) = (-3, 4)$. Hence $\frac{3}{2}x_2 = 4$ and therefore $x_2 = \frac{8}{3}$.

   (c) Require $\mathbf{w} + \mathbf{x} = \mathbf{u}$. Then $(w_1 + (-2), 4 + x_2) = (1, 2)$. Hence $w_1 - 2 = 1$, so $w_1 = 3$ and $x_2 + 4 = 2$, so $x_2 = -2$.

9. Use Equation (1) to find the length of a vector.

   (a) $\|(1, 2)\| = \sqrt{(1)^2 + (2)^2} = \sqrt{1 + 4} = \sqrt{5}$.

   (b) $\|(-3, -4)\| = \sqrt{(-3)^2 + (-4)^2} = \sqrt{9 + 16} = \sqrt{25} = 5$.

   (c) $\|(0, 2)\| = \sqrt{(0)^2 + (2)^2} = \sqrt{4} = 2$.

   (d) $\|(-4, 3)\| = \sqrt{(-4)^2 + (3)^2} = \sqrt{16 + 9} = \sqrt{25} = 5$.

11. The distance between points is determined using Equation (2).

   (a) $P_1(2, 3)$, $P_2(3, 4)$.
   Distance from P_1 to $P_2 = \sqrt{(3 - 2)^2 + (4 - 3)^2} = \sqrt{2}$.

   (b) $P_1(0, 0)$, $P_2(3, 4)$.
   Distance from P_1 to $P_2 = \sqrt{(3 - 0)^2 + (4 - 0)^2} = \sqrt{25} = 5$.

   (c) $P_1(-3, 2)$, $P_2(0, 1)$.
   Distance from P_1 to $P_2 = \sqrt{(0 - (-3))^2 + (1 - 2)^2} = \sqrt{10}$.

   (d) $P_1(0, 3)$, $P_2(2, 0)$.
   Distance from P_1 to $P_2 = \sqrt{(2 - 0)^2 + (0 - 3)^2} = \sqrt{13}$.

13. The vector $(-5, 6)$ is a linear combination of the vectors $(1, 2)$ and $(3, 4)$ if we can find c_1 and c_2 such that

$$(-5, 6) = c_1(1, 2) + c_2(3, 4).$$

Equating corresponding components leads to the linear system

$$c_1 + 3c_2 = -5$$
$$2c_1 + 4c_2 = 6.$$

The solution to this system is $c_1 = 19$, $c_2 = -8$, which means the given vector can be written as the linear combination

$$(-5, 6) = 19(1, 2) + (-8)(3, 4).$$

15. Using Equation (3), the area of the triangle is

$$\frac{1}{2}\left| \det\left(\begin{bmatrix} 3 & 3 & 1 \\ -1 & -1 & 1 \\ 4 & 1 & 1 \end{bmatrix} \right) \right| = \frac{1}{2}\left| (1) \begin{vmatrix} -1 & -1 \\ 4 & 1 \end{vmatrix} - (1) \begin{vmatrix} 3 & 3 \\ 4 & 1 \end{vmatrix} + (1) \begin{vmatrix} 3 & 3 \\ -1 & -1 \end{vmatrix} \right| = 6.$$

17. From the equation following Equation (3), we have

$$\text{area of } T = \left| \det\left(\begin{bmatrix} 2 & 3 & 1 \\ 5 & 3 & 1 \\ 4 & 5 & 1 \end{bmatrix} \right) \right| = 6.$$

19. (a) $\mathbf{x} = (3, 4)$. $\|\mathbf{x}\| = \sqrt{(3)^2 + (4)^2} = \sqrt{25} = 5$. Then a unit vector $\mathbf{u}$ in the direction of $\mathbf{x}$ is given by

$$\mathbf{u} = \frac{1}{\|\mathbf{x}\|}\mathbf{x} = \tfrac{1}{5}(3, 4) = \left(\tfrac{3}{5}, \tfrac{4}{5} \right).$$

(b) $\mathbf{x} = (-2, -3)$. $\|\mathbf{x}\| = \sqrt{(-2)^2 + (-3)^2} = \sqrt{13}$. Then a unit vector $\mathbf{u}$ in the direction of $\mathbf{x}$ is given by

$$\mathbf{u} = \frac{1}{\|\mathbf{x}\|}\mathbf{x} = \frac{1}{\sqrt{13}}(-2, -3) = \left(\frac{-2}{\sqrt{13}}, \frac{-3}{\sqrt{13}} \right).$$

(c) $\mathbf{x} = (5, 0)$. $\|\mathbf{x}\| = \sqrt{(5)^2 + (0)^2} = \sqrt{25} = 5$. Then a unit vector $\mathbf{u}$ in the direction of $\mathbf{x}$ is given by

$$\mathbf{u} = \frac{1}{\|\mathbf{x}\|}\mathbf{x} = \tfrac{1}{5}(5, 0) = (1, 0).$$

21. Use Equation (7) to find the cosine of the angle between vectors $\mathbf{u}$ and $\mathbf{v}$.

(a) $\cos\theta = \dfrac{(1)(2) + (2)(-3)}{\|\mathbf{u}\| \, \|\mathbf{v}\|} = \dfrac{-4}{\sqrt{5}\,\sqrt{13}} = \dfrac{-4}{\sqrt{65}}.$

(b) $\cos\theta = \dfrac{(1)(0) + (0)(1)}{\|\mathbf{u}\| \, \|\mathbf{v}\|} = 0.$

(c) $\cos\theta = \dfrac{(-3)(4) + (-4)(-3)}{\|\mathbf{u}\| \, \|\mathbf{v}\|} = 0.$

(d) $\cos\theta = \dfrac{(2)(-2) + (1)(-1)}{\|\mathbf{u}\|\,\|\mathbf{v}\|} = \dfrac{-5}{\sqrt{5}\,\sqrt{5}} = -1.$

23. $\mathbf{i} = (1,0)$ and $\mathbf{j} = (0,1)$.

    (a) $\mathbf{i}\cdot\mathbf{i} = (1,0)\cdot(1,0) = 1$ and $\mathbf{j}\cdot\mathbf{j} = (0,1)\cdot(0,1) = 1$.

    (b) $\mathbf{i}\cdot\mathbf{j} = (1,0)\cdot(0,1) = 0$.

25. Two vectors are parallel if one is a scalar multiple of the other. We determine a so that $k(a,4) = (2,5)$. We have

$$k(a,4) = (ka, 4k) = (2,5)$$

and equating corresponding entries, we have

$$ka = 2 \quad\text{and}\quad 4k = 5.$$

It follows that $k = \frac{5}{4}$ and hence $a = \frac{2}{k} = \frac{8}{5}$.

27. (a) $(1,3) = (1,0) + (0,3) = (1,0) + 3(0,1) = 1\mathbf{i} + 3\mathbf{j}$.

    (b) $(-2,-3) = (-2,0) + (0,-3) = -2(1,0) + (-3)(0,1) = -2\mathbf{i} + (-3)\mathbf{j}$.

    (c) $(-2,0) = -2(1,0) = -2\mathbf{i}$.

    (d) $(0,3) = 3(0,1) = 3\mathbf{j}$.

29. Represent the force of 300 pounds along the negative y-axis by the vector $-300\mathbf{j}$ and the force of 400 pounds along the negative x-axis by the vector $-400\mathbf{i}$. The resultant force is the direction of the vector $-400\mathbf{i} - 300\mathbf{j}$. The magnitude of the resultant force is

$$\|-400\mathbf{i} - 300\mathbf{j}\| = \sqrt{(-400)^2 + (-300)^2} = \sqrt{250000} = 500 \text{ pounds.}$$

T.1. Locate the point A on the x-axis which is x units from the origin. Construct a perpendicular to the x-axis through A. Locate B on the y-axis y units from the origin. Construct a perpendicular through B. The intersection of those two perpendiculars is the desired point in the plane.

T.3. $(x,y) + (-1)(x,y) = (x,y) + (-x,-y) = (x - x, y - y) = (0,0)$.

T.5. $\|\mathbf{u}\| = \left\|\dfrac{1}{\|\mathbf{x}\|}\mathbf{x}\right\| = \dfrac{1}{\|\mathbf{x}\|}\|\mathbf{x}\|\,(\text{by T.4 above}) = 1.$

T.7. (a) $\mathbf{u} \cdot \mathbf{u} = \|\mathbf{u}\|^2 = x^2 + y^2$; $\mathbf{u} \cdot \mathbf{u} = 0$ if and only if $x = 0$ and $y = 0$, that is, $\mathbf{u} = \mathbf{0}$.

(b) $(x_1, y_1) \cdot (x_2, y_2) = x_1 x_2 + y_1 y_2 = x_2 x_1 + y_2 y_1 = (x_2, y_2) \cdot (x_1, y_1)$.

(c) $[(x_1, y_1) + (x_2, y_2)] \cdot (x_3, y_3) = (x_1 + x_2) x_3 + (y_1 + y_2) y_3 = x_1 x_3 + y_1 y_3 + x_2 x_3 + y_2 y_3 = (x_1, y_1) \cdot (x_3, y_3) + (x_2, y_2) \cdot (x_3, y_3)$.

(d) $(cx_1, cy_1) \cdot (x_2, y_2) = cx_1 x_2 + cy_1 y_2 = (x_1, y_1) \cdot (cx_2, cy_2) = c(x_1 x_2 + y_1 y_2) = c[(x_1, y_1) \cdot (x_2, y_2)]$.

T.9. If $\mathbf{u}$ and $\mathbf{v}$ are parallel, then there exists a nonzero scalar k such that $\mathbf{v} = k\mathbf{u}$. Thus

$$\cos\theta = \frac{\mathbf{u} \cdot \mathbf{v}}{\|\mathbf{u}\| \|\mathbf{v}\|} = \frac{\mathbf{u} \cdot (k\mathbf{u})}{\|\mathbf{u}\| \|k\mathbf{u}\|} = \frac{k(\mathbf{u} \cdot \mathbf{u})}{\|\mathbf{u}\| \sqrt{(k\mathbf{u}) \cdot (k\mathbf{u})}} = \frac{k\|\mathbf{u}\|^2}{\|\mathbf{u}\| \sqrt{k^2} \|\mathbf{u}\|} = \frac{k}{\sqrt{k^2}} = \frac{k}{\pm k} = \pm 1.$$

Section 4.2, p. 244

1. (a) $\mathbf{u} = (1, 2, -3)$, $\mathbf{v} = (0, 1, -2)$.

$\mathbf{u} + \mathbf{v} = (1 + 0, 2 + 1, (-3) + (-2)) = (1, 3, -5)$.

$\mathbf{u} - \mathbf{v} = (1 - 0, 2 - 1, -3 - (-2)) = (1, 1, -1)$.

$2\mathbf{u} = (2(1), 2(2), 2(-3)) = (2, 4, -6)$.

$3\mathbf{u} - 2\mathbf{v} = (3(1), 3(2), 3(-3)) - (2(0), 2(1), 2(-2)) = (3, 6, -9) - (0, 2, -4) = (3, 4, -5)$.

(b) $\mathbf{u} = (4, -2, 1, 3)$, $\mathbf{v} = (-1, 2, 5, -4)$.

$\mathbf{u} + \mathbf{v} = (4 + (-1), (-2) + 2, 1 + 5, 3 + (-4)) = (3, 0, 6, -1)$.

$\mathbf{u} - \mathbf{v} = (4 - (-1), -2 - 2, 1 - 5, 3 - (-4)) = (5, -4, -4, 7)$.

$2\mathbf{u} = (2(4), 2(-2), 2(1), 2(3)) = (8, -4, 2, 6)$.

$3\mathbf{u} - 2\mathbf{v} = (3(4), 3(-2), 3(1), 3(3)) - (2(-1), 2(2), 2(5), 2(-4))$

$\qquad = (12, -6, 3, 9) - (-2, 4, 10, -8) = (14, -10, -7, 17)$.

3. $\mathbf{u} = \begin{bmatrix} 1 \\ -2 \\ 3 \end{bmatrix}$, $\mathbf{v} = \begin{bmatrix} -3 \\ -1 \\ 3 \end{bmatrix}$, $\mathbf{w} = \begin{bmatrix} a \\ -1 \\ b \end{bmatrix}$, $\mathbf{x} = \begin{bmatrix} 3 \\ c \\ 2 \end{bmatrix}$.

(a) Require $\mathbf{w} = \frac{1}{2}\mathbf{u}$. Then

$$\begin{bmatrix} a \\ -1 \\ b \end{bmatrix} = \frac{1}{2} \begin{bmatrix} 1 \\ -2 \\ 3 \end{bmatrix} = \begin{bmatrix} \frac{1}{2} \\ -1 \\ \frac{3}{2} \end{bmatrix}.$$

Hence $a = \frac{1}{2}$, and $b = \frac{3}{2}$.

(b) Require $\mathbf{w} + \mathbf{v} = \mathbf{u}$. Then

$$\begin{bmatrix} a - 3 \\ -2 \\ b + 3 \end{bmatrix} = \begin{bmatrix} 1 \\ -2 \\ 3 \end{bmatrix}.$$

Hence $a - 3 = 1$, $b + 3 = 3$, and it follows that $a = 4$ and $b = 0$.

(c) Require $\mathbf{w} + \mathbf{x} = \mathbf{v}$. Then

$$\begin{bmatrix} a + 3 \\ -1 + c \\ b + 2 \end{bmatrix} = \begin{bmatrix} -3 \\ -1 \\ 3 \end{bmatrix}.$$

Hence $a + 3 = -3$, $-1 + c = -1$, $b + 2 = 3$, and it follows that $a = -6$, $c = 0$, and $b = 1$.

5. $\mathbf{u} = (4, 5, -2, 3)$, $\mathbf{v} = (3, -2, 0, 1)$, $\mathbf{w} = (-3, 2, -5, 3)$, $c = 2$, and $d = 3$.

(a) Verify $\mathbf{u} + \mathbf{v} = \mathbf{v} + \mathbf{u}$.
$\mathbf{u} + \mathbf{v} = (4 + 3, 5 + (-2), -2 + 0, 3 + 1) = (7, 3, -2, 4)$.
$\mathbf{v} + \mathbf{u} = (3 + 4, -2 + 5, 0 + (-2), 1 + 3) = (7, 3, -2, 4)$.

(b) Verify $\mathbf{u} + (\mathbf{v} + \mathbf{w}) = (\mathbf{u} + \mathbf{v}) + \mathbf{w}$.
$\mathbf{u} + (\mathbf{v} + \mathbf{w}) = \mathbf{u} + (3 + (-3), -2 + 2, 0 + (-5), 1 + 3) = (4, 5, -2, 3) + (0, 0, -5, 4) = (4, 5, -7, 7)$.
$(\mathbf{u} + \mathbf{v}) + \mathbf{w} = (4 + 3, 5 + (-2), -2 + 0, 3 + 1) + \mathbf{w} = (7, 3, -2, 4) + (-3, 2, -5, 3) = (4, 5, -7, 7)$.

(c) Verify $\mathbf{u} + \mathbf{0} = \mathbf{0} + \mathbf{u} = \mathbf{u}$.
$\mathbf{u} + \mathbf{0} = (4 + 0, 5 + 0, -2 + 0, 3 + 0) = (4, 5, -2, 3) = \mathbf{u}$.
$\mathbf{0} + \mathbf{u} = (0 + 4, 0 + 5, 0 + (-2), 0 + 3) = (4, 5, -2, 3) = \mathbf{u}$.

(d) Verify $\mathbf{u} + (-\mathbf{u}) = \mathbf{0}$.
$\mathbf{u} + (-\mathbf{u}) = (4 - 4, 5 - 5, -2 - (-2), 3 - 3) = (0, 0, 0, 0) = \mathbf{0}$.

(e) Verify $c(\mathbf{u} + \mathbf{v}) = c\mathbf{u} + c\mathbf{v}$.
$c(\mathbf{u} + \mathbf{v}) = 2(4 + 3, 5 + (-2), -2 + 0, 3 + 1) = 2(7, 3, -2, 4) = (14, 6, -4, 8)$.
$c\mathbf{u} + c\mathbf{v} = 2(4, 5, -2, 3) + 2(3, -2, 0, 1) = (8, 10, -4, 6) + (6, -4, 0, 2) = (14, 6, -4, 8)$.

(f) Verify $(c + d)\mathbf{u} = c\mathbf{u} + d\mathbf{u}$.
$(c + d)\mathbf{u} = 5\mathbf{u} = 5(4, 5, -2, 3) = (20, 25, -10, 15)$.
$c\mathbf{u} + d\mathbf{u} = 2\mathbf{u} + 3\mathbf{u} = (8, 10, -4, 6) + (12, 15, -6, 9) = (20, 25, -10, 15)$.

(g) Verify $c(d\mathbf{u}) = (cd)\mathbf{u}$.
$c(d\mathbf{u}) = 2(3\mathbf{u}) = 2(12, 15, -6, 9) = (24, 30, -12, 18)$.
$(cd)\mathbf{u} = 6\mathbf{u} = (24, 30, -12, 18)$.

(h) Verify $1\mathbf{u} = \mathbf{u}$.
$1\mathbf{u} = (4, 5, -2, 3) = \mathbf{u}$.

7.

9. Let the head of the vector be (x, y, z). Then we have

$$(3, 4, -1) = (x - 1, y - (-2), z - 3).$$

Hence $x - 1 = 3$, $y - (-2) = 4$, $z - 3 = -1$, and it follows that $x = 4$, $y = 2$, and $z = 2$.

11. (a) $\mathbf{u} = (2, 3, 4)$.
$\|\mathbf{u}\| = \sqrt{2^2 + 3^2 + 4^2} = \sqrt{29}$.

(b) $\mathbf{u} = (0, -1, 2, 3)$.
$\|\mathbf{u}\| = \sqrt{0^2 + (-1)^2 + 2^2 + 3^2} = \sqrt{14}$.

(c) $\mathbf{u} = (-1, -2, 0)$.
$\|\mathbf{u}\| = \sqrt{(-1)^2 + (-2)^2 + 0^2} = \sqrt{5}$.

(d) $\mathbf{u} = (1, 2, -3, -4)$.

$\|\mathbf{u}\| = \sqrt{1^2 + 2^2 + (-3)^2 + (-4)^2} = \sqrt{30}$.

13. (a) Let $\mathbf{u} = (1, 1, 0)$ and $\mathbf{v} = (2, -3, 1)$.

$\|\mathbf{u} - \mathbf{v}\| = \sqrt{(1-2)^2 + (1-(-3))^2 + (0-1)^2} = \sqrt{18}$.

(b) Let $\mathbf{u} = (4, 2, -1, 6)$ and $\mathbf{v} = (4, 3, 1, 5)$.

$\|\mathbf{u} - \mathbf{v}\| = \sqrt{(4-4)^2 + (2-3)^2 + (-1-1)^2 + (6-5)^2} = \sqrt{6}$.

(c) Let $\mathbf{u} = (0, 2, 3)$ and $\mathbf{v} = (1, 2, -4)$.

$\|\mathbf{u} - \mathbf{v}\| = \sqrt{(0-1)^2 + (2-2)^2 + (3-(-4))^2} = \sqrt{50}$.

(d) Let $\mathbf{u} = (3, 4, 0, 1)$ and $\mathbf{v} = (2, 2, 1, -1)$.

$\|\mathbf{u} - \mathbf{v}\| = \sqrt{(3-2)^2 + (4-2)^2 + (0-1)^2 + (1+1)^2} = \sqrt{10}$.

15. Performing the indicated operations on the vectors, we have

$$c_1 \begin{bmatrix} 1 \\ 2 \\ -1 \end{bmatrix} + c_2 \begin{bmatrix} 1 \\ 3 \\ -2 \end{bmatrix} + c_3 \begin{bmatrix} 3 \\ 7 \\ -4 \end{bmatrix} = \begin{bmatrix} c_1 + c_2 + 3c_3 \\ 2c_1 + 3c_2 + 7c_3 \\ -c_1 - 2c_2 - 4c_3 \end{bmatrix} = \begin{bmatrix} 0 \\ 0 \\ 0 \end{bmatrix}.$$

Equating corresponding components gives the linear system

$$\begin{aligned} c_1 + c_2 + 3c_3 &= 0 \\ 2c_1 + 3c_2 + 7c_3 &= 0 \\ -c_1 - 2c_2 - 4c_3 &= 0 \end{aligned}$$

Next form the augmented matrix and use row operations to attempt to solve this system:

$$\begin{bmatrix} 1 & 1 & 3 & \vdots & 0 \\ 2 & 3 & 7 & \vdots & 0 \\ -1 & -2 & -4 & \vdots & 0 \end{bmatrix} \longrightarrow \begin{bmatrix} 1 & 1 & 3 & \vdots & 0 \\ 0 & 1 & 1 & \vdots & 0 \\ 0 & -1 & -1 & \vdots & 0 \end{bmatrix} \longrightarrow \begin{bmatrix} 1 & 0 & 2 & \vdots & 0 \\ 0 & 1 & 1 & \vdots & 0 \\ 0 & 0 & 0 & \vdots & 0 \end{bmatrix}$$

add (-2) row 1 to row 2   add row 2 to row 3

add row 1 to row 3   add (-1) row 2 to row 1

The augmented matrix represents the linear system

$$\begin{aligned} c_1 \phantom{{}+c_2} + 2c_3 &= 0 \\ c_2 + c_3 &= 0 \end{aligned}$$

It follows that $c_1 = -2c_3$ and $c_2 = -c_3$. Hence the value for c_3 can be chosen to be any real number. Let $c_3 = r$, then we have the solution $c_1 = -2r$, $c_2 = -r$, $c_3 = r$. One solution that is not all zero is, for $r = 1$, $c_1 = -2$, $c_2 = -1$, $c_3 = 1$.

17. $\mathbf{u} \cdot \mathbf{v} = 0$ provided $(a, 2, 1, a) \cdot (a, -1, -2, -3) = 0$. Computing the dot product, we find that

$$a^2 - 2 - 2 - 3a = 0 \implies a^2 - 3a - 4 = 0 \implies (a-4)(a+1) = 0.$$

Hence $a = 4$ or $a = -1$.

19. $\mathbf{u} = (1, 2, 3)$, $\mathbf{v} = (1, 2, -4)$.

Verify $|\mathbf{u} \cdot \mathbf{v}| \leq \|\mathbf{u}\| \|\mathbf{v}\|$.

$|\mathbf{u} \cdot \mathbf{v}| = |(1)(1) + (2)(2) + (3)(-4)| = |-7| = 7$.

$\|\mathbf{u}\| = \sqrt{1^2 + 2^2 + 3^2} = \sqrt{14}$.

$\|\mathbf{v}\| = \sqrt{1^2 + 2^2 + (-4)^2} = \sqrt{21}$.

Hence $\|\mathbf{u}\| \|\mathbf{v}\| = \sqrt{14} \sqrt{21} = \sqrt{294} = 7\sqrt{6}$. It follows that $|\mathbf{u} \cdot \mathbf{v}| = 7 \leq \|\mathbf{u}\| \|\mathbf{v}\| = 7\sqrt{6}$.

21. (a) $\mathbf{u} = (2, 3, 1)$ and $\mathbf{v} = (3, -2, 0)$.

$\mathbf{u} \cdot \mathbf{v} = 0$, $\|\mathbf{u}\| = \sqrt{14}$, $\|\mathbf{v}\| = \sqrt{13}$.

$\cos \theta = \dfrac{0}{\sqrt{14}\,\sqrt{13}} = 0$.

(b) $\mathbf{u} = (1, 2, -1, 3)$ and $\mathbf{v} = (0, 0, -1, -2)$.

$\mathbf{u} \cdot \mathbf{v} = -5$, $\|\mathbf{u}\| = \sqrt{15}$, $\|\mathbf{v}\| = \sqrt{5}$.

$\cos \theta = \dfrac{-5}{\sqrt{15}\,\sqrt{5}} = \dfrac{-5}{\sqrt{75}} = \dfrac{-5}{5\sqrt{3}} = \dfrac{-1}{\sqrt{3}}$.

(c) $\mathbf{u} = (2, 0, 1)$ and $\mathbf{v} = (2, 2, -1)$.

$\mathbf{u} \cdot \mathbf{v} = 3$, $\|\mathbf{u}\| = \sqrt{5}$, $\|\mathbf{v}\| = \sqrt{9} = 3$.

$\cos \theta = \dfrac{3}{\sqrt{5}\,(3)} = \dfrac{1}{\sqrt{5}}$.

(d) $\mathbf{u} = (0, 4, 2, 3)$ and $\mathbf{v} = (0, -1, 2, 0)$.

$\mathbf{u} \cdot \mathbf{v} = 0$, $\|\mathbf{u}\| = \sqrt{29}$, $\|\mathbf{v}\| = \sqrt{5}$.

$\cos \theta = \dfrac{0}{\sqrt{29}\,\sqrt{5}} = 0$.

23. $\mathbf{u}_1 = (4, 2, 6, -8)$, $\mathbf{u}_2 = (-2, 3, -1, -1)$, $\mathbf{u}_3 = (-2, -1, -3, 4)$, $\mathbf{u}_4 = (1, 0, 0, 2)$, $\mathbf{u}_5 = (1, 2, 3, -4)$, $\mathbf{u}_6 = (0, -3, 1, 0)$.

(a) To determine which vectors are orthogonal, we compute the dot products $\mathbf{u}_s \cdot \mathbf{u}_t$ for $s, t = 1, \ldots, 6$. If the dot product is zero, the vectors are orthogonal. We display the results in the following table.

$\cdot$	$\mathbf{u}_1$	$\mathbf{u}_2$	$\mathbf{u}_3$	$\mathbf{u}_4$	$\mathbf{u}_5$	$\mathbf{u}_6$
$\mathbf{u}_1$	120	0	-60	-12	58	0
$\mathbf{u}_2$	0	15	0	-4	5	-10
$\mathbf{u}_3$	-60	0	30	6	-29	0
$\mathbf{u}_4$	-12	-4	6	5	-7	0
$\mathbf{u}_5$	58	5	-29	-7	30	-3
$\mathbf{u}_6$	0	-10	0	0	-3	10

Thus we see that the following pairs of vectors are orthogonal: $\mathbf{u}_1$ and $\mathbf{u}_2$, $\mathbf{u}_1$ and $\mathbf{u}_6$, $\mathbf{u}_2$ and $\mathbf{u}_3$, $\mathbf{u}_3$ and $\mathbf{u}_6$, and $\mathbf{u}_4$ and $\mathbf{u}_6$.

(b) Vectors $\mathbf{u}_s$ and $\mathbf{u}_t$ are parallel, provided

$$|\mathbf{u}_s \cdot \mathbf{u}_t| = \|\mathbf{u}_s\|\,\|\mathbf{u}_t\|.$$

The products $\|\mathbf{u}_s\|\,\|\mathbf{u}_t\|$ are displayed in the following table

	$\|\mathbf{u}_1\|$	$\|\mathbf{u}_2\|$	$\|\mathbf{u}_3\|$	$\|\mathbf{u}_4\|$	$\|\mathbf{u}_5\|$	$\|\mathbf{u}_6\|$
$\|\mathbf{u}_1\|$	120	$30\sqrt{2}$	60	$10\sqrt{6}$	60	$20\sqrt{3}$
$\|\mathbf{u}_2\|$	$30\sqrt{2}$	15	$15\sqrt{2}$	$5\sqrt{3}$	$15\sqrt{2}$	$5\sqrt{6}$
$\|\mathbf{u}_3\|$	60	$15\sqrt{2}$	30	$5\sqrt{6}$	30	$10\sqrt{3}$
$\|\mathbf{u}_4\|$	$10\sqrt{6}$	$5\sqrt{3}$	$5\sqrt{6}$	5	$5\sqrt{6}$	$5\sqrt{2}$
$\|\mathbf{u}_5\|$	60	$15\sqrt{2}$	30	$5\sqrt{6}$	30	$10\sqrt{3}$
$\|\mathbf{u}_6\|$	$20\sqrt{3}$	$5\sqrt{6}$	$10\sqrt{3}$	$5\sqrt{2}$	$10\sqrt{3}$	10

Compare the entries of this table with the absolute value of the corresponding entries in the table in part (a). Naturally every vector is parallel to itself (see the diagonal entries). The only nondiagonal entries that are equal are in the $(1,3)$ and $(3,1)$ positions. Thus, $\mathbf{u}_1$ and $\mathbf{u}_3$ are parallel.

(c) Vectors $\mathbf{u}_s$ and $\mathbf{u}_t$ are in the same direction provided

$$\mathbf{u}_s \cdot \mathbf{u}_t = \|\mathbf{u}_s\| \|\mathbf{u}_t\|.$$

Compare corresponding entries from the tables in parts (a) and (b). Naturally a vector is in the same direction as itself (see the diagonal entries), but no pair of distinct vectors are in the same direction.

25. $\mathbf{v} = (a, b, c)$ is orthogonal to both $\mathbf{w} = (1, 2, 1)$ and $\mathbf{x} = (1, -1, 1)$ if both dot products $\mathbf{v} \cdot \mathbf{w}$ and $\mathbf{v} \cdot \mathbf{x}$ are equal to zero. We find that

$$\mathbf{v} \cdot \mathbf{w} = (a, b, c) \cdot (1, 2, 1) = a + 2b + c = 0$$
$$\mathbf{v} \cdot \mathbf{x} = (a, b, c) \cdot (1, -1, 1) = a - b + c = 0.$$

Adding (-1) times the first equation to the second equation, we obtain $-3b = 0$, hence $b = 0$. Substituting $b = 0$ into both equations, we obtain the same equation, namely $a + c = 0$, which has infinitely many solutions. Letting $a = 1$, we find one possible answer is: $a = 1$, $b = 0$, $c = -1$, and therefore $\mathbf{v} = (1, 0, -1)$.

27. (a) $\mathbf{x} = (2, -1, 3)$.

$\|\mathbf{x}\| = \sqrt{14}$, $\mathbf{u} = \dfrac{1}{\|\mathbf{x}\|}\mathbf{x} = \left(\dfrac{2}{\sqrt{14}}, -\dfrac{1}{\sqrt{14}}, \dfrac{3}{\sqrt{14}}\right)$.

(b) $\mathbf{x} = (1, 2, 3, 4)$.

$\|\mathbf{x}\| = \sqrt{30}$, $\mathbf{u} = \dfrac{1}{\|\mathbf{x}\|}\mathbf{x} = \left(\dfrac{1}{\sqrt{30}}, \dfrac{2}{\sqrt{30}}, \dfrac{3}{\sqrt{30}}, \dfrac{4}{\sqrt{30}}\right)$.

(c) $\mathbf{x} = (0, 1, -1)$.

$\|\mathbf{x}\| = \sqrt{2}$, $\mathbf{u} = \dfrac{1}{\|\mathbf{x}\|}\mathbf{x} = \left(0, \dfrac{1}{\sqrt{2}}, -\dfrac{1}{\sqrt{2}}\right)$.

(d) $\mathbf{x} = (0, -1, 2, -1)$.

$\|\mathbf{x}\| = \sqrt{6}$, $\mathbf{u} = \dfrac{1}{\|\mathbf{x}\|}\mathbf{x} = \left(0, -\dfrac{1}{\sqrt{6}}, \dfrac{2}{\sqrt{6}}, -\dfrac{1}{\sqrt{6}}\right)$.

29. (a) $(1, 2, -3) = (1, 0, 0), +(0, 2, 0) + (0, 0, -3) = (1, 0, 0) + 2(0, 1, 0) - 3(0, 0, 1) = \mathbf{i} + 2\mathbf{j} - 3\mathbf{k}$.

(b) $(2, 3, -1) = (2, 0, 0) + (0, 3, 0) + (0, 0, -1) = 2(1, 0, 0) + 3(0, 1, 0) - 1(0, 0, 1) = 2\mathbf{i} + 3\mathbf{j} - \mathbf{k}$.

(c) $(0, 1, 2) = (0, 0, 0) + (0, 1, 0) + (0, 0, 2) = 0(1, 0, 0) + (0, 1, 0) + 2(0, 0, 1) = 0\mathbf{i} + \mathbf{j} + 2\mathbf{k}$.

(d) $(0, 0, -2) = -2(0, 0, 1) = -2\mathbf{k}$.

31. A triangle with vertices $P_1(2, 3, -4)$, $P_2(3, 1, 2)$, $P_3(-3, 0, 4)$ is isosceles provided two of the sides are of equal length.

$$\left\|\overrightarrow{P_1 P_2}\right\| = \sqrt{(3-2)^2 + (1-3)^2 + (2-(-4))^2} = \sqrt{41}$$

$$\left\|\overrightarrow{P_1 P_3}\right\| = \sqrt{(-3-2)^2 + (0-3)^2 + (4-(-4))^2} = \sqrt{98}$$

$$\left\|\overrightarrow{P_2 P_3}\right\| = \sqrt{(-3-3)^2 + (0-1)^2 + (4-2)^2} = \sqrt{41}.$$

Thus, the side connecting P_1 to P_2 and the side connecting P_2 to P_3 have the same length.

33. Since "new salaries" = "old salaries" + "the increase", we have

$$\text{"new salaries"} = \mathbf{u} + .08\mathbf{u} = 1.08\mathbf{u}.$$

35. The average daily price of stock is $.5(\mathbf{t} + \mathbf{b})$.

37. $\mathbf{u} = (0, 1, 0, 1)$. Let $\mathbf{v} = (a, b, c, d)$. Then

$$\mathbf{u} + \mathbf{v} = (0 + a, 1 + b, 0 + c, 1 + d) = \mathbf{0}.$$

Hence

$$
\begin{aligned}
0 + a = 0 &\implies a = 0 \\
1 + b = 0 &\implies b = 1 \\
0 + c = 0 &\implies c = 0 \\
1 + d = 0 &\implies d = 1.
\end{aligned}
$$

Therefore $\mathbf{v} = (0, 1, 0, 1)$ is the only vector, since a vector has only one additive inverse.

39. $\mathbf{u} = (1, 0, 1)$. Let $\mathbf{v} = (a, b, c)$. Then

$$\mathbf{u} \cdot \mathbf{v} = (1)(a) + (0)(b) + (1)(c) = a + c = 0.$$

Then we have four possibilities:

$$
\begin{aligned}
a = 0, \ c = 0, \ b = 0 \\
a = 0, \ c = 0, \ b = 1 \\
a = 1, \ c = 1, \ b = 0 \\
a = 1, \ c = 1, \ b = 1
\end{aligned}
$$

Therefore the solutions for $\mathbf{v}$ are:

$$(0, 0, 0), (0, 0, 1), (1, 1, 0), (1, 1, 1).$$

T.1. (a) $\mathbf{u} + \mathbf{v} = \begin{bmatrix} u_1 + v_1 \\ \vdots \\ u_n + v_n \end{bmatrix} = \begin{bmatrix} v_1 + u_1 \\ \vdots \\ v_n + u_n \end{bmatrix} = \mathbf{v} + \mathbf{u}.$

(b) $\mathbf{u} + (\mathbf{v} + \mathbf{w}) = \begin{bmatrix} u_1 + (v_1 + w_1) \\ \vdots \\ u_n + (v_n + w_n) \end{bmatrix} = \begin{bmatrix} (u_1 + v_1) + w_1 \\ \vdots \\ (u_n + v_n) + w_n \end{bmatrix} = (\mathbf{u} + \mathbf{v}) + \mathbf{w}.$

(c) $\mathbf{u} + \mathbf{0} = \begin{bmatrix} u_1 + 0 \\ \vdots \\ u_n + 0 \end{bmatrix} = \begin{bmatrix} u_1 \\ \vdots \\ u_n \end{bmatrix} = \mathbf{u}.$

(d) $\mathbf{u} + (-\mathbf{u}) = \begin{bmatrix} u_1 + (-u_1) \\ \vdots \\ u_n + (-u_n) \end{bmatrix} = \begin{bmatrix} 0 \\ \vdots \\ 0 \end{bmatrix} = \mathbf{0}.$

(e) $c(\mathbf{u} + \mathbf{v}) = \begin{bmatrix} c(u_1 + v_1) \\ \vdots \\ c(u_n + v_n) \end{bmatrix} = \begin{bmatrix} cu_1 + cv_1 \\ \vdots \\ cu_n + cv_n \end{bmatrix} = c\mathbf{u} + c\mathbf{v}.$

(g) $c(d\mathbf{u}) = \begin{bmatrix} c(du_1) \\ \vdots \\ c(du_n) \end{bmatrix} = \begin{bmatrix} (cd)u_1 \\ \vdots \\ (cd)u_n \end{bmatrix} = (cd)\mathbf{u}.$

(h) $1\mathbf{u} = \begin{bmatrix} 1u_1 \\ \vdots \\ 1u_n \end{bmatrix} = \mathbf{u}$.

T.3. The origin O and the head of the vector $\mathbf{u} = (x_1, y_1, z_1)$, call it P, are opposite vertices of a parallelepiped with faces parallel to the coordinate planes (see Figure).

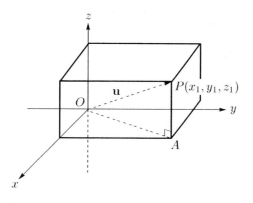

The face diagonal OA has length $\sqrt{x_1^2 + y_1^2}$ by one application of the Pythagorean Theorem. By a second application, the body diagonal has length

$$\|\mathbf{u}\| = \|OP\| = \sqrt{\left(\sqrt{x_1^2 + y_1^2}\right)^2 + z_1^2} = \sqrt{x_1^2 + y_1^2 + z_1^2}.$$

To establish (2), shift the parallelepiped so that the point representing the origin is at the head of the vector $\mathbf{u}$, and the point representing the head of the original vector $\mathbf{u}$ represents the head of vector $\mathbf{v}$. In this case the face diagonal OA has length $\sqrt{(u_1 - v_1)^2 + (u_2 - v_2)^2}$ and the body diagonal has length

$$\|\mathbf{u} - \mathbf{v}\| = \sqrt{\left(\sqrt{(u_1 - v_1)^2 + (u_2 - v_2)^2}\right)^2 + (u_3 - v_3)^2}$$

and the formula follows.

T.5. We know $\mathbf{u} \cdot \mathbf{v} = 0$ and $\mathbf{u} \cdot \mathbf{w} = 0$. We want to show $\mathbf{u}$ is perpendicular to any vector of the form $r\mathbf{v} + s\mathbf{w}$. Consider

$$\mathbf{u} \cdot (r\mathbf{v} + s\mathbf{w}) = \mathbf{u} \cdot (r\mathbf{v}) + \mathbf{u} \cdot (s\mathbf{w}) = r(\mathbf{u} \cdot \mathbf{v}) + s(\mathbf{u} \cdot \mathbf{w}) = r \cdot 0 + s \cdot 0 = 0.$$

(See Theorem 4.3.)

T.7. $\displaystyle\sum_{i=1}^{n} u_i(v_i + w_i) = \sum_{i=1}^{n} u_i v_i + \sum_{i=1}^{n} u_i w_i$.

T.9. $\|c\mathbf{u}\| = \left[\displaystyle\sum_{i=1}^{n}(cu_i)^2\right]^{\frac{1}{2}} = \left[c^2 \sum_{i=1}^{n} u_i^2\right]^{\frac{1}{2}} = |c|\,\|\mathbf{u}\|$.

T.11. By the remark following Example 12 in Section 1.3, we have

$$(A\mathbf{x}) \cdot \mathbf{y} = (A\mathbf{x})^T \mathbf{y} = \mathbf{x}^T(A^T \mathbf{y}) = \mathbf{x} \cdot (A^T \mathbf{y}).$$

T.13. Consider

$$\|\mathbf{u} + \mathbf{v}\|^2 = (u_1 + v_1)^2 + (u_2 + v_2)^2 + \cdots + (u_n + v_n)^2 = \|\mathbf{u}\|^2 + \|\mathbf{v}\|^2 + 2(\mathbf{u} \cdot \mathbf{v}).$$

Substitute $-\mathbf{v}$ for $\mathbf{v}$ to obtain

$$\|\mathbf{u} - \mathbf{v}\|^2 = \|\mathbf{u}\|^2 + \|\mathbf{v}\|^2 - 2(\mathbf{u} \cdot \mathbf{v}).$$

Adding these two equations, we find that

$$\|\mathbf{u} + \mathbf{v}\|^2 + \|\mathbf{u} - \mathbf{v}\|^2 = 2\|\mathbf{u}\|^2 + 2\|\mathbf{v}\|^2.$$

T.15. As in the solution to Exercise T.13 above,

$$\tfrac{1}{4}\|\mathbf{u} + \mathbf{v}\|^2 - \tfrac{1}{4}\|\mathbf{u} - \mathbf{v}\|^2 = \tfrac{1}{4}\left[\|\mathbf{u}\|^2 + \|\mathbf{v}\|^2 + 2(\mathbf{u} \cdot \mathbf{v})\right] - \tfrac{1}{4}\left[\|\mathbf{u}\|^2 + \|\mathbf{v}\|^2 - 2(\mathbf{u} \cdot \mathbf{v})\right] = \mathbf{u} \cdot \mathbf{v}.$$

T.17. Each element in B^3 is its own additive inverse. For example, $(1,0,0) + (1,0,0) = (0,0,0)$ or $(1,0,1) + (1,0,1) = (0,0,0)$. So the "negative" of $(1,0,0)$ is $(1,0,0)$.

T.19. The "negative" of each vector in B^4 is itself. For example, $(1,0,1,0) + (1,0,1,0) = (0,0,0,0)$.

T.21. (a) We want to find all $\mathbf{v}$ so that $\mathbf{v} \cdot (1,1,1,1) = 0$. Let $\mathbf{v} = (v_1, v_2, v_3, v_4)$. But

$$\mathbf{v} \cdot (1,1,1,1) = v_1 + v_2 + v_3 + v_4 = 0$$

whenever $\mathbf{v}$ has even parity, i.e., the sum of its entries is an even number.

(b) The vectors in $V_{\mathbf{u}}$ have odd parity.

(c) Yes, given a vector in B^4, its parity is well-defined, i.e., it can only be either even or odd.

ML.3. (a) $\mathbf{u} = [2 \ \ 0 \ \ 3]';\mathbf{v} = [2 \ \ -1 \ \ 1]';\mathbf{norm(u - v)}$
    ans =
                2.2361

(b) $\mathbf{u} = [2 \ \ 0 \ \ 0 \ \ 1];\mathbf{v} = [2 \ \ 5 \ \ -1 \ \ 3];\mathbf{norm(u - v)}$
    ans =
                5.4772

(c) $\mathbf{u} = [1 \ \ 0 \ \ 4 \ \ 3];\mathbf{v} = [-1 \ \ 1 \ \ 2 \ \ 2];\mathbf{norm(u - v)}$
    ans =
                3.1623

ML.5. (a) $\mathbf{u} = [5 \ \ 4 \ \ -4];\mathbf{v} = [3 \ \ 2 \ \ 1];$
    $\mathbf{dot(u,v)}$
    ans =
                19

(b) $\mathbf{u} = [3 \ \ -1 \ \ 0 \ \ 2];\mathbf{v} = [-1 \ \ 2 \ \ -5 \ \ -3];$
    $\mathbf{dot(u,v)}$
    ans =
                -11

(c) $\mathbf{u} = [1 \ \ 2 \ \ 3 \ \ 4 \ \ 5];$
    $\mathbf{dot(u, -u)}$
    ans =
                -55

ML.9. (a) $\mathbf{u} = [2 \quad 2 \quad -1]'$;
        unit = u/norm(u)
        unit =
                0.6667
                0.6667
                −0.3333
        format rat, unit
        unit =
                2/3
                2/3
                −1/3
        format

(b) $\mathbf{v} = [0 \quad 4 \quad -3 \quad 0]'$;
        unit = v/norm(v)
        unit =
                0
                0.8000
                −0.6000
                0
        format rat, unit
        unit =
                0
                4/5
                −3/5
                0
        format

(c) $\mathbf{w} = [1 \quad 0 \quad 1 \quad 0 \quad 3]'$;
        unit = w/norm(w)
        unit =
                0.3015
                0
                0.3015
                0
                0.9045
        format rat, unit
        unit =
                379/1257
                0
                379/1257
                0
                379/419
        format

Section 4.3, p. 255

In Exercises 1 and 3, we follow the techniques in Examples 1 and 2 to determine if L is a linear transformation.

1. (a) Let $\mathbf{u} = (x_1, y_1)$ and $\mathbf{v} = (x_2, y_2)$. Then

$$L(\mathbf{u} + \mathbf{v}) = L((x_1, y_1) + (x_2, y_2))$$
$$= L(x_1 + x_2, y_1 + y_2) = (x_1 + x_2 + 1, y_1 + y_2, x_1 + x_2 + y_1 + y_2).$$

On the other hand,

$$L(\mathbf{u}) + L(\mathbf{v}) = L(x_1, y_1) + L(x_2, y_2)$$
$$= (x_1 + 1, y_1, x_1 + y_1) + (x_2 + 1, y_2, x_2 + y_2)$$
$$= (x_1 + x_2 + 2, y_1 + y_2, x_1 + x_2 + y_1 + y_2).$$

Thus, $L(\mathbf{u} + \mathbf{v}) \neq L(\mathbf{u}) + L(\mathbf{v})$ and hence L is not a linear transformation.

(b) Let $\mathbf{u} = \begin{bmatrix} x_1 \\ y_1 \\ z_1 \end{bmatrix}$ and $\mathbf{v} = \begin{bmatrix} x_2 \\ y_2 \\ z_2 \end{bmatrix}$. Then

$$L(\mathbf{u} + \mathbf{v}) = L\left(\begin{bmatrix} x_1 + x_2 \\ y_1 + y_2 \\ z_1 + z_2 \end{bmatrix} \right) = \begin{bmatrix} x_1 + x_2 + y_1 + y_2 \\ y_1 + y_2 \\ x_1 + x_2 - z_1 - z_2 \end{bmatrix}.$$

On the other hand

$$L(\mathbf{u}) + L(\mathbf{v}) = \begin{bmatrix} x_1 + y_1 \\ y_1 \\ x_1 - z_1 \end{bmatrix} + \begin{bmatrix} x_2 + y_2 \\ y_2 \\ x_2 - z_2 \end{bmatrix} = \begin{bmatrix} x_1 + x_2 + y_1 + y_2 \\ y_1 + y_2 \\ x_1 + x_2 - z_1 - z_2 \end{bmatrix}.$$

Thus $L(\mathbf{u} + \mathbf{v}) = L(\mathbf{u}) + L(\mathbf{v})$. Next, let k be any real number. Then

$$L(k\mathbf{u}) = L\left(\begin{bmatrix} kx_1 \\ ky_1 \\ kz_1 \end{bmatrix} \right) = \begin{bmatrix} kx_1 + ky_1 \\ ky_1 \\ kx_1 - kz_1 \end{bmatrix} = \begin{bmatrix} k(x_1 + y_1) \\ ky_1 \\ k(x_1 - z_1) \end{bmatrix} = k \begin{bmatrix} x_1 + y_1 \\ y_1 \\ x_1 - z_1 \end{bmatrix} = kL(\mathbf{u}).$$

Thus L is a linear transformation.

(c) Let $\mathbf{u} = (x_1, y_1)$ and $\mathbf{v} = (x_2, y_2)$. Then

$$L(\mathbf{u} + \mathbf{v}) = L((x_1, y_1) + (x_2, y_2)) = L(x_1 + x_2, y_1 + y_2)$$
$$= ((x_1 + x_2)^2 + x_1 + x_2, y_1 + y_2 - (y_1 + y_2)^2)$$
$$= (x_1^2 + 2x_1 x_2 + x_2^2 + x_1 + x_2, y_1 + y_2 - y_1^2 - 2y_1 y_2 - y_2^2).$$

On the other hand

$$L(\mathbf{u}) + L(\mathbf{v}) = L(x_1, y_1) + L(x_2, y_2)$$
$$= (x_1^2 + x_1, y_1 - y_1^2) + (x_2^2 + x_2, y_2 - y_2^2)$$
$$= (x_1^2 + x_1 + x_2^2 + x_2, y_1 - y_1^2 + y_2 - y_2^2).$$

Thus, $L(\mathbf{u} + \mathbf{v}) \neq L(\mathbf{u}) + L(\mathbf{v})$ and hence L is not a linear transformation.

3. (a) Let $\mathbf{u} = (x_1, y_1, z_1)$ and $\mathbf{v} = (x_2, y_2, z_2)$. Then

$$L(\mathbf{u} + \mathbf{v}) = L((x_1, y_1, z_1) + (x_2, y_2, z_2)) = L(x_1 + x_2, y_1 + y_2, z_1 + z_2)$$
$$= (x_1 + x_2 + y_1 + y_2, 0, 2x_1 + 2x_2 - z_1 - z_2).$$

On the other hand

$$L(\mathbf{u}) + L(\mathbf{v}) = L(x_1, y_1, z_1) + L(x_2, y_2, z_2)$$
$$= (x_1 + y_1, 0, 2x_1 - z_1) + (x_2 + y_2, 0, 2x_2 - z_2)$$
$$= (x_1 + x_2 + y_1 + y_2, 0, 2x_1 + 2x_2 - z_1 - z_2).$$

Thus $L(\mathbf{u} + \mathbf{v}) = L(\mathbf{u}) + L(\mathbf{v})$. Next, let k be any real number. Then

$$L(k\mathbf{u}) = L(kx_1, ky_1, kz_1) = (kx_1 + ky_1, 0, 2kx_1 - kz_1)$$
$$= (k(x_1 + y_1), 0, k(2x_1 - z_1)) = k(x_1 + y_1, 0, 2x_1 - z_1) = kL(\mathbf{u}).$$

Hence L is a linear transformation.

(b) Let $\mathbf{u} = \begin{bmatrix} x_1 \\ y_1 \end{bmatrix}$ and $\mathbf{v} = \begin{bmatrix} x_2 \\ y_2 \end{bmatrix}$. Then

$$L(\mathbf{u} + \mathbf{v}) = L\left(\begin{bmatrix} x_1 + x_2 \\ y_1 + y_2 \end{bmatrix} \right) = \begin{bmatrix} (x_1 + x_2)^2 - (y_1 + y_2)^2 \\ (x_1 + x_2)^2 + (y_1 + y_2)^2 \end{bmatrix}$$

$$= \begin{bmatrix} x_1^2 + 2x_1 x_2 + x_2^2 - y_1^2 - 2y_1 y_2 - y_2^2 \\ x_1^2 + 2x_1 x_2 + x_2^2 + y_1^2 + 2y_1 y_2 + y_2^2 \end{bmatrix}.$$

On the other hand

$$L(\mathbf{u}) + L(\mathbf{v}) = \begin{bmatrix} x_1^2 - y_1^2 \\ x_1^2 + y_1^2 \end{bmatrix} + \begin{bmatrix} x_2^2 - y_2^2 \\ x_2^2 + y_2^2 \end{bmatrix} = \begin{bmatrix} x_1^2 + x_2^2 - y_1^2 - y_2^2 \\ x_1^2 + x_2^2 + y_1^2 + y_2^2 \end{bmatrix}.$$

Thus, $L(\mathbf{u} + \mathbf{v}) \neq L(\mathbf{u}) + L(\mathbf{v})$ and hence L is not a linear transformation.

5. Since $L \colon R^2 \to R^2$ is defined by $L(x, y) = (x, -y)$, we have $L(2, 3) = (2, -3)$.

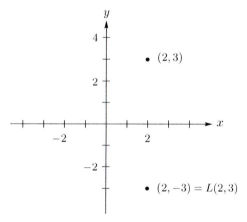

7. Since $L \colon R^2 \to R^2$ is defined by

$$L(x, y) = \begin{bmatrix} \cos \phi & -\sin \phi \\ \sin \phi & \cos \phi \end{bmatrix} \begin{bmatrix} x \\ y \end{bmatrix} \qquad \text{(see Example 10)}$$

we have

$$L(x, y) = \begin{bmatrix} \cos 30° & -\sin 30° \\ \sin 30° & \cos 30° \end{bmatrix} \begin{bmatrix} x \\ y \end{bmatrix} = \begin{bmatrix} \frac{\sqrt{3}}{2} & -\frac{1}{2} \\ \frac{1}{2} & \frac{\sqrt{3}}{2} \end{bmatrix} \begin{bmatrix} x \\ y \end{bmatrix}.$$

Thus

$$L(-1, 3) = \begin{bmatrix} \frac{\sqrt{3}}{2} & -\frac{1}{2} \\ \frac{1}{2} & \frac{\sqrt{3}}{2} \end{bmatrix} \begin{bmatrix} -1 \\ 3 \end{bmatrix} = \left(\frac{-3 - \sqrt{3}}{2}, \frac{-1 + 3\sqrt{3}}{2} \right) \approx (-2.366, 2.098).$$

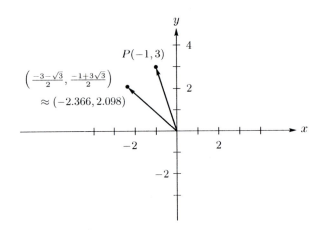

9. Since $L\colon R^2 \to R^2$ is defined by $L(\mathbf{u}) = -\mathbf{u}$, we have $L(3,2) = (-3,-2)$.

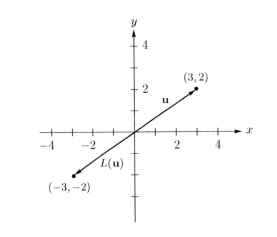

11. Since $L\colon R^3 \to R^2$ is defined by $L\left(\begin{bmatrix} x \\ y \\ z \end{bmatrix}\right) = \begin{bmatrix} x \\ x-y \end{bmatrix}$, we have $L\left(\begin{bmatrix} 2 \\ -1 \\ 3 \end{bmatrix}\right) = \begin{bmatrix} 2 \\ 3 \end{bmatrix}$.

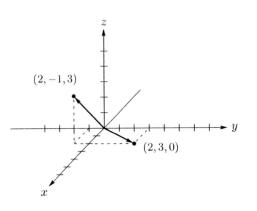

13. (a) We seek a vector $\mathbf{v} = \begin{bmatrix} x \\ y \\ z \end{bmatrix}$ such that $L(\mathbf{v}) = \mathbf{w}$:

$$L\left(\begin{bmatrix} x \\ y \\ z \end{bmatrix}\right) = \begin{bmatrix} x + z \\ y + z \\ x + 2y + 2z \end{bmatrix} = \begin{bmatrix} 1 \\ -1 \\ 0 \end{bmatrix}.$$

Hence we must solve the linear system

$$\begin{array}{rcr} x \quad\quad + \; z = & 1 \\ y + \; z = & -1 \\ x + 2y + 2z = & 0 \end{array}$$

with augmented matrix

$$\begin{bmatrix} 1 & 0 & 1 & \vdots & 1 \\ 0 & 1 & 1 & \vdots & -1 \\ 1 & 2 & 2 & \vdots & 0 \end{bmatrix}.$$

We find the reduced row echelon form of the augmented matrix to be

$$\begin{bmatrix} 1 & 0 & 0 & \vdots & 2 \\ 0 & 1 & 0 & \vdots & 0 \\ 0 & 0 & 1 & \vdots & -1 \end{bmatrix}.$$

Thus a solution to the system is $x = 2$, $y = 0$, $z = -1$, so $\mathbf{w}$ is in range L.

(b) We seek a vector $\mathbf{v} = \begin{bmatrix} x \\ y \\ z \end{bmatrix}$ such that $L(\mathbf{v}) = \mathbf{w}$:

$$L\left(\begin{bmatrix} x \\ y \\ z \end{bmatrix}\right) = \begin{bmatrix} x + z \\ y + z \\ x + 2y + 2z \end{bmatrix} = \begin{bmatrix} 2 \\ -1 \\ 3 \end{bmatrix}.$$

We find the reduced row echelon form of the augmented matrix to be

$$\begin{bmatrix} 1 & 0 & 0 & \vdots & 5 \\ 0 & 1 & 0 & \vdots & 2 \\ 0 & 0 & 1 & \vdots & -3 \end{bmatrix}.$$

Thus a solution to the system is $x = 5$, $y = 2$, $z = -3$, so $\mathbf{w}$ is in range L.

15. Given

$$\mathbf{w} = \begin{bmatrix} a \\ b \\ c \end{bmatrix},$$

where a, b, and c are any real numbers. Find

$$\mathbf{v} = \begin{bmatrix} x \\ y \\ z \end{bmatrix}$$

so that $L(\mathbf{v}) = \mathbf{w}$. We need to find a solution to the linear system

$$\begin{bmatrix} 4 & 1 & 3 \\ 2 & -1 & 3 \\ 2 & 2 & 0 \end{bmatrix} \begin{bmatrix} x \\ y \\ z \end{bmatrix} = \begin{bmatrix} a \\ b \\ c \end{bmatrix}.$$

The reduced row echelon form of the augmented matrix is

$$\begin{bmatrix} 1 & 0 & 1 & \vdots & \frac{1}{6}a + \frac{1}{6}b \\ 0 & 1 & -1 & \vdots & \frac{1}{3}a - \frac{2}{3}b \\ 0 & 0 & 0 & \vdots & -a + b + c \end{bmatrix}.$$

Thus a solution exists only for $-a + b + c = 0$.

17. Since

$$\begin{bmatrix} 4 \\ -3 \end{bmatrix} = 4\begin{bmatrix} 1 \\ 0 \end{bmatrix} - 3\begin{bmatrix} 0 \\ 1 \end{bmatrix} = 4\mathbf{i} - 3\mathbf{j}$$

we have

$$L\left(\begin{bmatrix} 4 \\ -3 \end{bmatrix}\right) = L(4\mathbf{i} - 3\mathbf{j}) = 4L(\mathbf{i}) - 3L(\mathbf{j}) = 4\begin{bmatrix} 2 \\ 3 \end{bmatrix} - 3\begin{bmatrix} -1 \\ 2 \end{bmatrix} = \begin{bmatrix} 8 \\ 12 \end{bmatrix} + \begin{bmatrix} 3 \\ -6 \end{bmatrix} = \begin{bmatrix} 11 \\ 6 \end{bmatrix}.$$

19. Since $L\colon R^3 \to R^2$ is as in Exercise 11, we have

$$L\left(\begin{bmatrix} x \\ y \\ z \end{bmatrix}\right) = \begin{bmatrix} x \\ x - y \end{bmatrix} = \begin{bmatrix} 0 \\ 0 \end{bmatrix}.$$

Equating corresponding components gives a linear system

$$\begin{aligned} x &= 0 \\ x - y &= 0 \end{aligned}$$

with augmented matrix

$$\begin{bmatrix} 1 & 0 & \vdots & 0 \\ 1 & -1 & \vdots & 0 \end{bmatrix}$$

which has reduced row echelon form

$$\begin{bmatrix} 1 & 0 & \vdots & 0 \\ 0 & 1 & \vdots & 0 \end{bmatrix}.$$

The solution is $x = 0$, $y = 0$ and $z = r$, where r is any real number. Thus $\mathbf{x} = \begin{bmatrix} 0 \\ 0 \\ r \end{bmatrix}$.

21. (a) Reflection about the y-axis. (b) Reflection about the origin. (c) Counterclockwise rotation through $\pi/2$ radians.

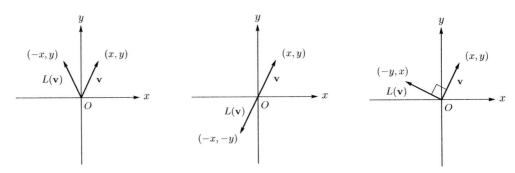

23. L is not a linear transformation. For, by Theorem 4.7(a), a linear transformation maps the zero vector to the zero vector. In this case, $L(\mathbf{0}) = L(0,0) = (1,0) \neq (0,0)$. Thus, L is not a linear transformation.

25. The standard matrix A representing

$$L\left(\begin{bmatrix} x \\ y \end{bmatrix}\right) = \begin{bmatrix} -x \\ y \end{bmatrix}$$

is the 2×2 matrix whose columns are $L(\mathbf{e}_1)$ and $L(\mathbf{e}_2)$, respectively.

$$L(\mathbf{e}_1) = L\left(\begin{bmatrix} 1 \\ 0 \end{bmatrix}\right) = \begin{bmatrix} -1 \\ 0 \end{bmatrix} = \text{col}_1(A), \qquad L(\mathbf{e}_2) = L\left(\begin{bmatrix} 0 \\ 1 \end{bmatrix}\right) = \begin{bmatrix} 0 \\ 1 \end{bmatrix} = \text{col}_2(A).$$

Hence

$$A = \begin{bmatrix} -1 & 0 \\ 0 & 1 \end{bmatrix} \quad \text{and} \quad A\mathbf{x} = \begin{bmatrix} -1 & 0 \\ 0 & 1 \end{bmatrix}\begin{bmatrix} x \\ y \end{bmatrix} = \begin{bmatrix} -x \\ y \end{bmatrix} = L(\mathbf{x}).$$

27. The standard matrix representing $L \colon R^2 \to R^2$ is (see Example 10)

$$A = \begin{bmatrix} \cos\frac{\pi}{4} & -\sin\frac{\pi}{4} \\ \sin\frac{\pi}{4} & \cos\frac{\pi}{4} \end{bmatrix} = \begin{bmatrix} \frac{\sqrt{2}}{2} & -\frac{\sqrt{2}}{2} \\ \frac{\sqrt{2}}{2} & \frac{\sqrt{2}}{2} \end{bmatrix}.$$

29. The standard matrix representing $L \colon R^3 \to R^3$ defined by

$$L\left(\begin{bmatrix} x \\ y \\ z \end{bmatrix}\right) = \begin{bmatrix} x - y \\ x + z \\ y - z \end{bmatrix}$$

is the 3×3 matrix whose columns are $L(\mathbf{e}_1)$, $L(\mathbf{e}_2)$, and $L(\mathbf{e}_3)$, respectively.

$$L(\mathbf{e}_1) = L\left(\begin{bmatrix} 1 \\ 0 \\ 0 \end{bmatrix}\right) = \begin{bmatrix} 1 \\ 1 \\ 0 \end{bmatrix}, \quad L(\mathbf{e}_2) = L\left(\begin{bmatrix} 0 \\ 1 \\ 0 \end{bmatrix}\right) = \begin{bmatrix} -1 \\ 0 \\ 1 \end{bmatrix}, \quad L(\mathbf{e}_3) = L\left(\begin{bmatrix} 0 \\ 0 \\ 1 \end{bmatrix}\right) = \begin{bmatrix} 0 \\ 1 \\ -1 \end{bmatrix}.$$

Hence

$$A = \begin{bmatrix} 1 & -1 & 0 \\ 1 & 0 & 1 \\ 0 & 1 & -1 \end{bmatrix} \quad \text{and} \quad A\mathbf{x} = \begin{bmatrix} 1 & -1 & 0 \\ 1 & 0 & 1 \\ 0 & 1 & -1 \end{bmatrix}\begin{bmatrix} x \\ y \\ z \end{bmatrix} = \begin{bmatrix} x - y \\ x + z \\ y - z \end{bmatrix} = L(\mathbf{x}).$$

31. (a) Following the technique of Example 9, we determine the numbers associated with the letters in the message.

S	E	N	D	H	I	M	M	O	N	E	Y
↕	↕	↕	↕	↕	↕	↕	↕	↕	↕	↕	↕
19	5	14	4	8	9	13	13	15	14	5	25

We break this string of numbers into four vectors in R^3 and multiply each vector by the matrix

$$A = \begin{bmatrix} 1 & 2 & 3 \\ 1 & 1 & 2 \\ 0 & 1 & 2 \end{bmatrix}.$$

We obtain

$$A\begin{bmatrix} 19 \\ 5 \\ 14 \end{bmatrix} = \begin{bmatrix} 71 \\ 52 \\ 33 \end{bmatrix}, \quad A\begin{bmatrix} 4 \\ 8 \\ 9 \end{bmatrix} = \begin{bmatrix} 47 \\ 30 \\ 26 \end{bmatrix}, \quad A\begin{bmatrix} 13 \\ 13 \\ 15 \end{bmatrix} = \begin{bmatrix} 84 \\ 56 \\ 43 \end{bmatrix}, \quad A\begin{bmatrix} 14 \\ 5 \\ 25 \end{bmatrix} = \begin{bmatrix} 99 \\ 69 \\ 55 \end{bmatrix}.$$

The final version of the code message is the string of numbers

$$71 \quad 52 \quad 33 \quad 47 \quad 30 \quad 26 \quad 84 \quad 56 \quad 43 \quad 99 \quad 69 \quad 55.$$

(b) We are given the coded message

$$67 \quad 44 \quad 41 \quad 49 \quad 39 \quad 19 \quad 113 \quad 76 \quad 62 \quad 104 \quad 69 \quad 55.$$

Breaking this into vectors in R^3, the decoded vector is obtained by multiplying each of the vectors by

$$A^{-1} = \begin{bmatrix} 0 & 1 & -1 \\ 2 & -2 & -1 \\ -1 & 1 & 1 \end{bmatrix}.$$

We obtain

$$A^{-1} \begin{bmatrix} 67 \\ 44 \\ 41 \end{bmatrix} = \begin{bmatrix} 3 \\ 5 \\ 18 \end{bmatrix}, \quad A^{-1} \begin{bmatrix} 49 \\ 39 \\ 19 \end{bmatrix} = \begin{bmatrix} 20 \\ 1 \\ 9 \end{bmatrix}, \quad A^{-1} \begin{bmatrix} 113 \\ 76 \\ 62 \end{bmatrix} = \begin{bmatrix} 14 \\ 12 \\ 25 \end{bmatrix}, \quad A^{-1} \begin{bmatrix} 104 \\ 69 \\ 55 \end{bmatrix} = \begin{bmatrix} 14 \\ 15 \\ 20 \end{bmatrix}.$$

Then the message is

$$\begin{array}{ccccccccccccc}
3 & 5 & 18 & 20 & 1 & 9 & 14 & 12 & 25 & & 14 & 15 & 20 \\
\updownarrow & \updownarrow & \updownarrow & \updownarrow & \updownarrow & \updownarrow & \updownarrow & \updownarrow & \updownarrow & & \updownarrow & \updownarrow & \updownarrow \\
C & E & R & T & A & I & N & L & Y & & N & O & T.
\end{array}$$

T.1. Using properties (a) and (b) in the definition of a linear transformation, we have

$$\begin{aligned}
L(c_1\mathbf{u}_1 + c_2\mathbf{u}_2 + \cdots + c_n\mathbf{u}_n) &= L((c_1\mathbf{u}_1 + c_2\mathbf{u}_2 + \cdots + c_{n-1}\mathbf{u}_{n-1}) + c_n\mathbf{u}_n) \\
&= L(c_1\mathbf{u}_1 + c_2\mathbf{u}_2 + \cdots + c_{n-1}\mathbf{u}_{n-1}) + L(c_n\mathbf{u}_n) \quad \text{[by property (a)]} \\
&= L(c_1\mathbf{u}_1 + c_2\mathbf{u}_2 + \cdots + c_{n-1}\mathbf{u}_{n-1}) + c_n L(\mathbf{u}_n) \quad \text{[by property (b)]}
\end{aligned}$$

Repeat with $L(c_1\mathbf{u}_1 + c_2\mathbf{u}_2 + \cdots + c_{n-1}\mathbf{u}_{n-1})$.

T.3. We have

$$L(\mathbf{u} + \mathbf{v}) = r(\mathbf{u} + \mathbf{v}) = r\mathbf{u} + r\mathbf{v} = L(\mathbf{u}) + L(\mathbf{v})$$

and

$$L(c\mathbf{u}) = r(c\mathbf{u}) = c(r\mathbf{u}) = cL(\mathbf{u}).$$

T.5. Given $L(\mathbf{u}) = a\mathbf{u} + b$, then

$$L(\mathbf{u} + \mathbf{v}) = a(\mathbf{u} + \mathbf{v}) + b = a\mathbf{u} + b + a\mathbf{v} + b - b = L(\mathbf{u}) + L(\mathbf{v}) - b.$$

So b must $= 0$. If $b = 0$, then $L(c\mathbf{u}) = a(c\mathbf{u}) = c(a\mathbf{u}) = cL(\mathbf{u})$. So a can be any real number. Check this also works for $L(\mathbf{u} + \mathbf{v})$.

T.7. We have $I(\mathbf{u} + \mathbf{v}) = \mathbf{u} + \mathbf{v} = I(\mathbf{u}) + I(\mathbf{v})$ and $I(c\mathbf{u}) = c\mathbf{u} = cI(\mathbf{u})$.

T.9. (a) $A^2 = \begin{bmatrix} \cos\phi & -\sin\phi \\ \sin\phi & \cos\phi \end{bmatrix} \begin{bmatrix} \cos\phi & -\sin\phi \\ \sin\phi & \cos\phi \end{bmatrix} = \begin{bmatrix} \cos^2\phi - \sin^2\phi & -2\sin\phi\cos\phi \\ 2\sin\phi\cos\phi & \cos^2\phi - \sin^2\phi \end{bmatrix}.$

$$= \begin{bmatrix} \cos 2\phi & -\sin 2\phi \\ \sin 2\phi & \cos 2\phi \end{bmatrix} = \begin{bmatrix} \cos 60° & -\sin 60° \\ \sin 60° & \cos 60° \end{bmatrix}$$

Thus, T_1 rotates $\mathbf{u}$ $60°$ counterclockwise.

(b) $A^{-1} = \begin{bmatrix} \cos\phi & \sin\phi \\ -\sin\phi & \cos\phi \end{bmatrix} = \begin{bmatrix} \cos(-\phi) & -\sin(-\phi) \\ \sin(-\phi) & \cos(-\phi) \end{bmatrix} = \begin{bmatrix} \cos(30°) & -\sin(-60°) \\ \sin(-30°) & \cos(-30°) \end{bmatrix}.$

Thus, T_2 rotates $\mathbf{u}$ $-30°$ counterclockwise or, equivalently, $30°$ clockwise.

(c) From the result in part (a), it is clear that $A^k\mathbf{u}$ rotates $\mathbf{u}$ $30k°$ counterclockwise. For this to equal $\mathbf{u}$, we must have $30k° = 360°$. Therefore, $k = 12$.

T.11. Let $\{\mathbf{e}_1, \ldots, \mathbf{e}_n\}$ be the natural basis for R^n. Then $I(\mathbf{e}_i) = \mathbf{e}_i$ for $i = 1, \ldots, n$. Hence the standard matrix representing I is the $n \times n$ identity matrix I_n.

ML.1. (a) $\mathbf{u} = [1 \quad 2]';\mathbf{v} = [0 \quad 3]';$
     $\mathbf{norm(u + v)}$
     ans $=$
        5.0990
     $\mathbf{norm(u) + norm(v)}$
     ans $=$
        5.2361

    (b) $\mathbf{u} = [1 \quad 2 \quad 3]';\mathbf{v} = [6 \quad 0 \quad 1]';$
     $\mathbf{norm(u + v)}$
     ans $=$
        8.3066
     $\mathbf{norm(u) + norm(v)}$
     ans $=$
        9.8244

Supplementary Exercises, p. 257

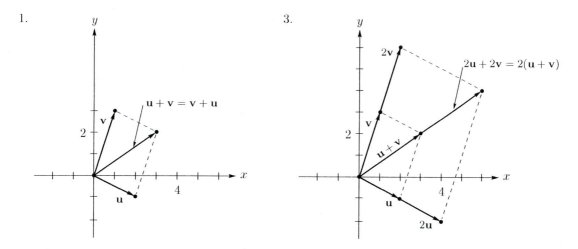

1.

3.

5. Let $\mathbf{x} = (a, b, c)$. Then

$$2(1, 2, -3) + 3(a, b, c) = (-2, 1, 1) - 5(a, b, c).$$

Equating corresponding components, we have

$$2 + 3a = -2 - 5a$$
$$4 + 3b = 1 - 5b$$
$$-6 + 3c = 1 - 5c.$$

Solving, we find that $a = -\frac{1}{2}$, $b = -\frac{3}{8}$, $c = \frac{7}{8}$. Hence $\mathbf{x} = \left(-\frac{1}{2}, -\frac{3}{8}, \frac{7}{8}\right)$.

7. Let $\mathbf{u} = (1, -1, 2, 3)$ and $\mathbf{v} = (2, 3, 1, -2)$.

   (a) $\|\mathbf{u}\| = \sqrt{(1)^2 + (-1)^2 + (2)^2 + (3)^2} = \sqrt{15}$

(b) $\|\mathbf{v}\| = \sqrt{(2)^2 + (3)^2 + (1)^2 + (-2)^2} = 3\sqrt{2}$.

(c) $\|\mathbf{u} - \mathbf{v}\| = \sqrt{(1-2)^2 + (-1-3)^2 + (2-1)^2 + (3-(-2))^2} = \sqrt{43}$.

(d) $\mathbf{u} \cdot \mathbf{v} = (1)(2) + (-1)(3) + (2)(1) + (3)(-2) = -5$.

(e) $\cos\theta = \dfrac{\mathbf{u} \cdot \mathbf{v}}{\|\mathbf{u}\|\,\|\mathbf{v}\|} = \dfrac{-5}{3\sqrt{30}} = \dfrac{-\sqrt{5}}{3\sqrt{6}}$.

9. $\|c(1,-2,2,0)\| = \sqrt{c^2[1^2 + (-2)^2 + 2^2 + 0^2]} = \sqrt{9c^2} = 3|c| = 9$. Thus $|c| = 3$, hence $c = 3$ or $c = -3$.

11. Let $L \colon R^2 \to R^2$ be defined by $L(x,y) = (x - 1, y - x)$. Also, let $\mathbf{u} = (x_1, y_1)$ and $\mathbf{v} = (x_2, y_2)$ be any vectors in R^2 and c any real scalar. Then

$$L(\mathbf{u} + \mathbf{v}) = L((x_1, y_1) + (x_2, y_2)) = L(x_1 + x_2, y_1 + y_2) = (x_1 + x_2 - 1, y_1 + y_2 - x_1 - x_2).$$

On the other hand,

$$L(\mathbf{u}) + L(\mathbf{v}) = (x_1 - 1, y_1 - x_1) + (x_2 - 1, y_2 - x_2) = (x_1 + x_2 - 2, y_1 + y_2 - x_1 - x_2).$$

Since $L(\mathbf{u} + \mathbf{v}) \neq L(\mathbf{u}) + L(\mathbf{v})$, L is not a linear transformation.

13. Let $\mathbf{v} = (-1, 2, 3)$. Then $\|\mathbf{v}\| = \sqrt{(-1)^2 + 2^2 + 3^2} = \sqrt{14}$. Hence the vector

$$\mathbf{u} = \frac{1}{\sqrt{14}}(-1, 2, 3) = \left(-\frac{1}{\sqrt{14}}, \frac{2}{\sqrt{14}}, \frac{3}{\sqrt{14}}\right)$$

is a unit vector parallel to $\mathbf{v}$.

15. Setting the dot products equal to zero we have

$$\mathbf{v} \cdot \mathbf{w} = 2a + b + 2 = 0$$
$$\mathbf{v} \cdot \mathbf{x} = a + 2 = 0.$$

Solving, we find that $a = -2$ and $b = 2$. Thus $\mathbf{v} = \begin{bmatrix} -2 \\ 2 \\ 2 \end{bmatrix}$.

17. We divide the region into two triangles, A_1 and A_2, and find the area of each.

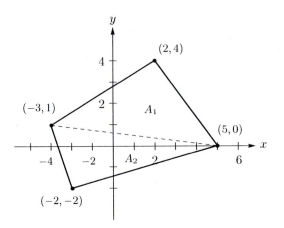

$$\text{area of } A_1 = \frac{1}{2}\left|\det\left(\begin{bmatrix} -3 & 1 & 1 \\ 2 & 4 & 1 \\ 5 & 0 & 1 \end{bmatrix}\right)\right| = \frac{1}{2}\left|-29\right| = \frac{29}{2}$$

$$\text{area of } A_2 = \frac{1}{2}\left|\det\left(\begin{bmatrix} -3 & 1 & 1 \\ -2 & -2 & 1 \\ 5 & 0 & 1 \end{bmatrix}\right)\right| = \frac{1}{2}\left|23\right| = \frac{23}{2}$$

Thus the total area is $A_1 + A_2 = \frac{29}{2} + \frac{23}{2} = 26$.

19. The standard matrix representing a clockwise rotation of R^2 through $\frac{\pi}{6}$ radians is

$$A = \begin{bmatrix} \cos\left(-\frac{\pi}{6}\right) & -\sin\left(-\frac{\pi}{6}\right) \\ \sin\left(-\frac{\pi}{6}\right) & \cos\left(\frac{-\pi}{6}\right) \end{bmatrix} = \begin{bmatrix} \frac{\sqrt{3}}{2} & \frac{1}{2} \\ -\frac{1}{2} & \frac{\sqrt{3}}{2} \end{bmatrix}.$$

21. (a) Find constants c_1, c_2, c_3 such that

$$(1, 3, -2) = c_1(1, 1, 0) + c_2(0, 1, 1) + c_3(0, 0, 1).$$

equating corresponding components, we have

$$\begin{aligned} c_1 &= 1 \\ c_1 + c_2 &= 3 \\ c_2 + c_3 &= -2. \end{aligned}$$

Solving this system we find $c_1 = 1$, $c_2 = 2$, $c_3 = -4$. thus

$$(1, 3, -2) = (1, 1, 0) + 2(0, 1, 1) - 4(0, 0, 1).$$

(b) $L(1, 3, -2) = L(1, 1, 0) + 2L(0, 1, 1) - 4L(0, 0, 1) = (2, -1) + 2(3, 2) - 4(1, -1) = (4, 7)$.

23. Let $L\colon R^n \to R^m$ be a linear transformation. Then there exists a unique $m \times n$ matrix A such that $L(\mathbf{x}) = A\mathbf{x}$. (See Section 4.3, Theorem 4.8). Thus, if A is 5×3, we have $n = 3$ and $m = 5$.

$$A\begin{bmatrix} a_1 \\ a_2 \\ a_3 \end{bmatrix} = \begin{bmatrix} c_1 \\ c_2 \\ c_3 \\ c_4 \\ c_5 \end{bmatrix}.$$

25. Let $\mathbf{u} = \begin{bmatrix} x \\ y \\ z \end{bmatrix}$. Since $L(\mathbf{u}) = A\mathbf{u}$, we have

$$\begin{bmatrix} 1 & 2 & 4 \\ 2 & 3 & 5 \\ -1 & -3 & -7 \end{bmatrix} \begin{bmatrix} x \\ y \\ z \end{bmatrix} = \begin{bmatrix} a \\ b \\ c \end{bmatrix}$$

with augmented matrix

$$\begin{bmatrix} 1 & 2 & 4 & \vdots & a \\ 2 & 3 & 5 & \vdots & b \\ -1 & -3 & -7 & \vdots & c \end{bmatrix}$$

which in reduced row echelon form is

$$\begin{bmatrix} 1 & 0 & -2 & \vdots & -3a + 2b \\ 0 & 1 & 3 & \vdots & 2a - b \\ 0 & 0 & 0 & \vdots & 3a - b + c \end{bmatrix}.$$

T.1. If $\mathbf{u} \cdot \mathbf{v} = 0$, then

$$\|\mathbf{u} + \mathbf{v}\| = \sqrt{(\mathbf{u}+\mathbf{v}) \cdot (\mathbf{u}+\mathbf{v})} = \sqrt{\mathbf{u}\cdot\mathbf{u} + 2(\mathbf{u}\cdot\mathbf{v}) + \mathbf{v}\cdot\mathbf{v}} = \sqrt{\mathbf{u}\cdot\mathbf{u} + \mathbf{v}\cdot\mathbf{v}}$$

and

$$\|\mathbf{u} - \mathbf{v}\| = \sqrt{(\mathbf{u}-\mathbf{v}) \cdot (\mathbf{u}-\mathbf{v})} = \sqrt{\mathbf{u}\cdot\mathbf{u} - 2(\mathbf{u}\cdot\mathbf{v}) + \mathbf{v}\cdot\mathbf{v}} = \sqrt{\mathbf{u}\cdot\mathbf{u} + \mathbf{v}\cdot\mathbf{v}}.$$

Hence $\|\mathbf{u} + \mathbf{v}\| = \|\mathbf{u} - \mathbf{v}\|$. On the other hand, if $\|\mathbf{u} + \mathbf{v}\| = \|\mathbf{u} - \mathbf{v}\|$, then

$$\|\mathbf{u} + \mathbf{v}\|^2 = \mathbf{u}\cdot\mathbf{u} + 2(\mathbf{u}\cdot\mathbf{v}) + \mathbf{v}\cdot\mathbf{v} = \mathbf{u}\cdot\mathbf{u} - 2(\mathbf{u}\cdot\mathbf{v}) + \mathbf{v}\cdot\mathbf{v} = \|\mathbf{u} - \mathbf{v}\|^2.$$

Simplifying, we have $2(\mathbf{u}\cdot\mathbf{v}) = -2(\mathbf{u}\cdot\mathbf{v})$, hence $\mathbf{u} \cdot \mathbf{v} = 0$.

T.3. Let $\mathbf{v} = (a, b, c)$ be a vector in R^3 that is orthogonal to every vector in R^3. Then $\mathbf{v} \cdot \mathbf{i} = 0$ so $(a, b, c) \cdot (1, 0, 0) = a = 0$. Similarly, $\mathbf{v} \cdot \mathbf{j} = 0$ and $\mathbf{v} \cdot \mathbf{k} = 0$ imply that $b = c = 0$. Therefore $\mathbf{v} = \mathbf{0}$.

T.5. If $\|\mathbf{u}\| = 0$, then $\|\mathbf{u}\| = \sqrt{\mathbf{u}\cdot\mathbf{u}} = 0$, so $\mathbf{u} \cdot \mathbf{u} = 0$. Part (a) of Theorem 4.3 implies that $\mathbf{u} = \mathbf{0}$.

Chapter 5

Applications of Vectors in R^2 and R^3 (Optional)

Section 5.1, p. 263

1. (a) $\mathbf{u} = 2\mathbf{i} + 3\mathbf{j} + 4\mathbf{k}$, $\mathbf{v} = -\mathbf{i} + 3\mathbf{j} - \mathbf{k}$.

$$\mathbf{u} \times \mathbf{v} = \begin{vmatrix} \mathbf{i} & \mathbf{j} & \mathbf{k} \\ 2 & 3 & 4 \\ -1 & 3 & -1 \end{vmatrix} = \begin{vmatrix} 3 & 4 \\ 3 & -1 \end{vmatrix} \mathbf{i} - \begin{vmatrix} 2 & 4 \\ -1 & -1 \end{vmatrix} \mathbf{j} + \begin{vmatrix} 2 & 3 \\ -1 & 3 \end{vmatrix} \mathbf{k} = -15\mathbf{i} - 2\mathbf{j} + 9\mathbf{k}.$$

(b) $\mathbf{u} = (1, 0, 1)$, $\mathbf{v} = (2, 3, -1)$.

$$\mathbf{u} \times \mathbf{v} = \begin{vmatrix} \mathbf{i} & \mathbf{j} & \mathbf{k} \\ 1 & 0 & 1 \\ 2 & 3 & -1 \end{vmatrix} = \begin{vmatrix} 0 & 1 \\ 3 & -1 \end{vmatrix} \mathbf{i} - \begin{vmatrix} 1 & 1 \\ 2 & -1 \end{vmatrix} \mathbf{j} + \begin{vmatrix} 1 & 0 \\ 2 & 3 \end{vmatrix} \mathbf{k} = -3\mathbf{i} + 3\mathbf{j} + 3\mathbf{k}.$$

(c) $\mathbf{u} = \mathbf{i} - \mathbf{j} + 2\mathbf{k}$, $\mathbf{v} = 3\mathbf{i} - 4\mathbf{j} + \mathbf{k}$.

$$\mathbf{u} \times \mathbf{v} = \begin{vmatrix} \mathbf{i} & \mathbf{j} & \mathbf{k} \\ 1 & -1 & 2 \\ 3 & -4 & 1 \end{vmatrix} = \begin{vmatrix} -1 & 2 \\ -4 & 1 \end{vmatrix} \mathbf{i} - \begin{vmatrix} 1 & 2 \\ 3 & 1 \end{vmatrix} \mathbf{j} + \begin{vmatrix} 1 & -1 \\ 3 & -4 \end{vmatrix} \mathbf{k} = 7\mathbf{i} + 5\mathbf{j} - \mathbf{k}.$$

(d) $\mathbf{u} = (2, -1, 1)$, $\mathbf{v} = -2\mathbf{u}$.

$$\begin{aligned} \mathbf{u} \times \mathbf{v} = \mathbf{u} \times (-2\mathbf{u}) &= -2(\mathbf{u} \times \mathbf{u}) \quad &\text{(by Theorem 5.1(d))} \\ &= -2(\mathbf{0}) = \mathbf{0} \quad &\text{(by Theorem 5.1(e))} \end{aligned}$$

3. Let $\mathbf{u} = \mathbf{i} + 2\mathbf{j} - 3\mathbf{k}$, $\mathbf{v} = 2\mathbf{i} + 3\mathbf{j} + \mathbf{k}$, $\mathbf{w} = 2\mathbf{i} - \mathbf{j} + 2\mathbf{k}$, $c = -3$.

(a) Verify $\mathbf{u} \times \mathbf{v} = -(\mathbf{v} \times \mathbf{u})$.

$$\mathbf{u} \times \mathbf{v} = \begin{vmatrix} \mathbf{i} & \mathbf{j} & \mathbf{k} \\ 1 & 2 & -3 \\ 2 & 3 & 1 \end{vmatrix} = \begin{vmatrix} 2 & -3 \\ 3 & 1 \end{vmatrix} \mathbf{i} - \begin{vmatrix} 1 & -3 \\ 2 & 1 \end{vmatrix} \mathbf{j} + \begin{vmatrix} 1 & 2 \\ 2 & 3 \end{vmatrix} \mathbf{k} = 11\mathbf{i} - 7\mathbf{j} - \mathbf{k}.$$

$$-(\mathbf{v} \times \mathbf{u}) = -\begin{vmatrix} \mathbf{i} & \mathbf{j} & \mathbf{k} \\ 2 & 3 & 1 \\ 1 & 2 & -3 \end{vmatrix} = -\begin{vmatrix} 3 & 1 \\ 2 & -3 \end{vmatrix} \mathbf{i} + \begin{vmatrix} 2 & 1 \\ 1 & -3 \end{vmatrix} \mathbf{j} - \begin{vmatrix} 2 & 3 \\ 1 & 2 \end{vmatrix} \mathbf{k} = 11\mathbf{i} - 7\mathbf{j} - \mathbf{k}.$$

(b) Verify $\mathbf{u} \times (\mathbf{v} + \mathbf{w}) = \mathbf{u} \times \mathbf{v} + \mathbf{u} \times \mathbf{w}$.

$$\mathbf{u} \times (\mathbf{v} + \mathbf{w}) = (\mathbf{i} + 2\mathbf{j} - 3\mathbf{k}) \times (4\mathbf{i} + 2\mathbf{j} + 3\mathbf{k})$$

$$= \begin{vmatrix} \mathbf{i} & \mathbf{j} & \mathbf{k} \\ 1 & 2 & -3 \\ 4 & 2 & 3 \end{vmatrix} = \begin{vmatrix} 2 & -3 \\ 2 & 3 \end{vmatrix} \mathbf{i} - \begin{vmatrix} 1 & -3 \\ 4 & 3 \end{vmatrix} \mathbf{j} + \begin{vmatrix} 1 & 2 \\ 4 & 2 \end{vmatrix} \mathbf{k} = 12\mathbf{i} - 15\mathbf{j} - 6\mathbf{k}$$

$$\mathbf{u} \times \mathbf{v} + \mathbf{u} \times \mathbf{w} = \begin{vmatrix} \mathbf{i} & \mathbf{j} & \mathbf{k} \\ 1 & 2 & -3 \\ 2 & 3 & 1 \end{vmatrix} + \begin{vmatrix} \mathbf{i} & \mathbf{j} & \mathbf{k} \\ 1 & 2 & -3 \\ 2 & -1 & 2 \end{vmatrix}$$

$$= (11\mathbf{i} - 7\mathbf{j} - \mathbf{k}) + \begin{vmatrix} 2 & -3 \\ -1 & 2 \end{vmatrix} \mathbf{i} - \begin{vmatrix} 1 & -3 \\ 2 & 2 \end{vmatrix} \mathbf{j} + \begin{vmatrix} 1 & 2 \\ 2 & -1 \end{vmatrix} \mathbf{k}$$

$$= (11\mathbf{i} - 7\mathbf{j} - \mathbf{k}) + (\mathbf{i} - 8\mathbf{j} - 5\mathbf{k}) = 12\mathbf{i} - 15\mathbf{j} - 6\mathbf{k}.$$

(c) Verify $(\mathbf{u} + \mathbf{v}) \times \mathbf{w} = \mathbf{u} \times \mathbf{w} + \mathbf{v} \times \mathbf{w}$.

$$(\mathbf{u} + \mathbf{v}) \times \mathbf{w} = (3\mathbf{i} + 5\mathbf{j} - 2\mathbf{k}) \times (2\mathbf{i} - \mathbf{j} + 2\mathbf{k})$$

$$= \begin{vmatrix} \mathbf{i} & \mathbf{j} & \mathbf{k} \\ 3 & 5 & -2 \\ 2 & -1 & 2 \end{vmatrix} = \begin{vmatrix} 5 & -2 \\ -1 & 2 \end{vmatrix} \mathbf{i} - \begin{vmatrix} 3 & -2 \\ 2 & 2 \end{vmatrix} \mathbf{j} + \begin{vmatrix} 3 & 5 \\ 2 & -1 \end{vmatrix} \mathbf{k}$$

$$= 8\mathbf{i} - 10\mathbf{j} - 13\mathbf{k}$$

$$\mathbf{u} \times \mathbf{w} + \mathbf{v} \times \mathbf{w} = \begin{vmatrix} \mathbf{i} & \mathbf{j} & \mathbf{k} \\ 1 & 2 & -3 \\ 2 & -1 & 2 \end{vmatrix} + \begin{vmatrix} \mathbf{i} & \mathbf{j} & \mathbf{k} \\ 2 & 3 & 1 \\ 2 & -1 & 2 \end{vmatrix}$$

$$= (\mathbf{i} - 8\mathbf{j} - 5\mathbf{k}) + \begin{vmatrix} 3 & 1 \\ -1 & 2 \end{vmatrix} \mathbf{i} - \begin{vmatrix} 2 & 1 \\ 2 & 2 \end{vmatrix} \mathbf{j} + \begin{vmatrix} 2 & 3 \\ 2 & -1 \end{vmatrix} \mathbf{k}$$

$$= (\mathbf{i} - 8\mathbf{j} - 5\mathbf{k}) + (7\mathbf{i} - 2\mathbf{j} - 8\mathbf{k}) = 8\mathbf{i} - 10\mathbf{j} - 13\mathbf{k}.$$

(d) Verify $c(\mathbf{u} \times \mathbf{v}) = (c\mathbf{u}) \times \mathbf{v} = \mathbf{u} \times (c\mathbf{v})$.

$$-3(\mathbf{u} \times \mathbf{v}) = -3(11\mathbf{i} - 7\mathbf{j} - \mathbf{k}) = -33\mathbf{i} + 21\mathbf{j} + 3\mathbf{k}$$

$$(-3\mathbf{u}) \times \mathbf{v} = (-3\mathbf{i} - 6\mathbf{j} + 9\mathbf{k}) \times (2\mathbf{i} + 3\mathbf{j} + \mathbf{k})$$

$$= \begin{vmatrix} \mathbf{i} & \mathbf{j} & \mathbf{k} \\ -3 & -6 & 9 \\ 2 & 3 & 1 \end{vmatrix} = \begin{vmatrix} -6 & 9 \\ 3 & 1 \end{vmatrix} \mathbf{i} - \begin{vmatrix} -3 & 9 \\ 2 & 1 \end{vmatrix} \mathbf{j} + \begin{vmatrix} -3 & -6 \\ 2 & 3 \end{vmatrix} \mathbf{k}$$

$$= -33\mathbf{i} + 21\mathbf{j} + 3\mathbf{k}$$

$$\mathbf{u} \times (c\mathbf{v}) = (\mathbf{i} + 2\mathbf{j} - 3\mathbf{k}) \times (-6\mathbf{i} - 9\mathbf{j} - 3\mathbf{k})$$

$$= \begin{vmatrix} \mathbf{i} & \mathbf{j} & \mathbf{k} \\ 1 & 2 & -3 \\ -6 & -9 & -3 \end{vmatrix} = \begin{vmatrix} 2 & -3 \\ -9 & -3 \end{vmatrix} \mathbf{i} - \begin{vmatrix} 1 & -3 \\ -6 & -3 \end{vmatrix} \mathbf{j} + \begin{vmatrix} 1 & 2 \\ -6 & -9 \end{vmatrix} \mathbf{k}$$

$$= -33\mathbf{i} + 21\mathbf{j} + 3\mathbf{k}.$$

5. $\mathbf{u} = \mathbf{i} - \mathbf{j} + 2\mathbf{k}$, $\mathbf{v} = 2\mathbf{i} + 2\mathbf{j} - \mathbf{k}$, $\mathbf{w} = \mathbf{i} + \mathbf{j} - \mathbf{k}$.

(a) Verify $(\mathbf{u} \times \mathbf{v}) \cdot \mathbf{w} = \mathbf{u} \cdot (\mathbf{v} \times \mathbf{w})$.

$$(\mathbf{u} \times \mathbf{v}) \cdot \mathbf{w} = \begin{vmatrix} \mathbf{i} & \mathbf{j} & \mathbf{k} \\ 1 & -1 & 2 \\ 2 & 2 & -1 \end{vmatrix} \cdot (\mathbf{i} + \mathbf{j} - \mathbf{k})$$

$$= \left(\begin{vmatrix} -1 & 2 \\ 2 & -1 \end{vmatrix} \mathbf{i} - \begin{vmatrix} 1 & 2 \\ 2 & -1 \end{vmatrix} \mathbf{j} + \begin{vmatrix} 1 & -1 \\ 2 & 2 \end{vmatrix} \mathbf{k} \right) \cdot (\mathbf{i} + \mathbf{j} - \mathbf{k})$$

$$= (-3\mathbf{i} + 5\mathbf{j} + 4\mathbf{k}) \cdot (\mathbf{i} + \mathbf{j} - \mathbf{k}) = -3 + 5 - 4 = -2$$

$$\mathbf{u} \cdot (\mathbf{v} \times \mathbf{w}) = (\mathbf{i} - \mathbf{j} + 2\mathbf{k}) \cdot \begin{vmatrix} \mathbf{i} & \mathbf{j} & \mathbf{k} \\ 2 & 2 & -1 \\ 1 & 1 & -1 \end{vmatrix}$$

$$= (\mathbf{i} - \mathbf{j} + 2\mathbf{k}) \cdot \left(\begin{vmatrix} 2 & -1 \\ 1 & -1 \end{vmatrix} \mathbf{i} - \begin{vmatrix} 2 & -1 \\ 1 & -1 \end{vmatrix} \mathbf{j} + \begin{vmatrix} 2 & 2 \\ 1 & 1 \end{vmatrix} \mathbf{k} \right)$$

$$= (\mathbf{i} - \mathbf{j} + 2\mathbf{k}) \cdot (-\mathbf{i} + \mathbf{j}) = -1 - 1 = -2.$$

(b) Verify $\mathbf{u} \times (\mathbf{v} \times \mathbf{w}) = (\mathbf{u} \cdot \mathbf{w})\mathbf{v} - (\mathbf{u} \cdot \mathbf{v})\mathbf{w}$.

$$\mathbf{u} \times (\mathbf{v} \times \mathbf{w}) = (\mathbf{i} - \mathbf{j} + 2\mathbf{k}) \times (-\mathbf{i} + \mathbf{j}) = \begin{vmatrix} \mathbf{i} & \mathbf{j} & \mathbf{k} \\ 1 & -1 & 2 \\ -1 & 1 & 0 \end{vmatrix}$$

$$= \begin{vmatrix} -1 & 2 \\ 1 & 0 \end{vmatrix} \mathbf{i} - \begin{vmatrix} 1 & 2 \\ -1 & 0 \end{vmatrix} \mathbf{j} + \begin{vmatrix} 1 & -1 \\ -1 & 1 \end{vmatrix} \mathbf{k} = -2\mathbf{i} - 2\mathbf{j}$$

$$(\mathbf{u} \cdot \mathbf{w})\mathbf{v} - (\mathbf{u} \cdot \mathbf{v})\mathbf{w} = -2\mathbf{v} + 2\mathbf{w} = -2\mathbf{i} - 2\mathbf{j}.$$

7. Use the results from Exercise 2 for $\mathbf{u} \times \mathbf{v}$.

(a) $\mathbf{u} \cdot (\mathbf{u} \times \mathbf{v}) = (\mathbf{i} - \mathbf{j} + 2\mathbf{k}) \cdot (-4\mathbf{i} + 4\mathbf{j} + 4\mathbf{k}) = -4 - 4 + 8 = 0.$
$\mathbf{v} \cdot (\mathbf{u} \times \mathbf{v}) = (3\mathbf{i} + \mathbf{j} + 2\mathbf{k}) \cdot (-4\mathbf{i} + 4\mathbf{j} + 4\mathbf{k}) = -12 + 4 + 8 = 0.$

(b) $\mathbf{u} \cdot (\mathbf{u} \times \mathbf{v}) = (2\mathbf{i} + \mathbf{j} - 2\mathbf{k}) \cdot (3\mathbf{i} - 8\mathbf{j} - \mathbf{k}) = 6 - 8 + 2 = 0.$
$\mathbf{v} \cdot (\mathbf{u} \times \mathbf{v}) = (\mathbf{i} + 3\mathbf{k}) \cdot (3\mathbf{i} - 8\mathbf{j} - \mathbf{k}) = 3 - 0 - 3 = 0.$

(c) $\mathbf{u} \cdot (\mathbf{u} \times \mathbf{v}) = (2\mathbf{j} + \mathbf{k}) \cdot (0\mathbf{i} + 0\mathbf{j} + 0\mathbf{k}) = 0.$
$\mathbf{v} \cdot (\mathbf{u} \times \mathbf{v}) = (6\mathbf{j} + 3\mathbf{k}) \cdot (0\mathbf{i} + 0\mathbf{j} + 0\mathbf{k}) = 0.$

(d) $\mathbf{u} \cdot (\mathbf{u} \times \mathbf{v}) = (4\mathbf{i} - 2\mathbf{k}) \cdot (4\mathbf{i} + 4\mathbf{j} + 8\mathbf{k}) = 16 + 0 - 16 = 0.$
$\mathbf{v} \cdot (\mathbf{u} \times \mathbf{v}) = (2\mathbf{j} - \mathbf{k}) \cdot (4\mathbf{i} + 4\mathbf{j} + 8\mathbf{k}) = 0 + 8 - 8 = 0.$

9. Let a triangle have vertices $\mathbf{v}_1 = (1, -2, 3)$, $\mathbf{v}_2 = (-3, 1, 4)$, $\mathbf{v}_3 = (0, 4, 3)$. Its area is given by

$$A_T = \tfrac{1}{2} \|(\mathbf{v_2} - \mathbf{v_1}) \times (\mathbf{v_3} - \mathbf{v_1})\|$$

$$= \tfrac{1}{2} \|(-4, 3, 1) \times (-1, 6, 0)\| = \tfrac{1}{2} \left\| \begin{vmatrix} \mathbf{i} & \mathbf{j} & \mathbf{k} \\ -4 & 3 & 1 \\ -1 & 6 & 0 \end{vmatrix} \right\|$$

$$= \tfrac{1}{2} \|-6\mathbf{i} + \mathbf{j} - 21\mathbf{k}\| = \tfrac{1}{2} \sqrt{(-6)^2 + 1^2 + (-21)^2} = \tfrac{1}{2} \sqrt{478}.$$

11. Area $= \|(\mathbf{i} + 3\mathbf{j} - 2\mathbf{k}) \times (3\mathbf{i} - \mathbf{j} - \mathbf{k})\| = \|-5\mathbf{i} - 5\mathbf{j} - 10\mathbf{k}\| = \sqrt{(-5)^2 + (-5)^2 + (-10)^2} = \sqrt{150} = 5\sqrt{6}.$

13. Volume $= |(\mathbf{i} - 2\mathbf{j} + 4\mathbf{k}) \cdot (3\mathbf{i} - 4\mathbf{j} + 7\mathbf{k})| = |3 + 8 + 28| = 39.$

T.1. (a) Interchange of the second and third rows of the determinant in (2) changes the sign of the determinant.

(b) $\begin{vmatrix} \mathbf{i} & \mathbf{j} & \mathbf{k} \\ u_1 & u_2 & u_3 \\ v_1 + w_1 & v_2 + w_2 & v_3 + w_3 \end{vmatrix} = \mathbf{i} \begin{vmatrix} u_2 & u_3 \\ v_2 + w_2 & v_3 + w_3 \end{vmatrix} - \mathbf{j} \begin{vmatrix} u_1 & u_3 \\ v_1 + w_1 & v_3 + w_3 \end{vmatrix}$

$$+ \mathbf{k} \begin{vmatrix} u_1 & u_2 \\ v_1 + w_1 & v_2 + w_2 \end{vmatrix}$$

$$= \mathbf{i}[u_2(v_3 + w_3) - u_3(v_2 + w_2)] - \mathbf{j}[u_1(v_3 + w_3) - u_3(v_1 + w_1)]$$
$$+ \mathbf{k}[u_1(v_2 + w_2) - u_2(v_1 + w_1)]$$

Within each bracket expand, separate, and collect those terms involving the v_i's and the w_i's, then rewrite as the sum of two determinants.

(c) Similar to the proof for (b).

(d) Follows from the homogeneity property for determinants: Theorem 3.5.

(e) Follows from Theorem 3.3.

(f) Follows from Theorem 3.4.

(g) $\mathbf{u} \times (\mathbf{v} \times \mathbf{w}) = \begin{vmatrix} \mathbf{i} & \mathbf{j} & \mathbf{k} \\ u_1 & u_2 & u_3 \\ v_2 w_3 - w_2 v_3 & -(v_1 w_3 - w_1 v_3) & v_1 w_2 - w_1 v_2 \end{vmatrix}$

$$= \mathbf{i}[u_2(v_1 w_2 - w_1 v_2) + u_3(v_1 w_3 - w_1 v_3)] - \mathbf{j}[u_1(v_1 w_2 - w_1 v_2) - u_3(v_2 w_3 - w_2 v_3)]$$
$$+ \mathbf{k}[-u_1(v_1 w_3 - w_1 v_3) - u_2(v_2 w_3 - w_2 v_3)]$$

$$= (u_1 w_1 + u_2 w_2 + u_3 w_3) \begin{bmatrix} v_1 \\ v_2 \\ v_3 \end{bmatrix} - (u_1 v_1 + u_2 v_2 + u_3 v_3) \begin{bmatrix} w_1 \\ w_2 \\ w_3 \end{bmatrix}$$

Hint: Verify this last step component by component, i.e., the first component of the last line is equal to the entries in the bracket indexed by $\mathbf{i}$ above it, etc.

(h) Similar to (g)

T.3. We have

$$\mathbf{j} \times \mathbf{i} = \begin{vmatrix} \mathbf{i} & \mathbf{j} & \mathbf{k} \\ 0 & 1 & 0 \\ 1 & 0 & 0 \end{vmatrix} = \mathbf{k} \begin{vmatrix} 0 & 1 \\ 1 & 0 \end{vmatrix} = -\mathbf{k}$$

$$\mathbf{k} \times \mathbf{j} = \begin{vmatrix} \mathbf{i} & \mathbf{j} & \mathbf{k} \\ 0 & 0 & 1 \\ 0 & 1 & 0 \end{vmatrix} = \mathbf{i} \begin{vmatrix} 0 & 1 \\ 1 & 0 \end{vmatrix} = -\mathbf{i}$$

$$\mathbf{i} \times \mathbf{k} = \begin{vmatrix} \mathbf{i} & \mathbf{j} & \mathbf{k} \\ 1 & 0 & 0 \\ 0 & 0 & 1 \end{vmatrix} = -\mathbf{j} \begin{vmatrix} 1 & 0 \\ 0 & 1 \end{vmatrix} = -\mathbf{j}.$$

T.5. If $\mathbf{v} = c\mathbf{u}$ for some c, then $\mathbf{u} \times \mathbf{v} = c(\mathbf{u} \times \mathbf{u}) = \mathbf{0}$. Conversely, if $\mathbf{u} \times \mathbf{v} = \mathbf{0}$, the area of the parallelogram with adjacent sides $\mathbf{u}$ and $\mathbf{v}$ is 0, and hence that parallelogram is degenerate: $\mathbf{u}$ and $\mathbf{v}$ are parallel.

T.7. Using Theorem 5.1(h),

$$(\mathbf{u} \times \mathbf{v}) \times \mathbf{w} + (\mathbf{v} \times \mathbf{w}) \times \mathbf{u} + (\mathbf{w} \times \mathbf{u}) \times \mathbf{v} = [(\mathbf{w} \cdot \mathbf{u})\mathbf{v} - (\mathbf{w} \cdot \mathbf{v})\mathbf{u}] + [(\mathbf{u} \cdot \mathbf{v})\mathbf{w} - (\mathbf{u} \cdot \mathbf{w})\mathbf{v}]$$
$$+ [(\mathbf{v} \cdot \mathbf{w})\mathbf{u} - (\mathbf{v} \cdot \mathbf{u})\mathbf{w}] = \mathbf{0}.$$

ML.1. (a) $\mathbf{u} = [1 \quad -2 \quad 3]; \mathbf{v} = [1 \quad 3 \quad 1]; \text{cross(u,v)}$
ans =
$\qquad -11 \quad 2 \quad 5$

(b) $\mathbf{u} = \begin{bmatrix} 1 & 0 & 3 \end{bmatrix}; \mathbf{v} = \begin{bmatrix} 1 & -1 & 2 \end{bmatrix}; \mathbf{cross(u,v)}$
   ans =
       3   1   -1

(c) $\mathbf{u} = \begin{bmatrix} 1 & 2 & -3 \end{bmatrix}; \mathbf{v} = \begin{bmatrix} 2 & -1 & 2 \end{bmatrix}; \mathbf{cross(u,v)}$
   ans =
       1   -8   -5

ML.5. Following Example 6 we proceed as follows in MATLAB.

   $\mathbf{u} = \begin{bmatrix} 3 & -2 & 1 \end{bmatrix}; \mathbf{v} = \begin{bmatrix} 1 & 2 & 3 \end{bmatrix}; \mathbf{w} = \begin{bmatrix} 2 & -1 & 2 \end{bmatrix};$
   $\mathbf{vol} = \mathbf{abs(dot(u,cross(v,w)))}$
   vol =
       8

Section 5.2, p. 269

1. In each of the following we substitute the coordinates of points P_1 and P_2 into Equation (5) and expand the determinant to find the equation of the line.

   (a) $\begin{vmatrix} x & y & 1 \\ -2 & -3 & 1 \\ 3 & 4 & 1 \end{vmatrix} = -7x + 5y + 1 = 0.$

   (b) $\begin{vmatrix} x & y & 1 \\ 2 & -5 & 1 \\ -3 & 4 & 1 \end{vmatrix} = -9x - 5y - 7 = 0.$

   (c) $\begin{vmatrix} x & y & 1 \\ 0 & 0 & 1 \\ -3 & 5 & 1 \end{vmatrix} = -5x - 3y = 0.$

   (d) $\begin{vmatrix} x & y & 1 \\ -3 & -5 & 1 \\ 0 & 2 & 1 \end{vmatrix} = -7x + 3y - 6 = 0.$

3. Convert the parametric form to the symmetric form by solving each equation for t and setting the results equal to one another. We obtain

$$\frac{x-3}{2} = \frac{y+2}{3} = \frac{z-4}{-3}.$$

A point is on the line provided the coordinates of the point satisfy the preceding string of equalities.

   (a) For $(1, 1, 1)$, we have

$$\frac{1-3}{2} = -1, \qquad \frac{1+2}{3} = 1, \qquad \frac{1-4}{-3} = 1.$$

   Thus $(1, 1, 1)$ is not on the line.

   (b) For $(1, -1, 0)$, we have

$$\frac{1-3}{2} = -1 \qquad \frac{-1+2}{3} = \frac{1}{3}, \qquad \frac{0-4}{-3} = \frac{4}{3}.$$

   Thus $(1, -1, 0)$ is not on the line.

   (c) For $(1, 0, -2)$, we have

$$\frac{1-3}{2} = -1, \qquad \frac{0+2}{3} = \frac{2}{3}, \qquad \frac{-2-4}{-3} = 2.$$

   Thus $(1, 0, -2)$ is not on the line.

(d) For $\left(4, -\frac{1}{2}, \frac{5}{2}\right)$, we have

$$\frac{4-3}{2} = \frac{1}{2}, \qquad \frac{-\frac{1}{2}+2}{3} = \frac{1}{2}, \qquad \frac{\frac{5}{2}-4}{-3} = \frac{1}{2}.$$

Thus $\left(4, -\frac{1}{2}, \frac{5}{2}\right)$ is on the line.

5. From Example 2 we have that the parametric equation of a line through $P_0(x_0, y_0, z_0)$ which is parallel to vector $\mathbf{u} = (a, b, c)$ is given by

$$\begin{aligned} x &= x_0 + ta \\ y &= y_0 + tb \qquad (-\infty < t < \infty) \\ z &= z_0 + tc. \end{aligned}$$

(a) Let $P_0 = (3, 4, -2)$ and let $\mathbf{u} = (4, -5, 2)$. Then the parametric equation is

$$\begin{aligned} x &= 3 + 4t \\ y &= 4 - 5t \qquad (-\infty < t < \infty) \\ z &= -2 + 2t. \end{aligned}$$

(b) Let $P_0 = (3, 2, 4)$ and let $\mathbf{u} = (-2, 5, 1)$. Then the parametric equation is

$$\begin{aligned} x &= 3 - 2t \\ y &= 2 + 5t \qquad (-\infty < t < \infty) \\ z &= 4 + t. \end{aligned}$$

(c) Let $P_0 = (0, 0, 0)$ and let $\mathbf{u} = (2, 2, 2)$. Then the parametric equation is

$$\begin{aligned} x &= 2t \\ y &= 2t \qquad (-\infty < t < \infty) \\ z &= 2t. \end{aligned}$$

Equivalently, $x = t$, $y = t$, $z = t$.

(d) Let $P_0 = (-2, -3, 1)$ and let $\mathbf{u} = (2, 3, 4)$. Then the parametric equation is

$$\begin{aligned} x &= -2 + 2t \\ y &= -3 + 3t \qquad (-\infty < t < \infty) \\ z &= 1 + 4t. \end{aligned}$$

7. Follow the method of Example 4.

(a) $\dfrac{x-2}{2} = \dfrac{y+3}{5} = \dfrac{z-1}{4}.$ (b) $\dfrac{x+3}{8} = \dfrac{y+2}{7} = \dfrac{z+2}{6}.$

(c) $\dfrac{x+2}{4} = \dfrac{y-3}{-6} = \dfrac{z-4}{1}.$ (d) $\dfrac{x}{4} = \dfrac{y}{5} = \dfrac{z}{2}.$

9. From Example 5 we have that the equation of a plane passing through point $P_0(x_0, y_0, z_0)$ and perpendicular to vector $\mathbf{n} = (a, b, c)$ is

$$a(x - x_0) + b(y - y_0) + c(z - z_0) = 0.$$

(a) Let $P_0 = (0, 2, -3)$ and vector $\mathbf{n} = (3, -2, 4)$. The equation of the plane is

$$3(x - 0) - 2(y - 2) + 4(z - (-3)) = 3x - 2y + 4z + 16 = 0.$$

(b) Let $P_0 = (-1, 3, 2)$ and vector $\mathbf{n} = (0, 1, -3)$. The equation of the plane is

$$0(x - (-1)) + 1(y - 3) - 3(z - 2) = y - 3z + 3 = 0.$$

(c) Let $P_0 = (-2, 3, 4)$ and vector $\mathbf{n} = (0, 0, -4)$. The equation of the plane is

$$0(x - (-2)) + 0(y - 3) - 4(z - 4) = -4z + 16 = 0 \quad \text{or equivalently} \quad -z + 4 = 0.$$

(d) Let $P_0 = (5, 2, 3)$ and vector $\mathbf{n} = (-1, -2, 4)$. The equation of the plane is

$$-1(x - 5) - 2(y - 2) + 4(z - 3) = -x - 2y + 4z - 3 = 0.$$

11. Following Example 10, we solve the pair of equations simultaneously.

(a) Form the augmented matrix of the linear system and row reduce it.

$$\begin{bmatrix} 2 & 3 & -4 & \vdots & -5 \\ -3 & 2 & 5 & \vdots & -6 \end{bmatrix} \begin{smallmatrix} \frac{1}{2}\mathbf{r}_1 \\ 3\mathbf{r}_1 + \mathbf{r}_2 \end{smallmatrix} \longrightarrow \begin{bmatrix} 1 & \frac{3}{2} & -2 & \vdots & -\frac{5}{2} \\ 0 & \frac{13}{2} & -1 & \vdots & -\frac{27}{2} \end{bmatrix} \begin{smallmatrix} \frac{2}{13}\mathbf{r}_2 \\ -\frac{3}{2}\mathbf{r}_2 + \mathbf{r}_1 \end{smallmatrix} \longrightarrow \begin{bmatrix} 1 & 0 & -\frac{23}{13} & \vdots & \frac{8}{13} \\ 0 & 1 & -\frac{2}{13} & \vdots & -\frac{27}{13} \end{bmatrix}$$

The solution of the linear system is

$$x = \tfrac{8}{13} + \tfrac{23}{13}r, \qquad y = -\tfrac{27}{13} + \tfrac{2}{13}r, \qquad z = r,$$

where r is any real number. Let $r = 13t$, where t is any real number. Then one way to write the parametric equation of the line of intersection is

$$\begin{aligned} x &= \tfrac{8}{13} + 23t \\ y &= -\tfrac{27}{13} + 2t \qquad (-\infty < t < \infty) \\ z &= 13t. \end{aligned}$$

(b) Form the augmented matrix of the linear system and row reduce it:

$$\begin{bmatrix} 3 & -2 & -5 & \vdots & -4 \\ 2 & 3 & 4 & \vdots & -8 \end{bmatrix} \begin{smallmatrix} \frac{1}{3}\mathbf{r}_1 \\ -2\mathbf{r}_1 + \mathbf{r}_2 \end{smallmatrix} \longrightarrow \begin{bmatrix} 1 & -\frac{2}{3} & -\frac{5}{3} & \vdots & -\frac{4}{3} \\ 0 & \frac{13}{3} & \frac{22}{3} & \vdots & -\frac{16}{3} \end{bmatrix} \begin{smallmatrix} \frac{3}{13}\mathbf{r}_2 \\ \frac{2}{3}\mathbf{r}_2 + \mathbf{r}_1 \end{smallmatrix} \longrightarrow \begin{bmatrix} 1 & 0 & -\frac{7}{13} & \vdots & -\frac{28}{13} \\ 0 & 1 & \frac{22}{13} & \vdots & -\frac{16}{13} \end{bmatrix}$$

The solution of the linear system is

$$x = -\tfrac{28}{13} + \tfrac{7}{13}r, \qquad y = -\tfrac{16}{13} - \tfrac{22}{13}r, \qquad z = r,$$

where r is any real number. Let $r = 13t$, where t is any real number. Then one way to write the parametric equation of the line of intersection is

$$\begin{aligned} x &= -\tfrac{28}{13} + 7t \\ y &= -\tfrac{16}{13} - 22t \qquad (-\infty < t < \infty) \\ z &= 13t \end{aligned}$$

(c) Form the augmented matrix of the linear system and row reduce it:

$$\begin{bmatrix} -1 & 2 & 1 & \vdots & 0 \\ 2 & -1 & 2 & \vdots & -8 \end{bmatrix} \begin{smallmatrix} -1\mathbf{r}_1 \\ -2\mathbf{r}_1 + \mathbf{r}_2 \end{smallmatrix} \longrightarrow \begin{bmatrix} 1 & -2 & -1 & \vdots & 0 \\ 0 & 3 & 4 & \vdots & -8 \end{bmatrix} \begin{smallmatrix} \frac{1}{3}\mathbf{r}_2 \\ 2\mathbf{r}_2 + \mathbf{r}_1 \end{smallmatrix} \longrightarrow \begin{bmatrix} 1 & 0 & \frac{5}{3} & \vdots & -\frac{16}{3} \\ 0 & 1 & \frac{4}{3} & \vdots & -\frac{8}{3} \end{bmatrix}$$

The solution of the linear system is

$$x = -\tfrac{16}{3} - \tfrac{5}{3}r, \qquad y = -\tfrac{8}{3} - \tfrac{4}{3}r, \qquad z = r,$$

where r is any real number. Let $r = -3t$, where t is any real number. Then one way to write the parametric equation of the line of intersection is

$$
\begin{aligned}
x &= -\tfrac{16}{3} + 5t \\
y &= -\tfrac{8}{3} + 4t \qquad (-\infty < t < \infty) \\
z &= -3t.
\end{aligned}
$$

13. Determine the equation of the line through $P_1 = (2, 3, -2)$ and $P_2 = (4, -2, -3)$. Then $P_3 = (0, 8, -1)$ is on the line through P_1 and P_2 if it satisfies an equation of the line. A vector parallel to the line through P_1 and P_2 is

$$\overrightarrow{P_1P_2} = (2, -5, -1).$$

Then a line through P_1 parallel to vector $(2, -5, -1)$ is

$$
\begin{aligned}
x &= 2 + 2t \\
y &= 3 - 5t \qquad (-\infty < t < \infty) \\
z &= -2 - t.
\end{aligned}
$$

The symmetric form for the equation of this line is

$$\frac{x-2}{2} = \frac{y-3}{-5} = \frac{z+2}{-1}.$$

Substituting the coordinates of P_3 into the equation, we have

$$\frac{0-2}{2} = -1, \qquad \frac{8-3}{-5} = -1, \qquad \frac{-1+2}{-1} = -1.$$

Thus, P_3 is on the line through P_1 and P_2. All three points are on the same line.

15. Equate the expressions for x, y, and z, respectively, to form a system of equations in s and t. We obtain

$$
\begin{array}{ll}
2 - 3s = 5 + 2t & \qquad -3s - 2t = 3 \\
3 + 2s = 1 - 3t \quad \text{which is equivalent to} & \qquad 2s + 3t = -2 \\
4 + 2s = 2 + t & \qquad 2s - t = -2.
\end{array}
$$

Form the associated augmented matrix and row reduce it:

$$
\left[\begin{array}{rr:r}
-3 & -2 & 3 \\
2 & 3 & -2 \\
2 & -1 & -2
\end{array}\right]
\longrightarrow
\left[\begin{array}{rr:r}
1 & 0 & -1 \\
0 & 1 & 0 \\
0 & 0 & 0
\end{array}\right].
$$

The solution is $s = -1$ and $t = 0$. Substituting $s = -1$ into the equation of the first line, we have that the point of intersection has coordinates

$$x = 5, \qquad y = 1, \qquad z = 2.$$

17. We show that the lines intersect in more than one point, hence they must be the same line. Rewrite the equation of the first line using the parameter s in place of t:

$$
\begin{array}{ll}
x = 2 + 3s & \qquad x = -1 - 9t \\
y = 3 - 2s \quad \text{and} & \qquad y = 5 + 6t\,. \\
z = -1 + 4s & \qquad z = -5 - 12t
\end{array}
$$

Equating the expressions for x, y, and z, respectively, we have a system of equations which has augmented matrix

$$\begin{bmatrix} 3 & 9 & \vdots & -3 \\ -2 & -6 & \vdots & 2 \\ 4 & 12 & \vdots & -4 \end{bmatrix} \longrightarrow \begin{bmatrix} 1 & 3 & \vdots & -1 \\ 0 & 0 & \vdots & 0 \\ 0 & 0 & \vdots & 0 \end{bmatrix}.$$

Thus there are infinitely many solutions, hence infinitely many intersections of the lines. It follows that the two lines are identical.

19. Let $P_1 = (-2, 3, 4)$, $P_2 = (4, -2, 5)$, $P_3 = (0, 2, 4)$. Also, let L_2 denote the line through P_2 and P_3. Let a plane through P_1 perpendicular to L_2 be denoted by π. Then a vector parallel to L_2 is $\mathbf{n} = (-4, 4, -1)$. It follows that $\mathbf{n}$ is a normal to π. Thus for any point $P = (x, y, z)$ in π, the equation of the plane is given by

$$\mathbf{n} \cdot \overrightarrow{P_1 P} = (-4, 4, -1) \cdot (x + 2, y - 3, z - 4) = -4x + 4y - z - 16 = 0,$$

or equivalently, $4x - 4y + z + 16 = 0$.

21. Let the lines be represented as follows:

$$L_1: \begin{array}{l} x = 3 + 2s \\ y = 4 - 3s \\ z = 5 + 4s \end{array} \quad \text{and} \quad L_2: \begin{array}{l} x = 1 - 2t \\ y = 7 + 4t \\ z = 1 - 3t. \end{array}$$

The vector $\mathbf{u} = (2, -3, 4)$ is parallel to L_1 and vector $\mathbf{v} = (-2, 4, -3)$ is parallel to L_2. The normal direction to a plane containing L_1 and L_2 is perpendicular to both $\mathbf{u}$ and $\mathbf{v}$. Thus $\mathbf{n} = \mathbf{u} \times \mathbf{v}$ is a normal to the plane. We have

$$\mathbf{n} = \mathbf{u} \times \mathbf{v} = \begin{vmatrix} \mathbf{i} & \mathbf{j} & \mathbf{k} \\ 2 & -3 & 4 \\ -2 & 4 & -3 \end{vmatrix} = -7\mathbf{i} - 2\mathbf{j} + 2\mathbf{k} = (-7, -2, 2).$$

We need only determine a point P_0 on one of the lines and then use Equation (8) to find the equation of the plane containing L_1 and L_2. If we set $s = 0$, then $P_0 = (3, 4, 5)$ and hence the equation of the plane is

$$\mathbf{n} \cdot \overrightarrow{P_0 P} = (-7, -2, 2) \cdot (x - 3, y - 4, z - 5) = -7x - 2y + 2z + 19 = 0,$$

or equivalently, $7x + 2y - 2z - 19 = 0$.

23. Let $P_0 = (-2, 5, -3)$ and let π denote the plane

$$2x - 3y + 4z + 7 = 0.$$

The vector $\mathbf{n} = (2, -3, 4)$ is normal to π and hence any line perpendicular to π must be parallel to $\mathbf{n}$. Thus the line through P_0 perpendicular to π in parametric form is

$$\begin{array}{ll} x = -2 + 2t & \\ y = 5 - 3t & (-\infty < t < \infty) \\ z = -3 + 4t. & \end{array}$$

T.1. Since by hypothesis a, b, and c are not all zero, take $a \neq 0$. Let $P_0 = \left(-\frac{d}{a}, 0, 0\right)$. Then from equations (8) and (9), the equation of the plane through P_0 with normal vector $\mathbf{n} = (a, b, c)$ is

$$a \left(x + \frac{d}{a} \right) + b(y - 0) + c(z - 0) = 0 \quad \text{or} \quad ax + by + cz + d = 0.$$

T.3. Possible solutions:

$$L_1: \quad x = s, y = z = 0 \quad \text{(the } x\text{-axis)}$$
$$L_2: \quad x = 0, y = 1, z = t.$$

T.5. Expand the determinant along the first row:

$$0 = \begin{vmatrix} x & y & z & 1 \\ a_1 & b_1 & c_1 & 1 \\ a_2 & b_2 & c_2 & 1 \\ a_3 & b_3 & c_3 & 1 \end{vmatrix} = xA_{11} + yA_{12} + zA_{13} + 1 \cdot A_{14} \tag{5.1}$$

where A_{1j} is the cofactor of the $1, j$th element, and (since it depends upon the second, third and fourth rows of the determinant) is a constant. Thus (5.1) is an equation of the form

$$ax + by + cz + d = 0$$

and so is the equation of some plane. The noncolinearity of the three points insures that the three cofactors A_{11}, A_{12}, A_{13} are not all zero. Next let $(x, y, z) = (a_i, b_i, c_i)$. The determinant has two equal rows, and so has the value zero. Thus the point P_i lies on the plane whose equation is (5.1). Thus (5.1) is an equation for the plane through P_1, P_2, P_3.

Supplementary Exercises, p. 271

1. $(x, y, z) \times (1, 2, 3) = (3y - 4, -3x + 2, 2x - y)$. Setting this cross product equal to $(0, 0, 0)$, we obtain the equations

$$3y - 4 = 0 \quad \Longrightarrow \quad y = \frac{4}{3}$$
$$-3x + 2 = 0 \quad \Longrightarrow \quad x = \frac{2}{3}$$
$$2x - y = 0 \quad \Longrightarrow \quad y = 2x$$

Therefore the solution is $x = \frac{2}{3}$ and $y = \frac{4}{3}$.

3. A point is on the line provided the coordinates of the point satisfy the given equations of the line in symmetric form
$$\frac{x - 3}{2} = \frac{y + 3}{4} = \frac{z + 5}{-4}.$$

(a) For $(1, 2, 3)$ we have
$$\frac{1 - 3}{2} = -1, \quad \frac{2 + 3}{4} = \frac{5}{4}, \quad \frac{3 + 5}{-4} = -2.$$

Thus $(1, 2, 3)$ is not on the line.

(b) For $(5, 1, -9)$ we have
$$\frac{5 - 3}{2} = 1, \quad \frac{1 + 3}{4} = 1, \quad \frac{-9 + 5}{-4} = 1.$$

Thus $(5, 1, -9)$ is on the line.

(c) For $(1, -7, -1)$ we have
$$\frac{1 - 3}{2} = -1, \quad \frac{-7 + 3}{4} = -1, \quad \frac{-1 + 5}{-4} = -1.$$

Thus $(1, -7, -1)$ is on the line.

T.1. Let $\mathbf{u}$, $\mathbf{v}$ be vectors in R^2. Then

$$(L_1 \circ L_2)(\mathbf{u} + \mathbf{v}) = L_1(L_2(\mathbf{u} + \mathbf{v})) = L_1(L_2(\mathbf{u}) + L_2(\mathbf{v})) \quad \text{since } L_2 \text{ is a linear operator}$$
$$= L_1(L_2(\mathbf{u})) + L_1(L_2(\mathbf{v})) \quad \text{since } L_1 \text{ is a linear operator}$$
$$= (L_1 \circ L_2)(\mathbf{u}) + (L_1 \circ L_2)(\mathbf{v}).$$

Moreover, for any scalar c,

$$(L_1 \circ L_2)(c\mathbf{u}) = L_1(L_2(c\mathbf{u})) = L_1(cL_2(\mathbf{u})) = cL_1(L_2(\mathbf{u})) = c(L_1 \circ L_2)(\mathbf{u}).$$

Therefore, $L_1 \circ L_2$ is a linear operator.

Chapter 6

Real Vector Spaces

Section 6.1, p. 278

1. No. If (x, y) and (x', y') are in V, then $x > 0$, $x' > 0$, $y > 0$, and $y' > 0$. Since $x + x' > 0$ and $y + y' > 0$, $(x, y) \oplus (x', y')$ is in V. However, if $c < 0$ and (x, y) is in V, then $c \odot (x, y)$ is not in V, since $cx < 0$ and $cy < 0$. For example, let $(2, 3)$ be in V, and let $c = -2$. Then $c \odot (x, y) = (-4, -6)$ is not in V.

3. No. If $a_1 t^2 + b_1 t + c_1$ and $a_2 t^2 + b_2 t + c_2$ are in V, then $b_1 = a_1 + 1$ and $b_2 = a_2 + 1$. Now

$$(a_1 t^2 + b_1 t + c_1) \oplus (a_2 t^2 + b_2 t + c_2) = (a_1 + a_2)t^2 + (b_1 + b_2)t + (c_1 + c_2),$$

and

$$b_1 + b_2 = (a_1 + 1) + (a_2 + 1) = a_1 + a_2 + 2,$$

so $(a_1 t^2 + b_1 t + c_1) \oplus (a_2 t^2 + b_2 t + c_2)$ is not in V.

5. Let $\mathbf{u} = \begin{bmatrix} x_1 \\ y_1 \end{bmatrix}$, $\mathbf{v} = \begin{bmatrix} x_2 \\ y_2 \end{bmatrix}$, and $\mathbf{w} = \begin{bmatrix} x_3 \\ y_3 \end{bmatrix}$ be any vectors in R^2; let c and d be any scalars.

Property (α): $\mathbf{u} + \mathbf{v} = \begin{bmatrix} x_1 \\ y_1 \end{bmatrix} + \begin{bmatrix} x_2 \\ y_2 \end{bmatrix} = \begin{bmatrix} x_1 + x_2 \\ y_1 + y_2 \end{bmatrix}$ which lies in R^2.
Thus R^2 is closed under addition of vectors.

Property (a): $\mathbf{u} + \mathbf{v} = \begin{bmatrix} x_1 \\ y_1 \end{bmatrix} + \begin{bmatrix} x_2 \\ y_2 \end{bmatrix} = \begin{bmatrix} x_1 + x_2 \\ y_1 + y_2 \end{bmatrix} = \begin{bmatrix} x_2 + x_1 \\ y_2 + y_1 \end{bmatrix} = \begin{bmatrix} x_2 \\ y_2 \end{bmatrix} + \begin{bmatrix} x_1 \\ y_1 \end{bmatrix} = \mathbf{v} + \mathbf{u}$.

Property (b): $\mathbf{u} + (\mathbf{v} + \mathbf{w}) = \begin{bmatrix} x_1 \\ y_1 \end{bmatrix} + \left(\begin{bmatrix} x_2 \\ y_2 \end{bmatrix} + \begin{bmatrix} x_3 \\ y_3 \end{bmatrix} \right) = \begin{bmatrix} x_1 \\ y_1 \end{bmatrix} + \begin{bmatrix} x_2 + x_3 \\ y_2 + y_3 \end{bmatrix}$

$$= \begin{bmatrix} x_1 + (x_2 + x_3) \\ y_1 + (y_2 + y_3) \end{bmatrix} = \begin{bmatrix} (x_1 + x_2) + x_3 \\ (y_1 + y_2) + y_3 \end{bmatrix}$$

$$= \begin{bmatrix} x_1 + x_2 \\ y_1 + y_2 \end{bmatrix} + \begin{bmatrix} x_3 \\ y_3 \end{bmatrix} = (\mathbf{u} + \mathbf{v}) + \mathbf{w}.$$

Property (c): Let $\mathbf{0} = \begin{bmatrix} 0 \\ 0 \end{bmatrix}$. Then

$$\mathbf{u} + \mathbf{0} = \begin{bmatrix} x_1 \\ y_1 \end{bmatrix} + \begin{bmatrix} 0 \\ 0 \end{bmatrix} = \begin{bmatrix} x_1 + 0 \\ y_1 + 0 \end{bmatrix} = \begin{bmatrix} x_1 \\ y_1 \end{bmatrix} = \mathbf{u}$$

$$\mathbf{0} + \mathbf{u} = \begin{bmatrix} 0 \\ 0 \end{bmatrix} + \begin{bmatrix} x_1 \\ y_1 \end{bmatrix} = \begin{bmatrix} 0 + x_1 \\ 0 + y_1 \end{bmatrix} = \begin{bmatrix} x_1 \\ y_1 \end{bmatrix} = \mathbf{u}.$$

Property (d): Let $-\mathbf{u} = \begin{bmatrix} -x_1 \\ -y_1 \end{bmatrix}$. Then $\mathbf{u} + (-\mathbf{u}) = \begin{bmatrix} x_1 \\ y_1 \end{bmatrix} + \begin{bmatrix} -x_1 \\ -y_1 \end{bmatrix} = \begin{bmatrix} x_1 - x_1 \\ y_1 - y_1 \end{bmatrix} = \begin{bmatrix} 0 \\ 0 \end{bmatrix} = \mathbf{0}$.

Property (β): $c\mathbf{u} = c\begin{bmatrix} x_1 \\ y_1 \end{bmatrix} = \begin{bmatrix} cx_1 \\ cy_1 \end{bmatrix}$ which is in R^2. Thus R^2 is closed under scalar multiplication.

Property (e): $c(\mathbf{u} + \mathbf{v}) = c\begin{bmatrix} x_1 + x_2 \\ y_1 + y_2 \end{bmatrix} = \begin{bmatrix} c(x_1 + x_2) \\ c(y_1 + y_2) \end{bmatrix} = \begin{bmatrix} cx_1 + cx_2 \\ cy_1 + cy_2 \end{bmatrix}$

$$= \begin{bmatrix} cx_1 \\ cy_1 \end{bmatrix} + \begin{bmatrix} cx_2 \\ cy_2 \end{bmatrix} = c\begin{bmatrix} x_1 \\ y_1 \end{bmatrix} + c\begin{bmatrix} x_2 \\ y_2 \end{bmatrix}.$$

Property (f): $(c + d)\mathbf{u} = \begin{bmatrix} (c+d)x_1 \\ (c+d)y_1 \end{bmatrix} = \begin{bmatrix} cx_1 + dx_1 \\ cy_1 + dy_1 \end{bmatrix} = \begin{bmatrix} cx_1 \\ cy_1 \end{bmatrix} + \begin{bmatrix} dx_1 \\ dy_1 \end{bmatrix} = c\mathbf{u} + d\mathbf{u}$.

Property (g): $c(d\mathbf{u}) = c\begin{bmatrix} dx_1 \\ dy_1 \end{bmatrix} = \begin{bmatrix} c(dx_1) \\ c(dy_1) \end{bmatrix} = \begin{bmatrix} (cd)x_1 \\ (cd)y_1 \end{bmatrix} = (cd)\mathbf{u}$.

Property (h): $1\mathbf{u} = 1\begin{bmatrix} x_1 \\ y_1 \end{bmatrix} = \begin{bmatrix} 1x_1 \\ 1y_1 \end{bmatrix} = \begin{bmatrix} x_1 \\ y_1 \end{bmatrix} = \mathbf{u}$.

7. Let V be the set of all ordered triples of real numbers of the form $(x, y, 0)$. Let $\mathbf{u} = (x, y, 0)$, $\mathbf{v} = (x', y', 0)$, and $\mathbf{w} = (x'', y'', 0)$ be arbitrary vectors in V, and let c and d be scalars. The operations $\oplus$ and $\odot$ are defined by

$$\mathbf{u} \oplus \mathbf{v} = (x, y, 0) \oplus (x', y', 0) = (x + x', y + y', 0)$$
$$c \odot \mathbf{u} = c \odot (x, y, 0) = (cx, cy, 0).$$

Property (α): $\mathbf{u} \oplus \mathbf{v} = (x + x', y + y', 0)$, which is an ordered triple of the correct form. Thus $\mathbf{u} \oplus \mathbf{v}$ is in V. That is, V is closed under the operation $\oplus$.

Property (a): $\mathbf{u} \oplus \mathbf{v} = (x + x', y + y', 0) = (x' + x, y' + y, 0) = \mathbf{v} + \mathbf{u}$.

Property (b): $\mathbf{u} \oplus (\mathbf{v} \oplus \mathbf{w}) = (x, y, 0) \oplus (x' + x'', y' + y'', 0) = (x + x' + x'', y + y' + y'', 0)$.
$(\mathbf{u} \oplus \mathbf{v}) \oplus \mathbf{w} = (x + x', y + y', 0) \oplus (x'', y'', 0) = (x + x' + x'', y + y' + y'', 0)$.
Thus, $\mathbf{u} \oplus (\mathbf{v} \oplus \mathbf{w}) = (\mathbf{u} \oplus \mathbf{v}) \oplus \mathbf{w}$.

Property (c): Let $\mathbf{0}$ be defined as the triple $(0, 0, 0)$ in V.
$\mathbf{u} \oplus \mathbf{0} = (x + 0, y + 0, 0) = (x, y, 0) = \mathbf{u}$
$\mathbf{0} \oplus \mathbf{u} = (0 + x, 0 + y, 0) = (x, y, 0) = \mathbf{u}$
Thus, $\mathbf{u} \oplus \mathbf{0} = \mathbf{0} \oplus \mathbf{u} = \mathbf{u}$.

Property (d): Let $-\mathbf{u}$ be defined to be the triple $(-x, -y, 0)$. Then, for any $\mathbf{u}$ in V, $-\mathbf{u}$ is also in V.
$\mathbf{u} \oplus (-\mathbf{u}) = (x + (-x), y + (-y), 0) = (0, 0, 0) = \mathbf{0}$.

Property (β): $(c \odot \mathbf{u}) = (cx, cy, 0)$, which is an ordered triple of the correct form. Thus $c \odot \mathbf{u}$ is in V. That is, V is closed under the operation $\odot$.

Property (e): $c \odot (\mathbf{u} \oplus \mathbf{v}) = c \odot (x + x', y + y', 0) = (c(x + x'), c(y + y'), 0)$
$$= (cx + cx', cy + cy', 0) = (cx, cy, 0) \oplus (cx', cy', 0)$$
$$= c \odot (x, y, 0) \oplus c \odot (x', y', 0) = c \odot \mathbf{u} \oplus c \odot \mathbf{v}.$$

Property (f): $(c + d) \odot \mathbf{u} = ((c + d)x, (c + d)y, 0) = (cx + dx, cy + dy, 0)$
$$= (cx, cy, 0) \oplus (dx, dy, 0)$$
$$= c \odot (x, y, 0) \oplus d \odot (x, y, 0)$$
$$= c \odot \mathbf{u} \oplus d \odot \mathbf{u}.$$

Property (g): $c \odot (d \odot \mathbf{u}) = c \odot (dx, dy, 0) = (cdx, cdy, 0) = (cd) \odot (x, y, 0) = (cd) \odot \mathbf{u}$.

Property (h): $1 \odot \mathbf{u} = (1x, 1y, 0) = (x, y, 0) = \mathbf{u}$.

Thus V with operations $\oplus$ and $\odot$ is a vector space.

9. Let V be the set of all real valued functions defined on R^1. Let $\mathbf{u} = f$, $\mathbf{v} = g$, and $\mathbf{w} = h$ denote real valued functions and let c and d be scalars. The operations $\oplus$ and $\odot$ are defined by

$$\mathbf{u} \oplus \mathbf{v} = (f \oplus g)(t) = f(t) + g(t)$$
$$c \odot \mathbf{u} = (c \odot f)(t) = cf(t).$$

Here, $\oplus$ is the addition of functions which is defined to be the point-by-point sum of their evaluations at t. Thus, $f(t) + g(t)$ is the addition of real numbers.

Property (α): $\mathbf{u} \oplus \mathbf{v} = (f \oplus g)(t) = f(t) + g(t)$.
     For each real number t, the value of the function $(f \oplus g)$ at t is the real number $f(t) + g(t)$. Hence, $(f \oplus g)$ is a real valued function on R^1. Thus $\mathbf{u} \oplus \mathbf{v}$ is in V. That is, V is closed under the operation $\oplus$.

Property (a): $\mathbf{u} \oplus \mathbf{v} = (f \oplus g)(t) = f(t) + g(t) = g(t) + f(t) = (g \oplus f)(t) = \mathbf{v} \oplus \mathbf{u}$.

Property (b): $\mathbf{u} \oplus (\mathbf{v} \oplus \mathbf{w}) = (f \oplus (g \oplus h))(t) = f(t) + (g \oplus h)(t)$
$$= f(t) + (g(t) + h(t)) = (f(t) + g(t)) + h(t)$$
$$= (f \oplus g)(t) + h(t) = ((f \oplus g) \oplus h)(t) = (\mathbf{u} \oplus \mathbf{v}) \oplus \mathbf{w}.$$

Property (c):   Let $\mathbf{0}$ be the element in V such that $\mathbf{0}(t) = 0$ for all real values t. Then

$$\mathbf{u} \oplus \mathbf{0} = (f \oplus \mathbf{0})(t) = f(t) + \mathbf{0}(t) = f(t) + 0 = f(t).$$

Thus function $\mathbf{u} \oplus \mathbf{0} = \mathbf{u}$. Similarly,

$$\mathbf{0} \oplus \mathbf{u} = (\mathbf{0} \oplus f)(t) = \mathbf{0}(t) + f(t) = 0 + f(t) = f(t).$$

Thus function $\mathbf{0} \oplus \mathbf{u} = \mathbf{u}$. Hence $\mathbf{u} \oplus \mathbf{0} = \mathbf{0} \oplus \mathbf{u} = \mathbf{u}$.

Property (d):   Let $-\mathbf{u} = -f$ be defined to be the function which when evaluated at t gives the real number $-f(t)$. For any $\mathbf{u}$ in V, $-\mathbf{u}$ is also in V. Moreover,

$$\mathbf{u} \oplus (-\mathbf{u}) = (f \oplus (-f))(t) = f(t) + (-f(t)) = 0.$$

Thus for any real number t, function $\mathbf{u} \oplus (-\mathbf{u})$ gives the value zero. From property (c), there is only one such function, $\mathbf{0}$. Hence, $\mathbf{u} \oplus (-\mathbf{u}) = \mathbf{0}$.

Property (β): $c \odot \mathbf{u} = (c \odot f)(t) = cf(t)$.
     For any real number t, function $c \odot f$ gives the real value $cf(t)$, hence $c \odot f$ is a real valued function on R^1. Thus $c \odot \mathbf{u}$ is in V. That is, V is closed under the operation $\odot$.

Property (e): $c \odot (\mathbf{u} \oplus \mathbf{v}) = c \odot (f \oplus g)(t) = c(f(t) + g(t))$
$$= cf(t) + cg(t) = (c \odot f)(t) + (c \odot g)(t)$$
$$= c \odot f \oplus c \odot g = c \odot \mathbf{u} \oplus c \odot \mathbf{v}.$$

Property (f): $(c + d) \odot \mathbf{u} = ((c + d) \odot f)(t) = (c + d)f(t) = cf(t) + df(t)$
$$= (c \odot f)(t) \oplus (d \odot f)(t) = c \odot f \oplus d \odot f = c \odot \mathbf{u} \oplus d \odot \mathbf{u}.$$

Property (g): $c \odot (d \odot \mathbf{u}) = c \odot (d \odot f)(t) = c(df(t)) = (cd)f(t) = ((cd) \odot f)(t) = (cd) \odot f = (cd) \odot \mathbf{u}.$

Property (h): $1 \odot \mathbf{u} = (1 \odot f) = (1 \odot f)(t) = 1f(t) = f(t) = f = \mathbf{u}.$

Thus V with operations $\oplus$ and $\odot$ is a vector space.

11. V is the set of ordered triples with operations

$$(x, y, z) \oplus (x', y', z') = (x', y + y', z')$$
$$c \odot (x, y, z) = (cx, cy, cz).$$

Let $\mathbf{u} = (x, y, z)$, $\mathbf{v} = (x', y', z')$, and $\mathbf{w} = (x'', y'', z'')$ be arbitrary vectors in V, and let c and d be scalars.

Property (α): $\mathbf{u} \oplus \mathbf{v} = (x', y + y', z')$ which is an ordered triple of real numbers. Thus V is closed with respect to $\oplus$.

Property (a): $\mathbf{u} \oplus \mathbf{v} = (x', y + y', z')$
$\mathbf{v} \oplus \mathbf{u} = (x, y' + y, z)$
Thus $\mathbf{u} \oplus \mathbf{v} \neq \mathbf{v} \oplus \mathbf{u}$, hence V with the operations of $\oplus$ and $\odot$ is not a vector space.

Property (b): $\mathbf{u} \oplus (\mathbf{v} \oplus \mathbf{w}) = (x, y, z) \oplus (x'', y' + y'', z'') = (x'', y + y' + y'', z'')$
$(\mathbf{u} \oplus \mathbf{v}) \oplus \mathbf{w} = (x', y + y', z') \oplus (x'', y'', z'') = (x'', y + y' + y'', z'')$
Thus $\mathbf{u} \oplus (\mathbf{v} \oplus \mathbf{w}) = (\mathbf{u} \oplus \mathbf{v}) \oplus \mathbf{w}$.

Property (c):   Let $\mathbf{0} = (0, 0, 0)$. Then $\mathbf{u} \oplus \mathbf{0} = (0, y + 0, 0) \neq \mathbf{u}$. This property fails to hold.

Property (d):   Let $-\mathbf{u} = (-x, -y, -z)$. Then $\mathbf{u} \oplus (-\mathbf{u}) = (-x, y - y, -z) = (-x, 0, -z) \neq \mathbf{0}$. This property fails to hold.

Property (β): $c \odot \mathbf{u} = (cx, cy, cz)$, which is an ordered triple of real numbers. Thus V is closed with respect to $\odot$.

Property (e): $c \odot (\mathbf{u} \oplus \mathbf{v}) = c \odot (x', y + y', z') = (cx', cy + cy', cz')$
$c \odot \mathbf{u} \oplus c \odot \mathbf{v} = (cx, cy, cz) \oplus (cx', cy', cz') = (cx', cy + cy', cz')$
Thus $c \odot (\mathbf{u} \oplus \mathbf{v}) = c \odot \mathbf{u} \oplus c \odot \mathbf{v}$.

Property (f): $(c + d) \odot \mathbf{u} = (cx + dx, cy + dy, cz + dz)$
$c \odot \mathbf{u} \oplus d \odot \mathbf{u} = (cx, cy, cz) \oplus (dx, dy, dz) = (dx, cy + dy, dz)$
Thus this property fails to hold.

Property (g): $c \odot (d \odot \mathbf{u}) = c \odot (dx, dy, dz) = (cdx, cdy, cdz)$
$(cd) \odot \mathbf{u} = (cdx, cdy, cdz)$
Thus $c \odot (d \odot \mathbf{u}) = (cd) \odot \mathbf{u}$.

Property (h): $1 \odot \mathbf{u} = (x, y, z) = \mathbf{u}$.

In summary: Properties (a), (c), (d), and (f) fail to hold.

13. V is the set of ordered triples of the form $(0, 0, z)$ with operations

$$(0, 0, z) \oplus (0, 0, z') = (0, 0, z + z')$$
$$c \odot (0, 0, z) = (0, 0, cz).$$

Let $\mathbf{u} = (0, 0, z)$, $\mathbf{v} = (0, 0, z')$, and $\mathbf{w} = (0, 0, z'')$ be arbitrary vectors in V, and let c and d be scalars.

Property (α): $\mathbf{u} \oplus \mathbf{v} = (0, 0, z + z')$ which is an ordered triple of the correct form to belong to V. Thus V is closed with respect to $\oplus$.

Property (a): $\mathbf{u} \oplus \mathbf{v} = (0, 0, z + z') = (0, 0, z' + z) = \mathbf{v} \oplus \mathbf{u}$
This property holds.

Property (b): $\mathbf{u} \oplus (\mathbf{v} \oplus \mathbf{w}) = (0, 0, z) \oplus (0, 0, z' + z'') = (0, 0, z + z' + z'')$
$(\mathbf{u} \oplus \mathbf{v}) \oplus \mathbf{w} = (0, 0, z + z') \oplus (0, 0, z'') = (0, 0, z + z' + z'')$
This property holds.

Property (c):   Let $\mathbf{0} = (0, 0, 0)$. Then $\mathbf{u} \oplus \mathbf{0} = (0, 0, z + 0) = (0, 0, z) = \mathbf{u}$
$\mathbf{0} \oplus \mathbf{u} = (0, 0, 0 + z) = (0, 0, z) = \mathbf{u}$
This property holds.

Property (d): Let $-\mathbf{u} = (0,0,-z)$. Then $\mathbf{u} \oplus (-\mathbf{u}) = (0,0,z-z) = (0,0,0) = \mathbf{0}$.
This property holds.

Property (β): $c \odot \mathbf{u} = (0,0,cz)$ which is an ordered triple of the correct form to belong to V. Thus V is closed with respect to $\odot$.

Property (e): $c \odot (\mathbf{u} \oplus \mathbf{v}) = c \odot (0,0,z+z') = (0,0,cz+cz')$
$c \odot \mathbf{u} \oplus c \odot \mathbf{v} = (0,0,cz) \oplus (0,0,cz') = (0,0,cz+cz')$
This property holds.

Property (f): $(c+d) \odot \mathbf{u} = (0,0,cz+dz)$
$c \odot \mathbf{u} \oplus d \odot \mathbf{u} = (0,0,cz) \oplus (0,0,dz) = (0,0,cz+dz)$
This property holds.

Property (g): $c \odot (d \odot \mathbf{u}) = c \odot (0,0,dz) = (0,0,cdz)$
$(cd) \odot \mathbf{u} = (0,0,cdz)$
This property holds.

Property (h): $1 \odot \mathbf{u} = (0,0,z) = \mathbf{u}$.
This property holds.

In summary: V is a vector space with operations $\oplus$ and $\odot$.

15. Let V be the set of all ordered pairs (x,y), where $x \le 0$, with operations

$$(x,y) \oplus (x',y') = (x+x', y+y')$$
$$c \odot (x,y) = (cx, cy).$$

Let $\mathbf{u} = (x,y)$, $\mathbf{v} = (x',y')$, and $\mathbf{w} = (x'',y'')$ be arbitrary vectors in V, and let c and d be scalars.

Property (α): $\mathbf{u} \oplus \mathbf{v} = (x+x', y+y')$. Since $x \le 0$ and $x' \le 0$, $x+x' \le 0$. Then $\mathbf{u} \oplus \mathbf{v}$ is an ordered pair with first entry ≤ 0. Thus V is closed with respect to $\oplus$.

Property (a): $\mathbf{u} \oplus \mathbf{v} = (x+x', y+y') = (x'+x, y'+y) = \mathbf{v} \oplus \mathbf{u}$.
This property holds.

Property (b): $\mathbf{u} \oplus (\mathbf{v} \oplus \mathbf{w}) = (x,y) \oplus (x'+x'', y'+y'') = (x+x'+x'', y+y'+y'')$
$(\mathbf{u} \oplus \mathbf{v}) \oplus \mathbf{w} = (x+x', y+y') \oplus (x'',y'') = (x+x'+x'', y+y'+y'')$
This property holds.

Property (c): Let $\mathbf{0} = (0,0)$. Since its first entry is ≤ 0, $\mathbf{0}$ is in V.
$\mathbf{u} \oplus \mathbf{0} = (x+0, y+0) = (x,y) = \mathbf{u}$
$\mathbf{0} \oplus \mathbf{u} = (0+x, 0+y) = (x,y) = \mathbf{u}$
This property holds.

Property (d): Let $-\mathbf{u} = (-x,-y)$. Since $x \le 0$, $-x \ge 0$. Hence $-\mathbf{u}$ is not in V for some $\mathbf{u}$ in V. Thus not every element in V has an additive inverse. It follows that V is not a vector space.

Property (β): $c \odot \mathbf{u} = (cx, cy)$. If $c < 0$ and $x \ne 0$, then $c \odot \mathbf{u}$ has its first component > 0 and hence is not in V. Thus V is not closed under $\odot$.

Property (e): $c \odot (\mathbf{u} \oplus \mathbf{v}) = c \odot (x+x', y+y') = (cx+cx', cy+cy')$ which may not be in V. Hence this property may fail.

Property (f): $(c+d) \odot \mathbf{u} = (cx+dx, cy+dy)$ which may not be in V. Hence this property may fail.

Property (g): $c \odot (d \odot \mathbf{u}) = c \odot (dx, dy) = (cdx, cdy)$ which may not be in V. Hence this property may fail.

Property (h): $1 \odot \mathbf{u} = (1x, 1y) = \mathbf{u}$
This property holds.

In summary: Properties (d), (β), (e), (f), and (g) fail to hold.

17. V is the set of all positive real numbers with operations

$$\mathbf{u} \oplus \mathbf{v} = \mathbf{u}\mathbf{v}$$
$$c \odot \mathbf{u} = \mathbf{u}^c.$$

Let $\mathbf{u}$ and $\mathbf{v}$ be real numbers and c and d be scalars.

Property (α): $\mathbf{u} \oplus \mathbf{v} = \mathbf{u}\mathbf{v}$ which is a positive real number. Thus V is closed with respect to $\oplus$.

Property (a): $\mathbf{u} \oplus \mathbf{v} = \mathbf{u}\mathbf{v} = \mathbf{v}\mathbf{u} = \mathbf{v} \oplus \mathbf{u}$.
This property holds.

Property (b): $\mathbf{u} \oplus (\mathbf{v} \oplus \mathbf{w}) = \mathbf{u} \oplus (\mathbf{v}\mathbf{w}) = \mathbf{u}(\mathbf{v}\mathbf{w}) = (\mathbf{u}\mathbf{v})\mathbf{w} = (\mathbf{u} \oplus \mathbf{v}) \oplus \mathbf{w}$
This property holds.

Property (c):   Let $\mathbf{0} = 1$. Then $\mathbf{0}$ is in V.
$\mathbf{u} \oplus \mathbf{0} = \mathbf{u}1 = \mathbf{u}$
$\mathbf{0} \oplus \mathbf{u} = 1\mathbf{u} = \mathbf{u}$
This property holds.

Property (d):   Let $-\mathbf{u} = \frac{1}{\mathbf{u}}$. Then $-\mathbf{u}$ is in V
$\mathbf{u} \oplus (-\mathbf{u}) = \mathbf{u}\left(\frac{1}{\mathbf{u}}\right) = 1 = \mathbf{0}$.
This property holds.

Property (β): $c \odot \mathbf{u} = \mathbf{u}^c$ which is a positive real number. Thus V is closed under $\odot$.

Property (e): $c \odot (\mathbf{u} \oplus \mathbf{v}) = c \odot (\mathbf{u}\mathbf{v}) = (\mathbf{u}\mathbf{v})^c$
$c \odot \mathbf{u} \oplus c \odot \mathbf{v} = \mathbf{u}^c \oplus \mathbf{v}^c = \mathbf{u}^c\mathbf{v}^c = (\mathbf{u}\mathbf{v})^c$
This property holds.

Property (f): $(c + d) \odot \mathbf{u} = \mathbf{u}^{c+d}$
$c \odot \mathbf{u} \oplus d \odot \mathbf{u} = \mathbf{u}^c \oplus \mathbf{u}^d = \mathbf{u}^c\mathbf{u}^d = \mathbf{u}^{c+d}$
This property holds.

Property (g): $c \odot (d \odot \mathbf{u}) = c \odot \mathbf{u}^d = (\mathbf{u}^d)^c = \mathbf{u}^{cd}$
$(cd) \odot \mathbf{u} = \mathbf{u}^{cd}$
This property holds.

Property (h): $1 \odot \mathbf{u} = \mathbf{u}^1 = \mathbf{u}$.
This property holds.

In summary: V with operations $\oplus$ and $\odot$ is a vector space.

19. Properties (α) and (β) hold since $\mathbf{0} \oplus \mathbf{0} = \mathbf{0}$ and $c \odot \mathbf{0} = \mathbf{0}$. We now verify the remaining properties in Definition 1:

Properties (a) and (b) hold vacuously, since there is only one vector in V.

Property (c) holds since $\mathbf{0}$ is the only vector in V and $\mathbf{0} \oplus \mathbf{0} = \mathbf{0}$, so $\mathbf{0}$ is the zero vector.

Property (d) holds, similarly, since $-\mathbf{0} = \mathbf{0}$.

Property (e) holds vacuously.

Property (f) holds since $(c + d) \odot \mathbf{0} = \mathbf{0}$ and $c \odot \mathbf{0} \oplus d \odot \mathbf{0} = \mathbf{0} \oplus \mathbf{0} = \mathbf{0}$.

Property (g) holds since $c \odot (d \odot \mathbf{u}) = c \odot \mathbf{0} = \mathbf{0}$ and $(cd) \odot \mathbf{0} = \mathbf{0}$.

Property (h) holds since $1 \odot \mathbf{0} = \mathbf{0}$.

T.1. $c\mathbf{u} = c(\mathbf{u} + \mathbf{0}) = c\mathbf{u} + c\mathbf{0} = c\mathbf{0} + c\mathbf{u}$ by Definition 1(c), (e) and (a). Add the negative of $c\mathbf{u}$ to both sides of this equation to get $\mathbf{0} = c\mathbf{0} + c\mathbf{u} + (-c\mathbf{u}) = c\mathbf{0} + \mathbf{0} = c\mathbf{0}$.

T.3. (cancellation): If $\mathbf{u} + \mathbf{v} = \mathbf{u} + \mathbf{w}$, then

$$(-\mathbf{u}) + (\mathbf{u} + \mathbf{v}) = (-\mathbf{u}) + (\mathbf{u} + \mathbf{w})$$
$$(-\mathbf{u} + \mathbf{u}) + \mathbf{v} = (-\mathbf{u} + \mathbf{u}) + \mathbf{w}$$
$$\mathbf{0} + \mathbf{v} = \mathbf{0} + \mathbf{w}$$
$$\mathbf{v} = \mathbf{w}$$

T.5. Let $\mathbf{0}_1$ and $\mathbf{0}_2$ be zero vectors. Then $\mathbf{0}_1 \oplus \mathbf{0}_2 = \mathbf{0}_1$ and $\mathbf{0}_1 \oplus \mathbf{0}_2 = \mathbf{0}_2$. So $\mathbf{0}_1 = \mathbf{0}_2$.

T.7. Show that if $\mathbf{u}, \mathbf{v} \in B^n$, then $\mathbf{u} + \mathbf{v} \in B^n$. Let

$$\mathbf{u} = \begin{bmatrix} u_1 \\ u_2 \\ \vdots \\ u_n \end{bmatrix} \quad \text{and} \quad \mathbf{v} = \begin{bmatrix} v_1 \\ v_2 \\ \vdots \\ v_n \end{bmatrix}.$$

then

$$\mathbf{u} + \mathbf{v} = \begin{bmatrix} u_1 + v_1 \\ u_2 + v_2 \\ \vdots \\ u_n + v_n \end{bmatrix}$$

and each component $u_i + v_i$ is either 0 or 1 under binary addition. So $\mathbf{u} + \mathbf{v} \in B^n$.

T.9. Under binary multiplication

$$1 \cdot u_i = \begin{cases} 0 & \text{if } u_i = 0 \\ 1 & \text{if } u_i = 1 \end{cases}.$$

So

$$1 \odot \mathbf{u} = 1 \cdot \begin{bmatrix} u_1 \\ u_2 \\ \vdots \\ u_n \end{bmatrix} = \begin{bmatrix} 1 \cdot u_1 \\ 1 \cdot u_2 \\ \vdots \\ 1 \cdot u_n \end{bmatrix} = \begin{bmatrix} u_1 \\ u_2 \\ \vdots \\ u_n \end{bmatrix} = \mathbf{u}$$

which is an element of B^n.

ML.1. Assume that $k + A$ means "add k to each element in A." Property (c) implies that the zero vector under $\oplus$ is the identity matrix $\begin{bmatrix} 1 & 0 \\ 0 & 1 \end{bmatrix}$. So if you choose any singular 2×2 matrix it will not satisfy property (d). It is also instructive to find counterexamples to property (h) or (e).

Section 6.2, p. 287

1. Let W be the subset of R^2 of all the points of the form (x, x), where x is a real number. W is a subspace of R^2 if conditions (α) and (β) of Theorem 6.2 hold.

   W is a subspace of R^2. To show this, let $\mathbf{u} = (a, a)$, $\mathbf{v} = (b, b)$, and let c be any real number.

   (α) $\mathbf{u} + \mathbf{v} = (a, a) + (b, b) = (a + b, a + b)$ is in W

   (β) $c \cdot \mathbf{u} = c(a, a) = (ca, ca)$ is in W

   W is closed under the operations of addition and scalar multiplication. Therefore W is a subspace.

3. Let V be the vector space R^2. Let W be the set of all vectors with tail at $(0, 0)$ whose head is a point inside or on the unit circle, $x^2 + y^2 = 1$. W is not a subspace of R^2. To show this, let $\mathbf{u} = (0, a)$ and $\mathbf{v} = (0, b)$, $0 < a, b \leq \sqrt{1 - x^2}$, be in W, and let c be any real number. Then

(α) $\mathbf{u} + \mathbf{v}$ may not be in W. For example, if $\mathbf{u} = \mathbf{v} = (0, \frac{3}{4})$, then $\mathbf{u} + vecv = (0, \frac{3}{2})$ is not in W.

(β) $c \cdot \mathbf{u}$ may not be in W. For example, if $\mathbf{u} = (0, \frac{3}{4})$ and $c = 4$, then $c \cdot \mathbf{u} = (0, 3)$ is not in W.

W is not closed under the operations of addition and scalar multiplication. Therefore W is not a subspace.

5. V is the vector space R^3. Let W be the given subset of V.

(a) W is not a subspace. To show this, let $\mathbf{u} = (a_1, b_1, c_1)$ and $\mathbf{v} = (a_2, b_2, c_2)$ be in W. Then $c_1 = c_2 = 2$. Now

$$\mathbf{u} + \mathbf{v} = (a_1 + a_2, b_1 + b_1, c_1 + c_2) = (a_1 + a_2, b_1 + b_1, 4)$$

which is not in W. Therefore Property (α) is not satisfied and hence W is not a subspace of V.

(b) W is a subspace of R^3. To show this, let $\mathbf{u} = (a_1, b_1, c_1)$ and $\mathbf{v} = (a_2, b_2, c_2)$ be in W. Then $c_1 = a_1 + b_1$ and $c_2 = a_2 + b_2$. Now

$$\mathbf{u} + \mathbf{v} = (a_1, b_1, c_1) + (a_2, b_2, c_2) = (a_1 + a_2, b_1 + b_2, c_1 + c_2)$$
$$r\mathbf{u} = r(a_1, b_1, c_1) = (ra_1, rb_1, rc_1).$$

Since

$$c_1 + c_2 = (a_1 + b_1) + (a_2 + b_2) = (a_1 + a_2) + (b_1 + b_2)$$

and

$$rc_1 = r(a_1 + b_1) = ra_1 + rb_1,$$

$\mathbf{u} + \mathbf{v}$ and $r\mathbf{u}$ are in W. Therefore W is a subspace of R^3.

(c) W is not a subspace of R^3. To show this, let $\mathbf{u} = (a_1, b_1, c_1)$ and $\mathbf{v} = (a_2, b_2, c_2)$ be in W. Then $c_1 > 0$ and $c_2 > 0$. Now

$$\mathbf{u} + \mathbf{v} = (a_1, b_1, c_1) + (a_2, b_2, c_2) = (a_1 + a_2, b_1 + b_2, c_1 + c_2).$$

Since $c_1 > 0$ and $c_2 > 0$, $c_1 + c_2 > 0$. Therefore $\mathbf{u} + \mathbf{v}$ is in W and Property (α) is satisfied. However, for any scalar r,

$$r\mathbf{u} = r(a_1, b_1, c_1) = (ra_1, rb_1, rc_1).$$

If $r < 0$, then $rc_1 < 0$ and therefore $r\mathbf{u}$ is not in W. Thus Property (β) is not satisfied, Hence W is not a subspace of R^3.

7. V is the vector space R^4. Let W be the given subset of R^4.

(a) W is not a subspace of R^4. To show this, let $\mathbf{u} = (a_1, b_1, c_1)$ and $\mathbf{v} = (a_2, b_2, c_2)$ be in W. Then $a_1 - b_1 = 2$ and $a_2 - b_2 = 2$. Now,

$$\mathbf{u} + \mathbf{v} = (a_1, b_1, c_1) + (a_2, b_2, c_2) = (a_1 + a_2, b_1 + b_2, c_1 + c_2).$$

Since

$$(a_1 + a_2) - (b_1 + b_2) = (a_1 - b_1) + (a_2 - b_2) = 2 - 2 = 0,$$

$\mathbf{u} + \mathbf{v}$ is not in W. Therefore Property (α) is not satisfied and hence W is not a subspace of R^4.

(b) W is a subspace of R^4. To show this, let $\mathbf{u} = (a_1, b_1, c_1, d_1)$ and $\mathbf{v} = (a_2, b_2, c_2, d_2)$ be in W. Then $c_1 = a_1 + 2b_1$, $d_1 = a_1 - 3b_1$ and $c_2 = a_2 + 2b_2$, $d_2 = a_2 - 3b_2$. Now,

$$\mathbf{u} + \mathbf{v} = (a_1, b_1, c_1, d_1) + (a_2, b_2, c_2, d_2) = (a_1 + a_2, b_1 + b_2, c_1 + c_2, d_1 + d_2)$$
$$r\mathbf{u} = r(a_1, b_1, c_1, d_1) = (ra_1, rb_1, rc_1, rd_1).$$

Since

$$c_1 + c_2 = (a_1 + 2b_1) + (a_2 + 2b_2) = (a_1 + a_2) + 2(b_1 + b_2)$$
$$d_1 + d_2 = (a_1 - 3b_1) + (a_2 - 3b_2) = (a_1 + a_2) - 3(b_1 + b_2)$$
$$rc_1 = r(a_1 + 2b_1) = ra_1 + 2(rb_1)$$
$$rd_1 = r(a_1 - 3b_1) = ra_1 - 3(rb_1),$$

$\mathbf{u} + \mathbf{v}$ and $r\mathbf{u}$ are in W. Therefore W is a subspace of R^4.

(c) W is a subspace of R^4. To show this, let $\mathbf{u} = (a_1, b_1, c_1, d_1)$ and $\mathbf{v} = (a_2, b_2, c_2, d_2)$ be in W. Then $a_1 = 0$, $b_1 = -d_1$, and $a_2 = 0$, $b_2 = -d_2$. Now,

$$\mathbf{u} + \mathbf{v} = (a_1, b_1, c_1, d_1) + (a_2, b_2, c_2, d_2) = (a_1 + a_2, b_1 + b_2, c_1 + c_2, d_1 + d_2).$$

Since

$$a_1 + a_2 = 0 \quad \text{and} \quad b_1 + b_2 = -d_1 - d_2 = -(d_1 + d_2),$$

$\mathbf{u} + \mathbf{v}$ is in W. Similarly,

$$r\mathbf{u} = r(a_1, b_1, c_1, d_1) = (ra_1, rb_1, rc_1, rd_1).$$

Since $ra_1 = r(0) = 0$ and $rb_1 = r(-d_1) = -(rd_1)$, $r\mathbf{u}$ is in W. Thus W is a subspace of R^4.

9. We are in vector space P_2. Let W be the given subset of polynomials.

(a) W is a subspace of P_2. To show this, let $\mathbf{u} = a_2 t^2 + a_1 t + a_0$ and $\mathbf{v} = b_2 t^2 + b_1 t + b_0$ be in W. Then $a_0 = 0$ and $b_0 = 0$. Now

$$\mathbf{u} + \mathbf{v} = (a_2 + b_2)t^2 + (a_1 + b_1)t + (a_0 + b_0)$$
$$r\mathbf{u} = (ra_2)t^2 + (ra_1)t + (ra_0).$$

Since $a_0 + b_0 = 0 + 0 = 0$ and $ra_0 = r(0) = 0$, $\mathbf{u} + \mathbf{v}$ and $r\mathbf{u}$ are in W. Therefore W is a subspace.

(b) W is not a subspace of P_2. To show this, let $\mathbf{u} = a_2 t^2 + a_1 t + a_0$ and $\mathbf{v} = b_2 t^2 + b_1 t + b_0$ be in W. Then $a_1 = 2$ and $b_1 = 2$. Now,

$$\mathbf{u} + \mathbf{v} = (a_2 + b_2)t^2 + (a_1 + b_1)t + (a_0 + b_0).$$

Since $a_0 + b_0 = 2 + 2 = 4 \neq 2$, $\mathbf{u} + \mathbf{v}$ is not in W. Therefore W is not a subspace of P_2.

(c) W is a subspace of P_2. To show this, let $\mathbf{u} = a_2 t^2 + a_1 t + a_0$ and $\mathbf{v} = b_2 t^2 + b_1 t + b_0$ be in W. Then $a_2 + a_1 = a_0$ and $b_2 + b_1 = b_0$. Now

$$\mathbf{u} + \mathbf{v} = (a_2 + b_2)t^2 + (a_1 + b_1)t + (a_0 + b_0)$$
$$r\mathbf{u} = (ra_2)t^2 + (ra_1)t + (ra_0).$$

Since

$$(a_2 + b_2) + (a_1 + b_1) = (a_2 + a_1) + (b_2 + b_1) = a_0 + b_0$$

and

$$ra_2 + ra_1 = r(a_2 + a_1) = ra_0,$$

$\mathbf{u} + \mathbf{v}$ and $r\mathbf{u}$ are in W. Therefore W is a subspace.

11. (a) P_2 is the set of all polynomials of degree 2 or less. That is, $P_2 = \{a_2t^2 + a_1t + a_0\}$.

   Property (α): $(a_2t^2 + a_1t + a_0) + (b_2t^2 + b_1t + b_0) = (a_2 + b_2)t^2 + (a_1 + b_1)t + (a_0 + b_0)$, which is a polynomial of degree 2 or less and hence is in W.

   Property (β): $c(a_2t^2 + a_1t + a_0) = ca_2t^2 + ca_1 + ca_0$, which is a polynomial of degree 2 or less and hence is in W.

   Therefore P_2 is a subspace of P_3.

   (b) $P_n = \{a_nt^n + a_{n-1}t^{n-1} + \cdots + a_1t + a_0\}$.

   Property (α): $(a_nt^n + a_{n-1}t^{n-1} + \cdots + a_1t + a_0) + (b_nt^n + b_{n-1}t^{n-1} + \cdots + b_1t + b_0)$
   $= (a_n + b_n)t^n + (a_{n-1} + b_{n-1})t^{n-1} + \cdots + (a_1 + b_1)t + (a_0 + b_0)$.
   which is a polynomial of degree $\leq n$, hence is in W.

   Property (β): $c(a_nt^n + a_{n-1}t^{n-1} + \cdots + a_1t + a_0) = ca_nt^n + ca_{n-1}t^{n-1} + \cdots + ca_1t + ca_0$, which is a polynomial of degree $\leq n$ and hence is in W.

   Thus, P_n is a subspace of P_{n+1}.

13. A polynomial determines a real valued function. Thus, P is a subset of the vector space of all real valued functions defined for all real numbers. Also, P is closed under vector addition and scalar multiplication. Thus it is a subspace of the function space.

15. Let

$$\mathbf{w}_1 = a\mathbf{u} + b\mathbf{v} = (2a + 4b, 2b, 3a - 5b, -4a + b)$$

and

$$\mathbf{w}_2 = a'\mathbf{u} + b'\mathbf{v} = (2a' + 4b', 2b', 3a' - 5b', -4a' + b')$$

lie in W. Then

$$\mathbf{w}_1 + \mathbf{w}_2 = (2(a + a') + 4(b + b'), 2(b + b'), 3(a + a') - 5(b + b') - 4(a + a') + (b + b'))$$

and

$$c\mathbf{w}_1 = (2ac + 4bc, 2bc, 3ac - 5bc, -4ac + bc)$$

both lie in W. By Theorem 6.2, W is a subspace.

17. (a) Not a subspace. Property (α) of Theorem 6.2 fails to hold:

$$\begin{bmatrix} a_1 & b_1 & c_1 \\ d_1 & e_1 & f_1 \end{bmatrix} + \begin{bmatrix} a_2 & b_2 & c_2 \\ d_2 & e_2 & f_2 \end{bmatrix} = \begin{bmatrix} a_1 + a_2 & b_1 + b_2 & c_1 + c_2 \\ d_1 + d_2 & e_1 + e_2 & f_1 + f_2 \end{bmatrix}$$

   is not in the subset W, since if $b_1 = 2c_1 + 1$ and $b_2 = 2c_2 + 1$, then $b_1 + b_2 = 2(c_1 + c_2) + 2$.

   (b) Not a subspace. Property (α) of Theorem 6.2 fails to hold:

$$\begin{bmatrix} 0 & 1 & a_1 \\ b_1 & c_1 & 0 \end{bmatrix} + \begin{bmatrix} 0 & 1 & a_2 \\ b_2 & c_2 & 0 \end{bmatrix} = \begin{bmatrix} 0 & 2 & a_1 + a_1 \\ b_1 + b_2 & c_1 + c_2 & 0 \end{bmatrix}$$

   is not in the subset W.

   (c) Subspace. Properties (α) and (β) of Theorem 6.2 are satisfied. Suppose

$$\begin{bmatrix} a_1 & b_1 & c_1 \\ d_1 & e_1 & f_1 \end{bmatrix} \quad \text{and} \quad \begin{bmatrix} a_2 & b_2 & c_2 \\ d_2 & e_2 & f_2 \end{bmatrix}$$

are in W, so that $a_1 + c_1 = 0$, $b_1 + d_1 + f_1 = 0$, $a_2 + c_2 = 0$, and $b_2 + d_2 + f_2 = 0$. Then

$$\begin{bmatrix} a_1 & b_1 & c_1 \\ d_1 & e_1 & f_1 \end{bmatrix} + \begin{bmatrix} a_2 & b_2 & c_2 \\ d_2 & e_2 & f_2 \end{bmatrix} = \begin{bmatrix} a_1 + a_2 & b_1 + b_2 & c_1 + c_2 \\ d_1 + d_2 & e_1 + e_2 & f_1 + f_2 \end{bmatrix}$$

is in W since

$$(a_1 + a_2) + (c_1 + c_2) = 0 \quad \text{and} \quad (b_1 + b_2) + (d_1 + d_2) + (f_1 + f_2) = 0.$$

Also, if k is any real number, then

$$k \begin{bmatrix} a_1 & b_1 & c_1 \\ d_1 & e_1 & f_1 \end{bmatrix} = \begin{bmatrix} ka_1 & kb_1 & kc_1 \\ kd_1 & ke_1 & kf_1 \end{bmatrix}$$

is in W since $ka_1 + kc_1 = 0$ and $kb_1 + kd_1 + kf_1 = 0$.

19. (a) Not a subspace. The sum of two singular matrices may be nonsingular. For example,

$$\begin{bmatrix} 1 & 0 \\ 0 & 0 \end{bmatrix} \quad \text{and} \quad \begin{bmatrix} 0 & 0 \\ 0 & 1 \end{bmatrix}$$

are singular matrices, but their sum

$$\begin{bmatrix} 1 & 0 \\ 0 & 1 \end{bmatrix}$$

is nonsingular.

(b) Subspace. By Exercise T.6 in Section 1.3, the product of two upper triangular matrices is upper triangular. Also, if k is any real number and $A = \begin{bmatrix} a_{ij} \end{bmatrix}$ is an upper triangular matrix ($a_{ij} = 0$ for $i > j$), then kA is also upper triangular ($ka_{ij} = 0$ for $i > j$). Thus properties (α) and (β) of Theorem 6.2 are satisfied.

(c) Not a subspace. Property (α) of Theorem 6.2 fails to hold: The matrices

$$\begin{bmatrix} 2 & 1 \\ 1 & 1 \end{bmatrix} \quad \text{and} \quad \begin{bmatrix} 1 & 0 \\ 0 & 1 \end{bmatrix}$$

have determinant equal to 1, but their sum

$$\begin{bmatrix} 3 & 1 \\ 1 & 2 \end{bmatrix}$$

is a matrix whose determinant is 5.

21. (a) Subspace. Properties (α) and (β) of Theorem 6.2 are satisfied: the sum of two integrable functions is integrable and the scalar multiple of an integrable function is integrable.

(b) Subspace. Properties (α) and (β) of Theorem 6.2 are satisfied. The sum of two bounded functions is bounded and the scalar multiple of a bounded function is bounded.

(c) Subspace. Properties (α) and (β) of Theorem 6.2 are satisfied. Then sum of two functions that are integrable on $[a, b]$ is integrable on $[a, b]$ and the scalar multiple of a function that is integrable on $[a, b]$ is also integrable on $[a, b]$.

(d) Subspace. Properties (α) and (β) of Theorem 6.2 are satisfied. The sum of two functions that are bounded on $[a, b]$ is bounded on $[a, b]$. The scalar multiple of a function that is bounded on $[a, b]$ is also bounded on $[a, b]$.

23. (a) The line segment shown is in the second quadrant and stops at the origin. Hence if (a, b) is a point on the line segment, $a \leq 0$ and $b \geq 0$. Thus vector

$$\mathbf{u} = \begin{bmatrix} a \\ b \end{bmatrix}$$

is in the set of vectors represented by the line segment. However,

$$-\mathbf{u} = (-1) \begin{bmatrix} a \\ b \end{bmatrix}$$

is not, since the point $(-a, -b)$ is not in the second quadrant and it follows that Property (β) is not satisfied. Hence this is not a subspace of R^2.

(b) The line shown goes through the origin. If (a, b) is any point on the line, the vector

$$\mathbf{u} = \begin{bmatrix} a \\ b \end{bmatrix}$$

is in the set of vectors represented by the line. One representation for the equation of the line is found using points (a, b) and $(0, 0)$. We have

$$y = \frac{b}{a} x \quad \Longleftrightarrow \quad ay = bx.$$

Similarly, if (c, d) is any other point on the line, then

$$\mathbf{v} = \begin{bmatrix} c \\ d \end{bmatrix}$$

is in the set of vectors represented by the line and $cy = dx$. Therefore

$$\mathbf{u} + \mathbf{v} = \begin{bmatrix} a \\ b \end{bmatrix} + \begin{bmatrix} c \\ d \end{bmatrix} = \begin{bmatrix} a + c \\ b + d \end{bmatrix}$$

and this vector is in the set represented by the line since

$$(a + c)y = ay + cy = bx + dx = (b + d)x.$$

Thus Property (α) is satisfied. Similarly, for any scalar r,

$$r\mathbf{u} = r \begin{bmatrix} a \\ b \end{bmatrix} = \begin{bmatrix} ra \\ rb \end{bmatrix}$$

and this vector is in the set represented by the line since

$$(ra)y = r(ay) = r(bx) = (rb)x.$$

Thus Property (β) is satisfied. Therefore this line represents a subspace of R^2.

25. A vector $\mathbf{v}$ belongs to span $\{\mathbf{v}_1, \mathbf{v}_2, \mathbf{v}_3\}$ provided $\mathbf{v}$ can be expressed as a linear combination of $\mathbf{v}_1$, $\mathbf{v}_2$, and $\mathbf{v}_3$. Following the procedure in Example 3, we form the expression

$$c_1 \mathbf{v}_1 + c_2 \mathbf{v}_2 + c_3 \mathbf{v}_3 = \mathbf{v},$$

and determine whether there exist values c_1, c_2, and c_3 for which this is true. For

$$\mathbf{v}_1 = (1, 0, 0, 1), \quad \mathbf{v}_2 = (1, -1, 0, 0), \quad \text{and} \quad \mathbf{v}_3 = (0, 1, 2, 1),$$

we have

$$(c_1 + c_2, -c_2 + c_3, 2c_3, c_1 + c_3) = \mathbf{v}.$$

We equate corresponding components to obtain a linear system. Construct the associated augmented matrix and compute its reduced row echelon form. If the system is consistent, then $\mathbf{v}$ belongs to span $\{\mathbf{v}_1, \mathbf{v}_2, \mathbf{v}_3\}$.

(a) For $\mathbf{v} = (-1, 4, 2, 2)$, the linear system is

$$
\begin{array}{rcr}
c_1 + c_2 & = & -1 \\
-c_2 + c_3 & = & 4 \\
2c_3 & = & 2 \\
c_1 \quad\;\; + c_3 & = & 2
\end{array}
$$

and the reduced row echelon form of the augmented matrix is

$$
\begin{bmatrix}
1 & 0 & 0 & \vdots & 0 \\
0 & 1 & 0 & \vdots & 0 \\
0 & 0 & 1 & \vdots & 0 \\
0 & 0 & 0 & \vdots & 1
\end{bmatrix}.
$$

The system is inconsistent. Hence $\mathbf{v}$ is not in span $\{\mathbf{v}_1, \mathbf{v}_2, \mathbf{v}_3\}$.

(b) For $\mathbf{v} = (1, 2, 0, 1)$, the linear system is

$$
\begin{array}{rcr}
c_1 + c_2 & = & 1 \\
-c_2 + c_3 & = & 2 \\
2c_3 & = & 0 \\
c_1 \quad\;\; + c_3 & = & 1
\end{array}
$$

and the reduced row echelon form of the augmented matrix is

$$
\begin{bmatrix}
1 & 0 & 0 & \vdots & 0 \\
0 & 1 & 0 & \vdots & 0 \\
0 & 0 & 1 & \vdots & 0 \\
0 & 0 & 0 & \vdots & 1
\end{bmatrix}.
$$

The system is inconsistent. Hence $\mathbf{v}$ is not in span $\{\mathbf{v}_1, \mathbf{v}_2, \mathbf{v}_3\}$.

(c) For $\mathbf{v} = (-1, 1, 4, 3)$, the linear system is

$$
\begin{array}{rcr}
c_1 + c_2 & = & -1 \\
-c_2 + c_3 & = & 1 \\
2c_3 & = & 4 \\
c_1 \quad\;\; + c_3 & = & 3
\end{array}
$$

and the reduced row echelon form of the augmented matrix is

$$
\begin{bmatrix}
1 & 0 & 0 & \vdots & 0 \\
0 & 1 & 0 & \vdots & 0 \\
0 & 0 & 1 & \vdots & 0 \\
0 & 0 & 0 & \vdots & 1
\end{bmatrix}.
$$

The system is inconsistent. Hence $\mathbf{v}$ is not in span $\{\mathbf{v}_1, \mathbf{v}_2, \mathbf{v}_3\}$.

(d) For $\mathbf{v} = (0, 1, 1, 0)$, the linear system is

$$
\begin{array}{rcr}
c_1 + c_2 & = & 0 \\
-c_2 + c_3 & = & 1 \\
2c_3 & = & 1 \\
c_1 \quad\;\; + c_3 & = & 0
\end{array}
$$

and the reduced row echelon form of the augmented matrix is

$$\begin{bmatrix} 1 & 0 & 0 & \vdots & 0 \\ 0 & 1 & 0 & \vdots & 0 \\ 0 & 0 & 1 & \vdots & 0 \\ 0 & 0 & 0 & \vdots & 1 \end{bmatrix}.$$

The system is inconsistent. Hence $\mathbf{v}$ is not in span $\{\mathbf{v}_1, \mathbf{v}_2, \mathbf{v}_3\}$.

27. We must determine if there are constants c_1, c_2, c_3 such that

$$c_1 p_1(t) + c_2 p_2(t) + c_3 p_3(t) = p(t).$$

(a) In this case

$$c_1(t^2 - t) + c_2(t^2 - 2t + 1) + c_3(-t^2 + 1) = 3t^2 - 3t + 1.$$

Expanding and equating coefficients of like powers of t, we obtain the linear system

$$\begin{aligned} c_1 + c_2 - c_3 &= 3 \\ -c_1 - 2c_2 &= -3 \\ c_2 + c_3 &= 1. \end{aligned}$$

The reduced row echelon form of the augmented matrix is

$$\begin{bmatrix} 1 & 0 & -2 & \vdots & 0 \\ 0 & 1 & 1 & \vdots & 0 \\ 0 & 0 & 0 & \vdots & 1 \end{bmatrix}.$$

The system is inconsistent and hence has no solution. Therefore $p(t) = 3t^2 - 3t + 1$ is not in the span of $p_1(t)$, $p_2(t)$, $p_3(t)$.

(b) In this case the augmented matrix of the system is

$$\begin{bmatrix} 1 & 1 & -1 & \vdots & 1 \\ -1 & -2 & 0 & \vdots & -1 \\ 0 & 1 & 1 & \vdots & 1 \end{bmatrix}$$

whose reduced row echelon form is

$$\begin{bmatrix} 1 & 0 & -2 & \vdots & 0 \\ 0 & 1 & 1 & \vdots & 0 \\ 0 & 0 & 0 & \vdots & 1 \end{bmatrix}.$$

The system is inconsistent and hence has no solution. Therefore $p(t) = t^2 - t + 1$ is not in the span of $p_1(t)$, $p_2(t)$, $p_3(t)$.

(c) In this case the augmented matrix of the system is

$$\begin{bmatrix} 1 & 1 & -1 & \vdots & 0 \\ -1 & -2 & 0 & \vdots & 1 \\ 0 & 1 & 1 & \vdots & 1 \end{bmatrix}$$

whose reduced row echelon form is

$$\begin{bmatrix} 1 & 0 & -2 & \vdots & 0 \\ 0 & 1 & 1 & \vdots & 0 \\ 0 & 0 & 0 & \vdots & 1 \end{bmatrix}.$$

The system is inconsistent and hence has no solution. Therefore $p(t) = t + 1$ is not in the span of $p_1(t)$, $p_2(t)$, $p_3(t)$.

(d) In this case the augmented matrix of the system is

$$\left[\begin{array}{ccc|c} 1 & 1 & -1 & 2 \\ -1 & -2 & 0 & -1 \\ 0 & 1 & 1 & -1 \end{array}\right].$$

whose reduced row echelon form is

$$\left[\begin{array}{ccc|c} 1 & 0 & -2 & 3 \\ 0 & 1 & 1 & -1 \\ 0 & 0 & 0 & 0 \end{array}\right].$$

The system is consistent and hence has solutions. Therefore $p(t) = 2t^2 - t - 1$ is in the span of $p_1(t)$, $p_2(t)$, $p_3(t)$.

29. W is a subspace of $V = B^3$. We must show that properties (α) and (β) of Theorem 6.2 are satisfied. Consider the vectors

$$\mathbf{u} = \begin{bmatrix} 1 \\ 0 \\ 0 \end{bmatrix}, \quad \mathbf{v} = \begin{bmatrix} 0 \\ 0 \\ 1 \end{bmatrix}, \quad \mathbf{w} = \begin{bmatrix} 1 \\ 0 \\ 1 \end{bmatrix}, \quad \mathbf{x} = \begin{bmatrix} 0 \\ 0 \\ 0 \end{bmatrix}.$$

Then $\mathbf{u} + \mathbf{v} = \mathbf{w}$, $\mathbf{u} + \mathbf{w} = \mathbf{v}$, and $\mathbf{v} + \mathbf{w} = \mathbf{u}$. Since $\mathbf{x}$ is the zero vector, $\mathbf{u} + \mathbf{x} = \mathbf{u}$, $\mathbf{v} + \mathbf{x} = \mathbf{v}$, and $\mathbf{w} + \mathbf{x} = \mathbf{w}$. Thus property ($\alpha$) is satisfied. In addition, if c is any scalar, then c is 0 or 1, so we have

$$0 \cdot \mathbf{u} = 0 \cdot \mathbf{v} = 0 \cdot \mathbf{w} = 0 \cdot \mathbf{x} = \begin{bmatrix} 0 \\ 0 \\ 0 \end{bmatrix} = \mathbf{x}$$

$$1 \cdot \mathbf{u} = \mathbf{u}, \ 1 \cdot \mathbf{v} = \mathbf{v}, \ 1 \cdot \mathbf{w} = \mathbf{w}, \ 1 \cdot \mathbf{x} = \mathbf{x}.$$

Thus property (β) is satisfied.

31. W is not a subspace. Property (α) of Theorem 6.2 is not satisfied. To show this, let

$$\mathbf{v} = \begin{bmatrix} a_1 \\ 1 \\ c_1 \\ d_1 \end{bmatrix} \quad \text{and} \quad \mathbf{u} = \begin{bmatrix} a_2 \\ 1 \\ c_2 \\ d_2 \end{bmatrix}.$$

Then

$$\mathbf{u} + \mathbf{v} = \begin{bmatrix} a_1 + a_2 \\ 0 \\ c_1 + c_2 \\ d_1 + d_2 \end{bmatrix} \quad \text{is not in } W.$$

Note that the second entry is always 0 since $1 + 1 = 0$.

33. We must determine if there are constants c_1, c_2, c_3 such that

$$c_1 \begin{bmatrix} 0 \\ 0 \\ 1 \end{bmatrix} + c_2 \begin{bmatrix} 1 \\ 0 \\ 1 \end{bmatrix} + c_3 \begin{bmatrix} 0 \\ 1 \\ 1 \end{bmatrix} = \begin{bmatrix} 1 \\ 1 \\ 1 \end{bmatrix}.$$

We obtain the linear system

$$\begin{aligned} c_2 &= 1 \\ c_3 &= 1 \\ c_1 + c_2 + c_3 &= 1 \implies c_1 = 1 \end{aligned}$$

Hence

$$\begin{bmatrix} 0 \\ 0 \\ 1 \end{bmatrix} + \begin{bmatrix} 1 \\ 0 \\ 1 \end{bmatrix} + \begin{bmatrix} 0 \\ 1 \\ 1 \end{bmatrix} = \begin{bmatrix} 1 \\ 1 \\ 1 \end{bmatrix}.$$

Therefore $\mathbf{u}$ belongs to span S.

T.1. If W is a subspace, then for $\mathbf{u}, \mathbf{v} \in W$, $\mathbf{u} + \mathbf{v}$ and $c\mathbf{u}$ lie in W by properties (α) and (β) of Definition 1. Conversely, assume (α) and (β) of Theorem 6.2 hold. We must show that properties (a)–(h) in Definition 1 hold.

Property (a) holds since, if $\mathbf{u}, \mathbf{v}$ are in W they are *a fortiori* in V, and therefore $\mathbf{u} + \mathbf{v} = \mathbf{v} + \mathbf{u}$ by property (a) for V. Similarly for (b). By (β), for $c = 0$, $\mathbf{0} = 0\mathbf{u}$ lies in W. Again by (β), for $c = -1$, $-\mathbf{u} = (-1)\mathbf{u}$ lies in W. Thus (d) holds. Finally, (e), (f), (g), (h) follow for W because those properties hold for any vectors in V and any scalars.

T.3. Since $\mathbf{b} \neq \mathbf{0}$, $A \neq O$. If $A\mathbf{x} = \mathbf{b}$ has no solutions, then that empty set of solutions is not a vector space. Otherwise, let $\mathbf{x}_0$ be a solution. Then $A(2\mathbf{x}_0) = 2(A\mathbf{x}_0) = 2\mathbf{b} \neq \mathbf{b}$ since $\mathbf{b} \neq \mathbf{0}$. Thus $\mathbf{x}_0 + \mathbf{x}_0 = 2\mathbf{x}_0$ is not a solution. Hence, the set of all solutions fails to be closed under either vector addition or scalar multiplication.

T.5. W must be closed under vector addition and under multiplication of a vector by an arbitrary scalar. Thus, along with $\mathbf{v}_1, \mathbf{v}_2, \ldots, \mathbf{v}_k$, W must contain $\displaystyle\sum_{i=1}^{k} a_i \mathbf{v}_i$ for any set of coefficients $a_1, \ldots, a_k$. Thus W contains span S.

T.7. We must show that $W = \{c\mathbf{x}_0 \mid c \text{ is a scalar}\}$ is closed under addition and scalar multiplication. We have $c\mathbf{x}_0 + d\mathbf{x}_0 = (c + d)\mathbf{x}_0$ is in W, and if r is a scalar then $r(c\mathbf{x}_0) = (rc)\mathbf{x}_0$ is in W. Hence W is a subspace.

T.9. Let V be a subspace of R^1 which is not the zero subspace and let $\mathbf{v} = \begin{bmatrix} v_1 \end{bmatrix} \neq \mathbf{0}$ be any vector in V. If $\mathbf{u}$ is any nonzero vector in R^1, then $\mathbf{u} = \left(\dfrac{u_1}{v_1}\right)\mathbf{v}$, a scalar multiple of $\mathbf{v}$, so R^1 is a subset of V. Hence, $V = R^1$.

T.11. Since $V = W_1 + W_2$, every vector $\mathbf{v}$ in W can be written as $\mathbf{w}_1 + \mathbf{w}_2$, $\mathbf{w}_1$ in W_1 and $\mathbf{w}_2$ in W_2. Suppose now that $\mathbf{v} = \mathbf{w}_1 + \mathbf{w}_2$ and $\mathbf{v} = \mathbf{w}_1' + \mathbf{w}_2'$. Then $\mathbf{w}_1 + \mathbf{w}_2 = \mathbf{w}_1' + \mathbf{w}_2'$, so

$$\mathbf{w}_1 - \mathbf{w}_1' = \mathbf{w}_2' - \mathbf{w}_2. \tag{6.1}$$

Call the vector $\mathbf{w}_1 - \mathbf{w}_1' = \mathbf{x}$. The left-hand side implies that $\mathbf{x}$ is in W_1, while the right-hand side implies $\mathbf{x}$ is in W_2. So $\mathbf{x}$ must be in the intersection $W_1 \cap W_2 = \{\mathbf{0}\}$ or $\mathbf{x} = \mathbf{0}$. Hence $\mathbf{w}_1 = \mathbf{w}_1'$ and $\mathbf{w}_2' = \mathbf{w}_2$, i.e., we can write $\mathbf{v}$ uniquely.

T.13. If W does not contain the zero vector, $\mathbf{0}$, then it is not closed under scalar multiplication (Let $r = 0$.) or vector addition (if $\mathbf{w} \in W$, then $-\mathbf{w} \in W$ and $\mathbf{w} + (-\mathbf{w}) = \mathbf{0}$).

T.15. We know from T.13 that any subspace must contain the zero vector, $\mathbf{0}$. Let $\mathbf{v}$ and $\mathbf{w}$ be any two vectors different from $\mathbf{0}$ and from each other. Suppose $W = \{\mathbf{0}, \mathbf{v}, \mathbf{w}\}$ forms a subspace of B^3. Then $\mathbf{v} + \mathbf{w}$ must be an element of W. But $\mathbf{v} + \mathbf{w} \neq \mathbf{v}$ since $\mathbf{w}$ is different from $\mathbf{0}$ and likewise $\mathbf{v} + \mathbf{w} \neq \mathbf{w}$. Binary addition implies that $\mathbf{v} = -\mathbf{v}$ and $\mathbf{w} = -\mathbf{w}$ so $\mathbf{v} + \mathbf{w} \neq \mathbf{0}$ because $\mathbf{v}$ is different from $\mathbf{w}$. So W is not closed under vector addition. Contradiction. So no three vectors can form a subspace in B^3.

T.17. Note that $\left\{ \begin{bmatrix} 0 \\ 0 \\ 0 \end{bmatrix}, \begin{bmatrix} 1 \\ 1 \\ 1 \end{bmatrix} \right\}$ is a subspace of B^3. To find more subspaces, knowing that they must contain the zero vector, $\mathbf{0}$, we form

$$W = \left\{ \begin{bmatrix} 0 \\ 0 \\ 0 \end{bmatrix}, \begin{bmatrix} 1 \\ 1 \\ 1 \end{bmatrix}, \mathbf{v} \right\}$$

where $\mathbf{v}$ is any other vector in B^3, different from $\begin{bmatrix} 0 \\ 0 \\ 0 \end{bmatrix}$ or $\begin{bmatrix} 1 \\ 1 \\ 1 \end{bmatrix}$. We know from T.15 that W must

also contain at least $\mathbf{v} + \begin{bmatrix} 1 \\ 1 \\ 1 \end{bmatrix}$ in order to be a subspace. There are three possibilities:

$$\left\{ \begin{bmatrix} 0 \\ 0 \\ 0 \end{bmatrix}, \begin{bmatrix} 1 \\ 1 \\ 1 \end{bmatrix}, \begin{bmatrix} 1 \\ 0 \\ 0 \end{bmatrix}, \begin{bmatrix} 0 \\ 1 \\ 1 \end{bmatrix} \right\}$$

$$\left\{ \begin{bmatrix} 0 \\ 0 \\ 0 \end{bmatrix}, \begin{bmatrix} 1 \\ 1 \\ 1 \end{bmatrix}, \begin{bmatrix} 0 \\ 1 \\ 0 \end{bmatrix}, \begin{bmatrix} 1 \\ 0 \\ 1 \end{bmatrix} \right\}$$

$$\left\{ \begin{bmatrix} 0 \\ 0 \\ 0 \end{bmatrix}, \begin{bmatrix} 1 \\ 1 \\ 1 \end{bmatrix}, \begin{bmatrix} 0 \\ 0 \\ 1 \end{bmatrix}, \begin{bmatrix} 1 \\ 1 \\ 0 \end{bmatrix} \right\}$$

We leave it to the reader to check that they are all closed under vector addition and scalar multiplication. To continue, adjoin any other vector $\mathbf{u}$ to the subspaces above. You will note that closure under vector addition requires that you include all the remaining vectors in B^3, hence B^3 itself is the last possible subspace that contains $\begin{bmatrix} 1 \\ 1 \\ 1 \end{bmatrix}$.

ML.3. (a) Following Example 11, we construct the augmented matrix that results from the expression $c_1\mathbf{v}_1 + c_2\mathbf{v}_2 + c_3\mathbf{v}_3 = \mathbf{v}$. Note that since the vectors are rows we need to convert them to columns to form this matrix. Next we obtain the reduced row echelon form of the associated linear system.

v1 = [1 0 0 1];v2 = [0 1 1 0];v3 = [1 1 1 1];v = [0 1 1 1];
rref([v1′ v2′ v3′ v′])

ans =

```
1  0  1  0
0  1  1  0
0  0  0  1
0  0  0  0
```

Since this represents an augmented matrix, the system is inconsistent and hence has no solution. Thus $\mathbf{v}$ is not a linear combination of $\{\mathbf{v}_1, \mathbf{v}_2, \mathbf{v}_3\}$.

(b) Here the strategy is similar to that in part (a) except that the vectors are already columns. We use the transpose operator to conveniently enter the vectors.

v1 = [1 2 − 1]′;v2 = [2 − 1 0]′;v3 = [− 1 8 − 3]′;v = [0 5 − 2]′;
rref([v1 v2 v3 v])

ans =

$$
\begin{array}{cccc}
1 & 0 & 3 & 2 \\
0 & 1 & -2 & -1 \\
0 & 0 & 0 & 0
\end{array}
$$

Since this matrix represents an augmented matrix, the system is consistent. It follows that $\mathbf{v}$ is a linear combination of $\{\mathbf{v}_1, \mathbf{v}_2, \mathbf{v}_3\}$. In fact, let $c_3 = r$, then $c_1 = 2 - 3r$ and $c_2 = 2r - 1$.

ML.5. (a) Follow the procedure in ML.3(b).
   v1 = [1 2 1 0 1]';v2 = [0 1 2 − 1 1]';
   v3 = [2 1 0 0 − 1]';v4 = [− 2 1 1 1 1]';
   v = [0 − 1 1 − 2 1]';
   rref([v1 v2 v3 v4 v])

   ans =

$$
\begin{array}{ccccc}
1 & 0 & 0 & 0 & 0 \\
0 & 1 & 0 & 0 & 1 \\
0 & 0 & 1 & 0 & -1 \\
0 & 0 & 0 & 1 & -1 \\
0 & 0 & 0 & 0 & 0
\end{array}
$$

   The system is consistent and it follows that $0\mathbf{v}_1 + \mathbf{v}_2 - \mathbf{v}_3 - \mathbf{v}_4 = \mathbf{v}$.

   (b) Associate a column vector of coefficients with each polynomial, then follow the method in part (a).
   v1 = [2 − 1 1]';v2 = [1 0 − 2]';v3 = [0 1 − 1]';v = [4 1 − 5]';
   rref([v1 v2 v3 v])

   ans =

$$
\begin{array}{cccc}
1 & 0 & 0 & 1 \\
0 & 1 & 0 & 2 \\
0 & 0 & 1 & 2
\end{array}
$$

   Since the system is consistent, we have that $p_1(t) + 2p_2(t) + 2p_3(t) = p(t)$.

ML.7. Associate a column vector with each polynomial as in ML.5(b).
   v1 = [0 1 − 1]';v2 = [0 1 1]';v3 = [1 1 1]';

   (a) v = [1 2 4]';
   rref([v1 v2 v3 v])

   ans =

$$
\begin{array}{cccc}
1 & 0 & 0 & -1 \\
0 & 1 & 0 & 2 \\
0 & 0 & 1 & 1
\end{array}
$$

   Since the system is consistent, $p(t)$ is in span S.

   (b) v = [2 1 − 2]';
   rref([v1 v2 v3 v])

   ans =

$$
\begin{array}{cccc}
1 & 0 & 0 & 1 \\
0 & 1 & 0 & -2 \\
0 & 0 & 1 & 2
\end{array}
$$

   Since the system is consistent, $p(t)$ is in span S.

(c) $\mathbf{v} = [-2 \quad 0 \quad 1]'$;
rref([v1 v2 v3 v])
ans =

1.0000	0	0	-0.5000
0	1.0000	0	2.5000
0	0	1.0000	-2.0000

Since the system is consistent, $p(t)$ is in span S.

Section 6.3, p. 301

1. We follow the method of Example 1. Let $\mathbf{v} = (a, b)$.

   (a) Determine whether constants c_1 and c_2 exist such that

   $$c_1(1, 2) + c_2(-1, 1) = \mathbf{v} = (a, b).$$

   The corresponding linear system is

   $$c_1 - c_2 = a$$
   $$2c_1 + c_2 = b$$

   which has solution

   $$c_1 = \frac{a + b}{3} \quad \text{and} \quad c_2 = \frac{-2a + b}{3}.$$

   Since there is a solution for every choice of a and b, $\{(1, 2), (-1, 1)\}$ spans R^2.

   (b) Determine whether constants c_1, c_2, and c_3 exist such that

   $$c_1(0, 0) + c_2(1, 1) + c_3(-2, -2) = \mathbf{v} = (a, b).$$

   The corresponding linear system is

   $$0c_1 + c_2 - 2c_3 = a$$
   $$0c_1 + c_2 - 2c_3 = b.$$

   The associated augmented matrix is

   $$\begin{bmatrix} 0 & 1 & -2 & \vdots & a \\ 0 & 1 & -2 & \vdots & b \end{bmatrix},$$

   which is row equivalent to

   $$\begin{bmatrix} 0 & 1 & -2 & \vdots & a \\ 0 & 0 & 0 & \vdots & b - a \end{bmatrix}.$$

   We see that this system is consistent only if $a = b$. Thus, not every vector (a, b) can be written as a linear combination of $\{(0, 0), (1, 1), (-2, -2)\}$. Hence these vectors do not span R^2.

   (c) Determine whether constants c_1, c_2, and c_3 exist such that

   $$c_1(1, 3) + c_2(2, -3) + c_3(0, 2) = \mathbf{v} = (a, b).$$

   The corresponding linear system is

   $$1c_1 + 2c_2 + 0c_3 = a$$
   $$3c_1 - 3c_2 + 2c_3 = b.$$

The associated augmented matrix is

$$\begin{bmatrix} 1 & 2 & 0 & \vdots & a \\ 3 & -3 & 2 & \vdots & b \end{bmatrix},$$

which is row equivalent to

$$\begin{bmatrix} 1 & 0 & \frac{4}{9} & \vdots & \frac{3a+2b}{9} \\ 0 & 1 & -\frac{2}{9} & \vdots & \frac{3a-b}{9} \end{bmatrix}.$$

This system is consistent for any choices of a and b. Thus $\{(1,3),(2,-3),(0,2)\}$ spans R^2.

(d) Determine whether constants c_1 and c_2 exist such that

$$c_1(2,4) + c_2(-1,2) = \mathbf{v} = (a,b).$$

The corresponding linear system is

$$2c_1 - c_2 = a$$
$$4c_1 + 2c_2 = b$$

which has solution

$$c_1 = \frac{2a+b}{8} \quad \text{and} \quad c_2 = \frac{b-2a}{4}.$$

Since there is a solution for every choice of a and b, $\{(2,4),(-1,2)\}$ spans R^2.

3. Following the technique of Example 1, let $\mathbf{v} = (a,b,c,d)$.

(a) Determine whether constants c_1, c_2, c_3, and c_4 exist such that

$$c_1(1,0,0,1) + c_2(0,1,0,0) + c_3(1,1,1,1) + c_4(1,1,1,0) = \mathbf{v} = (a,b,c,d).$$

The corresponding linear system is

$$c_1 \quad\;\; + c_3 + c_4 = a$$
$$c_2 + c_3 + c_4 = b$$
$$c_3 + c_4 = c$$
$$c_1 \quad\;\; + c_3 + \quad\;\; = d$$

which has augmented matrix

$$\begin{bmatrix} 1 & 0 & 1 & 1 & \vdots & a \\ 0 & 1 & 1 & 1 & \vdots & b \\ 0 & 0 & 1 & 1 & \vdots & c \\ 1 & 0 & 1 & 0 & \vdots & d \end{bmatrix}.$$

The reduced row echelon form is

$$\begin{bmatrix} 1 & 0 & 0 & 0 & \vdots & a-c \\ 0 & 1 & 0 & 0 & \vdots & b-c \\ 0 & 0 & 1 & 0 & \vdots & c+d-a \\ 0 & 0 & 0 & 1 & \vdots & a-d \end{bmatrix}.$$

The system is consistent for any choices of a, b, c, and d. Hence

$$\{(1,0,0,1),(0,1,0,0),(1,1,1,1),(1,1,1,0)\}$$

spans R^4.

(b) Determine whether constants c_1, c_2, and c_3, exist such that

$$c_1(1,2,1,0) + c_2(1,1,-1,0) + c_3(0,0,0,1) = \mathbf{v} = (a,b,c,d).$$

The corresponding linear system is

$$
\begin{aligned}
c_1 + c_2 \quad\;\; &= a \\
2c_1 + c_2 \quad\;\; &= b \\
c_1 - c_2 \quad\;\; &= c \\
c_3 &= d
\end{aligned}
$$

which has augmented matrix

$$
\begin{bmatrix}
1 & 1 & 0 & \vdots & a \\
2 & 1 & 0 & \vdots & b \\
1 & -1 & 0 & \vdots & c \\
0 & 0 & 1 & \vdots & d
\end{bmatrix}.
$$

Row operations $-2\mathbf{r}_1 + \mathbf{r}_2$, $-\mathbf{r}_1 + \mathbf{r}_3$, $-2\mathbf{r}_2 + \mathbf{r}_3$ give

$$
\begin{bmatrix}
1 & 1 & 0 & \vdots & a \\
0 & -1 & 0 & \vdots & b - 2a \\
0 & 0 & 0 & \vdots & 3a - 2b + c \\
0 & 0 & 1 & \vdots & d
\end{bmatrix}.
$$

We see that the system is inconsistent unless $3a - 2b + c = 0$. Thus there is not a solution for all choices of a, b, c and d and it follows that $\{(1,2,1,0),(1,1,-1,0),(0,0,0,1)\}$ does not span R^4.

(c) Determine whether constants c_1, c_2, c_3, c_4, and c_5 exist such that

$$c_1(6,4,-2,4) + c_2(2,0,0,1) + c_3(3,2,-1,2)$$
$$+ c_4(5,6,-3,2) + c_5(0,4,-2,-1) = \mathbf{v} = (a,b,c,d).$$

The corresponding linear system is

$$
\begin{aligned}
6c_1 + 2c_2 + 3c_3 + 5c_4 \quad\;\; &= a \\
4c_1 \quad\;\; + 2c_3 + 6c_4 + 4c_5 &= b \\
-2c_1 \quad\;\; - c_3 - 3c_4 - 2c_5 &= c \\
4c_1 + c_2 + 2c_3 + 2c_4 - c_5 &= d
\end{aligned}
$$

which has augmented matrix

$$
\begin{bmatrix}
6 & 2 & 3 & 5 & 0 & \vdots & a \\
4 & 0 & 2 & 6 & 4 & \vdots & b \\
-2 & 0 & -1 & -3 & -2 & \vdots & c \\
4 & 1 & 2 & 2 & -1 & \vdots & d
\end{bmatrix}.
$$

We see that rows 2 and 3 are proportional except in the augmented column. Thus applying row operation $2\mathbf{r}_3 + \mathbf{r}_2$ gives

$$
\begin{bmatrix}
6 & 2 & 3 & 5 & 0 & \vdots & a \\
0 & 0 & 0 & 0 & 0 & \vdots & b + 2c \\
-2 & 0 & -1 & -3 & -2 & \vdots & c \\
4 & 1 & 2 & 2 & -1 & \vdots & d
\end{bmatrix}.
$$

It follows that the system is inconsistent unless $b + 2c = 0$. Thus there is not a solution for all choices of a, b, c, and d. Hence $\{(6,4,-2,4),(2,0,0,1),(3,2,-1,2),(5,6,-3,2),(0,4,-2,-1)\}$ does not span R^4.

(d) Determine whether constants c_1, c_2, c_3, and c_4 exist such that

$$c_1(1,1,0,0) + c_2(1,2,-1,1) + c_3(0,0,1,1) + c_4(2,1,2,1) = \mathbf{v} = (a,b,c,d).$$

The corresponding linear system is

$$\begin{aligned}
c_1 + c_2 \quad\quad + 2c_4 &= a \\
c_1 + 2c_2 \quad\quad + c_4 &= b \\
- c_2 + c_3 + 2c_4 &= c \\
c_2 + c_3 + c_4 &= d
\end{aligned}$$

which has augmented matrix

$$\begin{bmatrix}
1 & 1 & 0 & 2 & \vdots & a \\
1 & 2 & 0 & 1 & \vdots & b \\
0 & -1 & 1 & 2 & \vdots & c \\
0 & 1 & 1 & 1 & \vdots & d
\end{bmatrix}.$$

The row operations $-\mathbf{r}_1 + \mathbf{r}_2$, $\mathbf{r}_2 + \mathbf{r}_3$, $-\mathbf{r}_2 + \mathbf{r}_4$, $-\mathbf{r}_3 + \mathbf{r}_4$, $-\mathbf{r}_4 + \mathbf{r}_3$, $\mathbf{r}_4 + \mathbf{r}_2$, $-3\mathbf{r}_4 + \mathbf{r}_1$, $-\mathbf{r}_2 + \mathbf{r}_1$, give an equivalent linear system with solution

$$\begin{aligned}
c_1 &= -4a + 5b + 3c - 3d, & c_2 &= a - b - c + d, \\
c_3 &= -3a + 3b + 2c - d, & c_4 &= 2a - 2b - c + d.
\end{aligned}$$

It follows that the system is consistent for all choices of a, b, c, and d. Hence

$$\{(1,1,0,0), (1,2,-1,1), (0,0,1,1), (2,1,2,1),\}$$

spans R^4.

5. Following the technique of Example 3, let $p(t) = at^3 + bt^2 + ct + d$. Determine whether constants c_1, c_2, c_3 and c_4 exist such that

$$c_1(t^3 + 2t + 1) + c_2(t^2 - t + 2) + c_3(t^3 + 2) + c_4(-t^3 + t^2 - 5t + 2) = p(t).$$

Expanding, grouping like terms, and equating coefficients of like powers of t gives the linear system

$$\begin{aligned}
c_1 \quad\quad + c_3 - c_4 &= a \\
c_2 \quad + c_4 &= b \\
2c_1 - c_2 \quad\quad - 5c_4 &= c \\
c_1 + 2c_2 + 2c_3 + 2c_4 &= d
\end{aligned}$$

which has augmented matrix

$$\begin{bmatrix}
1 & 0 & 1 & -1 & \vdots & a \\
0 & 1 & 0 & 1 & \vdots & b \\
2 & -1 & 0 & -5 & \vdots & c \\
1 & 2 & 2 & 2 & \vdots & d
\end{bmatrix}.$$

Row operations $-2\mathbf{r}_1 + \mathbf{r}_3$, $-\mathbf{r}_1 + \mathbf{r}_4$, $\mathbf{r}_2 + \mathbf{r}_3$, $-2\mathbf{r}_2 + \mathbf{r}_4$, $2\mathbf{r}_4 + \mathbf{r}_3$ give an equivalent system which contains a row of the form

$$\begin{bmatrix} 0 & 0 & 0 & 0 & \vdots & -4a - 3b + c + 2d \end{bmatrix}.$$

It follows that the system does not have a solution for all values of a, b, c, and d. Hence

$$\{t^3 + 2t + 1, t^2 - t + 2, t^3 + 2, -t^3 + t^2 - 5t + 2\}$$

does not span P_3.

7. Following the technique in Example 6 we find the reduced row echelon form of the augmented matrix $[A \vdots \mathbf{0}]$. We obtain

$$\begin{bmatrix} 1 & 0 & 0 & -1 & \vdots & 0 \\ 0 & 1 & 2 & 0 & \vdots & 0 \\ 0 & 0 & 0 & 0 & \vdots & 0 \\ 0 & 0 & 0 & 0 & \vdots & 0 \end{bmatrix}.$$

The general solution is then given by

$$x_4 = r, \qquad x_3 = s, \qquad x_2 = -2s, \qquad x_1 = r,$$

where r and s are any real numbers. In matrix form we have that any member of the solution space is given by

$$\mathbf{x} = r \begin{bmatrix} 1 \\ 0 \\ 0 \\ 1 \end{bmatrix} + s \begin{bmatrix} 0 \\ -2 \\ 1 \\ 0 \end{bmatrix}.$$

Hence $\left\{ \begin{bmatrix} 1 \\ 0 \\ 0 \\ 1 \end{bmatrix}, \begin{bmatrix} 0 \\ -2 \\ 1 \\ 0 \end{bmatrix} \right\}$ spans the null space of A.

9. To determine whether $\{\mathbf{x}_1, \mathbf{x}_2, \mathbf{x}_3\}$ is linearly independent, we form Equation (1):

$$c_1 \mathbf{x}_1 + c_2 \mathbf{x}_2 + c_3 \mathbf{x}_3 = \mathbf{0}$$

which is

$$\begin{aligned} c_1 + c_2 + \; c_3 &= 0 \\ 2c_1 \qquad + 6c_3 &= 0 \\ - c_2 + 2c_3 &= 0 \\ c_1 + c_2 \qquad &= 0. \end{aligned}$$

The reduced row echelon form of this matrix is

$$\begin{bmatrix} 1 & 0 & 0 & \vdots & 0 \\ 0 & 1 & 0 & \vdots & 0 \\ 0 & 0 & 1 & \vdots & 0 \\ 0 & 0 & 0 & \vdots & 0 \end{bmatrix}$$

so $c_1 = c_2 = c_3 = 0$. Hence $\{\mathbf{x}_1, \mathbf{x}_2, \mathbf{x}_3\}$ is linearly independent.

11. To determine if the given set of vectors is linearly dependent, we form Equation (1), which leads to a homogeneous system. If this homogeneous system has a nontrivial solution, then the given set of vectors is linearly dependent.

(a) $c_1(1, 1, 2, 1) + c_2(1, 0, 0, 2) + c_3(4, 6, 8, 6) + c_4(0, 3, 2, 1) = (0, 0, 0, 0)$
The corresponding homogeneous system is

$$\begin{aligned} c_1 + \; c_2 + 4c_3 \qquad &= 0 \\ c_1 \qquad + 6c_3 + 3c_4 &= 0 \\ 2c_1 \qquad + 8c_3 + 2c_4 &= 0 \\ c_1 + 2c_2 + 6c_3 + \; c_4 &= 0. \end{aligned}$$

The reduced row echelon form of the coefficient matrix is

$$\begin{bmatrix} 1 & 0 & 0 & -3 \\ 0 & 1 & 0 & -1 \\ 0 & 0 & 1 & 1 \\ 0 & 0 & 0 & 0 \end{bmatrix}.$$

Hence the solution of this homogeneous system is $c_1 = 3r$, $c_2 = r$, $c_3 = -r$, $c_4 = r$, where r is any real number. Thus the vectors are linearly dependent. Let $r = 1$. Then we have $3\mathbf{v}_1 + \mathbf{v}_2 - \mathbf{v}_3 + \mathbf{v}_4 = \mathbf{0}$. Hence $\mathbf{v}_3 = 3\mathbf{v}_1 + \mathbf{v}_2 + \mathbf{v}_4$.

(b) $c_1(1, -2, 3, -1) + c_2(-2, 4, -6, 2) = (0, 0, 0, 0)$

The corresponding homogeneous system is

$$\begin{aligned} c_1 - 2c_2 &= 0 \\ -2c_1 + 4c_2 &= 0 \\ 3c_1 - 6c_2 &= 0 \\ -c_1 + 2c_2 &= 0. \end{aligned}$$

The reduced row echelon form of the coefficient matrix is

$$\begin{bmatrix} 1 & -2 \\ 0 & 0 \\ 0 & 0 \\ 0 & 0 \end{bmatrix}.$$

Hence the general solution of this homogeneous system is $c_1 = 2r$, $c_2 = r$, where r is any real number. Thus the vectors are linearly dependent.

(c) $c_1(1, 1, 1, 1) + c_2(2, 3, 1, 2) + c_3(3, 1, 2, 1) + c_4(2, 2, 1, 1) = (0, 0, 0, 0)$

The corresponding homogeneous system is

$$\begin{aligned} c_1 + 2c_2 + 3c_3 + 2c_4 &= 0 \\ c_1 + 3c_2 + c_3 + 2c_4 &= 0 \\ c_1 + c_2 + 2c_3 + c_4 &= 0 \\ c_1 + 2c_2 + c_3 + c_4 &= 0. \end{aligned}$$

The reduced row echelon form of the coefficient matrix is I_4. Hence the only solution of this homogeneous system is the trivial solution $c_1 = c_2 = c_3 = c_4 = 0$. Thus the vectors are linearly independent.

(d) $c_1(4, 2, -1, 3) + c_2(6, 5, -5, 1) + c_3(2, -1, 3, 5) = (0, 0, 0, 0)$

The corresponding homogeneous system is

$$\begin{aligned} 4c_1 + 6c_2 + 2c_3 &= 0 \\ 2c_1 + 5c_2 - c_3 &= 0 \\ -c_1 - 5c_2 + 3c_3 &= 0 \\ 3c_1 + c_2 + 5c_3 &= 0. \end{aligned}$$

The reduced row echelon form of the coefficient matrix is

$$\begin{bmatrix} 1 & 0 & 2 \\ 0 & 1 & 1 \\ 0 & 0 & 0 \\ 0 & 0 & 0 \end{bmatrix}.$$

Hence the solution of this homogeneous system is $c_1 = -2r$, $c_2 = -r$, $c_3 = r$, where r is any real number. Thus the vectors are linearly dependent. Let $r = 1$. Then we have $-2\mathbf{v}_1 - \mathbf{v}_2 + \mathbf{v}_3 = \mathbf{0}$. Hence $\mathbf{v}_3 = 2\mathbf{v}_1 + \mathbf{v}_2$.

13. To determine if the given set of vectors is linearly dependent, we form Equation (1), which leads to a homogeneous system. If this homogeneous system has a nontrivial solution, then the given set of vectors is linearly dependent.

(a) Here we have

$$c_1 \begin{bmatrix} 1 & 1 \\ 1 & 2 \end{bmatrix} + c_2 \begin{bmatrix} 1 & 0 \\ 0 & 2 \end{bmatrix} + c_3 \begin{bmatrix} 0 & 3 \\ 1 & 2 \end{bmatrix} + c_4 \begin{bmatrix} 2 & 6 \\ 4 & 6 \end{bmatrix} = \begin{bmatrix} 0 & 0 \\ 0 & 0 \end{bmatrix}.$$

Carrying out the matrix operations and equating entries leads to the linear system

$$
\begin{aligned}
c_1 + c_2 \qquad\quad + 2c_4 &= 0 \\
c_1 \qquad\quad + 3c_3 + 6c_4 &= 0 \\
c_1 \qquad + c_3 + 4c_4 &= 0 \\
2c_1 + 2c_2 + 2c_3 + 6c_4 &= 0.
\end{aligned}
$$

The row reduced echelon form of the coefficient matrix is

$$\begin{bmatrix} 1 & 0 & 0 & 3 \\ 0 & 1 & 0 & -1 \\ 0 & 0 & 1 & 1 \\ 0 & 0 & 0 & 0 \end{bmatrix}.$$

The solution to the system is therefore $c_1 = -3r$, $c_2 = r$, $c_3 = -r$, $c_4 = r$, where r is any real number. Thus the matrices are linearly dependent. Setting $r = 1$ gives $c_1 = -3$, $c_2 = 1$, $c_3 = -1$, and $c_4 = 1$. Hence

$$-3 \begin{bmatrix} 1 & 1 \\ 1 & 2 \end{bmatrix} + \begin{bmatrix} 1 & 0 \\ 0 & 2 \end{bmatrix} - \begin{bmatrix} 0 & 3 \\ 1 & 2 \end{bmatrix} + \begin{bmatrix} 2 & 6 \\ 4 & 6 \end{bmatrix} = \begin{bmatrix} 0 & 0 \\ 0 & 0 \end{bmatrix}$$

from which it follows that

$$\begin{bmatrix} 2 & 6 \\ 4 & 6 \end{bmatrix} = 3 \begin{bmatrix} 1 & 1 \\ 1 & 2 \end{bmatrix} - \begin{bmatrix} 1 & 0 \\ 0 & 2 \end{bmatrix} + \begin{bmatrix} 0 & 3 \\ 1 & 2 \end{bmatrix}.$$

(b) Here we have

$$c_1 \begin{bmatrix} 1 & 1 \\ 1 & 1 \end{bmatrix} + c_2 \begin{bmatrix} 1 & 0 \\ 0 & 2 \end{bmatrix} + c_3 \begin{bmatrix} 0 & 1 \\ 0 & 2 \end{bmatrix} = \begin{bmatrix} 0 & 0 \\ 0 & 0 \end{bmatrix}.$$

Carrying out the matrix operations and equating entries leads to the linear system

$$
\begin{aligned}
c_1 + c_2 \qquad\quad &= 0 \\
c_1 \qquad + c_3 &= 0 \\
c_1 \qquad\qquad &= 0 \\
c_1 + 2c_2 + 2c_3 &= 0.
\end{aligned}
$$

The reduced row echelon form of the coefficient matrix is

$$\begin{bmatrix} 1 & 0 & 0 \\ 0 & 1 & 0 \\ 0 & 0 & 1 \\ 0 & 0 & 0 \end{bmatrix}.$$

Thus the system has only the trivial solution $c_1 = c_2 = c_3 = 0$ and therefore the given vectors are linearly independent.

(c) Here we have

$$c_1 \begin{bmatrix} 1 & 1 \\ 1 & 1 \end{bmatrix} + c_2 \begin{bmatrix} 2 & 3 \\ 1 & 2 \end{bmatrix} + c_3 \begin{bmatrix} 3 & 1 \\ 2 & 1 \end{bmatrix} + c_4 \begin{bmatrix} 2 & 2 \\ 1 & 1 \end{bmatrix} = \begin{bmatrix} 0 & 0 \\ 0 & 0 \end{bmatrix}.$$

Carrying out the matrix operations and equating entries leads to the linear system

$$c_1 + 2c_2 + 3c_3 + 2c_4 = 0$$
$$c_1 + 3c_2 + c_3 + 2c_4 = 0$$
$$c_1 + c_2 + 2c_3 + c_4 = 0$$
$$c_1 + 2c_2 + c_3 + c_4 = 0$$

The reduced row echelon form of the coefficient matrix is I_4. Thus the system has only the trivial solution and therefore the matrices are linearly independent.

15. We must determine conditions on c under which there are scalars a_1, a_2, and a_3, not all zero, such that

$$a_1(-1, 0, -1) + a_2(2, 1, 2) + a_3(1, 1, c) = (0, 0, 0).$$

This linear combination leads to the homogeneous system

$$-a_1 + 2a_2 + a_3 = 0$$
$$a_2 + a_3 = 0$$
$$-a_1 + 2a_2 + ca_3 = 0.$$

Carrying out row operations on this matrix leads to the matrix

$$\begin{bmatrix} 1 & 0 & 1 \\ 0 & 1 & 1 \\ 0 & 0 & c-1 \end{bmatrix}.$$

Clearly, the system has a nontrivial solution when $c - 1 = 0$, or equivalently, $c = 1$. Thus the vectors are linearly dependent when $c = 1$.

17. Following the technique of Example 1, determine whether constants c_1, c_2, c_3 exist such that

$$c_1 \begin{bmatrix} 1 \\ 1 \\ 0 \end{bmatrix} + c_2 \begin{bmatrix} 1 \\ 0 \\ 1 \end{bmatrix} + c_3 \begin{bmatrix} 1 \\ 1 \\ 1 \end{bmatrix} = \mathbf{v} = \begin{bmatrix} a \\ b \\ c \end{bmatrix}.$$

The corresponding linear system is

$$c_1 + c_2 + c_3 = a$$
$$c_1 + c_3 = b$$
$$c_2 + c_3 = c$$

which has augmented matrix

$$\begin{bmatrix} 1 & 1 & 1 & \vdots & a \\ 1 & 0 & 1 & \vdots & b \\ 0 & 1 & 1 & \vdots & c \end{bmatrix}.$$

The reduced row echelon form is

$$\begin{bmatrix} 1 & 0 & 0 & \vdots & a+c \\ 0 & 1 & 0 & \vdots & a+b+c+c \\ 0 & 0 & 1 & \vdots & a+b+c \end{bmatrix}.$$

The system is consistent for any choices of a, b, and c. Hence the given vectors space B^3.

19. Determine whether constants c_1, c_2, c_3, and c_4 exist such that

$$c_1 \begin{bmatrix} 1 \\ 1 \\ 0 \\ 0 \end{bmatrix} + c_2 \begin{bmatrix} 0 \\ 1 \\ 1 \\ 0 \end{bmatrix} + c_3 \begin{bmatrix} 0 \\ 0 \\ 1 \\ 1 \end{bmatrix} + c_4 \begin{bmatrix} 1 \\ 0 \\ 0 \\ 1 \end{bmatrix} = \mathbf{v} = \begin{bmatrix} a \\ b \\ c \\ d \end{bmatrix}.$$

The corresponding linear system is

$$\begin{aligned} c_1 \quad\quad\quad + c_4 &= a \\ c_1 + c_2 \quad\quad &= b \\ c_2 + c_3 \quad &= c \\ c_3 + c_4 &= d \end{aligned}$$

which has augmented matrix

$$\begin{bmatrix} 1 & 0 & 0 & 1 & \vdots & a \\ 1 & 1 & 0 & 0 & \vdots & b \\ 0 & 1 & 1 & 0 & \vdots & c \\ 0 & 0 & 1 & 1 & \vdots & d \end{bmatrix}.$$

This matrix is row equivalent to

$$\begin{bmatrix} 1 & 0 & 0 & 1 & \vdots & a \\ 0 & 1 & 0 & 1 & \vdots & a+b \\ 0 & 0 & 1 & 1 & \vdots & a+b+c \\ 0 & 0 & 0 & 0 & \vdots & a+b+c+d \end{bmatrix}.$$

We see that the system is inconsistent unless $a+b+c+d = 0$. There is not a solution for all choices of a, b, c, and d. Thus the vectors given do not span B^4.

21. Form the linear combination

$$c_1 \begin{bmatrix} 1 \\ 1 \\ 0 \\ 0 \end{bmatrix} + c_2 \begin{bmatrix} 0 \\ 1 \\ 1 \\ 0 \end{bmatrix} + c_3 \begin{bmatrix} 0 \\ 0 \\ 1 \\ 1 \end{bmatrix} + c_4 \begin{bmatrix} 1 \\ 0 \\ 0 \\ 1 \end{bmatrix} = \begin{bmatrix} 0 \\ 0 \\ 0 \\ 0 \end{bmatrix}.$$

The corresponding homogeneous system is

$$\begin{aligned} c_1 \quad\quad\quad + c_4 &= 0 \\ c_1 + c_2 \quad\quad &= 0 \\ c_2 + c_3 \quad &= 0 \\ c_3 + c_4 &= 0 \end{aligned}$$

The reduced row echelon form of the coefficient matrix is

$$\begin{bmatrix} 1 & 0 & 0 & 1 \\ 0 & 1 & 0 & 1 \\ 0 & 0 & 1 & 1 \\ 0 & 0 & 0 & 0 \end{bmatrix}.$$

Hence the solution of the homogeneous system is $c_1 = -r$, $c_2 = -r$, $c_3 = -r$, $c_4 = r$, where r is any real number. Thus the vectors are linearly dependent.

T.1. If $c_1\mathbf{e}_1 + c_2\mathbf{e}_2 + \cdots + c_n\mathbf{e}_n = (c_1, c_2, \ldots, c_n) = (0, 0, \ldots, 0) = \mathbf{0}$ in R^n, then $c_1 = c_2 = \cdots = c_n = 0$.

T.3. Assume that $S = \{\mathbf{v}_1, \mathbf{v}_2, \ldots, \mathbf{v}_k\}$ is linearly dependent. Then there are constants c_i, not all zero, such that

$$c_1\mathbf{v}_1 + c_2\mathbf{v}_2 + \cdots + c_k\mathbf{v}_k = \mathbf{0}.$$

Let c_j be a nonzero coefficient. Then, solving the equation for $\mathbf{v}_j$, we find that

$$\mathbf{v}_j = -\frac{c_1}{c_j}\mathbf{v}_1 - \frac{c_2}{c_j}\mathbf{v}_2 - \cdots - \frac{c_{j-1}}{c_j}\mathbf{v}_{j-1} - \frac{c_{j+1}}{c_j}\mathbf{v}_{j+1} - \cdots - \frac{c_k}{c_j}\mathbf{v}_k.$$

Conversely, if

$$\mathbf{v}_j = d_1\mathbf{v}_1 + d_2\mathbf{v}_2 + \cdots + d_{j-1}\mathbf{v}_{j-1} + d_{j+1}\mathbf{v}_{j+1} + \cdots + d_k\mathbf{v}_k$$

for some coefficients d_i, then

$$d_1\mathbf{v}_1 + d_2\mathbf{v}_2 + \cdots + (-1)\mathbf{v}_j + \cdots + d_k\mathbf{v}_k = \mathbf{0}$$

and the set S is linearly dependent.

T.5. Form the linear combination

$$c_1\mathbf{w}_1 + c_2\mathbf{w}_2 + c_3\mathbf{w}_3 = \mathbf{0}$$

which gives

$$c_1(\mathbf{v}_1 + \mathbf{v}_2) + c_2(\mathbf{v}_1 + \mathbf{v}_3) + c_3(\mathbf{v}_2 + \mathbf{v}_3) = (c_1 + c_2)\mathbf{v}_1 + (c_1 + c_3)\mathbf{v}_2 + (c_2 + c_3)\mathbf{v}_3 = \mathbf{0}.$$

Since S is linearly independent, we have

$$\begin{aligned} c_1 + c_2 \qquad\quad &= 0 \\ c_1 \qquad + c_3 &= 0 \\ c_2 + c_3 &= 0 \end{aligned}$$

a linear system whose augmented matrix is

$$\begin{bmatrix} 1 & 1 & 0 & \vdots & 0 \\ 1 & 0 & 1 & \vdots & 0 \\ 0 & 1 & 1 & \vdots & 0 \end{bmatrix}.$$

The reduced row echelon form is

$$\begin{bmatrix} 1 & 0 & 0 & \vdots & 0 \\ 0 & 1 & 0 & \vdots & 0 \\ 0 & 0 & 1 & \vdots & 0 \end{bmatrix}$$

thus $c_1 = c_2 = c_3 = 0$ which implies that $\{\mathbf{w}_1, \mathbf{w}_2, \mathbf{w}_3\}$ is linearly independent.

T.7. Suppose $\{\mathbf{v}_1, \mathbf{v}_2, \mathbf{v}_3\}$ is linearly dependent. Then one of the $\mathbf{v}_j$'s is a linear combination of the preceding vectors in the list (Theorem 6.4). It must be $\mathbf{v}_3$ since $\{\mathbf{v}_1, \mathbf{v}_2\}$ is linearly independent. Thus $\mathbf{v}_3$ belongs to span $\{\mathbf{v}_1, \mathbf{v}_2\}$. Contradiction.

T.9. Let $\mathbf{v}_i = \displaystyle\sum_{j=1}^{k} a_{ij}\mathbf{u}_j$ for $i = 1, 2, \ldots, m$. Then

$$\mathbf{w} = \sum_{i=1}^{m} b_i\mathbf{v}_i = \sum_{i=1}^{m} b_i \sum_{j=1}^{k} a_{ij}\mathbf{u}_j = \sum_{j=1}^{k} \left(\sum_{i=1}^{m} b_i a_{ij} \right) \mathbf{u}_j$$

is a linear combination of the vectors $\mathbf{u}_j$ in S.

T.11. Let $V = R^2$ and $S_2 = \{(0,0), (1,0), (0,1)\}$. Since S_2 contains the zero vector for R^2 it is linearly dependent. If $S_1 = \{(0,0), (1,0)\}$, it is linearly dependent for the same reason. If $S_1 = \{(1,0), (0,1)\}$, then

$$c_1(1,0) + c_2(0,1) = (0,0)$$

only if $c_1 = c_2 = 0$. Thus in this case S_1 is linearly independent.

T.13. If $\{\mathbf{u}, \mathbf{v}\}$ is linearly dependent, then there exist scalars c_1 and c_2, not both zero, such that $c_1\mathbf{u} + c_2\mathbf{v} = \mathbf{0}$. In fact, since neither $\mathbf{u}$ nor $\mathbf{v}$ is the zero vector, it follows that both c_1 and c_2 must be nonzero for $c_1\mathbf{u} + c_2\mathbf{v} = \mathbf{0}$. Hence we have $\mathbf{v} = -\dfrac{c_1}{c_2}\mathbf{u}$.

Alternatively, if $\mathbf{v} = k\mathbf{u}$, then $k \neq 0$ since $\mathbf{v} \neq \mathbf{0}$. Hence we have $\mathbf{v} - k\mathbf{u} = \mathbf{0}$ which implies that $\{\mathbf{u}, \mathbf{v}\}$ is linearly dependent.

T.15. Let $\mathbf{v} = a_1\mathbf{w}_1 + a_2\mathbf{w}_2 + \cdots + a_k\mathbf{w}_k$. Then

$$\begin{aligned}
\text{span}\{\mathbf{w}_1, \mathbf{w}_2, \ldots, \mathbf{w}_k, \mathbf{v}\} &= c_1\mathbf{w}_1 + c_2\mathbf{w}_2 + \cdots + c_k\mathbf{w}_k + c_{k+1}\mathbf{v} \\
&= c_1\mathbf{w}_1 + c_2\mathbf{w}_2 + \cdots + c_k\mathbf{w}_k + c_{k+1}(a_1\mathbf{w}_1 + a_2\mathbf{w}_2 + \cdots + a_k\mathbf{w}_k) \\
&= (c_1 + c_{k+1}a_1)\mathbf{w}_1 + (c_2 + c_{k+1}a_2)\mathbf{w}_2 + \cdots + (c_k + c_{k+1}a_k)\mathbf{w}_k \\
&= \text{span}\{\mathbf{w}_1, \mathbf{w}_2, \ldots, \mathbf{w}_k\} \\
&= W.
\end{aligned}$$

ML.1. In each case we form a linear combination of the vectors in A, set it equal to the zero vector, derive the associated linear system and find its reduced row echelon form.

(a) **v1 = [1 0 0 1];v2 = [0 1 1 0];v3 = [1 1 1 1];**
rref([v1′ v2′ v3′ zeros(4,1)])

ans =

```
    1   0   1   0
    0   1   1   0
    0   0   0   0
    0   0   0   0
```

This represents a homogeneous system with 2 equations in 3 unknowns, hence there is a nontrivial solution. Thus S is linearly dependent.

(b) **v1 = [1 2;1 0];v2 = [2 −1;1 2];v3 = [−3 1;0 1];**
rref([reshape(v1,4,1) reshape(v2,4,1) reshape(v3,4,1) zeros(4,1)])

ans =

```
    1   0   0   0
    0   1   0   0
    0   0   1   0
    0   0   0   0
```

The homogeneous system has only the trivial solution, hence S is linearly independent.

(c) **v1 = [1 2 1 0 1]; v2 = [0 1 2 −1 1]; v3 = [2 1 0 0 −1]; v4 = [−2 1 1 1 1];**
rref([v1′ v2′ v3′ v4′ zeros(5,1)])

```
ans =
    1   0   0   0   0
    0   1   0   0   0
    0   0   1   0   0
    0   0   0   1   0
    0   0   0   0   0
```

The homogeneous system has only the trivial solution, hence S is linearly independent.

Section 6.4, p. 314

1. Following the method of Example 2, we determine whether the sets are linearly independent and span R^2.

    (a) Let $\mathbf{v}_1 = (1, 3)$ and $\mathbf{v}_2 = (1, -1)$. Form the linear combination $c_1 \mathbf{v}_1 + c_2 \mathbf{v}_2 = \mathbf{0} = (0, 0)$. Adding vectors then equating corresponding components gives the linear system

    $$c_1 + c_2 = 0$$
    $$3c_1 - c_2 = 0.$$

    The coefficient matrix

    $$\begin{bmatrix} 1 & 1 \\ 3 & -1 \end{bmatrix}$$

    has reduced row echelon form I_2. Hence the only solution is the trivial solution $c_1 = c_2 = 0$ and $\{\mathbf{v}_1, \mathbf{v}_2\}$ is linearly independent. Let $\mathbf{v} = (a, b)$, where a and b are any real numbers. Form the linear combination $k_1 \mathbf{v}_1 + k_2 \mathbf{v}_2 = \mathbf{v}$. Adding vectors and equating corresponding components gives a linear system with augmented matrix

    $$\begin{bmatrix} 1 & 1 & \vdots & a \\ 3 & -1 & \vdots & b \end{bmatrix}$$

    which has reduced row echelon form

    $$\begin{bmatrix} 1 & 0 & \vdots & \frac{a+b}{4} \\ 0 & 1 & \vdots & \frac{3a-b}{4} \end{bmatrix}.$$

    It follows that there is a solution k_1, k_2 for any choices of a and b. Hence $\{\mathbf{v}_1, \mathbf{v}_2\}$ spans R^2. Thus $\{\mathbf{v}_1, \mathbf{v}_2\}$ is a basis for R^2.

    (b) Let $\mathbf{v}_1 = (0, 0)$, $\mathbf{v}_2 = (1, 2)$, and $\mathbf{v}_3 = (2, 4)$. Form the linear combination $c_1 \mathbf{v}_1 + c_2 \mathbf{v}_2 + c_3 \mathbf{v}_3 = \mathbf{0} = (0, 0)$. Adding vectors then equating corresponding components gives the linear system

    $$0c_1 + c_2 + 2c_3 = 0$$
    $$0c_1 + 2c_2 + 4c_3 = 0.$$

    The coefficient matrix

    $$\begin{bmatrix} 0 & 1 & 2 \\ 0 & 2 & 4 \end{bmatrix}$$

    has reduced row echelon form

    $$\begin{bmatrix} 0 & 1 & 2 \\ 0 & 0 & 0 \end{bmatrix}.$$

    Thus we have $c_1 = r$, $c_2 = -s$, $c_3 = s$, where r and s are arbitrary real numbers. Hence there are nontrivial solutions and $\{\mathbf{v}_1, \mathbf{v}_2, \mathbf{v}_3\}$ is linearly dependent. This set is not a basis for R^2.

(c) Let $\mathbf{v}_1 = (1,2)$, $\mathbf{v}_2 = (2,-3)$, and $\mathbf{v}_3 = (3,2)$. Form the linear combination $c_1\mathbf{v}_1+c_2\mathbf{v}_2+c_3\mathbf{v}_3 = \mathbf{0} = (0,0)$. Adding vectors then equating corresponding components gives the linear system

$$c_1 + 2c_2 + 3c_3 = 0$$
$$2c_1 - 3c_2 + 2c_3 = 0.$$

The coefficient matrix

$$\begin{bmatrix} 1 & 2 & 3 \\ 2 & -3 & 2 \end{bmatrix}$$

has reduced row echelon form

$$\begin{bmatrix} 1 & 0 & \frac{13}{7} \\ 0 & 1 & \frac{4}{7} \end{bmatrix}.$$

Hence $c_1 = -\frac{13}{7}r$, $c_2 = -\frac{4}{7}r$, $c_3 = r$, where r is any real number. Thus there are nontrivial solutions so the set is linearly dependent and is not a basis for R^2.

(d) Let $\mathbf{v}_1 = (1,3)$ and $\mathbf{v}_2 = (-2,6)$. Form the linear combination $c_1\mathbf{v}_1 + c_2\mathbf{v}_2 = \mathbf{0} = (0,0)$. Adding vectors then equating corresponding components gives the linear system

$$c_1 - 2c_2 = 0$$
$$3c_1 + 6c_2 = 0.$$

The coefficient matrix

$$\begin{bmatrix} 1 & -2 \\ 3 & 6 \end{bmatrix}$$

has reduced row echelon form I_2. Hence the only solution is the trivial solution $c_1 = c_2 = 0$ and $\{\mathbf{v}_1, \mathbf{v}_2\}$ is linearly independent. Let $\mathbf{v} = (a,b)$, where a and b are any real numbers. Form the linear combination $k_1\mathbf{v}_1 + k_2\mathbf{v}_2 = \mathbf{v}$. Adding vectors and equating corresponding components gives a linear system with augmented matrix

$$\begin{bmatrix} 1 & -2 & \vdots & a \\ 3 & 6 & \vdots & b \end{bmatrix}$$

which has reduced row echelon form

$$\begin{bmatrix} 1 & 0 & \vdots & \frac{3a+b}{6} \\ 0 & 1 & \vdots & \frac{b-3a}{12} \end{bmatrix}.$$

It follows that there is a solution k_1, k_2 for any choices of a and b. Hence $\{\mathbf{v}_1, \mathbf{v}_2\}$ spans R^2. Thus $\{\mathbf{v}_1, \mathbf{v}_2\}$ is a basis for R^2.

3. Following the method of Example 2, we determine whether the sets are linearly independent and span R^4.

(a) Let $\mathbf{v}_1 = (1,0,0,1)$, $\mathbf{v}_2 = (0,1,0,0)$, $\mathbf{v}_3 = (1,1,1,1)$, and $\mathbf{v}_4 = (0,1,1,1)$. Form the linear combination

$$c_1\mathbf{v}_1 + c_2\mathbf{v}_2 + c_3\mathbf{v}_3 + c_4\mathbf{v}_4 = \mathbf{0} = (0,0,0,0).$$

Adding vectors then equating corresponding components gives the linear system

$$c_1 \qquad + c_3 \qquad = 0$$
$$c_2 + c_3 + c_4 = 0$$
$$c_3 + c_4 = 0$$
$$c_1 \qquad + c_3 + c_4 = 0.$$

The coefficient matrix

$$\begin{bmatrix} 1 & 0 & 1 & 0 \\ 0 & 1 & 1 & 1 \\ 0 & 0 & 1 & 1 \\ 1 & 0 & 1 & 1 \end{bmatrix}$$

has reduced row echelon form I_4. Hence the only solution is the trivial solution $c_1 = c_2 = c_3 = c_4 = 0$ and $\{\mathbf{v}_1, \mathbf{v}_2, \mathbf{v}_3, \mathbf{v}_4\}$ is linearly independent. Let $\mathbf{v} = (a, b, c, d)$, where a, b, c, and d are any real numbers. Form the linear combination $k_1\mathbf{v}_1 + k_2\mathbf{v}_2 + k_3\mathbf{v}_3 + k_4\mathbf{v}_4 = \mathbf{v}$. Adding vectors and equating corresponding components gives a linear system with augmented matrix

$$\begin{bmatrix} 1 & 0 & 1 & 0 & \vdots & a \\ 0 & 1 & 1 & 1 & \vdots & b \\ 0 & 0 & 1 & 1 & \vdots & c \\ 1 & 0 & 1 & 1 & \vdots & d \end{bmatrix}$$

which has reduced row echelon form $\begin{bmatrix} I_4 & \vdots & \mathbf{q} \end{bmatrix}$, where the components of $\mathbf{q}$ are linear combinations of a, b, c, and d. It follows that there is a solution for any choice of a, b, c, and d. Hence $\{\mathbf{v}_1, \mathbf{v}_2, \mathbf{v}_3, \mathbf{v}_4\}$ spans R^4. Thus it follows that $\{\mathbf{v}_1, \mathbf{v}_2, \mathbf{v}_3, \mathbf{v}_4\}$ is a basis for R^4.

(b) Let $\mathbf{v}_1 = (1, -1, 0, 2)$, $\mathbf{v}_2 = (3, -1, 2, 1)$, and $\mathbf{v}_3 = (1, 0, 0, 1)$. Form the linear combination $c_1\mathbf{v}_1 + c_2\mathbf{v}_2 + c_3\mathbf{v}_3 = \mathbf{0} = (0, 0, 0, 0)$. Adding vectors then equating corresponding components gives the linear system

$$\begin{aligned} c_1 + 3c_2 + c_3 &= 0 \\ -c_1 - c_2 \phantom{{}+ c_3} &= 0 \\ 2c_2 \phantom{{}+ c_3} &= 0 \\ 2c_1 + c_2 + c_3 &= 0. \end{aligned}$$

The coefficient matrix

$$\begin{bmatrix} 1 & 3 & 1 \\ -1 & -1 & 0 \\ 0 & 2 & 0 \\ 2 & 1 & 1 \end{bmatrix}$$

has reduced row echelon form

$$\begin{bmatrix} 1 & 0 & 0 \\ 0 & 1 & 0 \\ 0 & 0 & 1 \\ 0 & 0 & 0 \end{bmatrix}.$$

Hence the only solution is the trivial solution $c_1 = c_2 = c_3 = 0$ and $\{\mathbf{v}_1, \mathbf{v}_2, \mathbf{v}_3\}$ is linearly independent. Let $\mathbf{v} = (a, b, c, d)$, where a, b, c, and d are any real numbers. Form the linear combination $k_1\mathbf{v}_1 + k_2\mathbf{v}_2 + k_3\mathbf{v}_3 = \mathbf{v}$. Adding vectors and equating corresponding components gives a linear system with augmented matrix

$$\begin{bmatrix} 1 & 3 & 1 & \vdots & a \\ -1 & -1 & 0 & \vdots & b \\ 0 & 2 & 0 & \vdots & c \\ 2 & 1 & 1 & \vdots & d \end{bmatrix}$$

which has reduced row echelon form in which the last row is

$$\begin{bmatrix} 0 & 0 & 0 & \vdots & -2a + 2b + 3c + 2d \end{bmatrix}.$$

It follows that there is not a solution k_1, k_2, and k_3 for all choices of a, b, c, and d. Hence $\{\mathbf{v}_1, \mathbf{v}_2, \mathbf{v}_3\}$ does not span R^4. Thus $\{\mathbf{v}_1, \mathbf{v}_2, \mathbf{v}_3\}$ is not a basis for R^4.

(c) Let

$$\mathbf{v}_1 = (-2, 4, 6, 4), \qquad \mathbf{v}_2 = (0, 1, 2, 0),$$
$$\mathbf{v}_3 = (-1, 2, 3, 2), \qquad \mathbf{v}_4 = (-3, 2, 5, 6), \quad \text{and} \quad \mathbf{v}_5 = (-2, -1, 0, 4).$$

Form the linear combination

$$c_1\mathbf{v}_1 + c_2\mathbf{v}_2 + c_3\mathbf{v}_3 + c_4\mathbf{v}_4 + c_5\mathbf{v}_5 = \mathbf{0} = (0, 0, 0, 0).$$

Adding vectors then equating corresponding components gives the linear system

$$\begin{array}{rcrcrcrcrcl}
-2c_1 & & & - & c_3 & - & 3c_4 & - & 2c_5 & = & 0 \\
4c_1 & + & c_2 & + & 2c_3 & + & 2c_4 & - & c_5 & = & 0 \\
6c_1 & + & 2c_2 & + & 3c_3 & + & 5c_4 & & & = & 0 \\
4c_1 & & & + & 2c_3 & + & 6c_4 & + & 4c_5 & = & 0.
\end{array}$$

By Theorem 1.8, this homogeneous system of 4 equations in 5 unknowns has a nontrivial solution. Thus $\{\mathbf{v}_1, \mathbf{v}_2, \mathbf{v}_3, \mathbf{v}_4, \mathbf{v}_5\}$ is linearly dependent and is not a basis for R^4.

(d) Let

$$\mathbf{v}_1 = (0, 0, 1, 1), \quad \mathbf{v}_2 = (-1, 1, 1, 2), \quad \mathbf{v}_3 = (1, 1, 0, 0), \quad \text{and} \quad \mathbf{v}_4 = (2, 1, 2, 1).$$

Form the linear combination

$$c_1\mathbf{v}_1 + c_2\mathbf{v}_2 + c_3\mathbf{v}_3 + c_4\mathbf{v}_4 = \mathbf{0} = (0, 0, 0, 0).$$

Adding vectors then equating corresponding components gives the linear system

$$\begin{array}{rcrcrcrcl}
 & - & c_2 & + & c_3 & + & 2c_4 & = & 0 \\
 & & c_2 & + & c_3 & + & c_4 & = & 0 \\
c_1 & + & c_2 & & & + & 2c_4 & = & 0 \\
c_1 & + & 2c_2 & & & + & c_4 & = & 0.
\end{array}$$

The coefficient matrix

$$\begin{bmatrix} 0 & -1 & 1 & 2 \\ 0 & 1 & 1 & 1 \\ 1 & 1 & 0 & 2 \\ 1 & 2 & 0 & 1 \end{bmatrix}$$

has reduced row echelon form I_4. Hence the only solution is the trivial solution $c_1 = c_2 = c_3 = c_4 = 0$ and $\{\mathbf{v}_1, \mathbf{v}_2, \mathbf{v}_3, \mathbf{v}_4\}$ is linearly independent. Let $\mathbf{v} = (a, b, c, d)$, where a, b, c, and d are any real numbers. Form the linear combination

$$k_1\mathbf{v}_+ k_2\mathbf{v}_2 + k_3\mathbf{v}_3 + k_4\mathbf{v}_4 = \mathbf{v}.$$

Adding vectors and equating corresponding components gives a linear system with augmented matrix

$$\left[\begin{array}{cccc:c} 0 & -1 & 1 & 2 & a \\ 0 & 1 & 1 & 1 & b \\ 1 & 1 & 0 & 2 & c \\ 1 & 2 & 0 & 1 & d \end{array}\right]$$

which has reduced row echelon form $\left[I_4 \vdots \mathbf{q}\right]$, where $\mathbf{q}$ is a linear combination of a, b, c, and d. It follows that there is a solution k_1, k_2, k_3, and k_4 for any choices of a, b, c, and d. Hence $\{\mathbf{v}_1, \mathbf{v}_2, \mathbf{v}_3, \mathbf{v}_4\}$ spans R^4 and is a basis for R^4.

5. We determine whether the sets are linearly independent and span P_3.

   (a) Let

   $$p_1(t) = t^3 + 2t^2 + 3t, \quad p_2(t) = 2t^3 + 1,$$
   $$p_3(t) = 6t^3 + 8t^2 + 6t + 4, \quad \text{and} \quad p_4(t) = t^3 + 2t^2 + t + 1.$$

   Form the linear combination

   $$c_1 p_1(t) + c_2 p_2(t) + c_3 p_3(t) + c_4 p_4(t) = \mathbf{0} = 0t^3 + 0t^2 + 0t + 0.$$

   Expanding and combining terms with like powers of t, we obtain

   $$(c_1 + 2c_2 + 6c_3 + c_4)t^3 + (2c_1 + 8c_3 + 2c_4)t^2 + (3c_1 + 6c_3 + c_4)t + (c_2 + 4c_3 + c_4) = \mathbf{0}.$$

   Equating coefficients of like powers of t from each side of the equation we obtain the linear system

   $$
   \begin{aligned}
   c_1 + 2c_2 + 6c_3 + \;\; c_4 &= 0 \\
   2c_1 \qquad\quad + 8c_3 + 2c_4 &= 0 \\
   3c_1 \qquad\quad + 6c_3 + \;\; c_4 &= 0 \\
   c_2 + 4c_3 + \;\; c_4 &= 0.
   \end{aligned}
   $$

   The coefficient matrix of this homogeneous system is

   $$
   \begin{bmatrix}
   1 & 2 & 6 & 1 \\
   2 & 0 & 8 & 2 \\
   3 & 0 & 6 & 1 \\
   0 & 1 & 4 & 1
   \end{bmatrix}.
   $$

   Applying row operations $-2\mathbf{r}_1 + \mathbf{r}_2$, $-3\mathbf{r}_1 + \mathbf{r}_3$, $\left(-\frac{1}{4}\right)\mathbf{r}_2$, $6\mathbf{r}_2 + \mathbf{r}_3$, $-\mathbf{r}_2 + \mathbf{r}_4$, $\left(-\frac{1}{6}\right)\mathbf{r}_3$, $-3\mathbf{r}_3 + \mathbf{r}_4$, $-\mathbf{r}_3 + \mathbf{r}_2$, $-6\mathbf{r}_3 + \mathbf{r}_1$, $-2\mathbf{r}_1 + \mathbf{r}_2$ gives the reduced row echelon form

   $$
   \begin{bmatrix}
   1 & 0 & 0 & -\frac{1}{3} \\
   0 & 1 & 0 & -\frac{1}{3} \\
   0 & 0 & 1 & \frac{1}{3} \\
   0 & 0 & 0 & 0
   \end{bmatrix}.
   $$

   Thus the general solution is $c_1 = \frac{1}{3}r$, $c_2 = \frac{1}{3}r$, $c_3 = -\frac{1}{3}r$, $c_4 = r$, where r is any real number. Hence the vectors are linearly dependent and are not a basis for P_3.

   (b) Let

   $$p_1(t) = t^3 + t^2 + 1, \quad p_2(t) = t^3 - 1, \quad \text{and} \quad p_3(t) = t^3 + t^2 + t.$$

   Form the linear combination

   $$c_1 p_1(t) + c_2 p_2(t) + c_3 p_3(t) = \mathbf{0} = 0t^3 + 0t^2 + 0t + 0.$$

   Expanding and combining terms with like powers of t we obtain

   $$(c_1 + c_2 + c_3)t^3 + (c_1 + c_3)t^2 + c_3 t + (c_1 - c_2) = \mathbf{0}.$$

Equating coefficients of like powers of t from each side of the equation we obtain the linear system

$$c_1 + c_2 + c_3 = 0$$
$$c_1 \quad + c_3 = 0$$
$$c_3 = 0$$
$$c_1 - c_2 \quad = 0.$$

The coefficient matrix of this homogeneous system is

$$\begin{bmatrix} 1 & 1 & 1 \\ 1 & 0 & 1 \\ 0 & 0 & 1 \\ 1 & -1 & 0 \end{bmatrix}.$$

Applying row operations $-\mathbf{r}_1 + \mathbf{r}_2$, $-\mathbf{r}_1 + \mathbf{r}_4$, $(-1)\mathbf{r}_2$, , $2\mathbf{r}_2 + \mathbf{r}_4$, $-\mathbf{r}_3 + \mathbf{r}_1$, $\mathbf{r}_3 + \mathbf{r}_4$, $-\mathbf{r}_2 + \mathbf{r}_1$ gives the reduced row echelon form

$$\begin{bmatrix} 1 & 0 & 0 \\ 0 & 1 & 0 \\ 0 & 0 & 1 \\ 0 & 0 & 0 \end{bmatrix}.$$

Hence the only solution is the trivial solution and the vectors are linearly independent. Let $p(t) = at^3 + bt^2 + ct + d$, where a, b, c, and d are arbitrary real numbers. Form the linear combination

$$k_1 p_1(t) + k_2 p_2(t) + k_3 p_3(t) = p(t).$$

Expanding, combining terms with like powers of t, and equating coefficients of like powers from both sides of the equation we obtain the linear system

$$k_1 + k_2 + k_3 = a$$
$$k_1 \quad + k_3 = b$$
$$k_3 = c$$
$$k_1 - k_2 \quad = d.$$

Applying the row operations given above, the reduced row echelon form is

$$\begin{bmatrix} 1 & 0 & 0 & \vdots & b - c \\ 0 & 1 & 0 & \vdots & a - b \\ 0 & 0 & 1 & \vdots & c \\ 0 & 0 & 0 & \vdots & a - 2b + c + d \end{bmatrix}.$$

It follows that the system is inconsistent for certain choices of a, b, c, and d. Hence the vectors do not span P_3 and are not a basis for P_3.

(c) Let

$$p_1(t) = t^3 + t^2 + t + 1, \quad p_2(t) = t^3 + 2t^2 + t + 3,$$
$$p_3(t) = 2t^3 + t^2 + 3t + 2, \quad \text{and} \quad p_4(t) = t^3 + t^2 + 2t + 2.$$

Form the linear combination

$$c_1 p_1(t) + c_2 p_2(t) + c_3 p_3(t) + c_4 p_4(t) = \mathbf{0} = 0t^3 + 0t^2 + 0t + 0.$$

Expanding and combining terms with like powers of t, we obtain

$$(c_1 + c_2 + 2c_3 + c_4)t^3 + (c_1 + 2c_2 + c_3 + c_4)t^2 + (c_1 + c_2 + 3c_3 + 2c_4)t + (c_1 + 3c_2 + 2c_3 + 2c_4) = \mathbf{0}.$$

Equating coefficients of like powers of t from each side of the equation we obtain the linear system

$$
\begin{aligned}
c_1 + c_2 + 2c_3 + c_4 &= 0 \\
c_1 + 2c_2 + c_3 + c_4 &= 0 \\
c_1 + c_2 + 3c_3 + 2c_4 &= 0 \\
c_1 + 3c_2 + 2c_3 + 2c_4 &= 0.
\end{aligned}
$$

The coefficient matrix of this homogeneous system has reduced row echelon form I_4. Hence the only solution is the trivial solution and the vectors are linearly independent. Let $p(t) = at^3 + bt^2 + ct + d$, where a, b, c, and d are arbitrary real numbers. Form the linear combination

$$k_1 p_1(t) + k_2 p_2(t) + k_3 p_3(t) + k_4 p_4(t) = p(t).$$

Expanding, combining terms with like powers of t, and equating coefficients of like powers from both sides of the equation we obtain the linear system

$$
\begin{aligned}
k_1 + k_2 + 2k_3 + k_4 &= a \\
k_1 + 2k_2 + k_3 + k_4 &= b \\
k_1 + k_2 + 3k_3 + 2k_4 &= c \\
k_1 + 3k_2 + 2k_3 + 2k_4 &= d.
\end{aligned}
$$

The reduced row echelon form of the coefficient matrix has the form $\begin{bmatrix} I_4 \vdots \mathbf{q} \end{bmatrix}$, where the entries of $\mathbf{q}$ are linear combinations of a, b, c, and d. Hence the system is consistent for all choices of a, b, c, and d. Thus the vectors span P_3. Since the vectors are both linearly independent and span, they are a basis for P_3.

(d) Let

$$p_1(t) = t^3 - t, \quad p_2(t) = t^3 + t^2 + 1, \quad \text{and} \quad p_3(t) = t - 1.$$

Form the linear combination

$$c_1 p_1(t) + c_2 p_2(t) + c_3 p_3(t) = \mathbf{0} = 0t^3 + 0t^2 + 0t + 0.$$

Expanding and combining terms with like powers of t, we obtain

$$(c_1 + c_2)t^3 + c_2 t^2 + (-c_1 + c_3)t + (c_2 - c_3) = \mathbf{0}.$$

Equating coefficients of like powers of t from both sides of the equation we obtain the linear system

$$
\begin{aligned}
c_1 + c_2 \quad\quad &= 0 \\
c_2 \quad\quad &= 0 \\
-c_1 \quad\quad + c_3 &= 0 \\
c_2 - c_3 &= 0.
\end{aligned}
$$

The coefficient matrix of this homogeneous system is

$$
\begin{bmatrix}
1 & 1 & 0 \\
0 & 1 & 0 \\
-1 & 0 & 1 \\
0 & 1 & -1
\end{bmatrix}.
$$

Applying row operations $\mathbf{r}_1 + \mathbf{r}_3$, $-\mathbf{r}_2 + \mathbf{r}_3$, $-\mathbf{r}_2 + \mathbf{r}_4$, $\mathbf{r}_3 + \mathbf{r}_4$, $-\mathbf{r}_2 + \mathbf{r}_1$ gives the reduced row echelon form

$$\begin{bmatrix} 1 & 0 & 0 \\ 0 & 1 & 0 \\ 0 & 0 & 1 \\ 0 & 0 & 0 \end{bmatrix}.$$

Hence the only solution is the trivial solution and the vectors are linearly independent. Let $p(t) = at^3 + bt^2 + ct + d$, where a, b, c, and d are arbitrary real numbers. Form the linear combination

$$k_1 p_1(t) + k_2 p_2(t) + k_3 p_3(t) = p(t).$$

Expanding, combining terms with like powers of t, and equating coefficients of like powers from both sides of the equation, we obtain the linear system

$$\begin{array}{rcl} k_1 + k_2 & = & a \\ k_2 & = & b \\ -k_1 \qquad + k_3 & = & c \\ k_2 - k_3 & = & d. \end{array}$$

Applying the row operations given above we obtain the reduced row echelon form of the associated augmented matrix as

$$\begin{bmatrix} 1 & 0 & 0 & \vdots & a - b \\ 0 & 1 & 0 & \vdots & b \\ 0 & 0 & 1 & \vdots & a - b + c \\ 0 & 0 & 0 & \vdots & a - 2b + c + d \end{bmatrix}.$$

It follows that there is not a solution for all choices of a, b, c, and d. Hence the vectors do not span P_3 and are not a basis for P_3.

7. Let $V = R^3$.

(a) Let $S = \{(1,1,1),(1,2,3),(0,1,0)\}$. Form a linear combination of the three vectors in S and set it equal to an arbitrary vector in V:

$$c_1(1,1,1) + c_2(1,2,3) + c_3(0,1,0) = (a,b,c).$$

Carrying out the operations on the left-hand side of this equation and equating corresponding entries leads to the linear system

$$\begin{array}{rcl} c_1 + c_2 & = & a \\ c_1 + 2c_2 + c_3 & = & b \\ c_1 + 3c_2 & = & c. \end{array}$$

The reduced row echelon form of the augmented matrix of this system has the form $\begin{bmatrix} I_3 & \vdots & \mathbf{q} \end{bmatrix}$, where the components of $\mathbf{q}$ are linear combinations of a, b, and c. Hence S spans V. If $a = b = c = 0$, it follows that the only solution is the trivial solution so S is linearly independent. Thus S is a basis for V. To write the vector $(2,1,3)$ in terms of the basis we set $a = 2$, $b = 1$ and $c = 3$ to get the augmented matrix

$$\begin{bmatrix} 1 & 1 & 0 & \vdots & 2 \\ 1 & 2 & 1 & \vdots & 1 \\ 1 & 3 & 0 & \vdots & 3 \end{bmatrix}$$

The reduced row echelon form of this matrix is

$$\begin{bmatrix} 1 & 0 & 0 & \vdots & \frac{3}{2} \\ 0 & 1 & 0 & \vdots & \frac{1}{2} \\ 0 & 0 & 1 & \vdots & -\frac{3}{2} \end{bmatrix}.$$

Thus the solution is $c_1 = \frac{3}{2}$, $c_2 = \frac{1}{2}$, $c_3 = -\frac{3}{2}$. It follows that

$$(2, 1, 3) = \tfrac{3}{2}(1, 1, 1) + \tfrac{1}{2}(1, 2, 3) - \tfrac{3}{2}(0, 1, 0).$$

(b) Let $S = \{(1, 2, 3), (2, 1, 3), (0, 0, 0)\}$. From Section 6.3, any set of vectors that contains the zero vector is linearly dependent. Hence S cannot be a basis.

9. Let $V = P_2$.

(a) Let $S = \{t^2 + t, t - 1, t + 1\}$. Form a linear combination of the vectors in S and set it equal to an arbitrary vector in V:

$$c_1(t^2 + t) + c_2(t - 1) + c_3(t + 1) = at^2 + bt + c.$$

Expanding, combining terms with like powers of t, and equating coefficients of like powers from both sides of the equation, we obtain the linear system

$$\begin{aligned} c_1 \qquad\qquad &= a \\ c_1 + c_2 + c_3 &= b \\ -c_2 + c_3 &= c. \end{aligned}$$

The reduced row echelon form of this matrix has the form $\begin{bmatrix} I_3 & \vdots & \mathbf{q} \end{bmatrix}$, where the entries of $\mathbf{q}$ are linear combinations of a, b, and c. Thus the system has a solution for any choice of a, b, and c, and therefore S spans P_2. If $a = b = c = 0$, the only solution of the system is the trivial solution so S is linearly independent. Thus S is a basis for P_2. To write $5t^2 - 3t + 8$ in terms of the basis, we set $a = 5$, $b = -3$, and $c = 8$ to get the matrix

$$\begin{bmatrix} 1 & 0 & 0 & \vdots & 5 \\ 1 & 1 & 1 & \vdots & -3 \\ 0 & -1 & 1 & \vdots & 8 \end{bmatrix}$$

whose reduced row echelon form is

$$\begin{bmatrix} 1 & 0 & 0 & \vdots & 5 \\ 0 & 1 & 0 & \vdots & -8 \\ 0 & 0 & 1 & \vdots & 0 \end{bmatrix}.$$

The solution is $c_1 = 5$, $c_2 = -8$, $c_3 = 0$. Therefore

$$5t^2 - 3t + 8 = 5(t^2 + t) - 8(t - 1).$$

(b) Let $S = \{t^2 + 1, t - 1\}$. Since a natural basis for V is $\{t^2, t, 1\}$, any basis must contain 3 vectors. Since S has only 2 elements, it cannot be a basis for V.

11. We follow the technique given in Example 5. Let $S = \{\mathbf{v}_1, \mathbf{v}_2, \mathbf{v}_3, \mathbf{v}_4\}$, where

$$\mathbf{v}_1 = (1, 2, 2), \quad \mathbf{v}_2 = (3, 2, 1), \quad \mathbf{v}_3 = (11, 10, 7), \quad \text{and} \quad \mathbf{v}_4 = (7, 6, 4).$$

Form the linear combination

$$c_1\mathbf{v}_1 + c_2\mathbf{v}_2 + c_3\mathbf{v}_3 + c_4\mathbf{v}_4 = \mathbf{0} = (0, 0, 0).$$

Expanding, adding vectors, and equating corresponding components gives the linear system

$$c_1 + 3c_2 + 11c_3 + 7c_4 = 0$$
$$2c_1 + 2c_2 + 10c_3 + 6c_4 = 0$$
$$2c_1 + c_2 + 7c_3 + 4c_4 = 0.$$

The coefficient matrix of this homogeneous system is

$$\begin{bmatrix} 1 & 3 & 11 & 7 \\ 2 & 2 & 10 & 6 \\ 2 & 1 & 7 & 4 \end{bmatrix}.$$

Applying row operations $-2\mathbf{r}_1 + \mathbf{r}_2$, $-2\mathbf{r}_1 + \mathbf{r}_3$, $\left(-\frac{1}{4}\right)\mathbf{r}_2$, $-5\mathbf{r}_2 + \mathbf{r}_3$, $-3\mathbf{r}_2 + \mathbf{r}_1$ gives reduced row echelon form

$$\begin{bmatrix} 1 & 0 & 2 & 1 \\ 0 & 1 & 3 & 2 \\ 0 & 0 & 0 & 0 \end{bmatrix}.$$

The leading 1's appear in columns 1 and 2, so $\{\mathbf{v}_1, \mathbf{v}_2\}$ is a basis for $W = \text{span } S$. Hence $\dim W = 2$.

13. We follow the technique given in Example 5. Let $S = \{p_1(t), p_2(t), p_3(t), p_4(t)\}$, where

$$p_1(t) = t^3 + t^2 - 2t + 1, \quad p_2(t) = t^2 + 1, \quad p_3(t) = t^3 - 2t, \quad \text{and} \quad p_4(t) = 2t^3 + t^2 - 4t + 3.$$

Form the linear combination

$$c_1 p_1(t) + c_2 p_2(t) + c_3 p_3(t) + c_4 p_4(t) = \mathbf{0}.$$

Expanding and adding like terms gives

$$(c_1 + c_3 + 2c_4)t^3 + (c_1 + c_2 + 3c_4)t^2 + (-2c_1 - 2c_3 - 4c_4)t + (c_1 + c_2 + 3c_4) = 0t^3 + 0t^2 + 0t + 0.$$

Equating coefficients of like powers of t on each side of the equation gives the linear system

$$c_1 + c_3 + 2c_4 = 0$$
$$c_1 + c_2 + 3c_4 = 0$$
$$-2c_1 - 2c_3 - 4c_4 = 0$$
$$c_1 + c_2 + 3c_4 = 0.$$

The coefficient matrix of this homogeneous system is

$$\begin{bmatrix} 1 & 0 & 1 & 2 \\ 1 & 1 & 0 & 3 \\ -2 & 0 & -2 & -4 \\ 1 & 1 & 0 & 3 \end{bmatrix}.$$

Applying row operations $-\mathbf{r}_1 + \mathbf{r}_2$, $2\mathbf{r}_1 + \mathbf{r}_3$, $-\mathbf{r}_1 + \mathbf{r}_4$, $-\mathbf{r}_2 + \mathbf{r}_4$, gives the reduced row echelon form

$$\begin{bmatrix} 1 & 0 & 1 & 2 \\ 0 & 1 & -1 & 1 \\ 0 & 0 & 0 & 0 \\ 0 & 0 & 0 & 0 \end{bmatrix}.$$

The leading 1's appear in columns 1 and 2, so $\{p_1(t), p_2(t)\}$ is a basis for $W = \text{span } S$. Hence $\dim W = 2$.

15. The general vector in M_{23} has the form

$$\begin{bmatrix} a & b & c \\ d & e & f \end{bmatrix}$$

which can be written as

$$\begin{bmatrix} a & b & c \\ d & e & f \end{bmatrix} = a\begin{bmatrix} 1 & 0 & 0 \\ 0 & 0 & 0 \end{bmatrix} + b\begin{bmatrix} 0 & 1 & 0 \\ 0 & 0 & 0 \end{bmatrix} + c\begin{bmatrix} 0 & 0 & 1 \\ 0 & 0 & 0 \end{bmatrix}$$
$$+ d\begin{bmatrix} 0 & 0 & 0 \\ 1 & 0 & 0 \end{bmatrix} + e\begin{bmatrix} 0 & 0 & 0 \\ 0 & 1 & 0 \end{bmatrix} + f\begin{bmatrix} 0 & 0 & 0 \\ 0 & 0 & 1 \end{bmatrix}.$$

Hence the vectors

$$\begin{bmatrix} 1 & 0 & 0 \\ 0 & 0 & 0 \end{bmatrix}, \begin{bmatrix} 0 & 1 & 0 \\ 0 & 0 & 0 \end{bmatrix}, \begin{bmatrix} 0 & 0 & 1 \\ 0 & 0 & 0 \end{bmatrix},$$
$$\begin{bmatrix} 0 & 0 & 0 \\ 1 & 0 & 0 \end{bmatrix}, \begin{bmatrix} 0 & 0 & 0 \\ 0 & 1 & 0 \end{bmatrix}, \begin{bmatrix} 0 & 0 & 0 \\ 0 & 0 & 1 \end{bmatrix}$$

span M_{23}. Forming the equation

$$c_1\begin{bmatrix} 1 & 0 & 0 \\ 0 & 0 & 0 \end{bmatrix} + c_2\begin{bmatrix} 0 & 1 & 0 \\ 0 & 0 & 0 \end{bmatrix} + c_3\begin{bmatrix} 0 & 0 & 1 \\ 0 & 0 & 0 \end{bmatrix}$$
$$+ c_4\begin{bmatrix} 0 & 0 & 0 \\ 1 & 0 & 0 \end{bmatrix} + c_5\begin{bmatrix} 0 & 0 & 0 \\ 0 & 1 & 0 \end{bmatrix} + c_6\begin{bmatrix} 0 & 0 & 0 \\ 0 & 0 & 1 \end{bmatrix} = \begin{bmatrix} 0 & 0 & 0 \\ 0 & 0 & 0 \end{bmatrix},$$

we obtain a homogeneous system of 6 equations in 6 unknowns, which is easily seen to have only the trivial solution. Hence the vectors given above are linearly independent and form a basis for M_{23}. Thus M_{23} has dimension 6. Generalizing, we find that M_{mn} has dimension mn.

17. Let W be the given subspace.

(a) Let (a, b, c) be an arbitrary element in W, where $b = a + c$. Then

$$(a, b, c) = (a, a + c, c) = a(1, 1, 0,) + c(0, 1, 1).$$

It follows that the set

$$S = \{(1, 1, 0), (0, 1, 1)\}$$

spans the subspace W. This set is linearly independent; for if $a = b = c = 0$ on the left side of the above equation, then $a = c = 0$. Thus S is linearly independent and spans W and is therefore a basis for W.

(b) Let (a, b, c) be an arbitrary element in W, where $b = a$. Then

$$(a, b, c) = (a, a, c) = a(1, 1, 0) + c(0, 0, 1).$$

It follows that the set

$$S = \{(1, 1, 0), (0, 0, 1)\}$$

spans the subspace W. This set is linearly independent; for if $a = b = c = 0$ on the left side of the above equation, then $a = c = 0$. Thus S is linearly independent and spans W and is therefore a basis for W.

(c) Let (a, b, c) be an arbitrary element in W, where $2a + b - c = 0$. Then $b = -2a + c$ and hence

$$(a, b, c) = (a, -2a + c, c) = a(1, -2, 0) + c(0, 1, 1).$$

It follows that the set

$$S = \{(1, -2, 0), (0, 1, 1)\}$$

spans the subspace W. This set is linearly independent; for if $a = b = c = 0$ on the left side of the above equation, then $a = c = 0$. Thus S is linearly independent and spans W and is therefore a basis for W.

19. We find a basis for each subspace W of R^4.

   (a) W consists of all vectors of the form $(a, b, c, a + b)$. Now,

$$(a, b, c, a + b) = a(1, 0, 0, 1) + b(0, 1, 0, 1) + c(0, 0, 1, 0).$$

   Let

$$S = \{(1, 0, 0, 1), (0, 1, 0, 1), (0, 0, 1, 0)\}.$$

   Then the preceding relation implies that S spans W. This set is linearly independent; for if

$$a(1, 0, 0, 1) + b(0, 1, 0, 1) + c(0, 0, 1, 0) = (a, b, c, a + b) = (0, 0, 0, 0),$$

   then $a = b = c = 0$. Thus S is a basis for W and therefore dim $W = 3$.

   (b) W consists of all vectors of the form $(a, b, a - b, a + b)$. Now

$$(a, b, a - b, a + b) = a(1, 0, 1, 1) + b(0, 1, 1, 1).$$

   Let

$$S = \{(1, 0, 1, 1), (0, 1, -1, 1)\}.$$

   Then the preceding relation implies that S spans W. This set is linearly independent; for if

$$a(1, 0, 1, 1) + b(0, 1, -1, 1) = (a, b, a - b, a + b) = (0, 0, 0, 0),$$

   then $a = b = 0$. thus S spans W and is linearly independent. Thus S is a basis for W and therefore dim $W = 2$.

21. The subspace W of P_2 consists of all vectors of the form $at^2 + bt + 2a - 3b$. Now,

$$at^2 + bt + 2a - 3b = a(t^2 + 2) + b(t - 3).$$

Thus $S = \{t^2 + 2, t - 3\}$ spans W. To see if S is linearly independent we form

$$c_1(t^2 + 2) + c_2(t - 3) = 0.$$

Equating like powers of t leads to the homogeneous system

$$
\begin{aligned}
c_1 \qquad\qquad &= 0 \\
c_2 &= 0 \\
2c_1 - 3c_2 &= 0
\end{aligned}
$$

whose solution is $c_1 = c_2 = 0$. Hence S is linearly independent and is then a basis for W.

23. (a) In Exercise 1(a) it is shown that $\{(1,3),(1,-1)\}$ is a basis for R^2. Thus

$$W = \text{span } \{(1,3),(1,-1)\} = R^2$$

and hence $\dim W = 2$.

(b) Let $\mathbf{v}_1 = (0,0)$, $\mathbf{v}_2 = (1,2)$, $\mathbf{v}_3 = (2,4)$, $S = \{\mathbf{v}_1,\mathbf{v}_2,\mathbf{v}_3\}$, and $W = \text{span } S$. Following the technique in Example 5, we determine a subset of S that is a basis for W. Form the linear combination

$$c_1\mathbf{v}_1 + c_2\mathbf{v}_2 + c_3\mathbf{v}_3 = \mathbf{0} = (0,0).$$

Expanding, adding vectors, and equating corresponding components gives the linear system

$$\begin{aligned} 0c_1 + c_2 + 2c_3 &= 0 \\ 0c_1 + 2c_2 + 4c_3 &= 0. \end{aligned}$$

The reduced row echelon form of the coefficient matrix of this homogeneous system is

$$\begin{bmatrix} 0 & 1 & 2 \\ 0 & 0 & 0 \end{bmatrix}.$$

The leading 1 appears in column 2, so $\{\mathbf{v}_2\}$ is a basis for W. Thus $\dim W = 1$.

(c) Let $\mathbf{v}_1 = (1,2)$, $\mathbf{v}_2 = (2,-3)$, $\mathbf{v}_3 = (3,2)$, $S = \{\mathbf{v}_1,\mathbf{v}_2,\mathbf{v}_3\}$, and $W = \text{span } S$. Following the technique in Example 5, we determine a subset of S that is a basis for W. Form the linear combination

$$c_1\mathbf{v}_1 + c_2\mathbf{v}_2 + c_3\mathbf{v}_3 = \mathbf{0} = (0,0).$$

Expanding, adding vectors, and equating corresponding components gives the linear system

$$\begin{aligned} c_1 + 2c_2 + 3c_3 &= 0 \\ 2c_1 - 3c_2 + 2c_3 &= 0. \end{aligned}$$

The reduced row echelon form of the coefficient matrix of this homogeneous system is

$$\begin{bmatrix} 1 & 0 & \frac{13}{7} \\ 0 & 1 & \frac{1}{7} \end{bmatrix}.$$

The leading 1's appear in columns 1 and 2, so $\{\mathbf{v}_1,\mathbf{v}_2\}$ is a basis for W. Thus $\dim W = 2$.

(d) In Exercise 1(d) it is shown that $\{(1,3),(-2,6)\}$ is a basis for R^2. Thus

$$W = \text{span } \{(1,3),(-2,6)\} = R^2 \quad \text{and} \quad \dim W = 2.$$

25. (a) Let $\mathbf{v}_1 = (1,0,0,1)$, $\mathbf{v}_2 = (0,1,0,0)$, $\mathbf{v}_3 = (1,1,1,1)$, $\mathbf{v}_4 = (0,1,1,1)$, $S = \{\mathbf{v}_1,\mathbf{v}_2,\mathbf{v}_3,\mathbf{v}_4\}$, and $W = \text{span } S$. In Exercise 3(a) it is shown that S is linearly independent. Thus $\dim W = 4$.

(b) Let $\mathbf{v}_1 = (1,-1,0,2)$, $\mathbf{v}_2 = (3,-1,2,1)$, $\mathbf{v}_3 = (1,0,0,1)$, $S = \{\mathbf{v}_1,\mathbf{v}_2,\mathbf{v}_3\}$, and $W = \text{span } S$. In Exercise 3(a) it is shown that S is linearly independent. Thus $\dim W = 3$.

(c) Let $\mathbf{v}_1 = (-2,4,6,4)$, $\mathbf{v}_2 = (0,1,2,0)$, $\mathbf{v}_3 = (-1,2,3,2)$, $\mathbf{v}_4 = (-3,2,5,6)$, $\mathbf{v}_5 = (-2,-1,0,4)$, $S = \{\mathbf{v}_1,\mathbf{v}_2,\mathbf{v}_3,\mathbf{v}_4,\mathbf{v}_5\}$, and $W = \text{span } S$. Following the technique in Example 5, we determine a subset of S that is a basis for W. Form the linear combination

$$c_1\mathbf{v}_1 + c_2\mathbf{v}_2 + c_3\mathbf{v}_3 + c_4\mathbf{v}_4 + c_5\mathbf{v}_5 = \mathbf{0} = (0,0,0,0).$$

Expanding, adding vectors, and equating corresponding components gives the linear system

$$-2c_1 \quad\quad -\ c_3 - 3c_4 - 2c_5 = 0$$
$$4c_1 +\ c_2 + 2c_3 + 2c_4 -\ c_5 = 0$$
$$6c_1 + 2c_2 + 3c_3 + 5c_4 \quad\quad = 0$$
$$4c_1 \quad\quad + 2c_3 + 6c_4 + 4c_5 = 0.$$

The reduced row echelon form of the coefficient matrix of this homogeneous system is

$$\begin{bmatrix} 1 & 0 & \frac{1}{2} & 0 & -\frac{1}{2} \\ 0 & 1 & 0 & 0 & -1 \\ 0 & 0 & 0 & 1 & 1 \\ 0 & 0 & 0 & 0 & 0 \end{bmatrix}.$$

The leading 1's appear in columns 1, 2, and 4, so $\{\mathbf{v}_1, \mathbf{v}_2, \mathbf{v}_4\}$ is a basis for W. Thus $\dim W = 3$.

(d) Let $\mathbf{v}_1 = (0, 0, 1, 1)$, $\mathbf{v}_2 = (-1, 1, 1, 2)$, $\mathbf{v}_3 = (1, 1, 0, 0)$, $\mathbf{v}_4 = (2, 1, 2, 1)$, $S = \{\mathbf{v}_1, \mathbf{v}_2, \mathbf{v}_3, \mathbf{v}_4\}$, and $W = \text{span } S$. In Exercise 3(d) it is shown that S is linearly independent. Thus $\dim W = 4$.

27. The subspace consists of all vectors of the form $at^3 + (3a - 5d)t^2 + (d + 4a)t + d$. Now,

$$at^3 + (3a - 5d)t^2 + (d + 4a)t + d = a(t^3 + 3t^2 + 4t) + d(-5t^2 + t + 1).$$

Thus $S = \{t^3 + 3t^2 + 4t, -5t^2 + t + 1\}$ spans the subspace. To see if S is linearly independent, we form

$$c_1(t^3 + 3t^2 + 4t) + c_2(-5t^2 + t + 1) = 0.$$

Equating like powers of t leads to the homogeneous system

$$c_1 \quad\quad = 0$$
$$3c_1 - 5c_2 = 0$$
$$4c_1 +\ c_2 = 0$$
$$c_2 = 0,$$

whose solution is $c_1 = c_2 = 0$. Hence S is linearly independent and is therefore a basis for the subspace. The dimension of the subspace is 2.

29. We follow the technique given in Example 9. Let $\mathbf{e}_1 = (1, 0, 0, 0)$, $\mathbf{e}_2 = (0, 1, 0, 0)$, $\mathbf{e}_3 = (0, 0, 1, 0)$, $\mathbf{e}_4 = (0, 0, 0, 1)$, $\mathbf{v}_1 = (1, 0, 1, 0)$, $\mathbf{v}_2 = (0, 1, -1, 0)$, and $S = \{\mathbf{v}_1, \mathbf{v}_2, \mathbf{e}_1, \mathbf{e}_2, \mathbf{e}_3, \mathbf{e}_4\}$. Then S spans R^4 since it contains a basis. We form the expression

$$c_1\mathbf{v}_1 + c_2\mathbf{v}_2 + c_3\mathbf{e}_1 + c_4\mathbf{e}_2 + c_5\mathbf{e}_3 + c_6\mathbf{e}_4 = (0, 0, 0, 0)$$

which leads to the homogeneous system with augmented matrix

$$\left[\begin{array}{cccccc:c} 1 & 0 & 1 & 0 & 0 & 0 & 0 \\ 0 & 1 & 0 & 1 & 0 & 0 & 0 \\ 1 & -1 & 0 & 0 & 1 & 0 & 0 \\ 0 & 0 & 0 & 0 & 0 & 1 & 0 \end{array}\right].$$

Transforming the augmented matrix to reduced row echelon form gives

$$\left[\begin{array}{cccccc:c} 1 & 0 & 0 & 1 & 1 & 0 & 0 \\ 0 & 1 & 0 & 1 & 0 & 0 & 0 \\ 0 & 0 & 1 & -1 & -1 & 0 & 0 \\ 0 & 0 & 0 & 0 & 0 & 1 & 0 \end{array}\right].$$

Since the leading 1's appear in columns 1, 2, 3, and 6, we conclude that $\{\mathbf{v}_1, \mathbf{v}_2, \mathbf{e}_1, \mathbf{e}_4\}$ is a basis for R^4 containing $\mathbf{v}_1$ and $\mathbf{v}_2$.

31. Let W be the subspace of M_{33} consisting of all symmetric matrices. A vector in W has the form

$$\begin{bmatrix} a & b & c \\ b & d & e \\ c & e & f \end{bmatrix}$$

which can be written as

$$\begin{bmatrix} a & b & c \\ b & d & e \\ c & e & f \end{bmatrix} = a \begin{bmatrix} 1 & 0 & 0 \\ 0 & 0 & 0 \\ 0 & 0 & 0 \end{bmatrix} + b \begin{bmatrix} 0 & 1 & 0 \\ 1 & 0 & 0 \\ 0 & 0 & 0 \end{bmatrix} + c \begin{bmatrix} 0 & 0 & 1 \\ 0 & 0 & 0 \\ 1 & 0 & 0 \end{bmatrix}$$

$$+ d \begin{bmatrix} 0 & 0 & 0 \\ 0 & 1 & 0 \\ 0 & 0 & 0 \end{bmatrix} + e \begin{bmatrix} 0 & 0 & 0 \\ 0 & 0 & 1 \\ 0 & 1 & 0 \end{bmatrix} + f \begin{bmatrix} 0 & 0 & 0 \\ 0 & 0 & 0 \\ 0 & 0 & 1 \end{bmatrix}.$$

Thus these six vectors span W. Writing the linear combination

$$c_1 \begin{bmatrix} 1 & 0 & 0 \\ 0 & 0 & 0 \\ 0 & 0 & 0 \end{bmatrix} + c_2 \begin{bmatrix} 0 & 1 & 0 \\ 1 & 0 & 0 \\ 0 & 0 & 0 \end{bmatrix} + c_3 \begin{bmatrix} 0 & 0 & 1 \\ 0 & 0 & 0 \\ 1 & 0 & 0 \end{bmatrix}$$

$$+ c_4 \begin{bmatrix} 0 & 0 & 0 \\ 0 & 1 & 0 \\ 0 & 0 & 0 \end{bmatrix} + c_5 \begin{bmatrix} 0 & 0 & 0 \\ 0 & 0 & 1 \\ 0 & 1 & 0 \end{bmatrix} + c_6 \begin{bmatrix} 0 & 0 & 0 \\ 0 & 0 & 0 \\ 0 & 0 & 1 \end{bmatrix} = \begin{bmatrix} 0 & 0 & 0 \\ 0 & 0 & 0 \\ 0 & 0 & 0 \end{bmatrix}.$$

we obtain a homogeneous system whose only solution is the trivial one. Hence these six vectors are also linearly independent and form a basis for W.

33. Let

$$S = \left\{ \begin{bmatrix} 1 \\ 0 \\ 0 \\ 0 \end{bmatrix}, \begin{bmatrix} 0 \\ 1 \\ 0 \\ 0 \end{bmatrix} \right\}.$$

Then the vectors in S are linearly independent and hence span a two-dimensional subspace of R^4.

35. Rewrite the equation of the plane as a homogeneous system. Solve for x and we have

$$\begin{aligned} x &= \tfrac{3}{2}y - 2z \\ y &= \phantom{\tfrac{3}{2}}y \\ z &= \phantom{\tfrac{3}{2}y - 2}z \end{aligned}$$

The column vector form of this system is

$$\begin{bmatrix} x \\ y \\ z \end{bmatrix} = y \begin{bmatrix} \tfrac{3}{2} \\ 1 \\ 0 \end{bmatrix} + z \begin{bmatrix} -2 \\ 0 \\ 1 \end{bmatrix}$$

which shows that each solution of the system must be a linear combination of

$$\mathbf{v}_1 = \begin{bmatrix} \tfrac{3}{2} \\ 1 \\ 0 \end{bmatrix} \quad \text{and} \quad \mathbf{v}_2 = \begin{bmatrix} -2 \\ 0 \\ 1 \end{bmatrix}.$$

This means that $\{\mathbf{v}_1, \mathbf{v}_2\}$ must span the solution set. $\mathbf{v}_1$ and $\mathbf{v}_2$ are also independent for if

$$c_1\mathbf{v}_1 + c_2\mathbf{v}_2 = c_1 \begin{bmatrix} \frac{3}{2} \\ 1 \\ 0 \end{bmatrix} + c_2 \begin{bmatrix} -2 \\ 0 \\ 1 \end{bmatrix} = \begin{bmatrix} \frac{3}{2}c_1 - 2c_2 \\ c_1 \\ c_2 \end{bmatrix} = \begin{bmatrix} 0 \\ 0 \\ 0 \end{bmatrix}$$

then $c_1 = c_2 = 0$. Therefore

$$\{\mathbf{v}_1, \mathbf{v}_2\} = \left\{ \begin{bmatrix} \frac{3}{2} \\ 1 \\ 0 \end{bmatrix}, \begin{bmatrix} -2 \\ 0 \\ 1 \end{bmatrix} \right\}$$

is a basis for the given plane. By Exercise T.9, we have

$$\left\{ \begin{bmatrix} 3 \\ 2 \\ 0 \end{bmatrix}, \begin{bmatrix} -2 \\ 0 \\ 1 \end{bmatrix} \right\}$$

is also a basis.

37. By Theorem 6.9, we only need to show that the vectors are linearly independent. Form the equation

$$c_1 \begin{bmatrix} 1 \\ 1 \\ 1 \end{bmatrix} + c_2 \begin{bmatrix} 1 \\ 1 \\ 0 \end{bmatrix} + c_3 \begin{bmatrix} 1 \\ 0 \\ 1 \end{bmatrix} = \begin{bmatrix} 0 \\ 0 \\ 0 \end{bmatrix}.$$

We obtain the linear system

$$\begin{aligned}
c_1 + c_2 + c_3 &= 0 \\
c_1 + c_2 \phantom{{}+ c_3} &= 0 \\
c_1 \phantom{{}+ c_2} + c_3 &= 0.
\end{aligned}$$

The reduced row echelon form of the coefficient matrix is

$$\begin{bmatrix} 1 & 0 & 0 \\ 0 & 1 & 0 \\ 0 & 0 & 1 \end{bmatrix}.$$

Hence the only solution to the linear system is $c_1 = c_2 = c_3 = 0$, which shows that the given vectors are linearly independent. Thus the vectors form a basis for B^3.

39. By Theorem 6.9, we only need to show that the vectors are linearly independent. Form the equation

$$c_1 \begin{bmatrix} 1 \\ 0 \\ 0 \\ 1 \end{bmatrix} + c_2 \begin{bmatrix} 0 \\ 0 \\ 1 \\ 1 \end{bmatrix} + c_3 \begin{bmatrix} 0 \\ 1 \\ 1 \\ 0 \end{bmatrix} + c_4 \begin{bmatrix} 1 \\ 1 \\ 0 \\ 0 \end{bmatrix} = \begin{bmatrix} 0 \\ 0 \\ 0 \\ 0 \end{bmatrix}.$$

We obtain the linear system

$$\begin{aligned}
c_1 \phantom{{}+ c_2 + c_3} + c_4 &= 0 \\
c_3 + c_4 &= 0 \\
c_2 + c_3 \phantom{{}+ c_4} &= 0 \\
c_1 + c_2 \phantom{{}+ c_3 + c_4} &= 0
\end{aligned}$$

The reduced row echelon form of the coefficient matrix is

$$\begin{bmatrix} 1 & 0 & 0 & 1 \\ 0 & 1 & 0 & 1 \\ 0 & 0 & 1 & 1 \\ 0 & 0 & 0 & 0 \end{bmatrix}.$$

Hence the solution of the homogeneous system is $c_1 = -r$, $c_2 = -r$, $c_3 = -r$, $c_4 = r$, where r is any real number. Thus the vectors are linearly dependent and do not form a basis for B^4.

T.1. Since the largest number of vectors in any linearly independent set is m, $\dim V = m$. The result follows from Theorem 6.9.

T.3. By Theorem 6.7, any linearly independent set T of vectors in V has at most n elements. Thus a set of $n+1$ vectors must be linearly dependent.

T.5. Let $S = \{\mathbf{w}_1, \mathbf{w}_2, \dots, \mathbf{w}_k\}$ be a linearly independent set of vectors in V and let $\{\mathbf{v}_1, \dots, \mathbf{v}_n\}$ be a basis for V (V is finite-dimensional). Let $S_1 = \{\mathbf{w}_1, \mathbf{w}_2, \dots, \mathbf{w}_k, \mathbf{v}_1, \mathbf{v}_2, \dots, \mathbf{v}_n\}$. S_1 spans V (since the subset $\{\mathbf{v}_1, \mathbf{v}_2, \dots, \mathbf{v}_n\}$ does). If S_1 is linearly independent, then it is a basis for V which contains S. Otherwise some vector in S_1 is a linear combination of the preceding vectors (Theorem 6.4). That vector cannot be one of the $\mathbf{w}_i$'s since S is linearly independent. So it is one of the $\mathbf{v}_j$'s. Delete it to form a new set S_2 with one fewer element than S_1 which also spans V. Either S_2 is a basis or else another $\mathbf{v}_j$ can be deleted. After a finite number of steps we arrive at a set S_p which is a basis for V and which contains the given set S.

T.7. Let $S = \{\mathbf{v}_1, \mathbf{v}_2, \dots, \mathbf{v}_m\}$ be a basis for W. By Theorem 6.8, there is a basis T for V which contains the linearly independent set S. Since $\dim W = m = \dim V$, T must have m elements. Thus $T = S$ and $V = W$.

T.9. If $\mathbf{v}$ is a linear combination of $\{\mathbf{v}_1, \mathbf{v}_2, \dots, \mathbf{v}_n\}$, then

$$\mathbf{v} = d_1\mathbf{v}_1 + d_2\mathbf{v}_2 + \cdots + d_n\mathbf{v}_n$$

and hence

$$\mathbf{v} = \frac{d_1}{c}(c\mathbf{v}_1) + d_2\mathbf{v}_2 + \cdots + d_n\mathbf{v}_n.$$

Therefore $\mathbf{v}$ is a linear combination of $\{c\mathbf{v}_1, \mathbf{v}_2, \dots, \mathbf{v}_n\}$. Similarly any vector which is a linear combination of the second set

$$\mathbf{v} = d_1(c\mathbf{v}_1) + d_2\mathbf{v}_2 + \cdots + d_n\mathbf{v}_n,$$

is a linear combination of the first set:

$$\mathbf{v} = (d_1c)\mathbf{v}_1 + \cdots + d_n\mathbf{v}_n.$$

Thus the two sets span V. Since the second set has n elements, it is also a basis for V.

T.11. Let $S = \{\mathbf{v}_1, \mathbf{v}_2, \dots, \mathbf{v}_n\}$. Since every vector in V can be written as a linear combination of the vectors in S, it follows that S spans V. Suppose now that

$$c_1\mathbf{v}_1 + c_2\mathbf{v}_2 + \cdots + c_n\mathbf{v}_n = \mathbf{0}.$$

We also have

$$0\mathbf{v}_1 + 0\mathbf{v}_2 + \cdots + 0\mathbf{v}_n = \mathbf{0}.$$

From the hypothesis it then follows that $c_1 = 0$, $c_2 = 0, \dots, c_n = 0$. Hence, S is a basis for V.

T.13. Since A is singular, Theorem 1.13 (Sec. 1.7) implies that the homogeneous system $A\mathbf{x} = \mathbf{0}$ has a nontrivial solution $\mathbf{w}$. Since $\{\mathbf{v}_1, \mathbf{v}_2, \dots, \mathbf{v}_n\}$ is a linearly independent set of vectors in R^n, it is a basis for R^n, so

$$\mathbf{w} = c_1\mathbf{v}_1 + c_2\mathbf{v}_2 + \cdots + c_n\mathbf{v}_n.$$

Observe that $\mathbf{w} \neq \mathbf{0}$, so $c_1, c_2, \dots, c_n$ are not all zero. Then

$$\mathbf{0} = A\mathbf{w} = A(c_1\mathbf{v}_1 + c_2\mathbf{v}_2 + \cdots + c_n\mathbf{v}_n) = c_1(A\mathbf{v}_1) + c_2(A\mathbf{v}_2) + \cdots + c_n(A\mathbf{v}_n).$$

Hence, $\{A\mathbf{v}_1, A\mathbf{v}_2, \dots, A\mathbf{v}_n\}$ is linearly dependent.

T.15. Suppose that $c_n, c_{n-1}, \ldots, c_1, c_0$ are scalars such that

$$c_n t^n + c_{n-1} t^{n-1} + \cdots + c_1 t + c_0 = 0.$$

Then

$$c_n t^n + c_{n-1} t^{n-1} + \cdots + c_1 t + c_0 = 0 = 0t^n + 0t^{n-1} + \cdots + 0t + 0(1).$$

Equating like powers of t, we obtain $c_n = c_{n-1} = \cdots = c_1 = c_0 = 0$. Hence the set $\{t^n, t^{n-1}, \ldots, t, 1\}$ is linear independent.

T.17. (a) $\mathbf{v}_1 = \begin{bmatrix} 1 \\ 0 \\ 0 \end{bmatrix}$, $\mathbf{v}_2 = \begin{bmatrix} 0 \\ 1 \\ 0 \end{bmatrix}$, $\mathbf{v}_3 = \begin{bmatrix} 0 \\ 0 \\ 1 \end{bmatrix}$.

(b) $\mathbf{v}_1 = \begin{bmatrix} 0 \\ 0 \\ 0 \end{bmatrix}$, $\mathbf{v}_2 = \begin{bmatrix} 1 \\ 0 \\ 0 \end{bmatrix}$, $\mathbf{v}_3 = \begin{bmatrix} 0 \\ 1 \\ 0 \end{bmatrix}$.

ML.1. Follow the procedure in Exercise ML.3 in Section 6.2.

v1 = [1 2 1]′;v2 = [2 1 1]′;v3 = [2 2 1]′;

rref([v1 v2 v3 zeros(size(v1))])

ans =

```
        1  0  0  0
        0  1  0  0
        0  0  1  0
```

It follows that the only solution is the trivial solution so S is linearly independent.

ML.3. Proceed as in ML.1.

v1 = [1 1 0 1]′;v2 = [2 1 1 −1]′;v3 = [0 0 1 1]′;v4 = [1 2 1 2]′;

rref([v1 v2 v3 v4 zeros(size(v1))])

ans =

```
        1  0  0  0  0
        0  1  0  0  0
        0  0  1  0  0
        0  0  0  1  0
```

It follows that S is linearly independent and since $\dim V = 4$, S is a basis for V.

ML.5. Here we do not know $\dim(\text{span } S)$, but $\dim(\text{span } S) =$ the number of linearly independent vectors in S. We proceed as we did in ML.1.

v1 = [1 2 1 0]′;v2 = [2 1 3 1]′;v3 = [2 2 1 2]′;

rref([v1 v2 v3 zeros(size(v1))])

ans =

```
        1  0  0  0
        0  1  0  0
        0  0  1  0
        0  0  0  0
```

The leading 1's imply that S is a linearly independent set hence $\dim(\text{span } S) = 3$ and S is a basis for V.

ML.7. $\mathbf{v1} = [1 \ \ 1 \ \ 0 \ \ 0]'; \mathbf{v2} = [-2 \ \ -2 \ \ 0 \ \ 0]'; \mathbf{v3} = [1 \ \ 0 \ \ 2 \ \ 1]'; \mathbf{v4} = [2 \ \ 1 \ \ 2 \ \ 1]'; \mathbf{v5} = [0 \ \ 1 \ \ 1 \ \ 1]';$

rref([v1 v2 v3 v4 v5 zeros(size(v1))])

ans =

```
    1  -2   0   1   0   0
    0   0   1   1   0   0
    0   0   0   0   1   0
    0   0   0   0   0   0
```

The leading 1's point to vectors $\mathbf{v}_1$, $\mathbf{v}_3$ and $\mathbf{v}_5$ and hence these vectors are a linearly independent set which also spans S. Thus $T = \{\mathbf{v}_1, \mathbf{v}_3, \mathbf{v}_5\}$ is a basis for span S. We have dim(span S) = 3 and span $S \neq R^4$.

ML.9. Proceed as in ML.2.

$\mathbf{v1} = [0 \ \ 1 \ \ -2]'; \mathbf{v2} = [0 \ \ 2 \ \ -1]'; \mathbf{v3} = [0 \ \ 4 \ \ -2]'; \mathbf{v4} = [1 \ \ -1 \ \ 1]'; \mathbf{v5} = [1 \ \ 2 \ \ 1]';$

rref([v1 v2 v3 v4 v5 zeros(size(v1))])

ans =

```
    1.0000        0   1.6000        0   0.6000   0
         0   1.0000   1.2000        0   1.2000   0
         0        0        0   1.0000   1.0000   0
```

It follows that $T = \{\mathbf{v}_1, \mathbf{v}_2, \mathbf{v}_4\}$ is a basis for span S. We have dim(span S) = 3 and it follows that span $S = P_2$.

ML.11. $\mathbf{v1} = [1 \ \ 0 \ \ -1 \ \ 1]'; \mathbf{v2} = [1 \ \ 0 \ \ 0 \ \ 2]';$

rref([v1 v2 eye(4) zeros(size(v1))])

ans =

```
    1.0000        0        0        0  -1.0000        0   0
         0   1.0000        0        0   0.5000   0.5000   0
         0        0   1.0000        0   0.5000  -0.5000   0
         0        0        0   1.0000        0        0   0
```

It follows that $\left\{\mathbf{v}_1, \mathbf{v}_2, \mathbf{e}_3 = [0 \ \ 0 \ \ 1 \ \ 0]', \mathbf{e}_4 = [0 \ \ 0 \ \ 0 \ \ 1]'\right\}$ is a basis for R^4. Hence, a basis for P_3 is $\{t^3 - t + 1, t^3 + 2, t, 1\}$.

Section 6.5, p. 327

1. Let W be the solution space to $A\mathbf{x} = \mathbf{0}$.

(a) To find W, we row reduce the augmented matrix of A:

$$\left[\begin{array}{ccc|c} 2 & -1 & -2 & 0 \\ -4 & 2 & 4 & 0 \\ -8 & 4 & 8 & 0 \end{array}\right] \rightarrow \left[\begin{array}{ccc|c} 2 & -1 & -2 & 0 \\ 0 & 0 & 0 & 0 \\ 0 & 0 & 0 & 0 \end{array}\right] \rightarrow \left[\begin{array}{ccc|c} 2 & -1 & -2 & 0 \\ 0 & 0 & 0 & 0 \\ 0 & 0 & 0 & 0 \end{array}\right]$$

add 2(row 1) to row 2   multiply row 1 by $\frac{1}{2}$
add 4(row 1) to row 3

The solutions to the system are therefore $x_1 = \frac{1}{2}r + s$, $x_2 = r$, $x_3 = s$, where r, s are arbitrary real numbers.

(b) Every solution has the form

$$\mathbf{x} = \begin{bmatrix} x_1 \\ x_2 \\ x_3 \end{bmatrix} = \begin{bmatrix} \frac{1}{2}r + s \\ r \\ s \end{bmatrix} = r \begin{bmatrix} \frac{1}{2} \\ 1 \\ 0 \end{bmatrix} + s \begin{bmatrix} 1 \\ 0 \\ 1 \end{bmatrix}.$$

Thus, every solution to $A\mathbf{x} = \mathbf{0}$ may be expressed as a linear combination of the two vectors

$$\mathbf{x}_1 = \begin{bmatrix} \frac{1}{2} \\ 1 \\ 0 \end{bmatrix} \quad \text{and} \quad \mathbf{x}_2 = \begin{bmatrix} 1 \\ 0 \\ 1 \end{bmatrix}.$$

(c)

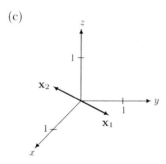

In Exercises 3–9, we form the augmented matrix associated with the homogeneous linear system, find its reduced row echelon form, and write out the solution, assigning arbitrary constants as needed. Following the method in Example 1 we determine a basis for the solution space and compute the dimension of the solution space.

3. Let $\begin{bmatrix} A \vdots \mathbf{0} \end{bmatrix}$ be the augmented matrix associated with the homogeneous system. We have

$$\begin{bmatrix} A \vdots \mathbf{0} \end{bmatrix} = \begin{bmatrix} 1 & 1 & 1 & 1 & \vdots & 0 \\ 2 & 1 & -1 & 1 & \vdots & 0 \end{bmatrix}.$$

Using row operations we find that the reduced row echelon form of $\begin{bmatrix} A \vdots \mathbf{0} \end{bmatrix}$ is

$$\begin{bmatrix} 1 & 0 & -2 & 0 & \vdots & 0 \\ 0 & 1 & 3 & 1 & \vdots & 0 \end{bmatrix}.$$

It follows that every solution is of the form

$$\mathbf{x} = \begin{bmatrix} 2s \\ -3s - t \\ s \\ t \end{bmatrix} = s \begin{bmatrix} 2 \\ -3 \\ 1 \\ 0 \end{bmatrix} + t \begin{bmatrix} 0 \\ -1 \\ 0 \\ 1 \end{bmatrix}.$$

Thus

$$\left\{ \begin{bmatrix} 2 \\ -3 \\ 1 \\ 0 \end{bmatrix}, \begin{bmatrix} 0 \\ -1 \\ 0 \\ 1 \end{bmatrix} \right\}$$

is a basis for the solution space and it has dimension 2.

5. Let $\begin{bmatrix} A \vdots \mathbf{0} \end{bmatrix}$ be the augmented matrix associated with the homogeneous system. We have

$$\begin{bmatrix} A \vdots \mathbf{0} \end{bmatrix} = \begin{bmatrix} 1 & 2 & -1 & 3 & \vdots & 0 \\ 2 & 2 & -1 & 2 & \vdots & 0 \\ 1 & 0 & 3 & 3 & \vdots & 0 \end{bmatrix}.$$

Using row operations we find that the reduced row echelon form of $\left[A \mathrel{\vdots} \mathbf{0}\right]$ is

$$
\begin{bmatrix}
1 & 0 & 0 & -1 & \vdots & 0 \\
0 & 1 & 0 & \frac{8}{3} & \vdots & 0 \\
0 & 0 & 1 & \frac{4}{3} & \vdots & 0
\end{bmatrix}.
$$

It follows that every solution is of the form

$$
\mathbf{x} =
\begin{bmatrix}
t \\
-\frac{8}{3}t \\
-\frac{4}{3}t \\
t
\end{bmatrix}
= t
\begin{bmatrix}
1 \\
-\frac{8}{3} \\
-\frac{4}{3} \\
1
\end{bmatrix}.
$$

Thus

$$
\left\{
\begin{bmatrix}
1 \\
-\frac{8}{3} \\
-\frac{4}{3} \\
1
\end{bmatrix}
\right\}
$$

is a basis for the solution space and it has dimension 1.

7. Let $\left[A \mathrel{\vdots} \mathbf{0}\right]$ be the augmented matrix associated with the homogeneous system. We have

$$
\left[A \mathrel{\vdots} \mathbf{0}\right] =
\begin{bmatrix}
1 & 2 & 1 & 2 & 1 & \vdots & 0 \\
1 & 2 & 2 & 1 & 2 & \vdots & 0 \\
2 & 4 & 3 & 3 & 3 & \vdots & 0 \\
0 & 0 & 1 & -1 & -1 & \vdots & 0
\end{bmatrix}.
$$

Using row operations we find that the reduced row echelon form of $\left[A \mathrel{\vdots} \mathbf{0}\right]$ is

$$
\begin{bmatrix}
1 & 2 & 0 & 3 & 0 & \vdots & 0 \\
0 & 0 & 1 & -1 & 0 & \vdots & 0 \\
0 & 0 & 0 & 0 & 1 & \vdots & 0 \\
0 & 0 & 0 & 0 & 0 & \vdots & 0
\end{bmatrix}.
$$

It follows that every solution is of the form

$$
\mathbf{x} =
\begin{bmatrix}
-2s - 3t \\
s \\
t \\
t \\
0
\end{bmatrix}
= s
\begin{bmatrix}
-2 \\
1 \\
0 \\
0 \\
0
\end{bmatrix}
+ t
\begin{bmatrix}
-3 \\
0 \\
1 \\
1 \\
0
\end{bmatrix}.
$$

Thus

$$
\left\{
\begin{bmatrix}
-2 \\
1 \\
0 \\
0 \\
0
\end{bmatrix},
\begin{bmatrix}
-3 \\
0 \\
1 \\
1 \\
0
\end{bmatrix}
\right\}
$$

is a basis for the solution space and it has dimension 2.

9. Let $[A \mid \mathbf{0}]$ be the augmented matrix associated with the homogeneous system. We have

$$[A \mid \mathbf{0}] = \begin{bmatrix} 1 & 2 & 2 & -1 & 1 & \vdots & 0 \\ 0 & 2 & 2 & -2 & -1 & \vdots & 0 \\ 2 & 6 & 2 & -4 & 1 & \vdots & 0 \\ 1 & 4 & 0 & -3 & 0 & \vdots & 0 \end{bmatrix}.$$

Using row operations we find that the reduced row echelon form of $[A \mid \mathbf{0}]$ is

$$\begin{bmatrix} 1 & 0 & 0 & 1 & 2 & \vdots & 0 \\ 0 & 1 & 0 & -1 & -\frac{1}{2} & \vdots & 0 \\ 0 & 0 & 1 & 0 & 0 & \vdots & 0 \\ 0 & 0 & 0 & 0 & 0 & \vdots & 0 \end{bmatrix}.$$

It follows that every solution is of the form

$$\mathbf{x} = \begin{bmatrix} -s - 2t \\ s + \frac{1}{2}t \\ 0 \\ s \\ t \end{bmatrix} = s \begin{bmatrix} -1 \\ 1 \\ 0 \\ 1 \\ 0 \end{bmatrix} + t \begin{bmatrix} -2 \\ \frac{1}{2} \\ 0 \\ 0 \\ 1 \end{bmatrix}.$$

Thus

$$\left\{ \begin{bmatrix} -1 \\ 1 \\ 0 \\ 1 \\ 0 \end{bmatrix}, \begin{bmatrix} -2 \\ \frac{1}{2} \\ 0 \\ 0 \\ 1 \end{bmatrix} \right\}$$

is a basis for the solution space and it has dimension 2.

11. The null space of matrix A is the same as the solution space of $A\mathbf{x} = \mathbf{0}$. To find a basis we follow the procedure used in Exercises 3–9. Let $[A \mid \mathbf{0}]$ be the augmented matrix associated with the homogeneous system. We have

$$[A \mid \mathbf{0}] = \begin{bmatrix} 1 & 2 & 3 & -1 & \vdots & 0 \\ 2 & 3 & 2 & 0 & \vdots & 0 \\ 3 & 4 & 1 & 1 & \vdots & 0 \\ 1 & 1 & -1 & 1 & \vdots & 0 \end{bmatrix}.$$

Using row operations we find that the reduced row echelon form of $[A \mid \mathbf{0}]$ is

$$\begin{bmatrix} 1 & 0 & -5 & 3 & \vdots & 0 \\ 0 & 1 & 4 & -2 & \vdots & 0 \\ 0 & 0 & 0 & 0 & \vdots & 0 \\ 0 & 0 & 0 & 0 & \vdots & 0 \end{bmatrix}.$$

It follows that every solution is of the form

$$\mathbf{x} = \begin{bmatrix} 5s - 3t \\ -4s + 2t \\ s \\ t \end{bmatrix} = s \begin{bmatrix} 5 \\ -4 \\ 1 \\ 0 \end{bmatrix} + t \begin{bmatrix} -3 \\ 2 \\ 0 \\ 1 \end{bmatrix}.$$

Thus

$$\left\{ \begin{bmatrix} 5 \\ -4 \\ 1 \\ 0 \end{bmatrix}, \begin{bmatrix} -3 \\ 2 \\ 0 \\ 1 \end{bmatrix} \right\}$$

is a basis for the null space of A.

13. Following Example 2, we form the matrix

$$1I_2 - A = \begin{bmatrix} 1 & 0 \\ 0 & 1 \end{bmatrix} - \begin{bmatrix} 3 & 2 \\ 1 & 2 \end{bmatrix} = \begin{bmatrix} -2 & -2 \\ -1 & -1 \end{bmatrix}.$$

Then the reduced row echelon form of matrix

$$\begin{bmatrix} -2 & -2 & \vdots & 0 \\ -1 & -1 & \vdots & 0 \end{bmatrix}$$

is

$$\begin{bmatrix} 1 & 1 & \vdots & 0 \\ 0 & 0 & \vdots & 0 \end{bmatrix}.$$

Hence every solution of the homogeneous system $(1I_2 - A)\mathbf{x} = \mathbf{0}$ is of the form

$$\mathbf{x} = \begin{bmatrix} -t \\ t \end{bmatrix},$$

where t is any real number. It follows that a basis for the solution space of $(1I_2 - A)\mathbf{x} = \mathbf{0}$ is

$$\left\{ \begin{bmatrix} -1 \\ 1 \end{bmatrix} \right\}.$$

15. Following Example 2, we form the matrix

$$1I_3 - A = \begin{bmatrix} 1 & 0 & -1 \\ -1 & 1 & 3 \\ 0 & -1 & -2 \end{bmatrix}.$$

Then the reduced row echelon form of the augmented matrix

$$\begin{bmatrix} 1 & 0 & -1 & \vdots & 0 \\ -1 & 1 & 3 & \vdots & 0 \\ 0 & -1 & -2 & \vdots & 0 \end{bmatrix}$$

is

$$\begin{bmatrix} 1 & 0 & -1 & \vdots & 0 \\ 0 & 1 & 2 & \vdots & 0 \\ 0 & 0 & 0 & \vdots & 0 \end{bmatrix}.$$

Hence every solution of the homogeneous system $(1I_3 - A)\mathbf{x} = \mathbf{0}$ is of the form

$$\mathbf{x} = \begin{bmatrix} t \\ -2t \\ t \end{bmatrix},$$

where t is any real number. It follows that a basis for the solution space of $(1I_3 - A)\mathbf{x} = \mathbf{0}$ is

$$\left\{ \begin{bmatrix} 1 \\ -2 \\ 1 \end{bmatrix} \right\}.$$

17. Follow the procedure in Example 3.

$$\lambda I_2 - A = \lambda \begin{bmatrix} 1 & 0 \\ 0 & 1 \end{bmatrix} - \begin{bmatrix} 2 & 3 \\ 2 & -3 \end{bmatrix} = \begin{bmatrix} \lambda - 2 & -3 \\ -2 & \lambda + 3 \end{bmatrix}.$$

The homogeneous system $(\lambda I_2 - A)\mathbf{x} = \mathbf{0}$ has a nontrivial solution if and only if $\det(\lambda I_2 - A) = 0$. We have

$$\det(\lambda I_2 - A) = \det\left(\begin{bmatrix} \lambda - 2 & -3 \\ -2 & \lambda + 3 \end{bmatrix} \right) = (\lambda - 2)(\lambda + 3) - (-2)(-3)$$

$$= \lambda^2 + \lambda - 12 = (\lambda - 3)(\lambda + 4) = 0$$

only if $\lambda = 3$ or -4. It follows that $(\lambda I_2 - A)\mathbf{x} = \mathbf{0}$ has a nontrivial solution when $\lambda = 3$ or -4.

19. Follow the procedure in Example 3.

$$\lambda I_3 - A = \lambda \begin{bmatrix} 1 & 0 & 0 \\ 0 & 1 & 0 \\ 0 & 0 & 1 \end{bmatrix} - \begin{bmatrix} 0 & 0 & 0 \\ 0 & 1 & -1 \\ 1 & 0 & 0 \end{bmatrix} = \begin{bmatrix} \lambda & 0 & 0 \\ 0 & \lambda - 1 & 1 \\ -1 & 0 & \lambda \end{bmatrix}.$$

The homogeneous system $(\lambda I_3 - A)\mathbf{x} = \mathbf{0}$ has a nontrivial solution if and only if $\det(\lambda I_3 - A) = 0$. We have

$$\det(\lambda I_3 - A) = \det\left(\begin{bmatrix} \lambda & 0 & 0 \\ 0 & \lambda - 1 & 1 \\ -1 & 0 & \lambda \end{bmatrix} \right) = \lambda^2(\lambda - 1) = 0$$

only if $\lambda = 0$ or 1. It follows that $(\lambda I_3 - A)\mathbf{x} = \mathbf{0}$ has a nontrivial solution when $\lambda = 0$ or 1.

21. The augmented matrix of the linear system is

$$\begin{bmatrix} 1 & 2 & -1 & -1 & \vdots & 3 \\ 1 & 1 & 3 & 2 & \vdots & -2 \\ 2 & -1 & 4 & 3 & \vdots & 1 \\ 2 & -2 & 8 & 6 & \vdots & -4 \end{bmatrix},$$

whose reduced echelon form is

$$\begin{bmatrix} 1 & 0 & 0 & 0 & \vdots & 3 \\ 0 & 1 & 0 & -\frac{1}{7} & \vdots & -\frac{5}{7} \\ 0 & 0 & 1 & \frac{5}{7} & \vdots & -\frac{10}{7} \\ 0 & 0 & 0 & 0 & \vdots & 0 \end{bmatrix}.$$

The solution to the system is therefore

$$\mathbf{x} = \begin{bmatrix} 3 \\ -\frac{5}{7} \\ -\frac{10}{7} \\ 0 \end{bmatrix} + \begin{bmatrix} 0 \\ \frac{1}{7}r \\ -\frac{5}{7}r \\ r \end{bmatrix},$$

where $\mathbf{x}_p = \begin{bmatrix} 3 \\ -\frac{5}{7} \\ -\frac{10}{7} \\ 0 \end{bmatrix}$ is a particular solution to the system and $\mathbf{x}_h = \begin{bmatrix} 0 \\ \frac{1}{7}r \\ -\frac{5}{7}r \\ r \end{bmatrix}$ is a solution to the

corresponding homogeneous system.

23. Let $\begin{bmatrix} A & \vdots & \mathbf{0} \end{bmatrix}$ be the augmented matrix associated with the homogeneous system. We have

$$\begin{bmatrix} A & \vdots & \mathbf{0} \end{bmatrix} = \begin{bmatrix} 1 & 1 & 0 & 1 & \vdots & 0 \\ 1 & 0 & 1 & 1 & \vdots & 0 \\ 1 & 1 & 1 & 0 & \vdots & 0 \end{bmatrix}.$$

Using row operations we find that the reduced row echelon form of $\begin{bmatrix} A & \vdots & \mathbf{0} \end{bmatrix}$ is

$$\begin{bmatrix} 1 & 0 & 0 & 0 & \vdots & 0 \\ 0 & 1 & 0 & 1 & \vdots & 0 \\ 0 & 0 & 1 & 1 & \vdots & 0 \end{bmatrix}.$$

It follows that every solution is of the form

$$\mathbf{x} = \begin{bmatrix} 0 \\ -t \\ -t \\ t \end{bmatrix} = t \begin{bmatrix} 0 \\ -1 \\ -1 \\ 1 \end{bmatrix} = t \begin{bmatrix} 0 \\ 1 \\ 1 \\ 1 \end{bmatrix}.$$

Thus $\left\{ \begin{bmatrix} 0 \\ 1 \\ 1 \\ 1 \end{bmatrix} \right\}$ is a basis for the solution space and it has dimension 1.

25. Let $\begin{bmatrix} A & \vdots & \mathbf{0} \end{bmatrix}$ be the augmented matrix associated with the homogeneous system. We have

$$\begin{bmatrix} 1 & 1 & 0 & 0 & 1 & \vdots & 0 \\ 1 & 0 & 1 & 1 & 1 & \vdots & 0 \\ 0 & 1 & 1 & 0 & 0 & \vdots & 0 \end{bmatrix}.$$

Using row operations we find that the reduced row echelon form of $\begin{bmatrix} A & \vdots & \mathbf{0} \end{bmatrix}$ is

$$\begin{bmatrix} 1 & 0 & 1 & 0 & 1 & \vdots & 0 \\ 0 & 1 & 1 & 0 & 0 & \vdots & 0 \\ 0 & 0 & 0 & 1 & 0 & \vdots & 0 \end{bmatrix}.$$

It follows that every solution is of the form

$$\mathbf{x} = \begin{bmatrix} -s-t \\ -s \\ s \\ 0 \\ t \end{bmatrix} = s \begin{bmatrix} -1 \\ -1 \\ 1 \\ 0 \\ 0 \end{bmatrix} + t \begin{bmatrix} -1 \\ 0 \\ 0 \\ 0 \\ 1 \end{bmatrix} = s \begin{bmatrix} 1 \\ 1 \\ 1 \\ 0 \\ 0 \end{bmatrix} + t \begin{bmatrix} 1 \\ 0 \\ 0 \\ 0 \\ 1 \end{bmatrix}.$$

Thus

$$
\left\{ \begin{bmatrix} 1 \\ 1 \\ 1 \\ 0 \\ 0 \end{bmatrix}, \begin{bmatrix} 1 \\ 0 \\ 0 \\ 0 \\ 1 \end{bmatrix} \right\}
$$

is a basis for the solution space and it has dimension 2.

27. The augmented matrix of the linear system is

$$
\begin{bmatrix} 1 & 1 & 0 & 1 & \vdots & 1 \\ 1 & 0 & 1 & 1 & \vdots & 1 \\ 0 & 1 & 1 & 0 & \vdots & 0 \\ 0 & 0 & 1 & 1 & \vdots & 0 \end{bmatrix}
$$

whose reduced row echelon form is

$$
\begin{bmatrix} 1 & 0 & 0 & 0 & \vdots & 1 \\ 0 & 1 & 0 & 1 & \vdots & 0 \\ 0 & 0 & 1 & 1 & \vdots & 0 \\ 0 & 0 & 0 & 0 & \vdots & 0 \end{bmatrix}.
$$

The solution to the system is therefore

$$
\mathbf{x} = \begin{bmatrix} 1 \\ 0 \\ 0 \\ 0 \end{bmatrix} + \begin{bmatrix} 0 \\ -b \\ -b \\ b \end{bmatrix} = \begin{bmatrix} 1 \\ 0 \\ 0 \\ 0 \end{bmatrix} + b \begin{bmatrix} 0 \\ 1 \\ 1 \\ 1 \end{bmatrix},
$$

where b is any bit. Hence $\mathbf{x}_p = \begin{bmatrix} 1 \\ 0 \\ 0 \\ 0 \end{bmatrix}$ is a particular solution to the system and $\mathbf{x}_h = b \begin{bmatrix} 0 \\ 1 \\ 1 \\ 1 \end{bmatrix}$ is a solution to the corresponding homogeneous system.

T.1. Since each vector in S is a solution to $A\mathbf{x} = \mathbf{0}$, we have $A\mathbf{x}_i = \mathbf{0}$ for $i = 1, 2, \ldots, k$. The span of S consists of all possible linear combinations of the vectors in S, hence

$$
\mathbf{y} = c_1\mathbf{x}_1 + c_2\mathbf{x}_2 + \cdots + c_k\mathbf{x}_k
$$

represents an arbitrary member of span S. We have

$$
A\mathbf{y} = c_1 A\mathbf{x}_1 + c_2 A\mathbf{x}_2 + \cdots + c_k A\mathbf{x}_k = c_1\mathbf{0} + c_2\mathbf{0} + \cdots + c_k\mathbf{0} = \mathbf{0}.
$$

Thus $\mathbf{y}$ is a solution to $A\mathbf{x} = \mathbf{0}$ and it follows that every member of span S is a solution to $A\mathbf{x} = \mathbf{0}$.

T.3. (a) Set $A = \begin{bmatrix} a_{ij} \end{bmatrix}$. Since the dimension of the null space of A is 3, the null space of A is R^3. Then the natural basis $\{\mathbf{e}_1, \mathbf{e}_2, \mathbf{e}_3\}$ is a basis for the null space of A. Forming $A\mathbf{e}_1 = \mathbf{0}$, $A\mathbf{e}_2 = \mathbf{0}$, $A\mathbf{e}_3 = \mathbf{0}$, we find that all the columns of A must be zero. Hence, $A = O$.

(b) Since $A\mathbf{x} = \mathbf{0}$ has a nontrivial solution, the null space of A contains a nonzero vector, so the dimension of the null space of A is not zero. If this dimension is 3, then by part (a), $A = O$, a contradiction. Hence, the dimension is either 1 or 2.

ML.1. Enter A into MATLAB and we find that

rref(A)

ans =

$$\begin{array}{rrrrr} 1 & 0 & 2 & 1 & 2 \\ 0 & 1 & 0 & 1 & -1 \\ 0 & 0 & 0 & 0 & 0 \end{array}$$

Write out the solution to the linear system $A\mathbf{x} = \mathbf{0}$ as

$$\mathbf{x} = r \begin{bmatrix} -2 \\ 0 \\ 1 \\ 0 \\ 0 \end{bmatrix} + s \begin{bmatrix} -1 \\ -1 \\ 0 \\ 1 \\ 0 \end{bmatrix} + t \begin{bmatrix} -2 \\ 1 \\ 0 \\ 0 \\ 1 \end{bmatrix}.$$

A basis for the null space of A consists of the three vectors above. We can compute such a basis directly using the command **homsoln** as shown next.

homsoln(A)

ans =

$$\begin{array}{rrr} -2 & -1 & -2 \\ 0 & -1 & 1 \\ 1 & 0 & 0 \\ 0 & 1 & 0 \\ 0 & 0 & 1 \end{array}$$

ML.3. Enter A into MATLAB and we find that

rref(A)

ans =

$$\begin{array}{rrrr} 1.0000 & 0 & -1.0000 & -1.3333 \\ 0 & 1.0000 & 2.0000 & 0.3333 \\ 0 & 0 & 0 & 0 \end{array}$$

format rat, ans

ans =

$$\begin{array}{rrrr} 1 & 0 & -1 & -4/3 \\ 0 & 1 & 2 & 1/3 \\ 0 & 0 & 0 & 0 \end{array}$$

format

Write out the solution to the linear system $A\mathbf{x} = \mathbf{0}$ as

$$\mathbf{x} = r \begin{bmatrix} 1 \\ -2 \\ 1 \\ 0 \end{bmatrix} + s \begin{bmatrix} \frac{4}{3} \\ -\frac{1}{3} \\ 0 \\ 1 \end{bmatrix}.$$

A basis for the null space of A consists of the two vectors above. We can compute such a basis directly using command **homsoln** as shown next.

homsoln(A)

```
ans =
        1.0000    1.3333
       -2.0000   -0.3333
        1.0000         0
             0    1.0000
```

format rat, ans

```
ans =
        1      4/3
       -2     -1/3
        1        0
        0        1
```

format

ML.5. Form the matrix $6I_3 - A$ in MATLAB as follows.

C = 6 * eye(3) − [1   2   3;3   2   1;2   1   3]

```
C =
        5   -2   -3
       -3    4   -1
       -2   -1    3
```

rref(C)

```
ans =
        1    0   -1
        0    1   -1
        0    0    0
```

The solution is $\mathbf{x} = \begin{bmatrix} t \\ t \\ t \end{bmatrix}$, for t any nonzero real number. Just choose $t \neq 0$ to obtain a nontrivial solution.

Section 6.6, p. 337

1. We follow the procedure used in Example 1. Let $S = \{\mathbf{v}_1, \mathbf{v}_2, \mathbf{v}_3, \mathbf{v}_4, \mathbf{v}_5\}$, where

$$\mathbf{v}_1 = (1, 2, 3), \quad \mathbf{v}_2 = (2, 1, 4), \quad \mathbf{v}_3 = (-1, -1, 2), \quad \mathbf{v}_4 = (0, 1, 2), \quad \mathbf{v}_5 = (1, 1, 1).$$

To find a basis for $V = \operatorname{span} S$, we form a matrix A with rows $\mathbf{v}_1, \mathbf{v}_2, \mathbf{v}_3, \mathbf{v}_4, \mathbf{v}_5$ and transform it to reduced row echelon form. The nonzero rows of the reduced row echelon form are a basis for V.

$$A = \begin{bmatrix} 1 & 2 & 3 \\ 2 & 1 & 4 \\ -1 & -1 & 2 \\ 0 & 1 & 2 \\ 1 & 1 & 1 \end{bmatrix}$$

and applying row operations we obtain the reduced row echelon form

$$\begin{bmatrix} 1 & 0 & 0 \\ 0 & 1 & 0 \\ 0 & 0 & 1 \\ 0 & 0 & 0 \\ 0 & 0 & 0 \end{bmatrix}.$$

Thus $\{(1,0,0),(0,1,0),(0,0,1)\}$ is a basis for V. (Hence, in this case $V = R^3$.)

3. We follow the procedure used in Example 3. Let $S = \{\mathbf{v}_1, \mathbf{v}_2, \mathbf{v}_3, \mathbf{v}_4, \mathbf{v}_5\}$, where

$$\mathbf{v}_1 = \begin{bmatrix} 1 \\ 2 \\ 1 \\ 2 \end{bmatrix}, \quad \mathbf{v}_2 = \begin{bmatrix} 2 \\ 1 \\ 2 \\ 1 \end{bmatrix}, \quad \mathbf{v}_3 = \begin{bmatrix} 3 \\ 2 \\ 3 \\ 2 \end{bmatrix}, \quad \mathbf{v}_4 = \begin{bmatrix} 3 \\ 3 \\ 3 \\ 3 \end{bmatrix}, \quad \mathbf{v}_5 = \begin{bmatrix} 5 \\ 3 \\ 5 \\ 3 \end{bmatrix}.$$

To find a basis for $V = \text{span } S$, we form a matrix A^T whose columns are the vectors $\mathbf{v}_1, \mathbf{v}_2, \mathbf{v}_3, \mathbf{v}_4, \mathbf{v}_5$ written in row form and transform it to reduced row echelon form. The nonzero rows of the reduced row echelon form written as columns are a basis for V.

$$A^T = \begin{bmatrix} 1 & 2 & 1 & 2 \\ 2 & 1 & 2 & 1 \\ 3 & 2 & 3 & 2 \\ 3 & 3 & 3 & 3 \\ 5 & 3 & 5 & 3 \end{bmatrix}.$$

Applying row operations we obtain the reduced row echelon form

$$\begin{bmatrix} 1 & 0 & 1 & 0 \\ 0 & 1 & 0 & 1 \\ 0 & 0 & 0 & 0 \\ 0 & 0 & 0 & 0 \\ 0 & 0 & 0 & 0 \end{bmatrix}.$$

Vectors

$$\left\{ \begin{bmatrix} 1 \\ 0 \\ 1 \\ 0 \end{bmatrix}, \begin{bmatrix} 0 \\ 1 \\ 0 \\ 1 \end{bmatrix} \right\}$$

are a basis for V.

Notation: We use the symbol $\text{rref}(A)$ to denote the reduced row echelon form of matrix A.

5. (a) To find a basis for the row space of A that are not row vectors of A we compute $\text{rref}(A)$ and select the nonzero rows as in Example 1.

$$\text{rref}(A) = \begin{bmatrix} 1 & 0 & -1 \\ 0 & 1 & 0 \\ 0 & 0 & 0 \\ 0 & 0 & 0 \end{bmatrix}$$

Thus $\{(1,0,-1),(0,1,0)\}$ is a basis for the row space of A.

(b) To find a basis for the row space of A that are row vectors of A we use the method described in Example 3. Compute $\text{rref}(A^T)$ and then use the leading 1's to point to the rows of A that form a basis for the row space of A.

$$\text{rref}(A^T) = \begin{bmatrix} 1 & 0 & -5 & -3 \\ 0 & 1 & 2 & 1 \\ 0 & 0 & 0 & 0 \end{bmatrix}$$

The leading 1's are in columns 1 and 2, hence rows 1 and 2 of A form a basis for the row space.

7. (a) To find a basis for the column space of A that does not consist of the columns of A we follow the procedure in Example 4. Compute $\text{rref}(A^T)$ and then take the transposes of the nonzero rows to get the desired basis.

$$\text{rref}(A^T) = \begin{bmatrix} 1 & 0 & 0 & 0 \\ 0 & 1 & 0 & \frac{1}{5} \\ 0 & 0 & 1 & \frac{3}{5} \\ 0 & 0 & 0 & 0 \end{bmatrix}$$

Thus

$$S = \left\{ \begin{bmatrix} 1 \\ 0 \\ 0 \\ 0 \end{bmatrix}, \begin{bmatrix} 0 \\ 1 \\ 0 \\ \frac{1}{5} \end{bmatrix}, \begin{bmatrix} 0 \\ 0 \\ 1 \\ \frac{3}{5} \end{bmatrix} \right\}$$

is a basis for the column space of A.

(b) To find a basis for the column space of A that consists of columns of A we follow the procedure in the second solution in Example 4. Compute $\text{rref}(A)$ and use the leading 1's to point to the columns of A that form the desired basis.

$$\text{rref}(A) = \begin{bmatrix} 1 & 0 & 1 & 0 \\ 0 & 1 & -3 & 0 \\ 0 & 0 & 0 & 1 \\ 0 & 0 & 0 & 0 \end{bmatrix}$$

The leading 1's are in columns 1, 2, and 4 and hence those columns of A form a basis for the column space of A.

9. (i) To find a basis for the row space of A that are not row vectors of A we compute $\text{rref}(A)$ and select the nonzero rows as in Example 1.

$$A = \begin{bmatrix} 1 & -2 & 5 \\ 2 & 3 & 2 \\ 0 & -7 & 8 \end{bmatrix}$$

$$\text{rref}(A) = \begin{bmatrix} 1 & 0 & \frac{19}{7} \\ 0 & 1 & -\frac{8}{7} \\ 0 & 0 & 0 \end{bmatrix}.$$

Thus $\left\{ \begin{bmatrix} 1 & 0 & \frac{19}{7} \end{bmatrix}, \begin{bmatrix} 0 & 1 & -\frac{8}{7} \end{bmatrix} \right\}$ is a basis for the row space of A.

(ii) To find a basis for the column space of A that does not consist of the columns of A we follow the procedure in Example 4. Compute $\text{rref}(A^T)$ and then take the transpose of the nonzero rows.

$$A^T = \begin{bmatrix} 1 & 2 & 0 \\ -2 & 3 & -7 \\ 5 & 2 & 8 \end{bmatrix}$$

$$\text{rref}(A^T) = \begin{bmatrix} 1 & 0 & 2 \\ 0 & 1 & -1 \\ 0 & 0 & 0 \end{bmatrix}$$

Thus $\left\{ \begin{bmatrix} 1 \\ 0 \\ 2 \end{bmatrix}, \begin{bmatrix} 0 \\ 1 \\ -1 \end{bmatrix} \right\}$ is a basis for the column space of A.

(iii) A basis for the row space of A^T consists of the transposes of a corresponding basis for the column space of A (see (ii)). Thus $\left\{ \begin{bmatrix} 1 & 0 & 2 \end{bmatrix}, \begin{bmatrix} 0 & 1 & -1 \end{bmatrix} \right\}$ is a basis for the row space of A^T.

(iv) A basis for the column space of A^T consists of the transposes of a corresponding basis for the row space of A (see (i)). Thus

$$\left\{ \begin{bmatrix} 1 \\ 0 \\ \frac{19}{7} \end{bmatrix}, \begin{bmatrix} 0 \\ 1 \\ -\frac{8}{7} \end{bmatrix} \right\}$$

is a basis for the column space of A^T.

11. The row rank of A is the number of nonzero rows in the row reduced echelon form of A. For

$$A = \begin{bmatrix} 1 & 2 & 3 & 2 & 1 \\ 3 & 1 & -5 & -2 & 1 \\ 7 & 8 & -1 & 2 & 5 \end{bmatrix}$$

the reduced row echelon form is

$$\begin{bmatrix} 1 & 0 & 0 & 0 & 0 \\ 0 & 1 & 0 & \frac{4}{13} & \frac{8}{13} \\ 0 & 0 & 1 & \frac{6}{13} & -\frac{1}{13} \end{bmatrix}$$

and hence A has row rank 3. The column rank of A is the number of nonzero rows in the reduced row echelon form of A^T. For matrix A this is

$$\begin{bmatrix} 1 & 0 & 0 \\ 0 & 1 & 0 \\ 0 & 0 & 1 \\ 0 & 0 & 0 \\ 0 & 0 & 0 \end{bmatrix}.$$

Thus the column rank of A is 3.

13. $\mathrm{rref}(A) = \begin{bmatrix} 1 & 0 & -3 & -\frac{13}{3} \\ 0 & 1 & 2 & \frac{11}{3} \\ 0 & 0 & 0 & 0 \\ 0 & 0 & 0 & 0 \end{bmatrix}$ and rank $A = 2$.

Since A represents the coefficients of a homogeneous system of four equations in four unknowns and $\mathrm{rref}(A)$ implies that the system is equivalent to two equations in four unknowns, it follows that two of the variables could be chosen arbitrarily so the nullity of A is 2. Hence

$$\text{rank } A + \text{nullity } A = 2 + 2 = n = 4.$$

15. $\mathrm{rref}(A) = \begin{bmatrix} 1 & 0 & 0 \\ 0 & 1 & 0 \\ 0 & 0 & 1 \end{bmatrix}$ and rank $A = 3$.

Since A represents the coefficients of a homogeneous system of three equations in three unknowns and $\mathrm{rref}(A)$ also represents a system of 3 equations in three unknowns, it follows that no variable can be chosen arbitrarily so the nullity of A is 0. Hence

$$\text{rank } A + \text{nullity } A = 3 + 0 = n = 3.$$

17. $\mathrm{rref}(A) = \begin{bmatrix} 1 & 0 & \frac{7}{3} \\ 0 & 1 & \frac{5}{3} \\ 0 & 0 & 0 \\ 0 & 0 & 0 \end{bmatrix}$ and rank $A = 2$.

Since A represents the coefficients of a homogeneous system of four equations in three unknowns and $\mathrm{rref}(A)$ implies that the system is equivalent to two equations in three unknowns, it follows that one of the variables could be chosen arbitrarily so the nullity of A is 1. Hence

$$\mathrm{rank}\ A + \mathrm{nullity}\ A = 2 + 1 = n = 3.$$

19. For a 4×6 matrix the largest possible rank is 4, since there can be at most 4 linearly independent rows. However, row rank = column rank; there can be at most 4 linearly independent columns. Thus the columns are linearly dependent.

21. Since A is 7×3 with rank 3 it follows that there are 3 linearly independent rows (rank = row rank); hence the rows of A are linearly dependent.

23. $\mathrm{rref}(A) = \begin{bmatrix} 1 & 0 & -3 \\ 0 & 1 & 0 \\ 0 & 0 & 0 \end{bmatrix}$ thus rank $A = 2$. By Theorem 6.13, A is singular.

25. $\mathrm{rref}(A) = I_4$, thus rank $A = 4$. By Theorem 6.13, A is nonsingular.

27. $\mathrm{rref}(A) = I_3$, thus rank $A = 3$. Hence Corollary 6.3 implies that $A\mathbf{x} = \mathbf{b}$ has a unique solution for every 3×1 matrix $\mathbf{b}$.

29. Form the matrix

$$A = \begin{bmatrix} 4 & 1 & 2 \\ 2 & 5 & -5 \\ 2 & -1 & 3 \end{bmatrix}.$$

Since $|A| = 0$, by Corollary 6.4 set S is linearly dependent in R^3.

31. $\mathrm{rref}(A) = \begin{bmatrix} 1 & 0 & 3 \\ 0 & 1 & 0 \\ 0 & 0 & 0 \end{bmatrix}$, thus rank $A = 2$.

By Corollary 6.5 the homogeneous linear system $A\mathbf{x} = \mathbf{0}$ has nontrivial solutions.

33. $\mathrm{rref}(A) = \begin{bmatrix} 1 & 0 & 0 & \frac{22}{83} \\ 0 & 1 & 0 & \frac{56}{83} \\ 0 & 0 & 1 & -\frac{60}{83} \end{bmatrix}$ and $\mathrm{rref}\left(\begin{bmatrix} A & \vdots & \mathbf{b} \end{bmatrix}\right)$ is the same but with a column of zeros attached.

Thus rank $A = \mathrm{rank}\begin{bmatrix} A & \vdots & \mathbf{b} \end{bmatrix}$, hence the linear system has a solution.

35. $\mathrm{rref}(A) = \begin{bmatrix} 1 & 0 & -1 & \frac{8}{7} \\ 0 & 1 & 1 & -\frac{10}{7} \\ 0 & 0 & 0 & 0 \end{bmatrix}$ and $\mathrm{rref}\left(\begin{bmatrix} A & \vdots & \mathbf{b} \end{bmatrix}\right) = \begin{bmatrix} 1 & 0 & -1 & \frac{8}{7} & \vdots & 0 \\ 0 & 1 & 1 & -\frac{10}{7} & \vdots & 0 \\ 0 & 0 & 0 & 0 & \vdots & 1 \end{bmatrix}$.

Hence rank $A \neq \mathrm{rank}\begin{bmatrix} A & \vdots & \mathbf{b} \end{bmatrix}$. Thus the linear system has no solution.

37. Let $A = \begin{bmatrix} 1 & 0 & 1 \\ 0 & 1 & 1 \\ 1 & 1 & 0 \end{bmatrix}$. Then $\mathrm{rref}(A) = \begin{bmatrix} 1 & 0 & 1 \\ 0 & 1 & 1 \\ 0 & 0 & 0 \end{bmatrix}$. Thus rank $A = 2$.

39. Let $A = \begin{bmatrix} 1 & 1 & 0 & 0 \\ 1 & 1 & 1 & 0 \\ 0 & 1 & 1 & 1 \\ 0 & 0 & 1 & 1 \end{bmatrix}$. Then $\mathrm{rref}(A) = \begin{bmatrix} 1 & 0 & 0 & 0 \\ 0 & 1 & 0 & 0 \\ 0 & 0 & 1 & 0 \\ 0 & 0 & 0 & 1 \end{bmatrix}$. Thus rank $A = 4$.

T.1. Rank $A = n$ if and only if A is nonsingular if and only if $\det A \neq 0$.

T.3. $S = \{\mathbf{v}_1, \mathbf{v}_2, \ldots, \mathbf{v}_n\}$ is linearly independent if and only if A has rank n if and only if $\det(A) \neq 0$.

T.5. Let rank $A = n$. Then $n = $ rank $A = $ column rank A implies that the n columns of A are linearly independent. Conversely, suppose the columns of A are linearly independent. Then $n = $ column rank $A = $ rank A.

T.7. Let $A\mathbf{x} = \mathbf{b}$ have a solution for every $m \times 1$ matrix $\mathbf{b}$. Then the columns of A span R^m. Thus there is a subset of m columns of A that is a basis for R^m and rank $A = m$. Conversely, if rank $A = m$, then column rank $A = m$. Thus m columns of A are a basis for R^m and hence all the columns of A span R^m. Since $\mathbf{b}$ is in R^m, it is a linear combination of the columns of A; that is, $A\mathbf{x} = \mathbf{b}$ has a solution for every $m \times 1$ matrix $\mathbf{b}$.

T.9. Suppose that the linear system $A\mathbf{x} = \mathbf{b}$ has at most one solution for every $m \times 1$ matrix $\mathbf{b}$. Since $A\mathbf{x} = \mathbf{0}$ always has the trivial solution, then $A\mathbf{x} = \mathbf{0}$ has only the trivial solution. Conversely, suppose that $A\mathbf{x} = \mathbf{0}$ has only the trivial solution. Then nullity $A = 0$, so by Theorem 6.12, rank $A = n$. Thus, $\dim($column space $A) = n$, so the n columns of A, which span its column space, form a basis for the column space. If $\mathbf{b}$ is an $m \times 1$ matrix then $\mathbf{b}$ is a vector in R^m. If $\mathbf{b}$ is in the column space of A, then B can be written as a linear combination of the columns of A in one and only one way. That is, $A\mathbf{x} = \mathbf{b}$ has exactly one solution. If $\mathbf{b}$ is not in the column space of A, then $A\mathbf{x} = \mathbf{b}$ has no solution. Thus, $A\mathbf{x} = \mathbf{b}$ has at most one solution.

T.11. We prove the contrapositive. Suppose that $A\mathbf{x} = \mathbf{b}$ is consistent. Assume that there are at least two different solutions $\mathbf{x}_1$ and $\mathbf{x}_2$. Then $A\mathbf{x}_1 = \mathbf{b}$ and $A\mathbf{x}_2 = \mathbf{b}$, so

$$A(\mathbf{x}_1 - \mathbf{x}_2) = A\mathbf{x}_1 - A\mathbf{x}_2 = \mathbf{b} - \mathbf{b} = \mathbf{0}.$$

That is, $A\mathbf{x} = \mathbf{0}$ has a nontrivial solution so nullity $A > 0$. By Theorem 6.12, rank $A < n$. Conversely, if rank $A < n$, then by Corollary 6.5, $A\mathbf{x} = \mathbf{0}$ has a nontrivial solution $\mathbf{y}$. Suppose that $\mathbf{x}_0$ is a solution to $A\mathbf{x} = \mathbf{b}$. Thus, $A\mathbf{y} = \mathbf{0}$ and $A\mathbf{x}_0 = \mathbf{b}$. Then $\mathbf{x}_0 + \mathbf{y}$ is a solution to $A\mathbf{x} = \mathbf{b}$, since

$$A(\mathbf{x}_0 + \mathbf{y}) = A\mathbf{x}_0 + A\mathbf{y} = \mathbf{b} + \mathbf{0} = \mathbf{b}.$$

Since $\mathbf{y} \neq \mathbf{0}$, $\mathbf{x}_0 + \mathbf{y} \neq \mathbf{x}_0$, so $A\mathbf{x} = \mathbf{b}$ has more than one solution.

ML.1. (a) Exercise 1

```
>>A = [1  2  3;2  1  4;−1  −1  2;0  1  2;1  1  1];
A =
       1     2    3
       2     1    4
      −1    −1    2
       0     1    2
       1     1    1
>>rref(A)
ans =
       1    0    0
       0    1    0
       0    0    1
       0    0    0
       0    0    0
```

So $\mathbf{w}_1 = (1,0,0)$, $\mathbf{w}_2 = (0,1,0)$, and $\mathbf{w}_3 = (0,0,1)$ form a basis for V.

(b) Exercise 2

```
>>A = [1  1  2  1;1  0  -3  1;0  1  1  2;0  0  1  1;1  0  0  1];
A =
     1   1    2   1
     1   0   -3   1
     0   1    1   2
     0   0    1   1
     1   0    0   1
>>rref(A)
A =
     1   0   0   0
     0   1   0   0
     0   0   1   0
     0   0   0   1
     0   0   0   0
```

So $\mathbf{w}_1 = (1,0,0,0)$, $\mathbf{w}_2 = (0,1,0,0)$, $\mathbf{w}_3 = (0,0,1,0)$, and $\mathbf{w}_4 = (0,0,0,1)$ form a basis for V.

(c) Exercise 3

```
>>A = [1  2  1  2;2  1  2  1;3  2  3  2;3  3  3  3;5  3  5  3];
A =
     1   2   1   2
     2   1   2   1
     3   2   3   2
     3   3   3   3
     5   3   5   3
>>rref(A)
ans =
     1   0   1   0
     0   1   0   1
     0   0   0   0
     0   0   0   0
     0   0   0   0
```

So $\mathbf{w}_1 = (1,0,1,0)$ and $\mathbf{w}_2 = (0,1,0,1)$ form a basis for V.

(d) Exercise 4

```
>>A = [1  2  1  1;2  1  3  1;0  2  1  2;3  2  1  4;5  0  0  -1];
A =
     1   2   1    1
     2   1   3    1
     0   2   1    2
     3   2   1    4
     5   0   0   -1
>>rref(A)
```

ans =

$$\begin{matrix} 1 & 0 & 0 & 0 \\ 0 & 1 & 0 & 0 \\ 0 & 0 & 1 & 0 \\ 0 & 0 & 0 & 1 \\ 0 & 0 & 0 & 0 \end{matrix}$$

So $\mathbf{w}_1 = (1,0,0,0)$, $\mathbf{w}_2 = (0,1,0,0)$, $\mathbf{w}_3 = (0,0,1,0)$, and $\mathbf{w}_4 = (0,0,0,1)$ form a basis for V.

ML.3. (a) The transposes of the nonzero rows of rref(A^T) give us one basis for the column space of A.

A = [1 3 1;2 5 0;4 11 2;6 9 1];

rref(A′)

ans =

$$\begin{matrix} 1 & 0 & 2 & 0 \\ 0 & 1 & 1 & 0 \\ 0 & 0 & 0 & 1 \end{matrix}$$

The leading ones of rref(A) point to the columns of A that form a basis for the column space of A.

rref(A)

ans =

$$\begin{matrix} 1 & 0 & 0 \\ 0 & 1 & 0 \\ 0 & 0 & 1 \\ 0 & 0 & 0 \end{matrix}$$

Thus columns 1, 2, and 3 of A are a basis for the column space of A.

(b) Follow the same procedure as in part (a).

A = [2 1 2 0;0 0 0 0;1 2 2 1;4 5 6 2;3 3 4 1];

rref(A′)

ans =

$$\begin{matrix} 1 & 0 & 0 & 1 & 1 \\ 0 & 0 & 1 & 2 & 1 \\ 0 & 0 & 0 & 0 & 0 \\ 0 & 0 & 0 & 0 & 0 \end{matrix}$$

rref(A)

ans =

$$\begin{matrix} 1.0000 & 0 & 0.6667 & -0.3333 \\ 0 & 1.0000 & 0.6667 & 0.6667 \\ 0 & 0 & 0 & 0 \\ 0 & 0 & 0 & 0 \end{matrix}$$

Thus columns 1 and 2 of A are a basis for the column space of A.

ML.5. Compare the rank of the coefficient matrix with the rank of the augmented matrix as in Theorem 6.14.

(a) **A = [1 2 4 − 1;0 1 2 0;3 1 1 − 2];b = [21 8 16]′;**

rank(A),rank([A b])

ans =

3

ans =

   3

The system is consistent.

(b) $\mathbf{A} = [1 \ \ 2 \ \ 1; 1 \ \ 1 \ \ 0; 2 \ \ 1 - 1]; \mathbf{b} = [3 \ \ 3 \ \ 3]';$
    rank(A),rank([A b])

  ans =

   2

  ans =

   3

The system is inconsistent.

(c) $\mathbf{A} = [1 \ \ 2; 2 \ \ 0; 2 \ \ 1; -1 \ \ 2]; \mathbf{b} = [3 \ \ 2 \ \ 3 \ \ 2]';$
    rank(A),rank([A b])

  ans =

   2

  ans =

   3

The system is inconsistent.

Section 6.7, p. 349

1. Since S is the natural basis for R^2 we have

$$\mathbf{v} = \begin{bmatrix} 3 \\ -2 \end{bmatrix} = 3 \begin{bmatrix} 1 \\ 0 \end{bmatrix} - 2 \begin{bmatrix} 0 \\ 1 \end{bmatrix}$$

and hence $[\mathbf{v}]_S = \begin{bmatrix} 3 \\ -2 \end{bmatrix}$.

3. Following the procedure in Example 3(b), we solve for c_1 and c_2 where

$$c_1(t+1) + c_2(t-2) = t+4.$$

Expanding the left side and collecting terms we have

$$(c_1 + c_2)t + (c_1 - 2c_2) = t + 4.$$

Equating the coefficients of like terms we obtain the system of equations

$$c_1 + c_2 = 1$$
$$c_1 - 2c_2 = 4.$$

Form the augmented matrix

$$\begin{bmatrix} 1 & 1 & \vdots & 1 \\ 1 & -2 & \vdots & 4 \end{bmatrix}.$$

Its reduced row echelon form is

$$\begin{bmatrix} 1 & 0 & \vdots & 2 \\ 0 & 1 & \vdots & -1 \end{bmatrix},$$

hence $[\mathbf{v}]_S = \begin{bmatrix} 2 \\ -1 \end{bmatrix}$.

5. Form the linear combination

$$c_1 \begin{bmatrix} 1 & 0 \\ 0 & 0 \end{bmatrix} + c_2 \begin{bmatrix} 0 & 0 \\ 1 & 0 \end{bmatrix} + c_3 \begin{bmatrix} 0 & 1 \\ 0 & 0 \end{bmatrix} + c_4 \begin{bmatrix} 0 & 0 \\ 0 & 1 \end{bmatrix} = \begin{bmatrix} 1 & 0 \\ -1 & 2 \end{bmatrix}.$$

Combining terms on the left side gives

$$\begin{bmatrix} c_1 & c_3 \\ c_2 & c_4 \end{bmatrix} = \begin{bmatrix} 1 & 0 \\ -1 & 2 \end{bmatrix}.$$

It follows that $\begin{bmatrix} \mathbf{v} \end{bmatrix}_S = \begin{bmatrix} 1 \\ -1 \\ 0 \\ 2 \end{bmatrix}.$

7. $\mathbf{v} = 1 \begin{bmatrix} 2 \\ 1 \end{bmatrix} + 2 \begin{bmatrix} -1 \\ 1 \end{bmatrix} = \begin{bmatrix} 0 \\ 3 \end{bmatrix}.$

9. $\mathbf{v} = -2t + 3(2t - 1) = 4t - 3.$

11. $\mathbf{v} = 2 \begin{bmatrix} -1 & 0 \\ 1 & 0 \end{bmatrix} + 1 \begin{bmatrix} 2 & 2 \\ 0 & 1 \end{bmatrix} - 1 \begin{bmatrix} 1 & 2 \\ -1 & 3 \end{bmatrix} + 3 \begin{bmatrix} 0 & 0 \\ 2 & 3 \end{bmatrix} = \begin{bmatrix} -1 & 0 \\ 9 & 7 \end{bmatrix}.$

13. Let

$$S = \{(1,2), (0,1)\} \quad \text{and} \quad T = \{(1,1), (2,3)\}$$

be ordered bases of R^2 and let $\mathbf{v} = (1,5)$ and $\mathbf{w} = (5,4)$ be vectors in R^2.

(a) Find $\begin{bmatrix} \mathbf{v} \end{bmatrix}_T$ and $\begin{bmatrix} \mathbf{w} \end{bmatrix}_T$. We express $\mathbf{v}$ and $\mathbf{w}$ in terms of the vectors in the T-basis.

$$c_1(1,1) + c_2(2,3) = \mathbf{v} = (1,5)$$

leads to the linear system with augmented matrix

$$\begin{bmatrix} 1 & 2 & \vdots & 1 \\ 1 & 3 & \vdots & 5 \end{bmatrix}.$$

Applying row operations $-\mathbf{r}_1 + \mathbf{r}_2$ and $-2\mathbf{r}_2 + \mathbf{r}_1$ gives the reduced row echelon form

$$\begin{bmatrix} 1 & 0 & \vdots & -7 \\ 0 & 1 & \vdots & 4 \end{bmatrix}.$$

Hence $c_1 = -7$ and $c_2 = 4$. Thus

$$\begin{bmatrix} \mathbf{v} \end{bmatrix}_T = \begin{bmatrix} -7 \\ 4 \end{bmatrix}.$$

Similarly,

$$c_1(1,1) + c_2(2,3) = \mathbf{w} = (5,4)$$

leads to the linear system with augmented matrix

$$\begin{bmatrix} 1 & 2 & \vdots & 5 \\ 1 & 3 & \vdots & 4 \end{bmatrix}.$$

Applying row operations $-\mathbf{r}_1 + \mathbf{r}_2$ and $-2\mathbf{r}_2 + \mathbf{r}_1$ gives the reduced row echelon form

$$\begin{bmatrix} 1 & 0 & \vdots & 7 \\ 0 & 1 & \vdots & -1 \end{bmatrix}.$$

Hence $c_1 = 7$ and $c_2 = -1$. Thus

$$[\mathbf{w}]_T = \begin{bmatrix} 7 \\ -1 \end{bmatrix}.$$

Note: In each of the linear systems above the coefficient matrix was the same. Only the right-hand side of the linear system changed. Hence we could compute the coordinates of $\mathbf{v}$ and $\mathbf{w}$ together by using the partitioned matrix

$$\begin{bmatrix} 1 & 2 & \vdots & 1 & \vdots & 5 \\ 1 & 3 & \vdots & 5 & \vdots & 4 \end{bmatrix}$$

with the same row operations. When put into reduced row echelon form the coordinates appear in the last two columns. For efficiency we usually drop all but the first set of vertical lines and write

$$\begin{bmatrix} 1 & 2 & \vdots & 1 & 5 \\ 1 & 3 & \vdots & 5 & 4 \end{bmatrix}.$$

This procedure generalizes to vector spaces other than R^2 and to more than two vectors.

(b) Find the transition matrix from the T-basis to the S-basis. We first find the coordinates of the vectors in the T-basis with respect to the S-basis. Hence we must solve the equations

$$c_1(1,2) + c_2(0,1) = (1,1)$$
$$c_1(1,2) + c_2(0,1) = (2,3).$$

Using the note in part (a), we must row reduce the matrix

$$\begin{bmatrix} 1 & 0 & \vdots & 1 & 2 \\ 2 & 1 & \vdots & 1 & 3 \end{bmatrix}.$$

Applying row operation $-2\mathbf{r}_1 + \mathbf{r}_2$ we have the reduced row echelon form

$$\begin{bmatrix} 1 & 0 & \vdots & 1 & 2 \\ 0 & 1 & \vdots & -1 & -1 \end{bmatrix}.$$

Thus the transition matrix from the T-basis to the S-basis is

$$P_{S \leftarrow T} = \begin{bmatrix} 1 & 2 \\ -1 & -1 \end{bmatrix}.$$

(c) We find the coordinates of $\mathbf{v}$ and $\mathbf{w}$ with respect to the S-basis by multiplying their coordinates with respect to the T-basis by the transition matrix. Using the results from parts (a) and (b) we have

$$[\mathbf{v}]_S = P_{S \leftarrow T}\, [\mathbf{v}]_T = \begin{bmatrix} 1 & 2 \\ -1 & -1 \end{bmatrix} \begin{bmatrix} -7 \\ 4 \end{bmatrix} = \begin{bmatrix} 1 \\ 3 \end{bmatrix}$$

$$[\mathbf{w}]_S = P_{S \leftarrow T}\, [\mathbf{w}]_T = \begin{bmatrix} 1 & 2 \\ -1 & -1 \end{bmatrix} \begin{bmatrix} 7 \\ -1 \end{bmatrix} = \begin{bmatrix} 5 \\ -6 \end{bmatrix}.$$

(d) To find the coordinate vectors of $\mathbf{v}$ and $\mathbf{w}$ with respect to the S-basis directly we proceed as in part (a). Here we find the coordinates using the partitioned matrix approach. Row reduce

$$\begin{bmatrix} 1 & 0 & \vdots & 1 & 5 \\ 2 & 1 & \vdots & 5 & 4 \end{bmatrix}.$$

Applying row operation $-2\mathbf{r}_1 + \mathbf{r}_2$ gives the reduced row echelon form

$$\begin{bmatrix} 1 & 0 & \vdots & 1 & 5 \\ 0 & 1 & \vdots & 3 & -6 \end{bmatrix}.$$

Hence

$$\left[\mathbf{v}\right]_S = \begin{bmatrix} 1 \\ 3 \end{bmatrix} \quad \text{and} \quad \left[\mathbf{w}\right]_S = \begin{bmatrix} 5 \\ -6 \end{bmatrix}.$$

(e) To find the transition matrix $Q_{T\leftarrow S}$ from the S-basis to the T-basis we find the coordinates of the vectors in S with respect to the T-basis. This leads to the linear systems

$$c_1(1,1) + c_2(2,3) = (1,2)$$
$$c_1(1,1) + c_2(2,3) = (0,1).$$

Combining these systems into the partitioned form discussed above, we row reduce the following matrix:

$$\begin{bmatrix} 1 & 2 & \vdots & 1 & 0 \\ 1 & 3 & \vdots & 2 & 1 \end{bmatrix}_{-1\mathbf{r}_1+\mathbf{r}_2} \longrightarrow \begin{bmatrix} 1 & 2 & \vdots & 1 & 0 \\ 0 & 1 & \vdots & 1 & 1 \end{bmatrix}_{-2\mathbf{r}_2+\mathbf{r}_1} \longrightarrow \begin{bmatrix} 1 & 0 & \vdots & -1 & -2 \\ 0 & 1 & \vdots & 1 & 1 \end{bmatrix}_{-1\mathbf{r}_1+\mathbf{r}_2}$$

Hence $Q_{T\leftarrow S} = \begin{bmatrix} -1 & -2 \\ 1 & 1 \end{bmatrix}$.

(f) From part (d) we have $\left[\mathbf{v}\right]_S = \begin{bmatrix} 1 \\ 3 \end{bmatrix}$ and $\left[\mathbf{w}\right]_S = \begin{bmatrix} 5 \\ -6 \end{bmatrix}$. Then

$$\left[\mathbf{v}\right]_T = Q_{T\leftarrow S}\left[\mathbf{v}\right]_S = \begin{bmatrix} -1 & -2 \\ 1 & 1 \end{bmatrix}\begin{bmatrix} 1 \\ 3 \end{bmatrix} = \begin{bmatrix} -7 \\ 4 \end{bmatrix}$$

$$\left[\mathbf{w}\right]_T = Q_{T\leftarrow S}\left[\mathbf{w}\right]_S = \begin{bmatrix} -1 & -2 \\ 1 & 1 \end{bmatrix}\begin{bmatrix} 5 \\ -6 \end{bmatrix} = \begin{bmatrix} 7 \\ -1 \end{bmatrix}.$$

15. Let $S = \{t^2+1, t-2, t+3\}$ and $T = \{2t^2+t, t^2+3, t\}$ be ordered bases of P_2 and let $\mathbf{v} = 8t^2 - 4t + 6$ and $\mathbf{w} = 7t^2 - t + 9$.

(a) Find $\left[\mathbf{v}\right]_T$ and $\left[\mathbf{w}\right]_T$. We express $\mathbf{v}$ and $\mathbf{w}$ in terms of the vectors in the T-basis:

$$c_1(2t^2+t) + c_2(t^2+3) + c_3(t) = 8t^2 - 4t + 6$$
$$c_1(2t^2+t) + c_2(t^2+3) + c_3(t) = 7t^2 - t + 9.$$

Combining like terms on the left and then equating like powers of t on the left and right leads to systems of equations with the same coefficient matrix but different right-hand sides. We express in the partition matrix form as used in Exercise 1 and row reduce it.

$$\begin{bmatrix} 2 & 1 & 0 & \vdots & 8 & 7 \\ 1 & 0 & 1 & \vdots & -4 & -1 \\ 0 & 3 & 0 & \vdots & 6 & 9 \end{bmatrix}_{\mathbf{r}_1 \leftrightarrow \mathbf{r}_2} \longrightarrow \begin{bmatrix} 2 & 1 & 0 & \vdots & 8 & 7 \\ 0 & 1 & -2 & \vdots & 16 & 9 \\ 0 & 3 & 0 & \vdots & 6 & 9 \end{bmatrix}_{-3\mathbf{r}_2+\mathbf{r}_3} \longrightarrow$$

$$\begin{bmatrix} 2 & 1 & 0 & \vdots & 8 & 7 \\ 0 & 1 & -2 & \vdots & 16 & 9 \\ 0 & 0 & 6 & \vdots & -42 & -18 \end{bmatrix}_{\substack{\frac{1}{6}\mathbf{r}_3 \\ 2\mathbf{r}_3 + \mathbf{r}_2 \\ -2\mathbf{r}_1 + \mathbf{r}_2}} \longrightarrow \begin{bmatrix} 1 & 0 & 0 & \vdots & 3 & 2 \\ 0 & 1 & 0 & \vdots & 2 & 3 \\ 0 & 0 & 1 & \vdots & -7 & -3 \end{bmatrix}.$$

Thus $\left[\mathbf{v}\right]_T = \begin{bmatrix} 3 \\ 2 \\ -7 \end{bmatrix}$ and $\left[\mathbf{w}\right]_T = \begin{bmatrix} 2 \\ 3 \\ -3 \end{bmatrix}.$

(b) Find the transition matrix from the T-basis to the S-basis. To do so, we first find the coordinates of the vectors in the T-basis with respect to the S-basis. Hence we must solve the equations

$$c_1(t^2 + 1) + c_2(t - 2) + c_3(t + 3) = 2t^2 + t, t^2 + 3, \text{ or } t.$$

Combining like terms on the left and then equating like powers of t on the left and right leads to systems of equations with the same coefficient matrix but different right-hand sides. We express the systems in the partition matrix form as used in Exercise 1 and row reduce it.

$$\begin{bmatrix} 1 & 0 & 0 & \vdots & 2 & 1 & 0 \\ 0 & 1 & 1 & \vdots & 1 & 0 & 1 \\ 1 & -2 & 3 & \vdots & 0 & 3 & 0 \end{bmatrix}_{-1r_1 + r_3} \longrightarrow \begin{bmatrix} 1 & 0 & 0 & \vdots & 2 & 1 & 0 \\ 0 & 1 & 1 & \vdots & 1 & 0 & 1 \\ 0 & -2 & 3 & \vdots & -2 & 2 & 0 \end{bmatrix}_{2r_2 + r_3} \longrightarrow$$

$$\begin{bmatrix} 1 & 0 & 0 & \vdots & 2 & 1 & 0 \\ 0 & 1 & 1 & \vdots & 1 & 0 & 1 \\ 0 & 0 & 5 & \vdots & 0 & 2 & 2 \end{bmatrix}_{\substack{\frac{1}{5}r_3 \\ -1r_3 + r_2}} \longrightarrow \begin{bmatrix} 1 & 0 & 0 & \vdots & 2 & 1 & 0 \\ 0 & 1 & 0 & \vdots & 1 & -\frac{2}{5} & \frac{3}{5} \\ 0 & 0 & 1 & \vdots & 0 & \frac{2}{5} & \frac{2}{5} \end{bmatrix}.$$

Thus $P_{S \leftarrow T} = \begin{bmatrix} 2 & 1 & 0 \\ 1 & -\frac{2}{5} & \frac{3}{5} \\ 0 & \frac{2}{5} & \frac{2}{5} \end{bmatrix}.$

(c) $[\mathbf{v}]_S = P_{S \leftarrow T} [\mathbf{v}]_T = \begin{bmatrix} 2 & 1 & 0 \\ 1 & -\frac{2}{5} & \frac{3}{5} \\ 0 & \frac{2}{5} & \frac{2}{5} \end{bmatrix} \begin{bmatrix} 3 \\ 2 \\ -7 \end{bmatrix} = \begin{bmatrix} 8 \\ -2 \\ -2 \end{bmatrix}.$

$[\mathbf{w}]_S = P_{S \leftarrow T} [\mathbf{w}]_T = \begin{bmatrix} 2 & 1 & 0 \\ 1 & -\frac{2}{5} & \frac{3}{5} \\ 0 & \frac{2}{5} & \frac{2}{5} \end{bmatrix} \begin{bmatrix} 2 \\ 3 \\ -3 \end{bmatrix} = \begin{bmatrix} 7 \\ -1 \\ 0 \end{bmatrix}.$

(d) To find the coordinate vectors of $\mathbf{v}$ and $\mathbf{w}$ with respect to the S-basis directly we proceed as in part (a). Here we find the coordinates using the partitioned matrix approach. Row reduce the following matrix:

$$\begin{bmatrix} 1 & 0 & 0 & \vdots & 8 & 7 \\ 0 & 1 & 1 & \vdots & -4 & -1 \\ 1 & -2 & 3 & \vdots & 6 & 9 \end{bmatrix}_{-1r_1 + r_3} \longrightarrow \begin{bmatrix} 1 & 0 & 0 & \vdots & 8 & 7 \\ 0 & 1 & 1 & \vdots & -4 & -1 \\ 0 & -2 & 3 & \vdots & -2 & 2 \end{bmatrix}_{2r_2 + r_3} \longrightarrow$$

$$\begin{bmatrix} 1 & 0 & 0 & \vdots & 8 & 7 \\ 0 & 1 & 1 & \vdots & -4 & -1 \\ 0 & 0 & 5 & \vdots & -10 & 0 \end{bmatrix}_{\substack{\frac{1}{5}r_3 \\ -1r_3 + r_2}} \longrightarrow \begin{bmatrix} 1 & 0 & 0 & \vdots & 8 & 7 \\ 0 & 1 & 0 & \vdots & -2 & -1 \\ 0 & 0 & 1 & \vdots & -2 & 0 \end{bmatrix}$$

Thus $[\mathbf{v}]_S = \begin{bmatrix} 8 \\ -2 \\ -2 \end{bmatrix}$ and $[\mathbf{w}]_S = \begin{bmatrix} 7 \\ -1 \\ 0 \end{bmatrix}.$

(e) To find the transition matrix $Q_{T \leftarrow S}$ from the S-basis to the T-basis we find the coordinates of the vectors in S with respect to the T-basis. This leads to the equations

$$c_1(2t^2 + t) + c_2(t^2 + 3) + c_3(t) = t^2 + 1, t - 2, \text{ or } t + 3.$$

Combining like terms on the left and then equating like powers of t on the left and right leads to systems of equations with the same coefficient matrix but different right-hand sides. We express the systems in the partitioned matrix form used in Exercise 1 and row reduce it. We have

$$\left[\begin{array}{ccc:ccc} 2 & 1 & 0 & 1 & 0 & 0 \\ 1 & 0 & 1 & 0 & 1 & 1 \\ 0 & 3 & 0 & 1 & -2 & 3 \end{array}\right]_{\mathbf{r}_1 \leftrightarrow \mathbf{r}_2} \longrightarrow \left[\begin{array}{ccc:ccc} 2 & 1 & 0 & 1 & 0 & 0 \\ 0 & 1 & -2 & 1 & -2 & -2 \\ 0 & 3 & 0 & 1 & -2 & 3 \end{array}\right]_{-3\mathbf{r}_2+\mathbf{r}_3} \longrightarrow$$

$$\left[\begin{array}{ccc:ccc} 2 & 1 & 0 & 1 & 0 & 0 \\ 0 & 1 & -2 & 1 & -2 & -2 \\ 0 & 0 & 6 & -2 & 4 & 9 \end{array}\right]_{\substack{\frac{1}{6}\mathbf{r}_3 \\ 2\mathbf{r}_3 + \mathbf{r}_2 \\ -2\mathbf{r}_1 + \mathbf{r}_2}} \longrightarrow \left[\begin{array}{ccc:ccc} 1 & 0 & 0 & \frac{1}{3} & \frac{1}{3} & -\frac{1}{2} \\ 0 & 1 & 0 & \frac{1}{3} & -\frac{2}{3} & 1 \\ 0 & 0 & 1 & -\frac{1}{3} & \frac{2}{3} & \frac{3}{2} \end{array}\right].$$

Thus $Q_{T \leftarrow S} = \begin{bmatrix} \frac{1}{3} & \frac{1}{3} & -\frac{1}{2} \\ \frac{1}{3} & -\frac{2}{3} & 1 \\ -\frac{1}{3} & \frac{2}{3} & \frac{3}{2} \end{bmatrix}.$

(f) $[\mathbf{v}]_T = Q_{T \leftarrow S}[\mathbf{v}]_S = \begin{bmatrix} \frac{1}{3} & \frac{1}{3} & -\frac{1}{2} \\ \frac{1}{3} & -\frac{2}{3} & 1 \\ -\frac{1}{3} & \frac{2}{3} & \frac{3}{2} \end{bmatrix} \begin{bmatrix} 8 \\ -2 \\ -2 \end{bmatrix} = \begin{bmatrix} 3 \\ 2 \\ -7 \end{bmatrix}$

$[\mathbf{w}]_T = Q_{T \leftarrow S}[\mathbf{w}]_S = \begin{bmatrix} \frac{1}{3} & \frac{1}{3} & -\frac{1}{2} \\ \frac{1}{3} & -\frac{2}{3} & 1 \\ -\frac{1}{3} & \frac{2}{3} & \frac{3}{2} \end{bmatrix} \begin{bmatrix} 7 \\ -1 \\ 0 \end{bmatrix} = \begin{bmatrix} 2 \\ 3 \\ -3 \end{bmatrix}$

17. Let

$$S = \{\mathbf{v}_1, \mathbf{v}_2, \mathbf{v}_3, \mathbf{v}_4\} = \left\{ \begin{bmatrix} 1 & 0 \\ 0 & 0 \end{bmatrix}, \begin{bmatrix} 0 & 1 \\ 1 & 0 \end{bmatrix}, \begin{bmatrix} 0 & 2 \\ 0 & 1 \end{bmatrix}, \begin{bmatrix} 0 & 0 \\ 1 & 1 \end{bmatrix} \right\}$$

and

$$T = \{\mathbf{w}_1, \mathbf{w}_2, \mathbf{w}_3, \mathbf{w}_4\} = \left\{ \begin{bmatrix} 1 & 1 \\ 0 & 0 \end{bmatrix}, \begin{bmatrix} 0 & 0 \\ 1 & 0 \end{bmatrix}, \begin{bmatrix} 0 & 0 \\ 0 & 1 \end{bmatrix}, \begin{bmatrix} 1 & 0 \\ 0 & 0 \end{bmatrix} \right\}$$

be ordered bases of M_{22} and let

$$\mathbf{v} = \begin{bmatrix} 1 & 1 \\ 1 & 1 \end{bmatrix} \quad \text{and} \quad \mathbf{w} = \begin{bmatrix} 1 & 2 \\ -2 & 1 \end{bmatrix}.$$

(a) Find $[\mathbf{v}]_T$ and $[\mathbf{w}]_T$. We express $\mathbf{v}$ and $\mathbf{w}$ in terms of the vectors in the T-basis. The equation

$$c_2\mathbf{w}_1 + c_2\mathbf{w}_2 + c_3\mathbf{w}_3 + c_4\mathbf{w}_4 = \mathbf{v}$$

leads to

$$\begin{bmatrix} c_1 + c_4 & c_1 \\ c_2 & c_3 \end{bmatrix} = \begin{bmatrix} 1 & 1 \\ 1 & 1 \end{bmatrix}.$$

Equating corresponding elements gives a linear system whose augmented matrix is

$$\left[\begin{array}{cccc:c} 1 & 0 & 0 & 1 & 1 \\ 1 & 0 & 0 & 0 & 1 \\ 0 & 1 & 0 & 0 & 1 \\ 0 & 0 & 1 & 0 & 1 \end{array}\right].$$

Applying row operation $-\mathbf{r}_1 + \mathbf{r}_2$ gives

$$\begin{bmatrix} 1 & 0 & 0 & 1 & \vdots & 1 \\ 0 & 0 & 0 & -1 & \vdots & 0 \\ 0 & 1 & 0 & 0 & \vdots & 1 \\ 0 & 0 & 1 & 0 & \vdots & 1 \end{bmatrix}.$$

It follows that $c_1 = c_2 = c_3 = 1$ and $c_4 = 0$. Thus

$$[\mathbf{v}]_T = \begin{bmatrix} 1 \\ 1 \\ 1 \\ 0 \end{bmatrix}.$$

To determine $[\mathbf{w}]_T$ we replace $\mathbf{v}$ in the previous steps by $\mathbf{w}$. The corresponding linear system is

$$\begin{bmatrix} c_1 + c_4 & c_1 \\ c_2 & c_3 \end{bmatrix} = \begin{bmatrix} 1 & 2 \\ -2 & 1 \end{bmatrix}.$$

Equating corresponding elements gives a linear system whose augmented matrix is

$$\begin{bmatrix} 1 & 0 & 0 & 1 & \vdots & 1 \\ 1 & 0 & 0 & 0 & \vdots & 2 \\ 0 & 1 & 0 & 0 & \vdots & -2 \\ 0 & 0 & 1 & 0 & \vdots & 1 \end{bmatrix}.$$

Row reducing this systems gives $c_1 = 2$, $c_2 = -2$, $c_3 = 1$, and $c_4 = -1$. Thus

$$[\mathbf{w}]_T = \begin{bmatrix} 2 \\ -2 \\ 1 \\ -1 \end{bmatrix}.$$

(b) Find the transition matrix from the T-basis to the S-basis. We first find the coordinates of the vectors in the T-basis with respect to S-basis. Hence we must solve the equations

$$c_1 \mathbf{v}_1 + c_2 \mathbf{v}_2 + c_3 \mathbf{v}_3 + c_4 \mathbf{v}_4 = \quad \mathbf{w}_1 \quad \vdots \quad \mathbf{w}_2 \quad \vdots \quad \mathbf{w}_3 \quad \vdots \quad \mathbf{w}_4 \ .$$

That is, there are really four equations to be solved. The right-hand side is the only thing that changes in each of the corresponding systems. Using the ideas from Exercise 13, we substitute in the vectors from M_{22}, equate corresponding components and obtain a set of linear systems with the same coefficient matrix but different right-hand sides. For efficiency we express this as in Exercise 13, using a partitioned matrix. We obtain the following matrix which we row reduce:

$$\begin{bmatrix} 1 & 0 & 0 & 0 & \vdots & 1 & 0 & 0 & 1 \\ 0 & 1 & 2 & 0 & \vdots & 1 & 0 & 0 & 0 \\ 0 & 1 & 0 & 1 & \vdots & 0 & 1 & 0 & 0 \\ 0 & 0 & 1 & 1 & \vdots & 0 & 0 & 1 & 0 \end{bmatrix}_{-1\mathbf{r}_2+\mathbf{r}_3}$$

$$\longrightarrow \begin{bmatrix} 1 & 0 & 0 & 0 & \vdots & 1 & 0 & 0 & 1 \\ 0 & 1 & 2 & 0 & \vdots & 1 & 0 & 0 & 0 \\ 0 & 0 & -2 & 1 & \vdots & -1 & 1 & 0 & 0 \\ 0 & 0 & 1 & 1 & \vdots & 0 & 0 & 1 & 0 \end{bmatrix}_{\mathbf{r}_3\leftrightarrow\mathbf{r}_4}$$

$$\longrightarrow \begin{bmatrix} 1 & 0 & 0 & 0 & \vdots & 1 & 0 & 0 & 1 \\ 0 & 1 & 2 & 0 & \vdots & 1 & 0 & 0 & 0 \\ 0 & 0 & 1 & 1 & \vdots & 0 & 0 & 1 & 0 \\ 0 & 0 & -2 & 1 & \vdots & -1 & 1 & 0 & 0 \end{bmatrix}_{\substack{-2\mathbf{r}_3+\mathbf{r}_2 \\ 2\mathbf{r}_3+\mathbf{r}_4}}$$

$$\longrightarrow \begin{bmatrix} 1 & 0 & 0 & 0 & \vdots & 1 & 0 & 0 & 1 \\ 0 & 1 & 0 & -2 & \vdots & 1 & 0 & -2 & 0 \\ 0 & 0 & 1 & 1 & \vdots & 0 & 0 & 1 & 0 \\ 0 & 0 & 0 & 3 & \vdots & -1 & 1 & 2 & 0 \end{bmatrix}_{\substack{(-\frac{1}{3})\,\mathbf{r}_4+\mathbf{r}_3 \\ \frac{2}{3}\mathbf{r}_4+\mathbf{r}_2 \\ \frac{1}{3}\mathbf{r}_4}}$$

$$\longrightarrow \begin{bmatrix} 1 & 0 & 0 & 0 & \vdots & 1 & 0 & 0 & 1 \\ 0 & 1 & 0 & 0 & \vdots & \frac{1}{3} & \frac{2}{3} & -\frac{2}{3} & 0 \\ 0 & 0 & 1 & 0 & \vdots & \frac{1}{3} & -\frac{1}{3} & \frac{1}{3} & 0 \\ 0 & 0 & 0 & 1 & \vdots & -\frac{1}{3} & \frac{1}{3} & \frac{2}{3} & 0 \end{bmatrix}$$

The coordinates of $\mathbf{w}_j$, $j = 1, 2, 3, 4$ are respectively the last four columns of the preceding matrix. Hence the transition matrix is the right-hand 4×4 block of the preceding matrix.

(c) We find the coordinates of $\mathbf{v}$ and $\mathbf{w}$ with respect to the S-basis by multiplying their coordinates with respect to the T-basis by the transition matrix. Using the results from parts (a) and (b), we have

$$[\mathbf{v}]_S = P_{S\leftarrow T}\,[\mathbf{v}]_T = \begin{bmatrix} 1 & 0 & 0 & 1 \\ \frac{1}{3} & \frac{2}{3} & -\frac{2}{3} & 0 \\ \frac{1}{3} & -\frac{1}{3} & \frac{1}{3} & 0 \\ -\frac{1}{3} & \frac{1}{3} & \frac{2}{3} & 0 \end{bmatrix} \begin{bmatrix} 1 \\ 1 \\ 1 \\ 0 \end{bmatrix} = \begin{bmatrix} 1 \\ \frac{1}{3} \\ \frac{1}{3} \\ \frac{2}{3} \end{bmatrix}$$

$$[\mathbf{w}]_S = P_{S\leftarrow T}\,[\mathbf{w}]_T = \begin{bmatrix} 1 & 0 & 0 & 1 \\ \frac{1}{3} & \frac{2}{3} & -\frac{2}{3} & 0 \\ \frac{1}{3} & -\frac{1}{3} & \frac{1}{3} & 0 \\ -\frac{1}{3} & \frac{1}{3} & \frac{2}{3} & 0 \end{bmatrix} \begin{bmatrix} 2 \\ -2 \\ 1 \\ -1 \end{bmatrix} = \begin{bmatrix} 1 \\ -\frac{4}{3} \\ \frac{5}{3} \\ -\frac{2}{3} \end{bmatrix}.$$

(d) To find the coordinate vectors of $\mathbf{v}$ and $\mathbf{w}$ with respect to the S-basis directly we proceed as in part (a). Here we find the coordinates using the partitioned matrix approach. From

$$c_1\mathbf{v}_1 + c_2\mathbf{v}_2 + c_3\mathbf{v}_3 + c_4\mathbf{v}_4 = \begin{bmatrix} \mathbf{v} & \vdots & \mathbf{w} \end{bmatrix}$$

we obtain linear systems whose augmented matrices written in partitioned form are represented by

$$\begin{bmatrix} 1 & 0 & 0 & 0 & \vdots & 1 & 1 \\ 0 & 1 & 2 & 0 & \vdots & 1 & 2 \\ 0 & 1 & 0 & 1 & \vdots & 1 & -2 \\ 0 & 0 & 1 & 1 & \vdots & 1 & 1 \end{bmatrix}.$$

Performing the same row operations as in part (b) we obtain

$$\begin{bmatrix} 1 & 0 & 0 & 0 & \vdots & 1 & 1 \\ 0 & 1 & 0 & 0 & \vdots & \frac{1}{3} & -\frac{4}{3} \\ 0 & 0 & 1 & 0 & \vdots & \frac{1}{3} & \frac{5}{3} \\ 0 & 0 & 0 & 1 & \vdots & \frac{2}{3} & -\frac{2}{3} \end{bmatrix}.$$

Thus $\left[\mathbf{v}\right]_S = \begin{bmatrix} 1 \\ \frac{1}{3} \\ \frac{1}{3} \\ \frac{2}{3} \end{bmatrix}$ and $\left[\mathbf{w}\right]_S = \begin{bmatrix} 1 \\ -\frac{4}{3} \\ \frac{5}{3} \\ -\frac{2}{3} \end{bmatrix}.$

(e) We could proceed directly as in part (b) reversing the roles of the S and T bases. However, if we call $Q_{T \leftarrow S}$ the transition matrix from the S- to the T-basis, then we have that $Q_{T \leftarrow S} = P_{S \leftarrow T}^{-1}$. We form the partitioned matrix $\left[P_{S \leftarrow T} \ \vdots \ I_4\right]$ and obtain its reduced row echelon form. The result is $\left[I_4 \ \vdots \ P_{S \leftarrow T}^{-1}\right]$. Hence

$$Q_{T \leftarrow S} = P_{S \leftarrow T}^{-1} = \begin{bmatrix} 0 & 1 & 2 & 0 \\ 0 & 1 & 0 & 1 \\ 0 & 0 & 1 & 1 \\ 1 & -1 & -2 & 0 \end{bmatrix}.$$

(f) From part (d) we have $\left[\mathbf{v}\right]_S$ and $\left[\mathbf{w}\right]_S$ and from part (e) we have $Q_{T \leftarrow S}$. Then

$$\left[\mathbf{v}\right]_T = Q_{T \leftarrow S} \left[\mathbf{v}\right]_S = \begin{bmatrix} 1 \\ 1 \\ 1 \\ 0 \end{bmatrix} \quad \text{and} \quad \left[\mathbf{w}\right]_T = Q_{T \leftarrow S} \left[\mathbf{w}\right]_S = \begin{bmatrix} 2 \\ -2 \\ 1 \\ -1 \end{bmatrix}.$$

19. We find the transition matrix $P_{S \leftarrow T}$ from the T-basis to the S-basis and then $\left[\mathbf{v}\right]_S = P_{S \leftarrow T} \left[\mathbf{v}\right]_T$. To find $P_{S \leftarrow T}$ we express the basis vectors in T in terms of those in S. The equations

$$c_1(1, -1) + c_2(2, 1) = (3, 0) \quad \text{and} \quad c_1(1, -1) + c_2(2, 1) = (4, -1)$$

lead to the partitioned matrix

$$\begin{bmatrix} 1 & 2 & \vdots & 3 & 4 \\ -1 & 1 & \vdots & 0 & -1 \end{bmatrix}$$

which we row reduce. We obtain

$$\begin{bmatrix} 1 & 0 & \vdots & 1 & 2 \\ 0 & 1 & \vdots & 1 & 1 \end{bmatrix}.$$

Thus it follows that $P_{S \leftarrow T} = \begin{bmatrix} 1 & 2 \\ 1 & 1 \end{bmatrix}$ and $\left[\mathbf{v}\right]_S = P_{S \leftarrow T} \left[\mathbf{v}\right]_T = \begin{bmatrix} 5 \\ 3 \end{bmatrix}.$

21. Following the procedure in Exercise 19, we are led to the partitioned matrix

$$\left[\begin{array}{ccc:ccc} -1 & 0 & 0 & -1 & 0 & -2 \\ 1 & 1 & 1 & 2 & 1 & 2 \\ 0 & 0 & 1 & 1 & 1 & 1 \end{array}\right].$$

Its reduced row echelon form is

$$\left[\begin{array}{ccc:ccc} 1 & 0 & 0 & 1 & 0 & 2 \\ 0 & 1 & 0 & 0 & 0 & -1 \\ 0 & 0 & 1 & 1 & 1 & 1 \end{array}\right]$$

and it follows that $Q_{T \leftarrow S}$, the transition matrix from S to T, is

$$Q_{T \leftarrow S} = \begin{bmatrix} 1 & 0 & 2 \\ 0 & 0 & -1 \\ 1 & 1 & 1 \end{bmatrix}.$$

Thus $\left[\mathbf{v}\right]_T = Q_{T \leftarrow S} \left[\mathbf{v}\right]_S = \begin{bmatrix} 1 & 0 & 2 \\ 0 & 0 & -1 \\ 1 & 1 & 1 \end{bmatrix} \begin{bmatrix} 2 \\ 0 \\ 1 \end{bmatrix} = \begin{bmatrix} 4 \\ -1 \\ 3 \end{bmatrix}.$

23. Denote the transition matrix from T to S by $P_{S \leftarrow T}$. Then from Equation (5) we have that the columns of $P_{S \leftarrow T}$ are the coordinates of the T-basis vectors with respect to the S-basis. Hence

$$\left[\mathbf{w}_1\right]_S = \begin{bmatrix} 1 \\ 2 \\ -1 \end{bmatrix} \implies \mathbf{w}_1 = 1\mathbf{v}_1 + 2\mathbf{v}_2 - \mathbf{v}_3 = \begin{bmatrix} 3 \\ 2 \\ 0 \end{bmatrix}$$

$$\left[\mathbf{w}_2\right]_S = \begin{bmatrix} 1 \\ 1 \\ -1 \end{bmatrix} \implies \mathbf{w}_2 = 1\mathbf{v}_1 + 1\mathbf{v}_2 - \mathbf{v}_3 = \begin{bmatrix} 2 \\ 1 \\ 0 \end{bmatrix}$$

$$\left[\mathbf{w}_3\right]_S = \begin{bmatrix} 2 \\ 1 \\ 1 \end{bmatrix} \implies \mathbf{w}_3 = 2\mathbf{v}_1 + 1\mathbf{v}_2 + 1\mathbf{v}_3 = \begin{bmatrix} 3 \\ 1 \\ 3 \end{bmatrix}.$$

25. Following the procedure used in Exercise 23, we have

$$\left[\mathbf{w}_1\right]_S = \begin{bmatrix} 2 \\ 1 \end{bmatrix} \implies \mathbf{w}_1 = 2\mathbf{v}_1 + 1\mathbf{v}_2 = \begin{bmatrix} 2 \\ 5 \end{bmatrix}$$

$$\left[\mathbf{w}_2\right]_S = \begin{bmatrix} 1 \\ 1 \end{bmatrix} \implies \mathbf{w}_2 = 1\mathbf{v}_1 + 1\mathbf{v}_2 = \begin{bmatrix} 1 \\ 3 \end{bmatrix}.$$

T.1. Let $\mathbf{v} = \mathbf{w}$. The coordinates of a vector relative to basis S are the coefficients used to express the vector in terms of the members of S. A vector has a unique expression in terms of the vectors of a basis, hence it follows that $\left[\mathbf{v}\right]_S$ must equal $\left[\mathbf{w}\right]_S$. Conversely, let

$$\left[\mathbf{v}\right]_S = \left[\mathbf{w}\right]_S = \begin{bmatrix} a_1 \\ a_2 \\ \vdots \\ a_n \end{bmatrix}.$$

Then

$$\mathbf{v} = a_1\mathbf{v}_1 + a_2\mathbf{v}_2 + \cdots + a_n\mathbf{v}_n \quad \text{and} \quad \mathbf{w} = a_1\mathbf{v}_1 + a_2\mathbf{v}_2 + \cdots + a_n\mathbf{v}_n.$$

Hence $\mathbf{v} = \mathbf{w}$.

T.3. Suppose that $\left\{ \left[\mathbf{w}_1 \right]_S, \left[\mathbf{w}_2 \right]_S, \ldots, \left[\mathbf{w}_k \right]_S \right\}$ is linearly dependent. Then there exist scalars a_i, $i = 1, 2, \ldots, k$, that are not all zero, such that

$$a_1 \left[\mathbf{w}_1 \right]_S + a_2 \left[\mathbf{w}_2 \right]_S + \cdots + a_n \left[\mathbf{w}_k \right]_S = \left[\mathbf{0}_V \right]_S.$$

Using Exercise T.2 we find that the preceding equation is equivalent to

$$\left[a_1 \mathbf{w}_1 + a_2 \mathbf{w}_2 + \cdots + a_k \mathbf{w}_k \right]_S = \left[\mathbf{0}_V \right]_S.$$

By Exercise T.1 we have

$$a_1 \mathbf{w}_1 + a_2 \mathbf{w}_2 + \cdots + a_k \mathbf{w}_k = \mathbf{0}_V.$$

Since the $\mathbf{w}$'s are linearly independent, the preceding equation is only true when all $a_i = 0$. Hence we have a contradiction and our assumption that the $\left[\mathbf{w}_i \right]_S$'s are linearly dependent must be false. It follows that $\left\{ \left[\mathbf{w}_1 \right]_S, \left[\mathbf{w}_2 \right]_S, \ldots, \left[\mathbf{w}_k \right]_S \right\}$ is linearly independent.

T.5. Consider the homogeneous system $M_S \mathbf{x} = \mathbf{0}$, where

$$\mathbf{x} = \begin{bmatrix} a_1 \\ a_2 \\ \vdots \\ a_n \end{bmatrix}.$$

This system can then be written in terms of the columns of M_S as

$$a_1 \mathbf{v}_1 + a_2 \mathbf{v}_2 + \cdots + a_n \mathbf{v}_n = \mathbf{0},$$

where $\mathbf{v}_j$ is the jth column of M_S. Since $\mathbf{v}_1, \mathbf{v}_2, \ldots, \mathbf{v}_n$ are linearly independent, we have $a_1 = a_2 = \cdots = a_n = 0$. Thus, $\mathbf{x} = \mathbf{0}$ is the only solution to $M_S \mathbf{x} = \mathbf{0}$, so by Theorem 1.13 (Sec. 1.7) we conclude that M_S is nonsingular.

T.7. (a) From Exercise T.6 we have

$$M_S \left[\mathbf{v} \right]_S = M_T \left[\mathbf{v} \right]_T.$$

From Exercise T.5 we know that M_S is nonsingular, so

$$\left[\mathbf{v} \right]_S = M_S^{-1} M_T \left[\mathbf{v} \right]_T.$$

Equation (2) is

$$\left[\mathbf{v} \right]_S = P_{S \leftarrow T} \left[\mathbf{v} \right]_T,$$

so

$$P_{S \leftarrow T} = M_S^{-1} M_T.$$

(b) Since M_S and M_T are nonsingular, M_S^{-1} is nonsingular, so $P_{S \leftarrow T}$, as the product of two nonsingular matrices, is nonsingular.

(c) $M_S = \begin{bmatrix} 2 & 1 & 1 \\ 0 & 2 & 1 \\ 1 & 0 & 1 \end{bmatrix}$, $M_T = \begin{bmatrix} 6 & 4 & 5 \\ 3 & -1 & 5 \\ 3 & 3 & 2 \end{bmatrix}$, $M_S^{-1} = \begin{bmatrix} \frac{2}{3} & -\frac{1}{3} & -\frac{1}{3} \\ \frac{1}{3} & \frac{1}{3} & -\frac{2}{3} \\ -\frac{2}{3} & \frac{1}{3} & \frac{4}{3} \end{bmatrix}$, $P_{S \leftarrow T} = \begin{bmatrix} 2 & 2 & 1 \\ 1 & -1 & 2 \\ 1 & 1 & 1 \end{bmatrix}.$

ML.1. Since S is a set consisting of three vectors in a 3-dimensional vector space, we can show that S is a basis by verifying that the vectors in S are linearly independent. It follows that if the reduced row echelon form of the three columns is I_3, they are linearly independent.

A = [1 2 1;2 1 0;1 0 2];

rref(A)

ans =

```
    1   0   0
    0   1   0
    0   0   1
```

To find the coordinates of **v** we solve the system $A\mathbf{c} = \mathbf{v}$. (Note that A is symmetric, so $A = A^T$.) We can do all three parts simultaneously as follows. Put the three columns whose coordinates we want to find into a matrix B.

B = [8 2 4;4 0 3;7 − 3 3];

rref([A B])

ans =

```
    1   0   0   1   −1   1
    0   1   0   2    2   1
    0   0   1   3   −1   1
```

The coordinates appear in the last three columns of the matrix above.

ML.3. Associate a column with each matrix and proceed as in ML.2.

A = [1 1 2 2;0 1 2 0;3 − 1 1 0;−1 0 0 0]′;

rref(A)

ans =

```
    1   0   0   0
    0   1   0   0
    0   0   1   0
    0   0   0   1
```

B = [1 0 0 1;2 7/6 10/3 2;1 1 1 1]′;

rref([A B])

ans =

1.0000	0	0	0	0.5000	1.0000	0.5000
0	1.0000	0	0	−0.5000	0.5000	0.1667
0	0	1.0000	0	0	0.3333	−0.3333
0	0	0	1.0000	−0.5000	0	−1.5000

The coordinates are the last three columns of the preceding matrix.

ML.5. **A = [0 0 1 − 1;0 0 1 1;0 1 1 0;1 0 − 1 0]′;**

B = [0 1 0 0;0 0 − 1 1;0 − 1 0 2;1 1 0 0]′;

rref([A B])

ans =

1.0000	0	0	0	−0.5000	−1.0000	−0.5000	0
0	1.0000	0	0	−0.5000	0	1.5000	0
0	0	1.0000	0	1.0000	0	−1.0000	1.0000
0	0	0	1.0000	0	0	0	1.0000

The transition matrix P is found in columns 5 through 8 of the preceding matrix.

ML.7. **S = [1 1 0;1 2 1;1 1 1];**
T = [1 1 0;0 1 1;1 0 2];
U = [2 − 1 1;1 2 − 2;1 1 1];

(a) The transition matrix from U to T will be the last 3 columns of rref([T U]).
rref([T U])

ans =

1.0000	0	0	1.0000	−1.6667	2.3333
0	1.0000	0	1.0000	0.6667	−1.3333
0	0	1.0000	0	1.3333	−0.6667

P = ans(:,4:6)

P =

1.0000	−1.6667	2.3333
1.0000	0.6667	−1.3333
0	1.3333	−0.6667

(b) The transition matrix from T to S will be the last 3 columns of rref([S T]).
rref([S T])

ans =

1	0	0	2	0	1
0	1	0	−1	1	−1
0	0	1	0	−1	2

Q = ans(:,4:6)

Q =

2	0	1
−1	1	−1
0	−1	2

(c) The transition matrix from U to S will be the last 3 columns of rref([S U]).
rref([S U])

ans =

1	0	0	2	−2	4
0	1	0	0	1	−3
0	0	1	−1	2	0

Z = ans(:,4:6)

Z =

2	−2	4
0	1	−3
−1	2	0

(d) **Q ∗ P** gives Z.

Section 6.8, p. 359

1. If the dot product of each pair of vectors is zero, then the set of vectors is orthogonal.

(a) $(1, -1, 2) \cdot (0, 2, -1) = -4$,
$(1, -1, 2) \cdot (-1, 1, 1) = 0$,
$(0, 2, -1) \cdot (-1, 1, 1) = 1$.
Thus the set is not orthogonal

(b) $(1, 2, -1, 1) \cdot (0, -1, -2, 0) = 0$
$(1, 2, -1, 1) \cdot (1, 0, 0, -1) = 0$
$(0, -1, -2, 0) \cdot (1, 0, 0, -1) = 0$
Thus the set is orthogonal.

(c) $(0, 1, 0, -1) \cdot (1, 0, 1, 1) = -1$
$(0, 1, 0, -1) \cdot (-1, 1, -1, 2) = -1$
$(1, 0, 1, 1) \cdot (-1, 1, -1, 2) = 0$
Thus the set is not orthogonal.

3. $\mathbf{u} \cdot \mathbf{v} = a - 1 - 4 = 0$ only if $a = 5$.

5. Let $\mathbf{u}_1 = (1, -1, 0)$ and $\mathbf{u}_2 = (2, 0, 1)$. Define $\mathbf{v}_1 = \mathbf{u}_1$. Compute

$$\mathbf{v}_2 = \mathbf{u}_2 - \left(\frac{\mathbf{u}_2 \cdot \mathbf{v}_1}{\mathbf{v}_1 \cdot \mathbf{v}_1} \right) \mathbf{v}_1 = (2, 0, 1) - \tfrac{2}{2}(1, -1, 0) = (1, 1, 1).$$

Then $\{\mathbf{v}_1, \mathbf{v}_2\}$ is an orthogonal basis for W. Let

$$\mathbf{w}_1 = \frac{1}{\|\mathbf{v}_1\|}\mathbf{v}_1 = \frac{1}{\sqrt{2}}(1, -1, 0) \quad \text{and} \quad \mathbf{w}_2 = \frac{1}{\|\mathbf{v}_2\|}\mathbf{v}_2 = \frac{1}{\sqrt{3}}(1, 1, 1).$$

Then $\{\mathbf{w}_1, \mathbf{w}_2\}$ is an orthonormal basis for W.

7. Let $\mathbf{u}_1 = (1, -1, 0, 1)$, $\mathbf{u}_2 = (2, 0, 0, -1)$, and $\mathbf{u}_3 = (0, 0, 1, 0)$. Define $\mathbf{v}_1 = \mathbf{u}_1$. Compute

$$\mathbf{v}_2 = \mathbf{u}_2 - \left(\frac{\mathbf{u}_2 \cdot \mathbf{v}_1}{\mathbf{v}_1 \cdot \mathbf{v}_1} \right) \mathbf{v}_1 = (2, 0, 0, -1) - \tfrac{1}{3}(1, -1, 0, 1) = \left(\tfrac{5}{3}, \tfrac{1}{3}, 0, -\tfrac{4}{3} \right),$$

$$\mathbf{v}_3 = \mathbf{u}_3 - \left(\frac{\mathbf{u}_3 \cdot \mathbf{v}_1}{\mathbf{v}_1 \cdot \mathbf{v}_1} \right) \mathbf{v}_1 - \left(\frac{\mathbf{u}_3 \cdot \mathbf{v}_2}{\mathbf{v}_2 \cdot \mathbf{v}_2} \right) \mathbf{v}_2$$
$$= (0, 0, 1, 0) - 0(1, -1, 0, 1) - 0 \left(\tfrac{5}{3}, \tfrac{1}{3}, 0, -\tfrac{4}{3} \right) = (0, 0, 1, 0).$$

Then $\{\mathbf{v}_1, \mathbf{v}_2, \mathbf{v}_3\}$ is an orthogonal basis for W. Clearing fractions in $\mathbf{v}_2$, we find that

$$\{\mathbf{v}_1', \mathbf{v}_2', \mathbf{v}_3'\} = \{(1, -1, 0, 1), (5, 1, 0, -4), (0, 0, 1, 0)\}$$

is also an orthogonal basis for W. Let

$$\mathbf{w}_1 = \frac{1}{\|\mathbf{v}_1'\|}\mathbf{v}_1' = \frac{1}{\sqrt{3}}(1, -1, 0, 1),$$

$$\mathbf{w}_2 = \frac{1}{\|\mathbf{v}_2'\|}\mathbf{v}_2' = \frac{1}{\sqrt{42}}(5, 1, 0, -4),$$

$$\mathbf{w}_3 = \frac{1}{\|\mathbf{v}_3'\|}\mathbf{v}_3' = (0, 0, 1, 0).$$

Then $\{\mathbf{w}_1, \mathbf{w}_2, \mathbf{w}_3\}$ is an orthonormal basis for W.

9. Let $\mathbf{u}_1 = (1, 2)$ and $\mathbf{u}_2 = (-3, 4)$.

(a) Define $\mathbf{v}_1 = \mathbf{u}_1$. Compute

$$\mathbf{v}_2 = \mathbf{u}_2 - \left(\frac{\mathbf{u}_2 \cdot \mathbf{v}_1}{\mathbf{v}_1 \cdot \mathbf{v}_1} \right) \mathbf{v}_1 = (-3, 4) - 1(1, 2) = (-4, 2).$$

Then $\{\mathbf{v}_1, \mathbf{v}_2\}$ is an orthogonal basis for R^2.

(b) Let

$$\mathbf{w}_1 = \frac{1}{\|\mathbf{v}_1\|}\mathbf{v}_1 = \frac{1}{\sqrt{5}}(1,2) \quad \text{and} \quad \mathbf{w}_2 = \frac{1}{\|\mathbf{v}_2\|}\mathbf{v}_2 = \frac{1}{2\sqrt{5}}(-4,2).$$

then $\{\mathbf{w}_1, \mathbf{w}_2\}$ is an orthonormal basis for R^2.

11. Let $\mathbf{u}_1 = \left(\frac{2}{3}, -\frac{2}{3}, \frac{1}{3}\right)$ and $\mathbf{u}_2 = \left(\frac{2}{3}, \frac{1}{3}, -\frac{2}{3}\right)$. We first find a basis for R^3 containing $\mathbf{u}_1$ and $\mathbf{u}_2$ following the technique of Example 9 of Section 6.4. Then we use the Gram–Schmidt process to transform it to an orthonormal basis. Let

$$\mathbf{e}_1 = (1,0,0), \quad \mathbf{e}_2 = (0,1,0), \quad \mathbf{e}_3 = (0,0,1) \quad \text{and} \quad S = \{\mathbf{u}_1, \mathbf{u}_2, \mathbf{e}_1, \mathbf{e}_2, \mathbf{e}_3\}.$$

S spans R^3 since it contains the basis $\{\mathbf{e}_1, \mathbf{e}_2, \mathbf{e}_3\}$. Form the equation

$$c_1\mathbf{u}_1 + c_2\mathbf{u}_2 + c_3\mathbf{e}_1 + c_4\mathbf{e}_2 + c_5\mathbf{e}_3 = (0,0,0)$$

which leads to the homogeneous system with augmented matrix

$$\begin{bmatrix} \frac{2}{3} & \frac{2}{3} & 1 & 0 & 0 & \vdots & 0 \\ -\frac{2}{3} & \frac{1}{3} & 0 & 1 & 0 & \vdots & 0 \\ \frac{1}{3} & -\frac{2}{3} & 0 & 0 & 1 & \vdots & 0 \end{bmatrix}.$$

Transforming this augmented matrix to reduced row echelon form we obtain

$$\begin{bmatrix} 1 & 0 & 0 & -2 & -1 & \vdots & 0 \\ 0 & 1 & 0 & -1 & -2 & \vdots & 0 \\ 0 & 0 & 1 & 2 & 2 & \vdots & 0 \end{bmatrix}.$$

The leading ones indicate that $\{\mathbf{u}_1, \mathbf{u}_2, \mathbf{e}_1\}$ is a basis for R^3.

Next we use the Gram–Schmidt process. Let $\mathbf{v}_1 = \mathbf{u}_1$. Compute

$$\mathbf{v}_2 = \mathbf{u}_2 - \left(\frac{\mathbf{u}_2 \cdot \mathbf{v}_1}{\mathbf{v}_1 \cdot \mathbf{v}_1}\right)\mathbf{v}_1 = \mathbf{u}_2 - 0\mathbf{u}_1 = \left(\frac{2}{3}, \frac{1}{3}, -\frac{2}{3}\right)$$

$$\mathbf{v}_3 = \mathbf{e}_1 - \left(\frac{\mathbf{e}_1 \cdot \mathbf{v}_1}{\mathbf{v}_1 \cdot \mathbf{v}_1}\right)\mathbf{v}_1 - \left(\frac{\mathbf{e}_1 \cdot \mathbf{v}_2}{\mathbf{v}_2 \cdot \mathbf{v}_2}\right)\mathbf{v}_2$$

$$= \mathbf{e}_1 - \frac{2}{3}\mathbf{v}_1 - \frac{2}{3}\mathbf{v}_2 = \left(\frac{1}{9}, \frac{2}{9}, \frac{2}{9}\right).$$

The set $\{\mathbf{v}_1, \mathbf{v}_2, \mathbf{v}_3\}$ is an orthogonal basis for R^3. Then $\{\mathbf{w}_1, \mathbf{w}_2, \mathbf{w}_3\}$ is an orthonormal basis, where

$$\mathbf{w}_1 = \frac{1}{\|\mathbf{v}_1\|}\mathbf{v}_1 = \left(\frac{2}{3}, -\frac{2}{3}, \frac{1}{3}\right),$$

$$\mathbf{w}_2 = \frac{1}{\|\mathbf{v}_2\|}\mathbf{v}_2 = \left(\frac{2}{3}, \frac{1}{3}, -\frac{2}{3}\right),$$

$$\mathbf{w}_3 = \frac{1}{\|\mathbf{v}_3\|}\mathbf{v}_3 = \left(\frac{1}{3}, \frac{2}{3}, \frac{2}{3}\right).$$

13. Let

$$\mathbf{u}_1 = (1,1,0,0), \quad \mathbf{u}_2 = (2,-1,0,1), \quad \mathbf{u}_3 = (3,-3,0,-2), \quad \mathbf{u}_4 = (1,-2,0,-3),$$

and

$$S = \{\mathbf{u}_1, \mathbf{u}_2, \mathbf{u}_3, \mathbf{u}_4\}.$$

We find a basis for $W = \text{span } S$ following the method in Example 5 of Section 6.4. To determine a basis for span S form the linear combination

$$c_1\mathbf{u}_1 + c_2\mathbf{u}_2 + c_3\mathbf{u}_3 + c_4\mathbf{u}_4 = \mathbf{0}.$$

Expanding, adding vectors, and equating corresponding components from each side of the equation we obtain a homogeneous system with coefficient matrix

$$\begin{bmatrix} 1 & 2 & 3 & 1 \\ 1 & -1 & -3 & -2 \\ 0 & 0 & 0 & 0 \\ 0 & 1 & -2 & -3 \end{bmatrix}$$

which has reduced row echelon form

$$\begin{bmatrix} 1 & 0 & 0 & 0 \\ 0 & 1 & 0 & -1 \\ 0 & 0 & 1 & 1 \\ 0 & 0 & 0 & 0 \end{bmatrix}.$$

The leading 1's are in columns 1, 2, and 3 so $S_1 = \{\mathbf{u}_1, \mathbf{u}_2, \mathbf{u}_3\}$ is a basis for span S. Next we use the Gram–Schmidt process. Let $\mathbf{v}_1 = \mathbf{u}_1$. Compute

$$\mathbf{v}_2 = \mathbf{u}_2 - \left(\frac{\mathbf{u}_2 \cdot \mathbf{v}_1}{\mathbf{v}_1 \cdot \mathbf{v}_1}\right)\mathbf{v}_1 = (2,-1,0,1) - \tfrac{1}{2}(1,1,0,0) = \left(\tfrac{3}{2}, -\tfrac{3}{2}, 0, 1\right)$$

$$\mathbf{v}_3 = \mathbf{u}_3 - \left(\frac{\mathbf{u}_3 \cdot \mathbf{v}_1}{\mathbf{v}_1 \cdot \mathbf{v}_1}\right)\mathbf{v}_1 - \left(\frac{\mathbf{u}_3 \cdot \mathbf{v}_2}{\mathbf{v}_2 \cdot \mathbf{v}_2}\right)\mathbf{v}_2$$

$$= (3,-3,0,-2) - 0(1,1,0,0) - \tfrac{14}{11}\left(\tfrac{3}{2}, -\tfrac{3}{2}, 0, 1\right) = \left(\tfrac{12}{11}, -\tfrac{12}{11}, 0, -\tfrac{36}{11}\right).$$

Then $\{\mathbf{v}_1, \mathbf{v}_2, \mathbf{v}_3\}$ is an orthogonal basis for W. Clearing fractions, we have

$$\{\mathbf{v}_1', \mathbf{v}_2', \mathbf{v}_3'\} = \{(1,1,0,0), (3,-3,0,2), (12,-12,0,-36)\}$$

is also an orthogonal basis for W. Let

$$\mathbf{w}_1 = \frac{1}{\|\mathbf{v}_1'\|}\mathbf{v}_1' = \frac{1}{\sqrt{2}}(1,1,0,0),$$

$$\mathbf{w}_2 = \frac{1}{\|\mathbf{v}_2'\|}\mathbf{v}_2' = \frac{1}{\sqrt{22}}(3,-3,0,2),$$

$$\mathbf{w}_3 = \frac{1}{\|\mathbf{v}_3'\|}\mathbf{v}_3' = \frac{1}{\sqrt{11}}(1,-1,0,-3).$$

It follows that $\{\mathbf{w}_1, \mathbf{w}_2, \mathbf{w}_3\}$ is an orthonormal basis for W.

15. Let W be the subspace of R^4 consisting of all vectors of the form $(a, a+b, c, b+c)$. Since

$$(a, a+b, c, b+c) = a(1,1,0,0) + b(0,1,0,1) + c(0,0,1,1)$$

it follows that $S = \{\mathbf{u}_1, \mathbf{u}_2, \mathbf{u}_3\} = \{(1,1,0,0), (0,1,0,1), (0,0,1,1)\}$ spans W. To show that S is a basis for W we show that S is linearly independent. Form the expression

$$c_1\mathbf{u}_1 + c_2\mathbf{u}_2 + c_3\mathbf{u}_3 = (0,0,0,0)$$

which has augmented matrix

$$\begin{bmatrix} 1 & 0 & 0 & \vdots & 0 \\ 1 & 1 & 0 & \vdots & 0 \\ 0 & 0 & 1 & \vdots & 0 \\ 0 & 1 & 1 & \vdots & 0 \end{bmatrix}.$$

The reduced row echelon form of this matrix is

$$\begin{bmatrix} 1 & 0 & 0 & \vdots & 0 \\ 0 & 1 & 0 & \vdots & 0 \\ 0 & 0 & 1 & \vdots & 0 \\ 0 & 0 & 0 & \vdots & 0 \end{bmatrix}$$

so $c_1 = c_2 = c_3 = 0$. Hence S is linearly independent. We next apply the Gram–Schmidt process to S. Let $\mathbf{v}_1 = \mathbf{u}_1$ and

$$\mathbf{v}_2 = \mathbf{u}_2 - \left(\frac{\mathbf{u}_2 \cdot \mathbf{v}_1}{\mathbf{v}_1 \cdot \mathbf{v}_1} \right) \mathbf{v}_1 = (0, 1, 0, 1) - \tfrac{1}{2}(1, 1, 0, 0) = \left(-\tfrac{1}{2}, \tfrac{1}{2}, 0, 1 \right)$$

$$\mathbf{v}_3 = \mathbf{u}_3 - \left(\frac{\mathbf{u}_3 \cdot \mathbf{v}_1}{\mathbf{v}_1 \cdot \mathbf{v}_1} \right) \mathbf{v}_1 - \left(\frac{\mathbf{u}_3 \cdot \mathbf{v}_2}{\mathbf{v}_2 \cdot \mathbf{v}_2} \right) \mathbf{v}_2$$

$$= (0, 0, 1, 1) - 0(1, 1, 0, 0) - \left(\frac{1}{\frac{3}{2}} \right) \left(-\tfrac{1}{2}, \tfrac{1}{2}, 0, 1 \right) = \left(\tfrac{1}{3}, -\tfrac{1}{3}, 1, \tfrac{1}{3} \right)$$

Then $\{\mathbf{w}_1, \mathbf{w}_2, \mathbf{w}_3\}$ is an orthonormal basis for W, where

$$\mathbf{w}_1 = \frac{1}{\|\mathbf{v}_1\|} \mathbf{v}_1 = \left(\frac{1}{\sqrt{2}}, \frac{1}{\sqrt{2}}, 0, 0 \right)$$

$$\mathbf{w}_2 = \frac{1}{\|\mathbf{v}_2\|} \mathbf{v}_2 = \left(-\frac{1}{\sqrt{6}}, \frac{1}{\sqrt{6}}, 0, \frac{2}{\sqrt{6}} \right)$$

$$\mathbf{w}_3 = \frac{1}{\|\mathbf{v}_3\|} \mathbf{v}_3 = \left(\frac{1}{\sqrt{12}}, -\frac{1}{\sqrt{12}}, \frac{3}{\sqrt{12}}, \frac{1}{\sqrt{12}} \right)$$

17. Let W be the subspace of R^4 consisting of all vectors of the form (a, b, c, d) such that $a - b - 2c + d = 0$. We have $a = b + 2c - d$ so W is all vectors of the form

$$(b + 2c - d, b, c, d) = b(1, 1, 0, 0) + c(2, 0, 1, 0) + d(-1, 0, 0, 1).$$

It follows that

$$S = \{\mathbf{u}_1, \mathbf{u}_2, \mathbf{u}_3\} = \{(1, 1, 0, 0), (2, 0, 1, 0), (-1, 0, 0, 1)\}$$

spans W. To show that S is a basis for W we show that S is linearly independent. Form the expression

$$c_1 \mathbf{u}_1 + c_2 \mathbf{u}_2 + c_3 \mathbf{u}_3 = (0, 0, 0, 0)$$

which has augmented matrix

$$\begin{bmatrix} 1 & 2 & -1 & \vdots & 0 \\ 1 & 0 & 0 & \vdots & 0 \\ 0 & 1 & 0 & \vdots & 0 \\ 0 & 0 & 1 & \vdots & 0 \end{bmatrix}.$$

The reduced row echelon form of this matrix is

$$\begin{bmatrix} 1 & 0 & 0 & \vdots & 0 \\ 0 & 1 & 0 & \vdots & 0 \\ 0 & 0 & 1 & \vdots & 0 \\ 0 & 0 & 0 & \vdots & 0 \end{bmatrix}$$

so $c_1 = c_2 = c_3 = 0$. Hence S is linearly independent. We next apply the Gram–Schmidt process to S. Let $\mathbf{v}_1 = \mathbf{u}_1$ and

$$\mathbf{v}_2 = \mathbf{u}_2 - \left(\frac{\mathbf{u}_2 \cdot \mathbf{v}_1}{\mathbf{v}_1 \cdot \mathbf{v}_1}\right)\mathbf{v}_1 = (2,0,1,0) - \tfrac{2}{2}(1,1,0,0) = (1,-1,1,0)$$

$$\mathbf{v}_3 = \mathbf{u}_3 - \left(\frac{\mathbf{u}_3 \cdot \mathbf{v}_1}{\mathbf{v}_1 \cdot \mathbf{v}_1}\right)\mathbf{v}_1 - \left(\frac{\mathbf{u}_3 \cdot \mathbf{v}_2}{\mathbf{v}_2 \cdot \mathbf{v}_2}\right)\mathbf{v}_2$$

$$= (-1,0,0,1) - \left(-\tfrac{1}{2}\right)(1,1,0,0) - \left(-\tfrac{1}{3}\right)(-1,0,0,1) = \left(-\tfrac{1}{6},\tfrac{1}{6},\tfrac{1}{3},1\right)$$

Then $\{\mathbf{w}_1, \mathbf{w}_2, \mathbf{w}_3\}$ is an orthonormal basis for W, where

$$\mathbf{w}_1 = \frac{1}{\|\mathbf{v}_1\|}\mathbf{v}_1 = \left(\frac{1}{\sqrt{2}}, \frac{1}{\sqrt{2}}, 0, 0\right)$$

$$\mathbf{w}_2 = \frac{1}{\|\mathbf{v}_2\|}\mathbf{v}_2 = \left(\frac{1}{\sqrt{3}}, -\frac{1}{\sqrt{3}}, \frac{1}{\sqrt{3}}, 0\right)$$

$$\mathbf{w}_3 = \frac{1}{\|\mathbf{v}_3\|}\mathbf{v}_3 = \left(-\frac{1}{\sqrt{42}}, \frac{1}{\sqrt{42}}, \frac{2}{\sqrt{42}}, \frac{6}{\sqrt{42}}\right)$$

19. Form the augmented matrix

$$\begin{bmatrix} 1 & 1 & -1 & \vdots & 0 \\ 2 & 1 & 3 & \vdots & 0 \\ 1 & 2 & -6 & \vdots & 0 \end{bmatrix}$$

and find its reduced row echelon form. We obtain

$$\begin{bmatrix} 1 & 0 & 4 & \vdots & 0 \\ 0 & 1 & -5 & \vdots & 0 \\ 0 & 0 & 0 & \vdots & 0 \end{bmatrix}$$

and it follows that the solution is

$$\begin{bmatrix} -4t \\ 5t \\ t \end{bmatrix}$$

for any real number t. Hence

$$\mathbf{u}_1 = \begin{bmatrix} -4 \\ 5 \\ 1 \end{bmatrix}$$

is a basis for the solution space. To find an orthonormal basis we compute

$$\frac{1}{\|\mathbf{u}_1\|}\mathbf{u}_1 = \frac{1}{\sqrt{42}}\begin{bmatrix} -4 \\ 5 \\ 1 \end{bmatrix}.$$

21. Let

$$\mathbf{w}_1 = \frac{1}{\sqrt{5}}(1,0,2), \quad \mathbf{w}_2 = \frac{1}{\sqrt{5}}(-2,0,1), \quad \mathbf{w}_3 = (0,1,0), \quad \text{and} \quad \mathbf{v} = (2,-3,1).$$

Following the procedure in Example 5, we compute

$$c_1 = \mathbf{v} \cdot \mathbf{w}_1 = \frac{4}{\sqrt{5}}, \quad c_2 = \mathbf{v} \cdot \mathbf{w}_2 = -\frac{3}{\sqrt{5}}, \quad c_3 = \mathbf{v} \cdot \mathbf{w}_3 = -3.$$

Then

$$\mathbf{v} = \frac{4}{\sqrt{5}}\mathbf{w}_1 - \frac{3}{\sqrt{5}}\mathbf{w}_2 - 3\mathbf{w}_3.$$

T.1. $\mathbf{e}_i \cdot \mathbf{e}_j = 0$ for $i \neq j$ and 1 for $i = j$.

T.3. Since an orthonormal set of vectors is an orthogonal set, we know by Theorem 6.16 they are linearly independent. Since there are n of them, they span R^n.

T.5. If $\mathbf{u}$ is orthogonal to $S = \{\mathbf{v}_1, \mathbf{v}_2, \ldots, \mathbf{v}_n\}$, then $\mathbf{u} \cdot \mathbf{v}_j = 0$ for $j = 1, \ldots, n$. Let $\mathbf{w}$ be in span S. Then $\mathbf{w}$ is a linear combination of the vectors in S:

$$\mathbf{w} = \sum_{j=1}^{n} c_j \mathbf{v}_j.$$

Thus

$$\mathbf{u} \cdot \mathbf{w} = \sum_{j=1}^{n} c_j(\mathbf{u} \cdot \mathbf{v}_j) = \sum_{j=1}^{n} c_j 0 = 0.$$

Hence $\mathbf{u}$ is orthogonal to every vector in span S.

T.7. If $\mathbf{u} \cdot \mathbf{v} = 0$, then $u_1 v_1 + u_2 v_2 + \cdots + u_n v_n = 0$. We have

$$\mathbf{u} \cdot (c\mathbf{v}) = u_1(cv_1) + u_2(cv_2) + \cdots + u_n(cv_n) = c(u_1 v_1 + u_2 v_2 + \cdots + u_n v_n) = c(0) = 0.$$

T.9. Since some of the vectors $\mathbf{v}_j$ can be zero, A can be singular.

T.11 Let $\mathbf{x}$ be in S. Then we can write $\mathbf{x} = \sum_{j=1}^{k} c_j \mathbf{u}_j$. Similarly if $\mathbf{y}$ is in T, we have $\mathbf{y} = \sum_{i=k+1}^{n} c_i \mathbf{u}_i$. Then

$$\mathbf{x} \cdot \mathbf{y} = \left(\sum_{j=1}^{k} c_j \mathbf{u}_j \right) \cdot \mathbf{y} = \sum_{j=1}^{k} c_j(\mathbf{u}_j \cdot \mathbf{y}) = \sum_{j=1}^{k} c_j \left(\mathbf{u}_j \cdot \sum_{i=k+1}^{n} c_i \mathbf{u}_i \right) = \sum_{j=1}^{k} c_j \left(\sum_{i=k+1}^{n} c_i(\mathbf{u}_j \cdot \mathbf{u}_i) \right).$$

Since $j \neq i$, $\mathbf{u}_j \cdot \mathbf{u}_i = 0$, hence $\mathbf{x} \cdot \mathbf{y} = 0$.

ML.1. Use the following MATLAB commands.

A = [1 1 0;1 0 1;0 0 1];

gschmidt(A)

```
ans =
      0.7071    0.7071         0
      0.7071   -0.7071         0
           0         0    1.0000
```

Write the columns in terms of $\sqrt{2}$. Note that $\frac{\sqrt{2}}{2} \approx 0.7071$.

ML.3. To find the orthonormal basis we proceed as follows in MATLAB.

A = [0 − 1 1;0 1 1;1 1 1]′;

G = gschmidt(A)

```
G =
           0         0    1.0000
     -0.7071    0.7071         0
      0.7071    0.7071         0
```

To find the coordinates of each vector with respect to the orthonormal basis T which consists of the columns of matrix G we express each vector as a linear combination of the columns of G. It follows that $[\mathbf{v}]_T$ is the solution to the linear system $G\mathbf{x} = \mathbf{v}$. We find the solution to all three systems at the same time as follows.

coord = rref([G [1 2 0;1 1 1; − 1 0 1]'])

coord =

1.0000	0	0	−1.4142	0	0.7071
0	1.0000	0	1.4142	1.4142	0.7071
0	0	1.0000	1.0000	1.0000	−1.0000

Columns 4, 5, and 6 are the solutions to parts (a), (b), and (c), respectively.

Section 6.9, p. 369

1. Let W be a subspace of R^3 with basis $\{\mathbf{w}_1, \mathbf{w}_2\}$, where $\mathbf{w}_1 = (1, 0, 1)$ and $\mathbf{w}_2 = (1, 1, 1)$. We first find the orthonormal basis of W using the Gram-Schmidt process.

Define $\mathbf{v}_1 = \mathbf{w}_1 = (1, 0, 1)$. Compute

$$\mathbf{v}_2 = \mathbf{w}_2 - \left(\frac{\mathbf{w}_2 \cdot \mathbf{v}_1}{\mathbf{v}_1 \cdot \mathbf{v}_1}\right)\mathbf{v}_1 = (1, 1, 1) - (1, 0, 1) = (0, 1, 0).$$

Then $\{\mathbf{u}_1, \mathbf{u}_2\}$ is an orthonormal basis for W, where

$$\mathbf{u}_1 = \frac{\mathbf{v}_1}{\|\mathbf{v}_1\|} = \left(\frac{1}{\sqrt{2}}, 0, \frac{1}{\sqrt{2}}\right)$$

$$\mathbf{u}_2 = \frac{\mathbf{v}_2}{\|\mathbf{v}_2\|} = (0, 1, 0)$$

The vector $\mathbf{w} = \text{proj}_W \mathbf{v}$, given by Equation (1) is the projection of $\mathbf{v}$ onto the subspace W.

$$\mathbf{w} = (\mathbf{v} \cdot \mathbf{u}_1)\mathbf{u}_1 + (\mathbf{v} \cdot \mathbf{u}_2)\mathbf{u}_2$$

We have $\mathbf{v} = (2, 2, 0)$ so

$$\mathbf{v} \cdot \mathbf{u}_1 = \frac{2}{\sqrt{2}}, \quad \mathbf{v} \cdot \mathbf{u}_2 = 2$$

Then

$$\mathbf{w} = \text{proj}_W \mathbf{v} = \frac{2}{\sqrt{2}}\left(\frac{1}{\sqrt{2}}, 0, \frac{1}{\sqrt{2}}\right) + 2(0, 1, 0) = (1, 2, 1)$$

We find

$$\mathbf{u} = \mathbf{v} - \mathbf{w} = (2, 2, 0) - (1, 2, 1) = (1, 0, -1)$$

Thus

$$\mathbf{v} = (2, 2, 0) = (1, 2, 1) + (1, 0, -1)$$

where $\mathbf{w} = (1, 2, 1)$ is in W and $\mathbf{u} = (1, 0, -1)$ is in $W^\perp$.

3. (a) Let $\mathbf{v} = (a, b, c)$ be a vector in $W^\perp$. Then

$$\mathbf{v} \cdot \mathbf{w} = (a, b, c) \cdot (2, -3, 1) = 2a - 3b + c = 0.$$

Therefore $a = \frac{3}{2}b - \frac{1}{2}c$ and hence the general vector $\mathbf{v}$ in $W^\perp$ has the form

$$\mathbf{v} = \left(\tfrac{3}{2}b - \tfrac{1}{2}c, b, c\right) = b\left(\tfrac{3}{2}, 1, 0\right) + c\left(-\tfrac{1}{2}, 0, 1\right).$$

where b and c are arbitrary scalars. Therefore the set

$$S = \left\{\left(\tfrac{3}{2}, 1, 0\right), \left(-\tfrac{1}{2}, 0, 1\right)\right\}$$

spans $W^\perp$. Since these two vectors are not scalar multiples of each other they are linearly independent and hence form a basis for $W^\perp$.

(b) $W^\perp$ is the plane through the origin determined by the vectors $\left(\frac{3}{2}, 1, 0\right)$ and $\left(-\frac{1}{2}, 0, 1\right)$. It consists of all points $P(x, y, z)$ such that

$$\mathbf{w} \cdot (x, y, z) = (2, -1, 3) \cdot (x, y, z) = 2x - y + 3z = 0.$$

5. Let $\mathbf{u} = (a_1, a_2, a_3, a_4, a_5)$ be a vector in $W^\perp$. Then $\mathbf{v}$ is orthogonal to each of the vectors $\mathbf{w}_1, \ldots, \mathbf{w}_5$. Setting $\mathbf{u} \cdot \mathbf{w}_i = 0$ for $i = 1, \ldots, 5$ leads to a homogeneous system. We row reduce the coefficient matrix of the system:

$$A = \begin{bmatrix} 2 & -1 & 1 & 3 & 0 \\ 1 & 2 & 0 & 1 & -2 \\ 4 & 3 & 1 & 5 & -4 \\ 3 & 1 & 2 & -1 & 1 \\ 2 & -1 & 2 & -2 & 3 \end{bmatrix} \longrightarrow \underset{\text{steps omitted}}{\cdots} \longrightarrow \begin{bmatrix} 1 & 0 & 0 & \frac{17}{5} & -\frac{8}{5} \\ 0 & 1 & 0 & -\frac{6}{6} & -\frac{1}{5} \\ 0 & 0 & 1 & -5 & 3 \\ 0 & 0 & 0 & 0 & 0 \\ 0 & 0 & 0 & 0 & 0 \end{bmatrix}.$$

The solution to the system is therefore

$$a_1 = -\tfrac{17}{5}r + \tfrac{8}{5}s, \quad a_2 = \tfrac{6}{5}r + \tfrac{1}{5}s, \quad a_3 = 5r - 3s, \quad a_4 = r, \quad a_5 = s,$$

where r and s are any real numbers. Hence

$$\mathbf{u} = \left(-\tfrac{17}{5}r + \tfrac{8}{5}s, \tfrac{6}{5}r + \tfrac{1}{5}s, 5r - 3s, r, s\right) = r\left(-\tfrac{17}{5}, \tfrac{6}{5}, 5, 1, 0\right) + s\left(\tfrac{8}{5}, \tfrac{1}{5}, -3, 0, 1\right).$$

Therefore the set

$$S = \left\{\left(-\tfrac{17}{5}, \tfrac{6}{5}, 5, 1, 0\right), \left(\tfrac{8}{5}, \tfrac{1}{5}, -3, 0, 1\right)\right\}$$

is a basis for $W^\perp$.

7. We follow the method used in Example 3. First transform the matrix A to its reduced row echelon form:

$$A = \begin{bmatrix} 2 & -1 & 3 & 4 \\ 0 & -3 & 7 & -2 \\ 1 & 1 & -2 & 3 \\ 1 & 4 & -9 & 5 \end{bmatrix} \longrightarrow \underset{\text{steps omitted}}{\cdots} \longrightarrow \begin{bmatrix} 1 & 0 & \frac{1}{3} & \frac{7}{3} \\ 0 & 1 & -\frac{7}{3} & \frac{2}{3} \\ 0 & 0 & 0 & 0 \\ 0 & 0 & 0 & 0 \end{bmatrix} = B.$$

To find the null space of A we solve the homogeneous system $B\mathbf{x} = \mathbf{0}$, obtaining

$$\mathbf{x} = \begin{bmatrix} -\frac{1}{3}r - \frac{7}{3}s \\ \frac{7}{3}r - \frac{2}{3}s \\ r \\ s \end{bmatrix} = r \begin{bmatrix} -\frac{1}{3} \\ \frac{7}{3} \\ 1 \\ 0 \end{bmatrix} + s \begin{bmatrix} -\frac{7}{3} \\ -\frac{2}{3} \\ 0 \\ 1 \end{bmatrix}$$

so

$$S = \left\{ \begin{bmatrix} -\frac{1}{3} \\ \frac{7}{3} \\ 1 \\ 0 \end{bmatrix}, \begin{bmatrix} -\frac{7}{3} \\ -\frac{2}{3} \\ 0 \\ 1 \end{bmatrix} \right\}$$

is a basis for the null space of A. The rows of B form a basis for the row space of A:

$$T = \left\{\left(1, 0, \tfrac{1}{3}, \tfrac{7}{3}\right), \left(0, 1, -\tfrac{7}{3}, \tfrac{2}{3}\right)\right\}.$$

Next, we have

$$A^T = \begin{bmatrix} 2 & 0 & 1 & 1 \\ -1 & -3 & 1 & 4 \\ 3 & 7 & -2 & -9 \\ 4 & -2 & 3 & 5 \end{bmatrix}.$$

We transform the matrix A^T to its reduced row echelon form:

$$A^T = \begin{bmatrix} 2 & 0 & 1 & 1 \\ -1 & -3 & 1 & 4 \\ 3 & 7 & -2 & -9 \\ 4 & -2 & 3 & 5 \end{bmatrix} \xrightarrow{\ \ \ } \underset{\text{steps omitted}}{\cdots} \xrightarrow{\ \ \ } \begin{bmatrix} 1 & 0 & \tfrac{1}{2} & \tfrac{1}{2} \\ 0 & 1 & -\tfrac{1}{2} & -\tfrac{3}{2} \\ 0 & 0 & 0 & 0 \\ 0 & 0 & 0 & 0 \end{bmatrix} = C$$

so

$$S' = \left\{ \begin{bmatrix} -\tfrac{1}{2} \\ \tfrac{1}{2} \\ 1 \\ 0 \end{bmatrix}, \begin{bmatrix} -\tfrac{1}{2} \\ \tfrac{3}{2} \\ 0 \\ 1 \end{bmatrix} \right\}$$

is a basis for the null space of A^T. The nonzero rows of C read vertically give a basis for the column space of A:

$$T' = \left\{ \begin{bmatrix} 1 \\ 0 \\ \tfrac{1}{2} \\ \tfrac{1}{2} \end{bmatrix}, \begin{bmatrix} 0 \\ 1 \\ -\tfrac{1}{2} \\ -\tfrac{3}{2} \end{bmatrix} \right\}.$$

9. Let W be the subspace of R^4 with basis $S = \{\mathbf{u}_1, \mathbf{u}_2, \mathbf{u}_3\}$, where $\mathbf{u}_1 = (1, 1, 0, 1)$, $\mathbf{u}_2 = (0, 1, 1, 0)$, $\mathbf{u}_3 = (-1, 0, 0, 1)$. We first transform S to an orthonormal basis for W by using the Gram–Schmidt process:

$$\mathbf{v}_1 = \mathbf{u}_1 = (1, 1, 0, 1)$$
$$\mathbf{v}_2 = \mathbf{u}_2 - \left(\frac{\mathbf{u}_2 \cdot \mathbf{v}_1}{\mathbf{v}_1 \cdot \mathbf{v}_1}\right)\mathbf{v}_1 = (0, 1, 1, 0) - \tfrac{1}{3}(1, 1, 0, 1) = \left(-\tfrac{1}{3}, \tfrac{2}{3}, 1, -\tfrac{1}{3}\right).$$

Multiplying $\mathbf{v}_2$ by 3 to clear fractions, we take

$$\mathbf{v}_2 = (-1, 2, 3, -1)$$
$$\mathbf{v}_3 = \mathbf{u}_3 - \left(\frac{\mathbf{u}_3 \cdot \mathbf{v}_1}{\mathbf{v}_1 \cdot \mathbf{v}_1}\right)\mathbf{v}_1 - \left(\frac{\mathbf{u}_3 \cdot \mathbf{v}_2}{\mathbf{v}_2 \cdot \mathbf{v}_2}\right)\mathbf{v}_2 = (-1, 0, 0, 1) - 0\mathbf{v}_1 - 0\mathbf{v}_3 = (-1, 0, 0, 1).$$

Normalizing $\mathbf{v}_1$, $\mathbf{v}_2$, and $\mathbf{v}_3$, we now have the orthonormal basis $\{\mathbf{w}_1, \mathbf{w}_2, \mathbf{w}_3\}$ for W, where

$$\mathbf{w}_1 = \frac{1}{\sqrt{3}}(1, 1, 0, 1), \qquad \mathbf{w}_2 = \frac{1}{\sqrt{15}}(-1, 2, 3, -1), \qquad \mathbf{w}_3 = \frac{1}{\sqrt{2}}(-1, 0, 0, 1).$$

If $\mathbf{v}$ is a vector in V, then

$$\text{proj}_W \mathbf{v} = (\mathbf{v} \cdot \mathbf{w}_1)\mathbf{w}_1 + (\mathbf{v} \cdot \mathbf{w}_2)\mathbf{w}_2 + (\mathbf{v} \cdot \mathbf{w}_3)\mathbf{w}_3.$$

(a) If $\mathbf{v} = (2, 1, 3, 0)$, then

$$\mathbf{v} \cdot \mathbf{w}_1 = \frac{3}{\sqrt{3}}, \qquad \mathbf{v} \cdot \mathbf{w}_2 = \frac{9}{\sqrt{15}}, \qquad \mathbf{v} \cdot \mathbf{w}_3 = -\frac{2}{\sqrt{2}},$$

so

$$\begin{aligned}
\operatorname{proj}_W \mathbf{v} &= \frac{3}{\sqrt{3}} \left(\frac{1}{\sqrt{3}}, \frac{1}{\sqrt{3}}, 0, \frac{1}{\sqrt{3}} \right) + \frac{9}{\sqrt{15}} \left(\frac{-1}{\sqrt{15}}, \frac{2}{\sqrt{15}}, \frac{3}{\sqrt{15}}, \frac{-1}{\sqrt{15}} \right) - \frac{2}{\sqrt{2}} \left(\frac{-1}{\sqrt{2}}, 0, 0, \frac{1}{\sqrt{2}} \right) \\
&= \left(\frac{7}{5}, \frac{11}{5}, \frac{9}{5}, -\frac{3}{5} \right).
\end{aligned}$$

(b) If $\mathbf{v} = (0, -1, 1, 0)$, then

$$\mathbf{v} \cdot \mathbf{w}_1 = -\frac{1}{\sqrt{3}}, \qquad \mathbf{v} \cdot \mathbf{w}_2 = \frac{1}{\sqrt{15}}, \qquad \mathbf{v} \cdot \mathbf{w}_3 = 0,$$

so

$$\operatorname{proj}_W \mathbf{v} = -\frac{1}{\sqrt{3}} \left(\frac{1}{\sqrt{3}}, \frac{1}{\sqrt{3}}, 0, \frac{1}{\sqrt{3}} \right) + \frac{1}{\sqrt{15}} \left(\frac{-1}{\sqrt{15}}, \frac{2}{\sqrt{15}}, \frac{3}{\sqrt{15}}, \frac{-1}{\sqrt{15}} \right) = \left(-\frac{2}{5}, -\frac{1}{5}, \frac{1}{5}, -\frac{2}{5} \right).$$

(c) If $\mathbf{v} = (0, 2, 0, 3)$, then

$$\mathbf{v} \cdot \mathbf{w}_1 = \frac{5}{\sqrt{3}}, \qquad \mathbf{v} \cdot \mathbf{w}_2 = \frac{1}{\sqrt{15}}, \qquad \mathbf{v} \cdot \mathbf{w}_3 = \frac{3}{\sqrt{2}},$$

so

$$\begin{aligned}
\operatorname{proj}_W \mathbf{v} &= \frac{5}{\sqrt{3}} \left(\frac{1}{\sqrt{3}}, \frac{1}{\sqrt{3}}, 0, \frac{1}{\sqrt{3}} \right) + \frac{1}{\sqrt{15}} \left(\frac{-1}{\sqrt{15}}, \frac{2}{\sqrt{15}}, \frac{3}{\sqrt{15}}, \frac{-1}{\sqrt{15}} \right) + \frac{3}{\sqrt{2}} \left(\frac{-1}{\sqrt{2}}, 0, 0, \frac{1}{\sqrt{2}} \right) \\
&= \left(\frac{1}{10}, \frac{9}{5}, \frac{1}{5}, \frac{31}{10} \right).
\end{aligned}$$

11. Let

$$\mathbf{w}_1 = \left(\frac{1}{\sqrt{2}}, 0, 0, -\frac{1}{\sqrt{2}} \right), \qquad \mathbf{w}_2 = (0, 0, 1, 0), \qquad \mathbf{w}_3 = \left(\frac{1}{\sqrt{2}}, 0, 0, \frac{1}{\sqrt{2}} \right).$$

Now $\mathbf{w} = \operatorname{proj}_W \mathbf{v}$ is given by Equation (1):

$$\mathbf{w} = \operatorname{proj}_W \mathbf{v} = (\mathbf{v} \cdot \mathbf{w}_1)\mathbf{w}_1 + (\mathbf{v} \cdot \mathbf{w}_2)\mathbf{w}_2 + (\mathbf{v} \cdot \mathbf{w}_3)\mathbf{w}_3.$$

We have $\mathbf{v} = (1, 0, 2, 3)$, so

$$\mathbf{v} \cdot \mathbf{w}_1 = -\frac{2}{\sqrt{2}}, \qquad \mathbf{v} \cdot \mathbf{w}_2 = 2, \qquad \mathbf{v} \cdot \mathbf{w}_3 = \frac{4}{\sqrt{2}}.$$

Then

$$\mathbf{w} = \operatorname{proj}_W \mathbf{v} = -\frac{2}{\sqrt{2}} \left(\frac{1}{\sqrt{2}}, 0, 0, -\frac{1}{\sqrt{2}} \right) + 2(0, 0, 1, 0) + \frac{4}{\sqrt{2}} \left(\frac{1}{\sqrt{2}}, 0, 0, \frac{1}{\sqrt{2}} \right) = (1, 0, 2, 3).$$

We find $\mathbf{u} = \mathbf{v} - \mathbf{w} = (1, 0, 2, 3) - (1, 0, 2, 3) = 0$.

13. As in Example 6, the distance from $\mathbf{v}$ to $\mathbf{w}$ is $\|\mathbf{v} - \operatorname{proj}_W \mathbf{v}\|$. We have $\mathbf{v} = (1, 2, -1, 0)$ so $\operatorname{proj}_W \mathbf{v}$ is given by Equation (1):

$$\mathbf{w} = \operatorname{proj}_W \mathbf{v} = (\mathbf{v} \cdot \mathbf{w}_1)\mathbf{w}_1 + (\mathbf{v} \cdot \mathbf{w}_2)\mathbf{w}_2 + (\mathbf{v} \cdot \mathbf{w}_3)\mathbf{w}_3.$$

We have

$$\mathbf{v} \cdot \mathbf{w}_1 = \frac{1}{\sqrt{2}}, \qquad \mathbf{v} \cdot \mathbf{w}_2 = -1, \qquad \mathbf{v} \cdot \mathbf{w}_3 = \frac{1}{\sqrt{2}}$$

so

$$\text{proj}_W \mathbf{v} = \frac{1}{\sqrt{2}} \left(\frac{1}{\sqrt{2}}, 0, 0, -\frac{1}{\sqrt{2}} \right) - (0, 0, 1, 0) + \frac{1}{\sqrt{2}} \left(\frac{1}{\sqrt{2}}, 0, 0, \frac{1}{\sqrt{2}} \right) = (1, 0, -1, 0).$$

Then

$$\mathbf{v} - \text{proj}_W \mathbf{v} = (1, 2, -1, 0) - (1, 0, -1, 0) = (0, 2, 0, 0)$$

so $\|\mathbf{v} - \text{proj}_W \mathbf{v}\| = 2$.

T.1. The zero vector is orthogonal to every vector in W.

T.3. Let $W = \text{span } S$, where $S = \{\mathbf{v}_1, \mathbf{v}_2, \ldots, \mathbf{v}_m\}$. If $\mathbf{u}$ is in $W^\perp$, then $\mathbf{u} \cdot \mathbf{w} = 0$ for any $\mathbf{w}$ in W. Hence $\mathbf{u} \cdot \mathbf{v}_i = 0$ for $i = 1, 2, \ldots, m$. Conversely, suppose that $\mathbf{u} \cdot \mathbf{v}_i = 0$ for $i = 1, 2, \ldots, m$. Let

$$\mathbf{w} = \sum_{i=1}^{m} c_i \mathbf{v}_i$$

be any vector in W. Then

$$\mathbf{u} \cdot \mathbf{w} = \sum_{i=1}^{m} c_i (\mathbf{u} \cdot \mathbf{v}_i) = 0.$$

Hence $\mathbf{u}$ is in $W^\perp$.

T.5. Let W be a subspace of R^n. By Theorem 6.20, we have $R^n = W \oplus W^\perp$. Let $\{\mathbf{w}_1, \mathbf{w}_2, \ldots, \mathbf{w}_r\}$ be a basis for W, so $\dim W = r$, and let $\{\mathbf{u}_1, \mathbf{u}_2, \ldots, \mathbf{u}_s\}$ be a basis for $W^\perp$, so $\dim W^\perp = s$. If $\mathbf{v}$ is in R^n, then $\mathbf{v} = \mathbf{w} + \mathbf{u}$, where $\mathbf{w}$ is in W and $\mathbf{u}$ is in $W^\perp$. Moreover, $\mathbf{w}$ and $\mathbf{u}$ are unique. Then

$$\mathbf{v} = \sum_{i=1}^{r} a_i \mathbf{w}_i + \sum_{j=1}^{s} b_j \mathbf{u}_j$$

so $S = \{\mathbf{w}_1, \mathbf{w}_2, \ldots, \mathbf{w}_r, \mathbf{v}_1, \mathbf{v}_2, \ldots, \mathbf{v}_s\}$ spans R^n. We now show that S is linearly independent. Suppose that

$$\sum_{i=1}^{r} a_i \mathbf{w}_i + \sum_{j=1}^{s} b_j \mathbf{u}_j = \mathbf{0}.$$

Then

$$\sum_{i=1}^{r} a_i \mathbf{w}_i = - \sum_{j=1}^{s} b_j \mathbf{u}_j,$$

so $\sum_{i=1}^{r} a_i \mathbf{w}_i$ lies in $W \cap W^\perp = \{\mathbf{0}\}$. Hence

$$\sum_{i=1}^{r} a_i \mathbf{w}_i = \mathbf{0},$$

and since $\mathbf{w}_1, \mathbf{w}_2, \ldots, \mathbf{w}_r$ are linearly independent, $a_1 = a_2 = \cdots = a_r = 0$. Similarly, $b_1 = b_2 = \cdots = b_r = 0$. Thus, S is also linearly independent and is then a basis for R^n. This means that

$$n = \dim R^n = r + s = \dim W + \dim W^\perp,$$

and $\mathbf{w}_1, \mathbf{w}_2, \ldots, \mathbf{w}_r, \mathbf{u}_1, \mathbf{u}_2, \ldots, \mathbf{u}_s$ is a basis for R^n.

ML.1. (a) $\mathbf{v} = [1 \quad 5 \quad -1 \quad 2]'$, $\mathbf{w} = [0 \quad 1 \quad 2 \quad 1]'$

   v =

     1
     5
     −1
     2

   w =

     0
     1
     2
     1

   proj = [dot(v,w)/norm(w)^2] * w

   proj =

      0
     0.8333
     1.6667
     0.8333

  format rat

  proj

  proj =

     0
     5/6
     5/3
     5/6

  format

(b) $\mathbf{v} = [1 \quad -2 \quad 3 \quad 0 \quad 1]'$, $\mathbf{w} = [1 \quad 1 \quad 1 \quad 1 \quad 1]'$

   v =

     1
     −2
     3
     0
     1

   w =

     1
     1
     1
     1
     1

   proj = [dot(v,w)/norm(w)^2] * w

proj =

       0.6000

       0.6000

       0.6000

       0.6000

       0.6000

format rat

proj

proj =

       3/5

       3/5

       3/5

       3/5

       3/5

format

ML.3. $\mathbf{w1} = [1 \quad 2 \quad 3]', \mathbf{w2} = [0 -3 \quad 2]'$

w1 =

    1

    2

    3

w2 =

     0

    −3

     2

(a) First note that w1 and w2 form an orthogonal basis for plane P.

$\mathbf{v} = [2 \quad 4 \quad 8]'$

v =

     2

     4

     8

$\mathbf{proj} = [\mathbf{dot(v,w1)/norm(w1)\char`^2}] * \mathbf{w1} + [\mathbf{dot(v,w2)/norm(w2)\char`^2}] * \mathbf{w2}$

proj =

     2.4286

     3.9341

     7.9011

(b) The distance from $\mathbf{v}$ to P is the length of the vector $-\mathrm{proj} + \mathbf{v}$.

$\mathbf{norm(- proj + v)}$

ans =

     0.4447

ML.5. $\mathbf{T} = [1 \quad 0;0 \quad 1;1 \quad 1;1 \quad 0;1 \quad 0]$

T =

1 0
0 1
1 1
1 0
1 0

b = [1 1 1 1 1]′

b =

1
1
1
1
1

(a) **rref([T b])**

ans =

1 0 0
0 1 0
0 0 1
0 0 0
0 0 0

Note that row [0 0 1] implies that the system is inconsistent.

(b) Note that the columns of T are not orthogonal, so we use the Gram–Schmidt process to find an orthonormal basis for the column space.

q = gschmidt(T)

q =

0.5000   −0.1890
      0   0.7559
0.5000   0.5669
0.5000   −0.1890
0.5000   −0.1890

Define the columns of q to be w1 and w2 which is an orthonormal basis for the column space.

w1 = q(:,1);w2 = q(:,2);
proj = dot(b,w1) ∗ w1 + dot(b,w2) ∗ w2

proj =

0.8571
0.5714
1.4286
0.8571
0.8571

Supplementary Exercises, p. 372

1. Let V be the set of all ordered pairs of real numbers. If (x, y) and (x', y') are in V, then $(x, y) \oplus (x', y') = (x - x', y - y')$ is also in V. Also, if c is a scalar, then $c \odot (x, y) = (cx, cy)$ is in V. Thus, properties (α) and (β) in Definition 1 in Section 6.1 hold. Next

Property (α): $(x, y) \oplus (x', y') = (x - x', y + y')$.

$(x', y') \oplus (x, y) = (x' - x, y' + y)$.

Since these are not the same in general (for example, $(2, 3) \oplus (4, 2) = (2 - 4, 3 + 2) = (-2, 5)$ and $(4, 2) \oplus (2, 3) = (4 - 2, 2 + 3) = (2, 5)$), property ($\alpha$) does not hold and we conclude that V is not a vector space.

3. Form the linear combination

$$c_1(1, 2, 1) + c_2(-1, 1, 2) + c_3(-3, -3, 0) = (4, 2, 1).$$

Expanding, adding vectors, and equating corresponding components on both sides of the equation gives a linear system with augmented matrix

$$\begin{bmatrix} 1 & -1 & -3 & \vdots & 4 \\ 2 & 1 & -3 & \vdots & 2 \\ 1 & 2 & 0 & \vdots & 1 \end{bmatrix}.$$

Applying row operations $-2\mathbf{r}_1 + \mathbf{r}_2$, $-\mathbf{r}_1 + \mathbf{r}_3$, $-\mathbf{r}_2 + \mathbf{r}_3$ gives an equivalent linear system with augmented matrix

$$\begin{bmatrix} 1 & -1 & -3 & \vdots & 4 \\ 0 & 3 & 3 & \vdots & -6 \\ 0 & 0 & 0 & \vdots & 3 \end{bmatrix}$$

which is inconsistent. Hence there are no solutions for any choice of constants c_1, c_2, c_3; thus $(4, 2, 1)$ is not a linear combination of these vectors.

5. Let $S = \{\mathbf{v}_1, \mathbf{v}_2, \mathbf{v}_3\}$, where

$$\mathbf{v}_1 = t^2 + 2t + 2, \qquad \mathbf{v}_2 = 2t^2 + 3t + 1, \qquad \mathbf{v}_3 = -t - 3.$$

Form the linear combination $c_1\mathbf{v}_1 + c_2\mathbf{v}_2 + c_3\mathbf{v}_3 = \mathbf{0}$. Expanding, adding like terms, and equating corresponding coefficients of like powers of t from both sides of the equation we obtain a homogeneous linear system whose coefficient matrix is

$$\begin{bmatrix} 1 & 2 & 0 \\ 2 & 3 & -1 \\ 2 & 1 & -3 \end{bmatrix}.$$

Applying row operations $-2\mathbf{r}_1 + \mathbf{r}_2$, $-2\mathbf{r}_1 + \mathbf{r}_3$, $(-1)\mathbf{r}_2$, $3\mathbf{r}_2 + \mathbf{r}_3$, $-2\mathbf{r}_2 + \mathbf{r}_1$ gives the reduced row echelon form

$$\begin{bmatrix} 1 & 0 & -2 \\ 0 & 1 & 1 \\ 0 & 0 & 0 \end{bmatrix}.$$

Thus we have $c_1 = 2r$, $c_2 = -r$, $c_3 = r$, where r is any real number. It follows that S is a linearly dependent set. For $r = 1$, we have $2\mathbf{v}_1 - \mathbf{v}_2 + \mathbf{v}_3 = \mathbf{0}$ which gives $\mathbf{v}_3 = -2\mathbf{v}_1 + \mathbf{v}_2$.

7. Let W be the subspace of R^4 of all vectors of the form $(a + b, b + c, a - b - 2c, b + c)$. Then

$$(a + b, b + c, a - b - 2c, b + c) = a(1, 0, 1, 0) + b(1, 1, -1, 1) + c(0, 1, -2, 1).$$

Let

$$\mathbf{v}_1 = (1, 0, 1, 0), \qquad \mathbf{v}_2 = (1, 1, -1, 1), \qquad \mathbf{v}_3 = (0, 1, -2, 1),$$

and $S = \{\mathbf{v}_1, \mathbf{v}_2, \mathbf{v}_3\}$. From the preceding equation we have that span $S = W$. Thus we need only determine if S is linearly independent. Form the linear combination

$$c_1\mathbf{v}_1 + c_2\mathbf{v}_2 + c_3\mathbf{v}_3 = \mathbf{0},$$

expand, add vectors, and equate corresponding components from both sides of the equation to obtain a linear system with coefficient matrix

$$\begin{bmatrix} 1 & 1 & 0 \\ 0 & 1 & 1 \\ 1 & -1 & -2 \\ 0 & 1 & 1 \end{bmatrix}.$$

applying row operations $-\mathbf{r}_1 + \mathbf{r}_3$, $2\mathbf{r}_2 + \mathbf{r}_3$, $-\mathbf{r}_2 + \mathbf{r}_4$, $-\mathbf{r}_2 + \mathbf{r}_1$ we have the reduced row echelon form

$$\begin{bmatrix} 1 & 0 & -1 \\ 0 & 1 & 1 \\ 0 & 0 & 0 \\ 0 & 0 & 0 \end{bmatrix}.$$

The solution is $c_1 = r$, $c_2 = -r$, $c_3 = r$, where r is any real number. Thus there are nontrivial solutions, hence S is linearly dependent. Let $r = 1$. Then $\mathbf{v}_1 - \mathbf{v}_2 + \mathbf{v}_3 = \mathbf{0}$ and it follows that $\mathbf{v}_3 = \mathbf{v}_2 - \mathbf{v}_1$. We drop $\mathbf{v}_3$ from S and hence $S_1 = \{\mathbf{v}_1, \mathbf{v}_2\}$ spans W. S_1 is linearly independent since $\mathbf{v}_1$ is not a multiple of $\mathbf{v}_2$. We have that S_1 is a basis for W and dim $W = 2$.

9. The coefficient matrix of the linear system is

$$\begin{bmatrix} 1 & 2 & -1 & 1 & 2 \\ 1 & 1 & 2 & -3 & 1 \end{bmatrix}.$$

Applying row operations $-\mathbf{r}_1 + \mathbf{r}_2$, $-1\mathbf{r}_2$, $-2\mathbf{r}_2 + \mathbf{r}_1$ gives the reduced row echelon form

$$\begin{bmatrix} 1 & 0 & 5 & -7 & 0 \\ 0 & 1 & -3 & 4 & 1 \end{bmatrix}.$$

Thus the solution is given by $x_1 = -5r + 7s$, $x_2 = 3r - 4s - t$, $x_3 = r$, $x_4 = s$, $x_5 = t$, where r, s, and t are any real numbers. Then the solution vector is

$$\mathbf{x} = \begin{bmatrix} -5r + 7s \\ 3r - 4s - t \\ r \\ s \\ t \end{bmatrix} = r\begin{bmatrix} -5 \\ 3 \\ 1 \\ 0 \\ 0 \end{bmatrix} + s\begin{bmatrix} 7 \\ -4 \\ 0 \\ 1 \\ 0 \end{bmatrix} + t\begin{bmatrix} 0 \\ -1 \\ 0 \\ 0 \\ 1 \end{bmatrix}.$$

It follows that

$$\left\{ \begin{bmatrix} -5 \\ 3 \\ 1 \\ 0 \\ 0 \end{bmatrix}, \begin{bmatrix} 7 \\ -4 \\ 0 \\ 1 \\ 0 \end{bmatrix}, \begin{bmatrix} 0 \\ -1 \\ 0 \\ 0 \\ 1 \end{bmatrix} \right\}$$

is a basis for the solution space which is of dimension 3.

11. Let $S = \{\mathbf{v}_1, \mathbf{v}_2\}$, where $\mathbf{v}_1 = t + 3$ and $\mathbf{v}_2 = 2t + \lambda^2 + 2$. Form the linear combination $c_1\mathbf{v}_1 + c_2\mathbf{v}_2 = \mathbf{0}$, expand, add like power terms, and equate coefficients of like powers of t from both sides of the equation to obtain the homogeneous linear system with coefficient matrix

$$\begin{bmatrix} 1 & 2 \\ 3 & \lambda^2 + 2 \end{bmatrix}.$$

Applying row operation $-3\mathbf{r}_1 + \mathbf{r}_2$ gives an equivalent linear system with coefficient matrix

$$\begin{bmatrix} 1 & 2 \\ 0 & \lambda^2 - 4 \end{bmatrix}.$$

This system has only the zero solution provided $\lambda^2 - 4 \neq 0$. Hence S will be linearly independent for all values of λ except $\lambda = 2$ or $\lambda = -2$.

13. Vector $(a^2, a, 1)$ is in span $\{(1, 2, 3), (1, 1, 1), (0, 1, 2)\}$ provided there exist constants c_1, c_2, c_3 such that

$$c_1(1, 2, 3) + c_2(1, 1, 1) + c_3(0, 1, 2) = (a^2, a, 1).$$

Adding vectors, and equating corresponding components gives the linear system

$$\begin{aligned} c_1 + c_2 \qquad &= a^2 \\ 2c_1 + c_2 + c_3 &= a \\ 3c_1 + c_2 + 2c_3 &= 1. \end{aligned}$$

Form the augmented matrix

$$\begin{bmatrix} 1 & 1 & 0 & \vdots & a^2 \\ 2 & 1 & 1 & \vdots & a \\ 3 & 1 & 2 & \vdots & 1 \end{bmatrix}$$

and use row operations $-2\mathbf{r}_1 + \mathbf{r}_2$, $-3\mathbf{r}_1 + \mathbf{r}_3$, $-\mathbf{r}_2 + \mathbf{r}_3$ to obtain an equivalent linear system represented by

$$\begin{bmatrix} 1 & 1 & 0 & \vdots & a^2 \\ 0 & -1 & 1 & \vdots & a - 2a^2 \\ 0 & 0 & 0 & \vdots & a^2 - 2a + 1 \end{bmatrix}.$$

This system is consistent provided $a^2 - 2a + 1 = (a - 1)^2 = 0$. That is, provided $a = 1$.

15. Let $\mathbf{u}$ be in span $\{\mathbf{v}_1, \mathbf{v}_2, \mathbf{v}_3\}$. Then there exist constants c_1, c_2, c_3 such that

$$\mathbf{u} = c_1\mathbf{v}_1 + c_2\mathbf{v}_2 + c_3\mathbf{v}_3.$$

$\mathbf{u}$ is in span $\{\mathbf{v}_1 + \mathbf{v}_2, \mathbf{v}_1 - \mathbf{v}_3, \mathbf{v}_1 + \mathbf{v}_3\}$ provided we can find constants k_1, k_2, k_3 such that

$$\begin{aligned} \mathbf{u} = c_1\mathbf{v}_1 + c_2\mathbf{v}_2 + c_3\mathbf{v}_3 &= k_1(\mathbf{v}_1 + \mathbf{v}_2) + k_2(\mathbf{v}_1 - \mathbf{v}_3) + k_3(\mathbf{v}_1 + \mathbf{v}_3) \\ &= (k_1 + k_2 + k_3)\mathbf{v}_1 + k_1\mathbf{v}_2 + (-k_2 + k_3)\mathbf{v}_3. \end{aligned}$$

Equating coefficients of corresponding vectors gives the linear system

$$\begin{aligned} k_1 + k_2 + k_3 &= c_1 \\ k_1 \qquad\qquad &= c_2 \\ -k_2 + k_3 &= c_3. \end{aligned}$$

Form the augmented matrix

$$\begin{bmatrix} 1 & 1 & 1 & \vdots & c_1 \\ 1 & 0 & 0 & \vdots & c_2 \\ 0 & -1 & 1 & \vdots & c_3 \end{bmatrix}$$

and use row operations to obtain its reduced row echelon form which is

$$\begin{bmatrix} 1 & 0 & 0 & \vdots & c_2 \\ 0 & 1 & 0 & \vdots & \frac{1}{2}(c_1 - c_2 - c_3) \\ 0 & 0 & 1 & \vdots & \frac{1}{2}(c_1 - c_2 + c_3) \end{bmatrix}.$$

Since the linear system is consistent, $\mathbf{u}$ is in span $\{\mathbf{v}_1 + \mathbf{v}_2, \mathbf{v}_1 - \mathbf{v}_3, \mathbf{v}_1 + \mathbf{v}_3\}$.

On the other hand, let $\mathbf{w}$ be in span $\{\mathbf{v}_1 + \mathbf{v}_2, \mathbf{v}_1 - \mathbf{v}_3, \mathbf{v}_1 + \mathbf{v}_3\}$. Then there exist constants k_1, k_2, k_3 such that

$$\mathbf{w} = k_1(\mathbf{v}_1 + \mathbf{v}_2) + k_2(\mathbf{v}_1 - \mathbf{v}_3) + k_3(\mathbf{v}_1 + \mathbf{v}_3).$$

However,

$$\mathbf{w} = (k_1 + k_2 + k_3)\mathbf{v}_1 + k_1\mathbf{v}_2 + (-k_2 + k_3)\mathbf{v}_3,$$

hence $\mathbf{w}$ is in span $\{\mathbf{v}_1, \mathbf{v}_2, \mathbf{v}_3\}$.

17. We check closure of vector addition and scalar multiplication.

   (a) Let $\mathbf{v}_1 = (x_1, mx_1 + b)$ and $\mathbf{v}_2 = (x_2, mx_2 + b)$. Then

   $$\mathbf{v}_1 + \mathbf{v}_2 = (x_1 + x_2, m(x_1 + x_2) + 2b)$$

   is of the form $(x, mx + b)$ provided $b = 0$. For any real scalar c,

   $$c\mathbf{v}_1 = (cx_1, cmx_1 + cb) = (cx_1, cmx_1).$$

   (Assuming $b = 0$.) Thus $c\mathbf{v}_1$ is of the form $(x, mx + b) = (x, mx)$ for all values of m. It follows that $\{(x, mx + b)\}$ is a subspace of R^2 for $b = 0$ and m any real number.

   (b) Let $\mathbf{v}_1 = (x_1, rx_1^2)$ and $\mathbf{v}_2 = (x_2, rx_2^2)$. Then

   $$\mathbf{v}_1 + \mathbf{v}_2 = (x_1 + x_2, r(x_1^2 + x_2^2))$$

   is of the form (x, rx^2) provided

   $$r(x_1^2 + x_2^2) = r(x_1 + x_2)^2 = r(x_1^2 + 2x_1x_2 + x_2^2).$$

   Thus, $2rx_1x_2$ must be zero for all x_1 and x_2. Hence $r = 0$. For any real scalar c,

   $$c\mathbf{v}_1 = (cx_1, crx_1^2) = (cx_1, 0)$$

   has the appropriate form for $r = 0$. It follows that $\{(x, rx^2)\}$ is a subspace of R^2 for $r = 0$.

19. Let $S = \{(1, 0, 0), (0, 1, 0), (0, 0, 1)\}$. If $W = R^3$, then W contains S. On the other hand, if W contains S, then W contains span $S = R^3$.

21. (a) The verification of Definition 1 in Section 6.1 follows from the properties of continuous functions and real numbers. In particular, in calculus it is shown that the sum of continuous functions is continuous and a real number times a continuous function is again a continuous function. This verifies (α) and (β) of Definition 1 in Section 6.1. We demonstrate that (a) and (e) hold and (b), (c), (d), (f), (g), (h) are shown in a similar way. To show (a), let f and g belong to $C[a, b]$. Then for t in $[a, b]$ we have

$$(f \oplus g)(t) = f(t) + g(t) = g(t) + f(t) = (g \oplus f)(t)$$

since $f(t)$ and $g(t)$ are real numbers and the addition of real numbers is commutative. To show (e), let c be any real number. Then

$$c \odot (f \oplus g)(t) = c(f(t) + g(t)) = cf(t) + cg(t)$$
$$= c \odot f(t) + c \odot g(t) = (c \odot f \oplus c \odot g)(t)$$

since c, $f(t)$, and $g(t)$ are real numbers and multiplication of real numbers distributes over addition of real numbers.

(b) Let f and g be in $W(k)$ and c be any real scalar. Then

$$(f \oplus g)(a) = f(a) + g(a) = k + k = 2k.$$

Hence $(f \oplus g)(a) = k$ provided $k = 0$. Also,

$$(c \odot f)(a) = cf(a) = ck.$$

Thus $(c \odot f)(a) = k$ for arbitrary c only if $k = 0$. $W(k)$ is a subspace of $C[a, b]$ for $k = 0$.

(c) Let f and g have roots at t_i, $i = 1, 2, \ldots, n$; that is, $f(t_i) = g(t_i) = 0$. It follows that $f \oplus g$ has roots at the t_i since

$$(f \oplus g)(t_i) = f(t_i) + g(t_i) = 0 + 0 = 0.$$

Similarly, $k \odot f$ has roots at the t_i since

$$(k \odot f)(t_i) = kf(t_i) = k(0) = 0.$$

23. The reduced row echelon form of the matrix is

$$\begin{bmatrix} 1 & 0 & 0 & \frac{2}{13} & \frac{15}{13} \\ 0 & 1 & 0 & \frac{17}{13} & -\frac{9}{13} \\ 0 & 0 & 1 & -\frac{3}{13} & -\frac{16}{13} \\ 0 & 0 & 0 & 0 & 0 \\ 0 & 0 & 0 & 0 & 0 \end{bmatrix}.$$

Hence the rank is 3.

25. Since the coefficient matrix is 8×10, rank $A \leq 8$. By Theorem 6.12 in Section 6.6,

$$\text{dim(sol space)} = \text{number of unknowns} - \text{rank } A.$$

So dim(sol space) ≥ 2.

27. $S = \{(1, 0, -1), (0, 1, 1), (0, 0, 1)\}$ and $T = \{(1, 0, 0), (0, 1, -1), (1, -1, 2)\}$ are bases for R^3. Let $\mathbf{v} = (2, 3, 5)$.

(a) To find $[\mathbf{v}]_T$ we solve a linear system obtained from the expression

$$c_1(1, 0, 0) + c_2(0, 1, -1) + c_3(1, -1, 2) = (2, 3, 5)$$

by finding the reduced row echelon form of the associated augmented matrix. The reduced row echelon form is

$$\begin{bmatrix} 1 & 0 & 0 & -6 \\ 0 & 1 & 0 & 11 \\ 0 & 0 & 1 & 8 \end{bmatrix}.$$

It follows that $[\mathbf{v}]_T = \begin{bmatrix} -6 \\ 11 \\ 8 \end{bmatrix}.$

(b) To find $\left[\mathbf{v}\right]_S$ we solve a linear system obtained from the expression

$$c_1(1, 0, -1) + c_2(0, 1, 1) + c_3(0, 0, 1) = (2, 3, 5)$$

by finding the reduced row echelon form of the associated augmented matrix. The reduced row echelon form is

$$\begin{bmatrix} 1 & 0 & 0 & \vdots & 2 \\ 0 & 1 & 0 & \vdots & 3 \\ 0 & 0 & 1 & \vdots & 4 \end{bmatrix}.$$

It follows that $\left[\mathbf{v}\right]_S = \begin{bmatrix} 2 \\ 3 \\ 4 \end{bmatrix}$.

(c) Following the procedure used in Exercise 13 in Section 6.7 we find the coordinates of the vectors of the members of T in terms of the vectors in S. Writing out the expressions leads to three systems of equations with the same coefficient matrix and the right sides are the vectors in T written as columns. We write these three systems in partitioned form and find the reduced row echelon form of the associated matrix. Specifically, the reduced row echelon form of

$$\begin{bmatrix} 1 & 0 & 0 & \vdots & 1 & 0 & 1 \\ 0 & 1 & 0 & \vdots & 0 & 1 & -1 \\ -1 & 1 & 1 & \vdots & 0 & -1 & 2 \end{bmatrix}$$

is

$$\begin{bmatrix} 1 & 0 & 0 & \vdots & 1 & 0 & 1 \\ 0 & 1 & 0 & \vdots & 0 & 1 & -1 \\ 0 & 0 & 1 & \vdots & 1 & -2 & 4 \end{bmatrix}.$$

Hence $P_{S \leftarrow T} = \begin{bmatrix} 1 & 0 & 1 \\ 0 & 1 & -1 \\ 1 & -2 & 4 \end{bmatrix}$.

(d) $\left[\mathbf{v}\right]_S = P_{S \leftarrow T}\left[\mathbf{v}\right]_T = \begin{bmatrix} 2 \\ 3 \\ 4 \end{bmatrix}$.

(e) Reversing the procedure in part (c), we find

$$Q_{T \leftarrow S} = P_{S \leftarrow T}^{-1} = \begin{bmatrix} 2 & -2 & -1 \\ -1 & 3 & 1 \\ -1 & 2 & 1 \end{bmatrix}.$$

(f) $\left[\mathbf{v}\right]_T = Q_{T \leftarrow S}\left[\mathbf{v}\right]_S = \begin{bmatrix} -6 \\ 11 \\ 8 \end{bmatrix}$.

29. We must have $\mathbf{u} \cdot \mathbf{v} = 0$ so

$$\frac{1}{\sqrt{2}}a + \frac{1}{\sqrt{2}}b = 0.$$

Also, if $\|\mathbf{v}\| = 1$, we must have $a^2 + 1 + b^2 = 0$, so $a = 0$ and $b = 0$.

31. (a) the solution to $A\mathbf{x} = \mathbf{0}$ is given by

$$\mathbf{x} = \begin{bmatrix} -5r + 2s \\ 2r + 5s \\ r \\ s \end{bmatrix} = r\begin{bmatrix} -5 \\ 2 \\ 1 \\ 0 \end{bmatrix} + s\begin{bmatrix} 2 \\ 5 \\ 0 \\ 1 \end{bmatrix}.$$

Then a basis for the solution space is

$$\begin{bmatrix} -5 \\ 2 \\ 1 \\ 0 \end{bmatrix}, \quad \begin{bmatrix} 2 \\ 5 \\ 0 \\ 1 \end{bmatrix}.$$

We now transform this basis to an orthonormal basis by using the Gram–Schmidt process:

$$\mathbf{u}_1 = \begin{bmatrix} -5 \\ 2 \\ 1 \\ 0 \end{bmatrix}, \quad \mathbf{u}_2 = \begin{bmatrix} 2 \\ 5 \\ 0 \\ 1 \end{bmatrix}.$$

Then

$$\mathbf{v}_1 = \mathbf{u}_1 = \begin{bmatrix} -5 \\ 2 \\ 1 \\ 0 \end{bmatrix}$$

$$\mathbf{v}_2 = \mathbf{u}_2 - \left(\frac{\mathbf{u}_2 \cdot \mathbf{v}_1}{\mathbf{v}_1 \cdot \mathbf{v}_1} \right) \mathbf{v}_1 = \begin{bmatrix} 2 \\ 5 \\ 0 \\ 1 \end{bmatrix} - 0\mathbf{v}_1 = \begin{bmatrix} 2 \\ 5 \\ 0 \\ 1 \end{bmatrix}.$$

Normalizing $\mathbf{v}_1$ and $\mathbf{v}_2$ we obtain

$$\mathbf{w}_1 = \frac{1}{\sqrt{30}} \begin{bmatrix} -5 \\ 2 \\ 1 \\ 0 \end{bmatrix}, \quad \mathbf{w}_2 = \frac{1}{\sqrt{30}} \begin{bmatrix} 2 \\ 5 \\ 0 \\ 1 \end{bmatrix}$$

as an orthonormal basis for the solution space of $A\mathbf{x} = \mathbf{0}$.

(b) The solution to $A\mathbf{x} = \mathbf{0}$ is given by

$$\mathbf{x} = \begin{bmatrix} -5r + 2s \\ 2r - 4s \\ r \\ s \end{bmatrix} = r \begin{bmatrix} -5 \\ 2 \\ 1 \\ 0 \end{bmatrix} + s \begin{bmatrix} 2 \\ -4 \\ 0 \\ 1 \end{bmatrix}.$$

Then a basis for the solution space is

$$\begin{bmatrix} -5 \\ 2 \\ 1 \\ 0 \end{bmatrix}, \quad \begin{bmatrix} 2 \\ -4 \\ 0 \\ 1 \end{bmatrix}.$$

We now transform this basis to an orthonormal basis by using the Gram–Schmidt process:

$$\mathbf{u}_1 = \begin{bmatrix} -5 \\ 2 \\ 1 \\ 0 \end{bmatrix}, \quad \mathbf{u}_2 = \begin{bmatrix} 2 \\ -4 \\ 0 \\ 1 \end{bmatrix}.$$

Then

$$\mathbf{v}_1 = \mathbf{u}_1 = \begin{bmatrix} -5 \\ 2 \\ 1 \\ 0 \end{bmatrix}$$

$$\mathbf{v}_2 = \mathbf{u}_2 - \left(\frac{\mathbf{u}_2 \cdot \mathbf{v}_1}{\mathbf{v}_1 \cdot \mathbf{v}_1} \right) \mathbf{v}_1 = \begin{bmatrix} 2 \\ -4 \\ 0 \\ 1 \end{bmatrix} - \left(\frac{-18}{30} \right) \begin{bmatrix} -5 \\ 2 \\ 0 \\ 1 \end{bmatrix}.$$

Normalizing $\mathbf{v}_1$ and $\mathbf{v}_2$ we obtain

$$\mathbf{w}_1 = \frac{1}{\sqrt{30}} \begin{bmatrix} -5 \\ 2 \\ 1 \\ 0 \end{bmatrix}, \qquad \mathbf{w}_2 = \frac{1}{\sqrt{255}} \begin{bmatrix} -5 \\ -14 \\ 3 \\ 5 \end{bmatrix}$$

as an orthonormal basis for the solution space of $A\mathbf{x} = \mathbf{0}$.

33. Let $\mathbf{u}_1 = (1, 0, 0, -1)$, $\mathbf{u}_2 = (1, -1, 0, 0)$, and $\mathbf{u}_3 = (0, 1, 0, 1)$. Let $\mathbf{v}_1 = (1, 0, 0, -1)$. Then

$$\mathbf{v}_2 = \mathbf{u}_2 - \left(\frac{\mathbf{u}_2 \cdot \mathbf{v}_1}{\mathbf{v}_1 \cdot \mathbf{v}_1} \right) = (1, -1, 0, 0) - \tfrac{1}{2}(1, 0, 0, -1) = \left(\tfrac{1}{2}, -1, 0, \tfrac{1}{2} \right).$$

Eliminating fractions, we take $\mathbf{v}_2 = (1, -2, 0, 1)$. Then

$$\mathbf{v}_3 = \mathbf{u}_3 - \left(\frac{\mathbf{u}_3 \cdot \mathbf{v}_1}{\mathbf{v}_1 \cdot \mathbf{v}_1} \right) \mathbf{v}_1 - \left(\frac{\mathbf{u}_3 \cdot \mathbf{v}_2}{\mathbf{v}_2 \cdot \mathbf{v}_2} \right) \mathbf{v}_2$$

$$= (0, 1, 0, 1) - \tfrac{-1}{2}(1, 0, 0, -1) - \tfrac{-1}{6}(1, -2, 0, 1)$$

$$= \left(\tfrac{2}{3}, \tfrac{2}{3}, 0, \tfrac{2}{3} \right).$$

Eliminating fractions, we take $\mathbf{v}_3 = (1, 1, 0, 1)$. Normalizing $\mathbf{v}_1$, $\mathbf{v}_2$, and $\mathbf{v}_3$ we obtain

$$\mathbf{w}_1 = \frac{1}{\sqrt{2}}(1, 0, 0, -1), \qquad \mathbf{w}_2 = \frac{1}{\sqrt{6}}(1, -2, 0, 1), \qquad \mathbf{w}_3 = \frac{1}{\sqrt{3}}(1, 1, 0, 1).$$

35. Let $W = \text{span } \{\mathbf{w}_1, \mathbf{w}_2\} = \text{span } \{(1, 0, 1), (0, 1, 0)\}$.

(a) If $\mathbf{u} = (u_1, u_2, u_3)$ is in $W^\perp$, then $\mathbf{u} \cdot \mathbf{w}_1 = 0$ and $\mathbf{u} \cdot \mathbf{w}_2 = 0$. Thus we have the linear system

$$\begin{aligned} u_1 \quad + u_3 &= 0 \\ u_2 \qquad &= 0 \end{aligned}$$

which has solution $u_1 = -r$, $u_2 = 0$, $u_3 = r$, where r is any real number. Hence all vectors in $W^\perp$ have the form $(-r, 0, r) = r(-1, 0, 1)$. It follows that $\{(-1, 0, 1)\}$ is a basis for $W^\perp$.

(b) We show that $T = \{(1, 0, 1), (0, 1, 0), (-1, 0, 1)\}$ is linearly independent; then Theorem 6.9 in Section 6.4 implies that T is a basis for R^3. Form

$$c_1(1, 0, 1) + c_2(0, 1, 0) + c_3(-1, 0, 1) = (0, 0, 0)$$

and construct the coefficient matrix of the corresponding homogeneous linear system. We obtain

$$\begin{bmatrix} 1 & 0 & -1 \\ 0 & 1 & 0 \\ 1 & 0 & 1 \end{bmatrix}$$

which has reduced row echelon form I_3. Hence the only solution is the trivial solution $c_1 = c_2 = c_3 = 0$. Thus T is linearly independent.

(c) Use Theorem 6.16 in Section 6.8. Let $S = \{\mathbf{w}_1, \mathbf{w}_2\} = \{(1, 0, 1), (0, 1, 0)\}$. S is an orthogonal basis for W, so we normalize the vectors to obtain an orthonormal basis. Let

$$\mathbf{w}_1' = \frac{1}{\|\mathbf{w}_1\|}\,\mathbf{w}_1 = \frac{1}{\sqrt{2}}\,\mathbf{w}_1 \quad \text{and} \quad \mathbf{w}_2' = \frac{1}{\|\mathbf{w}_2\|}\,\mathbf{w}_2 = \mathbf{w}_2.$$

(i) Let $\mathbf{v} = (1, 0, 0)$. From Equation (1) in Section 6.9,

$$\mathbf{w} = (\mathbf{v} \cdot \mathbf{w}_1')\mathbf{w}_1' + (\mathbf{v} \cdot \mathbf{w}_2')\mathbf{w}_2' + = \frac{1}{\sqrt{2}}\,\mathbf{w}_1' + 0\mathbf{w}_2' = \tfrac{1}{2}(1, 0, 1).$$

Then

$$\mathbf{u} = \mathbf{v} - \mathbf{w} = (1, 0, 0) - \tfrac{1}{2}(1, 0, 1) = \left(\tfrac{1}{2}, 0, -\tfrac{1}{2}\right).$$

Hence

$$\mathbf{v} = (1, 0, 0) = \left(\tfrac{1}{2}, 0, -\tfrac{1}{2}\right) + \left(\tfrac{1}{2}, 0, \tfrac{1}{2}\right) = \mathbf{u} + \mathbf{w}.$$

(ii) Let $\mathbf{v} = (1, 2, 3)$. From Equation (1) in Section 6.9,

$$\mathbf{w} = (\mathbf{v} \cdot \mathbf{w}_1')\mathbf{w}_1' + (\mathbf{v} \cdot \mathbf{w}_2')\mathbf{w}_2' + = \frac{4}{\sqrt{2}}\,\mathbf{w}_1' + 2\mathbf{w}_2' = (2, 2, 2).$$

Then

$$\mathbf{u} = \mathbf{v} - \mathbf{w} = (1, 2, 3) - (2, 2, 2) = (-1, 0, 1).$$

Hence

$$\mathbf{v} = (1, 2, 3) = (-1, 0, 1) + (2, 2, 2) = \mathbf{u} + \mathbf{w}.$$

37. We follow the method used in Example 3 in Section 6.9. First transform the matrix A to its reduced row echelon form:

$$A = \begin{bmatrix} 2 & 3 & -1 & 2 \\ 1 & 1 & -2 & 3 \\ 2 & 1 & 4 & 2 \end{bmatrix} \xrightarrow[\text{steps omitted}]{} \cdots \xrightarrow{} \begin{bmatrix} 1 & 0 & 0 & \frac{37}{11} \\ 0 & 1 & 0 & -\frac{20}{11} \\ 0 & 0 & 1 & -\frac{8}{11} \end{bmatrix} = B.$$

To find the null space of A we solve the homogeneous system $B\mathbf{x} = \mathbf{0}$, obtaining

$$\mathbf{x} = \begin{bmatrix} -\frac{37}{11}r \\ \frac{20}{11}r \\ -\frac{8}{11}r \\ r \end{bmatrix} = r \begin{bmatrix} -\frac{37}{11} \\ \frac{20}{11} \\ -\frac{8}{11} \\ 1 \end{bmatrix}$$

so

$$S = \left\{ \begin{bmatrix} -\frac{37}{11} \\ \frac{20}{11} \\ -\frac{8}{11} \\ 1 \end{bmatrix} \right\}$$

is a basis for the null space of A. The nonzero rows of B form a basis for the row space of A:

$$T = \left\{ \left(1, 0, 0, \tfrac{37}{11}\right), \left(0, 1, 0, -\tfrac{20}{11}\right), \left(0, 0, 1, -\tfrac{8}{11}\right) \right\}.$$

Next, we have

$$A^T = \begin{bmatrix} 2 & 1 & 2 \\ 3 & 1 & 1 \\ -1 & -2 & 4 \\ 2 & 3 & 2 \end{bmatrix}.$$

We transform the matrix A^T to its reduced row echelon form:

$$\begin{bmatrix} 2 & 1 & 2 \\ 3 & 1 & 1 \\ -1 & -2 & 4 \\ 2 & 3 & 2 \end{bmatrix} \xrightarrow[\text{steps omitted}]{\cdots} \begin{bmatrix} 1 & 0 & 0 \\ 0 & 1 & 0 \\ 0 & 0 & 1 \\ 0 & 0 & 0 \end{bmatrix} = C.$$

To find the null space of A^T we solve the homogeneous system $C\mathbf{x} = \mathbf{0}$ to obtain

$$\mathbf{x} = \begin{bmatrix} 0 \\ 0 \\ 0 \end{bmatrix}.$$

Thus the null space of A^T has no basis. The nonzero rows of C read vertically give a basis for the column space of A:

$$T' = \left\{ \begin{bmatrix} 1 \\ 0 \\ 0 \\ 0 \end{bmatrix}, \begin{bmatrix} 0 \\ 1 \\ 0 \\ 0 \end{bmatrix}, \begin{bmatrix} 0 \\ 0 \\ 1 \\ 0 \end{bmatrix} \right\}.$$

T.1. If A is nonsingular then $A\mathbf{x} = \mathbf{0}$ has only the trivial solution. Thus the dimension of the solution space is zero. Conversely, if $A\mathbf{x} = \mathbf{0}$ has a solution space of dimension zero, then $\mathbf{x} = \mathbf{0}$ is the only solution. Thus A is nonsingular.

T.3. Let $\mathbf{u} = (u_1, u_2, \ldots, u_n)$ and $\mathbf{e}_j$, $j = 1, 2, \ldots, n$, be the natural basis for R^n. Since $\{\mathbf{v}_1, \ldots, \mathbf{v}_n\}$ is a basis, there exist scalars $c_1, c_2, \ldots, c_n$ such that $\mathbf{e}_j = \sum\limits_{k=1}^{n} c_k \mathbf{v}_k$. Then

$$\mathbf{u} \cdot \mathbf{e}_j = u_j = \mathbf{u} \cdot \sum_{k=1}^{n} c_k \mathbf{v}_k = \sum_{k=1}^{n} c_k (\mathbf{u} \cdot \mathbf{v}_k) = \sum_{k=1}^{n} c_k (0) = 0$$

for each $j = 1, 2, \ldots, n$. Thus $\mathbf{u} = \mathbf{0}$.

T.5. (a) Theorem 6.10 (Sec. 6.6) implies that row space of A = row space of B. Thus rank A = row rank A = row rank B = rank B.

(b) Since A and B are row equivalent they have the same reduced row echelon form. It follows that the solutions of $A\mathbf{x} = \mathbf{0}$ and $B\mathbf{x} = \mathbf{0}$ are the same. Hence $A\mathbf{x} = \mathbf{0}$ if and only if $B\mathbf{x} = \mathbf{0}$.

T.7. $S = \{\mathbf{v}_1, \mathbf{v}_2, \ldots, \mathbf{v}_n\}$ is an orthonormal basis for R^n. Hence $\dim V = n$ and

$$\mathbf{v}_i \cdot \mathbf{v}_j = \begin{cases} 0 & \text{if } i \neq j \\ 1 & \text{if } i = j. \end{cases}$$

Let $T = \{a_1\mathbf{v}_1, a_2\mathbf{v}_2, \ldots, a_n\mathbf{v}_n\}$, where $a_j \neq 0$. To show that T is a basis we need only show that it spans R^n and then use Theorem 6.9(b) (Sec. 6.4). Let $\mathbf{v}$ belong to R^n. Then there exist scalars c_i, $i = 1, 2, \ldots, n$ such that

$$\mathbf{v} = c_1\mathbf{v}_1 + c_2\mathbf{v}_2 + \cdots + c_n\mathbf{v}_n.$$

Since $a_j \neq 0$, we have

$$\mathbf{v} = \frac{c_1}{a_1}\,a_1\mathbf{v}_1 + \frac{c_2}{a_2}\,a_2\mathbf{v}_2 + \cdots + \frac{c_n}{a_n}\,a_n\mathbf{v}_n$$

so span $T = R^n$. Next we show that the members of T are orthogonal. Since S is orthogonal, we have

$$(a_i\mathbf{v}_i) \cdot (a_j\mathbf{v}_j) = a_i a_j (\mathbf{v}_i \cdot \mathbf{v}_j) = \begin{cases} 0 & \text{if } i \neq j \\ a_i a_j & \text{if } i = j. \end{cases}$$

Hence T is an orthogonal set. In order for T to be an orthonormal set we must have $a_i a_j = 1$ for all i and j. This is only possible if all $a_i = 1$.

Chapter 7

Applications of Real Vector Spaces (Optional)

Section 7.1, p. 378

1. Let $\mathbf{u}_1$ and $\mathbf{u}_2$ be the column vectors of A:

$$\mathbf{u}_1 = \begin{bmatrix} 1 \\ -1 \end{bmatrix}, \qquad \mathbf{u}_2 = \begin{bmatrix} 2 \\ 3 \end{bmatrix}.$$

The vectors $\mathbf{u}_1$, $\mathbf{u}_2$ are linearly independent since neither is a multiple of the other. Hence $\{\mathbf{u}_1, \mathbf{u}_2\}$ is a basis for the column space W. We apply the Gram–Schmidt process to find an orthonormal basis. Let

$$\mathbf{v}_1 = \mathbf{u}_1 = \begin{bmatrix} 1 \\ -1 \end{bmatrix}$$

$$\mathbf{v}_2 = \mathbf{u}_2 - \frac{\mathbf{u}_2 \cdot \mathbf{v}_1}{\mathbf{v}_1 \cdot \mathbf{v}_1} \mathbf{v}_1 = \begin{bmatrix} 2 \\ 3 \end{bmatrix} - \frac{-1}{2} \begin{bmatrix} 1 \\ -1 \end{bmatrix} = \frac{5}{2} \begin{bmatrix} 1 \\ 1 \end{bmatrix}$$

The set $\{\mathbf{v}_1, \mathbf{v}_2\}$ is an orthogonal basis for W. To find an orthonormal basis, we determine unit vectors in the same direction as $\mathbf{v}_1$ and $\mathbf{v}_2$. Let

$$\mathbf{w}_1 = \frac{\mathbf{v}_1}{\|\mathbf{v}_1\|} = \frac{\mathbf{v}_1}{\sqrt{2}} = \frac{1}{\sqrt{2}} \begin{bmatrix} 1 \\ -1 \end{bmatrix}$$

$$\mathbf{w}_2 = \frac{\mathbf{v}_2}{\|\mathbf{v}_2\|} = \frac{\mathbf{v}_2}{\frac{5}{\sqrt{2}}} = \left(\frac{\sqrt{2}}{5} \right) \left(\frac{5}{2} \right) \begin{bmatrix} 1 \\ 1 \end{bmatrix} = \frac{1}{\sqrt{2}} \begin{bmatrix} 1 \\ 1 \end{bmatrix}$$

Then $\{\mathbf{w}_1, \mathbf{w}_2\}$ is an orthonormal basis for W. Therefore

$$Q = \begin{bmatrix} \mathbf{w}_1 & \mathbf{w}_2 \end{bmatrix} = \begin{bmatrix} \frac{1}{\sqrt{2}} & \frac{1}{\sqrt{2}} \\ -\frac{1}{\sqrt{2}} & \frac{1}{\sqrt{2}} \end{bmatrix} \approx \begin{bmatrix} 0.7071 & 0.7071 \\ -0.7071 & 0.7071 \end{bmatrix}.$$

To find the matrix $R = \begin{bmatrix} r_{ij} \end{bmatrix}$ we use $r_{ji} = \mathbf{u}_i \cdot \mathbf{w}_j$ to get:

$$r_{11} = \mathbf{u}_1 \cdot \mathbf{w}_1 = \frac{2}{\sqrt{2}} = \sqrt{2}, \qquad r_{12} = \mathbf{u}_2 \cdot \mathbf{w}_1 = -\frac{1}{\sqrt{2}}$$

$$r_{21} = \mathbf{u}_1 \cdot \mathbf{w}_2 = 0, \qquad r_{22} = \mathbf{u}_2 \cdot \mathbf{w}_2 = \frac{5\sqrt{2}}{2} = \frac{5}{\sqrt{2}}$$

Therefore

$$R = \begin{bmatrix} \sqrt{2} & -\frac{1}{\sqrt{2}} \\ 0 & \frac{5}{\sqrt{2}} \end{bmatrix} \approx \begin{bmatrix} 1.4142 & -0.7071 \\ 0 & 3.5355 \end{bmatrix}.$$

3. Let $\mathbf{u}_1$, $\mathbf{u}_2$, and $\mathbf{u}_3$ be the columns of A:

$$\mathbf{u}_1 = \begin{bmatrix} 1 \\ 2 \\ -1 \end{bmatrix}, \quad \mathbf{u}_2 = \begin{bmatrix} 0 \\ -3 \\ 2 \end{bmatrix}, \quad \mathbf{u}_3 = \begin{bmatrix} -1 \\ 3 \\ 4 \end{bmatrix}.$$

Row reducing the matrix A gives I_3. Therefore the vectors $\mathbf{u}_1$, $\mathbf{u}_2$, $\mathbf{u}_3$ are linearly independent and hence form a basis for the column space W. We apply the Gram–Schmidt process to find an orthogonal basis. Let

$$\mathbf{v}_1 = \mathbf{u}_1 = \begin{bmatrix} 1 \\ 2 \\ -1 \end{bmatrix}$$

$$\mathbf{v}_2 = \mathbf{u}_2 - \frac{\mathbf{u}_2 \cdot \mathbf{v}_1}{\mathbf{v}_1 \cdot \mathbf{v}_1} \mathbf{v}_1 = \begin{bmatrix} 0 \\ -3 \\ 2 \end{bmatrix} - \left(\frac{-8}{6} \right) \begin{bmatrix} 1 \\ 2 \\ -1 \end{bmatrix} = \begin{bmatrix} \frac{4}{3} \\ -\frac{1}{3} \\ \frac{2}{3} \end{bmatrix}$$

$$\mathbf{v}_3 = \mathbf{u}_3 - \frac{\mathbf{u}_3 \cdot \mathbf{v}_1}{\mathbf{v}_1 \cdot \mathbf{v}_1} \mathbf{v}_1 - \frac{\mathbf{u}_3 \cdot \mathbf{v}_2}{\mathbf{v}_2 \cdot \mathbf{v}_2} \mathbf{v}_2 = \begin{bmatrix} -1 \\ 3 \\ 4 \end{bmatrix} - \frac{1}{6} \begin{bmatrix} 1 \\ 2 \\ -1 \end{bmatrix} - \frac{\frac{1}{3}}{\frac{7}{3}} \begin{bmatrix} \frac{4}{3} \\ -\frac{1}{3} \\ \frac{2}{3} \end{bmatrix} = \begin{bmatrix} -\frac{19}{14} \\ \frac{38}{14} \\ \frac{57}{14} \end{bmatrix}.$$

The set $\{\mathbf{v}_1, \mathbf{v}_2, \mathbf{v}_3\}$ is an orthogonal basis for W. To find an orthonormal basis, we determine unit vectors in the same direction as $\mathbf{v}_1$, $\mathbf{v}_2$, and $\mathbf{v}_3$. Let

$$\mathbf{w}_1 = \frac{\mathbf{v}_1}{\|\mathbf{v}_1\|} = \frac{1}{\sqrt{6}} \begin{bmatrix} 1 \\ 2 \\ -1 \end{bmatrix}$$

$$\mathbf{w}_2 = \frac{\mathbf{v}_2}{\|\mathbf{v}_2\|} = \frac{1}{\sqrt{\frac{21}{9}}} \frac{1}{3} \begin{bmatrix} 4 \\ -1 \\ 2 \end{bmatrix} = \frac{1}{\sqrt{21}} \begin{bmatrix} 4 \\ -1 \\ 2 \end{bmatrix}$$

$$\mathbf{w}_3 = \frac{\mathbf{v}_3}{\|\mathbf{v}_3\|} = \frac{1}{\frac{19}{\sqrt{14}}} \frac{19}{14} \begin{bmatrix} -1 \\ 2 \\ 3 \end{bmatrix} = \frac{1}{\sqrt{14}} \begin{bmatrix} -1 \\ 2 \\ 3 \end{bmatrix}.$$

Therefore

$$Q = \begin{bmatrix} \mathbf{w}_1 & \mathbf{w}_2 & \mathbf{w}_3 \end{bmatrix} = \begin{bmatrix} \frac{1}{\sqrt{6}} & \frac{4}{\sqrt{21}} & -\frac{1}{\sqrt{14}} \\ \frac{2}{\sqrt{6}} & -\frac{1}{\sqrt{21}} & \frac{2}{\sqrt{14}} \\ -\frac{1}{\sqrt{6}} & \frac{2}{\sqrt{21}} & \frac{3}{\sqrt{14}} \end{bmatrix} \approx \begin{bmatrix} 0.4082 & 0.8729 & -0.2673 \\ 0.8165 & -0.2182 & 0.5345 \\ -0.4082 & 0.4364 & 0.8018 \end{bmatrix}$$

To find the matrix $R = [r_{ij}]$ we use $r_{ji} = \mathbf{u}_i \cdot \mathbf{w}_j$ to get:

$$r_{11} = \mathbf{u}_1 \cdot \mathbf{w}_1 = \frac{6}{\sqrt{6}}, \qquad r_{12} = \mathbf{u}_2 \cdot \mathbf{w}_1 = -\frac{8}{\sqrt{6}}, \qquad r_{13} = \mathbf{u}_3 \cdot \mathbf{w}_1 = \frac{1}{\sqrt{6}}$$

$$r_{21} = \mathbf{u}_1 \cdot \mathbf{w}_2 = 0, \qquad r_{22} = \mathbf{u}_2 \cdot \mathbf{w}_2 = \frac{7}{\sqrt{21}}, \qquad r_{23} = \mathbf{u}_3 \cdot \mathbf{w}_2 = \frac{1}{\sqrt{21}}$$

$$r_{31} = \mathbf{u}_1 \cdot \mathbf{w}_3 = 0, \qquad r_{32} = \mathbf{u}_2 \cdot \mathbf{w}_3 = 0, \qquad r_{33} = \mathbf{u}_3 \cdot \mathbf{w}_3 = \frac{19}{\sqrt{14}}.$$

Therefore

$$R = \begin{bmatrix} \frac{6}{\sqrt{6}} & -\frac{8}{\sqrt{6}} & \frac{1}{\sqrt{6}} \\ 0 & \frac{7}{\sqrt{21}} & \frac{1}{\sqrt{21}} \\ 0 & 0 & \frac{19}{\sqrt{14}} \end{bmatrix} \approx \begin{bmatrix} 2.4475 & -3.2660 & 0.4082 \\ 0 & 1.5275 & 0.2182 \\ 0 & 0 & 5.0780 \end{bmatrix}.$$

5. Let $\mathbf{u}_1$, $\mathbf{u}_2$, and $\mathbf{u}_3$ be the columns of A:

$$\mathbf{u}_1 = \begin{bmatrix} 1 \\ -1 \\ -1 \end{bmatrix}, \qquad \mathbf{u}_2 = \begin{bmatrix} 0 \\ 2 \\ -2 \end{bmatrix}, \qquad \mathbf{u}_3 = \begin{bmatrix} 2 \\ 0 \\ 2 \end{bmatrix}.$$

Row reducing the matrix A gives I_3. Therefore the vectors $\mathbf{u}_1$, $\mathbf{u}_2$, $\mathbf{u}_3$ are linearly independent and hence form a basis for the column space W. We apply the Gram–Schmidt process to find an orthogonal basis. Let

$$\mathbf{v}_1 = \mathbf{u}_1 = \begin{bmatrix} 1 \\ -1 \\ -1 \end{bmatrix}$$

$$\mathbf{v}_2 = \mathbf{u}_2 - \frac{\mathbf{u}_2 \cdot \mathbf{v}_1}{\mathbf{v}_1 \cdot \mathbf{v}_1} \mathbf{v}_1 = \begin{bmatrix} 0 \\ 2 \\ -2 \end{bmatrix} - \frac{0}{3} \begin{bmatrix} 1 \\ -1 \\ -1 \end{bmatrix} = \begin{bmatrix} 0 \\ 2 \\ -2 \end{bmatrix}$$

$$\mathbf{v}_3 = \mathbf{u}_3 - \frac{\mathbf{u}_3 \cdot \mathbf{v}_1}{\mathbf{v}_1 \cdot \mathbf{v}_1} \mathbf{v}_1 - \frac{\mathbf{u}_3 \cdot \mathbf{v}_2}{\mathbf{v}_2 \cdot \mathbf{v}_2} \mathbf{v}_2 = \begin{bmatrix} 2 \\ 0 \\ 2 \end{bmatrix} - \frac{0}{3} \begin{bmatrix} 1 \\ -1 \\ -1 \end{bmatrix} - \left(\frac{-4}{8} \right) \begin{bmatrix} 0 \\ 2 \\ -2 \end{bmatrix} = \begin{bmatrix} 2 \\ 1 \\ 1 \end{bmatrix}$$

The set $\{\mathbf{v}_1, \mathbf{v}_2, \mathbf{v}_3\}$ is an orthogonal basis for W. To find an orthonormal basis, we determine unit vectors in the same direction as $\mathbf{v}_1$, $\mathbf{v}_2$, and $\mathbf{v}_3$. Let

$$\mathbf{w}_1 = \frac{\mathbf{v}_1}{\|\mathbf{v}_1\|} = \frac{1}{\sqrt{3}} \begin{bmatrix} 1 \\ -1 \\ -1 \end{bmatrix}$$

$$\mathbf{w}_2 = \frac{\mathbf{v}_2}{\|\mathbf{v}_2\|} = \frac{1}{\sqrt{8}} \begin{bmatrix} 0 \\ 2 \\ -2 \end{bmatrix} = \frac{1}{\sqrt{2}} \begin{bmatrix} 0 \\ 1 \\ -1 \end{bmatrix}$$

$$\mathbf{w}_3 = \frac{\mathbf{v}_3}{\|\mathbf{v}_3\|} = \frac{1}{\sqrt{6}} \begin{bmatrix} 2 \\ 1 \\ 1 \end{bmatrix}$$

Therefore

$$Q = \begin{bmatrix} \mathbf{w}_1 & \mathbf{w}_2 & \mathbf{w}_3 \end{bmatrix} = \begin{bmatrix} \frac{1}{\sqrt{3}} & 0 & \frac{2}{\sqrt{6}} \\ -\frac{1}{\sqrt{3}} & \frac{1}{\sqrt{2}} & \frac{1}{\sqrt{6}} \\ -\frac{1}{\sqrt{3}} & -\frac{1}{\sqrt{2}} & \frac{1}{\sqrt{6}} \end{bmatrix} \approx \begin{bmatrix} 0.5774 & 0 & 0.8165 \\ -0.5774 & 0.7071 & 0.4082 \\ -0.5774 & -0.7071 & 0.4082 \end{bmatrix}.$$

To find the matrix $R = \begin{bmatrix} r_{ij} \end{bmatrix}$ we use $r_{ji} = \mathbf{u}_i \cdot \mathbf{w}_j$ to get:

$$r_{11} = \mathbf{u}_1 \cdot \mathbf{w}_1 = \sqrt{3}, \qquad r_{12} = \mathbf{u}_2 \cdot \mathbf{w}_1 = 0, \qquad r_{13} = \mathbf{u}_3 \cdot \mathbf{w}_1 = 0$$

$$r_{21} = \mathbf{u}_1 \cdot \mathbf{w}_2 = 0, \qquad r_{22} = \mathbf{u}_2 \cdot \mathbf{w}_2 = 2\sqrt{2}, \qquad r_{23} = \mathbf{u}_3 \cdot \mathbf{w}_2 = -\sqrt{2}$$

$$r_{31} = \mathbf{u}_1 \cdot \mathbf{w}_3 = 0, \qquad r_{32} = \mathbf{u}_2 \cdot \mathbf{w}_3 = 0, \qquad r_{33} = \mathbf{u}_3 \cdot \mathbf{w}_3 = \sqrt{6}$$

Therefore

$$R = \begin{bmatrix} \sqrt{3} & 0 & 0 \\ 0 & 2\sqrt{2} & -\sqrt{2} \\ 0 & 0 & \sqrt{6} \end{bmatrix} \approx \begin{bmatrix} 1.7321 & 0 & 0 \\ 0 & 2.8284 & -1.4142 \\ 0 & 0 & 2.4495 \end{bmatrix}.$$

T.1. We have

$$\mathbf{u}_i = \mathbf{v}_i + \frac{\mathbf{u}_i \cdot \mathbf{v}_1}{\mathbf{v}_1 \cdot \mathbf{v}_1} \mathbf{v}_1 + \frac{\mathbf{u}_i \cdot \mathbf{v}_2}{\mathbf{v}_2 \cdot \mathbf{v}_2} \mathbf{v}_2 + \cdots + \frac{\mathbf{u}_i \cdot \mathbf{v}_{i-1}}{\mathbf{v}_{i-1} \cdot \mathbf{v}_{i-1}} \mathbf{v}_{i-1}.$$

Then

$$r_{ii} = \mathbf{u}_i \cdot \mathbf{w}_i = \mathbf{v}_i \cdot \mathbf{w}_i + \frac{\mathbf{u}_i \cdot \mathbf{v}_1}{\mathbf{v}_1 \cdot \mathbf{v}_1} (\mathbf{v}_1 \cdot \mathbf{w}_i) + \frac{\mathbf{u}_i \cdot \mathbf{v}_2}{\mathbf{v}_2 \cdot \mathbf{v}_2} (\mathbf{v}_2 \cdot \mathbf{w}_i) + \cdots + \frac{\mathbf{u}_i \cdot \mathbf{v}_{i-1}}{\mathbf{v}_{i-1} \cdot \mathbf{v}_{i-1}} (\mathbf{v}_{i-1} \cdot \mathbf{w}_i)$$

$$= \mathbf{v}_i \cdot \mathbf{w}_i$$

because $\mathbf{v}_i \cdot \mathbf{w}_j = 0$ for $i \neq j$. Moreover,

$$\mathbf{w}_i = \frac{1}{\|\mathbf{v}_i\|} \mathbf{v}_i,$$

so $\mathbf{v}_i \cdot \mathbf{w}_i = \dfrac{1}{\|\mathbf{v}_i\|} (\mathbf{v}_i \cdot \mathbf{v}_i) = \|\mathbf{v}_i\|$.

Section 7.2, p. 388

1. In order to use Theorem 7.2 we must show that rank $A = 2$. We first obtain the reduced row echelon form of A:

$$\begin{bmatrix} 2 & 1 \\ 1 & 0 \\ 0 & -1 \\ -1 & 1 \end{bmatrix} \xrightarrow[\text{steps omitted}]{} \cdots \longrightarrow \begin{bmatrix} 1 & 0 \\ 0 & 1 \\ 0 & 0 \\ 0 & 0 \end{bmatrix}$$

Hence rank $A = 2$. Then using Theorem 7.2 we proceed as follows. Form the normal system:

$$A^T A \mathbf{x} = A^T \mathbf{b} \implies \begin{bmatrix} 6 & 1 \\ 1 & 3 \end{bmatrix} \mathbf{x} = \begin{bmatrix} 8 \\ 0 \end{bmatrix}.$$

Solving this linear system for $\mathbf{x}$ by Gauss–Jordan reduction, we obtain

$$\widehat{\mathbf{x}} \approx \begin{bmatrix} \frac{24}{17} \\ -\frac{8}{17} \end{bmatrix} \approx \begin{bmatrix} 1.4118 \\ -0.4706 \end{bmatrix}.$$

3. In order to use Theorem 7.2 we must show that rank $A = 3$. We first obtain the reduced row echelon form of A:

$$A = \begin{bmatrix} 1 & 2 & 1 \\ 1 & 3 & 2 \\ 2 & 5 & 3 \\ 2 & 0 & 1 \\ 3 & 1 & 1 \end{bmatrix} \longrightarrow \underset{\text{steps omitted}}{\cdots} \longrightarrow \begin{bmatrix} 1 & 0 & 0 \\ 0 & 1 & 0 \\ 0 & 0 & 1 \\ 0 & 0 & 0 \\ 0 & 0 & 0 \end{bmatrix}$$

Hence rank $A = 3$. Then using Theorem 7.2 we proceed as follows. Form the normal system:

$$A^T A \mathbf{x} = A^T \mathbf{b} \implies \begin{bmatrix} 19 & 18 & 14 \\ 18 & 39 & 24 \\ 14 & 24 & 16 \end{bmatrix} \mathbf{x} = \begin{bmatrix} -3 \\ 2 \\ 2 \end{bmatrix}.$$

Solving this linear system for $\mathbf{x}$ by Gauss–Jordan reduction, we obtain

$$\widehat{\mathbf{x}} \approx \begin{bmatrix} -\frac{9591}{6255} \\ -\frac{35028}{18765} \\ \frac{768}{180} \end{bmatrix} \approx \begin{bmatrix} -1.5333 \\ -1.8667 \\ 4.2667 \end{bmatrix}.$$

5. We first find the QR-factorization of

$$A = \begin{bmatrix} 2 & 1 \\ 1 & 0 \\ 0 & -1 \\ -1 & 1 \end{bmatrix}.$$

Let W denote the column space of A and let $\mathbf{u}_1$ and $\mathbf{u}_2$ be the column vectors of A:

$$\mathbf{u}_1 = \begin{bmatrix} 2 \\ 1 \\ 0 \\ -1 \end{bmatrix}, \qquad \mathbf{u}_2 = \begin{bmatrix} 1 \\ 0 \\ -1 \\ 1 \end{bmatrix}.$$

The vectors $\mathbf{u}_1$ and $\mathbf{u}_2$ are linearly independent since neither is a multiple of the other. We now apply the Gram–Schmidt process to find an orthonormal basis. Let

$$\mathbf{v}_1 = \mathbf{u}_1 = \begin{bmatrix} 2 \\ 1 \\ 0 \\ -1 \end{bmatrix}$$

$$\mathbf{v}_2 = \mathbf{u}_2 - \left(\frac{\mathbf{u}_2 \cdot \mathbf{v}_1}{\mathbf{v}_1 \cdot \mathbf{v}_1} \right) \mathbf{v}_1 = \begin{bmatrix} 1 \\ 0 \\ -1 \\ 1 \end{bmatrix} - \frac{1}{6} \begin{bmatrix} 2 \\ 1 \\ 0 \\ -1 \end{bmatrix} = \begin{bmatrix} \frac{4}{6} \\ -\frac{1}{6} \\ -1 \\ \frac{7}{6} \end{bmatrix}.$$

The set $\{\mathbf{v}_1, \mathbf{v}_2\}$ is an orthogonal basis for W. To find an orthonormal basis, we determine unit vectors in the same direction as $\mathbf{v}_1$ and $\mathbf{v}_2$. Let

$$\mathbf{w}_1 = \frac{1}{\|\mathbf{v}_1\|}\,\mathbf{v}_1 = \frac{1}{\sqrt{6}}\begin{bmatrix} 2 \\ 1 \\ 0 \\ -1 \end{bmatrix}$$

$$\mathbf{w}_2 = \frac{1}{\|\mathbf{v}_2\|}\,\mathbf{v}_2 = \frac{1}{\frac{\sqrt{102}}{6}}\begin{bmatrix} \frac{4}{6} \\ -\frac{1}{6} \\ -1 \\ \frac{7}{6} \end{bmatrix} = \frac{1}{\sqrt{102}}\begin{bmatrix} 4 \\ -1 \\ -6 \\ 7 \end{bmatrix}.$$

Then $\{\mathbf{w}_1, \mathbf{w}_2\}$ is an orthonormal basis for W. Therefore

$$Q = \begin{bmatrix} \mathbf{w}_1 & \mathbf{w}_2 \end{bmatrix} = \begin{bmatrix} \frac{2}{\sqrt{6}} & \frac{4}{\sqrt{102}} \\ \frac{1}{\sqrt{6}} & -\frac{1}{\sqrt{102}} \\ 0 & -\frac{6}{\sqrt{102}} \\ -\frac{1}{\sqrt{6}} & \frac{7}{\sqrt{102}} \end{bmatrix} \approx \begin{bmatrix} 0.8165 & 0.3961 \\ 0.4083 & -0.0990 \\ 0 & -0.5941 \\ -0.4083 & 0.6931 \end{bmatrix}.$$

To find the matrix $R = \begin{bmatrix} r_{ij} \end{bmatrix}$, we use $r_{ji} = \mathbf{u}_i \cdot \mathbf{w}_j$ to get

$$r_{11} = \mathbf{u}_1 \cdot \mathbf{w}_1 = \sqrt{6}, \qquad r_{12} = \mathbf{u}_2 \cdot \mathbf{w}_1 = \frac{1}{\sqrt{6}}$$

$$r_{21} = \mathbf{u}_1 \cdot \mathbf{w}_2 = 0 \qquad r_{22} = \mathbf{u}_2 \cdot \mathbf{w}_2 = \frac{\sqrt{17}}{\sqrt{6}}.$$

Therefore

$$R = \begin{bmatrix} \sqrt{6} & \frac{1}{\sqrt{6}} \\ 0 & \frac{\sqrt{17}}{\sqrt{6}} \end{bmatrix} \approx \begin{bmatrix} 2.4495 & 0.4083 \\ 0 & 1.6833 \end{bmatrix}.$$

We now solve $R\mathbf{x} = Q^T\mathbf{b}$:

$$\begin{bmatrix} 2.4495 & 0.4083 \\ 0 & 1.6833 \end{bmatrix}\mathbf{x} = \begin{bmatrix} 3.2661 \\ -0.7921 \end{bmatrix}$$

by Gauss–Jordan reduction, obtaining

$$\widehat{\mathbf{x}} \approx \begin{bmatrix} 1.4118 \\ -0.4706 \end{bmatrix}.$$

7. We use the procedure described just before Example 3. The given points are:

$$(x_1, y_1) = (2, 1); \qquad (x_2, y_2) = (3, 2); \qquad (x_3, y_3) = (4, 3); \qquad (x_4, y_4) = (5, 2).$$

Then let

$$A = \begin{bmatrix} x_1 & 1 \\ x_2 & 1 \\ x_3 & 1 \\ x_4 & 1 \end{bmatrix} = \begin{bmatrix} 2 & 1 \\ 3 & 1 \\ 4 & 1 \\ 5 & 1 \end{bmatrix}; \qquad \mathbf{x} = \begin{bmatrix} b_1 \\ b_0 \end{bmatrix}; \qquad \text{and} \quad \mathbf{b} = \begin{bmatrix} y_1 \\ y_2 \\ y_3 \\ y_4 \end{bmatrix} = \begin{bmatrix} 1 \\ 2 \\ 3 \\ 2 \end{bmatrix}.$$

Form the normal equations

$$A^T A\mathbf{x} = A^T\mathbf{b} \quad\Longrightarrow\quad \begin{bmatrix} 54 & 14 \\ 14 & 4 \end{bmatrix}\mathbf{x} = \begin{bmatrix} 30 \\ 8 \end{bmatrix}.$$

Solving this linear system by Gauss–Jordan reduction, we obtain

$$b_1 = 0.4 \quad\text{and}\quad b_0 = 0.6.$$

Hence the least squares line is

$$y = 0.4x + 0.6.$$

9. We use the procedure described just before Example 4. The given points are:

$$(x_1, y_1) = (2,3); \qquad (x_2, y_2) = (3,4); \qquad (x_3, y_3) = (4,3);$$
$$(x_4, y_4) = (5,4); \qquad (x_5, y_5) = (6,3); \qquad (x_6, y_6) = (7,4).$$

Then let

$$A = \begin{bmatrix} 2 & 1 \\ 3 & 1 \\ 4 & 1 \\ 5 & 1 \\ 6 & 1 \\ 7 & 1 \end{bmatrix}; \qquad \mathbf{x} = \begin{bmatrix} b_1 \\ b_0 \end{bmatrix}; \qquad \mathbf{b} = \begin{bmatrix} 3 \\ 4 \\ 3 \\ 4 \\ 3 \\ 4 \end{bmatrix}.$$

Form the normal equations

$$A^T A\mathbf{x} = A^T\mathbf{b} \quad\Longrightarrow\quad \begin{bmatrix} 139 & 27 \\ 27 & 6 \end{bmatrix}\mathbf{x} = \begin{bmatrix} 96 \\ 21 \end{bmatrix}.$$

Solving this linear system by Gauss–Jordan reduction, we obtain

$$\widehat{\mathbf{x}} \approx \begin{bmatrix} 0.086 \\ 3.114 \end{bmatrix}$$

Hence the least squares line is

$$y = 0.86x + 3.114$$

11. We use the procedure described just before Example 5. The given points are:

$$(x_1, y_1) = (0, 3.2); \qquad (x_2, y_2) = (0.5, 1.6); \qquad (x_3, y_3) = (1, 2);$$
$$(x_4, y_4) = (2, -0.4); \qquad (x_5, y_5) = (2.5, -0.8); \qquad (x_6, y_6) = (3, -1.6)$$
$$(x_7, y_7) = (4, 0.3); \qquad (x_8, y_8) = (5, 2.2)$$

Then let

$$A = \begin{bmatrix} 0 & 0 & 1 \\ 0.25 & 0.5 & 1 \\ 1 & 1 & 1 \\ 4 & 2 & 1 \\ 6.25 & 2.5 & 1 \\ 9 & 3 & 1 \\ 16 & 4 & 1 \\ 25 & 5 & 1 \end{bmatrix}; \qquad \mathbf{x} = \begin{bmatrix} a_2 \\ a_1 \\ a_0 \end{bmatrix}, \qquad \mathbf{b} = \begin{bmatrix} 3.2 \\ 1.6 \\ 2 \\ -0.4 \\ -0.8 \\ -1.6 \\ 0.3 \\ 2.2 \end{bmatrix}.$$

Form the normal equations

$$A^T A \mathbf{x} = A^T \mathbf{b} \implies \begin{bmatrix} 1018.1 & 240.8 & 61.5 \\ 240.8 & 61.5 & 18.0 \\ 61.5 & 18.0 & 8.0 \end{bmatrix} \mathbf{x} = \begin{bmatrix} 41.2 \\ 7.4 \\ 6.5 \end{bmatrix}.$$

Solving this linear system for $\mathbf{x}$ by Gauss–Jordan reduction, we obtain

$$\widehat{\mathbf{x}} \approx \begin{bmatrix} 0.5718 \\ -3.1314 \\ 3.4627 \end{bmatrix}$$

Hence the quadratic least squares polynomial for the given data is

$$y = 0.5718x^2 - 3.1314x + 3.4627.$$

13. (a) We use the procedure described just before Example 4. The given points are:

$$(x_1, y_1) = (1, 0.8); \qquad (x_2, y_2) = (2, 2.1); \qquad (x_3, y_3) = (3, 2.6);$$
$$(x_4, y_4) = (4, 2.0); \qquad (x_5, y_5) = (5, 3.1); \qquad (x_6, y_6) = (6, 3.3)$$

Then let

$$A = \begin{bmatrix} 1 & 1 \\ 2 & 1 \\ 3 & 1 \\ 4 & 1 \\ 5 & 1 \\ 6 & 1 \end{bmatrix}; \qquad \mathbf{x} = \begin{bmatrix} b_1 \\ b_0 \end{bmatrix}, \qquad \mathbf{b} = \begin{bmatrix} 0.8 \\ 2.1 \\ 2.6 \\ 2.0 \\ 3.1 \\ 3.3 \end{bmatrix}.$$

Form the normal equations

$$A^T A \mathbf{x} = A^T \mathbf{b} \implies \begin{bmatrix} 91 & 21 \\ 21 & 6 \end{bmatrix} \mathbf{x} = \begin{bmatrix} 56.1 \\ 13.9 \end{bmatrix}.$$

Solving this linear system $\mathbf{x}$ by Gauss–Jordan reduction, we obtain

$$\widehat{\mathbf{x}} \approx \begin{bmatrix} 0.426 \\ 0.827 \end{bmatrix}$$

Hence the least squares line is

$$y = 0.426x + 0.827.$$

(b) Letting $x = 10$ in the answer for part (a), we obtain $y = 5.087$. Thus, it will take the subject 5.087 hours to find his way out of the maze.

15. (a) We use the procedure described just before Example 4. The given points are:

$$(x_1, y_1) = (5, 2.3); \qquad (x_2, y_2) = (6, 3.2); \qquad (x_3, y_3) = (7, 4.1);$$
$$(x_4, y_4) = (8, 5.0); \qquad (x_5, y_5) = (9, 6.1); \qquad (x_6, y_6) = (10, 7.2)$$

Then let

$$A = \begin{bmatrix} 5 & 1 \\ 6 & 1 \\ 7 & 1 \\ 8 & 1 \\ 9 & 1 \\ 10 & 1 \end{bmatrix}; \qquad \mathbf{x} = \begin{bmatrix} b_1 \\ b_0 \end{bmatrix}, \qquad \text{and} \quad \mathbf{b} = \begin{bmatrix} 2.3 \\ 3.2 \\ 4.1 \\ 5.0 \\ 6.1 \\ 7.2 \end{bmatrix}.$$

Form the normal equations

$$A^T A\mathbf{x} = A^T \mathbf{b} \quad\Longrightarrow\quad \begin{bmatrix} 355 & 45 \\ 45 & 6 \end{bmatrix} \mathbf{x} = \begin{bmatrix} 226.3 \\ 27.9 \end{bmatrix}.$$

Solving this linear system $\mathbf{x}$ by Gauss–Jordan reduction, we obtain

$$\widehat{\mathbf{x}} \approx \begin{bmatrix} 0.974 \\ -2.657 \end{bmatrix}$$

Hence the least squares line is

$$y = 0.974x - 2.657.$$

(b) Letting $x = 14$ in the answer for part (a), we obtain $y = 10.979$. Thus, the annual sales is approximately 10.979 millions of dollars

17. (a) We row reduce the matrix, finding:

$$\begin{bmatrix} 1 & 3 & -3 \\ 2 & 4 & -2 \\ 0 & -1 & 2 \\ 1 & 2 & -1 \end{bmatrix} \xrightarrow[\text{steps omitted}]{\quad\cdots\quad} \begin{bmatrix} 1 & 0 & 3 \\ 0 & 1 & -2 \\ 0 & 0 & 0 \\ 0 & 0 & 0 \end{bmatrix}.$$

Therefore the rank of A is 2.

(b) To find a basis for the column space of A we determine the nonzero rows in the reduced row echelon form of A^T and convert them to columns. We find that

$$\begin{bmatrix} 1 & 2 & 0 & 1 \\ 3 & 4 & -1 & 2 \\ -3 & -2 & 2 & -1 \end{bmatrix} \xrightarrow[\text{steps omitted}]{\quad\cdots\quad} \begin{bmatrix} 1 & 0 & -1 & 0 \\ 0 & 1 & \frac{1}{2} & \frac{1}{2} \\ 0 & 0 & 0 & 0 \end{bmatrix}.$$

Then

$$\mathbf{u}_1 = \begin{bmatrix} 1 \\ 0 \\ -1 \\ 0 \end{bmatrix} \quad\text{and}\quad \mathbf{u}_2 = \begin{bmatrix} 0 \\ 1 \\ \frac{1}{2} \\ \frac{1}{2} \end{bmatrix}$$

form a basis for the column space of A. Applying the Gram–Schmidt process we obtain the orthonormal basis

$$\mathbf{w}_1 = \frac{1}{\sqrt{2}} \begin{bmatrix} 1 \\ 0 \\ -1 \\ 0 \end{bmatrix} \quad\text{and}\quad \mathbf{w}_2 = \frac{1}{\sqrt{22}} \begin{bmatrix} 1 \\ 4 \\ 1 \\ 2 \end{bmatrix}.$$

Hence we have

$$\text{proj}_W \mathbf{b} = (\mathbf{b} \cdot \mathbf{w}_1)\mathbf{w}_1 + (\mathbf{b} \cdot \mathbf{w}_2)\mathbf{w}_2 = \frac{1}{\sqrt{2}} \mathbf{w}_1 + \frac{3}{\sqrt{22}} \mathbf{w}_2 = \frac{1}{11} \begin{bmatrix} 7 \\ 6 \\ -4 \\ 3 \end{bmatrix}.$$

We now solve the system $A\widehat{\mathbf{x}} = \text{proj}_W \mathbf{b}$ by row reducing the matrix $\begin{bmatrix} A \ \vdots \ \text{proj}_W \mathbf{b} \end{bmatrix}$ to obtain

$$\begin{bmatrix} 1 & 0 & 3 & \vdots & -\frac{5}{11} \\ 0 & 1 & -2 & \vdots & \frac{4}{11} \\ 0 & 0 & 0 & \vdots & 0 \\ 0 & 0 & 0 & \vdots & 0 \end{bmatrix}.$$

Thus we see that there are infinitely many solutions to this least squares problem. The general solution is

$$\widehat{x}_1 = -3r - \tfrac{5}{11}, \qquad \widehat{x}_2 = 2r + \tfrac{4}{11}, \qquad \widehat{x}_3 = r,$$

where r is any real number. One possible answer is

$$\widehat{\mathbf{x}} = \begin{bmatrix} -\frac{5}{11} \\ \frac{4}{11} \\ 0 \end{bmatrix}.$$

T.1. We have

$$A = \begin{bmatrix} x_1 & 1 \\ x_2 & 1 \\ \vdots & \vdots \\ x_n & 1 \end{bmatrix}.$$

If at least two x-coordinates are unequal, then rank $A = 2$. Theorem 7.2 implies that $A^T A$ is nonsingular.

ML.1. Enter the data into MATLAB.

x = [2 3 4 5 6 7];y = [3 4 3 4 3 4];

c = lsqline(x,y)

We find that the least squares model is:

$$y = 0.08571 * x + 3.114.$$

ML.3. Enter the data into MATLAB.

x = [0 2 3 5 9];y = [185 170 166 152 110];

(a) Using command **c = lsqline(x,y)** we find that the least squares model is:

$$y = -8.278 * x + 188.1.$$

(b) Using the option to evaluate the model, we find that $x = 1$ gives 179.7778, $x = 6$ gives 138.3889, and $x = 8$ gives 121.8333.

(c) In the equation for the least squares line set $y = 160$ and solve for x. We find $x = 3.3893$ min.

ML.5. Data for quadratic least squares:

x	yy
−3.0000	0.5000
−2.5000	0
−2.0000	−1.1250
−1.5000	−1.1875
−1.0000	−1.0000
0	0.9375
0.5000	2.8750
1.0000	4.7500
1.5000	8.2500
2.0000	11.5000

v = polyfit(x,yy,2)

v =

    1.0204   3.1238   1.0507

Thus $y = 1.0204x^2 + 3.1238x + 1.0507$.

Section 7.3, p. 407

1. The code matrix of the parity $(m, m+1)$ code is

$$C = \begin{bmatrix} I_m \\ U \end{bmatrix}.$$

The code words are all the linear combinations of the columns of C which is the set of vectors in B^{m+1} given by

$$C\mathbf{b} = \begin{bmatrix} 1 & 0 & 0 & \cdots & 0 \\ 0 & 1 & 0 & \cdots & 0 \\ \vdots & \vdots & \vdots & \vdots & \vdots \\ 0 & 0 & 0 & \cdots & 1 \\ 1 & 1 & 1 & \cdots & 1 \end{bmatrix} \begin{bmatrix} b_1 \\ b_2 \\ \vdots \\ b_m \end{bmatrix} = \begin{bmatrix} b_1 \\ b_2 \\ \vdots \\ b_m \\ b_1 + b_2 + \cdots + b_m \end{bmatrix}.$$

These vectors form a subspace of B^{m+1}.

3. There are 2^3 code words in B^4 which are the images of the vectors $\mathbf{b}$ of B^3 by the matrix transformation $e \colon B^3 \to B^4$ defined by

$$e(\mathbf{b}) = C\mathbf{b} = \begin{bmatrix} 1 & 0 & 0 \\ 0 & 1 & 0 \\ 0 & 0 & 1 \\ 1 & 1 & 1 \end{bmatrix} \mathbf{b}.$$

The 2^3 possible messages are

$$\begin{bmatrix} 0 \\ 0 \\ 0 \end{bmatrix}, \begin{bmatrix} 0 \\ 0 \\ 1 \end{bmatrix}, \begin{bmatrix} 0 \\ 1 \\ 0 \end{bmatrix}, \begin{bmatrix} 1 \\ 0 \\ 0 \end{bmatrix}, \begin{bmatrix} 0 \\ 1 \\ 1 \end{bmatrix}, \begin{bmatrix} 1 \\ 1 \\ 0 \end{bmatrix}, \begin{bmatrix} 1 \\ 0 \\ 1 \end{bmatrix}, \begin{bmatrix} 1 \\ 1 \\ 1 \end{bmatrix}.$$

Thus the code words are

$$\begin{bmatrix} 0 \\ 0 \\ 0 \\ 0 \end{bmatrix}, \begin{bmatrix} 0 \\ 0 \\ 1 \\ 1 \end{bmatrix}, \begin{bmatrix} 0 \\ 1 \\ 0 \\ 1 \end{bmatrix}, \begin{bmatrix} 1 \\ 0 \\ 0 \\ 1 \end{bmatrix}, \begin{bmatrix} 0 \\ 1 \\ 1 \\ 0 \end{bmatrix}, \begin{bmatrix} 1 \\ 1 \\ 0 \\ 0 \end{bmatrix}, \begin{bmatrix} 1 \\ 0 \\ 1 \\ 0 \end{bmatrix}, \begin{bmatrix} 1 \\ 1 \\ 1 \\ 1 \end{bmatrix}.$$

5. The problem of finding a basis for the null space of the check matrix G reduces to the problem of finding a basis for the solution space of a homogeneous system with augmented matrix

$$\left[\begin{array}{ccccc:c} 1 & 1 & 0 & 1 & 0 & 0 \\ 1 & 0 & 1 & 0 & 1 & 0 \end{array}\right].$$

Its reduced row echelon form is

$$\left[\begin{array}{ccccc:c} 1 & 0 & 1 & 0 & 1 & 0 \\ 0 & 1 & 1 & 1 & 1 & 0 \end{array}\right].$$

It follows that

$$x_1 = x_3 + x_5$$
$$x_2 = x_3 + x_4 + x_5$$

If $x_3 = r$, $x_4 = s$, $x_5 = t$ are arbitrary bits we have

$$\mathbf{x} = \begin{bmatrix} x_1 \\ x_2 \\ x_3 \\ x_4 \\ x_5 \end{bmatrix} = \begin{bmatrix} r+t \\ r+s+t \\ r \\ s \\ t \end{bmatrix} = r\begin{bmatrix} 1 \\ 1 \\ 1 \\ 0 \\ 0 \end{bmatrix} + s\begin{bmatrix} 0 \\ 1 \\ 0 \\ 1 \\ 0 \end{bmatrix} + t\begin{bmatrix} 1 \\ 1 \\ 0 \\ 0 \\ 1 \end{bmatrix}.$$

Thus a basis for the null space of G is

$$\left\{ \begin{bmatrix} 1 \\ 1 \\ 1 \\ 0 \\ 0 \end{bmatrix}, \begin{bmatrix} 0 \\ 1 \\ 0 \\ 1 \\ 0 \end{bmatrix}, \begin{bmatrix} 1 \\ 1 \\ 0 \\ 0 \\ 1 \end{bmatrix} \right\}.$$

7. We have a $(3,5)$ code with code matrix

$$C = \begin{bmatrix} 1 & 0 & 0 \\ 0 & 1 & 0 \\ 0 & 0 & 1 \\ 0 & 1 & 1 \\ 1 & 0 & 0 \end{bmatrix}.$$

Using the result in Example 7 we find the check matrix is

$$G = \begin{bmatrix} 1 & 1 & 1 & 1 & 0 \\ 1 & 0 & 0 & 0 & 1 \end{bmatrix}.$$

Therefore

$$GC = \begin{bmatrix} 1 & 1 & 1 & 1 & 0 \\ 1 & 0 & 0 & 0 & 1 \end{bmatrix} \begin{bmatrix} 1 & 0 & 0 \\ 0 & 1 & 0 \\ 0 & 0 & 1 \\ 0 & 1 & 1 \\ 1 & 0 & 0 \end{bmatrix} = O.$$

9. The problem of finding a basis for the null space of the check matrix reduces to the problem of finding a basis for the solution space of a homogeneous system with augmented matrix

$$\left[\begin{array}{ccccc:c} 1 & 1 & 1 & 0 & 0 & 0 \\ 0 & 1 & 0 & 1 & 0 & 0 \\ 1 & 1 & 0 & 0 & 1 & 0 \end{array}\right].$$

Its reduced row echelon form is

$$\begin{bmatrix} 1 & 0 & 0 & 1 & 1 & | & 0 \\ 0 & 1 & 0 & 1 & 0 & | & 0 \\ 0 & 0 & 1 & 0 & 1 & | & 0 \end{bmatrix}.$$

It follows that

$$x_1 = x_4 + x_5$$
$$x_2 = x_4$$
$$x_3 = x_5.$$

If $x_4 = s$, $x_5 = t$ are arbitrary bits, the solution is

$$\mathbf{x} = \begin{bmatrix} x_1 \\ x_2 \\ x_3 \\ x_4 \\ x_5 \end{bmatrix} = \begin{bmatrix} s+t \\ s \\ t \\ s \\ t \end{bmatrix} = s \begin{bmatrix} 1 \\ 1 \\ 0 \\ 1 \\ 0 \end{bmatrix} + t \begin{bmatrix} 1 \\ 0 \\ 1 \\ 0 \\ 1 \end{bmatrix}.$$

Thus a basis for the null space of G is

$$\left\{ \begin{bmatrix} 1 \\ 1 \\ 0 \\ 1 \\ 0 \end{bmatrix}, \begin{bmatrix} 1 \\ 0 \\ 1 \\ 0 \\ 1 \end{bmatrix} \right\}.$$

11. The check matrix

$$G = \begin{bmatrix} 1 & 0 & 0 & 1 & 0 & 0 \\ 0 & 0 & 1 & 0 & 1 & 0 \\ 1 & 1 & 1 & 0 & 0 & 1 \end{bmatrix}$$

has code matrix (see Theorem 7.4)

$$C = \begin{bmatrix} 1 & 0 & 0 \\ 0 & 1 & 0 \\ 0 & 0 & 1 \\ 1 & 0 & 0 \\ 0 & 0 & 1 \\ 1 & 1 & 1 \end{bmatrix}.$$

The code words are all the linear combinations of the linearly independent columns of C, which is the set of all vectors in B^6 given by

$$C\mathbf{b} = \begin{bmatrix} 1 & 0 & 0 \\ 0 & 1 & 0 \\ 0 & 0 & 1 \\ 1 & 0 & 0 \\ 0 & 0 & 1 \\ 1 & 1 & 1 \end{bmatrix} \begin{bmatrix} b_1 \\ b_2 \\ b_3 \end{bmatrix}.$$

These vectors form a subspace of B^6 of dimension 3.

13. (a) Yes. No column of G is all zeros and the columns are distinct.

  (b) No. Not all columns are distinct.

15. (a) $G\mathbf{x}_t = \begin{bmatrix} 1 & 1 & 0 & 1 & 0 & 0 \\ 0 & 1 & 1 & 0 & 1 & 0 \\ 1 & 0 & 1 & 0 & 0 & 1 \end{bmatrix} \begin{bmatrix} 1 \\ 1 \\ 1 \\ 0 \\ 0 \\ 0 \end{bmatrix} = \begin{bmatrix} 0 \\ 0 \\ 0 \end{bmatrix}$ so no error was detected.

(b) $G\mathbf{x}_t = \begin{bmatrix} 1 & 1 & 0 & 1 & 0 & 0 \\ 0 & 1 & 1 & 0 & 1 & 0 \\ 1 & 0 & 1 & 0 & 0 & 1 \end{bmatrix} \begin{bmatrix} 1 \\ 0 \\ 1 \\ 0 \\ 1 \\ 1 \end{bmatrix} = \begin{bmatrix} 1 \\ 0 \\ 1 \end{bmatrix}$. By comparing this vector with the columns of matrix

G we find a single error was detected in the first bit. The corrected vector is $\begin{bmatrix} 0 \\ 0 \\ 1 \\ 0 \\ 1 \\ 1 \end{bmatrix}$.

(c) $G\mathbf{x}_t = \begin{bmatrix} 1 & 1 & 0 & 1 & 0 & 0 \\ 0 & 1 & 1 & 0 & 1 & 0 \\ 1 & 0 & 1 & 0 & 0 & 1 \end{bmatrix} \begin{bmatrix} 0 \\ 1 \\ 1 \\ 1 \\ 1 \\ 1 \end{bmatrix} = \begin{bmatrix} 0 \\ 1 \\ 0 \end{bmatrix}$. By comparing this vector with the columns of matrix

G a single error was detected in the fifth bit. The corrected vector is $\mathbf{x}_t = \begin{bmatrix} 0 \\ 1 \\ 1 \\ 1 \\ 0 \\ 1 \end{bmatrix}$.

17. When $n = 6$ the binary representations are

1	2	3	4	5	6
001	010	011	100	101	110

and the corresponding Hamming matrix is

$$H(6) = \begin{bmatrix} 0 & 0 & 0 & 1 & 1 & 1 \\ 0 & 1 & 1 & 0 & 0 & 1 \\ 1 & 0 & 1 & 0 & 1 & 0 \end{bmatrix}.$$

19. $H(6) = \begin{bmatrix} 0 & 0 & 0 & 1 & 1 & 1 \\ 0 & 1 & 1 & 0 & 0 & 1 \\ 1 & 0 & 1 & 0 & 1 & 0 \end{bmatrix}.$

Construct the augmented matrix

$$\begin{bmatrix} 0 & 0 & 0 & 1 & 1 & 1 & \vdots & 0 \\ 0 & 1 & 1 & 0 & 0 & 1 & \vdots & 0 \\ 1 & 0 & 1 & 0 & 1 & 0 & \vdots & 0 \end{bmatrix}.$$

Its reduced row echelon form is

$$\begin{bmatrix} 1 & 0 & 1 & 0 & 1 & 0 & \vdots & 0 \\ 0 & 1 & 1 & 0 & 0 & 1 & \vdots & 0 \\ 0 & 0 & 0 & 1 & 1 & 1 & \vdots & 0 \end{bmatrix}.$$

It follows that

$$x_1 = x_3 + x_5$$
$$x_2 = x_3 + x_6$$
$$x_4 = x_5 + x_6.$$

The general solution to this homogeneous system is

$$\mathbf{x} = \begin{bmatrix} x_1 \\ x_2 \\ x_3 \\ x_4 \\ x_5 \\ x_6 \end{bmatrix} = \begin{bmatrix} r+s \\ r+t \\ s+t \\ s \\ t \end{bmatrix} = r\begin{bmatrix} 1 \\ 1 \\ 1 \\ 0 \\ 0 \\ 0 \end{bmatrix} + s\begin{bmatrix} 1 \\ 0 \\ 0 \\ 1 \\ 1 \\ 0 \end{bmatrix} + t\begin{bmatrix} 0 \\ 1 \\ 0 \\ 1 \\ 0 \\ 1 \end{bmatrix},$$

where r, s, t are arbitrary bits. Therefore the code matrix corresponding to $H(6)$ is

$$\begin{bmatrix} 1 & 1 & 0 \\ 1 & 0 & 1 \\ 1 & 0 & 0 \\ 0 & 1 & 1 \\ 0 & 1 & 0 \\ 0 & 0 & 1 \end{bmatrix}.$$

21. $H(7) = \begin{bmatrix} 0 & 0 & 0 & 1 & 1 & 1 & 1 \\ 0 & 1 & 1 & 0 & 0 & 1 & 1 \\ 1 & 0 & 1 & 0 & 1 & 0 & 1 \end{bmatrix}.$

(a) $H(7)\mathbf{x}_t = \begin{bmatrix} 0 & 0 & 0 & 1 & 1 & 1 & 1 \\ 0 & 1 & 1 & 0 & 0 & 1 & 1 \\ 1 & 0 & 1 & 0 & 1 & 0 & 1 \end{bmatrix}\begin{bmatrix} 0 \\ 0 \\ 1 \\ 0 \\ 1 \\ 1 \\ 0 \end{bmatrix} = \begin{bmatrix} 0 \\ 0 \\ 0 \end{bmatrix}$ so no error was detected.

(b) $H(7)\mathbf{x}_t = \begin{bmatrix} 0 & 0 & 0 & 1 & 1 & 1 & 1 \\ 0 & 1 & 1 & 0 & 0 & 1 & 1 \\ 1 & 0 & 1 & 0 & 1 & 0 & 1 \end{bmatrix}\begin{bmatrix} 1 \\ 1 \\ 1 \\ 1 \\ 1 \\ 0 \\ 0 \end{bmatrix} = \begin{bmatrix} 0 \\ 0 \\ 1 \end{bmatrix}$. By comparing this vector with the columns of

matrix G we find a single error was detected in the first bit. The corrected vector is

$$\begin{bmatrix} 0 \\ 1 \\ 1 \\ 1 \\ 1 \\ 0 \\ 0 \end{bmatrix}.$$

(c) $H(7)\mathbf{x}_t = \begin{bmatrix} 0 & 0 & 0 & 1 & 1 & 1 & 1 \\ 0 & 1 & 1 & 0 & 0 & 1 & 1 \\ 1 & 0 & 1 & 0 & 1 & 0 & 1 \end{bmatrix} \begin{bmatrix} 1 \\ 0 \\ 0 \\ 0 \\ 1 \\ 1 \\ 1 \end{bmatrix} = \begin{bmatrix} 1 \\ 0 \\ 1 \end{bmatrix}$. By comparing this vector with the columns of

matrix G we find a single error was detected in the fifth bit. The corrected vector is

$$\begin{bmatrix} 1 \\ 0 \\ 0 \\ 0 \\ 0 \\ 1 \\ 1 \end{bmatrix}.$$

T.1. We want to show $GC = O$. The ijth element of $GC = ij$th element of

$$\begin{bmatrix} D & I_{n-m} \end{bmatrix} \begin{bmatrix} I_m \\ D \end{bmatrix} = \sum_{k=1}^{m} d_{ik}e_{kj} + \sum_{k=1}^{n-m} e_{ik}d_{kj}, \quad (i = 1, 2, \ldots, n - m, \ j = 1, 2, \ldots, m)$$

but $e_{kj} = 0$ whenever $k \neq j$ so

$$= d_{ij}e_{jj} + e_{ii}d_{ij} \quad \text{(but } e_{jj} = 1 \text{ for all } j)$$
$$= d_{ij} + d_{ij}$$
$$= 0 \quad \text{under binary arithmetic.}$$

T.3. Let $\mathbf{x}$ satisfy $Q\mathbf{x} = \mathbf{0}$, i.e.,

$$x_1\mathbf{q}_1 + x_2\mathbf{q}_2 + \cdots + x_n\mathbf{q}_n = \mathbf{0}_n \tag{7.1}$$

for some values of $x_1, x_2, \ldots, x_n$, where the $\mathbf{q}_i$'s are the columns of $\mathbf{q}$. Now rearrange the vectors on the left-hand side of (7.1), i.e., send the ith vector and corresponding scalar multiple to the $p(i)$th position, where p is a permutation on n letters. Rewrite this reordered equation (7.1) as a matrix product $Q_p\mathbf{x}_p$ where Q_p is the reordered set of column vectors $\mathbf{q}_j$, and x_p is their corresponding scalar multiplier. Note that x_p satisfies the homogeneous equation $Q_p\mathbf{x}_p = \mathbf{0}$ because it satisfies (7.1).

T.5. Note that $\mathbf{v} + \mathbf{w} = \mathbf{v} - \mathbf{w}$ in binary arithmetic since every vector is its own additive inverse. It is enough to show that $H(\mathbf{u}, \mathbf{v}) = $ number of places where $\mathbf{v}$ and $\mathbf{w}$ differ $= |\mathbf{v} + \mathbf{w}|$ but the ith entry of $\mathbf{v} + \mathbf{w}$ is

$$\begin{cases} 0 & \text{if } v_i = w_i \\ 1 & \text{if } v_i \neq w_i. \end{cases}$$

So $\mathbf{v} + \mathbf{w}$ records a 1 in the ith component only when $\mathbf{v}$ and $\mathbf{w}$ differ there. The weight, $|\mathbf{v} + \mathbf{w}|$, sums those ones.

T.7. Note that the message vectors in $H(5, 2)$ are all the bit vectors of length 2. So

$$C \cdot \begin{bmatrix} 0 & 1 & 0 & 1 \\ 0 & 0 & 1 & 1 \end{bmatrix} = \begin{bmatrix} 1 & 1 \\ 0 & 1 \\ 0 & 1 \\ 1 & 0 \\ 1 & 0 \end{bmatrix} \begin{bmatrix} 0 & 1 & 0 & 1 \\ 0 & 0 & 1 & 1 \end{bmatrix} = \begin{bmatrix} 0 & 1 & 1 & 0 \\ 0 & 0 & 1 & 1 \\ 0 & 0 & 1 & 1 \\ 0 & 1 & 0 & 1 \\ 0 & 1 & 0 & 1 \end{bmatrix} = \text{code words in } B^5$$

$\text{wt}(10011) = 3$, $\text{wt}(11100) = 3$, $\text{wt}(01111) = 4$.

T.9.   (a) The code words for the $(7,4)$ Hamming code are the columns in the matrix product

$$\begin{bmatrix} 1 & 1 & 0 & 1 \\ 1 & 0 & 1 & 1 \\ 1 & 0 & 0 & 0 \\ 0 & 1 & 1 & 1 \\ 0 & 1 & 0 & 0 \\ 0 & 0 & 1 & 0 \\ 0 & 0 & 0 & 1 \end{bmatrix} \begin{bmatrix} 0 & 0 & 0 & 0 & 0 & 0 & 0 & 0 & 1 & 1 & 1 & 1 & 1 & 1 & 1 & 1 \\ 0 & 0 & 0 & 0 & 1 & 1 & 1 & 1 & 0 & 0 & 0 & 0 & 1 & 1 & 1 & 1 \\ 0 & 0 & 1 & 1 & 0 & 0 & 1 & 1 & 0 & 0 & 1 & 1 & 0 & 0 & 1 & 1 \\ 0 & 1 & 0 & 1 & 0 & 1 & 0 & 1 & 0 & 1 & 0 & 1 & 0 & 1 & 0 & 1 \end{bmatrix}$$

$$= \begin{bmatrix} 0 & 1 & 0 & 1 & 1 & 0 & 1 & 0 & 1 & 0 & 1 & 0 & 0 & 1 & 0 & 1 \\ 0 & 1 & 1 & 0 & 0 & 1 & 1 & 0 & 1 & 0 & 0 & 1 & 1 & 0 & 0 & 1 \\ 0 & 0 & 0 & 0 & 0 & 0 & 0 & 0 & 1 & 1 & 1 & 1 & 1 & 1 & 1 & 1 \\ 0 & 1 & 1 & 0 & 1 & 0 & 0 & 1 & 0 & 1 & 1 & 0 & 1 & 0 & 0 & 1 \\ 0 & 0 & 0 & 0 & 1 & 1 & 1 & 1 & 0 & 0 & 0 & 0 & 1 & 1 & 1 & 1 \\ 0 & 0 & 1 & 1 & 0 & 0 & 1 & 1 & 0 & 0 & 1 & 1 & 0 & 0 & 1 & 1 \\ 0 & 1 & 0 & 1 & 0 & 1 & 0 & 1 & 0 & 1 & 0 & 1 & 0 & 1 & 0 & 1 \end{bmatrix}$$

Compare the minimum Hamming distance of each vector with the first column. It is 3. Compare the minimum Hamming distance of each vector with the 2nd column. It is 3. Do this for each column.

(b) The code words are 00000, 10011, 11100, and 01111. Their minimum Hamming distance is 3.

ML.1.   (a) >>**bingen(1,8,4)**

```
ans   =

      0  0  0  0  0  0  0  1
      0  0  0  1  1  1  1  0
      0  1  1  0  0  1  1  0
      1  0  1  0  1  0  1  0
```

(b) Augment this matrix with a column of zeros on the right and row reduce in binary arithmetic.

>>**aug=[ans [0 0 0 0]′]**

>>**binreduce(aug)**

```
ans   =

      1  0  1  0  1  0  1  0  0
      0  1  1  0  0  1  1  0  0
      0  0  0  1  1  1  1  0  0
      0  0  0  0  0  0  0  1  0
```

So the general solution is

$$x_8 = 0$$
$$x_7 = r$$
$$x_6 = s$$
$$x_5 = t$$
$$x_4 = -t - s - r$$
$$x_3 = u$$
$$x_2 = -u - s - r$$
$$x_1 = -u - t - r$$

or

$$r\begin{bmatrix}1\\1\\0\\1\\0\\0\\1\\0\end{bmatrix} + s\begin{bmatrix}0\\1\\0\\1\\0\\1\\0\\0\end{bmatrix} + t\begin{bmatrix}1\\0\\0\\1\\1\\0\\0\\0\end{bmatrix} + u\begin{bmatrix}1\\1\\1\\0\\0\\0\\0\\0\end{bmatrix}.$$

So

$$C = \begin{bmatrix} 1 & 0 & 1 & 1 \\ 1 & 1 & 0 & 1 \\ 0 & 0 & 0 & 1 \\ 1 & 1 & 1 & 0 \\ 0 & 0 & 1 & 0 \\ 0 & 1 & 0 & 0 \\ 1 & 0 & 0 & 0 \\ 0 & 0 & 0 & 0 \end{bmatrix}.$$

(c) Input C as a matrix and multiply with $H(1,8,4)$. The product should be O.

```
>>C=[1 0 1 1; 1 1 0 1; 0 0 0 1; 1 1 1 0; 0 0 1 0; 0 1 0 0; 1 0 0 0;
0 0 0 0]
>>H=bingen(1,8,4)

H   =
        0  0  0  0  0  0  0  1
        0  0  0  1  1  1  1  0
        0  1  1  0  0  1  1  0
        1  0  1  0  1  0  1  0
>>binprod(H,C)

ans  =
        0  0  0  0
        0  0  0  0
        0  0  0  0
        0  0  0  0
```

ML.3. (a) >>bingen(1,15,4)

```
ans  =
        0  0  0  0  0  0  0  1  1  1  1  1  1  1  1
        0  0  0  1  1  1  1  0  0  0  0  1  1  1  1
        0  1  1  0  0  1  1  0  0  1  1  0  0  1  1
        1  0  1  0  1  0  1  0  1  0  1  0  1  0  1
```

(b) Augment this matrix with a column of 0's on the right and row reduce using binary arithmetic.

```
>>aug=[ans [0 0 0 0]']
>>binreduce(aug)

ans  =
        1  0  1  0  1  0  1  0  1  0  1  0  1  0  1  0
        0  1  1  0  0  1  1  0  0  1  1  0  0  1  1  0
        0  0  0  1  1  1  1  0  0  0  0  1  1  1  1  0
        0  0  0  0  0  0  0  1  1  1  1  1  1  1  1  0
```

So the general solution is

$$x_{15} = r$$
$$x_{14} = s$$
$$x_{13} = t$$
$$x_{12} = u$$
$$x_{11} = v$$
$$x_{10} = w$$
$$x_9 = c$$
$$x_8 = -c - b - a - w - v - u - t - s - r$$
$$x_7 = d$$
$$x_6 = e$$
$$x_5 = f$$
$$x_4 = -f - e - d - u - t - s - r$$
$$x_3 = g$$
$$x_2 = -g - e - d - b - a - s - r$$
$$x_1 = -g - f - d - c - v - t - r$$

which we write as

$$
r\begin{bmatrix}1\\1\\0\\1\\0\\0\\0\\1\\0\\0\\0\\0\\0\\0\\1\end{bmatrix}
+ s\begin{bmatrix}0\\1\\0\\1\\0\\0\\0\\1\\0\\0\\0\\0\\0\\1\\0\end{bmatrix}
+ t\begin{bmatrix}1\\0\\0\\1\\0\\0\\0\\1\\0\\0\\0\\0\\0\\0\\0\end{bmatrix}
+ u\begin{bmatrix}0\\0\\0\\1\\0\\0\\0\\1\\0\\0\\0\\0\\1\\0\\0\end{bmatrix}
+ v\begin{bmatrix}1\\1\\0\\0\\0\\0\\0\\1\\0\\0\\0\\1\\0\\0\\0\end{bmatrix}
+ w\begin{bmatrix}0\\1\\0\\0\\0\\0\\0\\1\\0\\1\\0\\0\\0\\0\\0\end{bmatrix}
$$

$$
+ c\begin{bmatrix}1\\0\\0\\0\\0\\0\\0\\1\\1\\0\\0\\0\\0\\0\\0\end{bmatrix}
+ d\begin{bmatrix}1\\1\\0\\1\\0\\0\\1\\0\\0\\0\\0\\0\\0\\0\\0\end{bmatrix}
+ e\begin{bmatrix}0\\1\\0\\1\\0\\1\\0\\0\\0\\0\\0\\0\\0\\0\\0\end{bmatrix}
+ f\begin{bmatrix}1\\0\\0\\1\\1\\0\\0\\0\\0\\0\\0\\0\\0\\0\\0\end{bmatrix}
+ g\begin{bmatrix}1\\1\\1\\0\\0\\0\\0\\0\\0\\0\\0\\0\\0\\0\\0\end{bmatrix}
$$

Using the above vectors as columns, we construct C.

$$C = \begin{bmatrix} 1 & 0 & 1 & 0 & 1 & 0 & 1 & 1 & 0 & 1 & 1 \\ 1 & 1 & 0 & 0 & 1 & 1 & 0 & 1 & 1 & 0 & 1 \\ 0 & 0 & 0 & 0 & 0 & 0 & 0 & 0 & 0 & 0 & 1 \\ 1 & 1 & 1 & 1 & 0 & 0 & 0 & 1 & 1 & 1 & 0 \\ 0 & 0 & 0 & 0 & 0 & 0 & 0 & 0 & 0 & 1 & 0 \\ 0 & 0 & 0 & 0 & 0 & 0 & 0 & 0 & 1 & 0 & 0 \\ 0 & 0 & 0 & 0 & 0 & 0 & 0 & 1 & 0 & 0 & 0 \\ 1 & 1 & 1 & 1 & 1 & 1 & 1 & 0 & 0 & 0 & 0 \\ 0 & 0 & 0 & 0 & 0 & 0 & 1 & 0 & 0 & 0 & 0 \\ 0 & 0 & 0 & 0 & 0 & 1 & 0 & 0 & 0 & 0 & 0 \\ 0 & 0 & 0 & 0 & 1 & 0 & 0 & 0 & 0 & 0 & 0 \\ 0 & 0 & 0 & 1 & 0 & 0 & 0 & 0 & 0 & 0 & 0 \\ 0 & 0 & 1 & 0 & 0 & 0 & 0 & 0 & 0 & 0 & 0 \\ 0 & 1 & 0 & 0 & 0 & 0 & 0 & 0 & 0 & 0 & 0 \\ 1 & 0 & 0 & 0 & 0 & 0 & 0 & 0 & 0 & 0 & 0 \end{bmatrix}$$

Input C and multiply by $H(1,15,4)$ to see if the product is O.

```
>>C= [1 0 1 0 1 0 1 1 0 1 1;1 1 0 0 1 1 0 1 1 0 1;
0 0 0 0 0 0 0 0 0 0 1;1 1 1 1 0 0 0 1 1 1 0;
0 0 0 0 0 0 0 0 0 1 0;0 0 0 0 0 0 0 0 1 0 0;
0 0 0 0 0 0 0 1 0 0 0;1 1 1 1 1 1 1 0 0 0 0;
0 0 0 0 0 0 1 0 0 0 0;0 0 0 0 0 1 0 0 0 0 0;
0 0 0 0 1 0 0 0 0 0 0;0 0 0 1 0 0 0 0 0 0 0;
0 0 1 0 0 0 0 0 0 0 0;0 1 0 0 0 0 0 0 0 0 0;
1 0 0 0 0 0 0 0 0 0 0]
>>H=bingen(1,15,4)

H   =

     0  0  0  0  0  0  0  1  1  1  1  1  1  1  1
     0  0  0  1  1  1  1  0  0  0  0  1  1  1  1
     0  1  1  0  0  1  1  0  0  1  1  0  0  1  1
     1  0  1  0  1  0  1  0  1  0  1  0  1  0  1

>>binprod(H,C)

ans   =

     0  0  0  0  0  0  0  0  0  0  0
     0  0  0  0  0  0  0  0  0  0  0
     0  0  0  0  0  0  0  0  0  0  0
     0  0  0  0  0  0  0  0  0  0  0
```

Supplementary Exercises, p. 407

1. (a) Construct the augmented matrix

$$\begin{bmatrix} 0 & 0 & 0 & 0 & 0 & 0 & 0 & 1 & \vdots & 0 \\ 0 & 0 & 0 & 1 & 1 & 1 & 1 & 0 & \vdots & 0 \\ 0 & 1 & 1 & 0 & 0 & 1 & 1 & 0 & \vdots & 0 \\ 1 & 0 & 1 & 0 & 1 & 0 & 1 & 0 & \vdots & 0 \end{bmatrix}.$$

It reduced row echelon form is

$$
\begin{bmatrix}
1 & 0 & 1 & 0 & 1 & 0 & 1 & 0 & \vdots & 0 \\
0 & 1 & 1 & 0 & 0 & 1 & 1 & 0 & \vdots & 0 \\
0 & 0 & 0 & 1 & 1 & 1 & 1 & 0 & \vdots & 0 \\
0 & 0 & 0 & 0 & 0 & 0 & 0 & 1 & \vdots & 0
\end{bmatrix}.
$$

It follows that

$$x_1 = x_3 + x_5 + x_7$$
$$x_2 = x_3 + x_6 + x_7$$
$$x_4 = x_5 + x_6 + x_7$$
$$x_8 = 0.$$

The general solution to this homogeneous system is

$$
\mathbf{x} =
\begin{bmatrix} x_1 \\ x_2 \\ x_3 \\ x_4 \\ x_5 \\ x_6 \\ x_7 \\ x_8 \end{bmatrix}
=
\begin{bmatrix} r+s+u \\ r+t+u \\ r \\ s+t+u \\ s \\ t \\ u \\ 0 \end{bmatrix}
= r
\begin{bmatrix} 1 \\ 1 \\ 1 \\ 0 \\ 0 \\ 0 \\ 0 \\ 0 \end{bmatrix}
+ s
\begin{bmatrix} 1 \\ 0 \\ 0 \\ 1 \\ 1 \\ 0 \\ 0 \\ 0 \end{bmatrix}
+ t
\begin{bmatrix} 0 \\ 1 \\ 0 \\ 1 \\ 0 \\ 1 \\ 0 \\ 0 \end{bmatrix}
+ u
\begin{bmatrix} 1 \\ 1 \\ 0 \\ 1 \\ 0 \\ 0 \\ 1 \\ 0 \end{bmatrix},
$$

where r, s, t, and u are arbitrary bits. Therefore a basis for the corresponding subspace of code words is

$$
\left\{
\begin{bmatrix} 1 \\ 1 \\ 1 \\ 0 \\ 0 \\ 0 \\ 0 \\ 0 \end{bmatrix},
\begin{bmatrix} 1 \\ 0 \\ 0 \\ 1 \\ 1 \\ 0 \\ 0 \\ 0 \end{bmatrix},
\begin{bmatrix} 0 \\ 1 \\ 0 \\ 1 \\ 0 \\ 1 \\ 0 \\ 0 \end{bmatrix},
\begin{bmatrix} 1 \\ 1 \\ 0 \\ 1 \\ 0 \\ 0 \\ 1 \\ 0 \end{bmatrix}
\right\}.
$$

(b) $H(8)\mathbf{x}_t =
\begin{bmatrix}
0 & 0 & 0 & 0 & 0 & 0 & 0 & 1 \\
0 & 0 & 0 & 1 & 1 & 1 & 1 & 0 \\
0 & 1 & 1 & 0 & 0 & 1 & 1 & 0 \\
1 & 0 & 1 & 0 & 1 & 0 & 1 & 0
\end{bmatrix}
\begin{bmatrix} 1 \\ 1 \\ 1 \\ 1 \\ 1 \\ 1 \\ 1 \\ 0 \end{bmatrix}
=
\begin{bmatrix} 0 \\ 0 \\ 0 \\ 0 \end{bmatrix}$ so no error was detected. It is a code word.

3. Use the procedure described just before Example 4 in Section 7.2.

Let

$$
\mathbf{b} =
\begin{bmatrix} 1 \\ 2 \\ 4 \\ 10 \end{bmatrix}, \quad
A =
\begin{bmatrix} 0 & 1 \\ 3 & 1 \\ 5 & 1 \\ 8 & 1 \end{bmatrix}, \quad \text{and} \quad
\mathbf{x} =
\begin{bmatrix} b_1 \\ b_0 \end{bmatrix}.
$$

Then

$$
A^T A =
\begin{bmatrix} 0 & 3 & 5 & 8 \\ 1 & 1 & 1 & 1 \end{bmatrix}
\begin{bmatrix} 0 & 1 \\ 3 & 1 \\ 5 & 1 \\ 8 & 1 \end{bmatrix}
=
\begin{bmatrix} 98 & 16 \\ 16 & 4 \end{bmatrix}
$$

and

$$A^T \mathbf{b} = \begin{bmatrix} 0 & 3 & 5 & 8 \\ 1 & 1 & 1 & 1 \end{bmatrix} \begin{bmatrix} 1 \\ 2 \\ 4 \\ 10 \end{bmatrix} = \begin{bmatrix} 106 \\ 17 \end{bmatrix}.$$

Solving the normal system $A^T A \mathbf{x} = A^T \mathbf{b}$ for $\mathbf{x}$ by Gauss–Jordan reduction we obtain

$$\begin{bmatrix} 98 & 16 \\ 16 & 4 \end{bmatrix} \begin{bmatrix} b_1 \\ b_0 \end{bmatrix} = \begin{bmatrix} 106 \\ 17 \end{bmatrix}$$

with augmented matrix

$$\begin{bmatrix} 98 & 16 & \vdots & 106 \\ 16 & 4 & \vdots & 17 \end{bmatrix}$$

which has the solution

$$b_1 = \frac{19}{17}$$
$$b_0 = -\frac{15}{68}.$$

THerefore the least squares line is

$$y = \frac{19}{17}x - \frac{15}{68}.$$

Chapter 8

Eigenvectors, Eigenvalues, and Diagonalization

Section 8.1, p. 420

1. (a) $A\mathbf{x}_1 = \begin{bmatrix} 3 & -1 \\ -2 & 2 \end{bmatrix} \begin{bmatrix} r \\ 2r \end{bmatrix} = \begin{bmatrix} r \\ 2r \end{bmatrix} = 1 \begin{bmatrix} r \\ 2r \end{bmatrix}$.

   (b) $A\mathbf{x}_2 = \begin{bmatrix} 3 & -1 \\ -2 & 2 \end{bmatrix} \begin{bmatrix} r \\ -r \end{bmatrix} = \begin{bmatrix} 4r \\ -4r \end{bmatrix} = 4 \begin{bmatrix} r \\ -r \end{bmatrix}$.

3. $\det(\lambda I_3 - A) = \begin{vmatrix} \lambda - 1 & -2 & -1 \\ 0 & \lambda - 1 & -2 \\ 1 & -3 & \lambda - 2 \end{vmatrix}$

   $= (\lambda - 1)(\lambda - 1)(\lambda - 2) + 4 + 0 + (\lambda - 1) - 6(\lambda - 1) - 0$

   $= \lambda^3 - 4\lambda^2 + 7$.

5. $\det(\lambda I_3 - A) = \begin{vmatrix} \lambda - 4 & 1 & -3 \\ 0 & \lambda - 2 & -1 \\ 0 & 0 & \lambda - 3 \end{vmatrix} = (\lambda - 4)(\lambda - 2)(\lambda - 3) = \lambda^3 - 9\lambda^2 + 26\lambda - 24$.

7. $\det(\lambda I_3 - A) = \begin{vmatrix} \lambda - 2 & -1 & -2 \\ -2 & \lambda - 2 & 2 \\ -3 & -1 & \lambda - 1 \end{vmatrix} = \begin{vmatrix} \lambda - 4 & \lambda - 3 & 0 \\ -2 & \lambda - 2 & 2 \\ -3 & -1 & \lambda - 1 \end{vmatrix}$

   $= (\lambda - 4)\left[(\lambda - 2)(\lambda - 1) + 2\right] - (\lambda - 3)\left[-2(\lambda + 1) + 6\right]$

   $= (\lambda - 4)(\lambda^2 - 3\lambda + 4) + (\lambda - 4)(2\lambda - 6)$

   $= (\lambda - 4)(\lambda^2 - \lambda - 2) = (\lambda - 4)(\lambda - 2)(\lambda + 1) = \lambda^3 - 5\lambda^2 + 2\lambda + 8$

9. $\det(\lambda I_3 - A) = \begin{vmatrix} \lambda - 1 & 0 & 0 \\ 1 & \lambda - 3 & 0 \\ -3 & -2 & \lambda + 2 \end{vmatrix} = (\lambda - 1)(\lambda - 3)(\lambda + 2)$ is the characteristic polynomial.

   Thus the eigenvalues of A are $\lambda_1 = 1$, $\lambda_2 = 3$, and $\lambda_3 = -2$. To find the corresponding eigenvectors, solve the linear systems $(\lambda_j I_3 - A)\mathbf{x} = \mathbf{0}$. For $\lambda_1 = 1$, we have $(1I_3 - A)\mathbf{x} = \mathbf{0}$:

   $$\begin{bmatrix} 0 & 0 & 0 \\ 1 & -2 & 0 \\ -3 & -2 & 3 \end{bmatrix} \begin{bmatrix} x_1 \\ x_2 \\ x_3 \end{bmatrix} = \begin{bmatrix} 0 \\ 0 \\ 0 \end{bmatrix}.$$

The reduced row echelon form of the coefficient matrix is

$$\begin{bmatrix} 1 & 0 & -\frac{3}{4} \\ 0 & 1 & -\frac{3}{8} \\ 0 & 0 & 0 \end{bmatrix}.$$

Thus the solution is $x_1 = \frac{3}{4}r$, $x_2 = \frac{3}{8}r$, $x_3 = r$, where r is any real number, and therefore every vector of the form

$$\begin{bmatrix} \frac{3}{4}r \\ \frac{3}{8}r \\ r \end{bmatrix}, \quad r \neq 0,$$

is an eigenvector associated with eigenvalue $\lambda_1 = 1$. Let $r = 8$, and we have the eigenvector

$$\begin{bmatrix} 6 \\ 3 \\ 8 \end{bmatrix}.$$

For $\lambda_2 = 3$, we have $(3I_3 - A)\mathbf{x} = \mathbf{0}$:

$$\begin{bmatrix} 2 & 0 & 0 \\ 1 & 0 & 0 \\ -3 & -2 & 5 \end{bmatrix} \begin{bmatrix} x_1 \\ x_2 \\ x_3 \end{bmatrix} = \begin{bmatrix} 0 \\ 0 \\ 0 \end{bmatrix}.$$

The reduced row echelon form of the coefficient matrix is

$$\begin{bmatrix} 1 & 0 & 0 \\ 0 & 1 & -\frac{5}{2} \\ 0 & 0 & 0 \end{bmatrix}.$$

Thus the solution is $x_1 = 0$, $x_2 = \frac{5}{2}r$, $x_3 = r$, where r is any real number, and hence every vector of the form

$$\begin{bmatrix} 0 \\ \frac{5}{2}r \\ r \end{bmatrix}, \quad r \neq 0,$$

is an eigenvector associated with eigenvalue $\lambda_2 = 3$. Let $r = 2$, and we have eigenvector

$$\begin{bmatrix} 0 \\ 5 \\ 2 \end{bmatrix}.$$

For $\lambda_3 = -2$, we have $(-2I_3 - A)\mathbf{x} = \mathbf{0}$:

$$\begin{bmatrix} -3 & 0 & 0 \\ 1 & -5 & 0 \\ -3 & -2 & 0 \end{bmatrix} \begin{bmatrix} x_1 \\ x_2 \\ x_3 \end{bmatrix} = \begin{bmatrix} 0 \\ 0 \\ 0 \end{bmatrix}.$$

The reduced row echelon form of the coefficient matrix is

$$\begin{bmatrix} 1 & 0 & 0 \\ 0 & 1 & 0 \\ 0 & 0 & 0 \end{bmatrix}.$$

Thus the solution is $x_1 = 0$, $x_2 = 0$, $x_3 = r$, where r is any real number, and hence every vector of the form

$$\begin{bmatrix} 0 \\ 0 \\ r \end{bmatrix}, \quad r \neq 0,$$

is an eigenvector associated with eigenvalue $\lambda_3 = -2$. Let $r = 1$, and we have eigenvector

$$\begin{bmatrix} 0 \\ 0 \\ 1 \end{bmatrix}.$$

11. $\det(\lambda I_2 - A) = \begin{vmatrix} \lambda - 1 & 1 \\ -2 & \lambda - 4 \end{vmatrix} = (\lambda - 1)(\lambda - 4) + 2 = \lambda^2 - 5\lambda + 6 = (\lambda - 2)(\lambda - 3)$.

Thus the eigenvalues of A are $\lambda_1 = 2$, $\lambda_2 = 3$. To find the corresponding eigenvectors, solve the linear systems $(\lambda_j I_2 - A)\mathbf{x} = \mathbf{0}$. For $\lambda_1 = 2$, we have $(2I_2 - A)\mathbf{x} = \mathbf{0}$:

$$\begin{bmatrix} 1 & 1 \\ -2 & -2 \end{bmatrix} \begin{bmatrix} x_1 \\ x_2 \end{bmatrix} = \begin{bmatrix} 0 \\ 0 \end{bmatrix}.$$

The reduced row echelon form of the coefficient matrix is

$$\begin{bmatrix} 1 & 1 \\ 0 & 0 \end{bmatrix}.$$

Thus the solution is $x_1 = -r$, $x_2 = r$, where r is any real number, and hence every vector of the form

$$\begin{bmatrix} -r \\ r \end{bmatrix}, \quad r \neq 0,$$

is an eigenvector associated with eigenvalue $\lambda_1 = 2$. Let $r = -1$, and we have eigenvector

$$\begin{bmatrix} 1 \\ -1 \end{bmatrix}.$$

For $\lambda_2 = 3$, we have $(3I_2 - A)\mathbf{x} = \mathbf{0}$:

$$\begin{bmatrix} 2 & 1 \\ -2 & -1 \end{bmatrix} \begin{bmatrix} x_1 \\ x_2 \end{bmatrix} = \begin{bmatrix} 0 \\ 0 \end{bmatrix}.$$

The reduced row echelon form of the coefficient matrix is

$$\begin{bmatrix} 1 & \frac{1}{2} \\ 0 & 0 \end{bmatrix}.$$

Thus the solution is $x_1 = -\frac{1}{2}r$, $x_2 = r$, where r is any real number, and hence every vector of the form

$$\begin{bmatrix} -\frac{1}{2}r \\ r \end{bmatrix}, \quad r \neq 0,$$

is an eigenvector associated with eigenvalue $\lambda_2 = 3$. Let $r = -2$, and we have eigenvector

$$\begin{bmatrix} 1 \\ -2 \end{bmatrix}.$$

13. $\det(\lambda I_3 - A) = \begin{vmatrix} \lambda - 2 & -2 & -3 \\ -1 & \lambda - 2 & -1 \\ -2 & 2 & \lambda - 1 \end{vmatrix}$

$= (\lambda - 2)(\lambda - 2)(\lambda - 1) - 4 + 6 - 6(\lambda - 2) + 2(\lambda - 2) - 2(\lambda - 1)$

$= \lambda^3 - 5\lambda^2 + 2\lambda + 8 = (\lambda + 1)(\lambda - 2)(\lambda - 4)$

Thus the eigenvalues of A are $\lambda_1 = -1$, $\lambda_2 = 2$, $\lambda_3 = 4$. To find the corresponding eigenvectors, solve the linear systems $(\lambda_j I_3 - A)\mathbf{x} = \mathbf{0}$. For $\lambda_1 = -1$, we have $(-1 I_3 - A)\mathbf{x} = \mathbf{0}$:

$$\begin{bmatrix} -3 & -2 & -3 \\ -1 & -3 & -1 \\ -2 & 2 & -2 \end{bmatrix} \begin{bmatrix} x_1 \\ x_2 \\ x_3 \end{bmatrix} = \begin{bmatrix} 0 \\ 0 \\ 0 \end{bmatrix}.$$

The reduced row echelon form of the coefficient matrix is

$$\begin{bmatrix} 1 & 0 & 1 \\ 0 & 1 & 0 \\ 0 & 0 & 0 \end{bmatrix}.$$

Thus the solution is $x_1 = -r$, $x_2 = 0$, $x_3 = r$, where r is any real number, and hence every vector of the form

$$\begin{bmatrix} -r \\ 0 \\ r \end{bmatrix}, \quad r \neq 0,$$

is an eigenvector associated with eigenvalue $\lambda_1 = -1$. Let $r = -1$, and we have eigenvector

$$\begin{bmatrix} 1 \\ 0 \\ -1 \end{bmatrix}.$$

For $\lambda_2 = 2$, we have $(2 I_3 - A)\mathbf{x} = \mathbf{0}$:

$$\begin{bmatrix} 0 & -2 & -3 \\ -1 & 0 & -1 \\ -2 & 2 & 1 \end{bmatrix} \begin{bmatrix} x_1 \\ x_2 \\ x_3 \end{bmatrix} = \begin{bmatrix} 0 \\ 0 \\ 0 \end{bmatrix}.$$

The reduced row echelon form of the coefficient matrix is

$$\begin{bmatrix} 1 & 0 & 1 \\ 0 & 1 & \frac{3}{2} \\ 0 & 0 & 0 \end{bmatrix}.$$

Thus the solution is $x_1 = -r$, $x_2 = -\frac{3}{2}r$, $x_3 = r$, where r is any real number, and hence every vector of the form

$$\begin{bmatrix} -r \\ -\frac{3}{2}r \\ r \end{bmatrix}, \quad r \neq 0,$$

is an eigenvector associated with eigenvalue $\lambda_2 = 2$. Let $r = 2$, and we have eigenvector

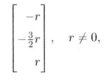

For $\lambda_3 = 4$, we have $(4I_3 - A)\mathbf{x} = \mathbf{0}$:

$$\begin{bmatrix} 2 & -2 & -3 \\ -1 & 2 & -1 \\ -2 & 2 & 3 \end{bmatrix} \begin{bmatrix} x_1 \\ x_2 \\ x_3 \end{bmatrix} = \begin{bmatrix} 0 \\ 0 \\ 0 \end{bmatrix}.$$

The reduced row echelon form of the coefficient matrix is

$$\begin{bmatrix} 1 & 0 & -4 \\ 0 & 1 & -\frac{5}{2} \\ 0 & 0 & 0 \end{bmatrix}.$$

Thus the solution is $x_1 = 4r$, $x_2 = \frac{5}{2}r$, $x_3 = r$, where r is any real number, and hence every vector of the form

$$\begin{bmatrix} 4r \\ \frac{5}{2}r \\ r \end{bmatrix}, \quad r \neq 0,$$

is an eigenvector associated with eigenvalue $\lambda_3 = 4$. Let $r = 2$, and we have eigenvector

$$\begin{bmatrix} 8 \\ 5 \\ 2 \end{bmatrix}.$$

15. $\det(\lambda I_4 - A) = \begin{vmatrix} \lambda - 1 & -2 & -3 & -4 \\ 0 & \lambda + 1 & -3 & -2 \\ 0 & 0 & \lambda - 3 & -3 \\ 0 & 0 & 0 & \lambda - 2 \end{vmatrix} = (\lambda - 1)(\lambda + 1)(\lambda - 3)(\lambda - 2)$

Thus the eigenvalues of A are $\lambda_1 = 1$, $\lambda_2 = -1$, $\lambda_3 = 3$, $\lambda_4 = 2$. To find the corresponding eigenvectors, solve the linear systems $(\lambda_j I_4 - A)\mathbf{x} = \mathbf{0}$. For $\lambda_1 = 1$, we have $(1I_4 - A)\mathbf{x} = \mathbf{0}$:

$$\begin{bmatrix} 0 & -2 & -3 & -4 \\ 0 & 2 & -3 & -2 \\ 0 & 0 & -2 & -3 \\ 0 & 0 & 0 & -1 \end{bmatrix} \begin{bmatrix} x_1 \\ x_2 \\ x_3 \\ x_4 \end{bmatrix} = \begin{bmatrix} 0 \\ 0 \\ 0 \\ 0 \end{bmatrix}.$$

The reduced row echelon form of the coefficient matrix is

$$\begin{bmatrix} 0 & 1 & 0 & 0 \\ 0 & 0 & 1 & 0 \\ 0 & 0 & 0 & 1 \\ 0 & 0 & 0 & 0 \end{bmatrix}.$$

Thus the solution is $x_1 = r$, $x_2 = 0$, $x_3 = 0$, $x_4 = 0$, where r is any real number, and hence every vector of the form

$$\begin{bmatrix} r \\ 0 \\ 0 \\ 0 \end{bmatrix}, \quad r \neq 0,$$

is an eigenvector associated with eigenvalue $\lambda_1 = 1$. Let $r = 1$, and we have eigenvector

$$\begin{bmatrix} 1 \\ 0 \\ 0 \\ 0 \end{bmatrix}.$$

For $\lambda_2 = -1$, we have $(-1I_4 - A)\mathbf{x} = \mathbf{0}$:

$$\begin{bmatrix} -2 & -2 & -3 & -4 \\ 0 & 0 & -3 & -2 \\ 0 & 0 & -4 & -3 \\ 0 & 0 & 0 & -3 \end{bmatrix} \begin{bmatrix} x_1 \\ x_2 \\ x_3 \\ x_4 \end{bmatrix} = \begin{bmatrix} 0 \\ 0 \\ 0 \\ 0 \end{bmatrix}.$$

The reduced row echelon form of the coefficient matrix is

$$\begin{bmatrix} 1 & 1 & 0 & 0 \\ 0 & 0 & 1 & 0 \\ 0 & 0 & 0 & 1 \\ 0 & 0 & 0 & 0 \end{bmatrix}.$$

Thus the solution is $x_1 = -r$, $x_2 = r$, $x_3 = 0$, $x_4 = 0$, where r is any real number, and hence every vector of the form

$$\begin{bmatrix} -r \\ r \\ 0 \\ 0 \end{bmatrix}, \quad r \neq 0,$$

is an eigenvector associated with eigenvalue $\lambda_2 = -1$. Let $r = -1$, and we have eigenvector

$$\begin{bmatrix} 1 \\ -1 \\ 0 \\ 0 \end{bmatrix}.$$

For $\lambda_3 = 3$, we have $(3I_4 - A)\mathbf{x} = \mathbf{0}$:

$$\begin{bmatrix} 2 & -2 & -3 & -4 \\ 0 & 4 & -3 & -2 \\ 0 & 0 & 0 & -3 \\ 0 & 0 & 0 & 1 \end{bmatrix} \begin{bmatrix} x_1 \\ x_2 \\ x_3 \\ x_4 \end{bmatrix} = \begin{bmatrix} 0 \\ 0 \\ 0 \\ 0 \end{bmatrix}.$$

The reduced row echelon form of the coefficient matrix is

$$\begin{bmatrix} 1 & 0 & -\frac{9}{4} & 0 \\ 0 & 1 & -\frac{3}{4} & 0 \\ 0 & 0 & 0 & 1 \\ 0 & 0 & 0 & 0 \end{bmatrix}.$$

Thus the solution is $x_1 = \frac{9}{4}r$, $x_2 = \frac{3}{4}r$, $x_3 = r$, $x_4 = 0$, where r is any real number, and hence every vector of the form

$$\begin{bmatrix} \frac{9}{4}r \\ \frac{3}{4}r \\ r \\ 0 \end{bmatrix}, \quad r \neq 0,$$

is an eigenvector associated with eigenvalue $\lambda_3 = 3$. Let $r = 4$, and we have eigenvector

$$\begin{bmatrix} 9 \\ 3 \\ 4 \\ 0 \end{bmatrix}.$$

For $\lambda_4 = 2$, we have $(2I_4 - A)\mathbf{x} = \mathbf{0}$:

$$\begin{bmatrix} 1 & -2 & -3 & -4 \\ 0 & 3 & -3 & -2 \\ 0 & 0 & -1 & -3 \\ 0 & 0 & 0 & 0 \end{bmatrix} \begin{bmatrix} x_1 \\ x_2 \\ x_3 \\ x_4 \end{bmatrix} = \begin{bmatrix} 0 \\ 0 \\ 0 \\ 0 \end{bmatrix}.$$

The reduced row echelon form of the coefficient matrix is

$$\begin{bmatrix} 1 & 0 & 0 & \frac{29}{3} \\ 0 & 1 & 0 & \frac{7}{3} \\ 0 & 0 & 1 & 3 \\ 0 & 0 & 0 & 0 \end{bmatrix}.$$

Thus the solution is $x_1 = -\frac{29}{3}r$, $x_2 = -\frac{7}{3}r$, $x_3 = -3r$, $x_4 = r$, where r is any real number, and hence every vector of the form

$$\begin{bmatrix} -\frac{29}{3}r \\ -\frac{7}{3}r \\ -3r \\ r \end{bmatrix}, \quad r \neq 0,$$

is an eigenvector associated with eigenvalue $\lambda_4 = 2$. Let $r = -3$, and we have eigenvector

$$\begin{bmatrix} 29 \\ 7 \\ 9 \\ -3 \end{bmatrix}.$$

17. (a) $\det(\lambda I_2 - A) = \begin{vmatrix} \lambda + 1 & 1 - i \\ -1 & \lambda \end{vmatrix} = (\lambda + 1)\lambda + (1 - i) = \lambda^2 + \lambda + 1 - i = (\lambda - i)(\lambda + 1 + i) = 0.$

Thus the eigenvalues of A are $\lambda_1 = i$ and $\lambda_2 = -1 - i$. To find the corresponding eigenvectors solve the linear systems $(\lambda_j I_2 - A)\mathbf{x} = \mathbf{0}$. For $\lambda_1 = i$ we have $(iI_2 - A)\mathbf{x} = \mathbf{0}$,

$$\begin{bmatrix} i + 1 & 1 - i \\ -1 & i \end{bmatrix} \begin{bmatrix} x_1 \\ x_2 \end{bmatrix} = \begin{bmatrix} 0 \\ 0 \end{bmatrix}.$$

The reduced form of the augmented matrix is

$$\begin{bmatrix} i & 1 & \vdots & 0 \\ 0 & 0 & \vdots & 0 \end{bmatrix}$$

and the solution is $x_1 = ir$, $x_2 = r$, where r is any nonzero number. Hence every vector of the form

$$\begin{bmatrix} ir \\ r \end{bmatrix}, \quad r \neq 0$$

is an eigenvector associated with eigenvalue $\lambda_1 = i$. Let $r = 1$ and we have eigenvector

$$\begin{bmatrix} i \\ 1 \end{bmatrix}.$$

For $\lambda_2 = -1 - i$ we have the system

$$\begin{bmatrix} -i & 1-i \\ -1 & -1-i \end{bmatrix} \begin{bmatrix} x_1 \\ x_2 \end{bmatrix} = \begin{bmatrix} 0 \\ 0 \end{bmatrix}.$$

The reduced form of the augmented matrix is

$$\begin{bmatrix} -i & 1-i & \vdots & 0 \\ 0 & 0 & \vdots & 0 \end{bmatrix}$$

and the solution is $x_1 = (-1 - i)r$, $x_2 = r$, where r is any nonzero number. Hence every vector of the form

$$\begin{bmatrix} (-1-i)r \\ r \end{bmatrix}, \quad r \neq 0$$

is an eigenvector associated with eigenvalue $\lambda_2 = -1 - i$. Let $r = 1$ and we have eigenvector

$$\begin{bmatrix} -1-i \\ 1 \end{bmatrix}.$$

(b) $\det(\lambda I_3 - A) = \begin{vmatrix} \lambda - i & -1 & 0 \\ -1 & \lambda - i & 0 \\ 0 & 0 & \lambda - 1 \end{vmatrix} = (\lambda - i)(\lambda - i)(\lambda - 1) - (\lambda - 1) = (\lambda - 1)(\lambda^2 - 2i\lambda - 2) =$

$(\lambda - 1)(\lambda - (1 + i))(\lambda - (-1 + i)) = 0$. Thus the eigenvalues of A are $\lambda_1 = 1$, $\lambda_2 = 1 + i$, and $\lambda_3 = -1 + i$. To find the corresponding eigenvectors solve the linear systems $(\lambda_j I_3 - A)\mathbf{x} = \mathbf{0}$. For $\lambda_1 = 1$ we have $(I_2 - A)\mathbf{x} = \mathbf{0}$,

$$\begin{bmatrix} 1-i & -1 & 0 \\ -1 & 1-i & 0 \\ 0 & 0 & 0 \end{bmatrix} \begin{bmatrix} x_1 \\ x_2 \\ x_3 \end{bmatrix} = \begin{bmatrix} 0 \\ 0 \\ 0 \end{bmatrix}.$$

The reduced form of the augmented matrix is

$$\begin{bmatrix} 1 & 0 & 0 & \vdots & 0 \\ 0 & 1 & 0 & \vdots & 0 \\ 0 & 0 & 0 & \vdots & 0 \end{bmatrix}$$

and the solution is $x_1 = 0$, $x_2 = 0$, $x_3 = r$, where r is any nonzero number. Hence every vector of the form

$$\begin{bmatrix} 0 \\ 0 \\ r \end{bmatrix}, \quad r \neq 0$$

is an eigenvector associated with eigenvalue $\lambda_1 = 1$. Let $r = 1$ and we have eigenvector

$$\begin{bmatrix} 0 \\ 0 \\ 1 \end{bmatrix}.$$

For $\lambda_2 = 1 + i$ we have the system

$$\begin{bmatrix} 1 & -1 & 0 \\ -1 & 1 & 0 \\ 0 & 0 & i \end{bmatrix} \begin{bmatrix} x_1 \\ x_2 \\ x_3 \end{bmatrix} = \begin{bmatrix} 0 \\ 0 \\ 0 \end{bmatrix}$$

which has solution $x_1 = r$, $x_2 = r$, $x_3 = 0$, where r is any nonzero number. Hence every vector of the form

$$\begin{bmatrix} r \\ r \\ 0 \end{bmatrix}, \quad r \neq 0$$

is an eigenvector associated with eigenvalue $\lambda_2 = 1 + i$. Let $r = 1$ and we have eigenvector

$$\begin{bmatrix} 1 \\ 1 \\ 0 \end{bmatrix}.$$

For $\lambda_3 = -1 + i$ we have the system

$$\begin{bmatrix} -1 & -1 & 0 \\ -1 & -1 & 0 \\ 0 & 0 & i - 2 \end{bmatrix} \begin{bmatrix} x_1 \\ x_2 \\ x_3 \end{bmatrix} = \begin{bmatrix} 0 \\ 0 \\ 0 \end{bmatrix},$$

which has solution $x_1 = -r$, $x_2 = r$, $x_3 = 0$, where r is any nonzero number. Hence every vector of the form

$$\begin{bmatrix} -r \\ r \\ 0 \end{bmatrix}, \quad r \neq 0$$

is an eigenvector associated with eigenvalue $\lambda_3 = -1 + i$. Let $r = 1$ and we have eigenvector

$$\begin{bmatrix} -1 \\ 1 \\ 0 \end{bmatrix}.$$

(c) $\det(\lambda I_3 - A) = \begin{vmatrix} \lambda & 1 & 0 \\ -1 & \lambda & 0 \\ 0 & -1 & \lambda \end{vmatrix} = \lambda(\lambda^2) - (-\lambda) = \lambda^3 + \lambda = \lambda(\lambda - i)(\lambda + i) = 0$

Thus the eigenvalues of A are $\lambda_1 = 0$, $\lambda_2 = i$, $\lambda_3 = -i$. To find the corresponding eigenvectors solve the linear systems $(\lambda_j I_3 - A)\mathbf{x} = \mathbf{0}$. For $\lambda_1 = 0$ we have the augmented matrix of the system $-A\mathbf{x} = \mathbf{0}$,

$$\begin{bmatrix} 0 & 1 & 0 & \vdots & 0 \\ -1 & 0 & 0 & \vdots & 0 \\ 0 & -1 & 0 & \vdots & 0 \end{bmatrix}$$

which has solution $x_1 = 0$, $x_2 = 0$, $x_3 = r$, where r is any nonzero number. Hence every vector of the form

$$\begin{bmatrix} 0 \\ 0 \\ r \end{bmatrix}, \quad r \neq 0$$

is an eigenvector associated with eigenvalue $\lambda_1 = 0$. Let $r = 1$ and we have eigenvector

$$\begin{bmatrix} 0 \\ 0 \\ 1 \end{bmatrix}.$$

For $\lambda_2 = i$ we have the augmented matrix

$$\begin{bmatrix} i & 1 & 0 & \vdots & 0 \\ -1 & i & 0 & \vdots & 0 \\ 0 & -1 & i & \vdots & 0 \end{bmatrix}$$

which has solution $x_1 = -r$, $x_2 = ir$, $x_3 = r$, where r is any nonzero number. Hence every vector of the form

$$\begin{bmatrix} -r \\ ir \\ r \end{bmatrix}, \quad r \neq 0$$

is an eigenvector associated with eigenvalue $\lambda_2 = i$. Let $r = 1$ and we have eigenvector

$$\begin{bmatrix} -1 \\ i \\ 1 \end{bmatrix}.$$

For $\lambda_3 = -i$ we have the augmented matrix of the system $(-iI_3 - A)\mathbf{x} = \mathbf{0}$,

$$\begin{bmatrix} -i & 1 & 0 & \vdots & 0 \\ -1 & -i & 0 & \vdots & 0 \\ 0 & -1 & -i & \vdots & 0 \end{bmatrix}$$

which has solution $x_1 = -r$, $x_2 = -ir$, $x_3 = r$, where r is any nonzero number. Hence every vector of the form

$$\begin{bmatrix} -r \\ -ir \\ r \end{bmatrix}, \quad r \neq 0$$

is an eigenvector associated with eigenvalue $\lambda_3 = -i$. Let $r = 1$ and we have eigenvector

$$\begin{bmatrix} -1 \\ -i \\ 1 \end{bmatrix}.$$

(d) $\det(\lambda I_3 - A) = \begin{vmatrix} \lambda & 0 & 9 \\ 0 & \lambda - 1 & 0 \\ -1 & 0 & \lambda \end{vmatrix} = \lambda^2(\lambda - 1) + 9(\lambda - 1) = (\lambda - 1)(\lambda - 3i)(\lambda + 3i) = 0$. Thus the eigenvalues of A are $\lambda_1 = 1$, $\lambda_2 = 3i$, $\lambda_3 = -3i$. To find the correspnding eigenvectors solve the linear systems $(\lambda_j I_3 - A)\mathbf{x} = \mathbf{0}$. For $\lambda_1 = 1$ we have the augmented matrix of the system $(I_3 - A)\mathbf{x} = \mathbf{0}$,

$$\begin{bmatrix} 1 & 0 & 9 & \vdots & 0 \\ 0 & 0 & 0 & \vdots & 0 \\ -1 & 0 & 1 & \vdots & 0 \end{bmatrix}$$

which has solution $x_1 = 0$, $x_2 = r$, $x_3 = 0$, where r is any nonzero number. Hence every vector of the form

$$\begin{bmatrix} 0 \\ r \\ 0 \end{bmatrix}, \quad r \neq 0$$

is an eigenvector associated with eigenvalue $\lambda_1 = 1$. Let $r = 1$ and we have eigenvector

$$\begin{bmatrix} 0 \\ 1 \\ 0 \end{bmatrix}.$$

For $\lambda_2 = 3i$ we have the augmented matrix of the system $((3i)I_3 - A)\mathbf{x} = \mathbf{0}$,

$$\begin{bmatrix} 3i & 0 & 9 & \vdots & 0 \\ 0 & 3i - 1 & 0 & \vdots & 0 \\ -1 & 0 & 3i & \vdots & 0 \end{bmatrix}$$

which has solution $x_1 = (3i)r$, $x_2 = 0$, $x_3 = r$, where r is any nonzero number. Hence every vector of the form

$$\begin{bmatrix} (3i)r \\ 0 \\ r \end{bmatrix}, \quad r \neq 0$$

is an eigenvector associated with eigenvalue $\lambda_2 = 3i$. Let $r = 1$ and we have eigenvector

$$\begin{bmatrix} 3i \\ 0 \\ 1 \end{bmatrix}.$$

For $\lambda_3 = -3i$ we have the augmented matrix of the system $((-3i)I_3 - A)\mathbf{x} = \mathbf{0}$,

$$\begin{bmatrix} -3i & 0 & 9 & \vdots & 0 \\ 0 & -3i-1 & 0 & \vdots & 0 \\ -1 & 0 & -3i & \vdots & 0 \end{bmatrix}$$

which has solution $x_1 = (-3i)r$, $x_2 = 0$, $x_3 = r$, where r is any nonzero number. Hence every vector of the form

$$\begin{bmatrix} (-3i)r \\ 0 \\ r \end{bmatrix}, \quad r \neq 0$$

is an eigenvector associated with eigenvalue $\lambda_3 = -3i$. Let $r = 1$ and we have eigenvector

$$\begin{bmatrix} -3i \\ 0 \\ 1 \end{bmatrix}.$$

19. Let $A = \begin{bmatrix} 2 & 2 & 3 & 4 \\ 0 & 2 & 3 & 2 \\ 0 & 0 & 1 & 1 \\ 0 & 0 & 0 & 1 \end{bmatrix}$. Since A is upper triangular, its eigenvalues are its diagonal entries: $\lambda_1 = 2$, $\lambda_2 = 2$, $\lambda_3 = 1$, $\lambda_4 = 1$. (See Exercise T.3.) To find the corresponding eigenvectors, solve $(2I_4 - A)\mathbf{x} = \mathbf{0}$:

$$\begin{bmatrix} 0 & -2 & -3 & -4 \\ 0 & 0 & -3 & -2 \\ 0 & 0 & 1 & -1 \\ 0 & 0 & 0 & 1 \end{bmatrix} \begin{bmatrix} x_1 \\ x_2 \\ x_3 \\ x_4 \end{bmatrix} = \begin{bmatrix} 0 \\ 0 \\ 0 \\ 0 \end{bmatrix}.$$

The reduced row echelon form of the coefficient matrix is

$$\begin{bmatrix} 0 & 1 & 0 & 0 \\ 0 & 0 & 1 & 0 \\ 0 & 0 & 0 & 1 \\ 0 & 0 & 0 & 0 \end{bmatrix}.$$

Thus the solution is $x_1 = r$, $x_2 = 0$, $x_3 = 0$, $x_4 = 0$, where r is any real number. The solution in vector form is

$$\begin{bmatrix} r \\ 0 \\ 0 \\ 0 \end{bmatrix} = r \begin{bmatrix} 1 \\ 0 \\ 0 \\ 0 \end{bmatrix}.$$

Thus

$$S = \{\mathbf{x}_1\} = \left\{ \begin{bmatrix} 1 \\ 0 \\ 0 \\ 0 \end{bmatrix} \right\}$$

is a basis for the eigenspace for eigenvalue $\lambda_1 = 2$. Next we solve $(1I_4 - A)\mathbf{x} = \mathbf{0}$:

$$\begin{bmatrix} -1 & -2 & -3 & -4 \\ 0 & -1 & -3 & -2 \\ 0 & 0 & 0 & -1 \\ 0 & 0 & 0 & 0 \end{bmatrix} \begin{bmatrix} x_1 \\ x_2 \\ x_3 \\ x_4 \end{bmatrix} = \begin{bmatrix} 0 \\ 0 \\ 0 \\ 0 \end{bmatrix}.$$

The reduced row echelon form of the coefficient matrix is

$$\begin{bmatrix} 1 & 0 & -3 & 0 \\ 0 & 1 & 3 & 0 \\ 0 & 0 & 0 & 1 \\ 0 & 0 & 0 & 0 \end{bmatrix}.$$

Thus the solution is $x_1 = 3r$, $x_2 = -3r$, $x_3 = r$, $x_4 = 0$, where r is any real number. In vector form the solution is

$$\begin{bmatrix} 3r \\ -3r \\ r \\ 0 \end{bmatrix} = r \begin{bmatrix} 3 \\ -3 \\ 1 \\ 0 \end{bmatrix}.$$

Thus we have

$$\mathbf{x}_2 = \begin{bmatrix} 3 \\ -3 \\ 1 \\ 0 \end{bmatrix}$$

as a basis for the eigenspace for eigenvalue $\lambda = 1$.

21. To find the eigenvectors associated with $\lambda = 2$, we solve

$$(2I_3 - A)\mathbf{x} = \begin{bmatrix} 0 & -1 & 0 \\ -1 & 0 & -1 \\ 0 & -1 & 0 \end{bmatrix} \begin{bmatrix} x_1 \\ x_2 \\ x_3 \end{bmatrix} = \begin{bmatrix} 0 \\ 0 \\ 0 \end{bmatrix}.$$

The general solution to this system is $x_1 = -r$, $x_2 = 0$, $x_3 = r$, or

$$\begin{bmatrix} -r \\ 0 \\ r \end{bmatrix} = r \begin{bmatrix} -1 \\ 0 \\ 1 \end{bmatrix}$$

where $r \neq 0$ is any real number. Thus a basis for the eigenspace of A associated with $\lambda = 2$ is

$$\left\{ \begin{bmatrix} -1 \\ 0 \\ 1 \end{bmatrix} \right\}.$$

23. To find the eigenvectors associated with $\lambda = 2$, we solve

$$(2I_4 - A)\mathbf{x} = \begin{bmatrix} -2 & -2 & 0 & 0 \\ -3 & -1 & 0 & 0 \\ 0 & 0 & 0 & -5 \\ 0 & 0 & 0 & 0 \end{bmatrix} \begin{bmatrix} x_1 \\ x_2 \\ x_3 \\ x_4 \end{bmatrix} = \begin{bmatrix} 0 \\ 0 \\ 0 \\ 0 \end{bmatrix}.$$

The general solution to this system is $x_1 = 0$, $x_2 = 0$, $x_3 = r$, $x_4 = 0$, or

$$\begin{bmatrix} 0 \\ 0 \\ r \\ 0 \end{bmatrix} = r \begin{bmatrix} 0 \\ 0 \\ 1 \\ 0 \end{bmatrix}$$

where $r \neq 0$ is any real number. Thus a basis for the eigenspace of A associated with $\lambda = 2$ is

$$\left\{ \begin{bmatrix} 0 \\ 0 \\ 1 \\ 0 \end{bmatrix} \right\}.$$

25. Let $A = \begin{bmatrix} 3 & 0 & 0 & 0 \\ 0 & 3 & 0 & 0 \\ 4 & 0 & 0 & 3 \\ 0 & 0 & -3 & 0 \end{bmatrix}$.

(a) To find the eigenvectors associated with $\lambda_1 = 3$ we solve the system $(3I_4 - A)\mathbf{x} = \mathbf{0}$ with augmented matrix

$$\left[\begin{array}{cccc:c} 0 & 0 & 0 & 0 & 0 \\ 0 & 0 & 0 & 0 & 0 \\ -4 & 0 & 3 & -3 & 0 \\ 0 & 0 & 3 & 3 & 0 \end{array} \right]$$

which has solution $x_1 = -\frac{3}{2}s$, $x_2 = r$, $x_3 = -s$, $x_4 = s$, where r and s are any nonzero numbers. Hence

$$\mathbf{x} = r \begin{bmatrix} 0 \\ 1 \\ 0 \\ 0 \end{bmatrix} + s \begin{bmatrix} -\frac{3}{2} \\ 0 \\ -1 \\ 1 \end{bmatrix}, \quad r \neq 0, \quad s \neq 0$$

and a basis for the eigenspace associated with the eigenvalue $\lambda_1 = 3$ is if $r = 1$, $s = -1$,

$$\left\{ \begin{bmatrix} 0 \\ 1 \\ 0 \\ 0 \end{bmatrix}, \begin{bmatrix} \frac{3}{2} \\ 0 \\ 1 \\ -1 \end{bmatrix} \right\}.$$

(b) To find the eigenvectors associated with $\lambda_2 = 3i$ we solve the system $((3i)I_4 - A)\mathbf{x} = \mathbf{0}$ with augmented matrix

$$\left[\begin{array}{cccc:c} 3i - 3 & 0 & 0 & 0 & 0 \\ 0 & 3i - 3 & 0 & 0 & 0 \\ -4 & 0 & 3i & -3 & 0 \\ 0 & 0 & 3 & 3i & 0 \end{array} \right]$$

which has solution $x_1 = 0$, $x_2 = 0$, $x_3 = -ir$, $x_4 = r$, where r is any nonzero number. Hence

$$\mathbf{x} = r \begin{bmatrix} 0 \\ 0 \\ -i \\ 1 \end{bmatrix}, \quad r \neq 0$$

and a basis for the eigenspace associated with the eigenvalue $\lambda_2 = 3i$ is if $r = i$

$$\left\{ \begin{bmatrix} 0 \\ 0 \\ 1 \\ i \end{bmatrix} \right\}.$$

27. Suppose that $\mathbf{u} = \begin{bmatrix} a_1 \\ a_2 \\ a_3 \end{bmatrix}$ is a stable age distribution. Then

$$A\mathbf{u} = \mathbf{u} \implies \begin{bmatrix} 0 & 0 & 8 \\ \frac{1}{4} & 0 & 0 \\ 0 & \frac{1}{2} & 0 \end{bmatrix} \begin{bmatrix} a_1 \\ a_2 \\ a_3 \end{bmatrix} = \begin{bmatrix} a_1 \\ a_2 \\ a_3 \end{bmatrix} \implies \begin{array}{l} 8a_3 = a_1 \\ \frac{1}{4}a_1 = a_2 \\ \frac{1}{2}a_2 = a_3 \end{array}$$

We find that $a_1 = 8r$, $a_2 = 2r$, $a_3 = r$, where $r > 0$ is a real number. Choosing $r = 1$, we obtain the stable age distribution

$$\begin{bmatrix} 8 \\ 2 \\ 1 \end{bmatrix}.$$

T.1. Let $\mathbf{u}$, $\mathbf{v}$ be vectors in S and c a scalar. Then

$$A(\mathbf{u} + \mathbf{v}) = A\mathbf{u} + A\mathbf{v} = \lambda_j\mathbf{u} + \lambda_j\mathbf{v} = \lambda_j(\mathbf{u} + \mathbf{v}) \quad \text{and} \quad A(c\mathbf{u}) = c(A\mathbf{u}) = c\lambda_j\mathbf{u} = \lambda_j(c\mathbf{u}).$$

Thus both $\mathbf{u} + \mathbf{v}$ and $c\mathbf{u}$ are either eigenvectors associated with λ_j or else $\mathbf{0}$. Hence both lie in S and S is a subspace.

T.3. $(\lambda I_n - A)$ is a triangular matrix whose determinant is the product of its diagonal elements, thus the characteristic polynomial of A is

$$f(\lambda) = (\lambda - a_{11})(\lambda - a_{22}) \cdots (\lambda - a_{nn}).$$

It follows that the eigenvalues of A are the diagonal elements of A.

T.5. $A^k\mathbf{x} = A^{k-1}(A\mathbf{x}) = A^{k-1}(\lambda\mathbf{x}) = \lambda A^{k-1}\mathbf{x} = \cdots = \lambda^k\mathbf{x}.$

T.7. (a) Let $\lambda_1, \lambda_2, \ldots, \lambda_n$ be the roots of the characteristic polynomial of A. Then

$$f(\lambda) = \det(\lambda I_n - A) = (\lambda - \lambda_1) \cdots (\lambda - \lambda_n).$$

Hence

$$f(0) = \det(-A) = (-\lambda_1)(-\lambda_2) \cdots (-\lambda_n) = (-1)^n \lambda_1 \lambda_2 \cdots \lambda_n.$$

Since $\det(-A) = (-1)^n \det(A)$ we have $\det(A) = \lambda_1 \cdots \lambda_n$.

(b) A is singular if and only if $\det(A) = 0$, if and only if [by (a)] 0 is a real root of the characteristic polynomial of A.

T.9. Let A be any $n \times n$ matrix.

(a) To find the coefficient of λ^{n-1} in the characteristic polynomial of A, first note that the characteristic polynomial is

$$\det(\lambda I_n - A) = \begin{vmatrix} \lambda - a_{11} & -a_{12} & \cdots & \cdots & -a_{1n} \\ -a_{21} & \lambda - a_{22} & \cdots & \cdots & -a_{2n} \\ \vdots & \vdots & \vdots & \vdots & \vdots \\ -a_{n1} & -a_{n2} & \cdots & -a_{n\,n-1} & \lambda - a_{nn} \end{vmatrix}.$$

Any product in $\det(\lambda I_n - A)$, other than the product of the diagonal entries, can contain at most $n - 2$ of the diagonal entries of $\lambda I_n - A$. This follows because at least two of the column indices must be out of natural order in every other product appearing in $\det(\lambda I_n - A)$. This implies that the coefficient of λ^{n-1} is formed by the expansion of the product of the diagonal entries. The coefficient of λ^{n-1} is the sum of the coefficients of λ^{n-1} from each of the products

$$-a_{ii}(\lambda - a_{11})\cdots(\lambda - a_{i-1\,i-1})(\lambda - a_{i+1\,i+1})\cdots(\lambda - a_{nn})$$

for $i = 1, 2, \ldots, n$. The coefficient of λ^{n-1} in each such term is $-a_{ii}$ so the coefficient of λ^{n-1} in the characteristic polynomial is

$$(-a_{11}) + (-a_{22}) + \cdots + (-a_{nn}) = -\operatorname{Tr}(A).$$

(b) If $\lambda_1, \lambda_2, \ldots, \lambda_n$ are the eigenvalues of A then $\lambda - \lambda_i$, for $i = 1, 2, \ldots, n$, are factors of the characteristic polynomial $\det(\lambda I_n - A)$. It follows that

$$\det(\lambda I_n - A) = (\lambda - \lambda_1)(\lambda - \lambda_2)\cdots(\lambda - \lambda_n).$$

Proceeding as in part (a), the coefficient of λ^{n-1} is the sum of the coefficients of λ^{n-1} from each of the products

$$-\lambda_i(\lambda - \lambda_1)(\lambda - \lambda_2)\cdots(\lambda - \lambda_{i-1})(\lambda - \lambda_{i+1})\cdots(\lambda - \lambda_n)$$

for $i = 1, 2, \ldots, n$. The coefficient of λ^{n-1} in each such term is $-\lambda_i$, so the coefficient of λ^{n-1} in the characteristic polynomial is

$$(-\lambda_1) + (-\lambda_2) + \cdots + (-\lambda_n) = -\operatorname{Tr}(A)$$

by (a). Thus, $\operatorname{Tr}(A)$ is the sum of the eigenvalues of A.

(c) We have
$$\det(\lambda I_n - A) = (\lambda - \lambda_1)(\lambda - \lambda_2)\cdots(\lambda - \lambda_n)$$
so the constant term is $\pm\lambda_1\lambda_2\cdots\lambda_n$.

(d) If $f(\lambda) = \det(\lambda I_n - A)$ is the characteristic polynomial of A, then $f(0) = \det(-A) = (-1)^n \det(A)$. Since $f(0) = a_n$, the constant term of $f(\lambda)$, $a_n = (-1)^n \det(A)$. The result follows from part (c).

T.11. If $A\mathbf{x} = \lambda\mathbf{x}$, then, for any scalar r,

$$(A + rI_n)\mathbf{x} = A\mathbf{x} + r\mathbf{x} = \lambda\mathbf{x} + r\mathbf{x} = (\lambda + r)\mathbf{x}.$$

Thus $\lambda + r$ is an eigenvalue of $A + rI_n$ with associated eigenvector $\mathbf{x}$.

T.13. We have

(a) $(A + B)\mathbf{x} = A\mathbf{x} + B\mathbf{x} = \lambda\mathbf{x} + \mu\mathbf{x} = (\lambda + \mu)\mathbf{x}$.

(b) $(AB)\mathbf{x} = A(B\mathbf{x}) = A(\mu\mathbf{x}) = \mu(A\mathbf{x}) = \mu(\lambda\mathbf{x})\mathbf{x} = (\lambda\mu)\mathbf{x}$.

T.15. Let S be the eigenspace of A associated with the eigenvalue λ_j. Let $\mathbf{s} \in S$. Then $L(\mathbf{s}) = A\mathbf{s} = \lambda_j \mathbf{s} \in S$ since S is a subspace (T.1) and hence closed under scalar multiplication. So S is invariant under L.

ML.1. Enter each matrix A into MATLAB and use command **poly(A)**.

(a) $\mathbf{A} = [1 \ \ 2; 2 \ \ -1];$
$\mathbf{v} = \mathbf{poly(A)}$

$\mathbf{v} =$

  1.0000 0 −5.0000

The characteristic polynomial is $\lambda^2 - 5$.

(b) $\mathbf{A} = [2 \ \ 4 \ \ 0; 1 \ \ 2 \ \ 1; 0 \ \ 4 \ \ 2];$
$\mathbf{v} = \mathbf{poly(A)}$

$\mathbf{v} =$

  1.0000 −6.0000 4.0000 8.0000

The characteristic polynomial is $\lambda^3 - 6\lambda^2 + 4\lambda + 8$.

(c) $\mathbf{A} = [1 \ \ 0 \ \ 0 \ \ 0; 2 \ \ -2 \ \ 0 \ \ 0; 0 \ \ 0 \ \ 2 \ \ -1; 0 \ \ 0 \ \ -1 \ \ 2];$
$\mathbf{v} = \mathbf{poly(A)}$

$\mathbf{v} =$

  1 −3 −3 11 −6

The characteristic polynomial is $\lambda^4 - 3\lambda^3 - 3\lambda^2 + 11\lambda - 6$.

ML.3. We solve the homogeneous system $(\lambda I_2 - A)\mathbf{x} = \mathbf{0}$ by finding the reduced row echelon form of the corresponding augmented matrix and then writing out the general solution.

(a) $\mathbf{A} = [1 \ \ 2; -1 \ \ 4];$
$\mathbf{M} = (3 * \mathbf{eye(size(A))}) - \mathbf{A})$
$\mathbf{rref}([\mathbf{M} \ [0 \ \ 0]'])$

ans $=$

  1 −1 0
  0   0 0

The general solution is $x_2 = r$, $x_1 = x_2 = r$. Let $r = 1$ and we have that $\begin{bmatrix} 1 & 1 \end{bmatrix}'$ is an eigenvector.

(b) $\mathbf{A} = [4 \ \ 0 \ \ 0; 1 \ \ 3 \ \ 0; 2 \ \ 1 \ \ -1];$
$\mathbf{M} = (-1 * \mathbf{eye(size(A))}) - \mathbf{A})$
$\mathbf{rref}([\mathbf{M} \ [0 \ \ 0 \ \ 0]'])$

ans $=$

  1 0 0 0
  0 1 0 0
  0 0 0 0

The general solution is $x_3 = r$, $x_2 = 0$, $x_1 = 0$. Let $r = 1$ and we have that $\begin{bmatrix} 0 & 0 & 1 \end{bmatrix}'$ is an eigenvector.

(c) $\mathbf{A} = [2 \ \ 1 \ \ 2; 2 \ \ 2 \ \ -2; 3 \ \ 1 \ \ 1];$
$\mathbf{M} = (2 * \mathbf{eye(size(A))}) - \mathbf{A})$
$\mathbf{rref}([\mathbf{M} \ [0 \ \ 0 \ \ 0]'])$

```
ans =
    1   0  -1   0
    0   1   2   0
    0   0   0   0
```

The general solution is $x_3 = r$, $x_2 = -2x_3 = -2r$, $x_1 = x_3 = r$. Let $r = 1$ and we have that $\begin{bmatrix} 1 & -2 & 1 \end{bmatrix}'$ is an eigenvector.

Section 8.2, p. 431

1. Diagonalizable. The eigenvalues are $\lambda_1 = -3$ and $\lambda_2 = 2$. The result follows by Theorem 8.5.

3. Diagonalizable. The eigenvalues are $\lambda_1 = 0$, $\lambda_2 = 2$, and $\lambda_3 = 3$. The result follows by Theorem 8.5.

5. Not diagonalizable. The characteristic polynomial is

$$p(\lambda) = \det(\lambda I_3 - A) = \begin{vmatrix} \lambda - 3 & -1 & 0 \\ 0 & \lambda - 3 & -1 \\ 0 & 0 & \lambda - 3 \end{vmatrix} = (\lambda - 3)^3.$$

Thus there is one eigenvalue of multiplicity 3: $\lambda_1 = \lambda_2 = \lambda_3 = 3$. To find the associated eigenvectors, we solve the system

$$(3I_3 - A)\mathbf{x} - \begin{bmatrix} 0 & -1 & 0 \\ 0 & 0 & -1 \\ 0 & 0 & 0 \end{bmatrix} \begin{bmatrix} x_1 \\ x_2 \\ x_3 \end{bmatrix} = \begin{bmatrix} 0 \\ 0 \\ 0 \end{bmatrix}$$

to obtain the eigenvectors

$$\begin{bmatrix} r \\ 0 \\ 0 \end{bmatrix} = r \begin{bmatrix} 1 \\ 0 \\ 0 \end{bmatrix}$$

where $r \neq 0$ is any real number. Since A does not have three linearly independent eigenvectors, A is not diagonalizable.

7. Not diagonalizable. The characteristic polynomial is

$$p(\lambda I_3 - A) = \begin{vmatrix} \lambda - 2 & 0 & -3 \\ 0 & \lambda - 1 & 0 \\ 0 & -1 & \lambda - 2 \end{vmatrix} = (\lambda - 2)(\lambda - 1)(\lambda - 2)$$

so the eigenvalues are $\lambda_1 = 2$, $\lambda_2 = 2$, $\lambda_3 = 1$. To find the eigenvectors associated with $\lambda_1 = \lambda_2 = 2$, we solve the system

$$(2I_3 - A)\mathbf{x} = \begin{bmatrix} 0 & 0 & -3 \\ 0 & 1 & 0 \\ 0 & -1 & 0 \end{bmatrix} \begin{bmatrix} x_1 \\ x_2 \\ x_3 \end{bmatrix} = \begin{bmatrix} 0 \\ 0 \\ 0 \end{bmatrix}$$

to obtain the eigenvectors

$$\begin{bmatrix} r \\ 0 \\ 0 \end{bmatrix} = r \begin{bmatrix} 1 \\ 0 \\ 0 \end{bmatrix}$$

where $r \neq 0$ is any scalar. Choose

$$\mathbf{x}_1 = \begin{bmatrix} 1 \\ 0 \\ 0 \end{bmatrix}$$

as an eigenvector associated with $\lambda_1 = \lambda_2 = 2$.

To find eigenvectors associated with $\lambda_3 = 1$, we solve the system

$$(1I_3 - A)\mathbf{x} = \begin{bmatrix} -1 & 0 & -3 \\ 0 & 0 & 0 \\ 0 & -1 & -1 \end{bmatrix} \begin{bmatrix} x_1 \\ x_2 \\ x_3 \end{bmatrix} = \begin{bmatrix} 0 \\ 0 \\ 0 \end{bmatrix}$$

to obtain the eigenvectors

$$\begin{bmatrix} -3r \\ r \\ r \end{bmatrix} = r \begin{bmatrix} -3 \\ 1 \\ 1 \end{bmatrix}$$

where $r \neq 0$ is any real number. Choose

$$x_2 = \begin{bmatrix} -3 \\ 1 \\ 1 \end{bmatrix}.$$

Since A does not have three linearly independent eigenvectors, A is not diagonalizable.

9. Let

$$D = \begin{bmatrix} 2 & 0 \\ 0 & -3 \end{bmatrix} \quad \text{and} \quad P = \begin{bmatrix} -1 & 1 \\ 2 & 1 \end{bmatrix}.$$

Then $P^{-1}AP = D$, so

$$A = PDP^{-1} = \frac{1}{3} \begin{bmatrix} -4 & -5 \\ -10 & 1 \end{bmatrix}$$

is a matrix whose eigenvalues and associated eigenvectors are as given.

11. To find the eigenvalues of A, we solve

$$p(\lambda) = \det(\lambda I_3 - A) = \begin{vmatrix} \lambda - 4 & -2 & -3 \\ -2 & \lambda - 1 & -2 \\ 1 & 2 & \lambda \end{vmatrix} = (\lambda - 1)^2(\lambda - 3) = 0.$$

The eigenvalues are $\lambda_1 = 1$, $\lambda_2 = 1$, $\lambda_3 = 3$. To find the eigenvectors associated with $\lambda_1 = \lambda_2 = 1$, we solve the system

$$(1I_3 - A)\mathbf{x} = \begin{bmatrix} -3 & -2 & -3 \\ -2 & 0 & -2 \\ 1 & 2 & 1 \end{bmatrix} \begin{bmatrix} x_1 \\ x_2 \\ x_3 \end{bmatrix} = \begin{bmatrix} 0 \\ 0 \\ 0 \end{bmatrix}$$

obtaining the solutions

$$\begin{bmatrix} r \\ 0 \\ -r \end{bmatrix} = r \begin{bmatrix} 1 \\ 0 \\ -1 \end{bmatrix},$$

where r is any nonzero real number. Choose $r = 1$ to obtain the eigenvector

$$\mathbf{x}_1 = \begin{bmatrix} 1 \\ 0 \\ -1 \end{bmatrix}.$$

To find the eigenvectors associated with $\lambda_3 = 3$, we solve the system

$$(3I_3 - A)\mathbf{x} = \begin{bmatrix} -1 & -2 & -3 \\ -2 & 2 & -2 \\ 1 & 2 & 3 \end{bmatrix} \begin{bmatrix} x_1 \\ x_2 \\ x_3 \end{bmatrix} = \begin{bmatrix} 0 \\ 0 \\ 0 \end{bmatrix}$$

obtaining the solutions

$$\begin{bmatrix} -\frac{5}{3}r \\ -\frac{2}{3}r \\ r \end{bmatrix} = r \begin{bmatrix} -\frac{5}{3} \\ -\frac{2}{3} \\ 1 \end{bmatrix},$$

where r is any nonzero real number. Choose $r = 3$ to obtain the eigenvector

$$\mathbf{x}_2 = \begin{bmatrix} -5 \\ -2 \\ 3 \end{bmatrix}.$$

Since there are only two linearly independent eigenvectors,

$$\left\{ \begin{bmatrix} 1 \\ 0 \\ -1 \end{bmatrix}, \begin{bmatrix} -5 \\ -2 \\ 3 \end{bmatrix} \right\},$$

A is not diagonalizable so no such matrix P exists.

13. To find the eigenvalues of A, we solve

$$p(\lambda) = \det(\lambda I_3 - A) = \begin{vmatrix} \lambda - 1 & -2 & -3 \\ 0 & \lambda - 1 & 0 \\ -2 & -1 & \lambda - 2 \end{vmatrix} = (\lambda - 4)(\lambda + 1)(\lambda - 1).$$

The eigenvalues are $\lambda_1 = 4$, $\lambda_2 = -1$, and $\lambda_3 = 1$. To find an eigenvector associated with $\lambda_1 = 4$, we solve the system

$$(4I_3 - A)\mathbf{x} = \begin{bmatrix} 3 & -2 & -3 \\ 0 & 3 & 0 \\ -2 & -1 & 2 \end{bmatrix} \begin{bmatrix} x_1 \\ x_2 \\ x_3 \end{bmatrix} = \begin{bmatrix} 0 \\ 0 \\ 0 \end{bmatrix}$$

obtaining the solutions

$$\begin{bmatrix} r \\ 0 \\ r \end{bmatrix} = r \begin{bmatrix} 1 \\ 0 \\ 1 \end{bmatrix},$$

where r is any nonzero real number. Choose $r = 1$ to obtain the eigenvector

$$\mathbf{x}_1 = \begin{bmatrix} 1 \\ 0 \\ 1 \end{bmatrix}.$$

To find an eigenvector associated with $\lambda_2 = -1$, we solve the system

$$(-1I_3 - A)\mathbf{x} = \begin{bmatrix} -2 & -2 & -3 \\ 0 & -2 & 0 \\ -2 & -1 & -3 \end{bmatrix} \begin{bmatrix} x_1 \\ x_2 \\ x_3 \end{bmatrix} = \begin{bmatrix} 0 \\ 0 \\ 0 \end{bmatrix}$$

obtaining the solutions

$$\begin{bmatrix} -\frac{3}{2}r \\ 0 \\ r \end{bmatrix} = r \begin{bmatrix} -\frac{3}{2} \\ 0 \\ 1 \end{bmatrix},$$

where r is any nonzero real number. Choose $r = 2$ to obtain the eigenvector

$$\mathbf{x}_2 = \begin{bmatrix} -3 \\ 0 \\ 2 \end{bmatrix}.$$

To find an eigenvector associated with $\lambda_3 = 1$, we solve the system

$$(1I_3 - A)\mathbf{x} = \begin{bmatrix} 0 & -2 & -3 \\ 0 & 0 & 0 \\ -2 & -1 & -1 \end{bmatrix} \begin{bmatrix} x_1 \\ x_2 \\ x_3 \end{bmatrix} = \begin{bmatrix} 0 \\ 0 \\ 0 \end{bmatrix}$$

obtaining the solutions

$$\begin{bmatrix} \frac{1}{4}r \\ -\frac{3}{2}r \\ r \end{bmatrix} = r \begin{bmatrix} \frac{1}{4} \\ -\frac{3}{2} \\ 1 \end{bmatrix},$$

where r is any nonzero real number. Choose $r = 4$ to obtain the eigenvector

$$\mathbf{x}_3 = \begin{bmatrix} 1 \\ -6 \\ 4 \end{bmatrix}.$$

There are three linearly independent eigenvectors

$$\left\{ \begin{bmatrix} 1 \\ 0 \\ 1 \end{bmatrix}, \begin{bmatrix} -3 \\ 0 \\ 2 \end{bmatrix}, \begin{bmatrix} 1 \\ -6 \\ 4 \end{bmatrix} \right\},$$

so A is diagonalizable and

$$P = \begin{bmatrix} 1 & -3 & 1 \\ 0 & 0 & -6 \\ 1 & 2 & 4 \end{bmatrix}.$$

15. To find the eigenvalues of matrix A we solve

$$p(\lambda) = \det(\lambda I_3 - A) = \begin{vmatrix} \lambda - 8 & -1 & 0 \\ 0 & \lambda - 8 & 0 \\ -8 & 0 & \lambda \end{vmatrix} = (\lambda - 8)^2 \lambda = 0$$

so the eigenvalues are $\lambda_1 = \lambda_2 = 8$, $\lambda_3 = 0$.

To find the eigenvectors associated with $\lambda_1 = \lambda_2 = 8$ we solve the system $(8I_3 - A)\mathbf{x} = \mathbf{0}$ with augmented matrix

$$\begin{bmatrix} 0 & -1 & 0 & \vdots & 0 \\ 0 & 0 & 0 & \vdots & 0 \\ -8 & 0 & 8 & \vdots & 0 \end{bmatrix}.$$

The general solution to this system is $x_1 = r$, $x_2 = 0$, $x_3 = r$, r any nonzero real number. Hence

$$\mathbf{x} = r \begin{bmatrix} 1 \\ 0 \\ 1 \end{bmatrix}, \quad r \neq 0$$

and an eigenvector associated with $\lambda_1 = \lambda_2 = 8$ is

$$\begin{bmatrix} 1 \\ 0 \\ 1 \end{bmatrix}.$$

To find the eigenvector associated with $\lambda_3 = 0$ we solve the system $(-A)\mathbf{x} = \mathbf{0}$ with augmented matrix

$$\begin{bmatrix} -8 & -1 & 0 & \vdots & 0 \\ 0 & -8 & 0 & \vdots & 0 \\ -8 & 0 & 0 & \vdots & 0 \end{bmatrix}.$$

The general solution is $x_1 = 0$, $x_2 = 0$, $x_3 = r$, r any nonzero real number. Hence

$$\mathbf{x} = r \begin{bmatrix} 0 \\ 0 \\ 1 \end{bmatrix}, \quad r \neq 0$$

and an eigenvector associated with $\lambda_3 = 0$ is

$$\begin{bmatrix} 0 \\ 0 \\ 1 \end{bmatrix}.$$

Since A does not have three linearly independent eigenvectors, A is not diagonalizable.

17. The matrix is upper triangular so its eigenvalues are $\lambda_1 = 3$, $\lambda_2 = 2$, and $\lambda_3 = 0$. To find the eigenvectors associated with $\lambda_1 = 3$, we solve the system

$$(3I_3 - A)\mathbf{x} = \begin{bmatrix} 0 & 2 & -1 \\ 0 & 1 & 0 \\ 0 & 0 & 3 \end{bmatrix} \begin{bmatrix} x_1 \\ x_2 \\ x_3 \end{bmatrix} = \begin{bmatrix} 0 \\ 0 \\ 0 \end{bmatrix}$$

obtaining the solution

$$\begin{bmatrix} r \\ 0 \\ 0 \end{bmatrix} = r \begin{bmatrix} 1 \\ 0 \\ 0 \end{bmatrix}$$

where $r \neq 0$ is any real number. Choose the eigenvector

$$\mathbf{x}_1 = \begin{bmatrix} 1 \\ 0 \\ 0 \end{bmatrix}.$$

For the eigenvectors associated with $\lambda_2 = 2$, we solve the system

$$(2I_3 - A)\mathbf{x} = \begin{bmatrix} -1 & 2 & -1 \\ 0 & 0 & 0 \\ 0 & 0 & 2 \end{bmatrix} \begin{bmatrix} xz_1 \\ x_2 \\ x_3 \end{bmatrix} = \begin{bmatrix} 0 \\ 0 \\ 0 \end{bmatrix}$$

obtaining the solution

$$\begin{bmatrix} 2r \\ r \\ 0 \end{bmatrix} = r \begin{bmatrix} 2 \\ 1 \\ 0 \end{bmatrix}$$

where $r \neq 0$ is any real number. Choose the eigenvector

$$\mathbf{x}_2 = \begin{bmatrix} 2 \\ 1 \\ 0 \end{bmatrix}.$$

For the eigenvectors associated with $\lambda_3 = 0$, we solve the system

$$(0I_3 - A)\mathbf{x} = \begin{bmatrix} 3 & -2 & 1 \\ 0 & 2 & 0 \\ 0 & 0 & 0 \end{bmatrix} \begin{bmatrix} x_1 \\ x_2 \\ x_3 \end{bmatrix} = \begin{bmatrix} 0 \\ 0 \\ 0 \end{bmatrix}$$

obtaining the solution

$$\begin{bmatrix} -\frac{1}{3}r \\ 0 \\ r \end{bmatrix} = r \begin{bmatrix} -\frac{1}{3} \\ 0 \\ 1 \end{bmatrix}$$

where $r \neq 0$ is any real number. Choose the eigenvector

$$x_3 = \begin{bmatrix} 1 \\ 0 \\ -3 \end{bmatrix}.$$

The eigenvectors

$$\left\{ \begin{bmatrix} 1 \\ 0 \\ 0 \end{bmatrix}, \begin{bmatrix} 2 \\ 1 \\ 0 \end{bmatrix}, \begin{bmatrix} 1 \\ 0 \\ -3 \end{bmatrix} \right\}$$

form a basis for R^3, so A is diagonalizable. The transition matrix P is

$$P = \begin{bmatrix} 1 & 2 & 1 \\ 0 & 1 & 0 \\ 0 & 0 & -3 \end{bmatrix}$$

and we have

$$P^{-1}AP = \begin{bmatrix} 3 & 0 & 0 \\ 0 & 2 & 0 \\ 0 & 0 & 0 \end{bmatrix}.$$

19. The characteristic polynomial is

$$p(\lambda) = \det(\lambda I_3 - A) = \begin{vmatrix} \lambda - 3 & 0 & 0 \\ -2 & \lambda - 3 & 0 \\ 0 & 0 & \lambda - 3 \end{vmatrix} = (\lambda - 3)^3$$

so the eigenvalues are $\lambda_1 = \lambda_2 = \lambda_3 = 3$. To find the eigenvectors associated with this eigenvalue, we solve

$$(3I_3 - A)\mathbf{x} = \begin{bmatrix} 0 & 0 & 0 \\ -2 & 0 & 0 \\ 0 & 0 & 0 \end{bmatrix} \begin{bmatrix} x_1 \\ x_2 \\ x_3 \end{bmatrix} = \begin{bmatrix} 0 \\ 0 \\ 0 \end{bmatrix}$$

to obtain the eigenvectors

$$\begin{bmatrix} 0 \\ r \\ s \end{bmatrix} = r \begin{bmatrix} 0 \\ 1 \\ 0 \end{bmatrix} + s \begin{bmatrix} 0 \\ 0 \\ 1 \end{bmatrix}$$

where r, s are any nonzero real numbers. Since there are only two linearly independent eigenvectors,

$$\left\{ \begin{bmatrix} 0 \\ 1 \\ 0 \end{bmatrix}, \begin{bmatrix} 0 \\ 0 \\ 1 \end{bmatrix} \right\},$$

A is not diagonalizable so no such matrix P exists.

21. $A = \begin{bmatrix} 4 & 1 & 0 & 0 \\ 0 & 4 & 0 & 0 \\ 0 & 0 & 4 & 0 \\ 1 & 1 & 0 & 0 \end{bmatrix}$. The characteristic polynomial is

$$p(\lambda) = \det(\lambda I_4 - A) = \begin{vmatrix} \lambda - 4 & -1 & 0 & 0 \\ 0 & \lambda - 4 & 0 & 0 \\ 0 & 0 & \lambda - 4 & 0 \\ -1 & -1 & 0 & \lambda \end{vmatrix} = (\lambda - 4)(\lambda - 4)(\lambda - 4)\lambda$$

so the eigenvalues are $\lambda_1 = \lambda_2 = \lambda_3 = 4$, $\lambda_4 = 0$.

To find the eigenvectors associated with eigenvalue $\lambda_1 = \lambda_2 = \lambda_3 = 4$ we solve the system $(4I_4 - A)\mathbf{x} = \mathbf{0}$ with augmented matrix

$$\left[\begin{array}{cccc:c} 0 & -1 & 0 & 0 & 0 \\ 0 & 0 & 0 & 0 & 0 \\ 0 & 0 & 0 & 0 & 0 \\ -1 & -1 & 0 & 4 & 0 \end{array}\right].$$

The general solution to this system is

$$x_1 = 4s, \quad x_2 = 0, \quad x_3 = r, \quad x_4 = s$$

where r and s are any nonzero real numbers. Hence

$$\mathbf{x} = r\begin{bmatrix} 0 \\ 0 \\ 1 \\ 0 \end{bmatrix} + s\begin{bmatrix} 4 \\ 0 \\ 0 \\ 1 \end{bmatrix}, \quad r \neq 0, \quad s \neq 0$$

and two linearly independent eigenvectors associated with $\lambda_1 = \lambda_2 = \lambda_3 = 4$ are

$$\begin{bmatrix} 0 \\ 0 \\ 1 \\ 0 \end{bmatrix}, \quad \begin{bmatrix} 4 \\ 0 \\ 0 \\ 1 \end{bmatrix}.$$

To find the eigenvectors associated with eigenvalue $\lambda_4 = 0$ we solve the system $(-A)\mathbf{x} = \mathbf{0}$ with augmented matrix

$$\left[\begin{array}{cccc:c} -4 & -1 & 0 & 0 & 0 \\ 0 & -4 & 0 & 0 & 0 \\ 0 & 0 & -4 & 0 & 0 \\ -1 & -1 & 0 & 0 & 0 \end{array}\right].$$

The general solution to this system is $x_1 = 0$, $x_2 = 0$, $x_3 = 0$, $x_4 = r$, where r is any nonzero real number. Hence

$$\mathbf{x} = r\begin{bmatrix} 0 \\ 0 \\ 0 \\ 1 \end{bmatrix},$$

and an eigenvector associated with $\lambda_4 = 0$ is

$$\begin{bmatrix} 0 \\ 0 \\ 0 \\ 1 \end{bmatrix}.$$

Since A does not have four linearly independent eigenvectors, A is not diagonalizable.

23. P is the matrix whose columns are the given eigenvectors:

$$P = \begin{bmatrix} -1 & 2 \\ 1 & 1 \end{bmatrix}, \quad D = P^{-1}AP = \begin{bmatrix} 3 & 0 \\ 0 & 4 \end{bmatrix}.$$

25. We choose two different nonsingular matrices P and compute $P^{-1}AP$: for example,

$$P = \begin{bmatrix} 1 & 4 \\ 0 & -3 \end{bmatrix}: \quad P^{-1}AP = \begin{bmatrix} 1 & \frac{4}{3} \\ 0 & -\frac{1}{3} \end{bmatrix}\begin{bmatrix} 3 & 4 \\ 0 & 0 \end{bmatrix}\begin{bmatrix} 1 & 4 \\ 0 & -3 \end{bmatrix} = \begin{bmatrix} 3 & 0 \\ 0 & 0 \end{bmatrix}$$

$$P = \begin{bmatrix} -4 & 1 \\ 3 & 0 \end{bmatrix}: \quad P^{-1}AP = \begin{bmatrix} 0 & \frac{1}{3} \\ 1 & \frac{4}{3} \end{bmatrix}\begin{bmatrix} 3 & 4 \\ 0 & 0 \end{bmatrix}\begin{bmatrix} -4 & 1 \\ 3 & 0 \end{bmatrix} = \begin{bmatrix} 0 & 0 \\ 0 & 3 \end{bmatrix}.$$

27. We choose two different nonsingular matrices P and compute $P^{-1}AP$: for example,

$$P = \begin{bmatrix} 0 & 0 & -1 \\ 0 & 1 & -1 \\ 1 & 0 & 1 \end{bmatrix} : \quad P^{-1}AP = \begin{bmatrix} 1 & 0 & 0 \\ 0 & 1 & 0 \\ 0 & 0 & 2 \end{bmatrix}$$

$$P = \begin{bmatrix} 0 & -1 & 0 \\ 0 & -1 & 1 \\ 1 & 1 & 0 \end{bmatrix} : \quad P^{-1}AP = \begin{bmatrix} 1 & 0 & 0 \\ 0 & 2 & 0 \\ 0 & 0 & 1 \end{bmatrix}.$$

29. Since A is upper triangular its eigenvalues are its diagonal entries. Since $\lambda = 2$ is an eigenvalue of multiplicity 2 we must show, by Theorem 8.4, that it has two linearly independent eigenvectors.

$$(2I_3 - A)\mathbf{x} = \begin{bmatrix} 0 & -3 & 0 \\ 0 & 1 & 0 \\ 0 & 0 & 0 \end{bmatrix} \begin{bmatrix} x_1 \\ x_2 \\ x_3 \end{bmatrix} = \begin{bmatrix} 0 \\ 0 \\ 0 \end{bmatrix}.$$

Row reducing the augmented matrix we obtain the equivalent linear system

$$\begin{bmatrix} 0 & 1 & 0 \\ 0 & 0 & 0 \\ 0 & 0 & 0 \end{bmatrix} \begin{bmatrix} x_1 \\ x_2 \\ x_3 \end{bmatrix} = \begin{bmatrix} 0 \\ 0 \\ 0 \end{bmatrix}.$$

It follows that there are two arbitrary constants in the general solution so there are two linearly independent eigenvectors. Hence the matrix is diagonalizable.

31. The matrix is lower triangular hence its eigenvalues are its diagonal entries. Since they are real and distinct the matrix is diagonalizable.

33. The characteristic polynomial is

$$p(\lambda) = \det(\lambda I_2 - A) = \begin{vmatrix} \lambda - 4 & -2 \\ -3 & \lambda - 3 \end{vmatrix} = \lambda^2 - 7\lambda + 6 = (\lambda - 6)(\lambda - 1)$$

so the eigenvalues are $\lambda_1 = 6$, $\lambda_2 = 1$. To find the eigenvectors associated with $\lambda_1 = 6$, we solve the system

$$(6I_2 - A)\mathbf{x} = \begin{bmatrix} 2 & -2 \\ -3 & 2 \end{bmatrix} \begin{bmatrix} x_1 \\ x_2 \end{bmatrix} = \begin{bmatrix} 0 \\ 0 \end{bmatrix}$$

to obtain the eigenvectors

$$\begin{bmatrix} r \\ r \end{bmatrix} = r \begin{bmatrix} 1 \\ 1 \end{bmatrix}.$$

Choose

$$\mathbf{x}_1 = \begin{bmatrix} 1 \\ 1 \end{bmatrix}.$$

To find the eigenvectors associated with $\lambda_2 = 1$, we solve the system

$$(1I_2 - A)\mathbf{x} = \begin{bmatrix} -3 & -2 \\ -3 & -2 \end{bmatrix} \begin{bmatrix} x_1 \\ x_2 \end{bmatrix} = \begin{bmatrix} 0 \\ 0 \end{bmatrix}$$

to obtain

$$\begin{bmatrix} -\frac{2}{3}r \\ r \end{bmatrix} = r \begin{bmatrix} -\frac{2}{3} \\ 1 \end{bmatrix}$$

where $r \neq 0$ is any real number. Choose

$$\mathbf{x}_2 = \begin{bmatrix} -2 \\ 3 \end{bmatrix}.$$

Since

$$\left\{ \begin{bmatrix} 1 \\ 1 \end{bmatrix}, \begin{bmatrix} -2 \\ 3 \end{bmatrix} \right\}$$

is a basis for R^2, A is diagonalizable. We have

$$P = \begin{bmatrix} 1 & -2 \\ 1 & 3 \end{bmatrix} \quad \text{and} \quad D = P^{-1}AP = \begin{bmatrix} 6 & 0 \\ 0 & 1 \end{bmatrix}.$$

35. The characteristic polynomial is

$$p(\lambda) = \det(\lambda I_3 - A) = \begin{vmatrix} \lambda - 2 & 2 & -3 \\ 0 & \lambda - 3 & 2 \\ 0 & 1 & \lambda - 2 \end{vmatrix} = \lambda^3 - 7\lambda^2 + 14\lambda - 8 = (\lambda - 2)(\lambda - 4)(\lambda - 1)$$

so the eigenvalues are $\lambda_1 = 2$, $\lambda_2 = 4$, $\lambda_3 = 1$. To find the eigenvectors associated with $\lambda_1 = 2$, we solve the system

$$(2I_3 - A)\mathbf{x} = \begin{bmatrix} 0 & 2 & -3 \\ 0 & -1 & 2 \\ 0 & 1 & 0 \end{bmatrix} \begin{bmatrix} x_1 \\ x_2 \\ x_3 \end{bmatrix} = \begin{bmatrix} 0 \\ 0 \\ 0 \end{bmatrix}$$

to obtain

$$\begin{bmatrix} r \\ 0 \\ 0 \end{bmatrix} = r \begin{bmatrix} 1 \\ 0 \\ 0 \end{bmatrix}$$

where $r \neq 0$ is any real number. Choose $\mathbf{x}_1 = \begin{bmatrix} 1 \\ 0 \\ 0 \end{bmatrix}$.

To find the eigenvectors associated with $\lambda_2 = 4$, we solve the system

$$(4I_3 - A)\mathbf{x} = \begin{bmatrix} 2 & 2 & -3 \\ 0 & 1 & 2 \\ 0 & 1 & 2 \end{bmatrix} \begin{bmatrix} x_1 \\ x_2 \\ x_3 \end{bmatrix} = \begin{bmatrix} 0 \\ 0 \\ 0 \end{bmatrix}$$

to obtain

$$\begin{bmatrix} \frac{7}{2}r \\ -2r \\ r \end{bmatrix} = r \begin{bmatrix} \frac{7}{2} \\ -2 \\ 1 \end{bmatrix}.$$

Choose $\mathbf{x}_3 = \begin{bmatrix} 7 \\ -4 \\ 2 \end{bmatrix}$.

To find the eigenvectors associated with $\lambda_3 = 1$, we solve the system

$$(1I_3 - A)\mathbf{x} = \begin{bmatrix} -1 & 2 & -3 \\ 0 & -2 & 2 \\ 0 & 1 & -1 \end{bmatrix} \begin{bmatrix} x_1 \\ x_2 \\ x_3 \end{bmatrix} = \begin{bmatrix} 0 \\ 0 \\ 0 \end{bmatrix}$$

to obtain the eigenvectors

$$\begin{bmatrix} -r \\ r \\ r \end{bmatrix} = r \begin{bmatrix} -1 \\ 1 \\ 1 \end{bmatrix}$$

where $r \neq 0$ is any real number. Choose $\mathbf{x}_3 = \begin{bmatrix} -1 \\ 1 \\ 1 \end{bmatrix}$. The eigenvectors $\{\mathbf{x}_1, \mathbf{x}_2, \mathbf{x}_3\}$ form a basis for R^3 so A is diagonalizable. We have

$$P = \begin{bmatrix} 1 & 7 & -1 \\ 0 & -4 & 1 \\ 0 & 2 & 1 \end{bmatrix} \quad \text{and} \quad D = P^{-1}AP = \begin{bmatrix} 2 & 0 & 0 \\ 0 & 4 & 0 \\ 0 & 0 & 1 \end{bmatrix}.$$

37. The matrix is upper triangluar with multiple eigenvalue $\lambda_1 = \lambda_2 = 1$ and associated eigenvector $\begin{bmatrix} 1 \\ 0 \end{bmatrix}$. Since there are not two linearly independent eigenvectors, the matrix is not diagonalizable.

39. The matrix has the multiple eigenvalue $\lambda_1 = \lambda_2 = -1$ with associated eigenvector $\begin{bmatrix} -1 \\ 1 \\ 0 \end{bmatrix}$. Since there are not two linearly independent eigenvectors, the matrix is not diagonalizable.

41. $8I_2 - A = \begin{bmatrix} 0 & -7 \\ 0 & 0 \end{bmatrix} \implies$ solution space has dimension 1 so A is defective.

43. $0I_3 - A = \begin{bmatrix} -3 & -3 & -3 \\ -3 & -3 & -3 \\ 3 & 3 & 3 \end{bmatrix} \implies$ solution space has dimension 2 so A is not defective since $\lambda = 0$ has multiplicity 2.

45. $D^9 = \begin{bmatrix} 2^9 & 0 \\ 0 & (-2)^9 \end{bmatrix} = \begin{bmatrix} 512 & 0 \\ 0 & -512 \end{bmatrix}$.

T.1. (a) $A = P^{-1}AP$ for $P = I_n$.

   (b) If $B = P^{-1}AP$, then $A = PBP^{-1}$ and so A is similar to B.

   (c) If $B = P^{-1}AP$ and $C = Q^{-1}BQ$ then $C = Q^{-1}P^{-1}APQ = (PQ)^{-1}A(PQ)$ with PQ nonsingular.

T.3. Necessary and sufficient conditions are: $(a-d)^2 + 4bc > 0$ for $b = c = 0$.

   For the characteristic polynomial of

   $$A = \begin{bmatrix} a & b \\ c & d \end{bmatrix} \quad \text{is} \quad f(\lambda) = \begin{vmatrix} \lambda - a & -b \\ -c & \lambda - d \end{vmatrix} = \lambda^2 + \lambda(-a - d) + ad - bc.$$

   Then $f(\lambda)$ has real roots if and only if $(a+d)^2 - 4(ad - bc) = (a-d)^2 + 4bc \geq 0$. If $(a-d)^2 + 4bc > 0$, then the eigenvalues are distinct and we can diagonalize. On the other hand, if $(a-d)^2 + 4bc = 0$, then the two eigenvalues λ_1 and λ_2 are equal and we have $\lambda_1 = \lambda_2 = \frac{a+d}{2}$. To find associated eigenvectors we solve the homogeneous system

   $$\begin{bmatrix} \frac{d-a}{2} & -b \\ -c & \frac{a-d}{2} \end{bmatrix} \begin{bmatrix} x_1 \\ x_2 \end{bmatrix} = \begin{bmatrix} 0 \\ 0 \end{bmatrix}.$$

   In this case A is diagonalizable if and only if the solution space has dimension $= 2$; that is, if and only if the rank of the coefficient matrix $= 0$, thus, if and only if $b = c = 0$ so that A is already diagonal.

T.5. False. As a counterexample, let $A = \begin{bmatrix} -3 & 2 \\ -2 & 1 \end{bmatrix}$. Then A has eigenvalues $\lambda_1 = -1$, $\lambda_2 = -1$, but all the eigenvectors are of the form $r \begin{bmatrix} 1 \\ 1 \end{bmatrix}$. Clearly A has only one linearly independent eigenvector and is not diagonalizable. However, $\det(A) \neq 0$, so A is nonsingular.

T.7. (a) If $P^{-1}AP = D$, a diagonal matrix, then $P^T A^T (P^{-1})^T = (P^{-1}AP)^T = D^T$ is diagonal, and $P^T = ((P^{-1})^T)^{-1}$, so A^T is similar to a diagonal matrix.

   (b) $P^{-1}A^k P = (P^{-1}AP)^k = D^k$ is diagonal.

T.9. Let $B = P^{-1}AP$. Then $\det(B) = \det(P^{-1}AP) = \det(P)^{-1} \det(A) \det(P) = \det(A)$.

T.11 The proof proceeds as in the proof of Theorem 8.5, with $k = n$.

ML.1. We find the eigenvalues and corresponding eigenvectors.

(a) **A = [0 2; − 1 3];**
r = roots(poly(A))

r =

$\qquad$ 2

$\qquad$ 1

The eigenvalues are distinct so A is diagonalizable. We find the corresponding eigenvectors.
M = (2 ∗ eye(size(A)) − A)
rref([M [0 0]′])

ans =

$\qquad$ 1   −1   0

$\qquad$ 0   0   0

The general solution is $x_1 = x_2$, $x_2 = r$. Let $r = 1$ and we have that $\begin{bmatrix} 1 & 1 \end{bmatrix}'$ is an eigenvector.
M = (1 ∗ eye(size(A)) − A)
rref([M [0 0]′])

ans =

$\qquad$ 1   −2   0

$\qquad$ 0   0   0

The general solution is $x_1 = 2x_2$, $x_2 = r$. Let $r = 1$ and we have that $\begin{bmatrix} 2 & 1 \end{bmatrix}'$ is an eigenvector.
P = [1 1;2 1]′

P =

$\qquad$ 1   2

$\qquad$ 1   1

invert(P) ∗ A ∗ P

ans =

$\qquad$ 2   0

$\qquad$ 0   1

(b) **A = [1 − 3;3 − 5];**
r = roots(poly(A))

r =

$\qquad$ −2

$\qquad$ −2

Next we determine eigenvectors corresponding to the eigenvalue -2.
M = (− 2 ∗ eye(size(A)) − A)
rref([M [0 0]′])

ans =

$\qquad$ 1   −1   0

$\qquad$ 0   0   0

The general solution is $x_1 = x_2$, $x_2 = r$. Let $r = 1$ and it follows that $\begin{bmatrix} 1 & 1 \end{bmatrix}'$ is an eigenvector, but there is only one linearly independent eigenvector. Hence A is not diagonalizable.

(c) **A = [0 0 4;5 3 6;6 0 5];**
r = roots(poly(A))

r =

      8.0000

      3.0000

     −3.0000

The eigenvalues are distinct, thus A is diagonalizable. We find corresponding eigenvectors.

M = (8 ∗ eye(size(A)) − A)

rref([M [0 0 0]′])

ans =

1.0000	0	−0.5000	0
0	1.0000	−1.7000	0
0	0	0	0

The general solution is $x_1 = 0.5x_3$, $x_2 = 1.7x_3$, $x_3 = r$. Let $r = 10$ and we have that $\begin{bmatrix} 2 & 17 & 10 \end{bmatrix}'$ is an eigenvector.

M = (3 ∗ eye(size(A)) − A)

rref([M [0 0 0]′])

ans =

1	0	0	0
0	0	1	0
0	0	0	0

The general solution is $x_1 = 0$, $x_3 = 0$, $x_2 = r$. Let $r = 1$ and we have that $\begin{bmatrix} 0 & 1 & 0 \end{bmatrix}'$ is an eigenvector.

M = (− 3 ∗ eye(size(A)) − A)

rref([M [0 0 0]′])

ans =

1.0000	0	1.3333	0
0	1.0000	−0.1111	0
0	0	0	0

The general solution is $x_1 = -\frac{4}{3}x_3$, $x_2 = \frac{1}{9}x_3$, $x_3 = r$. Let $r = 9$ and we have that $\begin{bmatrix} -12 & 1 & 9 \end{bmatrix}'$ is an eigenvector. Thus P is

$$P = \begin{bmatrix} 2 & 0 & -12 \\ 17 & 1 & 1 \\ 10 & 0 & 9 \end{bmatrix}.$$

invert(P) ∗ A ∗ P

ans =

8	0	0
0	3	0
0	0	−3

ML.3. **A = [− 1 1.5 − 1.5;− 2 2.5 − 1.5;− 2 2 − 1]**

   r = roots(poly(A))

r =

      1.0000

     −1.0000

      0.5000

The eigenvalues are distinct, hence A is diagonalizable.

M = (1 ∗ eye(size(A)) − A)

rref([M [0 0 0]′)

ans =

$$
\begin{array}{cccc}
1 & 0 & 0 & 0 \\
0 & 1 & -1 & 0 \\
0 & 0 & 0 & 0
\end{array}
$$

The general solution is $x_1 = 0$, $x_2 = x_3$, $x_3 = r$. Let $r = 1$ and we have that $\begin{bmatrix} 0 & 1 & 1 \end{bmatrix}'$ is an eigenvector.

M = (− 1 ∗ eye(size(A)) − A)

rref([M [0 0 0]′)

ans =

$$
\begin{array}{cccc}
1 & 0 & -1 & 0 \\
0 & 1 & -1 & 0 \\
0 & 0 & 0 & 0
\end{array}
$$

The general solution is $x_1 = x_3$, $x_2 = x_3$, $x_3 = r$. Let $r = 1$ and we have that $\begin{bmatrix} 1 & 1 & 1 \end{bmatrix}'$ is an eigenvector.

M = (.5 ∗ eye(size(A)) − A)

rref([M [0 0 0]′)

ans =

$$
\begin{array}{cccc}
1 & -1 & 0 & 0 \\
0 & 0 & 1 & 0 \\
0 & 0 & 0 & 0
\end{array}
$$

The general solution is $x_1 = x_2$, $x_3 = 0$, $x_2 = r$. Let $r = 1$ and we have that $\begin{bmatrix} 1 & 1 & 0 \end{bmatrix}'$ is an eigenvector. Hence let

P = [0 1 1;1 1 1;1 1 0]′

P =

$$
\begin{array}{ccc}
0 & 1 & 1 \\
1 & 1 & 1 \\
1 & 1 & 0
\end{array}
$$

then we have

A30 = P ∗ (diag([1 − 1 .5]))^30 ∗ invert(P)

A30 =

$$
\begin{array}{ccc}
1.0000 & -1.0000 & 1.0000 \\
0 & 0.0000 & 1.0000 \\
0 & 0 & 1.0000
\end{array}
$$

Since all the entries are not displayed as integers we set the format to long and redisplay the matrix to view its contents for more detail.

format long

A30

$\texttt{A30} =$

$$
\begin{array}{rrr}
1.0000000000000 & -0.99999999906868 & 0.99999999906868 \\
0 & 0.00000000093132 & 0.99999999906868 \\
0 & 0 & 1.00000000000000
\end{array}
$$

Section 8.3, p. 443

1. We show that $PP^T = I_3$:

$$
PP^T =
\begin{bmatrix}
\frac{2}{3} & -\frac{2}{3} & \frac{1}{3} \\
\frac{2}{3} & \frac{1}{3} & -\frac{2}{3} \\
\frac{1}{3} & \frac{2}{3} & \frac{2}{3}
\end{bmatrix}
\begin{bmatrix}
\frac{2}{3} & \frac{2}{3} & \frac{1}{3} \\
-\frac{2}{3} & \frac{1}{3} & \frac{2}{3} \\
\frac{1}{3} & -\frac{2}{3} & \frac{2}{3}
\end{bmatrix}
=
\begin{bmatrix}
1 & 0 & 0 \\
0 & 1 & 0 \\
0 & 0 & 1
\end{bmatrix}.
$$

3. (a) Let the columns of A be denoted

$$
\mathbf{x}_1 =
\begin{bmatrix} 1 \\ 0 \\ 0 \end{bmatrix}, \qquad
\mathbf{x}_2 =
\begin{bmatrix} 0 \\ \cos\theta \\ -\sin\theta \end{bmatrix}, \qquad
\mathbf{x}_3 =
\begin{bmatrix} 0 \\ \sin\theta \\ \cos\theta \end{bmatrix}.
$$

Then $\mathbf{x}_1 \cdot \mathbf{x}_2 = \mathbf{x}_1 \cdot \mathbf{x}_3 = \mathbf{x}_2 \cdot \mathbf{x}_3 = 0$ and $\mathbf{x}_1 \cdot \mathbf{x}_1 = \mathbf{x}_2 \cdot \mathbf{x}_2 = \mathbf{x}_3 \cdot \mathbf{x}_3 = 1$. Thus $\{\mathbf{x}_1, \mathbf{x}_2, \mathbf{x}_3\}$ is an orthonormal set.

(b) Let the columns of B be denoted by

$$
\mathbf{y}_1 =
\begin{bmatrix} 1 \\ 0 \\ 0 \end{bmatrix}, \qquad
\mathbf{y}_2 =
\begin{bmatrix} 0 \\ \frac{1}{\sqrt{2}} \\ -\frac{1}{\sqrt{2}} \end{bmatrix}, \qquad
\mathbf{y}_3 =
\begin{bmatrix} 0 \\ -\frac{1}{\sqrt{2}} \\ -\frac{1}{\sqrt{2}} \end{bmatrix}.
$$

Then $\mathbf{y}_1 \cdot \mathbf{y}_2 = \mathbf{y}_1 \cdot \mathbf{y}_3 = \mathbf{y}_2 \cdot \mathbf{y}_3 = 0$ and $\mathbf{y}_1 \cdot \mathbf{y}_1 = \mathbf{y}_2 \cdot \mathbf{y}_2 = \mathbf{y}_3 \cdot \mathbf{y}_3 = 1$. Thus $\{\mathbf{y}_1, \mathbf{y}_2, \mathbf{y}_3\}$ is an orthonormal set.

5. Let $A = \begin{bmatrix} 2 & 2 \\ 2 & 2 \end{bmatrix}$. Then

$$\det(\lambda I_2 - A) = \lambda(\lambda - 4)$$

and the eigenvalues of A are $\lambda_1 = 0$ and $\lambda_2 = 4$. To find an eigenvector $\mathbf{x}_1$ associated with $\lambda_1 = 0$, we solve $(0I_2 - A)\mathbf{x} = \mathbf{0}$:

$$\begin{bmatrix} -2 & -2 \\ -2 & -2 \end{bmatrix} \begin{bmatrix} x_1 \\ x_2 \end{bmatrix} = \begin{bmatrix} 0 \\ 0 \end{bmatrix}.$$

The reduced row echelon form of the coefficient matrix is

$$\begin{bmatrix} 1 & 1 \\ 0 & 0 \end{bmatrix}.$$

Thus we can take

$$\mathbf{x}_1 = \begin{bmatrix} 1 \\ -1 \end{bmatrix}.$$

To find an eigenvector $\mathbf{x}_2$ associated with $\lambda_2 = 4$, we solve $(4I_2 - A)\mathbf{x} = \mathbf{0}$:

$$\begin{bmatrix} 2 & -2 \\ -2 & 2 \end{bmatrix} \begin{bmatrix} x_1 \\ x_2 \end{bmatrix} = \begin{bmatrix} 0 \\ 0 \end{bmatrix}.$$

The reduced row echelon form of the coefficient matrix is

$$\begin{bmatrix} 1 & -1 \\ 0 & 0 \end{bmatrix}.$$

Thus we can take

$$\mathbf{x}_2 = \begin{bmatrix} 1 \\ 1 \end{bmatrix}.$$

We note that $\mathbf{x}_1 \cdot \mathbf{x}_2 = 0$; that is, $\mathbf{x}_1$ and $\mathbf{x}_2$ are orthogonal. To obtain an orthonormal basis of eigenvectors we normalize $\mathbf{x}_1$ and $\mathbf{x}_2$ as

$$\mathbf{z}_1 = \frac{1}{\|\mathbf{x}_1\|} \mathbf{x}_1 = \begin{bmatrix} \frac{1}{\sqrt{2}} \\ -\frac{1}{\sqrt{2}} \end{bmatrix}, \qquad \mathbf{z}_2 = \frac{1}{\|\mathbf{x}_2\|} \mathbf{x}_2 = \begin{bmatrix} \frac{1}{\sqrt{2}} \\ \frac{1}{\sqrt{2}} \end{bmatrix}.$$

Then set

$$P = \begin{bmatrix} \mathbf{z}_1 & \mathbf{z}_2 \end{bmatrix} = \begin{bmatrix} \frac{1}{\sqrt{2}} & \frac{1}{\sqrt{2}} \\ -\frac{1}{\sqrt{2}} & \frac{1}{\sqrt{2}} \end{bmatrix}$$

and hence

$$P^{-1}AP = \begin{bmatrix} 0 & 0 \\ 0 & 4 \end{bmatrix}.$$

7. Let $A = \begin{bmatrix} 0 & 0 & 0 \\ 0 & 2 & 2 \\ 0 & 2 & 2 \end{bmatrix}$. Then

$$\det(\lambda I_3 - A) = \lambda^2(\lambda - 4)$$

and the eigenvalues of A are $\lambda_1 = \lambda_2 = 0$, $\lambda_3 = 4$. To find an eigenvector $\mathbf{x}_1$ associated with $\lambda_1 = 0$, we solve $(0I_3 - A)\mathbf{x} = \mathbf{0}$:

$$\begin{bmatrix} 0 & 0 & 0 \\ 0 & -2 & -2 \\ 0 & -2 & -2 \end{bmatrix} \begin{bmatrix} x_1 \\ x_2 \\ x_3 \end{bmatrix} = \begin{bmatrix} 0 \\ 0 \\ 0 \end{bmatrix}.$$

The reduced row echelon form of the coefficient matrix is

$$\begin{bmatrix} 0 & 1 & 1 \\ 0 & 0 & 0 \\ 0 & 0 & 0 \end{bmatrix}.$$

The general solution is $x_1 = r$, $x_2 = -s$, $x_3 = s$, where r and s are any real numbers. In vector form the solution is

$$\begin{bmatrix} r \\ -s \\ s \end{bmatrix} = r \begin{bmatrix} 1 \\ 0 \\ 0 \end{bmatrix} + s \begin{bmatrix} 0 \\ -1 \\ 1 \end{bmatrix}.$$

Thus there are two linearly independent eigenvectors associated with $\lambda_1 = \lambda_2 = 0$;

$$\mathbf{x}_1 = \begin{bmatrix} 1 \\ 0 \\ 0 \end{bmatrix} \quad \text{and} \quad \mathbf{x}_2 = \begin{bmatrix} 0 \\ -1 \\ 1 \end{bmatrix}.$$

To find an eigenvector $\mathbf{x}_3$ associated with $\lambda_3 = 4$, we solve $(4I_3 - A)\mathbf{x} = \mathbf{0}$:

$$\begin{bmatrix} 4 & 0 & 0 \\ 0 & 2 & -2 \\ 0 & -2 & 2 \end{bmatrix} \begin{bmatrix} x_1 \\ x_2 \\ x_3 \end{bmatrix} = \begin{bmatrix} 0 \\ 0 \\ 0 \end{bmatrix}.$$

The reduced row echelon form of the coefficient matrix is

$$\begin{bmatrix} 1 & 0 & 0 \\ 0 & 1 & -1 \\ 0 & 0 & 0 \end{bmatrix}.$$

Thus we can take

$$\mathbf{x}_3 = \begin{bmatrix} 0 \\ 1 \\ 1 \end{bmatrix}.$$

Since A is symmetric, $\mathbf{x}_1$ and $\mathbf{x}_2$ are orthogonal to $\mathbf{x}_3$ (see Theorem 8.7). For this matrix A it is easily checked that $\mathbf{x}_1 \cdot \mathbf{x}_2 = 0$. (Note: this need not be true when the eigenvalue has multiplicity greater than 1.) Hence, $\{\mathbf{x}_1, \mathbf{x}_2, \mathbf{x}_3\}$ is orthogonal. To obtain an orthonormal basis of eigenvectors we normalize $\mathbf{x}_1$, $\mathbf{x}_2$, $\mathbf{x}_3$ to obtain

$$\mathbf{z}_1 = \frac{1}{\|\mathbf{x}_1\|} \mathbf{x}_1 = \mathbf{x}_1, \qquad \mathbf{z}_2 = \frac{1}{\|\mathbf{x}_2\|} \mathbf{x}_2 = \frac{1}{\sqrt{2}} \begin{bmatrix} 0 \\ -1 \\ 1 \end{bmatrix}, \qquad \mathbf{z}_3 = \frac{1}{\|\mathbf{x}_3\|} \mathbf{x}_3 = \frac{1}{\sqrt{2}} \begin{bmatrix} 0 \\ 1 \\ 1 \end{bmatrix}.$$

Then set $P = \begin{bmatrix} \mathbf{z}_1 & \mathbf{z}_2 & \mathbf{z}_3 \end{bmatrix}$ and hence

$$P^{-1}AP = \begin{bmatrix} 0 & 0 & 0 \\ 0 & 0 & 0 \\ 0 & 0 & 4 \end{bmatrix}.$$

9. Let $A = \begin{bmatrix} 0 & -1 & -1 \\ -1 & 0 & -1 \\ -1 & -1 & 0 \end{bmatrix}$. Then

$$\det(\lambda I_3 - A) = (\lambda + 2)(\lambda - 1)^2$$

and the eigenvalues of A are $\lambda_1 = \lambda_2 = 1$, $\lambda_3 = -2$. To find an eigenvector $\mathbf{x}_1$ associated with $\lambda_1 = 1$, we solve $(1I_3 - A)\mathbf{x} = \mathbf{0}$:

$$\begin{bmatrix} 1 & 1 & 1 \\ 1 & 1 & 1 \\ 1 & 1 & 1 \end{bmatrix} \begin{bmatrix} x_1 \\ x_2 \\ x_3 \end{bmatrix} = \begin{bmatrix} 0 \\ 0 \\ 0 \end{bmatrix}.$$

The reduced row echelon form of the coefficient matrix is

$$\begin{bmatrix} 1 & 1 & 1 \\ 0 & 0 & 0 \\ 0 & 0 & 0 \end{bmatrix}.$$

The solution is $x_1 = -r - s$, $x_2 = r$, $x_3 = s$, where r and s are any real numbers. In vector form the solution is

$$\begin{bmatrix} -r-s \\ r \\ s \end{bmatrix} = r \begin{bmatrix} -1 \\ 1 \\ 0 \end{bmatrix} + s \begin{bmatrix} -1 \\ 0 \\ 1 \end{bmatrix}.$$

Thus there are two linearly independent eigenvectors associated with $\lambda_1 = \lambda_2 = 1$;

$$\mathbf{x}_1 = \begin{bmatrix} -1 \\ 1 \\ 0 \end{bmatrix} \quad \text{and} \quad \mathbf{x}_2 = \begin{bmatrix} -1 \\ 0 \\ 1 \end{bmatrix}.$$

To find an eigenvector $\mathbf{x}_3$ associated with $\lambda_3 = -2$, we solve $(-2I_3 - A)\mathbf{x} = \mathbf{0}$:

$$\begin{bmatrix} -2 & 1 & 1 \\ 1 & -2 & 1 \\ 1 & 1 & -2 \end{bmatrix} \begin{bmatrix} x_1 \\ x_2 \\ x_3 \end{bmatrix} = \begin{bmatrix} 0 \\ 0 \\ 0 \end{bmatrix}.$$

The reduced row echelon form of the coefficient matrix is

$$\begin{bmatrix} 1 & 0 & -1 \\ 0 & 1 & -1 \\ 0 & 0 & 0 \end{bmatrix}.$$

Thus we can take

$$\mathbf{x}_3 = \begin{bmatrix} 1 \\ 1 \\ 1 \end{bmatrix}.$$

Since A is symmetric, $\mathbf{x}_1$ and $\mathbf{x}_2$ are orthogonal to $\mathbf{x}_3$ (see Theorem 8.7). For this matrix A, $\mathbf{x}_1 \cdot \mathbf{x}_2 = 1$, hence $\mathbf{x}_1$ and $\mathbf{x}_2$ are not orthogonal. We use the Gram–Schmidt process to obtain an orthogonal basis for the eigenspace associated with $\lambda_1 = 1$. Let

$$\mathbf{y}_1 = \mathbf{x}_1 = \begin{bmatrix} -1 \\ 1 \\ 0 \end{bmatrix} \quad \text{and} \quad \mathbf{y}_2 = \mathbf{x}_2 - \left(\frac{\mathbf{x}_2 \cdot \mathbf{y}_1}{\mathbf{y}_1 \cdot \mathbf{y}_1} \right) \mathbf{y}_1 = \begin{bmatrix} -\frac{1}{2} \\ -\frac{1}{2} \\ 1 \end{bmatrix}.$$

Let

$$\mathbf{y}_1' = \mathbf{y}_1 \quad \text{and} \quad \mathbf{y}_2' = 2\mathbf{y}_2 = \begin{bmatrix} -1 \\ -1 \\ 2 \end{bmatrix}.$$

The set $\{\mathbf{y}_1', \mathbf{y}_2'\}$ is an orthogonal basis for the eigenspace associated with $\lambda_1 = 1$. Then we normalize $\{\mathbf{y}_1', \mathbf{y}_2', \mathbf{x}_3\}$ to obtain

$$\mathbf{z}_1 = \frac{1}{\|\mathbf{y}_1'\|}\mathbf{y}_1' = \frac{1}{\sqrt{2}} \begin{bmatrix} -1 \\ 1 \\ 0 \end{bmatrix}, \qquad \mathbf{z}_2 = \frac{1}{\|\mathbf{y}_2'\|}\mathbf{y}_2' = \frac{1}{\sqrt{6}} \begin{bmatrix} -1 \\ -1 \\ 2 \end{bmatrix}, \qquad \mathbf{z}_3 = \frac{1}{\|\mathbf{x}_3'\|}\mathbf{x}_3' = \frac{1}{\sqrt{3}} \begin{bmatrix} 1 \\ 1 \\ 1 \end{bmatrix}.$$

Then set $P = \begin{bmatrix} \mathbf{z}_1 & \mathbf{z}_2 & \mathbf{z}_3 \end{bmatrix}$ and hence

$$P^{-1}AP = \begin{bmatrix} 1 & 0 & 0 \\ 0 & 1 & 0 \\ 0 & 0 & -2 \end{bmatrix}.$$

(The order of the columns of P and the order of the diagonal entries may vary.)

11. Let $A = \begin{bmatrix} 2 & 1 \\ 1 & 2 \end{bmatrix}$. Then

$$\det(\lambda I_2 - A) = (\lambda - 3)(\lambda - 1)$$

and the eigenvalues of A are $\lambda_1 = 3$ and $\lambda_2 = 1$. To find an eigenvector $\mathbf{x}_1$ associated with $\lambda_1 = 3$, we solve $(3I_2 - A)\mathbf{x} = \mathbf{0}$:

$$\begin{bmatrix} 1 & -1 \\ -1 & 1 \end{bmatrix} \begin{bmatrix} x_1 \\ x_2 \end{bmatrix} = \begin{bmatrix} 0 \\ 0 \end{bmatrix}.$$

The reduced row echelon form of the coefficient matrix is

$$\begin{bmatrix} 1 & -1 \\ 0 & 0 \end{bmatrix}.$$

Thus we can take

$$\mathbf{x}_1 = \begin{bmatrix} 1 \\ 1 \end{bmatrix}.$$

To find an eigenvector $\mathbf{x}_2$ associated with $\lambda_2 = 1$, we solve $(1I_2 - A)\mathbf{x} = \mathbf{0}$:

$$\begin{bmatrix} -1 & -1 \\ -1 & -1 \end{bmatrix} \begin{bmatrix} x_1 \\ x_2 \end{bmatrix} = \begin{bmatrix} 0 \\ 0 \end{bmatrix}.$$

The reduced row echelon form of the coefficient matrix is

$$\begin{bmatrix} 1 & 1 \\ 0 & 0 \end{bmatrix}.$$

Thus we can take

$$\mathbf{x}_2 = \begin{bmatrix} 1 \\ -1 \end{bmatrix}.$$

Since A is symmetric and $\lambda_1 \neq \lambda_2$, $\mathbf{x}_1$ and $\mathbf{x}_2$ are orthogonal. (See Theorem 8.7) To obtain an orthonormal basis of eigenvectors we normalize $\mathbf{x}_1$ and $\mathbf{x}_2$ as

$$\mathbf{z}_1 = \frac{1}{\|\mathbf{x}_1\|}\mathbf{x}_1 = \frac{1}{\sqrt{2}}\mathbf{x}_1 \quad \text{and} \quad \mathbf{z}_2 = \frac{1}{\|\mathbf{x}_2\|}\mathbf{x}_2 = \frac{1}{\sqrt{2}}\mathbf{x}_2.$$

Then set $P = \begin{bmatrix} \mathbf{z}_1 & \mathbf{z}_2 \end{bmatrix}$ and we have

$$P^{-1}AP = \begin{bmatrix} 3 & 0 \\ 0 & 1 \end{bmatrix}.$$

13. Let $A = \begin{bmatrix} 1 & 1 & 0 \\ 1 & 1 & 0 \\ 0 & 0 & 1 \end{bmatrix}$. Then

$$\det(\lambda I_3 - A) = (\lambda - 1)(\lambda - 2)\lambda$$

and the eigenvalues of A are $\lambda_1 = 1$, $\lambda_2 = 2$, and $\lambda_3 = 0$. To find an eigenvector $\mathbf{x}_1$ associated with $\lambda_1 = 1$, we solve $(1I_3 - A)\mathbf{x} = \mathbf{0}$:

$$\begin{bmatrix} 0 & -1 & 0 \\ -1 & 0 & 0 \\ 0 & 0 & 0 \end{bmatrix} \begin{bmatrix} x_1 \\ x_2 \\ x_3 \end{bmatrix} = \begin{bmatrix} 0 \\ 0 \\ 0 \end{bmatrix}.$$

The reduced row echelon form of the coefficient matrix is

$$\begin{bmatrix} 1 & 0 & 0 \\ 0 & 1 & 0 \\ 0 & 0 & 0 \end{bmatrix}.$$

Thus we can take

$$\mathbf{x}_1 = \begin{bmatrix} 0 \\ 0 \\ 1 \end{bmatrix}.$$

To find an eigenvector $\mathbf{x}_2$ associated with $\lambda_2 = 2$, we solve $(2I_3 - A)\mathbf{x} = \mathbf{0}$:

$$\begin{bmatrix} 1 & -1 & 0 \\ -1 & 1 & 0 \\ 0 & 0 & 1 \end{bmatrix} \begin{bmatrix} x_1 \\ x_2 \\ x_3 \end{bmatrix} = \begin{bmatrix} 0 \\ 0 \\ 0 \end{bmatrix}.$$

The reduced row echelon form of the coefficient matrix is

$$\begin{bmatrix} 1 & -1 & 0 \\ 0 & 0 & 1 \\ 0 & 0 & 0 \end{bmatrix}.$$

Thus we can take

$$\mathbf{x}_2 = \begin{bmatrix} 1 \\ 1 \\ 0 \end{bmatrix}.$$

To find an eigenvector $\mathbf{x}_3$ associated with $\lambda_3 = 0$, we solve $(0I_3 - A)\mathbf{x} = \mathbf{0}$:

$$\begin{bmatrix} -1 & -1 & 0 \\ -1 & -1 & 0 \\ 0 & 0 & -1 \end{bmatrix} \begin{bmatrix} x_1 \\ x_2 \\ x_3 \end{bmatrix} = \begin{bmatrix} 0 \\ 0 \\ 0 \end{bmatrix}.$$

The reduced row echelon form of the coefficient matrix is

$$\begin{bmatrix} 1 & 1 & 0 \\ 0 & 0 & 1 \\ 0 & 0 & 0 \end{bmatrix}.$$

Thus we can take

$$\mathbf{x}_3 = \begin{bmatrix} 1 \\ -1 \\ 0 \end{bmatrix}.$$

Since A is symmetric and the eigenvalues are distinct, Theorem 8.7 implies that $\{\mathbf{x}_1, \mathbf{x}_2, \mathbf{x}_3\}$ is orthogonal. To obtain an orthonormal basis of eigenvectors we normalize $\mathbf{x}_1$, $\mathbf{x}_2$, $\mathbf{x}_3$ to obtain

$$\mathbf{z}_1 = \frac{1}{\|\mathbf{x}_1\|}\mathbf{x}_1 = \mathbf{x}_1, \qquad \mathbf{z}_2 = \frac{1}{\|\mathbf{x}_2\|}\mathbf{x}_2 = \frac{1}{\sqrt{2}}\begin{bmatrix} 1 \\ 1 \\ 0 \end{bmatrix}, \qquad \mathbf{z}_3 = \frac{1}{\|\mathbf{x}_3\|}\mathbf{x}_3 = \frac{1}{\sqrt{2}}\begin{bmatrix} 1 \\ -1 \\ 0 \end{bmatrix}.$$

Then set $P = \begin{bmatrix} \mathbf{z}_1 & \mathbf{z}_2 & \mathbf{z}_3 \end{bmatrix}$ and hence

$$P^{-1}AP = \begin{bmatrix} 1 & 0 & 0 \\ 0 & 2 & 0 \\ 0 & 0 & 0 \end{bmatrix}.$$

15. Let $A = \begin{bmatrix} 1 & 0 & 0 \\ 0 & 1 & 1 \\ 0 & 1 & 1 \end{bmatrix}$. Then

$$\det(\lambda I_3 - A) = (\lambda - 1)\lambda(\lambda - 2)$$

and the eigenvalues of A are $\lambda_1 = 1$, $\lambda_2 = 0$, and $\lambda_3 = 2$. To find an eigenvector $\mathbf{x}_1$ associated with $\lambda_1 = 1$, we solve $(1I_3 - A)\mathbf{x} = \mathbf{0}$:

$$\begin{bmatrix} 0 & 0 & 0 \\ 0 & 0 & -1 \\ 0 & -1 & 0 \end{bmatrix}\begin{bmatrix} x_1 \\ x_2 \\ x_3 \end{bmatrix} = \begin{bmatrix} 0 \\ 0 \\ 0 \end{bmatrix}.$$

The reduced row echelon form of the coefficient matrix is

$$\begin{bmatrix} 0 & 1 & 0 \\ 0 & 0 & 1 \\ 0 & 0 & 0 \end{bmatrix}.$$

Thus we can take

$$\mathbf{x}_1 = \begin{bmatrix} 1 \\ 0 \\ 0 \end{bmatrix}.$$

To find an eigenvector $\mathbf{x}_2$ associated with $\lambda_2 = 0$, we solve $(0I_3 - A)\mathbf{x} = \mathbf{0}$:

$$\begin{bmatrix} -1 & 0 & 0 \\ 0 & -1 & -1 \\ 0 & -1 & -1 \end{bmatrix}\begin{bmatrix} x_1 \\ x_2 \\ x_3 \end{bmatrix} = \begin{bmatrix} 0 \\ 0 \\ 0 \end{bmatrix}.$$

The reduced row echelon form of the coefficient matrix is

$$\begin{bmatrix} 1 & 0 & 0 \\ 0 & 1 & 1 \\ 0 & 0 & 0 \end{bmatrix}.$$

Thus we can take

$$\mathbf{x}_2 = \begin{bmatrix} 0 \\ 1 \\ -1 \end{bmatrix}.$$

To find an eigenvector $\mathbf{x}_3$ associated with $\lambda_3 = 2$, we solve $(2I_3 - A)\mathbf{x} = \mathbf{0}$:

$$\begin{bmatrix} 1 & 0 & 0 \\ 0 & 1 & -1 \\ 0 & -1 & 1 \end{bmatrix} \begin{bmatrix} x_1 \\ x_2 \\ x_3 \end{bmatrix} = \begin{bmatrix} 0 \\ 0 \\ 0 \end{bmatrix}.$$

The reduced row echelon form of the coefficient matrix is

$$\begin{bmatrix} 1 & 0 & 0 \\ 0 & 1 & -1 \\ 0 & 0 & 0 \end{bmatrix}.$$

Thus we can take

$$\mathbf{x}_3 = \begin{bmatrix} 0 \\ 1 \\ 1 \end{bmatrix}.$$

Since A is symmetric and the eigenvalues are distinct, Theorem 8.7 implies that $\{\mathbf{x}_1, \mathbf{x}_2, \mathbf{x}_3\}$ is orthogonal. To obtain an orthonormal basis of eigenvectors we normalize $\mathbf{x}_1$, $\mathbf{x}_2$, $\mathbf{x}_3$ to obtain

$$\mathbf{z}_1 = \frac{1}{\|\mathbf{x}_1\|}\mathbf{x}_1 = \mathbf{x}_1, \qquad \mathbf{z}_2 = \frac{1}{\|\mathbf{x}_2\|}\mathbf{x}_2 = \frac{1}{\sqrt{2}}\begin{bmatrix} 0 \\ 1 \\ -1 \end{bmatrix}, \qquad \mathbf{z}_3 = \frac{1}{\|\mathbf{x}_3\|}\mathbf{x}_3 = \frac{1}{\sqrt{2}}\begin{bmatrix} 0 \\ 1 \\ 1 \end{bmatrix}.$$

Then set $P = \begin{bmatrix} \mathbf{z}_1 & \mathbf{z}_2 & \mathbf{z}_3 \end{bmatrix}$ and hence

$$P^{-1}AP = \begin{bmatrix} 1 & 0 & 0 \\ 0 & 0 & 0 \\ 0 & 0 & 2 \end{bmatrix}.$$

17. Let $A = \begin{bmatrix} 2 & 1 & 1 \\ 1 & 2 & 1 \\ 1 & 1 & 2 \end{bmatrix}$. Then

$$\det(\lambda I_3 - A) = (\lambda - 1)^2(\lambda - 4)$$

and the eigenvalues of A are $\lambda_1 = \lambda_2 = 1$, $\lambda_3 = 4$. To find an eigenvector $\mathbf{x}_1$ associated with $\lambda_1 = 1$, we solve $(1I_3 - A)\mathbf{x} = \mathbf{0}$:

$$\begin{bmatrix} -1 & -1 & -1 \\ -1 & -1 & -1 \\ -1 & -1 & -1 \end{bmatrix} \begin{bmatrix} x_1 \\ x_2 \\ x_3 \end{bmatrix} = \begin{bmatrix} 0 \\ 0 \\ 0 \end{bmatrix}.$$

The reduced row echelon form of the coefficient matrix is

$$\begin{bmatrix} 1 & 1 & 1 \\ 0 & 0 & 0 \\ 0 & 0 & 0 \end{bmatrix}.$$

The solution is $x_1 = -r - s$, $x_2 = r$, $x_3 = s$, where r and s are any real numbers. In vector form the solution is

$$\begin{bmatrix} -r - s \\ r \\ s \end{bmatrix} = r \begin{bmatrix} -1 \\ 1 \\ 0 \end{bmatrix} + s \begin{bmatrix} -1 \\ 0 \\ 1 \end{bmatrix}.$$

Thus there are two linearly independent eigenvectors,

$$\mathbf{x}_1 = \begin{bmatrix} -1 \\ 1 \\ 0 \end{bmatrix} \quad \text{and} \quad \mathbf{x}_2 = \begin{bmatrix} -1 \\ 0 \\ 1 \end{bmatrix}.$$

To find an eigenvector $\mathbf{x}_3$ associated with $\lambda_3 = 4$, we solve $(4I_3 - A)\mathbf{x} = \mathbf{0}$:

$$\begin{bmatrix} 2 & -1 & -1 \\ -1 & 2 & -1 \\ -1 & -1 & 2 \end{bmatrix} \begin{bmatrix} x_1 \\ x_2 \\ x_3 \end{bmatrix} = \begin{bmatrix} 0 \\ 0 \\ 0 \end{bmatrix}.$$

The reduced row echelon form of the coefficient matrix is

$$\begin{bmatrix} 1 & 0 & -1 \\ 0 & 1 & -1 \\ 0 & 0 & 0 \end{bmatrix}.$$

Thus we can take

$$\mathbf{x}_3 = \begin{bmatrix} 1 \\ 1 \\ 1 \end{bmatrix}.$$

Since A is symmetric $\mathbf{x}_1$ and $\mathbf{x}_2$ are orthogonal with $\mathbf{x}_3$ (see Theorem 8.7). For this matrix A, vectors $\mathbf{x}_1$ and $\mathbf{x}_2$ are not orthogonal. We use the Gram–Schmidt process to obtain an orthogonal basis for the eigenspace associated with $\lambda_1 = 1$ as follows. Let

$$\mathbf{y}_1 = \mathbf{x}_1 = \begin{bmatrix} -1 \\ 1 \\ 0 \end{bmatrix} \quad \text{and} \quad \mathbf{y}_2 = \mathbf{x}_2 - \left(\frac{\mathbf{x}_2 \cdot \mathbf{y}_1}{\mathbf{y}_1 \cdot \mathbf{y}_1} \right) \mathbf{y}_1 = \begin{bmatrix} -\frac{1}{2} \\ -\frac{1}{2} \\ 1 \end{bmatrix}.$$

Let

$$\mathbf{y}_1' = \mathbf{y}_1 \quad \text{and} \quad \mathbf{y}_2' = 2\mathbf{y}_2 = \begin{bmatrix} -1 \\ -1 \\ 2 \end{bmatrix}.$$

The set $\{\mathbf{y}_1', \mathbf{y}_2'\}$ is an orthogonal basis for the eigenspace associated with $\lambda_1 = 1$. Then we normalize $\{\mathbf{y}_1', \mathbf{y}_2', \mathbf{x}_3\}$ to obtain

$$\mathbf{z}_1 = \frac{1}{\|\mathbf{y}_1'\|} \mathbf{y}_1' = \frac{1}{\sqrt{2}} \begin{bmatrix} -1 \\ 1 \\ 0 \end{bmatrix}, \qquad \mathbf{z}_2 = \frac{1}{\|\mathbf{y}_2'\|} \mathbf{y}_2' = \frac{1}{\sqrt{6}} \begin{bmatrix} -1 \\ -1 \\ 2 \end{bmatrix}, \qquad \mathbf{z}_3 = \frac{1}{\|\mathbf{x}_3\|} \mathbf{x}_3 = \frac{1}{\sqrt{3}} \begin{bmatrix} 1 \\ 1 \\ 1 \end{bmatrix}.$$

The set $P = \begin{bmatrix} \mathbf{z}_1 & \mathbf{z}_2 & \mathbf{z}_3 \end{bmatrix}$ and hence

$$P^{-1}AP = \begin{bmatrix} 1 & 0 & 0 \\ 0 & 1 & 0 \\ 0 & 0 & 4 \end{bmatrix}.$$

T.1. In Exercise T.14 in Section 1.3, we showed that if $\mathbf{u}$ and $\mathbf{v}$ are vectors in R^n, then $\mathbf{u} \cdot \mathbf{v} = \mathbf{u}^T \mathbf{v}$. We now have

$$(A\mathbf{x}) \cdot \mathbf{y} = (A\mathbf{x})^T \mathbf{y} = \mathbf{x}^T A^T \mathbf{y} = \mathbf{x} \cdot (A^T \mathbf{y}).$$

T.3. The i, j entry of the matrix product $A^T A$ represents the ith row of A^T times the jth column of A. That is, the dot product of the ith and jth columns of A. If $A^T A = I_n$, then the dot product of the ith and jth columns of A is 1 if $i = j$ and 0 if $i \neq j$. Thus the columns of A form an orthonormal set in R^n. The converse is proved by reversing the steps in this argument.

T.5. Let

$$A = \begin{bmatrix} a & b \\ b & d \end{bmatrix}$$

be a 2×2 symmetric matrix. Then its characteristic polynomial is $\lambda^2 - (a+d)\lambda + (ad - b^2)$. The roots of this polynomial are

$$\lambda = \frac{a + d \pm \sqrt{(a+d)^2 - 4(ad - b^2)}}{2} = \frac{a + d \pm \sqrt{(a-d)^2 + 4b^2}}{2}.$$

If $b = 0$, A is already diagonal. If $b \neq 0$, the discriminant $(a - d)^2 + 4b^2$ is positive and there are two distinct real eigenvalues. Thus A is diagonalizable. By Theorem 8.4, there is a diagonalizing matrix P whose columns are linearly independent eigenvectors of A. We may assume further that those columns are unit vectors in R^2. By Theorem 8.7, the two columns are orthogonal. Thus P is an orthogonal matrix.

T.7. $(A^{-1})^T A^{-1} = (A^T)^{-1} A^{-1} = (AA^T)^{-1} = I_n^{-1} = I_n$.

T.9. For an $n \times n$ matrix A, if $A^T A \mathbf{y} = \mathbf{y} = I_n \mathbf{y}$ for all $\mathbf{y}$ in R^n, then $(A^T A - I_n)\mathbf{y} = \mathbf{0}$ for all $\mathbf{y}$ in R^n. Thus $A^T A - I_n$ is the zero matrix, so $A^T A = I_n$.

ML.1. (a) **A = [6 6;6 6];**
    [V,D] = eig(A)
    V =
        0.7071   0.7071
       −0.7071   0.7071

    D =
        0   0
        0   12
    Let $P = V$, then
    P = V;P′ * A * P
    ans =
        0   0
        0   12.0000

(b) **A = [1 2 2;2 1 2;2 2 1];**
    [V,D] = eig(A)
    V =
        0.7743   −0.2590   0.5774
       −0.6115   −0.5411   0.5774
       −0.1629   0.8001   0.5774

D =

$$\begin{array}{ccc} -1.0000 & 0 & 0 \\ 0 & -1.0000 & 0 \\ 0 & 0 & 5.0000 \end{array}$$

Let $P = V$, then

P = V;P′ * A * P

ans =

$$\begin{array}{ccc} -1.0000 & 0.0000 & 0.0000 \\ 0.0000 & -1.0000 & -0.0000 \\ 0.0000 & -0.0000 & 5.0000 \end{array}$$

(c) **A = [4 1 0;1 4 1;0 1 4];**
[V,D] = eig(A)

V =

$$\begin{array}{ccc} 0.5000 & -0.7071 & -0.5000 \\ 0.7071 & -0.0000 & 0.7071 \\ 0.5000 & 0.7071 & -0.5000 \end{array}$$

D =

$$\begin{array}{ccc} 5.4142 & 0 & 0 \\ 0 & 4.0000 & 0 \\ 0 & 0 & 2.5858 \end{array}$$

Let $P = V$, then

P = V;P′ * A * P

ans =

$$\begin{array}{ccc} 5.4142 & -0.0000 & -0.0000 \\ -0.0000 & 4.0000 & 0.0000 \\ -0.0000 & 0.0000 & 2.5858 \end{array}$$

Supplementary Exercises, p. 445

1. Let $A = \begin{bmatrix} -2 & 0 & 0 \\ 3 & 2 & 3 \\ 4 & -1 & 6 \end{bmatrix}$. The characteristic polynomial is given by

$$|\lambda I_3 - A| = \begin{vmatrix} \lambda + 2 & 0 & 0 \\ -3 & \lambda - 2 & -3 \\ -4 & 1 & \lambda - 6 \end{vmatrix} = (\lambda + 2)(\lambda - 3)(\lambda - 5).$$

Thus the eigenvalues are $\lambda_1 = -2$, $\lambda_2 = 3$, $\lambda_3 = 5$. To find an eigenvector $\mathbf{x}_1$ associated with $\lambda_1 = -2$, solve $(-2I_3 - A)\mathbf{x} = \mathbf{0}$:

$$\begin{bmatrix} 0 & 0 & 0 \\ -3 & -4 & -3 \\ -4 & 1 & -8 \end{bmatrix} \begin{bmatrix} x_1 \\ x_2 \\ x_3 \end{bmatrix} = \begin{bmatrix} 0 \\ 0 \\ 0 \end{bmatrix}.$$

The reduced row echelon form of the coefficient matrix is

$$\begin{bmatrix} 1 & 0 & \frac{35}{19} \\ 0 & 1 & -\frac{12}{19} \\ 0 & 0 & 0 \end{bmatrix}.$$

Thus the solution is $x_1 = -\frac{35}{19}r$, $x_2 = \frac{12}{19}r$, $x_3 = r$, where r is any real number. Let $r = 19$ and we have eigenvector

$$\mathbf{x}_1 = \begin{bmatrix} -35 \\ 12 \\ 19 \end{bmatrix}.$$

To find an eigenvector $\mathbf{x}_2$ associated with $\lambda_2 = 3$, solve $(3I_3 - A)\mathbf{x} = \mathbf{0}$:

$$\begin{bmatrix} 5 & 0 & 0 \\ -3 & 1 & -3 \\ -4 & 1 & -3 \end{bmatrix} \begin{bmatrix} x_1 \\ x_2 \\ x_3 \end{bmatrix} = \begin{bmatrix} 0 \\ 0 \\ 0 \end{bmatrix}.$$

The reduced row echelon form of the coefficient matrix is

$$\begin{bmatrix} 1 & 0 & 0 \\ 0 & 1 & -3 \\ 0 & 0 & 0 \end{bmatrix}.$$

Thus the solution is $x_1 = 0$, $x_2 = 3r$, $x_3 = r$, where r is any real number. Let $r = 1$ and we have eigenvector

$$\mathbf{x}_2 = \begin{bmatrix} 0 \\ 3 \\ 1 \end{bmatrix}.$$

To find an eigenvector $\mathbf{x}_3$ associated with $\lambda_3 = 5$, solve $(5I_3 - A)\mathbf{x} = \mathbf{0}$:

$$\begin{bmatrix} 7 & 0 & 0 \\ -3 & 3 & -3 \\ -4 & 1 & -1 \end{bmatrix} \begin{bmatrix} x_1 \\ x_2 \\ x_3 \end{bmatrix} = \begin{bmatrix} 0 \\ 0 \\ 0 \end{bmatrix}.$$

The reduced row echelon form of the coefficient matrix is

$$\begin{bmatrix} 1 & 0 & 0 \\ 0 & 1 & -1 \\ 0 & 0 & 0 \end{bmatrix}.$$

Thus the solution is $x_1 = 0$, $x_2 = r$, $x_3 = r$, where r is any real number. Let $r = 1$ and we have eigenvector

$$\mathbf{x}_3 = \begin{bmatrix} 0 \\ 1 \\ 1 \end{bmatrix}.$$

3. Let $A = \begin{bmatrix} 2 & 2 & 0 \\ 5 & -1 & 3 \\ 0 & 0 & 0 \end{bmatrix}$. We have

$$|\lambda I_3 - A| = \begin{vmatrix} \lambda - 2 & -2 & 0 \\ -5 & \lambda + 1 & -3 \\ 0 & 0 & \lambda \end{vmatrix} = \lambda(\lambda + 3)(\lambda - 4).$$

Thus the eigenvalues are $\lambda_1 = 0$, $\lambda_2 = -3$, and $\lambda_3 = 4$. Since the eigenvalues are real and distinct, A is diagonalizable. (See Theorem 8.5.)

5. Let $A = \begin{bmatrix} 1 & 0 & 1 \\ 0 & 1 & 0 \\ -1 & 0 & 1 \end{bmatrix}$. Then

$$|\lambda I_3 - A| = \begin{vmatrix} \lambda - 1 & 0 & -1 \\ 0 & \lambda - 1 & 0 \\ 1 & 0 & \lambda - 1 \end{vmatrix} = (\lambda - 1)(\lambda^2 - 2\lambda + 2).$$

Thus the eigenvalues are $\lambda_1 = 1$, $\lambda_2 = 1 + \sqrt{-1}$, $\lambda_3 = 1 - \sqrt{-1}$. Since not all of the roots of the characteristic polynomial are real, we say that A is not diagonalizable.

7. Let

$$A = \begin{bmatrix} 0 & 0 & 1 \\ 0 & 2 & 0 \\ 0 & 0 & 2 \end{bmatrix}.$$

Since A is upper triangular, its eigenvalues are $\lambda_1 = 0$, $\lambda_2 = \lambda_3 = 2$. To find an eigenvector associated with $\lambda_1 = 0$, solve $(0I_3 - A)\mathbf{x} = \mathbf{0}$:

$$\begin{bmatrix} 0 & 0 & -1 \\ 0 & -2 & 0 \\ 0 & 0 & -2 \end{bmatrix} \begin{bmatrix} x_1 \\ x_2 \\ x_3 \end{bmatrix} = \begin{bmatrix} 0 \\ 0 \\ 0 \end{bmatrix}.$$

The reduced row echelon form of the coefficient matrix is

$$\begin{bmatrix} 0 & 1 & 0 \\ 0 & 0 & 1 \\ 0 & 0 & 0 \end{bmatrix}.$$

The solution is $x_1 = r$, $x_2 = x_3 = 0$, where r is any real number. Let $r = 1$, then we have the eigenvector

$$\mathbf{x}_1 = \begin{bmatrix} 1 \\ 0 \\ 0 \end{bmatrix},$$

which is a basis for the eigenspace associated with $\lambda_1 = 0$. To find an eigenvector associated with $\lambda_2 = 2$, solve $(2I_3 - A)\mathbf{x} = \mathbf{0}$:

$$\begin{bmatrix} 2 & 0 & -1 \\ 0 & 0 & 0 \\ 0 & 0 & 0 \end{bmatrix} \begin{bmatrix} x_1 \\ x_2 \\ x_3 \end{bmatrix} = \begin{bmatrix} 0 \\ 0 \\ 0 \end{bmatrix}.$$

The reduced row echelon form of the coefficient matrix is

$$\begin{bmatrix} 1 & 0 & -\frac{1}{2} \\ 0 & 0 & 0 \\ 0 & 0 & 0 \end{bmatrix}.$$

The solution is $x_1 = \frac{1}{2}s$, $x_2 = r$, $x_3 = s$, where r and s are any real numbers. In vector form, the solution is

$$\begin{bmatrix} \frac{1}{2}s \\ r \\ s \end{bmatrix} = r \begin{bmatrix} 0 \\ 1 \\ 0 \end{bmatrix} + s \begin{bmatrix} \frac{1}{2} \\ 0 \\ 1 \end{bmatrix}.$$

Thus

$$\mathbf{x}_2 = \begin{bmatrix} 0 \\ 1 \\ 0 \end{bmatrix} \quad \text{and} \quad \mathbf{x}_3 = \begin{bmatrix} \frac{1}{2} \\ 0 \\ 1 \end{bmatrix}$$

are linearly independent eigenvectors associated with $\lambda_2 = 2$ and form a basis for the eigenspace. (Many other bases are possible.)

9. Let $A = \begin{bmatrix} 1 & 1 & 1 \\ 1 & 1 & 1 \\ 1 & 1 & 1 \end{bmatrix}$. We have

$$|\lambda I_3 - A| = \lambda^2 (\lambda - 3).$$

Thus the eigenvalues are $\lambda_1 = \lambda_2 = 0$ and $\lambda_3 = 3$. To find eigenvectors associated with eigenvalue λ_1, solve $(0I_3 - A)\mathbf{x} = \mathbf{0}$:

$$\begin{bmatrix} -1 & -1 & -1 \\ -1 & -1 & -1 \\ -1 & -1 & -1 \end{bmatrix} \begin{bmatrix} x_1 \\ x_2 \\ x_3 \end{bmatrix} = \begin{bmatrix} 0 \\ 0 \\ 0 \end{bmatrix}.$$

The reduced row echelon form of the coefficient matrix is

$$\begin{bmatrix} 1 & 1 & 1 \\ 0 & 0 & 0 \\ 0 & 0 & 0 \end{bmatrix}.$$

The solution is $x_1 = -r - s$, $x_2 = r$, $x_3 = s$, where r and s are any real numbers. In vector form the solution is

$$\begin{bmatrix} -r - s \\ r \\ s \end{bmatrix} = r \begin{bmatrix} -1 \\ 1 \\ 0 \end{bmatrix} + s \begin{bmatrix} -1 \\ 0 \\ 1 \end{bmatrix}.$$

Then

$$\mathbf{x}_1 = \begin{bmatrix} -1 \\ 1 \\ 0 \end{bmatrix} \quad \text{and} \quad \mathbf{x}_2 = \begin{bmatrix} -1 \\ 0 \\ 1 \end{bmatrix}$$

are linearly independent eigenvectors associated with eigenvalue $\lambda_1 = 0$. To find an eigenvector associated with $\lambda_3 = 3$, solve $(3I_3 - A)\mathbf{x} = \mathbf{0}$:

$$\begin{bmatrix} 2 & -1 & -1 \\ -1 & 2 & -1 \\ -1 & -1 & 2 \end{bmatrix} \begin{bmatrix} x_1 \\ x_2 \\ x_3 \end{bmatrix} = \begin{bmatrix} 0 \\ 0 \\ 0 \end{bmatrix}.$$

The reduced row echelon form of the coefficient matrix is

$$\begin{bmatrix} 1 & 0 & -1 \\ 0 & 1 & -1 \\ 0 & 0 & 0 \end{bmatrix}.$$

The solution is $x_1 = x_2 = x_3 = r$, where r is any real number. Let $r = 1$, then

$$\mathbf{x}_3 = \begin{bmatrix} 1 \\ 1 \\ 1 \end{bmatrix}$$

is an eigenvector associated with $\lambda_3 = 3$. Theorem 8.7 implies that $\mathbf{x}_3$ is orthogonal to both $\mathbf{x}_1$ and $\mathbf{x}_2$. However, $\mathbf{x}_1 \cdot \mathbf{x}_2 \neq 0$, thus we use the Gram–Schmidt process to determine an orthogonal pair of eigenvectors associated with $\lambda_1 = 0$. Let $\mathbf{y}_1 = \mathbf{x}_1$, and set

$$\mathbf{y}_2 = \mathbf{x}_2 - \left(\frac{\mathbf{x}_2 \cdot \mathbf{y}_1}{\mathbf{y}_1 \cdot \mathbf{y}_1} \right) \mathbf{y}_1 = \begin{bmatrix} -\frac{1}{2} \\ -\frac{1}{2} \\ 1 \end{bmatrix}.$$

To simplify the computations, let

$$\mathbf{y}_1' = \mathbf{y}_1 \quad \text{and} \quad \mathbf{y}_2' = 2\mathbf{y}_2 = \begin{bmatrix} -1 \\ -1 \\ 2 \end{bmatrix}.$$

We have that $\{\mathbf{y}_1', \mathbf{y}_2', \mathbf{x}_3\}$ is an orthogonal basis of eigenvectors. Next, we normalize these to obtain

$$\mathbf{z}_1 = \frac{1}{\|\mathbf{y}_1'\|} \mathbf{y}_1' = \begin{bmatrix} -\frac{1}{\sqrt{2}} \\ \frac{1}{\sqrt{2}} \\ 0 \end{bmatrix}, \quad \mathbf{z}_2 = \frac{1}{\|\mathbf{y}_2'\|} \mathbf{y}_2' = \begin{bmatrix} -\frac{1}{\sqrt{6}} \\ -\frac{1}{\sqrt{6}} \\ \frac{2}{\sqrt{6}} \end{bmatrix}, \quad \mathbf{z}_3 = \frac{1}{\|\mathbf{x}_3\|} \mathbf{x}_3 = \begin{bmatrix} \frac{1}{\sqrt{3}} \\ \frac{1}{\sqrt{3}} \\ \frac{1}{\sqrt{3}} \end{bmatrix}.$$

Set $P = \begin{bmatrix} \mathbf{z}_1 & \mathbf{z}_2 & \mathbf{z}_3 \end{bmatrix}$ to obtain

$$P^{-1}AP = D = \begin{bmatrix} 0 & 0 & 0 \\ 0 & 0 & 0 \\ 0 & 0 & 3 \end{bmatrix}.$$

11. If λ is an eigenvalue of A with corresponding eigenvector $\mathbf{x}$, then $A\mathbf{x} = \lambda\mathbf{x}$. Hence

$$A^2\mathbf{x} = A(A\mathbf{x}) = A(\lambda\mathbf{x}) = \lambda(A\mathbf{x}) = \lambda^2\mathbf{x}.$$

If $A^2 = A$, then $\lambda^2 = \lambda$. Thus $\lambda = 0$ or $\lambda = 1$.

T.1. Let P be a nonsingular matrix such that $P^{-1}AP = D$. Then

$$\mathrm{Tr}(D) = \mathrm{Tr}(P^{-1}AP) = \mathrm{Tr}(P^{-1}(AP)) = \mathrm{Tr}((AP)P^{-1}) = \mathrm{Tr}(APP^{-1}) = \mathrm{Tr}(AI_n) = \mathrm{Tr}(A).$$

T.3. Let P be such that $P^{-1}AP = B$.

(a) $B^T = (P^{-1}AP)^T = P^T A^T (P^{-1})^T = P^T A^T (P^T)^{-1}$; hence A^T and B^T are similar.

(b) $\mathrm{rank}(B) = \mathrm{rank}(P^{-1}AP) = \mathrm{rank}(P^{-1}A)$ (See Exercise T.6(c) in the Supplementary Exercises to Chapter 6.) $= \mathrm{rank}(A)$ (See Exercise T.6(d) in the Supplementary Exercises to Chapter 6.)

(c) $\det(B) = \det(P^{-1}AP) = \det(P^{-1})\det(A)\det(P) = (1/\det(P))\det(A)\det(P) = \det(A)$. Thus $\det(B) \neq 0$ if and only if $\det(A) \neq 0$.

(d) Since A and B are nonsingular and $B = P^{-1}AP$, $B^{-1} = (P^{-1}AP)^{-1} = P^{-1}A^{-1}P$. That is, A^{-1} and B^{-1} are similar.

(e) $\mathrm{Tr}(B) = \mathrm{Tr}(P^{-1}AP) = \mathrm{Tr}((P^{-1}A)P) = \mathrm{Tr}(P(P^{-1}A)) = \mathrm{Tr}(A)$. (See Supplementary Exercise T.1 in Chapter 1.)

T.5. $(cA)^T = (cA)^{-1}$ if and only if $cA^T = \frac{1}{c}A^{-1} = \frac{1}{c}A^T$. That is, $c = \frac{1}{c}$. Hence $c = \pm 1$.

T.7. (a) The characteristic polynomial is

$$f(\lambda) = \begin{vmatrix} 1 - \lambda & 2 & 3 \\ 2 & -1 - \lambda & 5 \\ 3 & 2 & 1 - \lambda \end{vmatrix} = -\lambda^3 + \lambda^2 + 24\lambda + 36.$$

(b) The characteristic polynomial is $f(\lambda) = (1 - \lambda)(2 - \lambda)(-3 - \lambda)$.

(c) The characteristic polynomial is

$$f(\lambda) = \begin{vmatrix} 3 - \lambda & 3 \\ 2 & 4 - \lambda \end{vmatrix} = \lambda^2 - 7\lambda + 6.$$

Chapter 9

Applications of Eigenvectors and Eigenvalues (Optional)

Section 9.1, p. 450

1. Let $A = \begin{bmatrix} 1 & 1 \\ 1 & 0 \end{bmatrix}$. The characteristic polynomial is

$$|\lambda I_2 - A| = \begin{vmatrix} \lambda - 1 & -1 \\ -1 & \lambda \end{vmatrix} = \lambda^2 - \lambda - 1.$$

Using the quadratic equation we find that the roots of $\lambda^2 - \lambda - 1 = 0$ are

$$\lambda_1 = \frac{1 + \sqrt{5}}{2} \quad \text{and} \quad \lambda_2 = \frac{1 - \sqrt{5}}{2}.$$

We find the corresponding eigenvectors as follows:

Case $\lambda = \lambda_1$. Solve

$$(\lambda_1 I_2 - A)\mathbf{x} = \begin{bmatrix} \frac{-1+\sqrt{5}}{2} & -1 \\ -1 & \frac{1+\sqrt{5}}{2} \end{bmatrix} \begin{bmatrix} x_1 \\ x_2 \end{bmatrix} = \mathbf{0}.$$

Row reduce the coefficient matrix to obtain

$$\begin{bmatrix} 1 & -\frac{1+\sqrt{5}}{2} \\ 0 & 0 \end{bmatrix}.$$

The solution is

$$x_1 = \frac{1 + \sqrt{5}}{2} r, \qquad x_2 = r,$$

where r is any real number. Let $r = 1$, then we have eigenvector

$$\mathbf{x}_1 = \begin{bmatrix} \frac{1+\sqrt{5}}{2} \\ 1 \end{bmatrix}.$$

Case $\lambda = \lambda_2$. Solve

$$(\lambda_2 I_2 - A)\mathbf{x} = \begin{bmatrix} \frac{-1-\sqrt{5}}{2} & -1 \\ -1 & \frac{1-\sqrt{5}}{2} \end{bmatrix} \begin{bmatrix} x_1 \\ x_2 \end{bmatrix} = \mathbf{0}.$$

Row reduce the coefficient matrix to obtain

$$\begin{bmatrix} 1 & -\frac{1-\sqrt{5}}{2} \\ 0 & 0 \end{bmatrix}.$$

The solution is

$$x_1 = \frac{1 - \sqrt{5}}{2} r \qquad x_2 = r,$$

where r is any real number. Let $r = 1$, then we have eigenvector

$$\mathbf{x}_2 = \begin{bmatrix} \frac{1-\sqrt{5}}{2} \\ 1 \end{bmatrix}.$$

3. The Fibonacci sequence is computed using the recursion relation

$$u_n = u_{n-1} + u_{n-2}, \quad n \geq 2, \quad u_0 = u_1 = 1.$$

From the discussion in the text, we have $u_4 = 5$ and $u_5 = 8$.

(a) $u_6 = u_5 + u_4 = 13$
$u_7 = u_6 + u_5 = 21$
$u_8 = u_7 + u_6 = 34.$

(b) $u_9 = u_8 + u_7 = 55$
$u_{10} = u_9 + u_8 = 89$
$u_{11} = u_{10} + u_9 = 144$
$u_{12} = u_{11} + u_{10} = 233.$

(c) $u_{13} = 377, u_{14} = 610, u_{15} = 987, u_{16} = 1597,$
$u_{17} = 2584, u_{18} = 4181, u_{19} = 6765, u_{20} = 10{,}946.$

T.1. Let us define u_{-1} to be 0. Then for $n = 0$,

$$A^1 = A = \begin{bmatrix} 1 & 1 \\ 1 & 0 \end{bmatrix} = \begin{bmatrix} u_1 & u_0 \\ u_0 & u_{-1} \end{bmatrix}$$

and, for $n = 1$,

$$A^2 = \begin{bmatrix} 2 & 1 \\ 1 & 1 \end{bmatrix} = \begin{bmatrix} u_2 & u_1 \\ u_1 & u_0 \end{bmatrix}.$$

Suppose that the formula

$$A^{n+1} = \begin{bmatrix} u_{n+1} & u_n \\ u_n & u_{n-1} \end{bmatrix} \tag{9.1}$$

holds for values up to and including n, $n \geq 1$. Then

$$A^{n+2} = A \cdot A^{n+1} = \begin{bmatrix} 1 & 1 \\ 1 & 0 \end{bmatrix} \begin{bmatrix} u_{n+1} & u_n \\ u_n & u_{n-1} \end{bmatrix}$$

$$= \begin{bmatrix} u_{n+1} + u_n & u_n + u_{n-1} \\ u_{n+1} & u_n \end{bmatrix} = \begin{bmatrix} u_{n+2} & u_{n+1} \\ u_{n+1} & u_n \end{bmatrix}.$$

Thus the formula (9.1) also holds for $n + 1$, so it holds for all natural numbers n. Using (9.1), we see that

$$u_{n+1} u_{n-1} - u_n^2 = \begin{vmatrix} u_{n+1} & u_n \\ u_n & u_{n-1} \end{vmatrix} = \det\left(A^{n+1}\right) = (\det(A))^{n+1} = (-1)^{n+1}.$$

Section 9.2, p. 460

1. Follow the steps in Example 2.

    (a) Since the system is diagonal,

    $$x_1 = b_1 e^{-3t}, \qquad x_2 = b_2 e^{4t}, \qquad x_3 = b_3 e^{2t},$$

    where b_1, b_2, b_3 are arbitrary constants. Thus the solution is

    $$\mathbf{x}(t) = \begin{bmatrix} b_1 e^{-3t} \\ b_2 e^{4t} \\ b_3 e^{2t} \end{bmatrix} = b_1 \begin{bmatrix} 1 \\ 0 \\ 0 \end{bmatrix} e^{-3t} + b_2 \begin{bmatrix} 0 \\ 1 \\ 0 \end{bmatrix} e^{4t} + b_3 \begin{bmatrix} 0 \\ 0 \\ 1 \end{bmatrix} e^{2t}.$$

    (b) For the initial conditions $x_1(0) = 3$, $x_2(0) = 4$, $x_3(0) = 5$ we have

    $$b_1 = 3, \qquad b_2 = 4, \qquad b_3 = 5.$$

    (Substitute $t = 0$ into the expressions for x_1, x_2, x_3.) The solution of the initial value problem is

    $$\mathbf{x}(t) = 3 \begin{bmatrix} 1 \\ 0 \\ 0 \end{bmatrix} e^{-3t} + 4 \begin{bmatrix} 0 \\ 1 \\ 0 \end{bmatrix} e^{4t} + 5 \begin{bmatrix} 0 \\ 0 \\ 1 \end{bmatrix} e^{2t}.$$

3. Let

$$A = \begin{bmatrix} 4 & 0 & 0 \\ 3 & -5 & 0 \\ 2 & 1 & 2 \end{bmatrix}.$$

Since A is lower triangular, its eigenvalues are the diagonal entries: $\lambda_1 = 4$, $\lambda_2 = -5$, $\lambda_3 = 2$. We find the corresponding eigenvectors by solving the linear systems $(\lambda_i I_3 - A)\mathbf{x} = \mathbf{0}$ with $i = 1, 2, 3$. We have:

For $\lambda_1 = 4$ the solution is $x_1 = \frac{6}{7}r$, $x_2 = \frac{2}{7}r$, and $x_3 = r$. Set $r = 7$ and we have the eigenvector

$$\mathbf{x}_1 = \begin{bmatrix} 6 \\ 2 \\ 7 \end{bmatrix}.$$

For $\lambda_2 = -5$ the solution is $x_1 = 0$, $x_2 = -7r$, and $x_3 = r$. Set $r = -1$ and we have the eigenvector

$$\mathbf{x}_2 = \begin{bmatrix} 0 \\ 7 \\ -1 \end{bmatrix}.$$

For $\lambda_3 = 2$ the solution is $x_1 = 0$, $x_2 = 0$, and $x_3 = r$. Set $r = 1$ and we have the eigenvector

$$\mathbf{x}_3 = \begin{bmatrix} 0 \\ 0 \\ 1 \end{bmatrix}.$$

From Equation (17), the solution is given by

$$\mathbf{x}(t) = b_1 \begin{bmatrix} 6 \\ 2 \\ 7 \end{bmatrix} e^{4t} + b_2 \begin{bmatrix} 0 \\ 7 \\ -1 \end{bmatrix} e^{-5t} + b_3 \begin{bmatrix} 0 \\ 0 \\ 1 \end{bmatrix} e^{2t}.$$

5. Let

$$A = \begin{bmatrix} 5 & 0 & 0 \\ 0 & -4 & 3 \\ 0 & 3 & 4 \end{bmatrix}.$$

We find the eigenvalues and associated eigenvectors of A. The characteristic polynomial is

$$|\lambda I_3 - A| = (\lambda - 5)^2 (\lambda + 5).$$

Hence the eigenvalues are $\lambda_1 = \lambda_2 = 5$, $\lambda_3 = -5$. Find associated eigenvectors:

Case $\lambda_1 = 5$. Solve the linear system

$$(5I_3 - A)\mathbf{x} = \begin{bmatrix} 0 & 0 & 0 \\ 0 & 9 & -3 \\ 0 & -3 & 1 \end{bmatrix} \begin{bmatrix} x_1 \\ x_2 \\ x_3 \end{bmatrix} = \mathbf{0}.$$

Row reduce the coefficient matrix to obtain

$$\begin{bmatrix} 0 & 1 & -\frac{1}{3} \\ 0 & 0 & 0 \\ 0 & 0 & 0 \end{bmatrix}.$$

The solution is $x_1 = r$, $x_2 = \frac{1}{3}s$, $x_3 = s$, where r and s are any real numbers. To find a pair of linearly independent eigenvectors, set $r = 1$ and $s = 0$; then set $r = 0$ and $s = 3$. This gives

$$\mathbf{x}_1 = \begin{bmatrix} 1 \\ 0 \\ 0 \end{bmatrix} \quad \text{and} \quad \mathbf{x}_2 = \begin{bmatrix} 0 \\ 1 \\ 3 \end{bmatrix}.$$

Case $\lambda_3 = -5$. Solve the linear system

$$(-5I_3 - A)\mathbf{x} = \begin{bmatrix} -10 & 0 & 0 \\ 0 & -1 & -3 \\ 0 & -3 & -9 \end{bmatrix} \begin{bmatrix} x_1 \\ x_2 \\ x_3 \end{bmatrix} = \mathbf{0}.$$

Row reduce the coefficient matrix to obtain

$$\begin{bmatrix} 1 & 0 & 0 \\ 0 & 1 & 3 \\ 0 & 0 & 0 \end{bmatrix}.$$

The solution is $x_1 = 0$, $x_2 = -3r$, $x_3 = r$, where r is any real number. Set $r = 1$ and we have eigenvector

$$\mathbf{x}_3 = \begin{bmatrix} 0 \\ -3 \\ 1 \end{bmatrix}.$$

The solution is

$$\mathbf{x}(t) = b_1 \begin{bmatrix} 1 \\ 0 \\ 0 \end{bmatrix} e^{5t} + b_2 \begin{bmatrix} 0 \\ 1 \\ 3 \end{bmatrix} e^{5t} + b_3 \begin{bmatrix} 0 \\ -3 \\ 1 \end{bmatrix} e^{-5t}.$$

7. Let

$$A = \begin{bmatrix} 1 & 2 & 3 \\ 0 & 1 & 0 \\ 2 & 1 & 2 \end{bmatrix}.$$

We find the eigenvalues and associated eigenvectors of A. The characteristic polynomial is

$$|\lambda I_3 - A| = (\lambda - 4)(\lambda + 1)(\lambda - 1).$$

Hence the eigenvalues are $\lambda_1 = 4$, $\lambda_2 = -1$, $\lambda_3 = 1$. Find associated eigenvectors:

Case $\lambda_1 = 4$. Solve the linear system

$$(4I_3 - A)\mathbf{x} = \begin{bmatrix} 3 & -2 & -3 \\ 0 & 3 & 0 \\ -2 & -1 & 2 \end{bmatrix} \begin{bmatrix} x_1 \\ x_2 \\ x_3 \end{bmatrix} = \mathbf{0}.$$

Row reduce the coefficient matrix to obtain

$$\begin{bmatrix} 1 & 0 & -1 \\ 0 & 1 & 0 \\ 0 & 0 & 0 \end{bmatrix}.$$

The solution is $x_1 = r$, $x_2 = 0$, $x_3 = r$, where r is any real number. Set $r = 1$. This gives eigenvector

$$\mathbf{x}_1 = \begin{bmatrix} 1 \\ 0 \\ 1 \end{bmatrix}.$$

Case $\lambda_2 = -1$. Solve the linear system

$$(-1I_3 - A)\,\mathbf{x} = \begin{bmatrix} -2 & -2 & -3 \\ 0 & -2 & 0 \\ -2 & -1 & -3 \end{bmatrix} \begin{bmatrix} x_1 \\ x_2 \\ x_3 \end{bmatrix} = \mathbf{0}.$$

Row reduce the coefficient matrix to obtain

$$\begin{bmatrix} 1 & 0 & \frac{3}{2} \\ 0 & 1 & 0 \\ 0 & 0 & 0 \end{bmatrix}.$$

The solution is $x_1 = -\frac{3}{2}r$, $x_2 = 0$, $x_3 = r$, where r is any real number. Set $r = 2$ and we have eigenvector

$$\mathbf{x}_2 = \begin{bmatrix} -3 \\ 0 \\ 2 \end{bmatrix}.$$

Case $\lambda_2 = 1$. Solve the linear system

$$(1I_3 - A)\mathbf{x} = \begin{bmatrix} 0 & -2 & -3 \\ 0 & 0 & 0 \\ -2 & -1 & -1 \end{bmatrix} \begin{bmatrix} x_1 \\ x_2 \\ x_3 \end{bmatrix} = \mathbf{0}.$$

Row reduce the coefficient matrix to obtain

$$\begin{bmatrix} 1 & 0 & -\frac{1}{4} \\ 0 & 1 & \frac{3}{2} \\ 0 & 0 & 0 \end{bmatrix}.$$

The solution is $x_1 = \frac{1}{4}r$, $x_2 = -\frac{3}{2}r$, $x_3 = r$, where r is any real number. Set $r = 4$ and we have eigenvector

$$\mathbf{x}_3 = \begin{bmatrix} 1 \\ -6 \\ 4 \end{bmatrix}.$$

The solution is

$$\mathbf{x}(t) = b_1 \begin{bmatrix} 1 \\ 0 \\ 1 \end{bmatrix} e^{4t} + b_2 \begin{bmatrix} -3 \\ 0 \\ 2 \end{bmatrix} e^{-t} + b_3 \begin{bmatrix} 1 \\ -6 \\ 4 \end{bmatrix} e^t.$$

9. Rewrite the system of differential equations in matrix form as

$$\begin{bmatrix} x_1'(t) \\ x_2'(t) \end{bmatrix} = \begin{bmatrix} -3 & 6 \\ 1 & -2 \end{bmatrix} \begin{bmatrix} x_1(t) \\ x_2(t) \end{bmatrix}.$$

Let

$$A = \begin{bmatrix} -3 & 6 \\ 1 & -2 \end{bmatrix}.$$

Find the eigenvalues and associated eigenvectors of A. The characteristic polynomial is

$$|\lambda I_2 - A| = \lambda(\lambda + 5).$$

The eigenvalues are $\lambda_1 = 0$ and $\lambda_2 = -5$. Find the associated eigenvectors

Case $\lambda_1 = 0$. Solve the linear system

$$(0I_2 - A)\mathbf{x} = \begin{bmatrix} 3 & -6 \\ -1 & 2 \end{bmatrix} \begin{bmatrix} x_1 \\ x_2 \end{bmatrix} = \mathbf{0}.$$

Row reduce the coefficient matrix to obtain

$$\begin{bmatrix} 1 & -2 \\ 0 & 0 \end{bmatrix}.$$

The solution is $x_1 = 2r$, $x_2 = r$, where r is any real number. Set $r = 1$ and we have eigenvector

$$\mathbf{x}_1 = \begin{bmatrix} 2 \\ 1 \end{bmatrix}.$$

Case $\lambda_2 = -5$. Solve the linear system

$$(-5I_2 - A)\,\mathbf{x} = \begin{bmatrix} -2 & -6 \\ -1 & -3 \end{bmatrix} \begin{bmatrix} x_1 \\ x_2 \end{bmatrix} = \mathbf{0}.$$

Row reduce the coefficient matrix to obtain

$$\begin{bmatrix} 1 & 3 \\ 0 & 0 \end{bmatrix}.$$

The solution is $x_1 = -3r$, $x_2 = r$, where r is any real number. Set $r = -1$ and we have eigenvector

$$\mathbf{x}_2 = \begin{bmatrix} 3 \\ -1 \end{bmatrix}.$$

The solution is

$$\mathbf{x}(t) = b_1 \begin{bmatrix} 2 \\ 1 \end{bmatrix} e^{0t} + b_2 \begin{bmatrix} 3 \\ -1 \end{bmatrix} e^{-5t} = b_1 \begin{bmatrix} 2 \\ 1 \end{bmatrix} + b_2 \begin{bmatrix} 3 \\ -1 \end{bmatrix} e^{-5t}.$$

Using the initial conditions $x_1(0) = 500$ and $x_2(0) = 200$, we have in matrix form

$$\mathbf{x}(0) = \begin{bmatrix} 500 \\ 200 \end{bmatrix}.$$

Set $t = 0$ in the solution and solve for b_1 and b_2. We have

$$\mathbf{x}(0) = \begin{bmatrix} 500 \\ 200 \end{bmatrix} = b_1 \begin{bmatrix} 2 \\ 1 \end{bmatrix} + b_2 \begin{bmatrix} 3 \\ -1 \end{bmatrix} = \begin{bmatrix} 2 & 3 \\ 1 & -1 \end{bmatrix} \begin{bmatrix} b_1 \\ b_2 \end{bmatrix} = A \begin{bmatrix} b_1 \\ b_2 \end{bmatrix}.$$

Solving this linear system we get $b_1 = 220$ and $b_2 = 20$. The solution of the initial value problem and the populations at time t are given by

$$\mathbf{x}(t) = 220 \begin{bmatrix} 2 \\ 1 \end{bmatrix} + 20 \begin{bmatrix} 3 \\ -1 \end{bmatrix} e^{-5t}.$$

T.1. Let $\mathbf{x}_1$ and $\mathbf{x}_2$ be solutions to the equation $\mathbf{x}' = A\mathbf{x}$, and let a and b be scalars. Then

$$\frac{d}{dt}(a\mathbf{x}_1 + b\mathbf{x}_2) = a\mathbf{x}_1' + b\mathbf{x}_2' = aA\mathbf{x}_1 + bA\mathbf{x}_2 = A(a\mathbf{x}_1 + b\mathbf{x}_2).$$

Thus $a\mathbf{x}_1 + b\mathbf{x}_2$ is also a solution to the given equation.

ML.1. **A = [1 1 0;0 0 1;8 − 14 7];**
[v,d] = eig(A)

v =

        −0.5774   0.2182   0.0605
        −0.5774   0.4364   0.2421
        −0.5774   0.8729   0.9684

d =

     1.0000   0   0
          0   2.0000   0
          0   0   4.0000

The general solution is given by

$$\mathbf{x}(t) = b_1 \begin{bmatrix} -0.5774 \\ -0.5774 \\ -0.5774 \end{bmatrix} e^t + b_2 \begin{bmatrix} 0.2182 \\ 0.4364 \\ 0.8729 \end{bmatrix} e^{2t} + b_3 \begin{bmatrix} 0.0605 \\ 0.2421 \\ 0.9684 \end{bmatrix} e^{4t}.$$

ML.3. $\mathbf{A} = [1 \ \ 2 \ \ 3; 0 \ \ 1 \ \ 0; 2 \ \ 1 \ \ 2];$

$[\mathbf{v}, \mathbf{d}] = \mathbf{eig}(\mathbf{A})$

$\mathbf{v} =$

−0.8321	−0.7071	−0.1374
0	0	0.8242
0.5547	−0.7071	−0.5494

$\mathbf{d} =$

−1	0	0
0	4	0
0	0	1

The general solution is given by

$$\mathbf{x}(t) = b_1 \begin{bmatrix} -0.8321 \\ 0 \\ 0.5547 \end{bmatrix} e^{-t} + b_2 \begin{bmatrix} -0.7071 \\ 0 \\ -0.7071 \end{bmatrix} e^{4t} + b_3 \begin{bmatrix} -0.1374 \\ 0.8242 \\ -0.5494 \end{bmatrix} e^t.$$

Section 9.3, p. 474

1. The eigenvalues of the coefficient matrix are $\lambda_1 = -1$ and $\lambda_2 = -3$ with associated eigenvectors

$$\mathbf{p}_1 = \begin{bmatrix} 1 \\ 0 \end{bmatrix} \quad \text{and} \quad \mathbf{p}_2 = \begin{bmatrix} 0 \\ 1 \end{bmatrix}.$$

Thus the origin is a stable equilibrium. The phase portrait shows all trajectories tending toward the origin.

3. The eigenvalues of the coefficient matrix are $\lambda_1 = -1$ and $\lambda_2 = -1$ with one linearly independent eigenvector $\mathbf{p}_1 = \begin{bmatrix} 1 \\ 0 \end{bmatrix}$. Thus the origin is a stable equilibrium. The phase portrait shows all trajectories tending toward the origin with those passing through points not on the eigenvector aligning themselves to be tangent to the eigenvector at the origin.

5. The eigenvalues of the coefficient matrix are $\lambda_1 = 2$ and $\lambda_2 = -2$ with associated eigenvectors

$$\mathbf{p}_1 = \begin{bmatrix} 1 \\ 1 \end{bmatrix} \quad \text{and} \quad \mathbf{p}_2 = \begin{bmatrix} -1 \\ 3 \end{bmatrix}.$$

Thus the origin is a saddle point. The phase portrait shows trajectories not in the direction of an eigenvector heading towards the origin, but bending away as $t \to \infty$.

7. The eigenvalues of the coefficient matrix are $\lambda_1 = -1$ and $\lambda_2 = -4$ with one linearly independent eigenvector

$$\mathbf{p}_1 = \begin{bmatrix} 1 \\ 1 \end{bmatrix} \quad \text{and} \quad \mathbf{p}_2 = \begin{bmatrix} -1 \\ 2 \end{bmatrix}.$$

Thus the origin is a stable equilibrium. The phase portrait shows all trajectories tending towards the origin.

9. The eigenvalues of the coefficient matrix are $\lambda_1 = 2i$ and $\lambda_2 = -2i$ with associated eigenvectors

$$\mathbf{p}_1 = \begin{bmatrix} -0.5345 + 0.8018i \\ 0.2673i \end{bmatrix} \quad \text{and} \quad \mathbf{p}_2 = \begin{bmatrix} -0.5345 - 0.8018i \\ -0.2673i \end{bmatrix}.$$

Since the eigenvalues have real part zero, the trajectories are ellipses whose major and minor axes are determined by the eigenvectors. The origin is called marginally stable in this case.

T.1.   (a) Let $x_1 = x$ and $x_2 = x'_1$. Then the differential equation $x'' + 2rx' + \dfrac{k}{m}x = 0$ can be written in the system form

$$x'_1 = x_2$$
$$x'_2 = x'' = -2rx' - \frac{k}{m}x = -\frac{k}{m}x_1 - 2rx_2.$$

In matrix form, this becomes

$$\mathbf{x}'(t) = \begin{bmatrix} x'_1 \\ x'_2 \end{bmatrix} = \begin{bmatrix} 0 & 1 \\ -\dfrac{k}{m} & -2r \end{bmatrix} \begin{bmatrix} x_1 \\ x_2 \end{bmatrix}.$$

(b) Here $m = 1$ and $r = 1$. The matrix A is

$$A = \begin{bmatrix} 0 & 1 \\ -k & -2 \end{bmatrix}.$$

(i)   If $k = 0.75$, we have

$$A = \begin{bmatrix} 0 & 1 \\ -0.75 & -2 \end{bmatrix}.$$

To find the eigenvalues of A, set

$$f(\lambda) = |A - \lambda I_2| = \begin{vmatrix} -\lambda & 1 \\ -0.75 & -2 - \lambda \end{vmatrix} = \lambda^2 + 2\lambda + 0.75 = 0.$$

Then
$$\lambda = -1.5, -0.5.$$

Thus there are two distinct negative eigenvalues. Hence all trajectories tend toward the equilibrium point at the origin.

(ii) If $k = 1$, we find that

$$f(\lambda) = \begin{vmatrix} -\lambda & 1 \\ -1 & -2 - \lambda \end{vmatrix} = \lambda^2 + 2\lambda + 1 = 0.$$

In this case $\lambda_1 = \lambda_2 = -1$. The eigenvalues are equal and negative, so the trajectories tend to a stable equilibrium point at the origin.

(iii) If $k = 2$, we find that

$$f(\lambda) = \begin{vmatrix} -\lambda & 1 \\ -2 & -2 - \lambda \end{vmatrix} = \lambda^2 + 2\lambda + 2 = 0.$$

In this case $\lambda = -1 \pm i$. Thus the eigenvalues are complex with negative real parts. Since the real part is negative, the trajectories are spirals that tend inward toward the origin. The origin is a stable equilibrium point.

(iv) If $k = 10$, we find that

$$f(\lambda) = \begin{vmatrix} -\lambda & 1 \\ -10 & -2 - \lambda \end{vmatrix} = \lambda^2 + 2\lambda + 10 = 0.$$

In this case $\lambda = -1 \pm 3i$. Thus the eigenvalues are complex with negative real parts. Since the real part is negative, the trajectories are spirals that tend inward toward the origin. The origin is a stable equilibrium point.

(c) For $m = 1$ and $r = 1$, $A = \begin{bmatrix} 0 & 1 \\ -k & -2 \end{bmatrix}$. Its characteristic polynomial is $\lambda^2 + 2\lambda + k$ and its eigenvalues are $\lambda = -1 \pm \sqrt{1 - k}$. The eigenvalues are complex for $k > 1$. (Recall that $k > 0$).

(d) Here $m = 1$ and $k = 1$, so

$$A = \begin{bmatrix} 0 & 1 \\ -1 & -2r \end{bmatrix}.$$

(i) If $r = 0$, $A = \begin{bmatrix} 0 & 1 \\ -1 & 0 \end{bmatrix}$. In this case the characteristic polynomial is $\lambda^2 + 1$, so the eigenvalues are $\pm i$. Thus, the trajectories are ellipses.

(ii) If $r = \frac{1}{2}$, $A = \begin{bmatrix} 0 & 1 \\ -1 & -1 \end{bmatrix}$. In this case the characteristic polynomial is $\lambda^2 + \lambda + 1$ and the eigenvalues are $-\frac{1}{2} \pm i\frac{\sqrt{3}}{2}$. The trajectories are spirals. Since the real part of the eigenvlaues are negative, the spirals goes inward toward the origin, which is a stable equilibrium point.

(iii) If $r = 1$, $A = \begin{bmatrix} 0 & 1 \\ -1 & -2 \end{bmatrix}$. In this case the characteristic polynomial is $\lambda^2 + 2\lambda + 1$, so the eigenvalues are $\lambda_1 = \lambda_2 = -1$. The trajectories tend to the stable equilibrium point at the origin.

(iv) If $r = \sqrt{2}$, $A = \begin{bmatrix} 0 & 1 \\ -1 & -2\sqrt{2} \end{bmatrix}$. In this case the characteristic polynomial is $\lambda^2 + 2\sqrt{2}\lambda + 1$, so the eigenvalues are $-\sqrt{2} \pm 1$. Both eigenvalues have negative real part so the trajectories tend toward the stable equilibrium point at the origin.

(e) The rider would experience an oscillatory up and down motion.

(f) Here $m = 1$ and $k = 1$ so $A = \begin{bmatrix} 0 & 1 \\ -1 & -2r \end{bmatrix}$. The characteristic polynomial is $\lambda^2 + 2r\lambda + 1$ so the eigenvlaues are $\lambda = -r \pm \sqrt{r^2 - 1}$. The eigenvalues are real when $r \geq 1$. (Note: $r \geq 0$).

Section 9.4, p. 483

1. We use Equations (1) and (2) along with Examples 1 and 2.

(a) $-3x^2 + 5xy - 2y^2 = \begin{bmatrix} x & y \end{bmatrix} \begin{bmatrix} -3 & \frac{5}{2} \\ \frac{5}{2} & -2 \end{bmatrix} \begin{bmatrix} x \\ y \end{bmatrix}$.

(b) $2x_1^2 + 3x_1x_2 - 5x_1x_3 + 7x_2x_3 = \begin{bmatrix} x_1 & x_2 & x_3 \end{bmatrix} \begin{bmatrix} 2 & \frac{3}{2} & -\frac{5}{2} \\ \frac{3}{2} & 0 & \frac{7}{2} \\ -\frac{5}{2} & \frac{7}{2} & 0 \end{bmatrix} \begin{bmatrix} x_1 \\ x_2 \\ x_3 \end{bmatrix}$.

(c) $3x_1^2 + x_2^2 - 2x_3^2 + x_1x_2 - x_1x_3 - 4x_2x_3 = \begin{bmatrix} x_1 & x_2 & x_3 \end{bmatrix} \begin{bmatrix} 3 & \frac{1}{2} & -\frac{1}{2} \\ \frac{1}{2} & 1 & -2 \\ -\frac{1}{2} & -2 & -2 \end{bmatrix} \begin{bmatrix} x_1 \\ x_2 \\ x_3 \end{bmatrix}$.

3. Since each of the matrices A is symmetric, we use Theorem 9.2. Hence the diagonal matrix D we seek has the eigenvalues of A as diagonal entries. Thus we compute the eigenvalues of A and form matrix D.

(a) Let $A = \begin{bmatrix} -1 & 0 & 0 \\ 0 & 1 & 1 \\ 0 & 1 & 1 \end{bmatrix}$. The characteristic polynomial of A is

$$\det(\lambda I_3 - A) = \begin{vmatrix} \lambda + 1 & 0 & 0 \\ 0 & \lambda - 1 & -1 \\ 0 & -1 & \lambda - 1 \end{vmatrix} = (\lambda + 1)\left[(\lambda - 1)^2 - 1\right] = \lambda(\lambda - 2)(\lambda + 1).$$

Hence the eigenvalues of A are $\lambda_1 = 0$, $\lambda_2 = 2$, and $\lambda_3 = -1$. Thus a diagonal matrix D congruent to A is

$$D = \begin{bmatrix} 0 & 0 & 0 \\ 0 & 2 & 0 \\ 0 & 0 & -1 \end{bmatrix}.$$

There is more than one diagonal matrix D congruent to A. Others are found by reordering the eigenvalues of A on the diagonal.

(b) Let $A = \begin{bmatrix} 1 & 1 & 1 \\ 1 & 1 & 1 \\ 1 & 1 & 1 \end{bmatrix}$. The characteristic polynomial of A is

$$\det(\lambda I_3 - A) = \begin{vmatrix} \lambda - 1 & -1 & -1 \\ -1 & \lambda - 1 & -1 \\ -1 & -1 & \lambda - 1 \end{vmatrix} = (\lambda - 1)^3 - 1 - 1 - 3(\lambda - 1)$$

$$= \lambda^3 - 3\lambda^2 = \lambda^2(\lambda - 3).$$

Hence the eigenvalues of A are $\lambda_1 = 0$, $\lambda_2 = 0$, and $\lambda_3 = 3$. Thus a diagonal matrix D congruent to A is

$$D = \begin{bmatrix} 0 & 0 & 0 \\ 0 & 0 & 0 \\ 0 & 0 & 3 \end{bmatrix}.$$

There is more than one diagonal matrix D congruent to A. Others are found by reordering the eigenvalues of A on the diagonal.

(c) Let $A = \begin{bmatrix} 0 & 2 & 2 \\ 2 & 0 & 2 \\ 2 & 2 & 0 \end{bmatrix}$. The characteristic polynomial of A is

$$\det(\lambda I_3 - A) = \begin{vmatrix} \lambda & -2 & -2 \\ -2 & \lambda & -2 \\ -2 & -2 & \lambda \end{vmatrix} = \lambda^3 - 8 - 8 - 12\lambda = \lambda^3 - 12\lambda - 16 = (\lambda - 4)(\lambda + 2)^2.$$

Hence the eigenvalues of A are $\lambda_1 = 4$, $\lambda_2 = -2$, and $\lambda_3 = -2$. Thus a diagonal matrix D congruent to A is

$$D = \begin{bmatrix} 4 & 0 & 0 \\ 0 & -2 & 0 \\ 0 & 0 & -2 \end{bmatrix}.$$

There is more than one diagonal matrix D congruent to A. Others are found by reordering the eigenvalues of A on the diagonal.

In Exercises 5–9, we form the matrix A of the quadratic form and find its eigenvalues $\lambda_1, \lambda_2, \ldots, \lambda_n$. Then Theorem 9.2 guarantees that the quadratic form

$$h(\mathbf{y}) = \lambda_1 y_1^2 + \lambda_2 y_2^2 + \cdots + \lambda_n y_n^2$$

is equivalent to the quadratic form $g(\mathbf{x}) = \mathbf{x}^T A \mathbf{x}$.

5. Let $g(\mathbf{x}) = 2x^2 - 4xy - y^2$. Then the matrix of the quadratic form is

$$A = \begin{bmatrix} 2 & -2 \\ -2 & -1 \end{bmatrix}.$$

The characteristic polynomial of A is

$$\det(\lambda I_2 - A) = \begin{vmatrix} \lambda - 2 & 2 \\ 2 & \lambda + 1 \end{vmatrix} = (\lambda - 2)(\lambda + 1) - 4 = \lambda^2 - \lambda - 6 = (\lambda - 3)(\lambda + 2).$$

Thus, the eigenvalues of A are $\lambda_1 = 3$, $\lambda_2 = -2$. Hence $g(\mathbf{x})$ is equivalent to $h(\mathbf{y}) = 3x'^2 - 2y'^2$, where $\mathbf{y} = \begin{bmatrix} x' & y' \end{bmatrix}^T$. (If the eigenvalues had been labeled $\lambda_1 = -2$, $\lambda_2 = 3$, then $h(\mathbf{y}) = -2x'^2 + 3y'^2$. Hence $h(\mathbf{y})$ is not unique.)

7. Let $g(\mathbf{x}) = 2x_1 x_3$. Then the matrix of the quadratic form is

$$A = \begin{bmatrix} 0 & 0 & 1 \\ 0 & 0 & 0 \\ 1 & 0 & 0 \end{bmatrix}.$$

The characteristic polynomial of A is

$$\det(\lambda I_3 - A) = \begin{bmatrix} \lambda & 0 & -1 \\ 0 & \lambda & 0 \\ -1 & 0 & \lambda \end{bmatrix} = \lambda^3 - \lambda = \lambda(\lambda^2 - 1) = \lambda(\lambda - 1)(\lambda + 1).$$

Thus the eigenvalues of A are $\lambda_1 = 0$, $\lambda_2 = 1$, $\lambda_3 = -1$. Hence $g(\mathbf{x})$ is equivalent to $h(\mathbf{y}) = y_2^2 - y_3^2$, where $\mathbf{y} = \begin{bmatrix} y_1 & y_2 & y_3 \end{bmatrix}^T$. (Using the eigenvalues in a different order produces another equivalent quadratic form of the desired type.)

9. Let $g(\mathbf{x}) = -2x_1^2 - 4x_2^2 + 4x_3^2 - 6x_2 x_3$. Then the matrix of the quadratic form is

$$A = \begin{bmatrix} -2 & 0 & 0 \\ 0 & -4 & -3 \\ 0 & -3 & 4 \end{bmatrix}.$$

The characteristic polynomial of A is

$$\det(\lambda I_3 - A) = \begin{vmatrix} \lambda + 2 & 0 & 0 \\ 0 & \lambda + 4 & 3 \\ 0 & 3 & \lambda - 4 \end{vmatrix} = (\lambda + 2)(\lambda + 4)(\lambda - 4) - 9(\lambda + 2)$$

$$= (\lambda + 2)(\lambda^2 - 25) = (\lambda + 2)(\lambda - 5)(\lambda + 5).$$

Thus the eigenvalues of A are $\lambda_1 = -2$, $\lambda_2 = 5$, $\lambda_3 = -5$. Hence $g(\mathbf{x})$ is equivalent to $h(\mathbf{y}) = -2y_1^2 + 5y_2^2 - 5y_3^2$, where $\mathbf{y} = \begin{bmatrix} y_1 & y_2 & y_3 \end{bmatrix}^T$. (Using the eigenvalues in a different order produces another equivalent quadratic form of the desired type.)

In Exercises 11–15, we form the matrix A of the quadratic form and find its eigenvalues $\lambda_1, \lambda_2, \ldots, \lambda_n$. We determine r, the number of nonzero eigenvalues, p, the number of positive eigenvalues, and $r - p$, the number of negative eigenvalues. Then Theorem 9.3 guarantees that the quadratic form

$$h(\mathbf{y}) = y_1^2 + y_2^2 + \cdots + y_p^2 - y_{p+1}^2 - \cdots - y_r^2$$

is equivalent to the quadratic form $g(\mathbf{x}) = \mathbf{x}^T A \mathbf{x}$.

11. Let $g(\mathbf{x}) = 2x^2 + 4xy + 2y^2$. Then the matrix of the quadratic form is

$$A = \begin{bmatrix} 2 & 2 \\ 2 & 2 \end{bmatrix}.$$

the characteristic polynomial of A is

$$\det(\lambda I_2 - A) = \begin{vmatrix} \lambda - 2 & -2 \\ -2 & \lambda - 2 \end{vmatrix} = (\lambda - 2)^2 - 4 = \lambda^2 - 4\lambda = \lambda(\lambda - 4).$$

Thus the eigenvalues of A are $\lambda_1 = 0$, $\lambda_2 = 4$. In this case $n = 2$, $r = 1$, $p = 1$, and $r - p = 0$. Hence $g(\mathbf{x})$ is equivalent to $h(\mathbf{y}) = y_1^2$, where $\mathbf{y} = \begin{bmatrix} y_1 & y_2 \end{bmatrix}^T$.

13. Let $g(\mathbf{x}) = 2x_1^2 + 4x_2^2 + 4x_3^2 + 10x_2x_3$. Then the matrix of the quadratic form is

$$A = \begin{bmatrix} 2 & 0 & 0 \\ 0 & 4 & 5 \\ 0 & 5 & 4 \end{bmatrix}.$$

The characteristic polynomial of A is

$$\det(\lambda I_3 - A) = \begin{vmatrix} \lambda - 2 & 0 & 0 \\ 0 & \lambda - 4 & -5 \\ 0 & -5 & \lambda - 4 \end{vmatrix} = (\lambda - 2)(\lambda - 4)^2 - 25(\lambda - 2)$$

$$= (\lambda - 2)(\lambda^2 - 8\lambda - 9) = (\lambda - 2)(\lambda - 9)(\lambda + 1).$$

Thus the eigenvalues are $\lambda_1 = 2$, $\lambda_2 = 9$, $\lambda_3 = -1$. In this case $n = 3$, $r = 3$, $p = 2$, and $r - p = 1$. Hence $g(\mathbf{x})$ is equivalent to $h(\mathbf{y}) = y_1^2 + y_2^2 - y_3^2$, where $\mathbf{y} = \begin{bmatrix} y_1 & y_2 & y_3 \end{bmatrix}^T$.

15. Let $g(\mathbf{x}) = -3x_1^2 + 2x_2^2 + 2x_3^2 + 4x_2x_3$. Then the matrix of the quadratic form is

$$A = \begin{bmatrix} -3 & 0 & 0 \\ 0 & 2 & 2 \\ 0 & 2 & 2 \end{bmatrix}.$$

The characteristic polynomial of A is

$$\det(\lambda I_3 - A) = \begin{vmatrix} \lambda + 3 & 0 & 0 \\ 0 & \lambda - 2 & 2 \\ 0 & 2 & \lambda - 2 \end{vmatrix} = (\lambda + 3)(\lambda - 2)^2 - 4(\lambda + 3)$$

$$= (\lambda + 3)(\lambda^2 - 4\lambda) = (\lambda + 3)\lambda(\lambda - 4).$$

Thus the eigenvalues of A are $\lambda_1 = -3$, $\lambda_2 = 0$, $\lambda_3 = 4$. In this case $n = 3$, $r = 2$, $p = 1$, and $r - p = 1$. Hence $g(\mathbf{x})$ is equivalent to $h(\mathbf{y}) = y_1^2 - y_2^2$, where $\mathbf{y} = \begin{bmatrix} y_1 & y_2 & y_3 \end{bmatrix}^T$.

17. Let $g(\mathbf{x}) = 4x_2^2 + 4x_3^2 - 10x_2x_3$. Then the matrix of the quadratic form is

$$A = \begin{bmatrix} 0 & 0 & 0 \\ 0 & 4 & -5 \\ 0 & -5 & 4 \end{bmatrix}.$$

The characteristic polynomial of A is

$$\det(\lambda I_3 - A) = \begin{vmatrix} \lambda & 0 & 0 \\ 0 & \lambda - 4 & 5 \\ 0 & 5 & \lambda - 4 \end{vmatrix} = \lambda(\lambda - 4)^2 - 25\lambda$$

$$= \lambda(\lambda^2 - 8\lambda - 9) = \lambda(\lambda - 9)(\lambda + 1).$$

Thus the eigenvalues of A are $\lambda_1 = 0$, $\lambda_2 = 9$, $\lambda_3 = -1$. In this case $n = 3$, $r = 2$, $p = 1$, and $r - p = 1$. Hence $g(\mathbf{x})$ is equivalent to $h(\mathbf{y}) = y_1^2 - y_2^2$, where $\mathbf{y} = \begin{bmatrix} y_1 & y_2 & y_3 \end{bmatrix}^T$. The rank of g is 2 and the signature of g is 0.

19. For a 2×2 matrix A of a quadratic form $g(\mathbf{x}) = \mathbf{x}^T A \mathbf{x}$, we have the following possibilities for its eigenvalues with the corresponding quadratic form of the type described in Theorem 9.3.

eigenvalues	quadratic form
both positive	$y_1^2 + y_2^2$
both negative	$-y_1^2 - y_2^2$
one positive, one negative	$y_1^2 - y_2^2$
one positive, one zero	y_1^2
one negative, one zero	$-y_1^2$

The conics for the equations $\mathbf{x}^T A \mathbf{x} = 1$ are given next.

$$y_1^2 + y_2^2 = 1 \quad \text{is a circle}$$
$$-y_1^2 - y_2^2 = 1 \quad \text{is empty; it represents no conic}$$
$$y_1^2 - y_2^2 = 1 \quad \text{is a hyperbola}$$
$$y_1^2 = 1 \quad \text{is a pair of lines; } y_1 = 1, \, y_1 = -1$$
$$-y_1^2 = 1 \quad \text{is empty; it represents no conic}$$

21. In the discussion preceding Example 7 it is stated that quadratic forms g and h are equivalent if and only if they have equal ranks and the same signature. We determine the rank and signature for each of the quadratic forms listed.

- Let $g_1(\mathbf{x}) = x_1^2 + x_2^2 + x_3^2 + 2x_1x_2$. Then the matrix of the quadratic form is

$$A = \begin{bmatrix} 1 & 1 & 0 \\ 1 & 1 & 0 \\ 0 & 0 & 1 \end{bmatrix}.$$

The characteristic polynomial of A is

$$\det(\lambda I_3 - A) = \begin{vmatrix} \lambda - 1 & -1 & 0 \\ -1 & \lambda - 1 & 0 \\ 0 & 0 & \lambda - 1 \end{vmatrix} = (\lambda - 1)^3 - (\lambda - 1)$$

$$= (\lambda - 1)(\lambda^2 - 2\lambda) = \lambda(\lambda - 1)(\lambda - 2).$$

Thus the eigenvalues of A are $\lambda_1 = 0$, $\lambda_2 = 1$, $\lambda_3 = 2$. In this case $n = 3$, $r = 2$, $p = 2$, and $r - p = 0$. The rank of g_1 is 2 and the signature of g_1 is 2.

- Let $g_2(\mathbf{x}) = 2x_2^2 + 2x_3^2 + 2x_2x_3$. Then the matrix of the quadratic form is

$$A = \begin{bmatrix} 0 & 0 & 0 \\ 0 & 2 & 1 \\ 0 & 1 & 2 \end{bmatrix}.$$

The characteristic polynomial of A is

$$\det(\lambda I_3 - A) = \begin{vmatrix} \lambda & 0 & 0 \\ 0 & \lambda - 2 & 1 \\ 0 & 1 & \lambda - 2 \end{vmatrix} = \lambda(\lambda - 2)^2 - \lambda$$

$$= \lambda(\lambda^2 - 4\lambda + 3) = \lambda(\lambda - 1)(\lambda - 3).$$

Thus the eigenvalues of A are $\lambda_1 = 0$, $\lambda_2 = 1$, $\lambda_3 = 3$. In this case $n = 3$, $r = 2$, $p = 2$, and $r - p = 0$. The rank of g_2 is 2 and the signature of g_2 is 2. Thus $g_1(\mathbf{x})$ and $g_2(\mathbf{x})$ are equivalent.

- Let $g_3(\mathbf{x}) = 3x_2^2 - 3x_3^2 + 8x_2x_3$. Then the matrix of the quadratic form is

$$A = \begin{bmatrix} 0 & 0 & 0 \\ 0 & 3 & 4 \\ 0 & 4 & -3 \end{bmatrix}.$$

The characteristic polynomial of A is

$$\det(\lambda I_3 - A) = \begin{vmatrix} \lambda & 0 & 0 \\ 0 & \lambda - 3 & -4 \\ 0 & -4 & \lambda + 3 \end{vmatrix} = \lambda(\lambda - 3)(\lambda + 3) - 16\lambda$$

$$= \lambda(\lambda^2 - 25) = \lambda(\lambda - 5)(\lambda + 5).$$

Thus the eigenvalues of A are $\lambda_1 = 0$, $\lambda_2 = 5$, $\lambda_3 = -5$. In this case $n = 3$, $r = 2$, $p = 1$, and $r - p = 1$. The rank of g_3 is 2 and the signature of g_3 is 0. Thus $g_3(\mathbf{x})$ is not equivalent to $g_1(\mathbf{x})$ or $g_2(\mathbf{x})$.

- Let $g_4(\mathbf{x}) = 3x_2^2 + 3x_3^2 - 4x_2x_3$. Then the matrix of the quadratic form is

$$A = \begin{bmatrix} 0 & 0 & 0 \\ 0 & 3 & -2 \\ 0 & -2 & 3 \end{bmatrix}.$$

The characteristic polynomial of A is

$$\det(\lambda I_3 - A) = \begin{vmatrix} \lambda & 0 & 0 \\ 0 & \lambda - 3 & 2 \\ 0 & 2 & \lambda - 3 \end{vmatrix} = \lambda(\lambda - 3)^2 - 4\lambda$$

$$= \lambda(\lambda^2 - 6\lambda + 5) = \lambda(\lambda - 5)(\lambda - 1).$$

Thus the eigenvalues of A are $\lambda_1 = 0$, $\lambda_2 = 5$, $\lambda_3 = 1$. In this case $n = 3$, $r = 2$, $p = 2$, and $r - p = 0$. The rank of g_4 is 2 and the signature of g_4 is 2. Thus $g_4(\mathbf{x})$ is equivalent to both $g_1(\mathbf{x})$ and $g_2(\mathbf{x})$.

23. Use Theorem 9.4 by computing the eigenvalues of each of the matrices.

(a) Let $A = \begin{bmatrix} 2 & -1 \\ -1 & 2 \end{bmatrix}$. The characteristic polynomial of A is

$$\det(\lambda I_2 - A) = \begin{vmatrix} \lambda - 2 & 1 \\ 1 & \lambda - 2 \end{vmatrix} = (\lambda - 2)^2 - 1 = \lambda^2 - 4\lambda + 3 = (\lambda - 1)(\lambda - 3).$$

Thus the eigenvalues of A are $\lambda_1 = 1$ and $\lambda_2 = 3$. Since all the eigenvalues are positive, A is positive definite.

(b) Let $A = \begin{bmatrix} 2 & 1 \\ 1 & 2 \end{bmatrix}$. The characteristic polynomial of A is

$$\det(\lambda I_2 - A) = \begin{vmatrix} \lambda - 2 & -1 \\ -1 & \lambda - 2 \end{vmatrix} = (\lambda - 2)^2 - 1 = \lambda^2 - 4\lambda + 3 = (\lambda - 1)(\lambda - 3) = 0.$$

Thus the eigenvalues of A are $\lambda_1 = 1$ and $\lambda_2 = 3$. Since all the eigenvalues are positive, A is positive definite.

(c) Let $A = \begin{bmatrix} 3 & 1 & 0 \\ 1 & 3 & 0 \\ 0 & 0 & 3 \end{bmatrix}$. The characteristic polynomial of A is

$$\det(\lambda I_3 - A) = \begin{vmatrix} \lambda - 3 & -1 & 0 \\ -1 & \lambda - 3 & 0 \\ 0 & 0 & \lambda - 3 \end{vmatrix} = (\lambda - 3)^3 - 1(\lambda - 3)$$

$$= (\lambda - 3)(\lambda^2 - 6\lambda + 8) = (\lambda - 3)(\lambda - 4)(\lambda - 2).$$

Thus the eigenvalues of A are $\lambda_1 = 3$, $\lambda_2 = 4$, and $\lambda_3 = 2$. Since all the eigenvalues are positive, A is positive definite.

(d) Let

$$A = \begin{bmatrix} 1 & 0 & 0 \\ 0 & 2 & 0 \\ 0 & 0 & -3 \end{bmatrix}.$$

Since A is diagonal, its eigenvalues are its diagonal entries. It follows that A is not positive definite.

(e) Let

$$A = \begin{bmatrix} 2 & 2 \\ 2 & 2 \end{bmatrix}.$$

Matrix A is singular, so one of its eigenvalues is zero. Thus A is not positive definite.

T.1. Let A be symmetric. Then $(P^T A P)^T = P^T A^T P = P^T A P$ since $A^T = A$. Hence $P^T A P$ is symmetric.

T.3. By Theorem 9.2, for the symmetric matrix A, there exists an orthogonal matrix P such that $P^{-1} A P = D$ is diagonal. Since P is orthogonal, $P^{-1} = P^T$. Thus A is congruent to D.

T.5. Let A be positive definite and $Q(\mathbf{x}) = \mathbf{x}^T A \mathbf{x}$. By Theorem 9.3, $Q(\mathbf{x})$ is a quadratic form which is equivalent to

$$Q'(\mathbf{y}) = y_1^2 + y_2^2 + \cdots + y_p^2 - y_{p+1}^2 - \cdots - y_r^2.$$

If Q and Q' are equivalent then $Q'(\mathbf{y}) > 0$ for each $\mathbf{y} \neq \mathbf{0}$. However, this can happen if and only if all terms in $Q'(\mathbf{y})$ are positive; that is, if and only if A is congruent to I_n, or if and only if $A = P^T I_n P = P^T P$.

ML.1. (a) **A = [− 1 0 0;0 1 1;0 1 1];**
eig(A)

ans =

   −1.0000

   2.0000

   0.0000

If we set the format to long e, then the eigenvalues are displayed as

ans =

   −1.000000000000000e + 000

   2.000000000000000e + 000

   2.220446049250313e − 016

Since the last value is extremely small, we will consider it zero. The **eig** command approximates the eigenvalues, hence errors due to using machine arithmetic can occur. Thus it follows that $\text{rank}(A) = 2$ and the signature of the quadratic form is 0.

(b) **A = ones(3);**
eig(A)

ans =

   0.0000

   −0.0000

   3.0000

If we set the format to long e, then the eigenvalues are displayed as

ans =

$$2.343881062810587e - 017$$
$$-7.011704839834072e - 016$$
$$2.999999999999999e + 000$$

We will consider the first two eigenvalues zero. Hence rank$(A) = 1$ and the signature of the quadratic form is 1.

(c) **A = [2 1 0 − 2;1 − 1 1 3;0 1 2 − 1;− 2 3 − 1 0];**

eig(A)

ans =

$$2.2896$$
$$1.6599$$
$$3.5596$$
$$-4.5091$$

It follows that rank$(A) = 4$ and the signature of the quadratic form is 2.

(d) **A = [2 − 1 0 0;− 1 2 − 1 0;0 − 1 2 − 1;0 0 − 1 2];**

eig(A)

ans =

$$1.3820$$
$$0.3820$$
$$2.6180$$
$$3.6180$$

It follows that rank$(A) = 4$ and the signature of the quadratic form is 4.

Section 9.5, p. 491

In Exercises 1–9, compare the given equations with those that appear in Figure 9.18. In some cases we may need to rearrange terms to properly identify the equation from the forms given in Figure 9.18.

1. Rewrite the equation $x^2 + 9y^2 - 9 = 0$ as

$$\frac{x^2}{9} + \frac{9y^2}{9} = \frac{9}{9} \quad \Longrightarrow \quad \frac{x^2}{9} + \frac{y^2}{1} = 1.$$

This is the equation of an ellipse in standard position with $a = 3$ and $b = 1$. The x-intercepts are $(-3, 0)$ and $(3, 0)$ and the y-intercepts are $(0, -1)$ and $(0, 1)$.

3. Rewrite the equation $25y^2 - 4x^2 = 100$ as

$$\frac{25y^2}{100} - \frac{4x^2}{100} = \frac{100}{100} \quad \Longrightarrow \quad \frac{y^2}{4} - \frac{x^2}{25} = 1.$$

This is the equation of a hyperbola in standard position with $a = 2$ and $b = 5$. The y-intercepts are $(0, -2)$ and $(0, 2)$.

5. Rewrite the equation $3x^2 - y^2 = 0$ as

$$y^2 = 3x^2 \quad \Longrightarrow \quad y = \sqrt{3}\,x \quad \text{and} \quad y = -\sqrt{3}\,x.$$

This represents the graph of a pair of intersecting lines which is a degenerate conic section.

7. Rewrite the equation $4x^2 + 4y^2 - 9 = 0$ as

$$\frac{4x^2}{9} + \frac{4y^2}{9} = \frac{9}{9} \quad \Longrightarrow \quad \frac{x^2}{\left(\frac{3}{2}\right)^2} + \frac{y^2}{\left(\frac{3}{2}\right)^2} = 1.$$

This is the equation of a circle in standard position with $a = \frac{3}{2}$.

9. Upon inspection, the equation $4x^2 + y^2 = 0$ is satisfied if and only if $x = y = 0$. Hence this is a degenerate conic section which represents the single point $(0,0)$.

In each of the Exercises 11–17, the equations do not contain cross-product terms. However, the equations do contain x^2 and x terms and/or y^2 and y terms. Thus we complete the square(s) and rewrite the equations as one of the forms in Figure 9.18.

11. In equation $x^2 + 2y^2 - 4x - 4y + 4 = 0$ we note that x^2 and x and y^2 and y terms appear. Hence we complete the square in both x and y. Rewrite the equation as

$$\underbrace{x^2 - 4x + 4}_{} + \underbrace{2y^2 - 4y +}_{} = 0$$

$$(x-2)^2 + 2\underbrace{(y^2 - 2y + }_{}) = 0$$

$$(x-2)^2 + 2\underbrace{(y^2 - 2y + 1)}_{} = 2$$

$$(x-2)^2 + 2(y-1)^2 = 2$$

Let $x' = x - 2$ and $y' = y - 1$ then the preceding equation is written as

$$x'^2 + 2y'^2 = 2.$$

Next we transform this equation to

$$\frac{x'^2}{2} + \frac{y'^2}{1} = 1.$$

If we translate the xy-coordinate system to the $x'y'$-coordinate system, whose origin is at $(2,1)$, then the graph is an ellipse in standard position with respect to the $x'y'$-coordinate system.

13. In equation $x^2 + y^2 - 8x - 6y = 0$ we note that x^2 and x and y^2 and y terms appear. Hence we complete the square in both x and y. Rewrite the equation as

$$\underbrace{x^2 - 8x + }_{} + \underbrace{y^2 - 6y + }_{} = 0$$

$$\underbrace{x^2 - 8x + 16}_{} + \underbrace{y^2 - 6y + 9}_{} = 16 + 9$$

$$(x-4)^2 + (y-3)^2 = 25$$

Let $x' = x - 4$ and $y' = y - 3$ then the preceding equation can be written as

$$x'^2 + y'^2 = 25.$$

Next we transform this equation to

$$\frac{x'^2}{25} + \frac{y'^2}{25} = 1 \quad \Longrightarrow \quad \frac{x'^2}{5^2} + \frac{y'^2}{5^2} = 1.$$

If we translate the xy-coordinate system to the $x'y'$-coordinate system, whose origin is at $(4,3)$, then the graph is a circle in standard position with respect to the $x'y'$-coordinate system.

15. In equation $y^2 - 4y = 0$ we note that only y^2 and y terms appear. Hence we complete the square only in y. Rewrite the equation as

$$y^2 - 4y + 4 = 4 \quad \Longrightarrow \quad (y - 2)^2 = 4.$$

Let $y' = y - 2$, then we have

$$y'^2 = 4 \quad \Longrightarrow \quad y' = 2 \quad \text{and} \quad y' = -2.$$

If we translate the xy-coordinate system to the $x'y'$-coordinate system, whose origin is at $(0, 2)$, then the graph is a pair of parallel lines.

17. In equation $x^2 + y^2 - 2x - 6y + 10 = 0$ we note that x^2 and x and y^2 and y terms appear. Hence we complete the square in both x and y. Rewrite the equation as

$$\underbrace{x^2 - 2x +}\ +\underbrace{y^2 - 6y +} = -10$$

$$\underbrace{x^2 - 2x + 1}+\underbrace{y^2 - 6y + 9} = -10 + 1 + 9$$

$$(x - 1)^2 + (y - 3)^2 = 0.$$

Let $x' = x - 1$ and $y' = y - 3$ then the preceding equation can be written as

$$x'^2 + y'^2 = 0.$$

If we translate the xy-coordinate system to the $x'y'$-coordinate system, whose origin is at $(1, 3)$, then the graph is a single point which is the origin of the $x'y'$-coordinate system.

In Exercises 19–23 each of the equations contains only an xy term. Thus a rotation will transform the graph to standard position in a new coordinate system. We determine the matrix A from Equation (5), find its eigenvalues, and construct the orthogonal matrix P that gives the correct rotation.

19. For equation $x^2 + xy + y^2 = 6$, we have

$$A = \begin{bmatrix} 1 & \frac{1}{2} \\ \frac{1}{2} & 1 \end{bmatrix}.$$

The characteristic equation is

$$\det(\lambda I_2 - A) = \begin{vmatrix} \lambda - 1 & -\frac{1}{2} \\ -\frac{1}{2} & \lambda - 1 \end{vmatrix} = \lambda^2 - 2\lambda + \frac{3}{4} = \left(\lambda - \frac{1}{2}\right)\left(\lambda - \frac{3}{2}\right) = 0.$$

Thus the eigenvalues of A are $\lambda_1 = \frac{1}{2}$ and $\lambda_2 = \frac{3}{2}$. Next we determine an eigenvector $\mathbf{x}_1$ associated with eigenvalue λ_1. Solving

$$(\lambda_1 I_2 - A)\mathbf{x} = \begin{bmatrix} -\frac{1}{2} & -\frac{1}{2} \\ -\frac{1}{2} & -\frac{1}{2} \end{bmatrix} \begin{bmatrix} x_1 \\ x_2 \end{bmatrix} = \begin{bmatrix} 0 \\ 0 \end{bmatrix}$$

we find

$$\mathbf{x} = \begin{bmatrix} r \\ -r \end{bmatrix}.$$

For $r = 1$, we have eigenvector

$$\mathbf{x}_1 = \begin{bmatrix} 1 \\ -1 \end{bmatrix}.$$

We determine an eigenvector $\mathbf{x}_2$ associated with eigenvalue λ_2 by solving

$$(\lambda_2 I_2 - A)\mathbf{x} = \begin{bmatrix} \frac{1}{2} & -\frac{1}{2} \\ -\frac{1}{2} & \frac{1}{2} \end{bmatrix} \begin{bmatrix} x_1 \\ x_2 \end{bmatrix} = \begin{bmatrix} 0 \\ 0 \end{bmatrix}.$$

We find

$$\mathbf{x} = \begin{bmatrix} r \\ r \end{bmatrix}.$$

For $r = 1$, we have eigenvector

$$\mathbf{x}_2 = \begin{bmatrix} 1 \\ 1 \end{bmatrix}.$$

Normalizing the eigenvectors and forming the matrix P, we have

$$P = \begin{bmatrix} \frac{1}{\sqrt{2}} & \frac{1}{\sqrt{2}} \\ -\frac{1}{\sqrt{2}} & \frac{1}{\sqrt{2}} \end{bmatrix} \qquad \text{(Note: } \det(P) = 1.\text{)}$$

Let $\mathbf{x} = P\mathbf{y}$, where

$$\mathbf{y} = \begin{bmatrix} x' \\ y' \end{bmatrix}.$$

Then we can rewrite the original equation as in (6) to obtain

$$\tfrac{1}{2}x'^2 + \tfrac{3}{2}y'^2 = 6 \quad \Longrightarrow \quad \frac{x'^2}{12} + \frac{y'^2}{4} = 1.$$

Thus this conic section is an ellipse. The preceding is only a possible answer. The roles of x' and y' would be reversed if the eigenvalues of A were used in a different order.

21. For equation $9x^2 + y^2 + 6xy = 4$, we have

$$A = \begin{bmatrix} 9 & 3 \\ 3 & 1 \end{bmatrix}.$$

The characteristic equation is

$$\det(\lambda I_2 - A) = \begin{vmatrix} \lambda - 9 & -3 \\ -3 & \lambda - 1 \end{vmatrix} = \lambda^2 - 10\lambda = \lambda(\lambda - 10) = 0.$$

Thus the eigenvalues of A are $\lambda_1 = 0$ and $\lambda_2 = 10$. Next we determine an eigenvector $\mathbf{x}_1$ associated with eigenvalue λ_1. Solving

$$(\lambda_1 I_2 - A)\mathbf{x} = \begin{bmatrix} -9 & -3 \\ -3 & -1 \end{bmatrix} \begin{bmatrix} x_1 \\ x_2 \end{bmatrix} = \begin{bmatrix} 0 \\ 0 \end{bmatrix},$$

we find

$$\mathbf{x} = \begin{bmatrix} -\frac{1}{3}r \\ r \end{bmatrix}.$$

For $r = 3$, we have eigenvector

$$\mathbf{x}_1 = \begin{bmatrix} -1 \\ 3 \end{bmatrix}.$$

We determine an eigenvector $\mathbf{x}_2$ associated with eigenvalue λ_2 by solving

$$(\lambda_2 I_2 - A)\mathbf{x} = \begin{bmatrix} 1 & -3 \\ -3 & 9 \end{bmatrix} \begin{bmatrix} x_1 \\ x_2 \end{bmatrix} = \begin{bmatrix} 0 \\ 0 \end{bmatrix}.$$

We find

$$\mathbf{x} = \begin{bmatrix} 3r \\ r \end{bmatrix}.$$

For $r = 1$, we have eigenvector

$$\mathbf{x}_2 = \begin{bmatrix} 3 \\ 1 \end{bmatrix}.$$

Normalizing the eigenvectors and forming the matrix P we have

$$P = \begin{bmatrix} -\frac{1}{\sqrt{10}} & \frac{3}{\sqrt{10}} \\ \frac{3}{\sqrt{10}} & \frac{1}{\sqrt{10}} \end{bmatrix}. \qquad \text{(Note that } \det(P) = -1.\text{)}$$

For a counterclockwise rotation we require $\det(P) = 1$. Since any nonzero multiple of an eigenvector is still an eigenvector for the same eigenvalue, we replace $\mathbf{x}_1$ by $-\mathbf{x}_1$ (an alternate procedure is to interchange columns of P) which results in redefining P as

$$P = \begin{bmatrix} \frac{1}{\sqrt{10}} & \frac{3}{\sqrt{10}} \\ -\frac{3}{\sqrt{10}} & \frac{1}{\sqrt{10}} \end{bmatrix}.$$

Let $\mathbf{x} = P\mathbf{y}$, where

$$\mathbf{y} = \begin{bmatrix} x' \\ y' \end{bmatrix}.$$

Then we can rewrite the original equation as in (6) to obtain

$$0x'^2 + 10y'^2 = 4 \implies y'^2 = \tfrac{4}{10} \implies y' = \pm \frac{2}{\sqrt{10}}.$$

Thus this conic section consists of two parallel lines. The preceding is only a possible answer. The roles of x' and y' would be reversed if the eigenvalues of A were used in a different order.

23. For equation $4x^2 + 4y^2 - 10xy = 0$, we have

$$A = \begin{bmatrix} 4 & -5 \\ -5 & 4 \end{bmatrix}.$$

The characteristic equation is

$$\det(\lambda I_2 - A) = \begin{vmatrix} \lambda - 4 & 5 \\ 5 & \lambda - 4 \end{vmatrix} = \lambda^2 - 8\lambda - 9 = (\lambda - 9)(\lambda + 1) = 0.$$

Thus the eigenvalues of A are $\lambda_1 = 9$ and $\lambda_2 = -1$. Next we determine an eigenvector $\mathbf{x}_1$ associated with eigenvalue λ_1. Solving

$$(\lambda_1 I_2 - A)\mathbf{x} = \begin{bmatrix} 5 & 5 \\ 5 & 5 \end{bmatrix} \begin{bmatrix} x_1 \\ x_2 \end{bmatrix} = \begin{bmatrix} 0 \\ 0 \end{bmatrix}$$

we find

$$\mathbf{x} = \begin{bmatrix} -r \\ r \end{bmatrix}.$$

For $r = 1$, we have eigenvector

$$\mathbf{x}_1 = \begin{bmatrix} -1 \\ 1 \end{bmatrix}.$$

We determine an eigenvector $\mathbf{x}_2$ associated with eigenvalue λ_2. Solving

$$(\lambda_2 I_2 - A)\mathbf{x} = \begin{bmatrix} -5 & 5 \\ 5 & -5 \end{bmatrix} \begin{bmatrix} x_1 \\ x_2 \end{bmatrix} = \begin{bmatrix} 0 \\ 0 \end{bmatrix}$$

we find

$$\mathbf{x} = \begin{bmatrix} r \\ r \end{bmatrix}.$$

For $r = 1$, we have eigenvector

$$\mathbf{x}_2 = \begin{bmatrix} 1 \\ 1 \end{bmatrix}.$$

Normalizing the eigenvectors and forming the matrix P we have

$$P = \begin{bmatrix} -\frac{1}{\sqrt{2}} & \frac{1}{\sqrt{2}} \\ \frac{1}{\sqrt{2}} & \frac{1}{\sqrt{2}} \end{bmatrix}. \qquad \text{(Note that } \det(P) = -1.\text{)}$$

For a counterclockwise rotation we require $\det(P) = 1$. Since any nonzero multiple of an eigenvector is still an eigenvector for the same eigenvalue, we replace $\mathbf{x}_1$ by $-\mathbf{x}_1$ which results in redefining P as

$$P = \begin{bmatrix} \frac{1}{\sqrt{2}} & \frac{1}{\sqrt{2}} \\ -\frac{1}{\sqrt{2}} & \frac{1}{\sqrt{2}} \end{bmatrix}. \qquad \text{(Note that } \det(P) = 1.\text{)}$$

Let $\mathbf{x} = P\mathbf{y}$, where

$$\mathbf{y} = \begin{bmatrix} x' \\ y' \end{bmatrix}.$$

Then we can rewrite the original equation as in (6) to obtain

$$9x'^2 - 1y'^2 = 0 \quad \Longrightarrow \quad y'^2 = 9x'^2 \quad \Longrightarrow \quad y' = \pm 3x'.$$

Thus this conic section is a pair of intersecting lines. The preceding is only a possible answer. The roles of x' and y' would be reversed if the eigenvalues of A were used in a different order.

In Exercises 25–29, we see that both cross product terms and x or y terms appear. Thus we must combine the rotation and completing the square techniques used earlier.

25. For equation $9x^2 + y^2 + 6xy - 10\sqrt{10}\,x + 10\sqrt{10}\,y + 90 = 0$, we have from Equation (5)

$$A = \begin{bmatrix} 9 & 3 \\ 3 & 1 \end{bmatrix}, \qquad B = \begin{bmatrix} -10\sqrt{10} & 10\sqrt{10} \end{bmatrix}, \qquad f = 90.$$

The characteristic equation for matrix A is

$$\det(\lambda I_2 - A) = \begin{vmatrix} \lambda - 9 & -3 \\ -3 & \lambda - 1 \end{vmatrix} = \lambda^2 - 10\lambda = \lambda(\lambda - 10) = 0.$$

Thus the eigenvalues are $\lambda_1 = 0$ and $\lambda_2 = 10$. Next we determine an eigenvector $\mathbf{x}_1$ associated with eigenvalue λ_1. Solving

$$(\lambda_1 I_2 - A)\mathbf{x} = \begin{bmatrix} -9 & -3 \\ -3 & -1 \end{bmatrix} \begin{bmatrix} x_1 \\ x_2 \end{bmatrix} = \begin{bmatrix} 0 \\ 0 \end{bmatrix}$$

we find

$$\mathbf{x} = \begin{bmatrix} -\frac{1}{3}r \\ r \end{bmatrix}.$$

For $r = 3$, we have eigenvector

$$\mathbf{x}_1 = \begin{bmatrix} -1 \\ 3 \end{bmatrix}.$$

We determine an eigenvector $\mathbf{x}_2$ associated with eigenvalue λ_2. Solving

$$(\lambda_2 I_2 - A)\mathbf{x} = \begin{bmatrix} 1 & -3 \\ -3 & 9 \end{bmatrix} \begin{bmatrix} x_1 \\ x_2 \end{bmatrix} = \begin{bmatrix} 0 \\ 0 \end{bmatrix}$$

we find

$$\mathbf{x} = \begin{bmatrix} 3r \\ r \end{bmatrix}.$$

For $r = 1$, we have eigenvector

$$\mathbf{x}_2 = \begin{bmatrix} 3 \\ 1 \end{bmatrix}.$$

Normalizing the eigenvectors and forming the matrix P we have

$$P = \begin{bmatrix} -\frac{1}{\sqrt{10}} & \frac{3}{\sqrt{10}} \\ \frac{3}{\sqrt{10}} & \frac{1}{\sqrt{10}} \end{bmatrix}. \qquad \text{(Note that } \det(P) = -1.\text{)}$$

For a counterclockwise rotation we require $\det(P) = 1$. Since any nonzero multiple of an eigenvector is still an eigenvector for the same eigenvalue, we replace $\mathbf{x}_1$ by $-\mathbf{x}_1$ which results in redefining P as

$$P = \begin{bmatrix} \frac{1}{\sqrt{10}} & \frac{3}{\sqrt{10}} \\ -\frac{3}{\sqrt{10}} & \frac{1}{\sqrt{10}} \end{bmatrix}$$

Let $\mathbf{x} = P\mathbf{y}$, where

$$\mathbf{y} = \begin{bmatrix} x' \\ y' \end{bmatrix}.$$

Then we can rewrite the original equation as in (6) to obtain

$$\begin{bmatrix} x' & y' \end{bmatrix} \begin{bmatrix} 0 & 0 \\ 0 & 10 \end{bmatrix} \begin{bmatrix} x' \\ y' \end{bmatrix} + \begin{bmatrix} -10\sqrt{10} & 10\sqrt{10} \end{bmatrix} \begin{bmatrix} \frac{1}{\sqrt{10}} & \frac{3}{\sqrt{10}} \\ -\frac{3}{\sqrt{10}} & \frac{1}{\sqrt{10}} \end{bmatrix} \begin{bmatrix} x' \\ y' \end{bmatrix} + 90 = 0.$$

Performing the matrix multiplications and simplifying gives

$$10y'^2 - 20y' + 40x' + 90 = 0.$$

Dividing by 10 and completing the square in y' gives

$$(y' - 1)^2 + 4(x' + 2) = 0.$$

Let $x'' = x' + 2$ and $y'' = y' - 1$; then we can write the equation as

$$y''^2 = -4x''.$$

Thus this conic section is a parabola. The preceding is only a possible answer. The roles of x'' and y'' would be reversed if the eigenvalues of A were used in a different order.

27. For equation $5x^2 + 12xy - 12\sqrt{13}\, x = 36$ we have from Equation (5)

$$A = \begin{bmatrix} 5 & 6 \\ 6 & 0 \end{bmatrix}, \qquad B = \begin{bmatrix} -12\sqrt{13} & 0 \end{bmatrix}, \qquad f = -36.$$

The characteristic equation for matrix A is

$$\det(\lambda I_2 - A) = \begin{vmatrix} \lambda - 5 & -6 \\ -6 & \lambda \end{vmatrix} = \lambda^2 - 5\lambda - 36 = (\lambda - 9)(\lambda + 4) = 0.$$

Thus the eigenvalues are $\lambda_1 = 9$ and $\lambda_2 = -4$. Next we determine an eigenvector $\mathbf{x}_1$ associated with eigenvalue λ_1. Solving

$$(\lambda_1 I_2 - A)\mathbf{x} = \begin{bmatrix} 4 & -6 \\ -6 & 9 \end{bmatrix} \begin{bmatrix} x_1 \\ x_2 \end{bmatrix} = \begin{bmatrix} 0 \\ 0 \end{bmatrix}$$

we find

$$\mathbf{x} = \begin{bmatrix} \frac{3}{2}r \\ r \end{bmatrix}.$$

For $r = 2$, we have eigenvector

$$\mathbf{x}_1 = \begin{bmatrix} 3 \\ 2 \end{bmatrix}.$$

We determine an eigenvector $\mathbf{x}_2$ associated with eigenvalue λ_2. Solving

$$(\lambda_2 I_2 - A)\mathbf{x} = \begin{bmatrix} -9 & -6 \\ -6 & -4 \end{bmatrix} \begin{bmatrix} x_1 \\ x_2 \end{bmatrix} = \begin{bmatrix} 0 \\ 0 \end{bmatrix}$$

we find

$$\mathbf{x} = \begin{bmatrix} -\frac{2}{3}r \\ r \end{bmatrix}.$$

For $r = 3$, we have eigenvector

$$\mathbf{x}_2 = \begin{bmatrix} -2 \\ 3 \end{bmatrix}.$$

Normalizing the eigenvectors and forming the matrix P we have

$$P = \begin{bmatrix} \frac{3}{\sqrt{13}} & -\frac{2}{\sqrt{13}} \\ \frac{2}{\sqrt{13}} & \frac{3}{\sqrt{13}} \end{bmatrix}. \qquad \text{(Note that } \det(P) = 1.)$$

Let $\mathbf{x} = P\mathbf{y}$, where

$$\mathbf{y} = \begin{bmatrix} x' \\ y' \end{bmatrix}.$$

Then we can rewrite the original equation as in (6) to obtain

$$\begin{bmatrix} x' & y' \end{bmatrix} \begin{bmatrix} 9 & 0 \\ 0 & -4 \end{bmatrix} \begin{bmatrix} x' \\ y' \end{bmatrix} + \begin{bmatrix} -12\sqrt{13} & 0 \end{bmatrix} \begin{bmatrix} \frac{3}{\sqrt{13}} & -\frac{2}{\sqrt{13}} \\ \frac{2}{\sqrt{13}} & \frac{3}{\sqrt{13}} \end{bmatrix} \begin{bmatrix} x' \\ y' \end{bmatrix} - 36 = 0.$$

Performing the matrix multiplications and simplifying gives

$$9x'^2 - 4y'^2 - 36x' + 24y' - 36 = 0.$$

Completing the square in x' and y' gives

$$9(x' - 2)^2 - 4(y' - 3)^2 = 36.$$

Let $x'' = x' - 2$ and $y'' = y' - 3$; then we can write the equation as

$$9x''^2 - 4y''^2 = 36$$

or equivalently

$$\frac{x''^2}{4} - \frac{y''^2}{9} = 1.$$

Thus this conic section is a hyperbola. The preceding is only a possible answer. The roles of x'' and y'' would be reversed if the eigenvalues of A were used in a different order.

29. For equation $x^2 - y^2 + 2\sqrt{3}\,xy + 6x = 0$ we have from Equation (5)

$$A = \begin{bmatrix} 1 & \sqrt{3} \\ \sqrt{3} & -1 \end{bmatrix}, \qquad B = \begin{bmatrix} 6 & 0 \end{bmatrix}, \qquad f = 0.$$

The characteristic equation for matrix A is

$$\det(\lambda I_2 - A) = \begin{vmatrix} \lambda - 1 & -\sqrt{3} \\ -\sqrt{3} & \lambda + 1 \end{vmatrix} = \lambda^2 - 4 = (\lambda - 2)(\lambda + 2) = 0.$$

Thus the eigenvalues are $\lambda_1 = 2$ and $\lambda_2 = -2$. Next we determine an eigenvector $\mathbf{x}_1$ associated with eigenvalue λ_1. Solving

$$(\lambda_1 I_2 - A)\mathbf{x} = \begin{bmatrix} 1 & -\sqrt{3} \\ -\sqrt{3} & 3 \end{bmatrix} \begin{bmatrix} x_1 \\ x_2 \end{bmatrix} = \begin{bmatrix} 0 \\ 0 \end{bmatrix}$$

we find

$$\mathbf{x} = \begin{bmatrix} \sqrt{3}\,r \\ r \end{bmatrix}.$$

For $r = 1$, we have eigenvector

$$\mathbf{x}_1 = \begin{bmatrix} \sqrt{3} \\ 1 \end{bmatrix}.$$

We determine an eigenvector $\mathbf{x}_2$ associated with eigenvalue λ_2. Solving

$$(\lambda_2 I_2 - A)\mathbf{x} = \begin{bmatrix} -3 & -\sqrt{3} \\ -\sqrt{3} & -1 \end{bmatrix} \begin{bmatrix} x_1 \\ x_2 \end{bmatrix} = \begin{bmatrix} 0 \\ 0 \end{bmatrix}$$

we find

$$\mathbf{x} = \begin{bmatrix} -\frac{\sqrt{3}}{3}r \\ r \end{bmatrix}.$$

For $r = 1$, we have eigenvector

$$\mathbf{x}_2 = \begin{bmatrix} -\frac{\sqrt{3}}{3} \\ 1 \end{bmatrix}.$$

Normalizing the eigenvectors and forming the matrix P we have

$$P = \begin{bmatrix} \frac{\sqrt{3}}{2} & -\frac{1}{2} \\ \frac{1}{2} & \frac{\sqrt{3}}{2} \end{bmatrix}. \qquad \text{(Note that } \det(P) = 1.)$$

Let $\mathbf{x} = P\mathbf{y}$, where

$$\mathbf{y} = \begin{bmatrix} x' \\ y' \end{bmatrix}.$$

The we can rewrite the original equation as in (6) to obtain

$$\begin{bmatrix} x' & y' \end{bmatrix} \begin{bmatrix} 2 & 0 \\ 0 & -2 \end{bmatrix} \begin{bmatrix} x' \\ y' \end{bmatrix} + \begin{bmatrix} 6 & 0 \end{bmatrix} \begin{bmatrix} \frac{\sqrt{3}}{2} & -\frac{1}{2} \\ \frac{1}{2} & \frac{\sqrt{3}}{2} \end{bmatrix} \begin{bmatrix} x' \\ y' \end{bmatrix} = 0.$$

Performing the matrix multiplication and simplifying gives

$$2x'^2 - 2y'^2 + 3\sqrt{3}\,x' - 3y' = 0.$$

Completing the square in x' and y' gives

$$2\left(x'^2 + \tfrac{3}{2}\sqrt{3}\,x' + \tfrac{27}{16}\right) - 2\left(y'^2 - \tfrac{3}{2}y' + \tfrac{9}{16}\right) = \tfrac{9}{4}$$

or

$$2\left(x' + \tfrac{3}{4}\sqrt{3}\right)^2 - 2\left(y' - \tfrac{3}{4}\right)^2 = \tfrac{9}{4}.$$

Let $x'' = x' + \frac{3\sqrt{3}}{4}$ and $y'' = y' - \frac{3}{4}$; then we can write the equation as

$$2x''^2 - 2y''^2 = \tfrac{9}{4}$$

or equivalently

$$\frac{x''^2}{\frac{9}{8}} - \frac{y''^2}{\frac{9}{8}} = 1.$$

Thus this conic section is a hyperbola. The preceding is only a possible answer. The roles of x'' and y'' would be reversed if the eigenvalues of A were used in a different order.

Section 9.6, p. 499

In Exercises 1–13, we determine the matrix A of the quadric surface, find its eigenvalues, and use the classification from Table 9.2.

1. For the quadric surface $x^2 + y^2 + 2z^2 - 2xy - 4xz - 4yz + 4x = 8$,

$$A = \begin{bmatrix} 1 & -1 & -2 \\ -1 & 1 & -2 \\ -2 & -2 & 2 \end{bmatrix}.$$

The characteristic equation is

$$\det(\lambda I_3 - A) = \begin{vmatrix} \lambda - 1 & 1 & 2 \\ 1 & \lambda - 1 & 2 \\ 2 & 2 & \lambda - 2 \end{vmatrix} = \lambda^3 - 4\lambda^2 - 4\lambda + 16$$

$$= \lambda^2(\lambda - 4) - 4(\lambda - 4) = (\lambda - 4)(\lambda^2 - 4)$$
$$= (\lambda - 4)(\lambda - 2)(\lambda + 2) = 0$$

and hence the eigenvalues are $\lambda = 4, 2, -2$. Thus the inertia of A is $(2, 1, 0)$ and it follows from Table 9.2 that this quadric surface is a hyperboloid of one sheet.

3. For the quadric surface $z = 4xy$ or equivalently $-4xy + z = 0$,

$$A = \begin{bmatrix} 0 & -2 & 0 \\ -2 & 0 & 0 \\ 0 & 0 & 0 \end{bmatrix}.$$

The characteristic equation is

$$\det(\lambda I_3 - A) = \begin{vmatrix} \lambda & 2 & 0 \\ 2 & \lambda & 0 \\ 0 & 0 & \lambda \end{vmatrix} = \lambda^3 - 4\lambda = \lambda(\lambda^2 - 4) = 0$$

and hence the eigenvalues are $\lambda = 2, -2, 0$. Thus the inertia of A is $(1, 1, 1)$ and it follows from Table 9.2 that this quadric surface is a hyperbolic paraboloid.

5. For the quadric surface $x^2 - y = 0$,

$$A = \begin{bmatrix} 1 & 0 & 0 \\ 0 & 0 & 0 \\ 0 & 0 & 0 \end{bmatrix}.$$

The characteristic equation is

$$\det(\lambda I_3 - A) = \begin{vmatrix} \lambda - 1 & 0 & 0 \\ 0 & \lambda & 0 \\ 0 & 0 & \lambda \end{vmatrix} = \lambda^2(\lambda - 1) = 0$$

and hence the eigenvalues are $\lambda = 1, 0, 0$. Thus the inertia of A is $(1, 0, 2)$ and it follows from Table 9.2 that this quadric surface is a parabolic cylinder.

7. For the quadric surface $5y^2 + 20y + z - 23 = 0$,

$$A = \begin{bmatrix} 0 & 0 & 0 \\ 0 & 5 & 0 \\ 0 & 0 & 0 \end{bmatrix}.$$

The characteristic equation is

$$\det(\lambda I_3 - A) = \begin{vmatrix} \lambda & 0 & 0 \\ 0 & \lambda - 5 & 0 \\ 0 & 0 & \lambda \end{vmatrix} = \lambda^2(\lambda - 5) = 0$$

and hence the eigenvalues are $\lambda = 5, 0, 0$. Thus the inertia of A is $(1, 0, 2)$ and it follows from Table 9.2 that this quadric surface is a parabolic cylinder.

9. For the quadric surface $4x^2 + 9y^2 + z^2 + 8x - 18y - 4x - 19 = 0$,

$$A = \begin{bmatrix} 4 & 0 & 0 \\ 0 & 9 & 0 \\ 0 & 0 & 1 \end{bmatrix}.$$

Since A is diagonal, its eigenvalues are $\lambda = 9, 4, 1$. Thus the inertia of A is $(3, 0, 0)$ and it follows from Table 9.2 that this quadric surface is an ellipsoid.

11. For the quadric surface $x^2 + 4y^2 + 4x + 16y - 16z - 4 = 0$,

$$A = \begin{bmatrix} 1 & 0 & 0 \\ 0 & 4 & 0 \\ 0 & 0 & 0 \end{bmatrix}.$$

Since A is diagonal, its eigenvalues are $\lambda = 4, 1, 0$. Thus the inertia of A is $(2, 0, 1)$ and it follows from Table 9.2 that this quadric surface is an elliptic paraboloid.

13. For the quadric surface $x^2 - 4z^2 - 4x + 8z = 0$,

$$A = \begin{bmatrix} 1 & 0 & 0 \\ 0 & 0 & 0 \\ 0 & 0 & -4 \end{bmatrix}.$$

Since A is diagonal, its eigenvalues are $\lambda = 1, -4, 0$. Thus the inertia of A is $(1, 1, 1)$ and it follows from Table 9.2 that this quadric surface is a hyperbolic paraboloid.

In Exercises 15–27 we proceed as above and then perform any rotations or translations that are required to obtain the standard form. If a rotation is required then we must find the eigenvectors of A.

15. For the quadric surface $x^2 + 2y^2 + 2z^2 + 2yz = 1$,

$$A = \begin{bmatrix} 1 & 0 & 0 \\ 0 & 2 & 1 \\ 0 & 1 & 2 \end{bmatrix}.$$

The characteristic equation of A is

$$\det(\lambda I_3 - A) = \begin{vmatrix} \lambda - 1 & 0 & 0 \\ 0 & \lambda - 2 & -1 \\ 0 & -1 & \lambda - 2 \end{vmatrix} = (\lambda - 1)(\lambda^2 - 4\lambda + 3) = (\lambda - 1)(\lambda - 1)(\lambda - 3) = 0$$

and hence the eigenvalues are $\lambda = 3, 1, 1$. Thus the inertia of A is $(3, 0, 0)$ and it follows from Table 9.2 that this quadric surface is an ellipsoid. Since there is a cross product term we must perform a rotation (and possibly a translation) to obtain the standard form. Hence we require the eigenvectors. Solving the appropriate homogeneous systems we have eigenvalue and eigenvector pairs

$$\lambda_1 = 3, \mathbf{v}_1 = \begin{bmatrix} 0 \\ 1 \\ 1 \end{bmatrix}; \quad \lambda_2 = 1, \mathbf{v}_2 = \begin{bmatrix} 1 \\ 0 \\ 0 \end{bmatrix}; \quad \lambda_3 = 1, \mathbf{v}_3 = \begin{bmatrix} 0 \\ -1 \\ 1 \end{bmatrix}.$$

Normalizing the eigenvectors we have

$$\mathbf{u}_1 = \begin{bmatrix} 0 \\ \frac{1}{\sqrt{2}} \\ \frac{1}{\sqrt{2}} \end{bmatrix}, \qquad \mathbf{u}_2 = \mathbf{v}_2 = \begin{bmatrix} 1 \\ 0 \\ 0 \end{bmatrix}, \qquad \mathbf{u}_3 = \begin{bmatrix} 0 \\ -\frac{1}{\sqrt{2}} \\ \frac{1}{\sqrt{2}} \end{bmatrix}.$$

Let $P = \begin{bmatrix} \mathbf{u}_1 & \mathbf{u}_2 & \mathbf{u}_3 \end{bmatrix}$. Then $\det(P) = -1$, so to have a counterclockwise rotation we will redefine the eigenvector $\mathbf{u}_2 = -\mathbf{v}_2$. (This is valid since any nonzero multiple of an eigenvector is another eigenvector.) Thus

$$P = \begin{bmatrix} 0 & -1 & 0 \\ \frac{1}{\sqrt{2}} & 0 & -\frac{1}{\sqrt{2}} \\ \frac{1}{\sqrt{2}} & 0 & \frac{1}{\sqrt{2}} \end{bmatrix}.$$

Let $\mathbf{x} = P\mathbf{y}$, where $\mathbf{y} = \begin{bmatrix} x' & y' & z' \end{bmatrix}^T$. Then the quadric surface can be written as

$$(P\mathbf{y})^T A (P\mathbf{y}) = x'^2 + y'^2 + 3z'^2 = 1.$$

Hence the standard form is

$$\frac{x'^2}{1} + \frac{y'^2}{1} + \frac{z'^2}{\frac{1}{3}} = 1.$$

Note that for this problem no translations were required. The expression for the standard form is not unique. The roles of x', y', and z' may be interchanged depending upon the order of the eigenvalues used as columns in P.

17. For the quadric surface $2xz - 2x - 4y - 4z + 8 = 0$,

$$A = \begin{bmatrix} 0 & 0 & 1 \\ 0 & 0 & 0 \\ 1 & 0 & 0 \end{bmatrix}.$$

The characteristic equation of A is

$$\det(\lambda I_3 - A) = \begin{bmatrix} \lambda & 0 & -1 \\ 0 & \lambda & 0 \\ -1 & 0 & \lambda \end{bmatrix} = \lambda^3 - \lambda = \lambda(\lambda^2 - 1) = 0$$

and hence the eigenvalues are $\lambda = 1, -1, 0$. Thus the inertia of A is $(1, 1, 1)$ and it follows from Table 9.2 that this quadric surface is a hyperbolic paraboloid. Since there is a cross product term we must perform a rotation (and possibly a translation) to obtain the standard form. Hence we require the eigenvectors. Solving the appropriate homogeneous systems we have eigenvalue and eigenvector pairs

$$\lambda_1 = 1, \mathbf{v}_1 = \begin{bmatrix} 1 \\ 0 \\ 1 \end{bmatrix}; \quad \lambda_2 = -1, \mathbf{v}_2 = \begin{bmatrix} 1 \\ 0 \\ -1 \end{bmatrix}; \quad \lambda_3 = 0, \mathbf{v}_3 = \begin{bmatrix} 0 \\ 1 \\ 0 \end{bmatrix}.$$

Normalizing the eigenvectors, we have

$$\mathbf{u}_1 = \begin{bmatrix} \frac{1}{\sqrt{2}} \\ 0 \\ \frac{1}{\sqrt{2}} \end{bmatrix}, \qquad \mathbf{u}_2 = \begin{bmatrix} \frac{1}{\sqrt{2}} \\ 0 \\ -\frac{1}{\sqrt{2}} \end{bmatrix}, \qquad \mathbf{u}_3 = \mathbf{v}_3 = \begin{bmatrix} 0 \\ 1 \\ 0 \end{bmatrix}.$$

Let $P = \begin{bmatrix} \mathbf{u}_1 & \mathbf{u}_2 & \mathbf{u}_3 \end{bmatrix}$. Then $\det(P) = 1$ so we have a counterclockwise rotation. Let $\mathbf{x} = P\mathbf{y}$, where $\mathbf{y} = \begin{bmatrix} x' & y' & z' \end{bmatrix}^T$. Then the quadric surface can be written as

$$(P\mathbf{y})^T A(P\mathbf{y}) + \begin{bmatrix} -2 & -4 & -4 \end{bmatrix} P\mathbf{y} + 8 = 0$$

which simplifies to

$$x'^2 - y'^2 - \frac{6}{\sqrt{2}}x' + \frac{2}{\sqrt{2}}y' - 4z' + 8 = 0.$$

Completing the square in x' and y', we have

$$\left(x'^2 - \frac{6}{\sqrt{2}}x' + \frac{9}{2} \right) - \left(y'^2 - \frac{2}{\sqrt{2}}y' + \frac{1}{2} \right) - 4z' + 8 = \frac{9}{2} - 1$$

or equivalently,

$$\left(x' - \frac{3}{\sqrt{2}} \right)^2 - \left(y' - \frac{1}{\sqrt{2}} \right)^2 - 4z' + 8 = 8.$$

Let $x'' = x' - \frac{3}{\sqrt{2}}$, $y'' = y' - \frac{1}{\sqrt{2}}$, and $z'' = z'$. Then we have

$$x''^2 - y''^2 - 4z'' = 0$$

and the standard form is

$$\frac{x''^2}{4} - \frac{y''^2}{4} = z''.$$

The expression for the standard form is not unique. The roles of x'', y'', and z'' may be interchanged depending upon the order of the eigenvectors used as columns in P.

19. For the quadric surface $x^2 + y^2 + z^2 + 2xy = 8$,

$$A = \begin{bmatrix} 1 & 1 & 0 \\ 1 & 1 & 0 \\ 0 & 0 & 1 \end{bmatrix}.$$

The characteristic equation of A is

$$\det(\lambda I_3 - A) = \begin{vmatrix} \lambda - 1 & -1 & 0 \\ -1 & \lambda - 1 & 0 \\ 0 & 0 & \lambda - 1 \end{vmatrix} = (\lambda - 1)^3 - (\lambda - 1)$$

$$= (\lambda - 1)(\lambda^2 - 2\lambda) = (\lambda - 1)(\lambda - 2)\lambda = 0$$

and hence the eigenvalues are $\lambda = 2, 1, 0$. Thus the inertia of A is $(2, 0, 1)$ and it follows from Table 9.2 that this quadric surface is an elliptic paraboloid. Since there is a cross product term we must perform a rotation to obtain the standard form. Hence we require the eigenvectors. Solving the appropriate homogeneous systems we have eigenvalue and eigenvector pairs

$$\lambda_1 = 2, \mathbf{v}_1 = \begin{bmatrix} 1 \\ 1 \\ 0 \end{bmatrix}; \quad \lambda_2 = 1, \mathbf{v}_2 = \begin{bmatrix} 0 \\ 0 \\ 1 \end{bmatrix}; \quad \lambda_3 = 0, \mathbf{v}_3 = \begin{bmatrix} -1 \\ 1 \\ 0 \end{bmatrix}.$$

Normalizing the eigenvectors we have

$$\mathbf{u}_1 = \begin{bmatrix} \frac{1}{\sqrt{2}} \\ \frac{1}{\sqrt{2}} \\ 0 \end{bmatrix}, \quad \mathbf{u}_2 = \mathbf{v}_2 = \begin{bmatrix} 0 \\ 0 \\ 1 \end{bmatrix}, \quad \mathbf{u}_3 = \begin{bmatrix} -\frac{1}{\sqrt{2}} \\ \frac{1}{\sqrt{2}} \\ 0 \end{bmatrix}.$$

Let $P = \begin{bmatrix} \mathbf{u}_1 & \mathbf{u}_2 & \mathbf{u}_3 \end{bmatrix}$. Then $\det(P) = -1$, so to have a counterclockwise rotation we will redefine the eigenvector $\mathbf{u}_2 = -\mathbf{v}_2$. (This is valid since any nonzero multiple of an eigenvector is another eigenvector.) Thus

$$P = \begin{bmatrix} \frac{1}{\sqrt{2}} & 0 & -\frac{1}{\sqrt{2}} \\ \frac{1}{\sqrt{2}} & 0 & \frac{1}{\sqrt{2}} \\ 0 & -1 & 0 \end{bmatrix}.$$

Let $\mathbf{x} = P\mathbf{y}$, where $\mathbf{y} = \begin{bmatrix} x' & y' & z' \end{bmatrix}^T$. Then the quadric surface can be written as

$$(P\mathbf{y})^T A (P\mathbf{y}) = 2x'^2 + y'^2 = 8.$$

Hence the standard form is

$$\frac{x'^2}{4} + \frac{y'^2}{8} = 1.$$

Note that for this problem no translations were required. The expression for the standard form is not unique. The roles of x', y', and z' may be interchanged depending upon the order of the eigenvectors used as columns in P.

21. For the quadric surface $2x^2 + 2y^2 + 4z^2 - 4xy - 8xz - 8yz + 8x = 15$,

$$A = \begin{bmatrix} 2 & -2 & -4 \\ -2 & 2 & -4 \\ -4 & -4 & 4 \end{bmatrix}.$$

The characteristic equation of A is

$$\det(\lambda I_3 - A) = \begin{vmatrix} \lambda - 2 & 2 & 4 \\ 2 & \lambda - 2 & 4 \\ 4 & 4 & \lambda - 4 \end{vmatrix} = \lambda^3 - 8\lambda^2 - 16\lambda + 128$$

$$= \lambda^2(\lambda - 8) - 16(\lambda - 8) = (\lambda - 8)(\lambda^2 - 16)$$
$$= (\lambda - 8)(\lambda - 4)(\lambda + 4) = 0$$

and hence the eigenvalues are $\lambda = 8, 4, -4$. Thus the inertia of A is $(2, 1, 0)$ and it follows from Table 9.2 that this quadric surface is a hyperboloid of one sheet. From the terms present we see that we must perform a rotation and a translation to obtain the standard form. Hence we require the eigenvectors. Solving the appropriate homogeneous systems, we have eigenvalue and eigenvector pairs

$$\lambda_1 = 8, \mathbf{v}_1 = \begin{bmatrix} -\frac{1}{2} \\ -\frac{1}{2} \\ 1 \end{bmatrix}; \quad \lambda_2 = 4, \mathbf{v}_2 = \begin{bmatrix} -1 \\ 1 \\ 0 \end{bmatrix}; \quad \lambda_3 = -4, \mathbf{v}_3 = \begin{bmatrix} 1 \\ 1 \\ 1 \end{bmatrix}.$$

Normalizing the eigenvectors we have

$$\mathbf{u}_1 = \begin{bmatrix} -\frac{1}{\sqrt{6}} \\ -\frac{1}{\sqrt{6}} \\ \frac{1}{\sqrt{\frac{3}{2}}} \end{bmatrix}, \quad \mathbf{u}_2 = \begin{bmatrix} -\frac{1}{\sqrt{2}} \\ \frac{1}{\sqrt{2}} \\ 0 \end{bmatrix}, \quad \mathbf{u}_3 = \begin{bmatrix} \frac{1}{\sqrt{3}} \\ \frac{1}{\sqrt{3}} \\ \frac{1}{\sqrt{3}} \end{bmatrix}.$$

Let $P = \begin{bmatrix} \mathbf{u}_1 & \mathbf{u}_2 & \mathbf{u}_3 \end{bmatrix}$. Then $\det(P) = -1$ so we do not have a counterclockwise rotation. Here we replace $\mathbf{u}_1$ by $-\mathbf{u}_1$ and redefine P to be

$$P = \begin{bmatrix} \frac{1}{\sqrt{6}} & -\frac{1}{\sqrt{2}} & \frac{1}{\sqrt{3}} \\ \frac{1}{\sqrt{6}} & \frac{1}{\sqrt{2}} & \frac{1}{\sqrt{3}} \\ -\frac{1}{\sqrt{\frac{3}{2}}} & 0 & \frac{1}{\sqrt{3}} \end{bmatrix}.$$

Then $\det(P) = 1$ (verify). Let $\mathbf{x} = P\mathbf{y}$, where $\mathbf{y} = \begin{bmatrix} x' & y' & z' \end{bmatrix}^T$; then the quadric surface can be written as

$$(P\mathbf{y})^T A (P\mathbf{y}) + \begin{bmatrix} 8 & 0 & 0 \end{bmatrix} P\mathbf{y} = 15$$

which simplifies to

$$8x'^2 + 4y'^2 - 4z'^2 + \frac{8}{\sqrt{6}} x' - \frac{8}{\sqrt{2}} y' + \frac{8}{\sqrt{3}} z' = 15.$$

Completing the square in x', y', and z' we have

$$8\left(x'^2 + \frac{1}{\sqrt{6}} x' + \frac{1}{24}\right) + 4\left(y'^2 - \frac{2}{\sqrt{2}} y' + \frac{1}{2}\right) - 4\left(z'^2 - \frac{2}{\sqrt{3}} z' + \frac{1}{3}\right) = \frac{8}{24} + 2 - \frac{4}{3} + 15,$$

or equivalently,

$$8\left(x' + \frac{1}{\sqrt{24}}\right)^2 + 4\left(y' - \frac{1}{\sqrt{2}}\right)^2 - 4\left(z' - \frac{1}{\sqrt{3}}\right)^2 = 16.$$

Let $x'' = x' + \frac{1}{\sqrt{24}}$, $y'' = y' - \frac{1}{\sqrt{2}}$, and $z'' = z' - \frac{1}{\sqrt{3}}$; then we have

$$8x''^2 + 4y''^2 - 4z''^2 = 16$$

and the standard form is

$$\frac{x''^2}{2} + \frac{y''^2}{4} - \frac{z''^2}{4} = 1.$$

The expression for the standard form is not unique. The roles of x'', y'', and z'' may be interchanged depending upon the order of the eigenvectors used as columns in P.

23. For the quadric surface $2y^2 + 2z^2 + 4yz + \frac{16}{\sqrt{2}} x + 4 = 0$,

$$A = \begin{bmatrix} 0 & 0 & 0 \\ 0 & 2 & 2 \\ 0 & 2 & 2 \end{bmatrix}.$$

The characteristic equation of A is

$$\det(\lambda I_3 - A) = \begin{vmatrix} \lambda & 0 & 0 \\ 0 & \lambda - 2 & -2 \\ 0 & -2 & \lambda - 2 \end{vmatrix} = \lambda[(\lambda - 2)^2 - 4] = \lambda^2(\lambda - 4) = 0$$

and thus the eigenvalues are $\lambda = 4, 0, 0$. Thus the inertia of A is $(1, 0, 2)$ and it follows from Table 9.2 that this quadric surface is a parabolic cylinder. Since there is a cross product term we must perform a rotation to obtain the standard form. Hence we require the eigenvectors. Solving the appropriate homogeneous systems we have eigenvalue and eigenvector pairs

$$\lambda_1 = 4, \mathbf{v}_1 = \begin{bmatrix} 0 \\ 1 \\ 1 \end{bmatrix}; \quad \lambda_2 = 0, \mathbf{v}_2 = \begin{bmatrix} 1 \\ 0 \\ 0 \end{bmatrix}; \quad \lambda_3 = 0, \mathbf{v}_3 = \begin{bmatrix} 0 \\ -1 \\ 1 \end{bmatrix}.$$

Normalizing the eigenvectors, we have

$$\mathbf{u}_1 = \begin{bmatrix} 0 \\ \frac{1}{\sqrt{2}} \\ \frac{1}{\sqrt{2}} \end{bmatrix}, \qquad \mathbf{u}_2 = \mathbf{v}_2 = \begin{bmatrix} 1 \\ 0 \\ 0 \end{bmatrix}, \qquad \mathbf{u}_3 = \begin{bmatrix} 0 \\ -\frac{1}{\sqrt{2}} \\ \frac{1}{\sqrt{2}} \end{bmatrix}.$$

Let $P = \begin{bmatrix} \mathbf{u}_1 & \mathbf{u}_2 & \mathbf{u}_3 \end{bmatrix}$. Then $\det(P) = -1$, so to have a counterclockwise rotation we will redefine the eigenvector $\mathbf{u}_2 = -\mathbf{v}_2$. (This is valid since any nonzero multiple of an eigenvector is another eigenvector.) Thus

$$P = \begin{bmatrix} 0 & -1 & 0 \\ \frac{1}{\sqrt{2}} & 0 & -\frac{1}{\sqrt{2}} \\ \frac{1}{\sqrt{2}} & 0 & \frac{1}{\sqrt{2}} \end{bmatrix}.$$

Let $\mathbf{x} = P\mathbf{y}$, where $\mathbf{y} = \begin{bmatrix} x' & y' & z' \end{bmatrix}^T$; then the quadric surface can be written as

$$(P\mathbf{y})^T A(P\mathbf{y}) + \begin{bmatrix} \frac{16}{\sqrt{2}} & 0 & 0 \end{bmatrix} P\mathbf{y} + 4 = 0,$$

or equivalently,

$$4x'^2 - \frac{16}{\sqrt{2}} y' + 4 = 0.$$

Next we rearrange the equation to determine the translation required. We have

$$4x'^2 - \frac{16}{\sqrt{2}} \left(y' - \frac{\sqrt{2}}{4} \right) = 0.$$

Let $x'' = x'$ and $y'' = y' - \frac{\sqrt{2}}{4}$. Then it follows that the equation is

$$4x''^2 - \frac{16}{\sqrt{2}} y'' = 0.$$

Hence the standard form is

$$x''^2 = \frac{4}{\sqrt{2}} y''.$$

The expression for the standard form is not unique. The roles of x'', y'', and z'' may be interchanged depending upon the order of the eigenvectors used as columns in P.

25. For the quadric surface $-x^2 - y^2 - z^2 + 4xy + 4xz + 4yz + \frac{3}{\sqrt{2}} x - \frac{3}{\sqrt{2}} y = 6$,

$$A = \begin{bmatrix} -1 & 2 & 2 \\ 2 & -1 & 2 \\ 2 & 2 & -1 \end{bmatrix}.$$

The characteristic equation of A is

$$\det(\lambda I_3 - A) = \begin{vmatrix} \lambda + 1 & -2 & -2 \\ -2 & \lambda + 1 & -2 \\ -2 & -2 & \lambda + 1 \end{vmatrix} = \lambda^3 + 3\lambda^2 - 9\lambda - 27 = (\lambda - 3)(\lambda + 3)^2 = 0$$

and hence the eigenvalues are $\lambda = 3, -3, -3$. Thus the inertia of A is $(1, 2, 0)$ and it follows from Table 9.2 that this quadric surface is a hyperboloid of two sheets. From the terms present we see that we must perform a rotation and a translation to obtain the standard form. Hence we require the eigenvectors. Solving the appropriate homogeneous systems we have eigenvalue and eigenvector pairs

$$\lambda_1 = 3, \mathbf{v}_1 = \begin{bmatrix} 1 \\ 1 \\ 1 \end{bmatrix}; \quad \lambda_2 = -3, \mathbf{v}_2 = \begin{bmatrix} -1 \\ 1 \\ 0 \end{bmatrix}; \quad \lambda_3 = -3, \mathbf{v}_3 = \begin{bmatrix} -1 \\ 0 \\ 1 \end{bmatrix}.$$

Unfortunately the eigenvectors corresponding to the eigenvalue $\lambda = -3$, which has multiplicity two, are not orthogonal. Thus we need to use the Gram–Schmidt process. For consistency of notation, let $\mathbf{y}_1 = \mathbf{v}_1$ and proceed as follows: let $\mathbf{y}_2 = \mathbf{v}_2$ (which is orthogonal to $\mathbf{y}_1$), then define

$$\mathbf{y}_3 = \mathbf{v}_3 - \frac{\mathbf{x}_3 \cdot \mathbf{y}_2}{\mathbf{y}_2 \cdot \mathbf{y}_2} \mathbf{y}_2 = \begin{bmatrix} -\frac{1}{2} \\ -\frac{1}{2} \\ 1 \end{bmatrix}.$$

Normalizing the eigenvectors $\mathbf{y}_j$, we have

$$\mathbf{u}_1 = \begin{bmatrix} \frac{1}{\sqrt{3}} \\ \frac{1}{\sqrt{3}} \\ \frac{1}{\sqrt{3}} \end{bmatrix}, \quad \mathbf{u}_2 = \begin{bmatrix} -\frac{1}{\sqrt{2}} \\ \frac{1}{\sqrt{2}} \\ 0 \end{bmatrix}, \quad \mathbf{u}_3 = \begin{bmatrix} -\frac{1}{\sqrt{6}} \\ -\frac{1}{\sqrt{6}} \\ \sqrt{\frac{3}{2}} \end{bmatrix}.$$

Let $P = \begin{bmatrix} \mathbf{u}_1 & \mathbf{u}_2 & \mathbf{u}_3 \end{bmatrix}$. Then $\det(P) = 1$ so we do have a counterclockwise rotation. Let $\mathbf{x} = P\mathbf{y}$, where $\mathbf{y} = \begin{bmatrix} x' & y' & z' \end{bmatrix}^T$; then the quadric surface can be written as

$$(P\mathbf{y})^T A(P\mathbf{y}) + \begin{bmatrix} \frac{3}{\sqrt{2}} & -\frac{3}{\sqrt{2}} & 0 \end{bmatrix} P\mathbf{y} = 6$$

which simplifies to

$$3x'^2 - 3y'^2 - 3z'^2 - 3y' = 6.$$

Completing the square in y' we have

$$3x'^2 - 3\left(y'^2 + y' + \frac{1}{4}\right) - 3z'^2 = 6 - \frac{3}{4}$$

or equivalently

$$3x'^2 - 3\left(y' + \frac{1}{2}\right)^2 - 3z'^2 = \frac{21}{4}.$$

Let $x'' = x'$, $y'' = y' + \frac{1}{2}$, and $z'' = z'$. Then we have

$$3x''^2 - 3y''^2 - 3z''^2 = \frac{21}{4}$$

and the standard form is

$$\frac{x''^2}{\frac{7}{4}} - \frac{y''^2}{\frac{7}{4}} - \frac{z''^2}{\frac{7}{4}} = 1.$$

The expression for the standard form is not unique. The roles of x'', y'', and z'' may be interchanged depending upon the order of the eigenvectors used as columns in P.

27. For the quadric surface $x^2 + y^2 - z^2 - 2x - 4y - 4z + 1 = 0$,

$$A = \begin{bmatrix} 1 & 0 & 0 \\ 0 & 1 & 0 \\ 0 & 0 & -1 \end{bmatrix}.$$

The characteristic equation is

$$\det(\lambda I_3 - A) = \begin{vmatrix} \lambda - 1 & 0 & 0 \\ 0 & \lambda - 1 & 0 \\ 0 & 0 & \lambda + 1 \end{vmatrix} = (\lambda - 1)^2(\lambda + 1) = 0$$

and hence the eigenvalues are $\lambda = 1, 1, -1$. Thus the inertia of A is $(2, 1, 0)$ and it follows from Table 9.2 that this quadric surface is a hyperboloid of one sheet. Since there are no cross product terms we do not need the eigenvectors. We merely complete the square in each of the variables to get

$$(x^2 - 2x + 1) + (y^2 - 4y + 4) - (z^2 + 4z + 4) = -1 + 1 + 4 - 4$$
$$(x - 1)^2 + (y - 2)^2 - (z + 2)^2 = 0.$$

Let $x' = x - 1$, $y' = y - 2$, and $z' = z + 2$. Then we have the standard form

$$x''^2 + y''^2 - z''^2 = 0.$$

This is really a cone which is a special case of a hyperboloid of one sheet.

Supplementary Exercises, p. 500

1. (a) For $A(t) = \begin{bmatrix} t^2 & \dfrac{1}{t+1} \\ 4 & e^{-t} \end{bmatrix}$ we have

$$\frac{d}{dt}[A(t)] = \begin{bmatrix} 2t & \dfrac{-1}{(t+1)^2} \\ 0 & -e^{-t} \end{bmatrix} \quad \text{and} \quad \int_0^t A(s)\,ds = \begin{bmatrix} \frac{1}{3}t^3 & \ln|t+1| \\ 4t & -e^{-t} + 1 \end{bmatrix}.$$

(b) For $A(t) = \begin{bmatrix} \sin 2t & 0 & 0 \\ 0 & 1 & -t \\ 0 & te^{t^2} & \dfrac{t}{t^2+1} \end{bmatrix}$ we have

$$\frac{d}{dt}[A(t)] = \begin{bmatrix} 2\cos 2t & 0 & 0 \\ 0 & 0 & -1 \\ 0 & e^{t^2} + 2t^2 e^{t^2} & \dfrac{1 - t^2}{(t^2+1)^2} \end{bmatrix}$$

and

$$\int_0^t A(s)\,ds = \begin{bmatrix} -\frac{1}{2}\cos 2t + \frac{1}{2} & 0 & 0 \\ 0 & t & -\frac{1}{2}t^2 \\ 0 & \frac{1}{2}e^{t^2} - \frac{1}{2} & \frac{1}{2}\ln(t^2+1) \end{bmatrix}.$$

3. (a) Using the methods of Chapter 8 we find the eigenvalue and eigenvector pairs for A are

$$\lambda_1 = 0, \; \mathbf{x}_1 = \begin{bmatrix} 4 \\ 4 \\ 1 \end{bmatrix}; \quad \lambda_2 = 5, \; \mathbf{x}_2 = \begin{bmatrix} -1 \\ -6 \\ 1 \end{bmatrix}; \quad \lambda_3 = -3, \; \mathbf{x}_3 = \begin{bmatrix} -1 \\ 2 \\ 1 \end{bmatrix}.$$

Let

$$D = \begin{bmatrix} 0 & 0 & 0 \\ 0 & 5 & 0 \\ 0 & 0 & -3 \end{bmatrix} \quad \text{and} \quad P = \begin{bmatrix} \mathbf{x}_1 & \mathbf{x}_2 & \mathbf{x}_3 \end{bmatrix}.$$

Then we solve the system $P\mathbf{b} = \mathbf{x}_0$ and obtain $\mathbf{b} = \begin{bmatrix} \frac{2}{5} \\ \frac{7}{20} \\ \frac{1}{4} \end{bmatrix}$. Hence

$$\mathbf{x}(t) = Pe^{Dt}\mathbf{b} = \frac{2}{5}\begin{bmatrix} 4 \\ 4 \\ 1 \end{bmatrix} + \frac{7}{20}\begin{bmatrix} -1 \\ -6 \\ 1 \end{bmatrix}e^{5t} + \frac{1}{4}\begin{bmatrix} -1 \\ 2 \\ 1 \end{bmatrix}e^{-3t}.$$

(b) Using the methods of Chapter 8 we find the eigenvalue and eigenvector pairs for A are

$$\lambda_1 = 0, \; \mathbf{x}_1 = \begin{bmatrix} 1 \\ 0 \\ 0 \end{bmatrix}; \quad \lambda_2 = 2, \; \mathbf{x}_2 = \begin{bmatrix} 1 \\ 2 \\ 4 \end{bmatrix}; \quad \lambda_3 = -4, \; \mathbf{x}_3 = \begin{bmatrix} 1 \\ -4 \\ 16 \end{bmatrix}.$$

Let

$$D = \begin{bmatrix} 0 & 0 & 0 \\ 0 & 2 & 0 \\ 0 & 0 & -4 \end{bmatrix} \quad \text{and} \quad P = \begin{bmatrix} \mathbf{x}_1 & \mathbf{x}_2 & \mathbf{x}_3 \end{bmatrix}.$$

Then we solve the system $P\mathbf{b} = \mathbf{x}_0$ and obtain $\mathbf{b} = \begin{bmatrix} \frac{7}{8} \\ \frac{1}{12} \\ \frac{1}{24} \end{bmatrix}$. Hence

$$\mathbf{x}(t) = Pe^{Dt}\mathbf{b} = \frac{7}{8}\begin{bmatrix} 1 \\ 0 \\ 0 \end{bmatrix} + \frac{1}{12}\begin{bmatrix} 1 \\ 2 \\ 4 \end{bmatrix}e^{2t} + \frac{1}{24}\begin{bmatrix} 1 \\ -4 \\ 16 \end{bmatrix}e^{-4t}.$$

5. (a) Let

$$A = \begin{bmatrix} 1 & 1 \\ 3 & -1 \end{bmatrix}.$$

We find the eigenvalues and associated eigenvectors of A. The characteristic polynomial is

$$|\lambda I_2 - A| = \lambda^2 - 4 = (\lambda - 2)(\lambda + 2).$$

Hence the eigenvalues of A are $\lambda_1 = 2$ and $\lambda_2 = -2$. Find associated eigenvectors:
Case $\lambda = 2$. Solve the linear system

$$(2I_2 - A)\mathbf{x} = \begin{bmatrix} 1 & -1 \\ -3 & 3 \end{bmatrix}\begin{bmatrix} x_1 \\ x_2 \end{bmatrix} = \mathbf{0}.$$

A solution is easily seen to be $x_1 = r$, $x_2 = r$. Thus,

$$\mathbf{x}_1 = \begin{bmatrix} 1 \\ 1 \end{bmatrix}$$

is an associated eigenvector.

Case $\lambda = -2$. Solve the linear system

$$\begin{bmatrix} -3 & -1 \\ -3 & -1 \end{bmatrix} \begin{bmatrix} x_1 \\ x_2 \end{bmatrix} = \mathbf{0}.$$

A solution is easily seen to be $x_1 = -\frac{1}{3}r$, $x_2 = r$. Thus,

$$\mathbf{x}_2 = \begin{bmatrix} -1 \\ 3 \end{bmatrix}$$

is an associated eigenvector.

The solution is

$$\mathbf{x}(t) = b_1 \begin{bmatrix} 1 \\ 1 \end{bmatrix} e^{2t} + b_2 \begin{bmatrix} -1 \\ 3 \end{bmatrix} e^{-2t}.$$

(b) For the initial conditions $x_1(0) = 4$ and $x_2(0) = 6$, we have

$$b_1 \begin{bmatrix} 1 \\ 1 \end{bmatrix} e^{2(0)} + b_2 \begin{bmatrix} -1 \\ 3 \end{bmatrix} e^{-2(0)} = \begin{bmatrix} 4 \\ 6 \end{bmatrix}$$

which leads to the linear system

$$b_1 \begin{bmatrix} 1 \\ 1 \end{bmatrix} + b_2 \begin{bmatrix} -1 \\ 3 \end{bmatrix} = \begin{bmatrix} 4 \\ 6 \end{bmatrix}$$

whose augmented matrix is

$$\begin{bmatrix} 1 & -1 & \vdots & 4 \\ 1 & 3 & \vdots & 6 \end{bmatrix}.$$

Transforming this matrix to reduced row echelon form, we have

$$\begin{bmatrix} 1 & 0 & \vdots & \frac{9}{2} \\ 0 & 1 & \vdots & \frac{1}{2} \end{bmatrix}$$

so $b_1 = \frac{9}{2}$ and $b_2 = \frac{1}{2}$. the solution to the initial value problem is

$$\mathbf{x} = \frac{9}{2} \begin{bmatrix} 1 \\ 1 \end{bmatrix} e^{2t} + \frac{1}{2} \begin{bmatrix} -1 \\ 3 \end{bmatrix} e^{-2t}.$$

7. The characteristic polynomial of the matrix A in Exercise 5 is $\lambda^2 - 4$. Hence the eigenvalues are ± 2. Therefore, the trajectory moves toward the origin. The origin is a saddle point.

T.1. Proceed by showing corresponding entries of the matrices involved are equal.

(a)
$$\left[\frac{d}{dt} [c_1 A(t) + c_2 B(t)] \right]_{ij} = \frac{d}{dt} [c_1 A(t) + c_2 B(t)]_{ij}$$

$$= \frac{d}{dt} [c_1 a_{ij}(t) + c_2 b_{ij}(t)] = c_1 \frac{d}{dt} a_{ij} + c_2 \frac{d}{dt} b_{ij}(t)$$

$$= c_1 \left[\frac{d}{dt} A(t) \right]_{ij} + c_2 \left[\frac{d}{dt} B(t) \right]_{ij} = \left[c_1 \frac{d}{dt} A(t) + c_2 \frac{d}{dt} B(t) \right]_{ij}.$$

(b)
$$\left[\int_a^t (c_1 A(s) + c_2 B(s))\, ds\right]_{ij} = \int_a^t [c_1 A(s) + c_2 B(s)]_{ij}\, ds$$

$$= \int_a^t [c_1 a_{ij}(s) + c_2 b_{ij}(s)]\, ds = c_1 \int_a^t a_{ij}\, ds + c_2 \int_a^t b_{ij}\, ds$$

$$= c_1 \left[\int_a^t A(s)\, ds\right]_{ij} + c_2 \left[\int_a^t B(s)\, ds\right]_{ij}$$

$$= \left[c_1 \int_a^t A(s)\, ds + c_2 \int_a^t B(s)\, ds\right]_{ij}.$$

(c)
$$\left[\frac{d}{dt}[A(t)B(t)]\right]_{ij} = \frac{d}{dt}[A(t)B(t)]_{ij} = \frac{d}{dt}\left(\sum_{k=1}^n a_{ik}(t)b_{kj}(t)\right)$$

$$= \sum_{k=1}^n \frac{d}{dt}[a_{ik}(t)b_{kj}(t)] = \sum_{k=1}^n \left(\frac{d}{dt}[a_{ik}(t)]b_{kj}(t) + a_{ik}(t)\frac{d}{dt}[b_{kj}(t)]\right)$$

$$= \sum_{k=1}^n \left(\frac{d}{dt}[a_{ik}(t)]b_{kj}(t)\right) + \sum_{k=1}^n \left(a_{ik}(t)\frac{d}{dt}[b_{kj}(t)]\right)$$

$$= \left[\frac{d}{dt}[A(t)B(t)]\right]_{ij} + \left[A(t)\frac{d}{dt}[B(t)]\right]_{ij}$$

$$= \left[\frac{d}{dt}[A(t)B(t) + A(t)\frac{d}{dt}[B(t)]\right]_{ij}.$$

T.3. If $AB = BA$, then $(A+B)^2 = A^2 + 2AB + B^2$, $(A+B)^3 = A^3 + 3A^2B + 3AB^2 + B^3$, and in general for $k = 2, 3, \ldots$

$$(A + B)^k = A^k + \binom{k}{1}A^{k-1}B + \binom{k}{2}A^{k-2}B^2 + \cdots + \binom{k}{k-1}AB^{k-1} + B^k.$$

Applying these results to the product of the series in the following, we see that

$$e^A e^B = \left[I_n + A + A^2\frac{1}{2!} + A^3\frac{1}{3!} + \cdots\right]\left[I_n + B + B^2\frac{1}{2!} + B^3\frac{1}{3!} + \cdots\right]$$

$$= I_n + (A + B) + \frac{1}{2!}(A^2 + 2AB + B^2) + \frac{1}{3!}(A^3 + 3A^2B + 3AB^2 + B^3) + \cdots$$

$$= e^{A+B}$$

T.5. Recall that if C is $n \times n$ and R is $n \times 1$ then $CR = r_1 \text{col}_1(C) + r_2 \text{col}_2(C) + \cdots + r_n \text{col}_n(C)$. It follows that

$$\mathbf{x}(t) = \begin{bmatrix} p_1 & p_2 & \cdots & p_n \end{bmatrix} \begin{bmatrix} e^{c_1 t} b_1 \\ e^{c_2 t} b_2 \\ \vdots \\ e^{c_n t} b_n \end{bmatrix} = P \begin{bmatrix} e^{c_1 t} & 0 & \cdots & 0 \\ 0 & e^{c_2 t} & \cdots & \vdots \\ 0 & 0 & \cdot & \cdot \\ \vdots & \vdots & \ddots & \vdots \\ 0 & 0 & \cdots & e^{c_n t} \end{bmatrix} \begin{bmatrix} b_1 \\ b_2 \\ \vdots \\ b_n \end{bmatrix} = Pe^{Dt}B.$$

Chapter 10

Linear Transformations and Matrices

Section 10.1, p. 507

1. (a) Let $\mathbf{u} = (x_1, y_1)$ and $\mathbf{v} = (x_2, y_2)$. Then

$$L(\mathbf{u} + \mathbf{v}) = L((x_1, y_1) + (x_2, y_2)) = L(x_1 + x_2, y_1 + y_2)$$
$$= ((x_1 + x_2 + y_1 + y_2, x_1 + x_2 - (y_1 + y_2)).$$

On the other hand,

$$L(\mathbf{u}) + L(\mathbf{v}) = L((x_1, y_1)) + L((x_2, y_2)) = (x_1 + y_1, x_1 - y_1) + (x_2 + y_2, x_2 - y_2)$$
$$= (x_1 + y_1 + x_2 + y_2, x_1 - y_1 + x_2 - y_2)$$
$$= (x_1 + x_2 + y_1 + y_2, x_1 + x_2 - (y_1 + y_2)).$$

Thus, $L(\mathbf{u} + \mathbf{v}) = L(\mathbf{u}) + L(\mathbf{v})$. Next, let c be any real number. Then

$$L(c\mathbf{u}) = L(cx, cy) = (cx_1 + cy_1, cx_1 - cy_1) = c(x_1 + y_1, x_1 - y_1) = cL(\mathbf{u}).$$

Thus L is a linear transformation.

(b) Let

$$\mathbf{u} = \begin{bmatrix} x_1 \\ y_1 \\ z_1 \end{bmatrix} \quad \text{and} \quad \mathbf{v} = \begin{bmatrix} x_2 \\ y_2 \\ z_2 \end{bmatrix}.$$

Then

$$L(\mathbf{u} + \mathbf{v}) = L\left(\begin{bmatrix} x_1 \\ y_1 \\ z_1 \end{bmatrix} + \begin{bmatrix} x_2 \\ y_2 \\ z_2 \end{bmatrix}\right) = L\left(\begin{bmatrix} x_1 + x_2 \\ y_1 + y_2 \\ z_1 + z_2 \end{bmatrix}\right) = \begin{bmatrix} x_1 + x_2 + 1 \\ y_1 + y_2 - (z_1 + z_2) \end{bmatrix}.$$

On the other hand,

$$L(\mathbf{u}) + L(\mathbf{v}) = L\left(\begin{bmatrix} x_1 \\ y_1 \\ z_1 \end{bmatrix}\right) + L\left(\begin{bmatrix} x_2 \\ y_2 \\ z_2 \end{bmatrix}\right) = \begin{bmatrix} x_1 + 1 \\ y_1 - z_1 \end{bmatrix} + \begin{bmatrix} x_2 + 1 \\ y_2 - z_2 \end{bmatrix} = \begin{bmatrix} x_1 + x_2 + 2 \\ y_1 + y_2 - (z_1 + z_2) \end{bmatrix}.$$

Since $L(\mathbf{u} + \mathbf{v}) \neq L(\mathbf{u}) + L(\mathbf{v})$, L is not a linear transformation.

(c) Let

$$\mathbf{u} = \begin{bmatrix} x_1 \\ y_1 \\ z_1 \end{bmatrix} \quad \text{and} \quad \mathbf{v} = \begin{bmatrix} x_2 \\ y_2 \\ z_2 \end{bmatrix}.$$

Then

$$L(\mathbf{u} + \mathbf{v}) = L\left(\begin{bmatrix} x_1 \\ y_1 \\ z_1 \end{bmatrix} + \begin{bmatrix} x_2 \\ y_2 \\ z_2 \end{bmatrix}\right) = L\left(\begin{bmatrix} x_1 + x_2 \\ y_1 + y_2 \\ z_1 + z_2 \end{bmatrix}\right) = \begin{bmatrix} 1 & 2 & 3 \\ -1 & 2 & 4 \end{bmatrix} \begin{bmatrix} x_1 + x_2 \\ y_1 + y_2 \\ z_1 + z_2 \end{bmatrix}.$$

On the other hand, using the properties of matrix addition and multiplication, we have

$$L(\mathbf{u}) + L(\mathbf{v}) = \begin{bmatrix} 1 & 2 & 3 \\ -1 & 2 & 4 \end{bmatrix} \begin{bmatrix} x_1 \\ y_1 \\ z_1 \end{bmatrix} + \begin{bmatrix} 1 & 2 & 3 \\ -1 & 2 & 4 \end{bmatrix} \begin{bmatrix} x_2 \\ y_2 \\ z_2 \end{bmatrix}$$

$$= \begin{bmatrix} 1 & 2 & 3 \\ -1 & 2 & 4 \end{bmatrix} \left(\begin{bmatrix} x_1 \\ y_1 \\ z_1 \end{bmatrix} + \begin{bmatrix} x_2 \\ y_2 \\ z_2 \end{bmatrix}\right)$$

$$= \begin{bmatrix} 1 & 2 & 3 \\ -1 & 2 & 4 \end{bmatrix} \begin{bmatrix} x_1 + x_2 \\ y_1 + y_2 \\ z_1 + z_2 \end{bmatrix},$$

so $L(\mathbf{u} + \mathbf{v}) = L(\mathbf{u}) + L(\mathbf{v})$. Next, let c be any real number. Then using the properties of scalar multiplication, we have

$$L(c\mathbf{u}) = L\left(\begin{bmatrix} cx_1 \\ cy_1 \\ cz_1 \end{bmatrix}\right) = \begin{bmatrix} 1 & 2 & 3 \\ -1 & 2 & 4 \end{bmatrix} \begin{bmatrix} cx_1 \\ cy_1 \\ cz_1 \end{bmatrix} = c\left(\begin{bmatrix} 1 & 2 & 3 \\ -1 & 2 & 4 \end{bmatrix} \begin{bmatrix} x_1 \\ y_1 \\ z_1 \end{bmatrix}\right) = cL(\mathbf{u}).$$

Hence L is a linear transformation.

3. (a) $L(p(t) + q(t)) = t[p(t) + q(t)] + [p(0) + q(0)] = [tp(t) + p(0)] + [tq(t) + q(0)] = L(p(t)) + L(q(t))$. If c is any real number, then

$$L(cp(t)) = tcp(t) + cp(0) = c(tp(t) + p(0)) = cL(p(t)).$$

Hence L is a linear transformation.

(b) $L(p(t) + q(t)) = t(p(t) + q(t)) + t^2 + 1$. On the other hand,

$$L(p(t)) + L(q(t)) = [tp(t) + t^2 + 1] + [tq(t) + t^2 + 1] = t[p(t) + q(t)] + 2t^2 + 2.$$

Since $L(p(t) + q(t)) \neq L(p(t)) + L(q(t))$, L is not a linear transformation.

(c) $L((a_1 t + b_1) + (a_2 t + b_2)) = L((a_1 + a_2)t + b_1 + b_2) = (a_1 + a_2)t^2 + (a_1 + a_2) - (b_1 + b_2)$. On the other hand,

$$\begin{aligned} L(a_1 t + b_1) + L(a_2 t + b_2) &= a_1 t^2 + (a_1 - b_1) + a_2 t^2 + (a_2 - b_2) \\ &= (a_1 + a_2)t^2 + (a_1 + a_2) - (b_1 + b_2) \\ &= L((a_1 t + b_1) + (a_2 t + b_2)). \end{aligned}$$

Next, if k is any real number, then

$$L(k(at + b)) = L((ka)t + kb) = kat^2 + (ka - kb) = k(at^2 + (a - b)) = kL(at + b).$$

Hence L is a linear transformation.

5. (a) No:

$$\begin{aligned} L((a_1 t^2 + b_1 t + c_1) + (a_2 t^2 + b_2 t + c_2)) &= L((a_1 + b_1)t^2 + (b_1 + b_2)t + (c_1 + c_2)) \\ &= (a_1 + a_2 + 1)t^2 + [(b_1 + b_2) - (c_1 + c_2)]t + [(a_1 + a_2) + (c_1 + c_2)] \end{aligned}$$

while

$$L(a_1t^2 + b_1t + c_1) + L(a_2t^2 + b_2t + c_2)$$
$$= (a_1 + 1)t^2 + (b_1 - c_1)t + (a_1 + c_1) + (a_2 + 1)t^2 + (b_2 - c_2)t + (a_2 - c_2)$$
$$= (a_1 + a_2 + 2)t^2 + [(b_1 + b_2) - (c_1 + c_2)]t + [(a_1 + a_2) + (c_1 + c_2)].$$

(b) Yes:

$$L((a_1t^2 + b_1t + c_1) + (a_2t^2 + b_2t + c_2))$$
$$= L((a_1 + a_2)t^2 + (b_1 + b_2)t + (c_1 + c_2))$$
$$= (a_1 + a_2)t^2 + ((b_1 + b_2) - (c_1 + c_2))t + (a_1 + a_2) - (b_1 + b_2)$$
$$= [a_1t^2 + (b_1 - c_1)t + (a_1 - b_1)] + [a_2t^2 + (b_2 - c_2)t + (a_2 - b_2)]$$
$$= L(a_1t^2 + b_1t + c_1) + L(a_2t^2 + b_2t + c_2)$$

and if k is any real number

$$L(k(at^2 + bt + c)) = L(kat^2 + kbt + kc) = kat^2 + (kb - kc)t + (ka - kc)$$
$$= k[at^2 + (b - c)t + (a - c)] = kL(at^2 + bt + c).$$

(c) Yes:

$$L((a_1t^2 + b_1t + c_1) + (a_2t^2 + b_2t + c_2)) = L((a_1 + a_2)t^2 + (b_1 + b_2)t + (c_1 + c_2)) = 0$$

and

$$L(a_1t^2 + b_1t + c_1) + L(a_2t^2 + b_2t + c_2) = 0 + 0 = 0.$$

also, if k is any real number,

$$L(k(a_1t^2 + b_1t + c_1)) = L(ka_1t^2 + kb_1t + kc_1) = 0$$

and

$$kL(a_1t^2 + b_1t + c_1) = k(0) = 0.$$

7. Yes:

$$L\left(\begin{bmatrix} a_1 & b_1 \\ c_1 & d_1 \end{bmatrix} + \begin{bmatrix} a_2 & b_2 \\ c_2 & d_2 \end{bmatrix}\right) = L\left(\begin{bmatrix} a_1 + a_2 & b_1 + b_2 \\ c_1 + c_2 & d_1 + d_2 \end{bmatrix}\right)$$

$$= \begin{bmatrix} b_1 + b_2 & (c_1 + c_2) - (d_1 + d_2) \\ (c_1 + c_2) + (d_1 + d_2) & 2(a_1 + a_2) \end{bmatrix}$$

$$= \begin{bmatrix} b_1 & c_1 - d_1 \\ c_1 + d_1 & 2a_1 \end{bmatrix} + \begin{bmatrix} b_2 & c_2 - d_2 \\ c_2 + d_2 & 2a_2 \end{bmatrix}$$

$$= L\left(\begin{bmatrix} a_1 & b_1 \\ c_1 & d_1 \end{bmatrix}\right) + L\left(\begin{bmatrix} a_2 & b_2 \\ c_2 & d_2 \end{bmatrix}\right).$$

Also, if k is any real number,

$$L\left(k\begin{bmatrix} a & b \\ c & d \end{bmatrix}\right) = L\left(\begin{bmatrix} ka & kb \\ kc & kd \end{bmatrix}\right) = \begin{bmatrix} kb & kc - kd \\ kc + kd & 2ka \end{bmatrix}$$

$$= k\begin{bmatrix} b & c - d \\ c + d & 2a \end{bmatrix} = kL\left(\begin{bmatrix} a & b \\ c & d \end{bmatrix}\right).$$

9. Yes:

$$L\left(\begin{bmatrix} a_1 & b_1 \\ c_1 & d_1 \end{bmatrix} + \begin{bmatrix} a_2 & b_2 \\ c_2 & d_2 \end{bmatrix}\right) = L\left(\begin{bmatrix} a_1 + a_2 & b_1 + b_2 \\ c_1 + c_2 & d_1 + d_2 \end{bmatrix}\right)$$
$$= (a_1 + a_1) + (d_1 + d_2)$$
$$= (a_1 + d_1) + (a_2 + d_2)$$
$$= L\left(\begin{bmatrix} a_1 & b_1 \\ c_1 & d_1 \end{bmatrix}\right) + L\left(\begin{bmatrix} a_2 & b_2 \\ c_2 & d_2 \end{bmatrix}\right).$$

also, if k is any real number

$$L\left(k\begin{bmatrix} a & b \\ c & d \end{bmatrix}\right) = L\left(\begin{bmatrix} ka & kb \\ kc & kd \end{bmatrix}\right) = ka + kd = k(a + d) = kL\left(\begin{bmatrix} a & b \\ c & d \end{bmatrix}\right).$$

11. (a) $L\left(\begin{bmatrix} 1 & 2 & 0 & -1 \\ 3 & 0 & 2 & 3 \\ 4 & 1 & -2 & 1 \end{bmatrix}\right) = \begin{bmatrix} 2 & 3 & 1 \\ 1 & 2 & -3 \end{bmatrix}\begin{bmatrix} 1 & 2 & 0 & -1 \\ 3 & 0 & 2 & 3 \\ 4 & 1 & -2 & 1 \end{bmatrix} = \begin{bmatrix} 15 & 5 & 4 & 8 \\ -5 & -1 & 10 & 2 \end{bmatrix}.$

(b) Let $M = \begin{bmatrix} 2 & 3 & 1 \\ 1 & 2 & -3 \end{bmatrix}$. Then $L(A) = MA$. Therefore

$$L(A + B) = M(A + B) = MA + MB = L(A) + L(B)$$

and, for any real number c,

$$L(cA) = M(cA) = c(MA) = cL(A).$$

Therefore L is a linear transformation.

13. Let f and g be differentiable functions and k be any real number. From Example 4 we have $L(f) = f'$, the derivative of f. We assume that the properties of derivatives from calculus are familiar. Then

$$L(f + g) = (f + g)' = f' + g' = L(f) + L(g)$$

and

$$L(kf) = (kf)' = kf' = kL(f).$$

Thus L is a linear transformation.

15. $L: V \to R^n$ is the function that associates to vector $\mathbf{v}$ in V its coordinates with respect to the S-basis; $L(\mathbf{v}) = \begin{bmatrix} \mathbf{v} \end{bmatrix}_S$. We find $\begin{bmatrix} \mathbf{v} \end{bmatrix}_S$ by solving the linear equation

$$c_1\mathbf{v}_1 + c_2\mathbf{v}_2 + \cdots + c_n\mathbf{v}_n = \mathbf{v}.$$

Let $A = \begin{bmatrix} \mathbf{v}_1 & \mathbf{v}_2 & \cdots & \mathbf{v}_n \end{bmatrix}$ and $\mathbf{c} = \begin{bmatrix} c_1 & c_2 & \cdots & c_n \end{bmatrix}^T$. Since $\{\mathbf{v}_1, \mathbf{v}_2, \ldots, \mathbf{v}_n\}$ is a basis for V, A is a nonsingular matrix. Then the preceding equation can be expressed in matrix form as

$$A\mathbf{c} = \mathbf{v}$$

and it follows that

$$\begin{bmatrix} \mathbf{v} \end{bmatrix}_S = \mathbf{c} = A^{-1}\mathbf{v} = L(\mathbf{v}).$$

Let $\mathbf{v}$ and $\mathbf{w}$ be in V and let c be any real number. Then

$$L(\mathbf{v} + \mathbf{w}) = A^{-1}(\mathbf{v} + \mathbf{w}) = A^{-1}\mathbf{v} + A^{-1}\mathbf{w} = \begin{bmatrix} \mathbf{v} \end{bmatrix}_S + \begin{bmatrix} \mathbf{w} \end{bmatrix}_S = L(\mathbf{v}) + L(\mathbf{w})$$
$$L(k\mathbf{v}) = A^{-1}(k\mathbf{v}) = k(A^{-1}\mathbf{v}) = kL(\mathbf{v}).$$

Hence L is a linear transformation.

17. We observe that the vectors $(1, 1)$ and $(-1, 1)$ form a basis for R^2.

   (a) We first express $(-1, 5)$ as a linear combination of $(1, 1)$ and $(-1, 1)$ obtaining

$$(-1, 5) = 2(1, 1) + 3(-1, 1).$$

   Then

$$L(-1, 5) = L(2(1, 1) + 3(-1, 1)) = 2L(1, 1) + 3L(-1, 1) = 2(1, -2) + 3(2, 3) = (8, 5).$$

   (b) We first express (a_1, a_2) as a linear combination of $(1, 1)$ and $(-1, 1)$, obtaining

$$(a_1, a_2) = \left(\frac{a_1 + a_2}{2}\right)(1, 1) + \left(\frac{-a_1 + a_2}{2}\right)(-1, 1).$$

   Then

$$L(a_1, a_2) = L\left(\left(\frac{a_1 + a_2}{2}\right)(1, 1)\right) + L\left(\left(\frac{-a_1 + a_2}{2}\right)(-1, 1)\right)$$

$$= \left(\frac{a_1 + a_2}{2}\right)L(1, 1) + \left(\frac{-a_1 + a_2}{2}\right)L(-1, 1)$$

$$= \left(\frac{a_1 + a_2}{2}\right)(1, -2) + \left(\frac{-a_1 + a_2}{2}\right)(2, 3)$$

$$= \left(\frac{-a_1 + 3a_2}{2}, \frac{-5a_1 + a_2}{2}\right).$$

19. First, note that the vectors $t + 1$ and $t - 1$ form a basis for P_1.

   (a) We first express $6t - 4$ as a linear combination of $t + 1$ and $t - 1$, obtaining

$$6t - 4 = 1(t + 1) + 5(t - 1).$$

   Then

$$L(6t - 4) = L(1(t + 1) + 5(t - 1)) = 1L(t + 1) + 5L(t - 1)$$
$$= 1(2t + 3) + 5(3t - 2) = 17t - 7.$$

   (b) We first express $at + b$ as a linear combination of $t + 1$ and $t - 1$, obtaining

$$at + b = \left(\frac{a + b}{2}\right)(t + 1) + \left(\frac{a - b}{2}\right)(t - 1).$$

   Then

$$L(at + b) = L\left[\left(\frac{a + b}{2}\right)(t + 1) + \left(\frac{a - b}{2}\right)(t - 1)\right]$$

$$= \left(\frac{a + b}{2}\right)L(t + 1) + \left(\frac{a - b}{2}\right)L(t - 1)$$

$$= \left(\frac{a + b}{2}\right)(2t + 3) + \left(\frac{a - b}{2}\right)(3t - 2)$$

$$= \left(\frac{5a - b}{2}\right)t + \frac{a + 5b}{2}.$$

T.1. $L(c_1\mathbf{v}_1 + c_2\mathbf{v}_2 + \cdots + c_k\mathbf{v}_k) = L(c_1\mathbf{v}_1 + + L(c_2\mathbf{v}_2 + \cdots + c_k\mathbf{v}_k)$
$$= c_1L(\mathbf{v}_1) + L(c_2\mathbf{v}_2) + L(c_3\mathbf{v}_3 + \cdots + c_k\mathbf{v}_k)$$
$$= \cdots = c_1L(\mathbf{v}_1) + c_2L(\mathbf{v}_2) + \cdots + c_kL(\mathbf{v}_k).$$

T.3. $L(\mathbf{u} - \mathbf{v}) = L(\mathbf{u} + (-1)\mathbf{v}) = L(\mathbf{u}) + (-1)L(\mathbf{v}) = L(\mathbf{u}) - L(\mathbf{v}).$

T.5. Let A and B be in M_{nn}. Then

$$\mathrm{Tr}(A + B) = \sum_{i=1}^{n}(a_{ii} + b_{ii}) = \sum_{i=1}^{n}a_{ii} + \sum_{i=1}^{n}b_{ii} = \mathrm{Tr}(A) + \mathrm{Tr}(B)$$

and

$$\mathrm{Tr}(cA) = \sum_{i=1}^{n}(ca_{ii}) = c\sum_{i=1}^{n}a_{ii} = c\,\mathrm{Tr}(A).$$

Hence Tr is a linear transformation.

T.7. We have $L(\mathbf{u} + \mathbf{v}) = \mathbf{0}_W = \mathbf{0}_W + \mathbf{0}_W = L(\mathbf{u}) + L(\mathbf{v})$ and $L(c\mathbf{u}) = \mathbf{0}_W = c\mathbf{0}_W = cL(\mathbf{u}).$

T.9. Let $\mathbf{w}_1$ and $\mathbf{w}_2$ be in $L(V_1)$ and let c be a scalar. Then $\mathbf{w}_1 = L(\mathbf{v}_1)$ and $\mathbf{w}_2 = L(\mathbf{v}_2)$, where $\mathbf{v}_1$ and $\mathbf{v}_2$ are in V_1. Then

$$\mathbf{w}_1 + \mathbf{w}_2 = L(\mathbf{v}_1) + L(\mathbf{v}_2) = L(\mathbf{v}_1 + \mathbf{v}_2) \quad \text{and} \quad c\mathbf{w}_1 = cL(\mathbf{v}_1) = L(c\mathbf{v}_1).$$

Since $\mathbf{v}_1 + \mathbf{v}_2$ and $c\mathbf{v}_1$ are in V_1, we conclude that $\mathbf{w}_1 + \mathbf{w}_2$ and $c\mathbf{w}_1$ lie in $L(V_1)$. Hence $L(V_1)$ is a subspace of V.

T.11. Let $\mathbf{v}_1$ and $\mathbf{v}_2$ be in $L^{-1}(W_1)$ and let c be a scalar. Then $L(\mathbf{v}_1 + \mathbf{v}_2) = L(\mathbf{v}_1) + L(\mathbf{v}_2)$ is in W_1 since $L(\mathbf{v}_1)$ and $L(\mathbf{v}_2)$ are in W_1 and W_1 is a subspace of V. Hence $\mathbf{v}_1 + \mathbf{v}_2$ is in $L^{-1}(W_1)$. Similarly, $L(c\mathbf{v}_1) = cL(\mathbf{v}_1)$ is in W_1 so $c\mathbf{v}_1$ is in $L^{-1}(W_1)$. Hence, $L^{-1}(W_1)$ is a subspace of V.

T.13. Let U be the eigenspace of A with associated eigenvalue λ. Let $\mathbf{w}$ be in U. Then $L(\mathbf{w}) = A\mathbf{w} = \lambda\mathbf{w}$. Therefore $L(\mathbf{w})$ is in U since U is closed under scalar multiplication.

ML.1. (a) $\mathbf{A} = [1 \quad 0;0 \quad 0]; \mathbf{B} = [0 \quad 0;0 \quad 1];$
  $\det(\mathbf{A} + \mathbf{B})$
  `ans =`
  $\qquad$ 1
  $\det(\mathbf{A}) + \det(\mathbf{B})$
  `ans =`
  $\qquad$ 0

(b) $\mathbf{A} = \mathbf{eye}(3); \mathbf{B} = -\mathbf{ones}(3);$
  $\det(\mathbf{A} + \mathbf{B})$
  `ans =`
  $\qquad$ -2
  $\det(\mathbf{A}) + \det(\mathbf{B})$
  `ans =`
  $\qquad$ 1

Section 10.2, p.519

1. Let $L: R^2 \to R^2$ be the linear transformation defined by $L(a_1, a_2) = (a_1, 0)$.

   (a) $L(0, 2) = (0, 0)$, thus $(0, 2)$ is in ker L.

   (b) $L(2, 2) = (2, 0)$, thus $(2, 2)$ is not in ker L.

   (c) Since $L(3, 0) = (3, 0)$, $(3, 0)$ is in range L.

   (d) Vector $(3, 2)$ is not in range L since it is not of the form $(a_1, 0)$.

   (e) Vector (a_1, a_2) is in ker L provided $L(a_1, a_2) = (0, 0)$. Since $L(a_1, a_2) = (a_1, 0)$, it follows that (a_1, a_2) is in ker L only if $a_1 = 0$. Thus

   $$\ker L = \{(0, r) \mid r \text{ is any real number}\}.$$

   (f) Range $L = \{(r, 0) \mid r \text{ is any real number}\}$.

3. Let $L: R^2 \to R^3$ be defined by $L(x, y) = (x, x + y, y)$.

   (a) ker L is the set of all vectors whose image is $(0, 0, 0)$. Thus we set $L(x, y) = (x, x + y, y) = (0, 0, 0)$. equating corresponding components gives $x = y = 0$, and it follows that ker $L = \{(0, 0)\}$.

   (b) From (a), ker $L = \{\mathbf{0}\}$ hence Theorem 10.5 implies that L is one-to-one.

   (c) Let $\mathbf{y} = (y_1, y_2, y_3)$. To determine if L is onto, we ask if there is a vector $\mathbf{x} = (x_1, x_2)$ such that $L(\mathbf{x}) = \mathbf{y}$ for an arbitrary vector $\mathbf{y}$. We have $L(\mathbf{x}) = (x_1, x_1 + x_2, x_2) = (y_1, y_2, y_3)$. Equating corresponding components gives a linear system with augmented matrix

   $$\begin{bmatrix} 1 & 0 & \vdots & y_1 \\ 1 & 1 & \vdots & y_2 \\ 0 & 1 & \vdots & y_3 \end{bmatrix}$$

   which has reduced row echelon form

   $$\begin{bmatrix} 1 & 0 & \vdots & y_1 \\ 0 & 1 & \vdots & y_3 \\ 0 & 0 & \vdots & y_2 - y_1 - y_3 \end{bmatrix}.$$

   The system is inconsistent unless $y_2 - y_1 - y_3 = 0$, hence L is not onto.

5. Let $A = \begin{bmatrix} 1 & 0 & -1 & 3 & -1 \\ 1 & 0 & 0 & 2 & -1 \\ 2 & 0 & -1 & 5 & -1 \\ 0 & 0 & -1 & 1 & 0 \end{bmatrix}$ and $L(\mathbf{x}) = A\mathbf{x}$ for $\mathbf{x}$ in R^5.

   (a) $\mathbf{x}$ is in ker L provided $A\mathbf{x} = \mathbf{0}$. Thus we have a homogeneous linear system with coefficient matrix A. The reduced row echelon form of A is

   $$\begin{bmatrix} 1 & 0 & 0 & 2 & 0 \\ 0 & 0 & 1 & -1 & 0 \\ 0 & 0 & 0 & 0 & 1 \\ 0 & 0 & 0 & 0 & 0 \end{bmatrix}.$$

   Thus the solution is $x_1 = -2r$, $x_2 = s$, $x_3 = r$, $x_4 = r$, $x_5 = 0$, where r and s are any real numbers. In vector form the solution is

   $$\begin{bmatrix} -2r \\ s \\ r \\ r \\ 0 \end{bmatrix} = r \begin{bmatrix} -2 \\ 0 \\ 1 \\ 1 \\ 0 \end{bmatrix} + s \begin{bmatrix} 0 \\ 1 \\ 0 \\ 0 \\ 0 \end{bmatrix}.$$

Thus $\left\{ \begin{bmatrix} -2 \\ 0 \\ 1 \\ 1 \\ 0 \end{bmatrix}, \begin{bmatrix} 0 \\ 1 \\ 0 \\ 0 \\ 0 \end{bmatrix} \right\}$ is a basis for ker L.

(b) Let $S = \{\mathbf{v}_1, \mathbf{v}_2, \mathbf{v}_3, \mathbf{v}_4, \mathbf{v}_5\}$, where $\mathbf{v}_j =$ column j of A. Then span $S =$ range L. However, S is linearly dependent since $\mathbf{v}_2$ is the zero vector. Thus we can drop $\mathbf{v}_2$ and $S_1 = \{\mathbf{v}_1, \mathbf{v}_3, \mathbf{v}_4, \mathbf{v}_5\}$ spans range L. Following Example 4 of Section 6.6, we form a matrix with rows corresponding to the vectors in S_1:

$$\begin{bmatrix} 1 & 1 & 2 & 0 \\ -1 & 0 & -1 & -1 \\ 3 & 2 & 5 & 1 \\ -1 & -1 & -1 & 0 \end{bmatrix}.$$

The reduced row echelon form is

$$\begin{bmatrix} 1 & 0 & 0 & 1 \\ 0 & 1 & 0 & -1 \\ 0 & 0 & 1 & 0 \\ 0 & 0 & 0 & 0 \end{bmatrix}.$$

The nonzero rows written as columns are

$$\left\{ \begin{bmatrix} 1 \\ 0 \\ 0 \\ 1 \end{bmatrix}, \begin{bmatrix} 0 \\ 1 \\ 0 \\ -1 \end{bmatrix}, \begin{bmatrix} 0 \\ 0 \\ 1 \\ 0 \end{bmatrix} \right\}$$

and these form a basis for range L. Alternatively, a linearly independent subset of S can be determined using the method of Example 5 in Section 6.4. The reduced row echelon form of A is

$$\begin{bmatrix} 1 & 0 & 0 & 2 & 0 \\ 0 & 0 & 1 & -1 & 0 \\ 0 & 0 & 0 & 0 & 1 \\ 0 & 0 & 0 & 0 & 0 \end{bmatrix}.$$

Since the leading 1's are in columns 1, 3, and 5,

$$\{\mathbf{v}_1, \mathbf{v}_3, \mathbf{v}_5\} = \left\{ \begin{bmatrix} 1 \\ 1 \\ 2 \\ 0 \end{bmatrix}, \begin{bmatrix} -1 \\ 0 \\ -1 \\ -1 \end{bmatrix}, \begin{bmatrix} -1 \\ -1 \\ -1 \\ 0 \end{bmatrix} \right\}$$

is a basis for span S.

(c) $\dim(\ker L) = 2$, $\dim(\text{range } L) = 3$, $\dim R^5 = 5$, thus

$$\dim(\ker L) + \dim(\text{range } L) = \dim R^5.$$

7. Let $L: R^4 \to R^3$ be defined by $L\left(\begin{bmatrix} x \\ y \\ z \\ w \end{bmatrix}\right) = \begin{bmatrix} x + y \\ y - z \\ z - w \end{bmatrix}.$

(a) For arbitrary real numbers y_1, y_2, y_3 we determine if there is a vector in R^4 whose image is

$$\begin{bmatrix} y_1 \\ y_2 \\ y_3 \end{bmatrix}.$$

We have

$$\begin{bmatrix} x + y \\ y - z \\ z - w \end{bmatrix} = \begin{bmatrix} y_1 \\ y_2 \\ y_3 \end{bmatrix}.$$

Equating corresponding components gives a linear system with augmented matrix

$$\begin{bmatrix} 1 & 1 & 0 & 0 & \vdots & y_1 \\ 0 & 1 & -1 & 0 & \vdots & y_2 \\ 0 & 0 & 1 & -1 & \vdots & y_3 \end{bmatrix}.$$

The reduced row echelon form is

$$\begin{bmatrix} 1 & 0 & 0 & 1 & \vdots & y_1 - y_2 - y_3 \\ 0 & 1 & 0 & -1 & \vdots & y_2 + y_3 \\ 0 & 0 & 1 & -1 & \vdots & y_3 \end{bmatrix}.$$

The system is consistent for any values of y_1, y_2, y_3, and hence L is onto.

(b) From part (a) L is onto, thus range $L = R^3$. Then from Theorem 10.7 we have

$$\dim(\ker L) = \dim R^4 - \dim(\text{range } L) = 4 - 3 = 1.$$

(c) To verify Theorem 10.7, we find $\dim(\ker L)$ directly by finding a basis for $\ker L$. Vector $\mathbf{x}$ in R^4 is in $\ker L$ provided

$$L(\mathbf{x}) = \begin{bmatrix} x + y \\ y - z \\ z - w \end{bmatrix} = \begin{bmatrix} 0 \\ 0 \\ 0 \end{bmatrix} = \mathbf{0}.$$

Equating corresponding components gives a homogeneous system with coefficient matrix

$$\begin{bmatrix} 1 & 1 & 0 & 0 \\ 0 & 1 & -1 & 0 \\ 0 & 0 & 1 & -1 \end{bmatrix}$$

whose reduced row echelon form is

$$\begin{bmatrix} 1 & 0 & 0 & 1 \\ 0 & 1 & 0 & -1 \\ 0 & 0 & 1 & -1 \end{bmatrix}.$$

The solution is $x = -r$, $y = r$, $z = r$, $w = r$, where r is any real number. Thus every vector in $\ker L$ is of the form

$$\begin{bmatrix} -r \\ r \\ r \\ r \end{bmatrix} = r \begin{bmatrix} -1 \\ 1 \\ 1 \\ 1 \end{bmatrix}$$

and $\dim(\ker L) = 1$. Theorem 10.7 follows.

9. (a) Let $L(x, y) = (x + y, y)$. then $L: R^2 \to R^2$. We determine $\ker L$ and range L. Set $L(x, y) = (x + y, y) = (0, 0)$. Equating corresponding components gives the homogeneous linear system

$$x + y = 0$$
$$y = 0$$

whose only solution is $x = y = 0$. Thus $\ker L = \{(0, 0)\}$ and $\dim(\ker L) = 0$. By Theorem 10.5, L is one-to-one and by Corollary 10.3, L is onto. Thus range $L = R^2$ and $\dim(\text{range } L) = 2$. Hence

$$\dim(\ker L) + \dim(\text{range } L) = 0 + 2 = \dim(R^2).$$

(b) Let

$$L(\mathbf{x}) = L\left(\begin{bmatrix} x \\ y \\ z \end{bmatrix}\right) = A\mathbf{x} = \begin{bmatrix} 4 & -1 & -1 \\ 2 & 2 & 3 \\ 2 & -3 & -4 \end{bmatrix} \begin{bmatrix} x \\ y \\ z \end{bmatrix}.$$

Then $\ker L$ is the solution space of $A\mathbf{x} = \mathbf{0}$. The reduced row echelon form of A is

$$\begin{bmatrix} 1 & 0 & \frac{1}{10} \\ 0 & 1 & \frac{7}{5} \\ 0 & 0 & 0 \end{bmatrix}$$

and it follows that the solution is $x = -\frac{1}{10}r$, $y = -\frac{7}{5}r$, $z = r$, where r is any real number. Hence $\ker L$ is the set of all vectors of the form

$$r \begin{bmatrix} -\frac{1}{10} \\ -\frac{7}{5} \\ 1 \end{bmatrix}$$

and $\dim(\ker L) = 1$. Next we determine a basis for range L.

$$L(\mathbf{x}) = x \begin{bmatrix} 4 \\ 2 \\ 2 \end{bmatrix} + y \begin{bmatrix} -1 \\ 2 \\ -3 \end{bmatrix} + z \begin{bmatrix} -1 \\ 3 \\ -4 \end{bmatrix}.$$

Let

$$S = \left\{ \begin{bmatrix} 4 \\ 2 \\ 2 \end{bmatrix}, \begin{bmatrix} -1 \\ 2 \\ -3 \end{bmatrix}, \begin{bmatrix} -1 \\ 3 \\ -4 \end{bmatrix} \right\}.$$

Then span $S = $ range L. Following Example 4 of Section 6.6 we form the matrix with rows that are the vectors in S:

$$\begin{bmatrix} 4 & 2 & 2 \\ -1 & 2 & -3 \\ -1 & 3 & -4 \end{bmatrix}.$$

The reduced row echelon form of this matrix is

$$\begin{bmatrix} 1 & 0 & 1 \\ 0 & 1 & -1 \\ 0 & 0 & 0 \end{bmatrix}.$$

The nonzero rows written as columns form a basis for range L, thus $\dim(\text{range } L) = 2$. Hence

$$\dim(\ker L) + \dim(\text{range } L) = 1 + 2 = 3 = \dim(R^3).$$

(c) Let $L(x, y, z) = (x + y - z, x + y, y + z)$. To find a basis for $\ker L$, set

$$L(x, y, z) = (x + y - z, x + y, y + z) = (0, 0, 0)$$

and equate corresponding components. The resulting homogeneous system is

$$\begin{aligned} x + y - z &= 0 \\ x + y \quad\;\;\, &= 0 \\ y + z &= 0. \end{aligned}$$

The reduced row echelon form of the coefficient matrix is I_3, hence the only solution is $x = y = z = 0$. Thus $\ker L = \{\mathbf{0}\}$ and $\dim(\ker L) = 0$. Then L is one-to-one by Theorem 10.5 and Corollary 10.3 implies that L is onto. It follows that range $L = R^3$ and $\dim(\text{range } L) = 3$. We have

$$\dim(\ker L) + \dim(\text{range } L) = 0 + 3 = 3 = \dim(R^3).$$

11. Let $L\colon P_2 \to P_2$ be defined by $L(at^2 + bt + c) = (a + c)t^2 + (b + c)t$.

(a) $L(t^2 - t - 1) = 0t^2 - 2t \neq 0$, thus $t^2 - t - 1$ is not in $\ker L$.

(b) $L(t^2 + t - 1) = 0t^2 + 0t = 0$, thus $t^2 + t - 1$ is in $\ker L$.

(c) Set

$$L(at^2 + bt + c) = (a + c)t^2 + (b + c)t = 2t^2 - t.$$

Equating coefficients of like powers of t on both sides of the equation gives the linear system

$$\begin{aligned} a \quad\;\; + c &= \;\;\, 2 \\ b + c &= -1. \end{aligned}$$

Form the corresponding augmented matrix

$$\begin{bmatrix} 1 & 0 & 1 & \vdots & 2 \\ 0 & 1 & 1 & \vdots & -1 \end{bmatrix}.$$

It is already in reduced row echelon form, thus the system is consistent and $2t^2 - t$ is in range L.

(d) Set

$$L(at^2 + bt + c) = (a + c)t^2 + (b + c)t = t^2 - t - 2.$$

Equating coefficients of like powers of t on both sides of the equation gives the linear system

$$\begin{aligned} a \quad\;\; + c &= \;\;\; 1 \\ b + c &= -1 \\ 0 &= -2. \end{aligned}$$

The system is inconsistent, thus $t^2 - t - 2$ is not in range L.

(e) Set

$$L(at^2 + bt + c) = (a + c)t^2 + (b + c)t = 0t^2 + 0t + 0.$$

Equating coefficients of like powers of t on both sides of the equation gives the linear system

$$\begin{aligned} a \quad\;\; + c &= 0 \\ b + c &= 0 \end{aligned}$$

whose solution is $a = -r$, $b = -r$, $c = r$, where r is any real number. Thus $\ker L$ is the set of all elements in P_2 of the form $-rt^2 - rt + r$ and $\{-t^2 - t + 1\}$ is a basis.

(f) Let $S = \{t^2, t\}$. From the definition of L we see that span $S = $ range L. Since S is linearly independent, it is a basis for range L.

13. (a) From $L\left(\begin{bmatrix} a & b \\ c & d \end{bmatrix}\right) = \begin{bmatrix} a+b & b+c \\ a+d & b+d \end{bmatrix} = \begin{bmatrix} 0 & 0 \\ 0 & 0 \end{bmatrix}$ it follows that

$$
\begin{aligned}
a + b & & & = 0 \\
& b + c & & = 0 \\
a & & + d & = 0 \\
& b & + d & = 0.
\end{aligned}
$$

This linear system has augmented matrix

$$
\begin{bmatrix}
1 & 1 & 0 & 0 & \vdots & 0 \\
0 & 1 & 1 & 0 & \vdots & 0 \\
1 & 0 & 0 & 1 & \vdots & 0 \\
0 & 1 & 0 & 1 & \vdots & 0
\end{bmatrix}.
$$

The reduced row echelon form of the coefficient matrix of this system is I_4. Thus $a = b = c = d = 0$ and hence the kernel consists only of the zero matrix.

(b) $L\left(\begin{bmatrix} a & b \\ c & d \end{bmatrix}\right) = \begin{bmatrix} a+b & b+c \\ a+d & b+d \end{bmatrix} = a\begin{bmatrix} 1 & 0 \\ 1 & 0 \end{bmatrix} + b\begin{bmatrix} 1 & 1 \\ 0 & 1 \end{bmatrix} + c\begin{bmatrix} 0 & 1 \\ 0 & 0 \end{bmatrix} + d\begin{bmatrix} 0 & 0 \\ 1 & 1 \end{bmatrix}$.

Hence range L is spanned by the four 2×2 matrices in the preceding expression. Since $\dim M_{22} = 4$, Theorem 4.9b implies these four matrices are linearly independent. Thus

$$
\left\{ \begin{bmatrix} 1 & 0 \\ 1 & 0 \end{bmatrix}, \begin{bmatrix} 1 & 1 \\ 0 & 1 \end{bmatrix}, \begin{bmatrix} 0 & 1 \\ 0 & 0 \end{bmatrix}, \begin{bmatrix} 0 & 0 \\ 1 & 1 \end{bmatrix} \right\}
$$

is a basis for range L.

15. (a) If $\mathbf{v} = \begin{bmatrix} a & b \\ c & d \end{bmatrix}$, then

$$
L(\mathbf{v}) = L\left(\begin{bmatrix} a & b \\ c & d \end{bmatrix}\right) = \begin{bmatrix} 1 & 2 \\ 1 & 1 \end{bmatrix}\begin{bmatrix} a & b \\ c & d \end{bmatrix} - \begin{bmatrix} a & b \\ c & d \end{bmatrix}\begin{bmatrix} 1 & 2 \\ 1 & 1 \end{bmatrix} = \begin{bmatrix} 2c - b & 2d - 2a \\ a - d & b - 2c \end{bmatrix}.
$$

Then $\mathbf{v}$ is in ker L if $L(\mathbf{v}) = \mathbf{0}$, so we have

$$
\begin{aligned}
-b + 2c & & = 0 \\
-2a & + 2d & = 0 \\
a & - d & = 0 \\
b - 2c & & = 0.
\end{aligned}
$$

Solving this homogeneous system, we obtain

$$
\mathbf{v} = \begin{bmatrix} s & r \\ \frac{1}{2}r & s \end{bmatrix},
$$

which can be written as

$$
\begin{bmatrix} s & r \\ \frac{1}{2}r & s \end{bmatrix} = s\begin{bmatrix} 1 & 0 \\ 0 & 1 \end{bmatrix} + r\begin{bmatrix} 0 & 1 \\ \frac{1}{2} & 0 \end{bmatrix}.
$$

Hence $\left\{ \begin{bmatrix} 1 & 0 \\ 0 & 1 \end{bmatrix}, \begin{bmatrix} 0 & 1 \\ \frac{1}{2} & 0 \end{bmatrix} \right\}$ is a basis for ker L.

(b) Every vector in range L is of the form

$$\begin{bmatrix} 2c - b & 2d - 2a \\ a - d & b - 2c \end{bmatrix}$$

which can be written as

$$\begin{bmatrix} 2c - b & 2d - 2a \\ a - d & b - 2c \end{bmatrix} = a \begin{bmatrix} 0 & -2 \\ 1 & 0 \end{bmatrix} + b \begin{bmatrix} -1 & 0 \\ 0 & 1 \end{bmatrix} + c \begin{bmatrix} 2 & 0 \\ 0 & -2 \end{bmatrix} + d \begin{bmatrix} 0 & 2 \\ -1 & 0 \end{bmatrix}.$$

Hence $\left\{ \begin{bmatrix} 0 & -2 \\ 1 & 0 \end{bmatrix}, \begin{bmatrix} -1 & 0 \\ 0 & 1 \end{bmatrix} \right\}$ is a basis for range L.

17. Let $L: P_2 \to P_1$ be defined by $L(p(t)) = L(at^2 + bt + c) = p'(t) = 2at + b$.

(a) Set $L(p(t)) = 2at + b = \mathbf{0}$. Equating coefficients of like powers of t from both sides of the equation gives $a = 0$, $b = 0$. Hence, c can be chosen arbitrarily. Let $c = r$, any real number. Then $\ker L$ is the set of all polynomials of the form $0t^2 + 0t + r$, that is, the set of all constant polynomials. Thus $\{1\}$ is a basis for $\ker L$.

(b) From the definition of L we have that

$$\text{span } S = \text{span } \{t, 1\} = \text{range } L.$$

Since S is linearly independent, $\{t, 1\}$ is a basis for range L.

19. Let $L: R^4 \to R^6$ be a linear transformation.

(a) If $\dim(\ker L) = 2$, then from Theorem 10.7

$$\dim(\text{range } L) = \dim R^4 - \dim(\ker L) = 4 - 2 = 2.$$

(b) If $\dim(\text{range } L) = 3$, then from Theorem 10.7

$$\dim(\ker L) = \dim R^4 - \dim(\text{range } L) = 4 - 3 = 1.$$

T.1. Let $\mathbf{x}$ and $\mathbf{y}$ be solutions to $A\mathbf{x} = \mathbf{b}$, so $L(\mathbf{x}) = \mathbf{b}$ and $L(\mathbf{y}) = \mathbf{b}$. Then

$$L(\mathbf{x} - \mathbf{y}) = L(\mathbf{x}) - L(\mathbf{y}) = \mathbf{b} - \mathbf{b} = \mathbf{0}.$$

Hence $\mathbf{x} - \mathbf{y} = \mathbf{z}$ is in $\ker L$.

T.3. If $\mathbf{w}$ lies in the range of L, then $\mathbf{w} = A\mathbf{v}$ for some $\mathbf{v}$ in R^n, and $\mathbf{w}$ is a linear combination of columns of A. Thus $\mathbf{w}$ lies in the column space of A. Conversely, if $\mathbf{w}$ is in the column space of A, then $\mathbf{w}$ is a linear combination of the columns of A. Hence $\mathbf{w} = A\mathbf{v}$ for some $\mathbf{v}$ in R^n, which implies that $\mathbf{w}$ is in the range of L.

T.5. Let $\mathbf{w}$ be any vector in range L. Then there exists a vector $\mathbf{v}$ in V such that $L(\mathbf{v}) = \mathbf{w}$. Next, there exist scalars $c_1, \ldots, c_k$ such that $\mathbf{v} = c_1 \mathbf{v}_1 + \cdots + c_k \mathbf{v}_k$. Thus

$$\mathbf{w} = L(c_1 \mathbf{v}_1 + \cdots + c_k \mathbf{v}_k) = c_1 L(\mathbf{v}_1) + \cdots + c_k L(\mathbf{v}_k).$$

Hence $\{L(\mathbf{v}_1), L(\mathbf{v}_2), \ldots, L(\mathbf{v}_k)\}$ spans range L.

T.7. Suppose that S is linearly dependent. Then there exist constants $c_1, c_2, \ldots, c_n$, not all zero, such that

$$c_1 \mathbf{v}_1 + c_2 \mathbf{v}_2 + \cdots + c_n \mathbf{v}_n = \mathbf{0}.$$

Then

$$L(c_1\mathbf{v}_1 + c_2\mathbf{v}_2 + \cdots + c_n\mathbf{v}_n) = L(\mathbf{0}) = \mathbf{0}$$

or

$$c_1L(\mathbf{v}_1) + c_2L(\mathbf{v}_2) + \cdots + c_nL(\mathbf{v}_n) = \mathbf{0}$$

which implies that T is linearly dependent, a contradiction.

T.9. Let L be one-to-one and let $S = \{\mathbf{v}_1, \ldots, \mathbf{v}_k\}$ be a linearly independent set of vectors in V. Suppose that $\{L(\mathbf{v}_1), L(\mathbf{v}_2), \ldots, L(\mathbf{v}_k)\}$ is linearly dependent. Then there exist constants $c_1, c_2, \ldots, c_k$, not all zero, such that

$$c_1L(\mathbf{v}_1) + c_2L(\mathbf{v}_2) + \cdots + c_kL(\mathbf{v}_k) = \mathbf{0} \quad \text{or} \quad L(c_1\mathbf{v}_1 + c_2\mathbf{v}_2 + \cdots + c_k\mathbf{v}_k) = \mathbf{0} = L(\mathbf{0}).$$

Since L is one-to-one, we have $c_1\mathbf{v}_1 + c_2\mathbf{v}_2 + \cdots + c_k\mathbf{v}_k = \mathbf{0}$, which implies that S is linearly dependent, a contradiction.

T.11.   (a) Let $\mathbf{v}$ and $\mathbf{w}$ be vectors in V. From Equations (2) and (3) in Section 6.7, we have

$$L(\mathbf{v} + \mathbf{w}) = \big[\mathbf{v} + \mathbf{w}\big]_S = \big[\mathbf{v}\big]_S + \big[\mathbf{w}\big]_S = L(\mathbf{v}) + L(\mathbf{w})$$

and

$$L(c\mathbf{v}) = \big[c\mathbf{v}\big]_S = c\big[\mathbf{v}\big]_S = cL(\mathbf{v}).$$

(b) Let

$$L(\mathbf{v}) = \begin{bmatrix} a_1 \\ a_2 \\ \vdots \\ a_n \end{bmatrix} = \big[\mathbf{v}\big]_S \quad \text{and} \quad L(\mathbf{w}) = \begin{bmatrix} b_1 \\ b_2 \\ \vdots \\ b_n \end{bmatrix} = \big[\mathbf{w}\big]_S$$

and assume that $L(\mathbf{v}) = L(\mathbf{w})$. Then $a_i = b_i$, $1 \le i \le n$. This implies that

$$\mathbf{v} = a_1\mathbf{v}_1 + a_2\mathbf{v}_2 + \cdots + a_n\mathbf{v}_n = b_1\mathbf{v}_1 + b_2\mathbf{v}_2 + \cdots + b_n\mathbf{v}_n = \mathbf{w}.$$

Hence, L is one-to-one.

(c) Let

$$\mathbf{w} = \begin{bmatrix} c_1 \\ c_2 \\ \vdots \\ c_n \end{bmatrix}$$

be an arbitrary vector in R^n. If we let

$$\mathbf{v} = c_1\mathbf{v}_1 + c_2\mathbf{v}_2 + \cdots + c_n\mathbf{v}_n$$

then $\mathbf{v}$ is in V and $L(\mathbf{v}) = \big[\mathbf{v}\big]_S = \mathbf{w}$. Hence, L is onto.

ML.1. $\mathbf{A} = [1 \ \ 2 \ \ 5 \ \ 5; -2 \ \ -3 \ \ -8 \ \ -7];$

rref(A)

```
ans =

    1   0   1   -1
    0   1   2    3
```

It follows that the general solution to $A\mathbf{x} = \mathbf{0}$ is obtained from

$$x_1 \quad + \quad x_3 - \quad x_4 = 0$$
$$x_2 + 2x_3 + 3x_4 = 0.$$

Let $x_3 = r$ and $x_4 = s$, then $x_2 = -2r - 3s$ and $x_1 = -r + s$. Thus

$$\mathbf{x} = \begin{bmatrix} -r + s \\ -2r - 3s \\ r \\ s \end{bmatrix} = r \begin{bmatrix} -1 \\ -2 \\ 1 \\ 0 \end{bmatrix} + s \begin{bmatrix} 1 \\ -3 \\ 0 \\ 1 \end{bmatrix}$$

and

$$\left\{ \begin{bmatrix} -1 \\ -2 \\ 1 \\ 0 \end{bmatrix}, \begin{bmatrix} 1 \\ -3 \\ 0 \\ 1 \end{bmatrix} \right\}$$

is a basis for ker L.

To find a basis for range L proceed as follows.

rref(A′)′

ans =

```
1  0  0  0
0  1  0  0
```

Then $\left\{ \begin{bmatrix} 1 \\ 0 \end{bmatrix}, \begin{bmatrix} 0 \\ 1 \end{bmatrix} \right\}$ is a basis for range L.

ML.3. $[3 \ \ 3 \ \ -3 \ \ 1 \ \ 11; -4 \ \ -4 \ \ 7 \ \ -2 \ \ -19; 2 \ \ 2 \ \ -3 \ \ 1 \ \ 9];$

rref(A)

ans =

```
1  1  0  0   2
0  0  1  0  -1
0  0  0  1   2
```

It follows that the general solution to $A\mathbf{x} = \mathbf{0}$ is obtained from

$$x_1 + x_2 \qquad + 2x_5 = 0$$
$$x_3 \quad - \quad x_5 = 0$$
$$x_4 + 2x_5 = 0.$$

Let $x_5 = r$ and $x_2 = s$, then $x_4 = -2r$ and $x_3 = r$, $x_1 = -s - 2r$. Thus

$$\mathbf{x} = \begin{bmatrix} -2r - s \\ s \\ r \\ -2r \\ r \end{bmatrix} = r \begin{bmatrix} -2 \\ 0 \\ 1 \\ -2 \\ 1 \end{bmatrix} + s \begin{bmatrix} -1 \\ 1 \\ 0 \\ 0 \\ 0 \end{bmatrix}$$

and $\left\{ \begin{bmatrix} -2 \\ 0 \\ 1 \\ -2 \\ 1 \end{bmatrix}, \begin{bmatrix} -1 \\ 1 \\ 0 \\ 0 \\ 0 \end{bmatrix} \right\}$ is a basis for ker L. To find a basis for range L proceed as follows.

rref(A')'

ans =

```
        1  0  0  0  0
        0  1  0  0  0
        0  0  1  0  0
```

Thus the columns of I_3 are a basis for range L.

Section 10.3, p. 532

1. Let $L\colon R^2 \to R^2$ be defined by $L(x, y) = (x - 2y, x + 2y)$,

$$S = \{\mathbf{v}_1, \mathbf{v}_2\} = \{(1, -1), (0, 1)\} \quad \text{and} \quad T = \{\mathbf{w}_1, \mathbf{w}_2\} = \{(1, 0), (0, 1)\}.$$

We follow the steps given in the summary of the procedure in Theorem 10.8.

(a) $L(\mathbf{v}_1) = L((1, -1)) = (3, -1)$, $L(\mathbf{v}_2) = L((0, 1)) = (-2, 2)$.
To find $\big[L(\mathbf{v}_i)\big]_S$ we express $L(\mathbf{v}_i)$ as linear combinations of the vectors in S. Thus we solve for c_1 and c_2 in

$$c_1\mathbf{v}_1 + c_2\mathbf{v}_2 = L(\mathbf{v}_1)$$

and k_1 and k_2 in

$$k_1\mathbf{v}_1 + k_2\mathbf{v}_2 = L(\mathbf{v}_2).$$

Substituting in the expression for the vectors and combining terms leads to the linear systems

$$
\begin{aligned}
c_1 &= 3 & k_1 &= -2 \\
-c_1 + c_2 &= -1 & -k_1 + k_2 &= 2
\end{aligned}
$$

respectively. These systems have the same coefficient matrix but different right-hand sides. Hence form the partitioned matrix

$$
\left[\begin{array}{cc:c:c}
1 & 0 & 3 & -2 \\
-1 & 1 & -1 & 2
\end{array}\right]
$$

and find its reduced row echelon form which is

$$
\left[\begin{array}{cc:c:c}
1 & 0 & 3 & -2 \\
0 & 1 & 2 & 0
\end{array}\right].
$$

It follows that $c_1 = 3$, $c_2 = 2$, and $k_1 = -2$, $k_2 = 0$, so

$$
\big[L(\mathbf{v}_1)\big]_S = \begin{bmatrix} 3 \\ 2 \end{bmatrix} \quad \text{and} \quad \big[L(\mathbf{v}_2)\big]_S = \begin{bmatrix} -2 \\ 0 \end{bmatrix}.
$$

Hence the matrix of L with respect to S is

$$
\left[\big[L(\mathbf{v}_1)\big]_S \quad \big[L(\mathbf{v}_2)\big]_S \right] = \begin{bmatrix} 3 & -2 \\ 2 & 0 \end{bmatrix}.
$$

(b) From part (a), $L(\mathbf{v}_1) = (3, -1)$ and $L(\mathbf{v}_2) = (-2, 2)$. Here we must solve

$$c_1\mathbf{w}_1 + c_2\mathbf{w}_2 = L(\mathbf{v}_1) \quad \text{and} \quad k_1\mathbf{w}_1 + k_2\mathbf{w}_2 = L(\mathbf{v}_2).$$

Since T is the natural basis for R^2, we have $\big[L(\mathbf{v}_i)\big]_T = (L(\mathbf{v}_i))^T$, for $i = 1, 2$. Thus the matrix representing L with respect to S and T is

$$
\left[\big[L(\mathbf{v}_1)\big]_T \quad \big[L(\mathbf{v}_2)_T\big] \right] = \begin{bmatrix} 3 & -2 \\ -1 & 2 \end{bmatrix}.
$$

(c) $L(\mathbf{w}_1) = (1, 1)$, $L(\mathbf{w}_2) = (-2, 2)$.

To find $\left[L(\mathbf{w}_i)\right]_S$ we express $L(\mathbf{w}_i)$ as linear combinations of the vectors in S. Thus we solve for c_1, c_2, and for k_1, k_2 in

$$c_1 \mathbf{v}_1 + c_2 \mathbf{v}_2 = L(\mathbf{w}_1) \quad \text{and} \quad k_1 \mathbf{v}_1 + k_2 \mathbf{v}_2 = L(\mathbf{w}_2).$$

Substituting in the expressions for the vectors and combining terms leads to the linear systems

$$\begin{aligned} c_1 \quad &= 1 & k_1 \quad &= -2 \\ -c_1 + c_2 &= 1 & -k_1 + k_2 &= 2 \end{aligned}$$

respectively. These systems have the same coefficient matrix but different right-hand sides. Hence form the partitioned matrix

$$\left[\begin{array}{cc:c:c} 1 & 0 & 1 & -2 \\ -1 & 1 & 1 & 2 \end{array}\right]$$

and find its reduced row echelon form which is

$$\left[\begin{array}{cc:c:c} 1 & 0 & 1 & -2 \\ 0 & 1 & 2 & 0 \end{array}\right].$$

It follows that $c_1 = 1$, $c_2 = 2$, and $k_1 = -2$, $k_2 = 0$, so

$$\left[L(\mathbf{w}_1)\right]_S = \begin{bmatrix} 1 \\ 2 \end{bmatrix} \quad \text{and} \quad \left[L(\mathbf{w}_2)\right]_S = \begin{bmatrix} -2 \\ 0 \end{bmatrix}.$$

Hence the matrix of L with respect to S is

$$\left[\begin{array}{cc} \left[L(\mathbf{w}_1)\right]_S & \left[L(\mathbf{w}_2)\right]_S \end{array}\right] = \begin{bmatrix} 1 & -2 \\ 2 & 0 \end{bmatrix}.$$

(d) From part (c), $L(\mathbf{w}_1) = (1, 1)$, $L(\mathbf{w}_2) = (-2, 2)$. Here we must solve

$$c_1 \mathbf{w}_1 + c_2 \mathbf{w}_2 = L(\mathbf{w}_1) \quad \text{and} \quad k_1 \mathbf{w}_1 + k_2 \mathbf{w}_2 = L(\mathbf{w}_2).$$

Since T is the natural basis for R^2, we have $\left[L(\mathbf{w}_i)\right]_T = (L(\mathbf{w}_i))^T$, for $i = 1, 2$. Thus the matrix representing L with respect to T is

$$\left[\begin{array}{cc} \left[L(\mathbf{w}_1)\right]_T & \left[L(\mathbf{w}_2)\right]_T \end{array}\right] = \begin{bmatrix} 1 & -2 \\ 1 & 2 \end{bmatrix}.$$

(e) Let $\mathbf{u} = (2, -1)$. Then $L(\mathbf{u}) = L((2, -1)) = (4, 0)$. We find that

$$\left[\mathbf{u}\right]_S = \begin{bmatrix} 2 \\ 1 \end{bmatrix} \quad \text{and} \quad \left[\mathbf{u}\right]_T = \begin{bmatrix} 2 \\ -1 \end{bmatrix}.$$

We proceed as follows.

From part (a): $[L(\mathbf{u})]_S = \begin{bmatrix} 3 & -2 \\ 2 & 0 \end{bmatrix} [\mathbf{u}]_S = \begin{bmatrix} 3 & -2 \\ 2 & 0 \end{bmatrix} \begin{bmatrix} 2 \\ 1 \end{bmatrix} = \begin{bmatrix} 4 \\ 4 \end{bmatrix}$

Thus $L(\mathbf{u}) = 4\mathbf{v}_1 + 4\mathbf{v}_2 = (4, 0)$.

From part (b): $[L(\mathbf{u})]_T = \begin{bmatrix} 3 & -2 \\ -1 & 2 \end{bmatrix} [\mathbf{u}]_S = \begin{bmatrix} 3 & -2 \\ -1 & 2 \end{bmatrix} \begin{bmatrix} 2 \\ 1 \end{bmatrix} = \begin{bmatrix} 4 \\ 0 \end{bmatrix}$

Thus $L(\mathbf{u}) = 4\mathbf{w}_1 + 0\mathbf{w}_2 = (4, 0)$.

From part (c): $[L(\mathbf{u})]_S = \begin{bmatrix} 1 & -2 \\ 2 & 0 \end{bmatrix} [\mathbf{u}]_T = \begin{bmatrix} 1 & -2 \\ 2 & 0 \end{bmatrix} \begin{bmatrix} 2 \\ -1 \end{bmatrix} = \begin{bmatrix} 4 \\ 4 \end{bmatrix}$

Thus $L(\mathbf{u}) = 4\mathbf{v}_1 + 4\mathbf{v}_2 = (4, 0)$.

From part (d): $[L(\mathbf{u})]_T = \begin{bmatrix} 1 & -2 \\ 1 & 2 \end{bmatrix} [\mathbf{u}]_T = \begin{bmatrix} 1 & -2 \\ 1 & 2 \end{bmatrix} \begin{bmatrix} 2 \\ -1 \end{bmatrix} = \begin{bmatrix} 4 \\ 0 \end{bmatrix}$

Thus $L(\mathbf{u}) = 4\mathbf{w}_1 + 0\mathbf{w}_2 = (4, 0)$.

3. Let $L: R^2 \to R^3$ be defined by

$$L\left(\begin{bmatrix} x \\ y \end{bmatrix}\right) = \begin{bmatrix} x - 2y \\ 2x + y \\ x + y \end{bmatrix},$$

$S = \{\mathbf{v}_1, \mathbf{v}_2\}$ be the natural basis for R^2, and $T = \{\mathbf{w}_1, \mathbf{w}_2, \mathbf{w}_3\}$ be the natural basis for R^3. Also, let

$$S' = \{\mathbf{v}_1', \mathbf{v}_2'\} = \left\{ \begin{bmatrix} 1 \\ -1 \end{bmatrix}, \begin{bmatrix} 0 \\ 1 \end{bmatrix} \right\}$$

be another basis for R^2 and

$$T' = \{\mathbf{w}_1', \mathbf{w}_2', \mathbf{w}_3'\} = \left\{ \begin{bmatrix} 1 \\ 1 \\ 0 \end{bmatrix}, \begin{bmatrix} 0 \\ 1 \\ 1 \end{bmatrix}, \begin{bmatrix} 1 \\ -1 \\ 1 \end{bmatrix} \right\}$$

be another basis for R^3. We follow the steps given in the summary of the procedure in Theorem 10.8.

(a) $L(\mathbf{v}_1) = L\left(\begin{bmatrix} 1 \\ 0 \end{bmatrix}\right) = \begin{bmatrix} 1 \\ 2 \\ 1 \end{bmatrix}$, $L(\mathbf{v}_2) = L\left(\begin{bmatrix} 0 \\ 1 \end{bmatrix}\right) = \begin{bmatrix} -2 \\ 1 \\ 1 \end{bmatrix}$.

Next we determine the coordinates of $L(\mathbf{v}_1)$ and $L(\mathbf{v}_2)$ with respect to T. Since T is the natural basis we have

$$[L(\mathbf{v}_1)]_T = \begin{bmatrix} 1 \\ 2 \\ 1 \end{bmatrix}, \qquad [L(\mathbf{v}_2)]_T = \begin{bmatrix} -2 \\ 1 \\ 1 \end{bmatrix}.$$

The matrix A of L with respect to S and T is

$$A = \begin{bmatrix} [L(\mathbf{v}_1)]_T & [L(\mathbf{v}_2)]_T \end{bmatrix} = \begin{bmatrix} 1 & -2 \\ 2 & 1 \\ 1 & 1 \end{bmatrix}.$$

(b) $L(\mathbf{v}_1') = L\left(\begin{bmatrix} 1 \\ -1 \\ 0 \end{bmatrix}\right) = \begin{bmatrix} 3 \\ 1 \\ 0 \end{bmatrix}$, $L(\mathbf{v}_2') = L\left(\begin{bmatrix} 0 \\ 1 \end{bmatrix}\right) = \begin{bmatrix} -2 \\ 1 \\ 1 \end{bmatrix}$.

Next we determine the coordinates of $L(\mathbf{v}_1')$ and $L(\mathbf{v}_2')$ with respect to T'. Since T' is not the natural basis we must solve a system of equations to obtain the coordinates.

$$\left[L(\mathbf{v}_1')\right] = \begin{bmatrix} 3 \\ 1 \\ 0 \end{bmatrix} = c_1\mathbf{w}_1' + c_2\mathbf{w}_2' + c_3\mathbf{w}_3'.$$

Adding vectors and equating corresponding components from both sides of the equation gives the linear system

$$\begin{bmatrix} 1 & 0 & 1 \\ 1 & 1 & -1 \\ 0 & 1 & 1 \end{bmatrix} \begin{bmatrix} c_1 \\ c_2 \\ c_3 \end{bmatrix} = \begin{bmatrix} 3 \\ 1 \\ 0 \end{bmatrix}.$$

Forming the augmented matrix and applying row operations we obtain the reduced row echelon form

$$\begin{bmatrix} 1 & 0 & 0 & \vdots & \frac{7}{3} \\ 0 & 1 & 0 & \vdots & -\frac{2}{3} \\ 0 & 0 & 1 & \vdots & \frac{2}{3} \end{bmatrix}.$$

Thus $c_1 = \frac{7}{3}$, $c_2 = -\frac{2}{3}$, $c_3 = \frac{2}{3}$ and hence

$$\left[L(\mathbf{v}_1')\right]_{T'} = \begin{bmatrix} \frac{7}{3} \\ -\frac{2}{3} \\ \frac{2}{3} \end{bmatrix}.$$

Next, set

$$L(\mathbf{v}_2') = \begin{bmatrix} -2 \\ 1 \\ 1 \end{bmatrix} = c_1\mathbf{w}_1' + c_2\mathbf{w}_2' + c_3\mathbf{w}_3'.$$

Adding vectors and equating corresponding components from both sides of the equation gives the linear system

$$\begin{bmatrix} 1 & 0 & 1 \\ 1 & 1 & -1 \\ 0 & 1 & 1 \end{bmatrix} \begin{bmatrix} c_1 \\ c_2 \\ c_3 \end{bmatrix} = \begin{bmatrix} -2 \\ 1 \\ 1 \end{bmatrix}.$$

Forming the augmented matrix and applying row operations we obtain the reduced row echelon form

$$\begin{bmatrix} 1 & 0 & 0 & \vdots & -\frac{4}{3} \\ 0 & 1 & 0 & \vdots & \frac{5}{3} \\ 0 & 0 & 1 & \vdots & -\frac{2}{3} \end{bmatrix}.$$

Thus $c_1 = -\frac{4}{3}$, $c_2 = \frac{5}{3}$, $c_3 = -\frac{2}{3}$ and hence

$$\left[L(\mathbf{v}_2')\right]_{T'} = \begin{bmatrix} -\frac{4}{3} \\ \frac{5}{3} \\ -\frac{2}{3} \end{bmatrix}.$$

The matrix A of L with respect to S' and T' is

$$A = \begin{bmatrix} [L(\mathbf{v}_1)]_{T'} & [L(\mathbf{v}_2)]_{T'} \end{bmatrix} = \begin{bmatrix} \frac{7}{3} & -\frac{4}{3} \\ -\frac{2}{3} & \frac{5}{3} \\ \frac{2}{3} & -\frac{2}{3} \end{bmatrix}.$$

(c) Let $\mathbf{x} = \begin{bmatrix} 1 \\ 2 \end{bmatrix}$. Then from the definition of L,

$$L(\mathbf{x}) = L\left(\begin{bmatrix} 1 \\ 2 \end{bmatrix}\right) = \begin{bmatrix} -3 \\ 4 \\ 3 \end{bmatrix}.$$

Next we use Theorem 10.8 to express the relationship between coordinates of vectors in V with respect to a basis for V and coordinates of vectors in W with respect to a basis for W. From part (a),

$$[L(\mathbf{x})]_T = A\,[\mathbf{x}]_S = \begin{bmatrix} 1 & -2 \\ 2 & 1 \\ 1 & 1 \end{bmatrix} \begin{bmatrix} 1 \\ 2 \end{bmatrix}_S = \begin{bmatrix} 1 & -2 \\ 2 & 1 \\ 1 & 1 \end{bmatrix} \begin{bmatrix} 1 \\ 2 \end{bmatrix} = \begin{bmatrix} -3 \\ 4 \\ 3 \end{bmatrix}_T.$$

Hence

$$L(\mathbf{x}) = -3\mathbf{w}_1 + 4\mathbf{w}_2 + 3\mathbf{w}_3 = \begin{bmatrix} -3 \\ 4 \\ 3 \end{bmatrix}.$$

From part (b)

$$L(\mathbf{x}) = A\,[\mathbf{x}]_{S'} = \begin{bmatrix} \frac{7}{3} & -\frac{4}{3} \\ -\frac{2}{3} & \frac{5}{3} \\ \frac{2}{3} & -\frac{2}{3} \end{bmatrix} \begin{bmatrix} 1 \\ 2 \end{bmatrix}_{S'} = \begin{bmatrix} \frac{7}{3} & -\frac{4}{3} \\ -\frac{2}{3} & \frac{5}{3} \\ \frac{2}{3} & -\frac{2}{3} \end{bmatrix} \begin{bmatrix} 1 \\ 3 \end{bmatrix} = \begin{bmatrix} -\frac{5}{3} \\ \frac{13}{3} \\ -\frac{4}{3} \end{bmatrix}_{T'}$$

where we used $\mathbf{x} = 1\mathbf{v}'_1 + 3\mathbf{v}'_2$. Hence

$$L(\mathbf{x}) = \left(-\tfrac{5}{3}\right)\mathbf{w}'_1 + \left(\tfrac{13}{3}\right)\mathbf{w}'_2 + \left(-\tfrac{4}{3}\right)\mathbf{w}'_3 = \begin{bmatrix} -3 \\ 4 \\ 3 \end{bmatrix}.$$

5. Let $L\colon R^3 \to R^2$ be defined by

$$L\left(\begin{bmatrix} x \\ y \\ z \end{bmatrix}\right) = \begin{bmatrix} x + y \\ y - z \end{bmatrix},$$

$S = \{\mathbf{v}_1, \mathbf{v}_2, \mathbf{v}_3\}$ be the natural basis for R^3, and $T = \{\mathbf{w}_1, \mathbf{w}_2\}$ be the natural basis for R^2. Also, let

$$S' = \{\mathbf{v}'_1, \mathbf{v}'_2, \mathbf{v}'_3\} = \left\{ \begin{bmatrix} 1 \\ 1 \\ 0 \end{bmatrix}, \begin{bmatrix} 0 \\ 1 \\ 0 \end{bmatrix}, \begin{bmatrix} -1 \\ 1 \\ 1 \end{bmatrix} \right\}$$

be another basis for R^3 and

$$T' = \{\mathbf{w}'_1, \mathbf{w}'_2\} = \left\{ \begin{bmatrix} -1 \\ 1 \end{bmatrix}, \begin{bmatrix} 1 \\ 2 \end{bmatrix} \right\}$$

be another basis for R^2. We follow the steps given in the summary of the procedure in Theorem 10.8.

(a) $L(\mathbf{v}_1) = L\left(\begin{bmatrix} 1 \\ 0 \\ 0 \end{bmatrix}\right) = \begin{bmatrix} 1 \\ 0 \end{bmatrix}$, $L(\mathbf{v}_2) = L\left(\begin{bmatrix} 0 \\ 1 \\ 0 \end{bmatrix}\right) = \begin{bmatrix} 1 \\ 1 \end{bmatrix}$, $L(\mathbf{v}_3) = L\left(\begin{bmatrix} 0 \\ 0 \\ 1 \end{bmatrix}\right) = \begin{bmatrix} 0 \\ -1 \end{bmatrix}$.

Next we determine the coordinates of $L(\mathbf{v}_1)$, $L(\mathbf{v}_2)$, and $L(\mathbf{v}_3)$ with respect to T. Since T is the natural basis we have

$$\left[L(\mathbf{v}_1)\right]_T = \begin{bmatrix} 1 \\ 0 \end{bmatrix}, \quad \left[L(\mathbf{v}_2)\right]_T = \begin{bmatrix} 1 \\ 1 \end{bmatrix}, \quad \left[L(\mathbf{v}_3)\right]_T = \begin{bmatrix} 0 \\ -1 \end{bmatrix}.$$

The matrix A of L with respect to S and T is

$$A = \begin{bmatrix} \left[L(\mathbf{v}_1)\right]_T & \left[L(\mathbf{v}_2)\right]_T & \left[L(\mathbf{v}_3)\right]_T \end{bmatrix} = \begin{bmatrix} 1 & 1 & 0 \\ 0 & 1 & -1 \end{bmatrix}.$$

(b) $L(\mathbf{v}_1') = L\left(\begin{bmatrix} 1 \\ 1 \\ 0 \end{bmatrix}\right) = \begin{bmatrix} 2 \\ 1 \end{bmatrix}$, $L(\mathbf{v}_2') = L\left(\begin{bmatrix} 0 \\ 1 \\ 0 \end{bmatrix}\right) = \begin{bmatrix} 1 \\ 1 \end{bmatrix}$ $L(\mathbf{v}_3') = L\left(\begin{bmatrix} -1 \\ 1 \\ 1 \end{bmatrix}\right) = \begin{bmatrix} 0 \\ 0 \end{bmatrix}$

Next we determine the coordinates of $L(\mathbf{v}_1')$, $L(\mathbf{v}_2')$, and $L(\mathbf{v}_3')$ with respect to T'. Since T' is not the natural basis we must solve a system of equations to obtain the coordinates.

$$L(\mathbf{v}_1') = \begin{bmatrix} 2 \\ 1 \end{bmatrix} = c_1 \mathbf{w}_1' + c_2 \mathbf{w}_2'.$$

Adding vectors and equating corresponding components from both sides of the equation gives the linear system

$$\begin{bmatrix} -1 & 1 \\ 1 & 2 \end{bmatrix} \begin{bmatrix} c_1 \\ c_2 \end{bmatrix} = \begin{bmatrix} 2 \\ 1 \end{bmatrix}.$$

Forming the augmented matrix and applying row operations we obtain the reduced row echelon form

$$\begin{bmatrix} 1 & 0 & \vdots & -1 \\ 0 & 1 & \vdots & 1 \end{bmatrix}.$$

Thus $c_1 = -1$, $c_2 = 1$ and hence

$$\left[L(\mathbf{v}_1')\right]_{T'} = \begin{bmatrix} -1 \\ 1 \end{bmatrix}.$$

Next, set

$$L(\mathbf{v}_2') = \begin{bmatrix} 1 \\ 1 \end{bmatrix} = c_1 \mathbf{w}_1' + c_2 \mathbf{w}_2'.$$

Adding vectors and equating corresponding components from both sides of the equation gives the linear system

$$\begin{bmatrix} -1 & 1 \\ 1 & 2 \end{bmatrix} \begin{bmatrix} c_1 \\ c_2 \end{bmatrix} = \begin{bmatrix} 1 \\ 1 \end{bmatrix}.$$

Forming the augmented matrix and applying row operations we obtain the reduced row echelon form

$$\begin{bmatrix} 1 & 0 & \vdots & -\frac{1}{3} \\ 0 & 1 & \vdots & \frac{2}{3} \end{bmatrix}.$$

Thus $c_1 = -\frac{1}{3}$, $c_2 = \frac{2}{3}$ and

$$\left[L(\mathbf{v}_2')\right]_{T'} = \begin{bmatrix} -\frac{1}{3} \\ \frac{2}{3} \end{bmatrix}.$$

Finally, set

$$L(\mathbf{v}_3') = \begin{bmatrix} 0 \\ 0 \end{bmatrix} = c_1\mathbf{w}_1' + c_2\mathbf{w}_2'.$$

Then $c_1 = 0$ and $c_2 = 0$, so that

$$\left[L(\mathbf{v}_3')\right]_{T'} = \begin{bmatrix} 0 \\ 0 \end{bmatrix}.$$

The matrix A of L with respect to S' and T' is

$$A = \left[\left[L(\mathbf{v}_1')\right]_{T'} \quad \left[L(\mathbf{v}_2')\right]_{T'} \quad \left[L(\mathbf{v}_3')\right]_{T'}\right] = \begin{bmatrix} -1 & -\frac{1}{3} & 0 \\ 1 & \frac{2}{3} & 0 \end{bmatrix}.$$

(c) Let $\mathbf{x} = \begin{bmatrix} 1 \\ 2 \\ 3 \end{bmatrix}$ and $\mathbf{y} = L(\mathbf{x})$. Then from the definition of L,

$$L(\mathbf{x}) = L\left(\begin{bmatrix} 1 \\ 2 \\ 3 \end{bmatrix}\right) = \begin{bmatrix} 3 \\ -1 \end{bmatrix}.$$

Next we use Theorem 10.8 to express the relationship between coordinates of vectors in V with respect to a basis for V and coordinates of vectors in W with respect to a basis for W. From part (a),

$$\left[L(\mathbf{x})\right]_T = A\left[\mathbf{x}\right]_S = \begin{bmatrix} 1 & 1 & 0 \\ 0 & 1 & -1 \end{bmatrix}\begin{bmatrix} 1 \\ 2 \\ 3 \end{bmatrix}_S = \begin{bmatrix} 1 & 1 & 0 \\ 0 & 1 & -1 \end{bmatrix}\begin{bmatrix} 1 \\ 2 \\ 3 \end{bmatrix} = \begin{bmatrix} 3 \\ -1 \end{bmatrix}_T.$$

Hence $L(\mathbf{x}) = 3\mathbf{w}_1 - 1\mathbf{w}_2 = \begin{bmatrix} 3 \\ -1 \end{bmatrix}$. From part (b),

$$\left[L(\mathbf{x})\right]_{T'} = A\left[\mathbf{x}\right]_{S'} = \begin{bmatrix} -1 & -\frac{1}{3} & 0 \\ 1 & \frac{2}{3} & 0 \end{bmatrix}\begin{bmatrix} 1 \\ 2 \\ 3 \end{bmatrix}_{S'} = \begin{bmatrix} -1 & -\frac{1}{3} & 0 \\ 1 & \frac{2}{3} & 0 \end{bmatrix}\begin{bmatrix} 4 \\ -5 \\ 3 \end{bmatrix} = \begin{bmatrix} -\frac{7}{3} \\ \frac{2}{3} \end{bmatrix}_{T'}$$

where we have used $\mathbf{x} = 4\mathbf{v}_1' - 5\mathbf{v}_2' + 3\mathbf{v}_3'$. Hence

$$L(\mathbf{x}) = \left(-\frac{7}{3}\right)\mathbf{w}_1' + \left(\frac{2}{3}\right)\mathbf{w}_2' = \begin{bmatrix} 3 \\ -1 \end{bmatrix}.$$

7. Let $L\colon P_1 \to P_3$ be defined by $L(p(t)) = t^2 p(t)$. Let

$$S = \{p_1(t), p_2(t)\} = \{t, 1\} \quad \text{and} \quad S' = \{p_1'(t), p_2'(t)\} = \{t, t+1\}$$

be bases for P_1. Let

$$T = \{q_1(t), q_2(t), q_3(t), q_4(t)\} = \{t^3, t^2, t, 1\}$$

and

$$T' = \{q_1'(t), q_2'(t), q_3'(t), q_4'(t)\} = \{t^3, t^2 - 1, t, t+1\}$$

be bases for P_3. We proceed as in Example 6.

(a) $L(p_1(t)) = t^2(t) = t^3$, $L(p_2(t)) = t^2(1) = t^2$.

Next we determine the coordinates of $L(p_j(t))$ with respect to the T basis. Since T is the natural basis we have

$$\left[L(p_1(t))\right]_T = \begin{bmatrix} 1 \\ 0 \\ 0 \\ 0 \end{bmatrix} \quad \text{and} \quad \left[L(p_2(t))\right]_T = \begin{bmatrix} 0 \\ 1 \\ 0 \\ 0 \end{bmatrix}.$$

Then the matrix of L with respect to S and T is

$$A = \begin{bmatrix} \left[L(p_1(t))\right]_T & \left[L(p_2(t))\right]_T \end{bmatrix} = \begin{bmatrix} 1 & 0 \\ 0 & 1 \\ 0 & 0 \\ 0 & 0 \end{bmatrix}.$$

(b) $L(p_1'(t)) = t^2(t) = t^3$, $L(p_2'(t)) = t^2(t+1) = t^3 + t^2$.

Next we determine the coordinates of $L(p_j'(t))$ with respect to the T' basis.

$$L(p_1'(t)) = 1q_1(t) \quad \Longrightarrow \quad \left[L(p_1'(t))\right]_{T'} = \begin{bmatrix} 1 \\ 0 \\ 0 \\ 0 \end{bmatrix}$$

$$L(p_2'(t)) = 1q_1'(t) + 1q_2'(t) - 1q_3'(t) + 1q_4'(t) \quad \Longrightarrow \quad \left[L(p_2'(t))\right]_{T'} = \begin{bmatrix} 1 \\ 1 \\ -1 \\ 1 \end{bmatrix}.$$

Then the matrix of L with respect to S' and T' is

$$A = \begin{bmatrix} \left[L(p_1'(t))\right]_{T'} & \left[L(p_2'(t))\right]_{T'} \end{bmatrix} = \begin{bmatrix} 1 & 1 \\ 0 & 1 \\ 0 & -1 \\ 0 & 1 \end{bmatrix}.$$

9. Let $L: M_{22} \to M_{22}$ be defined by $L(A) = A^T$,

$$S = \{\mathbf{v}_1, \mathbf{v}_2, \mathbf{v}_3, \mathbf{v}_4\} = \left\{ \begin{bmatrix} 1 & 0 \\ 0 & 0 \end{bmatrix}, \begin{bmatrix} 0 & 1 \\ 0 & 0 \end{bmatrix}, \begin{bmatrix} 0 & 0 \\ 1 & 0 \end{bmatrix}, \begin{bmatrix} 0 & 0 \\ 0 & 1 \end{bmatrix} \right\},$$

and

$$T = \{\mathbf{w}_1, \mathbf{w}_2, \mathbf{w}_3, \mathbf{w}_4\} = \left\{ \begin{bmatrix} 1 & 1 \\ 0 & 0 \end{bmatrix}, \begin{bmatrix} 0 & 1 \\ 0 & 0 \end{bmatrix}, \begin{bmatrix} 0 & 0 \\ 1 & 1 \end{bmatrix}, \begin{bmatrix} 1 & 0 \\ 0 & 1 \end{bmatrix} \right\}.$$

(a) $L(\mathbf{v}_1) = \mathbf{v}_1$, $L(\mathbf{v}_2) = \mathbf{v}_3$, $L(\mathbf{v}_3) = \mathbf{v}_2$, $L(\mathbf{v}_4) = \mathbf{v}_4$.

Thus it follows that

$$\left[L(\mathbf{v}_1)\right]_S = \begin{bmatrix} 1 \\ 0 \\ 0 \\ 0 \end{bmatrix}, \quad \left[L(\mathbf{v}_2)\right]_S = \begin{bmatrix} 0 \\ 0 \\ 1 \\ 0 \end{bmatrix}, \quad \left[L(\mathbf{v}_3)\right]_S = \begin{bmatrix} 0 \\ 1 \\ 0 \\ 0 \end{bmatrix}, \quad \left[L(\mathbf{v}_4)\right]_S = \begin{bmatrix} 0 \\ 0 \\ 0 \\ 1 \end{bmatrix}$$

and hence the matrix of L with respect to S is

$$\begin{bmatrix} \left[L(\mathbf{v}_1)\right]_S & \left[L(\mathbf{v}_2)\right]_S & \left[L(\mathbf{v}_3)\right]_S & \left[L(\mathbf{v}_4)\right]_S \end{bmatrix} = \begin{bmatrix} 1 & 0 & 0 & 0 \\ 0 & 0 & 1 & 0 \\ 0 & 1 & 0 & 0 \\ 0 & 0 & 0 & 1 \end{bmatrix}.$$

(b) Use the computations of $L(\mathbf{v}_i)$, $i = 1, 2, 3, 4$ in part (a). Here each $L(\mathbf{v}_i)$ must be expressed in terms of the T-basis. Hence we consider four equations

$$c_1 \mathbf{w}_1 + c_2 \mathbf{w}_2 + c_3 \mathbf{w}_3 + c_4 \mathbf{w}_4 = L(\mathbf{v}_i), \quad i = 1, 2, 3, 4. \tag{10.1}$$

We form the linear system associated with (10.1). In each case the coefficient matrix is the same but the right-hand side is different. We illustrate the construction in the case that $i = 1$ and then show how it can be used to find $\left[L(\mathbf{v}_i)\right]_T$ for $i = 1, 2, 3, 4$. Combining the terms on the left side of (10.1) gives

$$\begin{bmatrix} c_1 + c_4 & c_1 + c_2 \\ c_3 & c_3 + c_4 \end{bmatrix} = L(\mathbf{v}_1) = \begin{bmatrix} 1 & 0 \\ 0 & 0 \end{bmatrix}$$

and equating corresponding entries gives the linear system

$$
\begin{aligned}
c_1 +   + c_4 &= 1 \\
c_1 + c_2 &= 0 \\
c_3 &= 0 \\
c_3 + c_4 &= 0.
\end{aligned}
$$

In matrix form we have

$$\begin{bmatrix} 1 & 0 & 0 & 1 \\ 1 & 1 & 0 & 0 \\ 0 & 0 & 1 & 0 \\ 0 & 0 & 1 & 1 \end{bmatrix} \begin{bmatrix} c_1 \\ c_2 \\ c_3 \\ c_4 \end{bmatrix} = \begin{bmatrix} 1 \\ 0 \\ 0 \\ 0 \end{bmatrix}.$$

For $L(\mathbf{v}_2)$, $L(\mathbf{v}_3)$, and $L(\mathbf{v}_4)$ the right-hand sides are respectively

$$\begin{bmatrix} 0 \\ 0 \\ 1 \\ 0 \end{bmatrix}, \quad \begin{bmatrix} 0 \\ 1 \\ 0 \\ 0 \end{bmatrix}, \quad \text{and} \quad \begin{bmatrix} 0 \\ 0 \\ 0 \\ 1 \end{bmatrix}.$$

To solve the four systems we partition the augmented matrix

$$\left[\begin{array}{cccc:c:c:c:c} 1 & 0 & 0 & 1 & 1 & 0 & 0 & 0 \\ 1 & 1 & 0 & 0 & 0 & 0 & 1 & 0 \\ 0 & 0 & 1 & 0 & 0 & 1 & 0 & 0 \\ 0 & 0 & 1 & 1 & 0 & 0 & 0 & 1 \end{array}\right]$$

and row reduce it to obtain

$$\left[\begin{array}{cccc:c:c:c:c} 1 & 0 & 0 & 0 & 1 & 1 & 0 & -1 \\ 0 & 1 & 0 & 0 & -1 & -1 & 1 & 1 \\ 0 & 0 & 1 & 0 & 0 & 1 & 0 & 0 \\ 0 & 0 & 0 & 1 & 0 & -1 & 0 & 1 \end{array}\right]$$

$$= \left[I_4 \; \vdots \; \left[L(\mathbf{v}_1)\right]_T \; \vdots \; \left[L(\mathbf{v}_2)\right]_T \; \vdots \; \left[L(\mathbf{v}_3)\right]_T \; \vdots \; \left[L(\mathbf{v}_4)\right]_T \right].$$

Thus it follows that the matrix of L with respect to S and T is

$$\begin{bmatrix} 1 & 1 & 0 & -1 \\ -1 & -1 & 1 & 1 \\ 0 & 1 & 0 & 0 \\ 0 & -1 & 0 & 1 \end{bmatrix}.$$

(c) Here we compute $L(\mathbf{w}_i)$, $i = 1, 2, 3, 4$ and then find $\left[L(\mathbf{w}_i)\right]_S$ to determine the columns of the matrix representing L with respect to T and S.

$$L(\mathbf{w}_1) = \begin{bmatrix} 1 & 0 \\ 1 & 0 \end{bmatrix}, \quad L(\mathbf{w}_2) = \begin{bmatrix} 0 & 0 \\ 1 & 0 \end{bmatrix}, \quad L(\mathbf{w}_3) = \begin{bmatrix} 0 & 1 \\ 0 & 1 \end{bmatrix}, \quad L(\mathbf{w}_4) = \begin{bmatrix} 1 & 0 \\ 0 & 1 \end{bmatrix}.$$

Following part (b) we consider the four equations

$$c_1\mathbf{v}_1 + c_2\mathbf{v}_2 + c_3\mathbf{v}_3 + c_4\mathbf{v}_4 = L(\mathbf{w}_i), \quad i = 1, 2, 3, 4.$$

Combining terms on the left gives

$$\begin{bmatrix} c_1 & c_2 \\ c_3 & c_4 \end{bmatrix} = L(\mathbf{w}_i), \quad i = 1, 2, 3, 4.$$

It follows that we are led to the partitioned matrix

$$\begin{bmatrix} 1 & 0 & 0 & 0 & \vdots & 1 & \vdots & 0 & \vdots & 0 & \vdots & 1 \\ 0 & 1 & 0 & 0 & \vdots & 0 & \vdots & 0 & \vdots & 1 & \vdots & 0 \\ 0 & 0 & 1 & 0 & \vdots & 1 & \vdots & 1 & \vdots & 0 & \vdots & 0 \\ 0 & 0 & 0 & 1 & \vdots & 0 & \vdots & 0 & \vdots & 1 & \vdots & 1 \end{bmatrix}$$

which is already in reduced row echelon form. Hence the matrix representing L with respect to T and S is

$$\begin{bmatrix} 1 & 0 & 0 & 1 \\ 0 & 0 & 1 & 0 \\ 1 & 1 & 0 & 0 \\ 0 & 0 & 1 & 1 \end{bmatrix}.$$

(d) Using the computations of $L(\mathbf{w}_i)$ from part (c), we consider the four equations

$$c_1\mathbf{w}_1 + c_2\mathbf{w}_2 + + c_3\mathbf{w}_3 + c_4\mathbf{w}_4 = L(\mathbf{w}_i), \quad i = 1, 2, 3, 4.$$

Combining the terms of the left side gives

$$\begin{bmatrix} c_1 + c_4 & c_1 + c_2 \\ c_3 & c_4 \end{bmatrix} = L(\mathbf{w}_i), \quad i = 1, 2, 3, 4.$$

It follows that we are led to the partitioned matrix

$$\begin{bmatrix} 1 & 0 & 0 & 1 & \vdots & 1 & \vdots & 0 & \vdots & 0 & \vdots & 1 \\ 1 & 1 & 0 & 0 & \vdots & 0 & \vdots & 0 & \vdots & 1 & \vdots & 0 \\ 0 & 0 & 1 & 0 & \vdots & 1 & \vdots & 1 & \vdots & 0 & \vdots & 0 \\ 0 & 0 & 0 & 1 & \vdots & 0 & \vdots & 0 & \vdots & 1 & \vdots & 1 \end{bmatrix}$$

whose reduced row echelon form is

$$\begin{bmatrix} 1 & 0 & 0 & 0 & \vdots & 1 & \vdots & 0 & \vdots & -1 & \vdots & 0 \\ 0 & 1 & 0 & 0 & \vdots & -1 & \vdots & 0 & \vdots & 2 & \vdots & 0 \\ 0 & 0 & 1 & 0 & \vdots & 1 & \vdots & 1 & \vdots & 0 & \vdots & 0 \\ 0 & 0 & 0 & 1 & \vdots & 0 & \vdots & 0 & \vdots & 1 & \vdots & 1 \end{bmatrix}.$$

Hence the matrix of L with respect to T is

$$\begin{bmatrix} 1 & 0 & -1 & 0 \\ -1 & 0 & 2 & 0 \\ 1 & 1 & 0 & 0 \\ 0 & 0 & 1 & 1 \end{bmatrix}.$$

11. For $L: R^3 \rightarrow R^3$, the matrix representing L with respect to the natural basis is

$$A = \begin{bmatrix} 1 & 3 & 1 \\ 1 & 2 & 0 \\ 0 & 1 & 1 \end{bmatrix}.$$

Since we are using the natural basis, the coordinates of a vector are just the components of the vector. Hence the image of a vector under L is obtain by multiplying the vector by A.

(a) $L\left(\begin{bmatrix} 1 \\ 2 \\ 3 \end{bmatrix} \right) = A \begin{bmatrix} 1 \\ 2 \\ 3 \end{bmatrix} = \begin{bmatrix} 10 \\ 5 \\ 5 \end{bmatrix}.$

(b) $L\left(\begin{bmatrix} 0 \\ 1 \\ 1 \end{bmatrix} \right) = A \begin{bmatrix} 0 \\ 1 \\ 1 \end{bmatrix} = \begin{bmatrix} 4 \\ 2 \\ 2 \end{bmatrix}.$

13. (a) By Theorem 10.8,

$$\left[L(\mathbf{v}_1) \right]_S = A \left[\mathbf{v}_1 \right]_S = A \begin{bmatrix} 1 \\ 2 \end{bmatrix}_S = \begin{bmatrix} 2 & -3 \\ -1 & 4 \end{bmatrix} \begin{bmatrix} 1 \\ 0 \end{bmatrix} = \begin{bmatrix} 2 \\ -1 \end{bmatrix}$$

$$\left[L(\mathbf{v}_2) \right]_S = A \left[\mathbf{v}_2 \right]_S = A \begin{bmatrix} 1 \\ -1 \end{bmatrix}_S = \begin{bmatrix} 2 & -3 \\ -1 & 4 \end{bmatrix} \begin{bmatrix} 0 \\ 1 \end{bmatrix} = \begin{bmatrix} -3 \\ 4 \end{bmatrix}.$$

(b) Using the results of (a),

$$L(\mathbf{v}_1) = 2\mathbf{v}_1 - \mathbf{v}_2 = \begin{bmatrix} 1 \\ 5 \end{bmatrix} \quad \text{and} \quad L(\mathbf{v}_2) = -3\mathbf{v}_1 + 4\mathbf{v}_2 = \begin{bmatrix} 1 \\ -10 \end{bmatrix}.$$

(c) We first compute the coordinates of

$$\mathbf{x} = \begin{bmatrix} -2 \\ 3 \end{bmatrix}$$

with respect to S. Set $c_1\mathbf{v}_1 + c_2\mathbf{v}_2 = \mathbf{x}$. Adding vectors and equating corresponding components from both sides of the equation gives the linear system

$$\begin{bmatrix} 1 & 1 \\ 2 & -1 \end{bmatrix} \begin{bmatrix} c_1 \\ c_2 \end{bmatrix} = \begin{bmatrix} -2 \\ 3 \end{bmatrix}.$$

Form the augmented matrix and row reduce it to obtain $c_1 = \frac{1}{3}$ and $c_2 = -\frac{7}{3}$. Thus

$$\left[\mathbf{x} \right]_S = \begin{bmatrix} -2 \\ 3 \end{bmatrix}_S = \begin{bmatrix} \frac{1}{3} \\ -\frac{7}{3} \end{bmatrix}.$$

By Theorem 10.8,

$$\left[L\left(\begin{bmatrix} -2 \\ 3 \end{bmatrix} \right) \right]_S = A \begin{bmatrix} -2 \\ 3 \end{bmatrix}_S = \begin{bmatrix} 2 & -3 \\ -1 & 4 \end{bmatrix} \begin{bmatrix} \frac{1}{3} \\ -\frac{7}{3} \end{bmatrix} = \begin{bmatrix} \frac{23}{3} \\ -\frac{29}{3} \end{bmatrix}.$$

Hence

$$L\left(\begin{bmatrix} -2 \\ 3 \end{bmatrix} \right) = \left(\frac{23}{3} \right)\mathbf{v}_1 - \left(\frac{29}{3} \right)\mathbf{v}_2 = \begin{bmatrix} -2 \\ 25 \end{bmatrix}.$$

15. Let $L: P_1 \to P_2$ be represented by

$$A = \begin{bmatrix} 1 & 0 \\ 2 & 1 \\ -1 & -2 \end{bmatrix}$$

with respect to bases

$$S = \{\mathbf{v}_1, \mathbf{v}_2\} = \{t + 1, t - 1\} \quad \text{and} \quad T = \{\mathbf{w}_1, \mathbf{w}_2, \mathbf{w}_3\} = \{t^2 + 1, t, t - 1\}.$$

(a) $\left[L(\mathbf{v}_i)\right]_T = A \left[\mathbf{v}_i\right]_S$ for $i = 1, 2$, so we first find $\left[\mathbf{v}_i\right]_S$. It follows that

$$\left[\mathbf{v}_1\right]_S = \begin{bmatrix} 1 \\ 0 \end{bmatrix} \quad \text{and} \quad \left[\mathbf{v}_2\right]_S = \begin{bmatrix} 0 \\ 1 \end{bmatrix}.$$

Thus

$$\left[L(\mathbf{v}_1)\right]_T = \begin{bmatrix} 1 & 0 \\ 2 & 1 \\ -1 & -2 \end{bmatrix} \begin{bmatrix} 1 \\ 0 \end{bmatrix} = \begin{bmatrix} 1 \\ 2 \\ -1 \end{bmatrix} \quad \text{and} \quad \left[L(\mathbf{v}_2)\right]_T = \begin{bmatrix} 1 & 0 \\ 2 & 1 \\ -1 & -2 \end{bmatrix} \begin{bmatrix} 0 \\ 1 \end{bmatrix} = \begin{bmatrix} 0 \\ 1 \\ -2 \end{bmatrix}.$$

(b) $L(\mathbf{v}_1) = 1\mathbf{w}_1 + 2\mathbf{w}_2 - 1\mathbf{w}_3 = (t^2 + 1) + 2t - (t - 1) = t^2 + t + 2$

$L(\mathbf{v}_2) = 0\mathbf{w}_1 + 1\mathbf{w}_2 - 2\mathbf{w}_3 = t - 2(t - 1) = -t + 2.$

(c) To compute $L(2t + 1)$ we first find $\left[2t + 1\right]_S$. Hence we solve

$$c_1 \mathbf{v}_1 + c_2 \mathbf{v}_2 = c_1(t + 1) + c_2(t - 1) = 2t + 1. \tag{10.2}$$

We have

$$(c_1 + c_2)t + (c_1 - c_2) = 2t + 1$$

which gives the linear system

$$c_1 + c_2 = 2$$
$$c_1 - c_2 = 1.$$

It follows that $c_1 = \frac{3}{2}$ and $c_2 = \frac{1}{2}$, so

$$\left[2t + 1\right]_S = \begin{bmatrix} \frac{3}{2} \\ \frac{1}{2} \end{bmatrix}.$$

Hence

$$\left[L(2t + 1)\right]_T = A \left[2t + 1\right]_S = \begin{bmatrix} 1 & 0 \\ 2 & 1 \\ -1 & -2 \end{bmatrix} \begin{bmatrix} \frac{3}{2} \\ \frac{1}{2} \end{bmatrix} = \begin{bmatrix} \frac{3}{2} \\ \frac{7}{2} \\ -\frac{5}{2} \end{bmatrix}$$

so

$$L(2t + 1) = \left(\tfrac{3}{2}\right) \mathbf{w}_1 + \left(\tfrac{7}{2}\right) \mathbf{w}_2 + \left(-\tfrac{5}{2}\right) \mathbf{w}_3$$
$$= \left(\tfrac{3}{2}\right) (t^2 + 1) + \left(\tfrac{7}{2}\right) t + \left(-\tfrac{5}{2}\right) (t - 1) = \left(\tfrac{3}{2}\right) t^2 + t + 4$$

(d) In part (c) the right side of (10.2) is replaced by $at + b$ so we are led to the linear system

$$c_1 + c_2 = a$$
$$c_1 - c_2 = b.$$

whose solution is

$$c_1 = \frac{a + b}{2}, \qquad c_2 = \frac{a - b}{2}.$$

Thus

$$\left[L(at + b)\right]_T = A \begin{bmatrix} \frac{a+b}{2} \\ \frac{a-b}{2} \end{bmatrix} = \begin{bmatrix} \frac{a+b}{2} \\ \frac{3a+b}{2} \\ \frac{-3a+b}{2} \end{bmatrix}$$

and

$$L(at + b) = \frac{a + b}{2} \mathbf{w}_1 + \frac{3a + b}{2} \mathbf{w}_2 + \frac{-3a + b}{2} \mathbf{w}_3$$
$$= \frac{a + b}{2}(t^2 + 1) + \frac{3a + b}{2} t + \frac{-3a + b}{2}(t - 1)$$
$$= \frac{a + b}{2} t^2 + bt + 2a.$$

17. Let $L: P_1 \to P_1$, where $L(t + 1) = t - 1$ and $L(t - 1) = 2t + 1$.

(a) $S = \{\mathbf{v}_1, \mathbf{v}_2\} = \{t + 1, t - 1\}$ is a basis for P_1. The matrix representing L with respect to S is

$$\left[\begin{bmatrix} L(\mathbf{v}_1) \end{bmatrix}_S \quad \begin{bmatrix} L(\mathbf{v}_2) \end{bmatrix}_S \right].$$

Thus we must solve the two systems

$$c_1 \mathbf{v}_1 + c_2 \mathbf{v}_2 = L(\mathbf{v}_i), \quad i = 1, 2.$$

We are lead to linear systems where the corresponding partitioned matrix is

$$\begin{bmatrix} 1 & 1 & \vdots & 1 & \vdots & 2 \\ 1 & -1 & \vdots & -1 & \vdots & 1 \end{bmatrix}.$$

The reduced row echelon form of this matrix is

$$\begin{bmatrix} 1 & 0 & \vdots & 0 & \vdots & \frac{3}{2} \\ 0 & 1 & \vdots & 1 & \vdots & \frac{1}{2} \end{bmatrix}.$$

Thus the matrix representing L with respect to S is

$$\begin{bmatrix} 0 & \frac{3}{2} \\ 1 & \frac{1}{2} \end{bmatrix}.$$

(b) To compute $L(2t + 3)$ from the definition of L we first express $2t + 3$ in terms of S. We have

$$2t + 3 = k_1 \mathbf{v}_1 + k_2 \mathbf{v}_2 = k_1(t + 1) + k_2(t - 1) = (k_1 + k_2)t + (k_1 - k_2)$$

which leads to the system

$$k_1 + k_2 = 2$$
$$k_1 - k_2 = 3$$

whose solution is $k_1 = \frac{5}{2}$, $k_2 = -\frac{1}{2}$. Then since L is a linear transformation we have

$$
\begin{aligned}
L(2t + 3) &= L\left(\left(\tfrac{5}{2}\right)\mathbf{v}_1 - \left(\tfrac{1}{2}\right)\mathbf{v}_2\right) \\
&= \left(\tfrac{5}{2}\right)L(\mathbf{v}_1) - \left(\tfrac{1}{2}\right)L(\mathbf{v}_2) \\
&= \left(\tfrac{5}{2}\right)(t - 1) - \left(\tfrac{1}{2}\right)(2t + 1) \\
&= \left(\tfrac{3}{2}\right)t - 3.
\end{aligned}
$$

To use the matrix from part (a) we use the fact that

$$
\left[2t + 3\right]_S = \begin{bmatrix} \frac{5}{2} \\ -\frac{1}{2} \end{bmatrix},
$$

so that

$$
\left[L(2t + 3)\right]_S = \left(-\tfrac{3}{4}\right)\mathbf{v}_1 + \left(\tfrac{9}{4}\right)\mathbf{v}_2 = \left(\tfrac{3}{2}\right)t - 3.
$$

(c) Using part (b), we replace $2t + 3$ by $at + b$ and we are led to the linear system

$$k_1 + k_2 = a$$
$$k_1 - k_2 = b$$

whose solution is

$$
k_1 = \frac{a + b}{2}, \qquad k_2 = \frac{a - b}{2}.
$$

then

$$
\left[L(at + b)\right]_S = A \begin{bmatrix} \frac{a+b}{2} \\ \frac{a-b}{2} \end{bmatrix} = \begin{bmatrix} 0 & \frac{3}{2} \\ 1 & \frac{1}{2} \end{bmatrix} \begin{bmatrix} \frac{a+b}{2} \\ \frac{a-b}{2} \end{bmatrix} = \begin{bmatrix} \frac{3(a-b)}{4} \\ \frac{3a+b}{4} \end{bmatrix}
$$

and it follows that

$$
L(at + b) = \left(\frac{3(a - b)}{4}\right)(t + 1) + \left(\frac{3a + b}{4}\right)(t - 1) = \left(\frac{3a - b}{2}\right)t - b.
$$

19. $L\colon P_1 \to P_1$ is represented by

$$
A = \begin{bmatrix} 2 & 3 \\ -1 & -2 \end{bmatrix}
$$

with respect to basis $S = \{\mathbf{v}_1, \mathbf{v}_2\} = \{t + 1, t - 1\}$. Let $T = \{t, 1\}$ be the natural basis for P_1. To find the matrix representing L with respect to T we determine P, the transition matrix from T to S and then compute $P^{-1}AP$. We have that

$$
P = \begin{bmatrix} [t]_S & [1]_S \end{bmatrix}.
$$

To find $[t]_S$ we solve $c_1\mathbf{v}_1 + c_2\mathbf{v}_2 = t$ and to find $[1]_S$ we solve $k_1\mathbf{v}_1 + k_2\mathbf{v}_2 = 1$. Each of these leads to a linear system with the same coefficient matrix but a different right-hand side. Hence we are led to the partitioned matrix

$$
\begin{bmatrix} 1 & 1 & \vdots & 1 & \vdots & 0 \\ 1 & -1 & \vdots & 0 & \vdots & 1 \end{bmatrix}.
$$

The reduced row echelon form is

$$\begin{bmatrix} 1 & 0 & \vdots & \frac{1}{2} & \vdots & \frac{1}{2} \\ 0 & 1 & \vdots & \frac{1}{2} & \vdots & -\frac{1}{2} \end{bmatrix}.$$

Thus $P = \begin{bmatrix} \frac{1}{2} & \frac{1}{2} \\ \frac{1}{2} & -\frac{1}{2} \end{bmatrix}$ and $P^{-1}AP = \begin{bmatrix} 1 & 0 \\ 4 & -1 \end{bmatrix}.$

21. $L \colon P_3 \to P_3$ is defined by $L(p(t)) = p''(t) + p(0)$. Let

$$S = \{\mathbf{v}_1, \mathbf{v}_2, \mathbf{v}_3, \mathbf{v}_4\} = \{1, t, t^2, t^3\} \quad \text{and} \quad T = \{\mathbf{w}_1, \mathbf{w}_2, \mathbf{w}_3, \mathbf{w}_4\} = \{t^3, t^2 - 1, t, 1\}$$

be bases for P_3.

(a) $L(\mathbf{v}_1) = L(1) = (1)'' + 1 = 1$
$L(\mathbf{v}_2) = L(t) = (t)'' + 0 = 0$
$L(\mathbf{v}_3) = L(t^2) = (t^2)'' + 0 = 2$
$L(\mathbf{v}_4) = L(t^3) = (t^3)'' + 0 = 6t$
Since S is the natural basis for P_3, $\left[L(\mathbf{v}_i)\right]_S$ for $i = 1, 2, 3, 4$ is easily computed:

$$\left[L(\mathbf{v}_1)\right]_S = \begin{bmatrix} 1 \\ 0 \\ 0 \\ 0 \end{bmatrix}, \quad \left[L(\mathbf{v}_2)\right]_S = \begin{bmatrix} 0 \\ 0 \\ 0 \\ 0 \end{bmatrix}, \quad \left[L(\mathbf{v}_3)\right]_S = \begin{bmatrix} 2 \\ 0 \\ 0 \\ 0 \end{bmatrix}, \quad \left[L(\mathbf{v}_4)\right]_S = \begin{bmatrix} 0 \\ 6 \\ 0 \\ 0 \end{bmatrix}.$$

Hence the matrix representing L with respect to S is

$$A = \begin{bmatrix} 1 & 0 & 2 & 0 \\ 0 & 0 & 0 & 6 \\ 0 & 0 & 0 & 0 \\ 0 & 0 & 0 & 0 \end{bmatrix}.$$

(b) $L(\mathbf{w}_1) = L(t^3) = (t^3)'' + 0 = 6t$
$L(\mathbf{w}_2) = L(t^2 - 1) = (t^2 - 1)'' + (-1) = 1$
$L(\mathbf{w}_3) = L(t) = (t)'' + 0 = 0$
$L(\mathbf{w}_4) = L(1) = (1)'' + 1 = 1$
Next we express $L(\mathbf{w}_i)$ in terms of the vectors in T. This can be done here without solving any linear systems because of the simplicity of the vectors in T. We have

$$\left[L(\mathbf{w}_1)\right]_T = \begin{bmatrix} 0 \\ 0 \\ 6 \\ 0 \end{bmatrix}, \quad \left[L(\mathbf{w}_2)\right]_T = \begin{bmatrix} 0 \\ 0 \\ 0 \\ 1 \end{bmatrix}, \quad \left[L(\mathbf{w}_3)\right]_T = \begin{bmatrix} 0 \\ 0 \\ 0 \\ 0 \end{bmatrix}, \quad \left[L(\mathbf{w}_4)\right]_T = \begin{bmatrix} 0 \\ 0 \\ 0 \\ 1 \end{bmatrix}.$$

It follows that the matrix representing L with respect to T is

$$B = \begin{bmatrix} 0 & 0 & 0 & 0 \\ 0 & 0 & 0 & 0 \\ 6 & 0 & 0 & 0 \\ 0 & 1 & 0 & 1 \end{bmatrix}.$$

(c) Given that A from part (a) represents L with respect to the S-basis we compute the matrix L with respect to the T-basis indirectly by determining the transition matrix P from T to S and proceeding as in Figure 10.4. From Section 6.6 we have that

$$P = \begin{bmatrix} \left[\mathbf{w}_1\right]_S & \left[\mathbf{w}_2\right]_S & \left[\mathbf{w}_3\right]_S & \left[\mathbf{w}_4\right]_S \end{bmatrix}.$$

Because S is the natural basis the $\left[\mathbf{w}_i\right]_S$ are easy to determine. We have

$$P = \begin{bmatrix} 0 & -1 & 0 & 1 \\ 0 & 0 & 1 & 0 \\ 0 & 1 & 0 & 0 \\ 1 & 0 & 0 & 0 \end{bmatrix}.$$

Computing P^{-1} we have that the matrix representing L with respect to T is

$$P^{-1}AP = \begin{bmatrix} 0 & 0 & 0 & 1 \\ 0 & 0 & 1 & 0 \\ 0 & 1 & 0 & 0 \\ 1 & 0 & 1 & 0 \end{bmatrix} \begin{bmatrix} 1 & 0 & 2 & 0 \\ 0 & 0 & 0 & 6 \\ 0 & 0 & 0 & 0 \\ 0 & 0 & 0 & 0 \end{bmatrix} \begin{bmatrix} 0 & -1 & 0 & 1 \\ 0 & 0 & 1 & 0 \\ 0 & 1 & 0 & 0 \\ 1 & 0 & 0 & 0 \end{bmatrix} = B.$$

23. $L(e^t) = (e^t)' = e^t \implies \left[L(e^t)\right]_S = \begin{bmatrix} 1 \\ 0 \end{bmatrix}$

$L(e^{-t}) = (e^{-t})' = -e^{-t} \implies \left[L(e^{-t})\right]_S = \begin{bmatrix} 0 \\ -1 \end{bmatrix}$

Thus the matrix representing L with respect to S is $\begin{bmatrix} 1 & 0 \\ 0 & -1 \end{bmatrix}$.

25. From Example 10 in Section 4.3 (see Equation (6)), it follows that the matrix of L is

$$A = \begin{bmatrix} \cos 45° & -\sin 45° \\ \sin 45° & \cos 45° \end{bmatrix} = \begin{bmatrix} \frac{1}{\sqrt{2}} & -\frac{1}{\sqrt{2}} \\ \frac{1}{\sqrt{2}} & \frac{1}{\sqrt{2}} \end{bmatrix}.$$

Since $A^T A = I_2$, it follows that A is orthogonal.

T.1. If $\mathbf{x} = a_1\mathbf{v}_1 + \cdots + a_n\mathbf{v}_n$ is a vector in V, then

$$L(\mathbf{x}) = L\left(\sum_{j=1}^{n} a_j\mathbf{v}_j\right) = \sum_{j=1}^{n} a_j L(\mathbf{v}_j).$$

Then

$$\left[L(\mathbf{x})\right]_T = \left[L\left(\sum_{j=1}^{n} a_j\mathbf{v}_j\right)\right]_T = \sum_{j=1}^{n} a_j \left[L(\mathbf{v}_j)\right]_T$$

$$= \begin{bmatrix} c_{11} \\ c_{21} \\ \vdots \\ c_{m1} \end{bmatrix} a_1 + \begin{bmatrix} c_{12} \\ c_{22} \\ \vdots \\ c_{m2} \end{bmatrix} a_2 + \cdots + \begin{bmatrix} c_{1n} \\ c_{2n} \\ \vdots \\ c_{mn} \end{bmatrix} a_n$$

$$= \begin{bmatrix} c_{11} & c_{12} & \cdots & c_{1n} \\ c_{21} & c_{22} & \cdots & c_{2n} \\ \vdots & \vdots & & \vdots \\ c_{m1} & c_{m2} & \cdots & c_{mn} \end{bmatrix} \begin{bmatrix} a_1 \\ a_2 \\ \vdots \\ a_n \end{bmatrix} = A\left[\mathbf{v}\right]_S.$$

We now show that A is unique. Assume that $A^* = \left[c_{ij}^*\right]$ is another matrix having the same properties as A does, with $A \neq A^*$. Since all the elements of A and A^* are not equal, say the kth columns of these matrices are unequal. Now $\left[\mathbf{v}_k\right]_S = \mathbf{e}_k$. Then

$$\left[L(\mathbf{v}_k)\right]_T = A\left[\mathbf{v}_k\right]_S = A\mathbf{e}_k = k\text{th column of } A,$$

and

$$\left[L(\mathbf{v}_k)\right]_T = A^* \left[\mathbf{v}_k\right]_S = A^*\mathbf{e}_k = k\text{th column of } A^*.$$

Thus, $L(\mathbf{v}_k)$ has two different coordinate vectors with respect to T, which is impossible. Hence A is unique.

T.3. Let $S = \{\mathbf{v}_1,\ldots,\mathbf{v}_n\}$ be a basis for V, $T = \{\mathbf{w}_1,\ldots,\mathbf{w}_m\}$ a basis for W. Then $O(\mathbf{v}_j) = \mathbf{0}_W = 0\cdot\mathbf{w}_1 + \cdots + 0\cdot\mathbf{w}_m$. If A is the matrix of the zero transformation with respect to these bases, then the jth column of A is $\mathbf{0}$. Thus A is the $m \times n$ zero matrix.

T.5. Let $S = \{\mathbf{v}_1,\mathbf{v}_2,\ldots,\mathbf{v}_n\}$ and $T = \{\mathbf{w}_1,\mathbf{w}_2,\ldots,\mathbf{w}_n\}$. The matrix of I with respect to S and T is the matrix whose jth column is $\left[I(\mathbf{v}_j)\right]_T = \left[\mathbf{v}_j\right]_T$. This is precisely the transition matrix $P_{T\leftarrow S}$ from the S-basis to the T-basis. (See Section 6.7).

T.7. Suppose that L is one-to-one and onto. Then $\dim(\ker L) = 0$. Since $\ker L$ is the solution space of the homogeneous system $A\mathbf{x} = \mathbf{0}$, this homogeneous system has only the trivial solution. Theorem 1.13 implies that A is nonsingular. Conversely, if A is nonsingular, then Theorem 1.12 implies that the only solution to $A\mathbf{x} = \mathbf{0}$ is the trivial one.

T.9. Suppose that L is an isometry. Then $L(\mathbf{v}_i)\cdot L(\mathbf{v}_j) = \mathbf{v}_i \cdot \mathbf{v}_j$ so $L(\mathbf{v}_i)\cdot L(\mathbf{v}_j) = 1$ if $i = j$ and 0 if $i \neq j$. Hence, $T = \{L(\mathbf{v}_1), L(\mathbf{v}_2),\ldots,L(\mathbf{v}_n)\}$ is an orthonormal basis for R^n. Conversely, suppose that T is an orthonormal basis for R^n. Then $L(\mathbf{v}_i)\cdot L(\mathbf{v}_j) = 1$ if $i = j$ and 0 if $i \neq j$. Thus, $L(\mathbf{v}_i)\cdot L(\mathbf{v}_j) = \mathbf{v}_i \cdot \mathbf{v}_j$, so L is an isometry.

ML.1. From the definition of L, note that we can compute images under L using matrix multiplication: $L(\mathbf{v}) = C\mathbf{v}$, where

C = [2 −1 0;1 1 − 3]

```
C =
    2   −1    0
    1    1   −3
```

This observation makes it easy to compute $L(\mathbf{v}_i)$ in MATLAB. Entering the vectors in set S and computing their images, we have

v1 = [1 1 1]′; v2 = [1 2 1]′; v3 = [0 1 − 1]′;

Denote the images as **Lvi**:

Lv1 = C ∗ v1

```
Lv1 =
        1
       −1
```

Lv2 = C ∗ v2

```
Lv2 =
        0
        0
```

Lv3 = C ∗ v3

```
Lv3 =
       −1
        4
```

To find the coordinates of **Lvi** with respect to the T basis we solve the three systems involved all at once using the **rref** command.

rref([[1 2;2 1] Lv1 Lv2 Lv3])

ans =

```
    1   0   −1   0    3
    0   1    1   0   −2
```

The last 3 columns give the matrix A representing L with respect to bases S and T.

A = ans(:,3:5)

A =

```
   −1   0    3
    1   0   −2
```

Section 10.4, p. 547

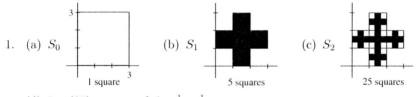

1. (a) S_0 (b) S_1 (c) S_2

1 square   5 squares   25 squares

(d) $5 \times (25)$ squares of size $\frac{1}{9} \times \frac{1}{9}$.

(e) S_1, S_2, and S_3 are composed of crosses made up of 5 squares of the same size.

(f) area $(S_0) = 3^2 = 9$, area $(S_1) = 5(\frac{1}{3}(3))^2 = 5$, area $(S_2) = 5^2(\frac{1}{3^2}(3))^2 = \frac{25}{9}$; area $(S_3) = \frac{125}{81}$;

$\dfrac{\text{area}\,(S_1)}{\text{area}\,(S_0)} = \dfrac{5}{9}; \ \dfrac{\text{area}\,(S_2)}{\text{area}\,(S_1)} = \dfrac{\frac{25}{9}}{9} = \dfrac{5}{9}; \ \dfrac{\text{area}\,(S_3)}{\text{area}\,(S_2)} = \dfrac{\frac{125}{81}}{\frac{25}{9}} = \dfrac{5}{9}$.

3. (a) $\frac{1}{8}$ (b) [figure]   (c) 2^{-6}.

(d) For 10.8(a) length $= 2 = 1 + 2(\frac{1}{2})$.

For 10.8(b) length $= 3 = 1 + 2(\frac{1}{2}) + 4(\frac{1}{4})$;

For 10.8(c) length $= 4 = 1 + 2(\frac{1}{2}) + 4(\frac{1}{4}) + 8(\frac{1}{8})$.

For 10.9   length $= 7$.

5. $T(T(\mathbf{v})) = \mathbf{v} + 2\mathbf{b}$, $T(T(T(\mathbf{v}))) = \mathbf{v} + 3\mathbf{b}$,
$T^k(\mathbf{v}) = \mathbf{v} + k\mathbf{b}$; the vector $\mathbf{v}$ is translated by $k\mathbf{b}$.

7. $T(\mathbf{v}) = \begin{bmatrix} 3 & -1 \\ 4 & 0 \end{bmatrix} \begin{bmatrix} v_1 \\ v_2 \end{bmatrix} + \begin{bmatrix} 1 \\ -5 \end{bmatrix}$; $A = \begin{bmatrix} 3 & -1 \\ 4 & 0 \end{bmatrix}$, $\mathbf{b} = \begin{bmatrix} 1 \\ -5 \end{bmatrix}$.

9. $A = \begin{bmatrix} 1 & 4 \\ -1 & 3 \end{bmatrix}$, $\mathbf{b} = \begin{bmatrix} -2 \\ 1 \end{bmatrix}$.

11. $S = \begin{bmatrix} 0 & 0 & 1 & 2 & 3 & 3 & 0 \\ 0 & 1 & 1 & 3 & 1 & 0 & 0 \end{bmatrix}$. Recall from Example 1 that to compute $T(S)$ we compute AS, then add the vector $\mathbf{b}$ to each column of the result of AS.

(a) $T(S) = AS + \mathbf{b} = \begin{bmatrix} 2 & -2 \\ 1 & 2 \end{bmatrix} \begin{bmatrix} 0 & 0 & 1 & 2 & 3 & 3 & 0 \\ 0 & 1 & 1 & 3 & 1 & 0 & 0 \end{bmatrix} + \begin{bmatrix} -2 & -2 & -2 & -2 & -2 & -2 & -2 \\ 1 & 1 & 1 & 1 & 1 & 1 & 1 \end{bmatrix}$

$= \begin{bmatrix} -2 & -4 & -2 & -4 & 2 & 4 & -2 \\ 1 & 2 & 4 & 8 & 8 & 7 & 1 \end{bmatrix}$.

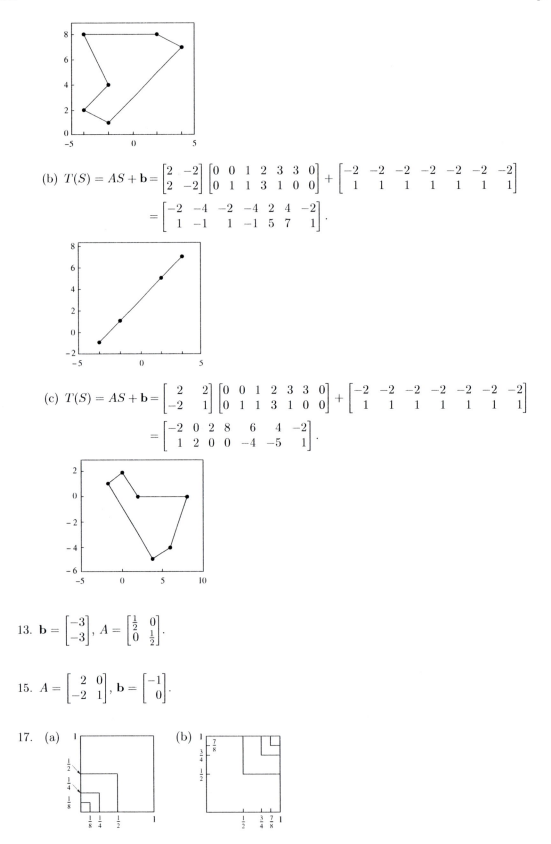

(b) $T(S) = AS + \mathbf{b} = \begin{bmatrix} 2 & -2 \\ 2 & -2 \end{bmatrix} \begin{bmatrix} 0 & 0 & 1 & 2 & 3 & 3 & 0 \\ 0 & 1 & 1 & 3 & 1 & 0 & 0 \end{bmatrix} + \begin{bmatrix} -2 & -2 & -2 & -2 & -2 & -2 & -2 \\ 1 & 1 & 1 & 1 & 1 & 1 & 1 \end{bmatrix}$

$\qquad = \begin{bmatrix} -2 & -4 & -2 & -4 & 2 & 4 & -2 \\ 1 & -1 & 1 & -1 & 5 & 7 & 1 \end{bmatrix}.$

(c) $T(S) = AS + \mathbf{b} = \begin{bmatrix} 2 & 2 \\ -2 & 1 \end{bmatrix} \begin{bmatrix} 0 & 0 & 1 & 2 & 3 & 3 & 0 \\ 0 & 1 & 1 & 3 & 1 & 0 & 0 \end{bmatrix} + \begin{bmatrix} -2 & -2 & -2 & -2 & -2 & -2 & -2 \\ 1 & 1 & 1 & 1 & 1 & 1 & 1 \end{bmatrix}$

$\qquad = \begin{bmatrix} -2 & 0 & 2 & 8 & 6 & 4 & -2 \\ 1 & 2 & 0 & 0 & -4 & -5 & 1 \end{bmatrix}.$

13. $\mathbf{b} = \begin{bmatrix} -3 \\ -3 \end{bmatrix}$, $A = \begin{bmatrix} \frac{1}{2} & 0 \\ 0 & \frac{1}{2} \end{bmatrix}.$

15. $A = \begin{bmatrix} 2 & 0 \\ -2 & 1 \end{bmatrix}$, $\mathbf{b} = \begin{bmatrix} -1 \\ 0 \end{bmatrix}.$

17. (a)   (b)

(c)   (d)

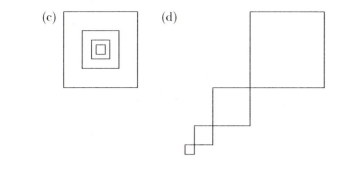

T.1.   (a) $A = \begin{bmatrix} -1 & 0 \\ 0 & 1 \end{bmatrix}$, $\mathbf{b} = \begin{bmatrix} 0 \\ 2 \end{bmatrix}$.

(b) $A = \begin{bmatrix} 0 & 1 \\ -1 & 0 \end{bmatrix}$, $\mathbf{b} = \begin{bmatrix} 0 \\ -1 \end{bmatrix}$.

T.3.   (a) $A = \begin{bmatrix} \frac{1}{3} & 0 \\ 0 & \frac{1}{3} \end{bmatrix}$.

(b) $\mathbf{b}_1 = \begin{bmatrix} 1 \\ 1 \end{bmatrix}$, $\mathbf{b}_2 = \begin{bmatrix} 0 \\ 1 \end{bmatrix}$, $\mathbf{b}_3 = \begin{bmatrix} 1 \\ 2 \end{bmatrix}$, $\mathbf{b}_4 = \begin{bmatrix} 2 \\ 1 \end{bmatrix}$, $\mathbf{b}_5 = \begin{bmatrix} 1 \\ 0 \end{bmatrix}$.

$$T_1(\mathbf{v}) = \begin{bmatrix} \frac{1}{3} & 0 \\ 0 & \frac{1}{3} \end{bmatrix} \mathbf{v} + \begin{bmatrix} 1 \\ 1 \end{bmatrix}$$

$$T_2(\mathbf{v}) = \begin{bmatrix} \frac{1}{3} & 0 \\ 0 & \frac{1}{3} \end{bmatrix} \mathbf{v} + \begin{bmatrix} 0 \\ 1 \end{bmatrix}$$

$$T_3(\mathbf{v}) = \begin{bmatrix} \frac{1}{3} & 0 \\ 0 & \frac{1}{3} \end{bmatrix} \mathbf{v} + \begin{bmatrix} 1 \\ 2 \end{bmatrix}$$

$$T_4(\mathbf{v}) = \begin{bmatrix} \frac{1}{3} & 0 \\ 0 & \frac{1}{3} \end{bmatrix} \mathbf{v} + \begin{bmatrix} 2 \\ 1 \end{bmatrix}$$

$$T_5(\mathbf{v}) = \begin{bmatrix} \frac{1}{3} & 0 \\ 0 & \frac{1}{3} \end{bmatrix} \mathbf{v} + \begin{bmatrix} 1 \\ 0 \end{bmatrix}$$

T.5. Here we consider solutions $\mathbf{v}$ of the matrix equation $A\mathbf{v} = \mathbf{v}$. The equation $A\mathbf{v} = \mathbf{v}$ implies that $\lambda = 1$ is an eigenvalue of A. Using this observation we can state the following. T has a fixed point when 1 is an eigenvalue of A. In this case there will be infinitely many fixed points since any nonzero multiple of an eigenvector is another eigenvector.

ML.1. Command **fernifs([0 .2],30000)** produces the following figure.

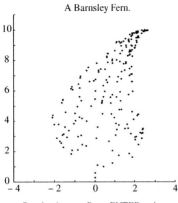

A Barnsley Fern.

Routine is over. Press ENTER twice.

Supplementary Exercises, p. 552

1. Yes. Let $\mathbf{u} = (x_1, y_1)$ and $\mathbf{v} = (x_2, y_2)$. Then

$$L(\mathbf{u} + \mathbf{v}) = L((x_1, y_1) + (x_2, y_2)) = L(x_1 + x_2, y_1 + y_2)$$
$$= ((x_1 + x_2) + (y_1 + y_2), 2(x_1 + x_2) - (y_1 + y_2))$$
$$= (x_1 + y_1, 2x_1 - y_1) + (x_2 + y_2, 2x_2 - y_2) = L(\mathbf{u}) + L(\mathbf{v}).$$

Also, if c is any real number, then

$$L(c\mathbf{u}) = L(c(x_1, y_1)) = L(cx_1, cy_1) = (cx_1 + cy_1, 2cx_1 - cy_1) = c(x_1 + y_1, 2x_1 - y_1) = cL(\mathbf{u}).$$

3. Since $S = \{t - 1, t + 1\}$ is a basis for P_1 we determine coefficients c_1 and c_2 such that

$$c_1(t - 1) + c_2(t + 1) = 5t + 1.$$

Expanding, collecting like terms, and equating coefficients of like powers of t from both sides of the equation we obtain the linear system

$$\begin{bmatrix} 1 & 1 \\ -1 & 1 \end{bmatrix} \begin{bmatrix} c_1 \\ c_2 \end{bmatrix} = \begin{bmatrix} 5 \\ 1 \end{bmatrix}$$

whose solution is $c_1 = 2$ and $c_2 = 3$. Then

$$L(5t + 1) = L(2(t - 1) + 3(t + 1)) = 2L(t - 1) + 3L(t + 1)$$
$$= 2(t + 2) + 3(2t + 1) = 8t + 7.$$

5. Let $L \colon R^2 \to R^3$ be defined by $L(x, y) = (x + y, x - y, x + 2y)$.

   (a) Following the technique in Example 12 in Section 10.2, we have

   $$L(x, y) = (x + y, x - y, x + 2y) = x(1, 1, 1) + y(1, -1, 2).$$

   Thus $S = \{\mathbf{v}_1, \mathbf{v}_2\} = \{(1, 1, 1), (1, -1, 2)\}$ spans range L. Since $\mathbf{v}_2$ is not a scalar multiple of $\mathbf{v}_1$, S is also linearly independent. Thus S is a basis for range L.

   (b) L is not onto since range L is not identically R^3. This follows since $\dim(\text{range } L) = 2$ but $\dim(R^3) = 3$.

7. Let $S = \{p_1(t), p_2(t), p_3(t)\} = \{t^2 - 1, t + 2, t - 1\}$ be a basis for P_2.

(a) Following Example 4 of Section 10.3, we determine coefficients c_1, c_2, and c_3 such that

$$p(t) = 2t^2 - 2t + 6 = c_1 p_1(t) + c_2 p_2(t) + c_3 p_3(t).$$

Expanding, collecting like terms, and equating coefficients of like powers of t from both sides of the equation yields the linear system

$$\begin{aligned} c_1 &&=& \ \ 2 \\ c_2 + c_3 &=& -2 \\ -c_1 + 2c_2 - c_3 &=& \ \ 6. \end{aligned}$$

Form the augmented matrix and determine its reduced row echelon form. We obtain the solution as $c_1 = 2$, $c_2 = 2$, $c_3 = -4$. Then

$$\left[p(t)\right]_S = \begin{bmatrix} 2 \\ 2 \\ -4 \end{bmatrix}.$$

(b) We have $\left[p(t)\right]_S = \begin{bmatrix} 2 \\ -1 \\ 3 \end{bmatrix}$. Then

$$p(t) = 2p_1(t) - p_2(t) + 3p_3(t) = 2t^2 + 2t - 7.$$

9. Let $L: P_1 \to P_1$ be a linear transformation which is represented by the matrix

$$A = \begin{bmatrix} 2 & -3 \\ 1 & 2 \end{bmatrix}$$

with respect to the basis $S = \{p_1(t), p_2(t)\} = \{t - 2, t + 1\}$.

(a) $\left[p_1(t)\right]_S = \begin{bmatrix} 1 \\ 0 \end{bmatrix}$, hence $\left[L(p_1(t))\right]_S = A \left[p_1(t)\right]_S = \begin{bmatrix} 2 \\ 1 \end{bmatrix}$.

$\left[p_2(t)\right]_S = \begin{bmatrix} 0 \\ 1 \end{bmatrix}$, hence $\left[L(p_2(t))\right]_S = A \left[p_2(t)\right]_S = \begin{bmatrix} -3 \\ 2 \end{bmatrix}$.

(b) Using the coordinates from part (a),

$$L(p_1(t)) = 2p_1(t) + 1p_2(t) = 3t - 3 \quad \text{and} \quad L(p_2(t)) = -3p_1(t) + 2p_2(t) = -t + 8.$$

(c) We first compute the coordinates of $p(t) = t + 2$ with respect to S. We have

$$t + 2 = \left(-\tfrac{1}{3}\right)(t - 2) + \tfrac{4}{3}(t + 1) \quad \Longrightarrow \quad \left[p(t)\right]_S = \begin{bmatrix} -\tfrac{1}{3} \\ \tfrac{4}{3} \end{bmatrix}.$$

Then

$$\begin{aligned} L(p(t)) = L\left(\left(-\tfrac{1}{3}\right)(t - 2) + \tfrac{4}{3}(t + 1)\right) &= \left(-\tfrac{1}{3}\right)L(t - 2) + \tfrac{4}{3}L(t + 1) \\ &= \left(-\tfrac{1}{3}\right)(3t - 3) + \tfrac{4}{3}(-t + 8) = -\tfrac{7}{3}t + \tfrac{35}{3}. \end{aligned}$$

11. Let the natural basis $S = \{\mathbf{e}_1, \mathbf{e}_2, \mathbf{e}_3\}$ and let

$$T = \{\mathbf{w}_1, \mathbf{w}_2, \mathbf{w}_3\} = \left\{ \begin{bmatrix} 0 \\ 1 \\ 1 \end{bmatrix}, \begin{bmatrix} 1 \\ 0 \\ 1 \end{bmatrix}, \begin{bmatrix} 1 \\ 1 \\ 0 \end{bmatrix} \right\}.$$

Then

$$L(\mathbf{e}_1) = \begin{bmatrix} 0 \\ 1 \\ 1 \end{bmatrix} = \mathbf{w}_1 \implies \left[L(\mathbf{e}_1)\right]_T = \begin{bmatrix} 1 \\ 0 \\ 0 \end{bmatrix}$$

$$L(\mathbf{e}_2) = \begin{bmatrix} 1 \\ 0 \\ 1 \end{bmatrix} = \mathbf{w}_2 \implies \left[L(\mathbf{e}_2)\right]_T = \begin{bmatrix} 0 \\ 1 \\ 0 \end{bmatrix}$$

$$L(\mathbf{e}_3) = \begin{bmatrix} 1 \\ 1 \\ 0 \end{bmatrix} = \mathbf{w}_3 \implies \left[L(\mathbf{e}_3)\right]_T = \begin{bmatrix} 0 \\ 0 \\ 1 \end{bmatrix}$$

The matrix of L with respect to S and T is I_3.

13. Yes. We have

$$L(B_1 + B_2) = A(B_1 + B_2) - (B_1 + B_2)A = (AB_1 - B_1A) + (AB_2 - B_2A) = L(B_1) + L(B_2)$$

and

$$L(cB) = A(cB) - (cB)A = c(AB - BA) = cL(B).$$

15. (a) We have

$$L(f + g) = (f + g)(0) = f(0) + g(0) = L(f) + L(g)$$
$$L(cf) = (cf)(0) = cf(0) = cL(f)$$

(b) The kernel of L consists of any continuous function f such that $L(f) = f(0) = 0$. That is, f is in ker L provided the value of f at $x = 0$ is zero. The following functions are in ker L:

$$x, \quad x^2, \quad x\cos x, \quad \sin x, \quad \frac{x}{x^2 + 1}, \quad xe^x.$$

(c) Yes. In this case

$$L(f + g) = (f + g)\left(\tfrac{1}{2}\right) = f\left(\tfrac{1}{2}\right) + g\left(\tfrac{1}{2}\right) = L(f) + L(g)$$
$$L(cf) = (cf)\left(\tfrac{1}{2}\right) = cf\left(\tfrac{1}{2}\right) = cL(f).$$

17. (a)   (b) A rectangular spiral.

T.1. Since $\left[\mathbf{v}_j\right]_S = \mathbf{e}_j = \begin{bmatrix} 0 \\ \vdots \\ 0 \\ 1 \\ 0 \\ \vdots \\ 0 \end{bmatrix} \leftarrow$ jth position , $\left\{\left[\mathbf{v}_1\right]_S, \ldots, \left[\mathbf{v}_n\right]_S\right\}$ is the standard basis for R^n.

T.3. (a) $(L_1 \boxplus L_2)(\mathbf{u} + \mathbf{v}) = L_1(\mathbf{u} + \mathbf{v}) + L_2(\mathbf{u} + \mathbf{v})$
$$= L_1(\mathbf{u}) + L_1(\mathbf{v}) + L_2(\mathbf{u}) + L_2(\mathbf{v}) = (L_1 \boxplus L_2)(\mathbf{u}) + (L_1 \boxplus L_2)(\mathbf{v})$$

$(L_1 \boxplus L_2)(k\mathbf{u}) = L_1(k\mathbf{u}) + L_2(k\mathbf{u})$
$$= kL_1(\mathbf{u}) + kL_2(\mathbf{u}) = k(L_1(\mathbf{u}) + L_2(\mathbf{u})) = k(L_1 \boxplus L_2)(\mathbf{u})$$

(b) $(c \boxdot L)(\mathbf{u} + \mathbf{v}) = cL(\mathbf{u} + \mathbf{v}) = cL(\mathbf{u}) + cL(\mathbf{v}) = (c \boxdot L)(\mathbf{u}) + (c \boxdot L)(\mathbf{v})$
$(c \boxdot L)(k\mathbf{u}) = cL(k\mathbf{u}) = ckL(\mathbf{u}) = kcL(\mathbf{u}) = k(c \boxdot L)(\mathbf{u})$

(c) $(L_1 \boxplus L_2)(\mathbf{v}) = L_1(\mathbf{v}) + L_2(\mathbf{v}) = (v_1 + v_2, v_2 + v_3) + (v_1 + v_3, v_2) = (2v_1 + v_2 + v_3, 2v_2 + v_3)$.
$(-2 \boxdot L_1)(\mathbf{v}) = -2L(\mathbf{v}) = -2(v_1 + v_2, v_2 + v_3) = (-2v_1 - 2v_2, -2v_2 - 2v_3)$.

T.5. (a) By Theorem 6.5, $L(\mathbf{x}) = A\mathbf{x}$ is one-to-one if and only if $\ker(L)$, which is the solution space of $A\mathbf{x} = \mathbf{0}$, is equal to $\{\mathbf{0}_V\}$. The solution space of $A\mathbf{x} = \mathbf{0}$ is $\{\mathbf{0}_V\}$ if and only if the only linear combination of the columns of A that gives the zero vector is the one in which all the coefficients are zero. This is the case if and only if the columns of A are linearly independent, which is equivalent to rank $A = n$.

(b) $L(A) = A\mathbf{x}$ is onto if and only if range L is R^m. But range L is equal to the column space of A and

$$m = \dim(\text{range } L) = \dim(\text{column space of } A) = \text{column rank of } A = \text{rank } A.$$

Chapter 11

Linear Programming (Optional)

Section 11.1, p. 572

1. Let x be the number of tons of regular steel to be made and let y be the number of tons of special steel to be made. Since each ton of regular steel requires 2 hours in the open-hearth furnace and each ton of special steel also requires 2 hours in the open-hearth furnace, the total amount of time all the steel produced is in the open-hearth furnace is

$$2x + 2y.$$

Similarly, each ton of regular steel requires 5 hours in the soaking pit and each ton of special steel requires 3 hours in the soaking pit, thus the total amount of time all the steel produced is in the soaking pit is

$$5x + 3y.$$

The open-hearth furnace is available 8 hours per day and the soaking pit is available 15 hours per day, hence we must have

$$2x + 2y \le 8$$
$$5x + 3y \le 15.$$

Since x and y cannot be negative, it follows that

$$x \ge 0 \quad \text{and} \quad y \ge 0.$$

The profit on a ton of regular steel is \$120 and it is \$100 on a ton of special steel, thus the total profit (in dollars) is

$$z = 120x + 100y.$$

The problem in mathematical form is: Find values of x and y that will maximize

$$z = 120x + 100y$$

subject to the following restrictions that must be satisfied by x and y:

$$2x + 2y \le 8$$
$$5x + 3y \le 15$$
$$x \ge 0 \quad \text{and} \quad y \ge 0.$$

3. Referring to Exercise 2, the additional rule would impose the further restriction

$$y \leq \tfrac{1}{2}x.$$

5. Let x be the number of minutes of advertising and let y be the number of minutes of comedy. The total time of the program is

$$x + y,$$

which cannot exceed 30 minutes, hence

$$x + y \leq 30$$

and of course

$$x \geq 0 \quad \text{and} \quad y \geq 0.$$

The advertiser requires at least 2 minutes of advertising time, while the station insists on no more than 4 minutes of advertising time, hence

$$x \geq 2 \quad \text{and} \quad x \leq 4.$$

Similarly, the comedian insists on at least 24 minutes for the comedy portion of the show, so

$$y \geq 24.$$

Since each minute of advertising attracts 40,000 viewers and each minute of comedy attracts 45,000 viewers, the number of viewer-minutes is

$$z = 40{,}000x + 45{,}000y.$$

The problem in mathematical form is: Find values of x and y that will maximize

$$z = 40{,}000x + 45{,}000y$$

subject to the following restrictions that must be satisfied by x and y:

$$x + y \leq 30$$
$$y \geq 24$$
$$x \geq 2$$
$$x \leq 4$$
$$x \geq 0 \quad \text{and} \quad y \geq 0.$$

7. Let x be the number of units of A and let y be the number of units of B. Of course, x and y cannot be negative, thus

$$x \geq 0 \quad \text{and} \quad y \geq 0.$$

Each unit of A has 1 gram of fat and each unit of B has 2 grams of fat, thus the total amount of fat provided is

$$x + 2y.$$

Since there is to be no more than 10 grams of fat, we have

$$x + 2y \leq 10.$$

Similarly, each unit of A has 1 gram of carbohydrates and the same for a unit of B, hence the total amount of carbohydrates is

$$x + y.$$

There is to be no more than 7 grams of carbohydrates, thus we have

$$x + y \leq 7.$$

Each unit of A has 4 grams of protein and each unit of B has 6 grams of protein, thus the total amount of protein provided is

$$z = 4x + 6y.$$

The problem in mathematical form is: Find values of x and y that will maximize

$$z = 4x + 6y$$

subject to the following restrictions that must be satisfied by x and y:

$$x + 2y \leq 10$$
$$x + y \leq 7$$
$$x \geq 0 \quad \text{and} \quad y \geq 0.$$

9. Let x be the number of units of type A grain and let y be the number of units of type B grain. Of course, x and y cannot be negative, so

$$x \geq 0 \quad \text{and} \quad y \geq 0.$$

Each unit of type A contains 2 grams of fat and each unit of type B contains 3 grams of fat, thus the total number of grams of fat is

$$2x + 3y.$$

Since at least 18 grams of fat is required, we have

$$2x + 3y \geq 18.$$

Similarly, each unit of type A contains 1 gram of protein and each unit of type B contains 3 grams of protein, hence the total number of grams of protein is

$$x + 3y.$$

Since there is to be at least 12 grams of protein, we have

$$x + 3y \geq 12.$$

Each unit of type A contains 80 calories and each unit of type B contains 60 calories, thus the total number of calories is

$$80x + 60y.$$

Since there is to be at least 480 calories, we have

$$80x + 60y \geq 480.$$

Each unit of type A costs 10 cents and each unit of type B costs 12 cents, then the total cost (in cents) is

$$z = 10x + 12y.$$

The problem in mathematical form is: Find values of x and y that will minimize

$$z = 10x + 12y$$

subject to the following restrictions that must be satisfied by x and y:

$$
\begin{aligned}
2x + 3y &\geq 18 \\
x + 3y &\geq 12 \\
80x + 60y &\geq 480 \\
x \geq 0 \quad \text{and} \quad y &\geq 0.
\end{aligned}
$$

11.

13.

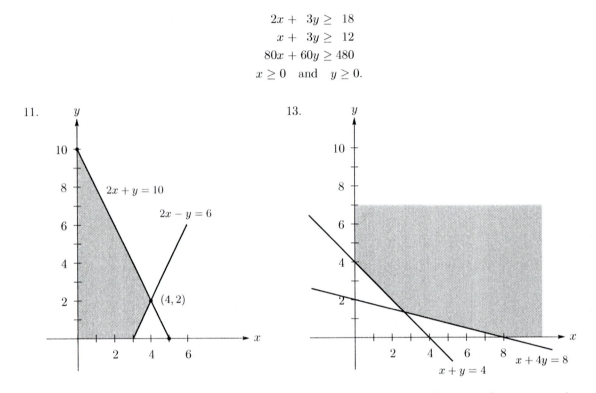

15. We follow the four step procedure given after Theorem 11.1. The feasible region S appears in the accompanying sketch.

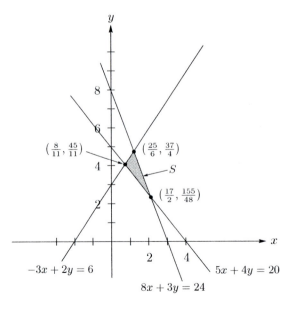

We see that S is bounded and it has extreme points $\left(\frac{8}{11}, \frac{45}{11}\right)$, $\left(\frac{25}{6}, \frac{37}{4}\right)$, and $\left(\frac{17}{2}, \frac{155}{48}\right)$. The values of the objective function are shown in the table:

point	$z = 3x - y$
$\left(\frac{8}{11}, \frac{45}{11}\right)$	$-\frac{21}{11}$
$\left(\frac{25}{6}, \frac{37}{4}\right)$	$\frac{39}{12}$
$\left(\frac{17}{2}, \frac{155}{48}\right)$	$\frac{1069}{48}$

Thus the minimum occurs at extreme point $\left(\frac{8}{11}, \frac{45}{11}\right)$. Hence the optimal solution is

$$x = \frac{8}{11}, \qquad y = \frac{45}{11}.$$

17. We follow the four step procedure given after Theorem 11.1. The problem is to maximize

$$z = 0.08x + 0.10y$$

subject to the following restrictions that must be satisfied by x and y:

$$x + y \leq 6000$$
$$y \leq 4000$$
$$x \geq 1500$$
$$x \geq 0 \quad \text{and} \quad y \geq 0.$$

The feasible region S appears in the accompanying sketch. We see that S is bounded and it has extreme points $(1500, 0)$, $(1500, 4000)$, $(2000, 4000)$, and $(6000, 0)$. The values of the objective function are shown in the table:

point	$z = 0.08x + 0.10y$
$(1500, 0)$	120
$(1500, 4000)$	520
$(2000, 4000)$	560
$(6000, 0)$	480

Thus the maximum occurs at extreme point $(2000, 4000)$. Hence the optimal solution is

$$x = 2000, \qquad y = 4000.$$

The trust fund should invest \$2000 in bond A and \$4000 in bond B to give a maximum return on the investment of \$560.

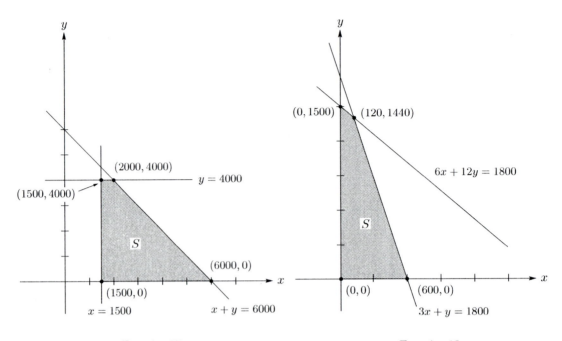

Exercise 17   Exercise 19

19. We follow the four step procedure given after Theorem 11.1. The problem is to maximize the revenue (in cents)

$$z = 30x + 60y$$

subject to the following restrictions that must be satisfied by x and y:

$$6x + 12y \leq 18000$$
$$3x + y \leq 1800$$
$$x \geq 0 \quad \text{and} \quad y \geq 0.$$

The feasible region S appears in the sketch above. We see that S is bounded and it has extreme points $(0,0)$, $(0,1500)$, $(120,1440)$, and $(600,0)$. The values of the objective function are shown in the table:

point	$z = 30x + 60y$
$(0,0)$	0
$(0,1500)$	90000
$(120,1440)$	90000
$(600,0)$	18000

Thus a maximum occurs at extreme points $(0,1500)$ and $(120,1440)$. Hence there is an optimal solution at

$$x = 0, \quad y = 1500$$

and another at

$$x = 120, \quad y = 1440.$$

In each case the maximum revenue is $900.

21. We follow the four step procedure given after Theorem 11.1. The problem is to minimize the cost (in cents)

$$z = 60x + 50y$$

subject to the following restrictions that must be satisfied by x and y:

$$3x + 5y \leq 15$$
$$4x + 4y \geq 16$$
$$x \geq 0 \quad \text{and} \quad y \geq 0.$$

The feasible region S appears in the sketch below.

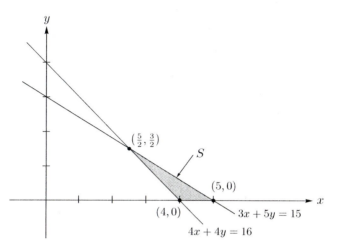

We see that S is bounded and it has extreme points $(4, 0)$, $\left(\frac{5}{2}, \frac{3}{2}\right)$, and $(5, 0)$. The values of the objective function are shown in the table:

point	$x = 60x + 50y$
$(4, 0)$	240
$\left(\frac{5}{2}, \frac{3}{2}\right)$	225
$(5, 0)$	300

Thus the minimum occurs at extreme point $\left(\frac{5}{2}, \frac{3}{2}\right)$. Hence the optimal solution is

$$x = \tfrac{5}{2}, \qquad y = \tfrac{3}{2}.$$

The generators should burn $\frac{5}{2}$ gallons of fuel L and $\frac{3}{2}$ gallons of fuel H to achieve a minimum cost of \$2.25 an hour.

23. We follow the four step procedure given after Theorem 11.1. The problem is to minimize

$$z = 20{,}000x + 25{,}000y$$

subject to the following restrictions that must be satisfied by x and y:

$$40x + 60y \geq 300$$
$$2x + 3y \leq 12$$
$$x \geq 0 \quad \text{and} \quad y \geq 0.$$

From the accompanying sketch we see that there is no feasible set of solutions. Hence there is no pair (x, y) which minimizes the objective function subject to the constraints.

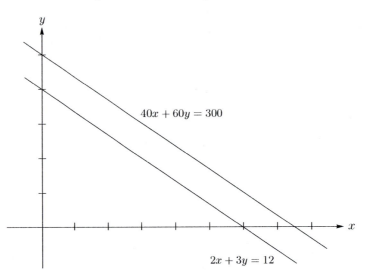

25. (a) This is a standard linear programming problem.

   (b) This is not a standard linear programming problem because we are asked to minimize the objective function.

   (c) This is not a standard linear programming problem because x_3 may be negative.

   (d) This is not a standard linear programming problem because of the equality constraint $3x_1 - 2x_2 + 2x_3 = 4$.

27. The only change needed is to convert the constraint

$$-3x_1 + 2x_2 - 3x_3 \geq -4$$

to a $\leq$ constraint. We replace the preceding constraint with

$$3x_1 - 2x_2 + 3x_3 \leq 4.$$

29. We introduce slack variable x_4 to convert constraint

$$3x_1 + x_2 - 4x_3 \leq 3$$

to the form

$$3x_1 + x_2 - 4x_3 + x_4 = 3.$$

Next introduce slack variable x_5 to convert constraint

$$x_1 - 2x_2 + 6x_3 \leq 21$$

to the form

$$x_1 - 2x_2 + 6x_3 + x_5 = 21.$$

Finally introduce slack variable x_6 to convert constraint

$$x_1 - x_2 - x_3 \leq 9$$

to the form

$$x_1 - x_2 - x_3 + x_6 = 9.$$

Then the new problem is:

Maximize $z = 2x_1 + 3x_2 + 7x_3$

subject to

$$\begin{aligned}
3x_1 + x_2 - 4x_3 + x_4 \quad\quad\quad &= 3 \\
x_1 - 2x_2 + 6x_3 \quad\quad + x_5 \quad\quad &= 21 \\
x_1 - x_2 - x_3 \quad\quad\quad + x_6 &= 9 \\
x_1 \geq 0, x_2 \geq 0, x_3 \geq 0, x_4 \geq 0, x_5 \geq 0, x_6 \geq 0.
\end{aligned}$$

Section 11.2, p. 589

1. We first rewrite this standard linear programming problem using slack variables. We have:

Maximize $z = 3x + 7y$

subject to

$$\begin{aligned}
3x - 2y + u \quad\quad\quad &= 7 \\
2x + 5y \quad + v \quad\quad &= 6 \\
2x + 3y \quad\quad\quad + w &= 8 \\
x \geq 0, y \geq 0, u \geq 0, v \geq 0, w \geq 0.
\end{aligned}$$

To form the initial tableau we head the columns with the variable names and enter the coefficients of the constraints into the rows with the right-hand sides of the constraints in the rightmost column of the tableau. Next we rewrite the objective function in the form

$$-3x - 7y + z = 0$$

and place its coefficient into the bottom row (the objective row) of the tableau. Finally, along the left side of the tableau we list the basic variable of the corresponding equation.

	x	y	u	v	w	z	
u	3	-2	1	0	0	0	7
v	2	5	0	1	0	0	6
w	2	3	0	0	1	0	8
	-3	-7	0	0	0	1	0

3. We first rewrite this standard linear programming problem using slack variables. We have:

Maximize $z = 2x_1 + 2x_2 + 3x_3 + x_4$

subject to

$$\begin{aligned}
3x_1 - 2x_2 + x_3 + x_4 + x_5 \quad\quad\quad\quad &= 6 \\
x_1 + x_2 + x_3 + x_4 \quad\quad + x_6 \quad\quad &= 8 \\
2x_1 - 3x_2 - x_3 + 2x_4 \quad\quad\quad\quad + x_7 &= 10 \\
x_1 \geq 0, x_2 \geq 0, x_3 \geq 0, x_4 \geq 0, x_5 \geq 0, x_6 \geq 0, x_7 \geq 0.
\end{aligned}$$

To form the initial tableau we head the columns with the variable names and enter the coefficients of the constraints into the rows with the right-hand sides of the constraints in the rightmost column of the tableau. Next, we rewrite the objective function in the form

$$-2x_1 - 2x_2 - 3x_3 - x_4 + z = 0$$

and place its coefficients into the bottom row (the objective row) of the tableau. Finally along the left side of the tableau we list the basic variables of the corresponding equation.

	x_1	x_2	x_3	x_4	x_5	x_6	x_7	z	
x_5	3	-2	1	1	1	0	0	0	6
x_6	1	1	1	1	0	1	0	0	8
x_7	2	-3	-1	2	0	0	1	0	10
	-2	-2	-3	-1	0	0	0	1	0

5. We follow the method used in Example 2. First reformulate the standard linear programming problem with slack variables:

Maximize $z = 2x + 3y$

subject to

$$3x + 5y + u \quad = 6$$
$$2x + 3y \quad + v = 7$$
$$x \geq 0, y \geq 0, u \geq 0, v \geq 0.$$

Form the initial tableau:

	x	y	u	v	z	
u	3	5	1	0	0	6
v	2	3	0	1	0	7
	-2	-3	0	0	1	0

Next we determine the pivotal column and pivotal row. The most negative entry in the objective row is -3, hence the y-column is the pivotal column. Thus the θ-ratios are $\left\{\frac{6}{5}, \frac{7}{3}\right\}$ and $\frac{6}{5}$ is the smallest. Hence the pivotal row is the u-row. It follows that the pivot has value 5. This information is marked in the next tableau.

		$\downarrow$				
	x	y	u	v	z	
$\leftarrow u$	3	5	1	0	0	6
v	2	3	0	1	0	7
	-2	-3	0	0	1	0

Next perform pivotal elimination. Take $\frac{1}{5}\mathbf{r}_1$, $-3\mathbf{r}_1 + \mathbf{r}_2$, $3\mathbf{r}_1 + \mathbf{r}_3$, and relabel the departing variable with the entering variable name. The resulting tableau is:

	x	y	u	v	z	
y	$\frac{3}{5}$	1	$\frac{1}{5}$	0	0	$\frac{6}{5}$
v	$\frac{1}{5}$	0	$-\frac{3}{5}$	1	0	$\frac{17}{5}$
	$-\frac{1}{5}$	0	$\frac{3}{5}$	0	1	$\frac{18}{5}$

Since there is a negative entry in the objective row, we do not have the optimal solution. We repeat the process. Determine the pivotal column and pivotal row. The most negative entry in the objective row is $-\frac{1}{5}$, hence the x-column is the pivotal column. Thus the θ-ratios are $\{2, 17\}$ and 2 is the smallest. Thus the pivotal row is the y-row. It follows that the pivot has value $\frac{3}{5}$. This information is marked in the next tableau.

	x	y	u	v	z	
$\leftarrow y$	$\frac{3}{5}$	1	$\frac{1}{5}$	0	0	$\frac{6}{5}$
v	$\frac{1}{5}$	0	$-\frac{3}{5}$	1	0	$\frac{17}{5}$
	$-\frac{1}{5}$	0	$\frac{3}{5}$	0	1	$\frac{18}{5}$

Next perform pivotal elimination. Take $\frac{5}{3}\mathbf{r}_1$, $-\frac{1}{5}\mathbf{r}_1 + \mathbf{r}_2$, $\frac{1}{5}\mathbf{r}_1 + \mathbf{r}_3$, and relabel the departing variable with the entering variable name. The resulting tableau is:

	x	y	u	v	z	
x	1	$\frac{5}{3}$	$\frac{1}{3}$	0	0	2
v	0	$-\frac{1}{3}$	$-\frac{2}{3}$	1	0	3
	0	$\frac{1}{3}$	$\frac{2}{3}$	0	1	4

Since all the entries in the objective row are nonnegative, we have the optimal solution

$$x = 2, \qquad y = 0.$$

The slack variables are

$$u = 0, \qquad v = 3,$$

and the optimal value of z is 4.

7. We follow the method used in Example 2. First reformulate the standard linear programming problem with slack variables:

Maximize $z = 2x + 5y$

subject to

$$2x - 3y + u \qquad = 4$$
$$x - 2y \qquad + v = 6$$
$$x \geq 0, y \geq 0, u \geq 0, v \geq 0.$$

Form the initial tableau:

	x	y	u	v	z	
u	2	-3	1	0	0	4
v	1	-2	0	1	0	6
	-2	-5	0	0	1	0

Next we determine the pivotal column and pivotal row. The most negative entry in the objective row is -5, hence the y-column is the pivotal column. Since none of the entries in the pivotal column above the objective row is nonnegative, the problem has no finite optimum.

9. We follow the method used in Example 2. First reformulate the standard linear programming problem with slack variables:

Maximize $z = 2x_1 - 4x_2 + 5x_3$

subject to

$$3x_1 + 2x_2 + x_3 + x_4 \qquad = 6$$
$$3x_1 - 6x_2 + 7x_3 \qquad + x_5 = 9$$
$$x_1 \geq 0, x_2 \geq 0, x_3 \geq 0, x_4 \geq 0, x_5 \geq 0.$$

Form the initial tableau:

	x_1	x_2	x_3	x_4	x_5	z	
x_4	3	2	1	1	0	0	6
x_5	3	−6	7	0	1	0	9
	−2	4	−5	0	0	1	0

Next we determine the pivotal column and pivotal row. The most negative entry in the objective row is −5, hence the x_3-column is the pivotal column. Thus the θ-ratios are $\left\{6, \frac{9}{7}\right\}$ and $\frac{9}{7}$ is the smallest. Hence the pivotal row is the x_5-row. It follows that the pivot has value 7. This information is marked in the next tableau.

			↓				
	x_1	x_2	x_3	x_4	x_5	z	
x_4	3	2	1	1	0	0	6
← x_5	3	−6	7	0	1	0	9
	−2	4	−5	0	0	1	0

Next perform pivotal elimination. Take $\frac{1}{7}\mathbf{r}_2$, $-\mathbf{r}_2 + \mathbf{r}_1$, $5\mathbf{r}_2 + \mathbf{r}_3$, and relabel the departing variable with the entering variable name. The resulting tableau is:

	x_1	x_2	x_3	x_4	x_5	z	
x_4	$\frac{18}{7}$	$\frac{20}{7}$	0	1	$-\frac{1}{7}$	0	$\frac{33}{7}$
x_3	$\frac{3}{7}$	$-\frac{6}{7}$	1	0	$\frac{1}{7}$	0	$\frac{9}{7}$
	$\frac{1}{7}$	$-\frac{2}{7}$	0	0	$\frac{5}{7}$	1	$\frac{45}{7}$

There is a negative entry in the objective row, thus we do not have an optimal solution. Determine the next pivotal column and pivotal row. The most negative entry in the objective row is $-\frac{2}{7}$ hence the x_2-column is the pivotal column. The θ-ratio is $\left\{\frac{33}{20}\right\}$ and $\frac{33}{20}$ is the smallest. (Only the positive entries of the pivotal column are used in forming the θ-ratios.) Thus the pivot has value $\frac{20}{7}$. This information is marked in the next tableau.

		↓					
	x_1	x_2	x_3	x_4	x_5	z	
← x_4	$\frac{18}{7}$	$\frac{20}{7}$	0	1	$-\frac{1}{7}$	0	$\frac{33}{7}$
x_3	$\frac{3}{7}$	$-\frac{6}{7}$	1	0	$\frac{1}{7}$	0	$\frac{9}{7}$
	$\frac{1}{7}$	$-\frac{2}{7}$	0	0	$\frac{5}{7}$	1	$\frac{45}{7}$

Next perform pivotal elimination. Take $\frac{7}{20}\mathbf{r}_1$, $\frac{6}{7}\mathbf{r}_1 + \mathbf{r}_2$, $\frac{2}{7}\mathbf{r}_1 + \mathbf{r}_3$, and relabel the departing variable with the entering variable name. The resulting tableau is:

	x_1	x_2	x_3	x_4	x_5	z	
x_2	$\frac{9}{10}$	1	0	$\frac{7}{20}$	$-\frac{1}{20}$	0	$\frac{33}{20}$
x_3	$\frac{6}{5}$	0	1	$\frac{3}{10}$	$\frac{1}{10}$	0	$\frac{27}{10}$
	$\frac{2}{5}$	0	0	$\frac{1}{10}$	$\frac{7}{10}$	1	$\frac{69}{10}$

Since all the entries of the objective row are nonnegative, we have an optimal solution,

$$x_1 = 0, \qquad x_2 = \tfrac{33}{20}, \qquad x_3 = \tfrac{27}{10}.$$

The slack variables are

$$x_4 = 0, \qquad x_5 = 0,$$

and the optimal value of z is $\frac{69}{10}$.

11. We follow the method of Example 2. First reformulate the standard linear programming problem with slack variables:

Maximize $z = x_1 + 2x_2 - x_3 + 5x_4$

subject to

$$
\begin{aligned}
2x_1 + 3x_2 + \ x_3 - \ x_4 + x_5 \qquad\quad &= 8 \\
3x_1 + \ x_2 - 4x_3 + 5x_4 \qquad + x_6 &= 9 \\
x_1 \geq 0, x_2 \geq 0, x_3 \geq 0, x_4 \geq 0, x_5 \geq 0, x_6 &\geq 0.
\end{aligned}
$$

Form the initial tableau:

	x_1	x_2	x_3	x_4	x_5	x_6	z	
x_5	2	3	1	-1	1	0	0	8
x_6	3	1	-4	5	0	1	0	9
	-1	-2	1	-5	0	0	1	0

Next we determine the pivotal column and pivotal row. The most negative entry in the objective row is -5 hence the x_4-column is the pivotal column. Thus the θ-ratio is $\left\{\frac{9}{5}\right\}$ and $\frac{9}{5}$ is the smallest. (Only the positive entries of the pivotal column are used in forming the θ-ratios.) Hence the pivotal row is the x_6-row. It follows that the pivot has value 5. This information is marked in the next tableau.

	x_1	x_2	x_3	$\downarrow$ x_4	x_5	x_6	z	
x_5	2	3	1	-1	1	0	0	8
$\leftarrow x_6$	3	1	-4	5	0	1	0	9
	-1	-2	1	-5	0	0	1	0

Next perform pivotal elimination. Take $\frac{1}{5}\mathbf{r}_2$, $\mathbf{r}_2 + \mathbf{r}_1$, $5\mathbf{r}_2 + \mathbf{r}_3$, and relabel the departing variable with the entering variable name. The resulting tableau is:

	x_1	x_2	x_3	x_4	x_5	x_6	z	
x_5	$\frac{13}{5}$	$\frac{16}{5}$	$\frac{1}{5}$	0	1	$\frac{1}{5}$	0	$\frac{49}{5}$
x_4	$\frac{3}{5}$	$\frac{1}{5}$	$-\frac{4}{5}$	1	0	$\frac{1}{5}$	0	$\frac{9}{5}$
	2	-1	-3	0	0	1	1	9

Since an entry in the objective row is negative, we do not have an optimal solution. Determine the pivotal column and pivotal row. The most negative entry in the objective row is -3 hence the x_3-column is the pivotal column. Thus the θ-ratio is $\{49\}$ and 49 is the smallest. (Only the positive entries of the pivotal column are used in forming the θ-ratios.) Hence the pivotal row is the x_5-row. It follows that the pivot has value $\frac{1}{5}$. This information is marked in the next tableau.

	x_1	x_2	x_3	x_4	x_5	x_6	z	
$\leftarrow x_5$	$\frac{13}{5}$	$\frac{16}{5}$	$\frac{1}{5}$	0	1	$\frac{1}{5}$	0	$\frac{49}{5}$
x_4	$\frac{3}{5}$	$\frac{1}{5}$	$-\frac{4}{5}$	1	0	$\frac{1}{5}$	0	$\frac{9}{5}$
	2	-1	-3	0	0	1	1	9

Next perform pivotal elimination. Take $5\mathbf{r}_1$, $\frac{4}{5}\mathbf{r}_1 + \mathbf{r}_2$, $3\mathbf{r}_1 + \mathbf{r}_3$, and relabel the departing variable with the entering variable name. The resulting tableau is:

	x_1	x_2	x_3	x_4	x_5	x_6	z	
x_3	13	16	1	0	5	1	0	49
x_4	11	13	0	1	4	1	0	41
	41	47	0	0	15	4	1	156

Since all the entries of the objective row are nonnegative, we have an optimal solution,

$$x_1 = 0, \qquad x_2 = 0, \qquad x_3 = 49, \qquad x_4 = 41.$$

The slack variables are

$$x_5 = 0, \qquad x_6 = 0$$

and the optimal value of z is 156.

13. The standard linear programming problem from Exercise 4 of Section 11.1 is:

Maximize $z = 30x + 60y$

subject to

$$6x + 12y \leq 18000$$
$$3x + y \leq 1800$$
$$x \geq 0 \quad \text{and} \quad y \geq 0.$$

We first reformulate the problem introducing slack variables. In this form we have:

Maximize $z = 30x + 60y$

subject to

$$6x + 12y + u \quad\quad = 18000$$
$$3x + \quad y \quad\quad + v = \quad 1800$$
$$x \geq 0, y \geq 0, u \geq 0, v \geq 0.$$

Form the initial tableau:

	x	y	u	v	z	
u	6	12	1	0	0	18000
v	3	1	0	1	0	1800
	-30	-60	0	0	1	0

Next we determine the pivotal column and pivotal row. The most negative entry in the objective row is -60 hence the y-column is the pivotal column. Thus the θ-ratios are $\left\{ \frac{18000}{12}, 1800 \right\}$ and $\frac{18000}{12}$ is the smallest. Hence the pivotal row is the u-row. It follows that the pivot has value 12. This information is marked in the next tableau.

		x	y	u	v	z	
$\leftarrow u$		6	12	1	0	0	18000
v		3	1	0	1	0	1800
		-30	-60	0	0	1	0

Next perform pivotal elimination. Take $\frac{1}{12}\mathbf{r}_1$, $-\mathbf{r}_1 + \mathbf{r}_2$, $60\mathbf{r}_1 + \mathbf{r}_3$, and relabel the departing variable with the entering variable name. The resulting tableau is:

	x	y	u	v	z	
y	$\frac{1}{2}$	1	$\frac{1}{12}$	0	0	1500
v	-3	0	$-\frac{1}{12}$	1	0	300
	0	0	5	0	1	90000

Since all the entries of the objective row are nonnegative, we have an optimal solution,

$$x = 0, \quad\quad y = 1500.$$

The slack variables are

$$u = 0, \quad\quad v = 300$$

and the optimal value of z is 90,000. Thus this optimal solution implies that no containers from the Smith Corporation and 1500 containers from the Johnson Corporation per truckload will give a maximum profit of $900. This problem has another optimal solution. See Exercise 19, Section 11.1.

15. Let x_1 be the number of tons of coal, x_2 be the number of tons of oil, and x_3 be the number of tons of gas. The problem is:

Maximize $z = 600x_1 + 550x_2 + 500x_3$

subject to

$$20x_1 + \quad 18x_2 + \quad 15x_3 \leq \quad 60$$
$$15x_1 + \quad 12x_2 + \quad 10x_3 \leq \quad 75$$
$$200x_1 + 220x_2 + 250x_3 \leq 2000$$
$$x_1 \geq 0, x_2 \geq 0, x_3 \geq 0.$$

Reformulating the problem using slack variables we have:

Maximize $z = 600x_1 + 550x_2 + 500x_3$

subject to

$$
\begin{array}{rcl}
20x_1 + 18x_2 + 15x_3 + x_4 & = & 60 \\
15x_1 + 12x_2 + 10x_3 + x_5 & = & 75 \\
200x_1 + 220x_2 + 250x_3 + x_6 & = & 2000 \\
\end{array}
$$
$$x_1 \geq 0, x_2 \geq 0, x_3 \geq 0, x_4 \geq 0, x_5 \geq 0, x_6 \geq 0.$$

Form the initial tableau:

	x_1	x_2	x_3	x_4	x_5	x_6	z	
x_4	20	18	15	1	0	0	0	60
x_5	15	12	10	0	1	0	0	75
x_6	200	220	250	0	0	1	0	2000
	-600	-550	-500	0	0	0	1	0

Next we determine the pivotal column and pivotal row. The most negative entry in the objective row is -600, hence the x_1-column is the pivotal column. Thus the θ-ratios are $\left\{ \frac{60}{20}, \frac{75}{15}, \frac{2000}{200} \right\}$ and $\frac{60}{20}$ is the smallest. Hence the pivotal row is the x_4-row. It follows that the pivot has value 20. This information is marked in the next tableau.

	$\downarrow$							
	x_1	x_2	x_3	x_4	x_5	x_6	z	
$\leftarrow x_4$	20	18	15	1	0	0	0	60
x_5	15	12	10	0	1	0	0	75
x_6	200	220	250	0	0	1	0	2000
	-600	-550	-500	0	0	0	1	0

Next perform pivotal elimination. Take $\frac{1}{20}\mathbf{r}_1$, $-15\mathbf{r}_1 + \mathbf{r}_2$, $-200\mathbf{r}_1 + \mathbf{r}_3$, $600\mathbf{r}_1 + \mathbf{r}_4$ and relabel the departing variable with the entering variable name. The resulting tableau is:

	x_1	x_2	x_3	x_4	x_5	x_6	z	
x_1	1	$\frac{9}{10}$	$\frac{3}{4}$	$\frac{1}{20}$	0	0	0	3
x_5	0	$-\frac{3}{2}$	$-\frac{5}{4}$	$-\frac{3}{4}$	1	0	0	30
x_6	0	40	100	-10	0	1	0	1400
	0	-10	-50	30	0	0	1	1800

Since an entry in the objective row is negative, we do not have an optimal solution. Determine the pivotal column and pivotal row. The most negative entry in the objective row is -50 hence the x_3-column is the pivotal column. Thus the θ-ratios are $\{4, 14\}$ and 4 is the smallest. Hence the pivotal row is the x_1-row. It follows that the pivot has value $\frac{3}{4}$. This information is marked in the next tableau.

	x_1	x_2	x_3	x_4	x_5	x_6	z	
$\leftarrow x_1$	1	$\frac{9}{10}$	$\frac{3}{4}$	$\frac{1}{20}$	0	0	0	3
x_5	0	$-\frac{3}{2}$	$-\frac{5}{4}$	$-\frac{3}{4}$	1	0	0	30
x_6	0	40	100	-10	0	1	0	1400
	0	-10	-50	30	0	0	1	1800

Next perform pivotal elimination. Take $\frac{4}{3}\mathbf{r}_1$, $\frac{5}{4}\mathbf{r}_1 + \mathbf{r}_2$, $-100\mathbf{r}_1 + \mathbf{r}_3$, $50\mathbf{r}_1 + \mathbf{r}_4$ and relabel the departing variable with the entering variable name. The resulting tableau is:

	x_1	x_2	x_3	x_4	x_5	x_6	z	
x_2	$\frac{4}{3}$	$\frac{6}{5}$	1	$\frac{1}{15}$	0	0	0	4
x_5	$\frac{5}{3}$	0	0	$-\frac{2}{3}$	1	0	0	35
x_6	$-\frac{400}{3}$	-80	0	$-\frac{50}{3}$	0	1	0	1000
	$\frac{200}{3}$	50	0	$\frac{65}{2}$	0	0	1	2000

Since all the entries of the objective row nonnegative, we have an optimal solution,

$$x_1 = 0, \qquad x_2 = 0, \qquad x_3.$$

The slack variables are

$$x_4 = 0, \qquad x_5 = 35, \qquad x_6 = 1000$$

and the optimal value of z is 2000. Thus, 4 tons of gas and no oil and no coal should be used to have a maximum of 2000 kilowatts generated.

T.1. We must show that if $\mathbf{x}$ and $\mathbf{y}$ are any two feasible solutions, then for any $0 \le r \le 1$, the vector $r\mathbf{x} + (1 - r)\mathbf{y}$ is also a feasible solution. First, since $r \ge 0$ and $(1 - r) \ge 0$, and $A\mathbf{x} \le \mathbf{b}$, $A\mathbf{y} \le \mathbf{b}$,

$$A[r\mathbf{x} + (1 - r)\mathbf{y}] = rA\mathbf{x} + (1 - r)A\mathbf{y} \le r\mathbf{b} + (1 - r)\mathbf{b} = \mathbf{b}.$$

Also, since $\mathbf{x} \ge \mathbf{0}$, $\mathbf{y} \ge \mathbf{0}$,

$$r\mathbf{x} + (1 - r)\mathbf{y} \ge r \cdot \mathbf{0} + (1 - r) \cdot \mathbf{0} = \mathbf{0}.$$

Thus $r\mathbf{x} + (1 - r)\mathbf{y}$ is a feasible solution.

ML.5. As a check, the solution is $x_4 = \frac{2}{3}$, $x_3 = 0$, $x_2 = \frac{1}{3}$, $x_1 = 1$, all other variables zero and the optimal value of z is 11. The final tableau is as follows:

0	0	1.6667	1	0.0000	-0.1667	0	0.6667
0	1	3.3333	0	1.0000	-0.3333	0	0.3333
1	0	-3.0000	0	-1.0000	0.5000	0	1.0000
0	0	1.0000	0	1.0000	1.0000	1	11.0000

ML.6. As a check, the solution is $x_3 = \frac{4}{3}$, $x_2 = 4$, $x_7 = \frac{22}{3}$, all other variables zero, and the optimal value of z is $\frac{28}{3}$. The final tableau is as follows:

0.4444	0	1	−0.1111	0.3333	−0.1111	0	0	1.3333
0.6667	1	0	1.3333	0	0.3333	0	0	4.0000
0.7778	0	0	−2.4444	−0.6667	−0.4444	1	0	7.3333
0.7778	0	0	1.5556	0.3333	0.5556	0	1	9.3333

Section 11.3, p. 598

1. Let

$$\mathbf{c} = \begin{bmatrix} 3 \\ 2 \end{bmatrix}, \qquad A = \begin{bmatrix} 4 & 3 \\ 5 & -2 \\ 6 & 8 \end{bmatrix}, \qquad \mathbf{b} = \begin{bmatrix} 7 \\ 6 \\ 9 \end{bmatrix}, \qquad \mathbf{x} = \begin{bmatrix} x_1 \\ x_2 \end{bmatrix}.$$

Define

$$\mathbf{y} = \begin{bmatrix} y_1 \\ y_2 \\ y_3 \end{bmatrix}.$$

Then the primal problem is

Maximize $z = \mathbf{c}^T \mathbf{x}$

subject to

$$A\mathbf{x} \leq \mathbf{b}$$
$$\mathbf{x} \geq \mathbf{0}.$$

The dual problem is

Minimize $z' = \mathbf{b}^T \mathbf{y}$

subject to

$$A^T \mathbf{y} \geq \mathbf{c}$$
$$\mathbf{y} \geq \mathbf{0}.$$

That is,

Minimize $z' = 7y_1 + 6y_2 + 9y_3$

subject to

$$4y_1 + 5y_2 + 6y_3 \geq 3$$
$$3y_1 - 2y_2 + 8y_3 \geq 2$$
$$y_1 \geq 0, y_2 \geq 0, y_3 \geq 0.$$

3. Let

$$\mathbf{c} = \begin{bmatrix} 3 \\ 5 \end{bmatrix}, \qquad A = \begin{bmatrix} 2 & 3 \\ 8 & -9 \\ 10 & 15 \end{bmatrix}, \qquad \mathbf{b} = \begin{bmatrix} 7 \\ 12 \\ 18 \end{bmatrix}, \qquad \mathbf{x} = \begin{bmatrix} x_1 \\ x_2 \end{bmatrix}.$$

Define

$$\mathbf{y} = \begin{bmatrix} y_1 \\ y_2 \\ y_3 \end{bmatrix}.$$

Then the primal problem is

Minimize $z = \mathbf{c}^T\mathbf{x}$

subject to

$$A\mathbf{x} \geq \mathbf{b}$$
$$\mathbf{x} \geq \mathbf{0}.$$

The dual problem is

Maximize $z' = \mathbf{b}^T\mathbf{y}$

subject to

$$A^T\mathbf{y} \leq \mathbf{c}$$
$$\mathbf{y} \geq \mathbf{0}.$$

That is,

Maximize $z' = 7y_1 + 12y_2 + 18y_3$

subject to

$$2y_1 + 8y_2 + 10y_3 \leq 3$$
$$3y_1 - 9y_2 + 15y_3 \leq 5$$
$$y_1 \geq 0, y_2 \geq 0, y_3 \geq 0.$$

5. The primal problem is

Maximize $z = 3x_1 + 6x_2 + 9x_3$

subject to

$$3x_1 + 2x_2 - 3x_3 \leq 12$$
$$5x_1 + 4x_2 + 7x_3 \leq 18$$
$$x_1 \geq 0, x_2 \geq 0, x_3 \geq 0.$$

In matrix form we have

Maximize $z = \begin{bmatrix} 3 & 6 & 9 \end{bmatrix} \begin{bmatrix} x_1 \\ x_2 \\ x_3 \end{bmatrix}$

subject to

$$\begin{bmatrix} 3 & 2 & -3 \\ 5 & 4 & 7 \end{bmatrix} \begin{bmatrix} x_1 \\ x_2 \\ x_3 \end{bmatrix} \leq \begin{bmatrix} 12 \\ 18 \end{bmatrix} \quad \text{and} \quad \begin{bmatrix} x_1 \\ x_2 \\ x_3 \end{bmatrix} \geq \mathbf{0}.$$

Then the dual problem is

Minimize $z' = \begin{bmatrix} 12 & 18 \end{bmatrix} \begin{bmatrix} y_1 \\ y_2 \end{bmatrix}$

subject to

$$\begin{bmatrix} 3 & 5 \\ 2 & 4 \\ -3 & 7 \end{bmatrix} \begin{bmatrix} y_1 \\ y_2 \end{bmatrix} \geq \begin{bmatrix} 3 \\ 6 \\ 9 \end{bmatrix} \quad \text{and} \quad \begin{bmatrix} y_1 \\ y_2 \end{bmatrix} \geq \mathbf{0}.$$

Then the dual of the dual problem is

Maximize $z'' = \begin{bmatrix} 3 & 6 & 9 \end{bmatrix} \begin{bmatrix} w_1 \\ w_2 \\ w_3 \end{bmatrix}$

subject to

$$\begin{bmatrix} 3 & 2 & -3 \\ 5 & 4 & 7 \end{bmatrix} \begin{bmatrix} w_1 \\ w_2 \\ w_3 \end{bmatrix} \leq \begin{bmatrix} 12 \\ 18 \end{bmatrix} \quad \text{and} \quad \begin{bmatrix} w_1 \\ w_2 \\ w_3 \end{bmatrix} \geq \mathbf{0}.$$

Let $z'' = z$ and $w_j = x_j$, $j = 1, 2, 3$ and we have the original problem.

7. From Exercise 6 of Section 11.2, we have the standard linear programming problem

Maximize $z = 2x + 5y$

subject to

$$3x + 7y \leq 6$$
$$2x + 6y \leq 7$$
$$3x + 2y \leq 5$$
$$x \geq 0, y \geq 0.$$

This problem has an optimal solution and the final tableau is

	x	y	u	v	w	z	
y	$\frac{3}{7}$	1	$\frac{1}{7}$	0	0	0	$\frac{6}{7}$
v	$-\frac{4}{7}$	0	$-\frac{6}{7}$	1	0	0	$\frac{13}{7}$
w	$\frac{15}{7}$	0	$-\frac{2}{7}$	0	1	0	$\frac{23}{7}$
	$\frac{1}{7}$	0	$\frac{5}{7}$	0	0	1	$\frac{30}{7}$

Theorem 11.4 implies that the dual problem also has an optimal solution with the same optimal value. This final tableau contains the optimal solution to the dual problem in the objective row under the columns of the slack variables. Thus the optimal solution of the dual problem is

$$y_1 = \tfrac{5}{7}, \qquad y_2 = 0, \qquad y_3 = 0.$$

The optimal value of the dual problem is

$$z' = \tfrac{30}{7}.$$

9. From Exercise 10 of Section 11.2, we have the standard linear programming problem

Maximize $z = 2x_1 + 4x_2 - 3x_3$

subject to

$$5x_1 + 2x_2 + x_3 \leq 5$$
$$3x_1 - 2x_2 + 3x_3 \leq 10$$
$$4x_1 + 5x_2 - x_3 \leq 20$$
$$x_1 \geq 0, x_2 \geq 0, x_3 \geq 0.$$

This problem has an optimal solution and the final tableau is

	x_1	x_2	x_3	x_4	x_5	x_6	z	
x_2	$\frac{5}{2}$	1	$\frac{1}{2}$	$\frac{1}{2}$	0	0	0	$\frac{5}{2}$
x_5	8	0	4	1	1	0	0	15
x_6	$-\frac{17}{2}$	0	$-\frac{7}{2}$	$-\frac{5}{2}$	0	1	0	$\frac{15}{2}$
	8	0	5	2	0	0	1	10

Theorem 11.4 implies that the dual problem also has an optimal solution with the same optimal value. This final tableau contains the optimal solution to the dual problem in the objective row under the columns of the slack variables. Thus the optimal solution of the dual problem is

$$y_1 = 2, \qquad y_2 = 0, \qquad y_3 = 0.$$

The optimal value of the dual problem is

$$z' = 10.$$

Section 11.4, p. 612

1.

$$C$$

		2 fingers shown	3 fingers shown
R	2 fingers shown	-4	5
	3 fingers shown	5	-6

3.

$$B$$

		Abington	Wyncote
A	Abington	50	60
	Wyncote	25	50

5. As in Example 3, we determine the row minima and column maxima.

(a)

		Row minima
5	4	4
3	-2	-2

Column maxima   5   4

Since 4 is both a row minimum and a column maximum, it is a saddle point.

(b)

			Row minima
2	1	0	0
3	1	-2	-2
4	2	-4	-4

Column maxima   4   2   0

Since 0 is both a row minimum and a column maximum, it is a saddle point.

(c)

			Row minima
3	4	5	3
-2	5	1	-2
-1	0	1	-1

Column maxima   3   5   5

Since 3 is both a row minimum and a column maximum, it is a saddle point.

$$\text{(d)} \quad \begin{bmatrix} 5 & 2 & 4 & 2 \\ 0 & -1 & 2 & 0 \\ 3 & 2 & 3 & 2 \\ 1 & 0 & -1 & -1 \end{bmatrix} \quad \begin{array}{c} \textbf{Row minima} \\ 2 \\ -1 \\ 2 \\ -1 \end{array}$$

Column maxima   5   2   4   2

Since the 2's (there are four of them) are both a row minimum and a column maximum, each is a saddle point.

7. For a strictly determined game the optimal strategy for each player is to choose the move represented by the saddle point. Then both players choose with probability 1 the row and column representing their respective moves.

(a) Determine the saddle point v.

$$\begin{bmatrix} 2 & 1 & 3 \\ -2 & 0 & 2 \end{bmatrix} \quad \begin{array}{c} \textbf{Row minima} \\ 1 \\ -2 \end{array}$$

Column maxima   2   1   3

Since 1 is both a row minimum and a column maximum it is a saddle point. Thus R should make the first move and C the second move ($v = 1$ is the $(1, 2)$ element). It follows that the optimal strategy for R is $\mathbf{p} = \begin{bmatrix} 1 & 0 \end{bmatrix}$ and the optimal strategy for C is

$$\mathbf{q} = \begin{bmatrix} 0 \\ 1 \\ 0 \end{bmatrix}.$$

The payoff for R is the value of the saddle point, $v = 1$.

(b) Determine the saddle point v.

$$\begin{bmatrix} -2 & -2 & 4 & 5 \\ -2 & -2 & 1 & 0 \\ 0 & 1 & 1 & 2 \end{bmatrix} \quad \begin{array}{c} \textbf{Row minima} \\ -2 \\ -2 \\ 0 \end{array}$$

Column maxima   0   1   4   5

Since the 0 in the $(3, 1)$ element is both a row minimum and a column maximum, it is a saddle point. Thus R should make the third move and C the first move. It follows that the optimal strategy for R is $\mathbf{p} = \begin{bmatrix} 0 & 0 & 1 \end{bmatrix}$ and the optimal strategy for C is

$$\mathbf{q} = \begin{bmatrix} 1 \\ 0 \\ 0 \\ 0 \end{bmatrix}.$$

The payoff for R is the value of the saddle point, $v = 0$. This game is a fair game.

(c) Determine the saddle point.

$$\begin{bmatrix} 6 & 4 \\ 7 & 4 \end{bmatrix} \quad \begin{array}{c} \textbf{Row minima} \\ 4 \\ 4 \end{array}$$

Column maxima   7   4

Since the 4 in the $(1, 2)$ element and the 4 in the $(2, 2)$ element are both a row minimum and a column maximum each is a saddle point. There are two optimal strategies:

(i) R should make the first move and C the second move. Then

$$\mathbf{p} = \begin{bmatrix} 1 & 0 \end{bmatrix} \quad \text{and} \quad \mathbf{q} = \begin{bmatrix} 0 \\ 1 \end{bmatrix}.$$

(ii) R should make the second move and C the second move. Then

$$\mathbf{p} = \begin{bmatrix} 0 & 1 \end{bmatrix} \quad \text{and} \quad \mathbf{q} = \begin{bmatrix} 0 \\ 1 \end{bmatrix}.$$

In either case the payoff is $v = 4$.

9. Let $A = \begin{bmatrix} 3 & -3 \\ 2 & 5 \\ 1 & 0 \end{bmatrix}$.

(a) $E(\mathbf{p}, \mathbf{q}) = \mathbf{p}A\mathbf{q} = \begin{bmatrix} \frac{1}{2} & \frac{1}{3} & \frac{1}{6} \end{bmatrix} \begin{bmatrix} 3 & -3 \\ 2 & 5 \\ 1 & 0 \end{bmatrix} \begin{bmatrix} \frac{1}{6} \\ \frac{5}{6} \end{bmatrix} = \frac{19}{36}$.

(b) Let

$$\mathbf{p} = \begin{bmatrix} 0 & 0 & 1 \end{bmatrix} \quad \text{and} \quad \mathbf{q} = \begin{bmatrix} \frac{1}{7} \\ \frac{6}{7} \end{bmatrix}.$$

Then $E(\mathbf{p}, \mathbf{q}) = \mathbf{p}A\mathbf{q} = \frac{1}{7}$.

11. Let $A = \begin{bmatrix} -3 & 2 \\ 4 & -5 \end{bmatrix}$. Then the optimal strategy for R is obtained from Equation (8):

$$p_1 = \frac{-5 - 4}{-3 + (-5) - 2 - 4} = \frac{9}{14} \quad \text{and} \quad p_2 = \frac{-3 - 2}{-3 + (-5) - 2 - 4} = \frac{5}{14}.$$

The value of the game is obtained from Equation (9):

$$v = \frac{(-3)(-5) - (2)(4)}{-3 + (-5) - 2 - 4} = -\frac{1}{2}.$$

The optimal strategy for C is obtained from Equation (13):

$$q_1 = \frac{-5 - 2}{-3 + (-5) - 2 - 4} = \frac{1}{2} \quad \text{and} \quad q_2 = \frac{-3 - 4}{-3 + (-5) - 2 - 4} = \frac{1}{2}.$$

13. Follow Example 10. Here we add 4 to each entry of the payoff matrix

$$\begin{bmatrix} 2 & -3 & 4 \\ 4 & 0 & 1 \\ 3 & 2 & -2 \end{bmatrix}$$

to obtain

$$A = \begin{bmatrix} 6 & 1 & 8 \\ 8 & 4 & 5 \\ 7 & 6 & 2 \end{bmatrix},$$

which has positive entries. From (18), the linear programming problem for C is

Maximize $x_1 + x_2 + x_3$
subject to

$$6x_1 + x_2 + 8x_3 \leq 1$$
$$8x_1 + 4x_2 + 5x_3 \leq 1$$
$$7x_1 + 6x_2 + 2x_3 \leq 1$$
$$x_1 \geq 0, x_2 \geq 0, x_3 \geq 0$$

where

$$x_i = \frac{q_i}{v} \quad \text{and} \quad x_1 + x_2 + x_3 = \frac{1}{v}.$$

Reformulating the problem using slack variables, we have

Maximize $x_1 + x_2 + x_3$
subject to

$$6x_1 + x_2 + 8x_3 + x_4 \qquad\qquad = 1$$
$$8x_1 + 4x_2 + 5x_3 \qquad + x_5 \qquad = 1$$
$$7x_1 + 6x_2 + 2x_3 \qquad\qquad + x_6 = 1$$
$$x_1 \geq 0, x_2 \geq 0, x_3 \geq 0, x_4 \geq 0, x_5 \geq 0, x_6 \geq 0.$$

Form the initial tableau

	x_1	x_2	x_3	x_4	x_5	x_6	z	
$\leftarrow x_4$	6	1	8	1	0	0	0	1
x_5	8	4	5	0	1	0	0	1
x_6	7	6	2	0	0	1	0	1
	-1	-1	-1	0	0	0	1	0

Determine the pivotal column and pivotal row. The most negative entry in the objective row is -1. We choose the x_3-column as the pivotal column. Thus the θ-ratios are $\left\{\frac{1}{8}, \frac{1}{5}, \frac{1}{2}\right\}$ and $\frac{1}{8}$ is the smallest. Thus the pivotal row is the x_4-row. It follows that the pivot has value 8. This information is marked on the preceding tableau. Next perform pivotal elimination. Take $\frac{1}{8}\mathbf{r}_1$, $-5\mathbf{r}_1 + \mathbf{r}_2$, $-2\mathbf{r}_1 + \mathbf{r}_3$, $\mathbf{r}_1 + \mathbf{r}_4$, and relabel the departing variable with the entering variable name. The resulting tableau is:

	x_1	x_2	x_3	x_4	x_5	x_6	z	
x_3	$\frac{3}{4}$	$\frac{1}{8}$	1	$\frac{1}{8}$	0	0	0	$\frac{1}{8}$
$\leftarrow x_5$	$\frac{17}{4}$	$\frac{27}{8}$	0	$-\frac{5}{8}$	1	0	0	$\frac{3}{8}$
x_6	$\frac{11}{2}$	$\frac{23}{4}$	0	$-\frac{1}{4}$	0	1	0	$\frac{3}{4}$
	$-\frac{1}{4}$	$-\frac{7}{8}$	0	$\frac{1}{8}$	0	0	1	$\frac{1}{8}$

Since there is a negative entry in the objective row, we do not have an optimal solution. We repeat the process.

Determine the pivotal column and pivotal row. The most negative entry in the objective row is $-\frac{7}{8}$, hence the x_2-column is the pivotal column. Thus the θ-ratios are $\left\{1, \frac{1}{9}, \frac{3}{23}\right\}$ and $\frac{1}{9}$ is the smallest.

Hence the pivotal row is the x_5-row. It follows that the pivot has value $\frac{27}{8}$. This information is marked on the preceding tableau. Next perform pivotal elimination. Take $\frac{8}{27}\mathbf{r}_2$, $\left(-\frac{1}{8}\right)\mathbf{r}_2 + \mathbf{r}_1$, $\left(-\frac{23}{4}\right)\mathbf{r}_2 + \mathbf{r}_3$, $\frac{7}{8}\mathbf{r}_2 + \mathbf{r}_4$, and relabel the departing variable with the entering variable name. The resulting tableau is:

	x_1	x_2	x_3	x_4 ↓	x_5	x_6	z	
x_3	$\frac{16}{27}$	0	1	$\frac{4}{27}$	$-\frac{1}{27}$	0	0	$\frac{1}{9}$
x_2	$\frac{34}{27}$	1	0	$-\frac{5}{27}$	$\frac{8}{27}$	0	0	$\frac{1}{9}$
← x_6	$-\frac{47}{27}$	0	0	$\frac{22}{27}$	$-\frac{46}{27}$	1	0	$\frac{1}{9}$
	$\frac{23}{27}$	0	0	$-\frac{1}{27}$	$\frac{7}{27}$	0	1	$\frac{2}{9}$

Since there is a negative entry in the objective row, we do not have the optimal solution. We repeat the process.

Determine the pivotal column and pivotal row. The most negative entry in the objective row is $-\frac{1}{27}$, hence we choose the x_4-column as the pivotal column. Thus the θ-ratios are $\left\{\frac{3}{4}, \frac{3}{22}\right\}$ and $\frac{3}{22}$ is the smallest. Hence the pivotal row is the x_6-row. It follows that the pivot has value $\frac{22}{27}$. This information is marked on the preceding tableau. Next perform pivotal elimination. Take $\frac{27}{22}\mathbf{r}_3$, $\left(-\frac{4}{27}\right)\mathbf{r}_3 + \mathbf{r}_1$, $\frac{5}{27}\mathbf{r}_3 + \mathbf{r}_2$, $\frac{1}{27}\mathbf{r}_3 + \mathbf{r}_4$, and relabel the departing variable with the entering variable name. The resulting tableau is:

	x_1	x_2	x_3	x_4	x_5	x_6	z	
x_3	$\frac{82}{297}$	0	1	0	$\frac{3}{11}$	$-\frac{2}{11}$	0	$\frac{1}{11}$
x_2	$\frac{19}{22}$	1	0	0	$-\frac{1}{11}$	$\frac{5}{22}$	0	$\frac{3}{22}$
x_4	$-\frac{47}{22}$	0	0	1	$-\frac{23}{11}$	$\frac{27}{22}$	0	$\frac{3}{22}$
	$\frac{17}{22}$	0	0	0	$\frac{2}{11}$	$\frac{1}{22}$	1	$\frac{5}{22}$

Since all the entries of the objective row are nonnegative, we have the optimal solution,

$$x_1 = 0, \qquad x_2 = \frac{3}{22}, \qquad x_3 = \frac{1}{11}.$$

The maximum value of v is

$$v = \frac{1}{\frac{5}{22}} = \frac{22}{5}.$$

Hence

$$\begin{aligned} q_1 &= x_1 \cdot v = 0, \\ q_2 &= x_2 \cdot v = \frac{3}{22}\frac{22}{5} = \frac{3}{5} \\ q_3 &= x_3 \cdot v = \frac{1}{11}\frac{22}{5} = \frac{2}{5}. \end{aligned}$$

Thus an optimal strategy for C is

$$\mathbf{q} = \begin{bmatrix} 0 \\ \frac{3}{5} \\ \frac{2}{5} \end{bmatrix}.$$

An optimal strategy for R is found in the objective row under the columns of the slack variables. Thus,

$$y_1 = 0, \qquad y_2 = \tfrac{2}{11}, \qquad y_3 = \tfrac{1}{22}$$

and

$$p_1 = y_1 \cdot v = 0,$$
$$p_2 = y_2 \cdot v = \tfrac{4}{5},$$
$$p_3 = y_3 \cdot v = \tfrac{1}{5}.$$

Hence an optimal strategy for R is $\mathbf{p} = \begin{bmatrix} 0 & \tfrac{4}{5} & \tfrac{1}{5} \end{bmatrix}$.

15. Let

$$A = \begin{bmatrix} 0 & -4 & 3 & 0 \\ 2 & -3 & 4 & 1 \\ -1 & 2 & 2 & 2 \\ 1 & -4 & 3 & 0 \end{bmatrix}.$$

Examine A for a recessive row or a recessive column. Each element of row 1 is less than or equal to the corresponding element of row 2. Thus we can drop row 1. The new matrix is

$$A_1 = \begin{bmatrix} 2 & -3 & 4 & 1 \\ -1 & 2 & 2 & 2 \\ 1 & -4 & 3 & 0 \end{bmatrix}.$$

Each element of column 3 is greater than or equal to the corresponding element in column 4. Thus we can drop column 3. The new matrix is

$$A_2 = \begin{bmatrix} 2 & -3 & 1 \\ -1 & 2 & 2 \\ 1 & -4 & 0 \end{bmatrix}.$$

Each element of column 3 is greater than or equal to the corresponding element in column 2. Thus we can drop column 3. The new matrix is

$$A_3 = \begin{bmatrix} 2 & -3 \\ -1 & 2 \\ 1 & -4 \end{bmatrix}.$$

Each element of row 3 is less than the corresponding element in row 1. Thus we can drop row 3. The new matrix is

$$A_4 = \begin{bmatrix} 2 & -3 \\ -1 & 2 \end{bmatrix}.$$

We inspect A_4 for a saddle point:

		Row minima
2	−3	−3
−1	2	−1
Column maxima 2	2	

A_4 has no saddle point, but we can use Equations (8), (9), and (13). We obtain

$$p_1 = \tfrac{3}{8}, \qquad p_2 = \tfrac{5}{8},$$

thus the strategy for R is $\mathbf{p} = \begin{bmatrix} \frac{3}{8} & \frac{5}{8} \end{bmatrix}$. Also,

$$q_1 = \tfrac{5}{8}, \qquad q_2 = \tfrac{3}{8},$$

thus the strategy for C is

$$\mathbf{q} = \begin{bmatrix} \frac{5}{8} \\ \frac{3}{8} \end{bmatrix}.$$

The value of the game is $v = \frac{1}{8}$. To form A_4 we dropped rows 1 and 4, hence for the original game $\mathbf{p} = \begin{bmatrix} 0 & \frac{3}{8} & \frac{5}{8} & 0 \end{bmatrix}$. Similarly, since columns 3 and 4 were dropped,

$$\mathbf{q} = \begin{bmatrix} \frac{5}{8} \\ \frac{3}{8} \\ 0 \\ 0 \end{bmatrix}.$$

The value of the original game is the same as the game using matrix A_4.

17. The payoff matrix from Exercise 2 is

$$\begin{bmatrix} 0 & 1 & -1 \\ -1 & 0 & 1 \\ 1 & -1 & 0 \end{bmatrix}.$$

Proceed as in Example 10. Add 2 to each entry so that all the entries are positive. We have

$$A = \begin{bmatrix} 2 & 3 & 1 \\ 1 & 2 & 3 \\ 3 & 1 & 2 \end{bmatrix}.$$

The linear programming problem for C is

Maximize $x_1 + x_2 + x_3$
subject to

$$\begin{aligned}
2x_1 + 3x_2 + x_3 &\le 1 \\
x_1 + 2x_2 + 3x_3 &\le 1 \\
3x_1 + x_2 + 2x_2 &\le 1 \\
x_1 \ge 0, x_2 \ge 0, x_3 &\ge 0
\end{aligned}$$

where

$$x_i = \frac{q_i}{v} \quad \text{and} \quad x_1 + x_2 + x_3 = \frac{1}{v}.$$

Reformulate the problem using slack variables and we have

Maximize $x_1 + x_2 + x_3$
subject to

$$\begin{aligned}
2x_1 + 3x_2 + x_3 + x_4 \qquad\qquad &= 1 \\
x_1 + 2x_2 + 3x_3 \qquad + x_5 \qquad &= 1 \\
3x_1 + x_2 + 2x_3 \qquad\qquad + x_6 &= 1 \\
x_1 \ge 0, x_2 \ge 0, x_3 \ge 0, x_4 \ge 0, x_5 \ge 0, x_6 &\ge 0
\end{aligned}$$

Form the initial tableau:

	x_1	x_2	x_3	x_4	x_5	x_6	z	
x_4	2	3	1	1	0	0	0	1
$\leftarrow x_5$	1	2	3	0	1	0	0	1
x_6	3	1	2	0	0	1	0	1
	-1	-1	-1	0	0	0	1	0

Determine the pivotal column and pivotal row. The most negative entry in the objective row is -1. We choose the x_3-column as the pivotal column. Thus the θ-ratios are $\left\{1, \frac{1}{3}, \frac{1}{2}\right\}$ and $\frac{1}{3}$ is the smallest. Hence the pivotal row is the x_5-row. It follows that the pivot has value 3. This information is marked on the preceding tableau. Next perform pivotal elimination. Take $\frac{1}{3}\mathbf{r}_2$, $-\mathbf{r}_2+\mathbf{r}_1$, $-2\mathbf{r}_2+\mathbf{r}_3$, $\mathbf{r}_2 + \mathbf{r}_4$, and relabel the departing variable with the entering variable name. The resulting tableau is:

	x_1	x_2	x_3	x_4	x_5	x_6	z	
x_4	$\frac{5}{3}$	$\frac{7}{3}$	0	1	$-\frac{1}{3}$	0	0	$\frac{2}{3}$
x_3	$\frac{1}{3}$	$\frac{2}{3}$	1	0	$\frac{1}{3}$	0	0	$\frac{1}{3}$
$\leftarrow x_6$	$\frac{7}{3}$	$-\frac{1}{3}$	0	0	$-\frac{2}{3}$	1	0	$\frac{1}{3}$
	$-\frac{2}{3}$	$-\frac{1}{3}$	0	0	$\frac{1}{3}$	0	1	$\frac{1}{3}$

Since there is a negative entry in the objective row, we do not have the optimal solution. We repeat the process.

Determine the pivotal column and pivotal row. The most negative entry in the objective row is $-\frac{2}{3}$. We choose the x_1-column as the pivotal column. Thus the θ-ratios are $\left\{\frac{2}{5}, 1, \frac{1}{7}\right\}$ and $\frac{1}{7}$ is the smallest.

Hence the pivotal row is the x_6-row. It follows that the pivot has value $\frac{7}{3}$. This information is marked on the preceding tableau. Next perform pivotal elimination. Take $\frac{3}{7}\mathbf{r}_3$, $\left(-\frac{5}{3}\right)\mathbf{r}_3 + \mathbf{r}_1$, $\left(-\frac{1}{3}\right)\mathbf{r}_3 + \mathbf{r}_2$, $\frac{2}{3}\mathbf{r}_3 + \mathbf{r}_4$, and relabel the departing variable with the entering variable name. The resulting tableau is:

	x_1	x_2	x_3	x_4	x_5	x_6	z	
$\leftarrow x_4$	0	$\frac{18}{7}$	0	1	$\frac{1}{7}$	$-\frac{5}{7}$	0	$\frac{3}{7}$
x_3	0	$\frac{5}{7}$	1	0	$\frac{3}{7}$	$-\frac{1}{7}$	0	$\frac{2}{7}$
x_1	1	$-\frac{1}{7}$	0	0	$-\frac{2}{7}$	$\frac{3}{7}$	0	$\frac{1}{7}$
	0	$-\frac{3}{7}$	0	0	$\frac{1}{7}$	$\frac{2}{7}$	1	$\frac{3}{7}$

Since there is a negative entry in the objective row, we do not have the optimal solution. We repeat the process.

Determine the pivotal column and pivotal row. The most negative entry in the objective row is $-\frac{3}{7}$. We choose the x_2-column as the pivotal column. Thus the θ-ratios are $\left\{\frac{1}{6}, \frac{2}{5}\right\}$ and $\frac{1}{6}$ is the smallest. Hence the pivotal row is the x_4-row. It follows that the pivot has value $\frac{18}{7}$. This information is marked on the preceding tableau. Next perform pivotal elimination. Take $\frac{7}{18}\mathbf{r}_1$, $\left(-\frac{5}{7}\right)\mathbf{r}_1 + \mathbf{r}_2$, $\frac{1}{7}\mathbf{r}_1 + \mathbf{r}_3$, $\frac{3}{7}\mathbf{r}_1 + \mathbf{r}_4$, and relabel the departing variable with the entering variable name. The resulting tableau is:

	x_1	x_2	x_3	x_4	x_5	x_6	z	
x_2	0	1	0	$\frac{7}{18}$	$\frac{1}{18}$	$-\frac{5}{18}$	0	$\frac{1}{6}$
x_3	0	0	1	$-\frac{5}{18}$	$\frac{7}{18}$	$\frac{1}{18}$	0	$\frac{1}{6}$
x_1	1	0	0	$\frac{1}{18}$	$\frac{37}{126}$	$\frac{7}{18}$	0	$\frac{1}{6}$
	0	0	0	$\frac{1}{6}$	$\frac{1}{6}$	$\frac{1}{6}$	1	$\frac{1}{2}$

We have the optimal solution to the linear programming problem:

$$x_1 = \tfrac{1}{6}, \qquad x_2 = \tfrac{1}{6}, \qquad x_3 = \tfrac{1}{6}$$

and hence the maximum value of $x_1 + x_2 + x_3$ is $\frac{1}{2}$. It follows that $v = 2$ and

$$q_1 = x_1 \cdot v = \tfrac{1}{3},$$
$$q_2 = x_2 \cdot v = \tfrac{1}{3},$$
$$q_3 = x_3 \cdot v = \tfrac{1}{3}.$$

Hence an optimal strategy for C is

$$\mathbf{q} = \begin{bmatrix} \frac{1}{3} \\ \frac{1}{3} \\ \frac{1}{3} \end{bmatrix}.$$

An optimal strategy for R is found in the objective row under the columns corresponding to the slack variables. We have

$$y_1 = \tfrac{1}{6}, \qquad y_2 = \tfrac{1}{6}, \qquad y_3 = \tfrac{1}{6},$$

hence

$$p_1 = y_1 \cdot v = \tfrac{1}{3},$$
$$p_2 = y_2 \cdot v = \tfrac{1}{3},$$
$$p_3 = y_3 \cdot v = \tfrac{1}{3}.$$

An optimal strategy for R is $\mathbf{p} = \begin{bmatrix} \frac{1}{3} & \frac{1}{3} & \frac{1}{3} \end{bmatrix}$. The value of the original game is $v - 2 = 0$.

19. The payoff matrix for Exercise 4 is

$$\begin{bmatrix} -5 & 5 \\ 10 & -10 \end{bmatrix}.$$

We attempt to find a saddle point:

			Row minima
	-5	5	-5
	10	-10	-10
Column maxima	10	5	

Hence there is no saddle point, so we use Equations (8), (9), and (13). We obtain

$$p_1 = \tfrac{2}{3} \quad \text{and} \quad p_2 = \tfrac{1}{3}.$$

Thus the optimal strategy for R is $\mathbf{p} = \begin{bmatrix} \frac{2}{3} & \frac{1}{3} \end{bmatrix}$. Similarly,

$$q_1 = \tfrac{1}{2} \quad \text{and} \quad q_2 = \tfrac{1}{2}$$

which implies that the optimal strategy for C is

$$\mathbf{q} = \begin{bmatrix} \frac{1}{2} \\ \frac{1}{2} \end{bmatrix}.$$

The value of the game is $v = 0$.

T.1. The expected payoff to R is the sum of terms of the form (Probability that R plays row i and C plays column j) × (Payoff to R when R plays i and C plays j) $= p_i q_j a_{ij}$. Summing over all $1 \leq i \leq m$ and $1 \leq j \leq n$, we get

$$\text{Expected payoff to } R = \sum_{i=1}^{m} \sum_{j=1}^{n} p_i a_{ij} q_j = \mathbf{p} A \mathbf{q}.$$

Supplementary Exercises, p. 614

1. The feasible region S appears in the accompanying figure.

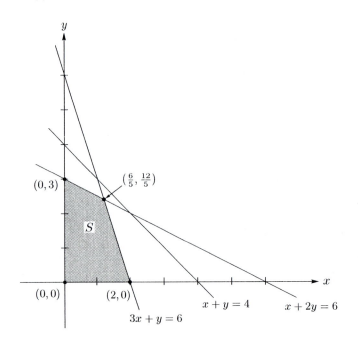

There are 4 extreme points, $(0,0)$, $(0,3)$, $\left(\frac{6}{5}, \frac{12}{5}\right)$, and $(2,0)$. The region S is bounded, thus by Theorem 11.1, the objective function $z = 2x + 3y$ has a maximum at one of these points. The values of the objective function at the extreme points are shown in the following table.

point	$z = 2x + 3y$
$(0,0)$	0
$(0,3)$	9
$\left(\frac{6}{5}, \frac{12}{5}\right)$	$\frac{48}{5}$
$(2,0)$	4

There is a maximum at extreme point $\left(\frac{6}{5}, \frac{12}{5}\right)$. The maximum is $z = \frac{48}{5}$.

3. We first reformulate this standard linear programming problem using slack variables. We have

Maximize $z = 50x + 100y$

subject to

$$\begin{aligned} x + 2y + u \quad\quad\quad\quad &= 16 \\ 3x + 2y \quad + v \quad\quad &= 24 \\ 2x + 2y \quad\quad\quad + w &= 18 \end{aligned}$$
$$x \geq 0, y \geq 0, u \geq 0, v \geq 0, w \geq 0.$$

Form the initial tableau.

	x	y	u	v	w	z	
u	1	2	1	0	0	0	16
v	3	2	0	1	0	0	24
w	2	2	0	0	1	0	18
	-50	-100	0	0	0	1	0

Next we determine the pivotal column and pivotal row. The most negative entry in the objective row is -100, hence the y-column is the pivotal column. Thus the θ-ratios are $\left\{\frac{16}{2}, \frac{24}{2}, \frac{18}{2}\right\}$ and $\frac{16}{2}$ is the smallest. Thus the pivotal row is the u-row. It follows that the pivot has the value 2. This information is marked in the next tableau.

			$\downarrow$					
	x	y	u	v	w	z	
$\leftarrow u$	1	2	1	0	0	0	16
v	3	2	0	1	0	0	24
w	2	2	0	0	1	0	18
	-50	-100	0	0	0	1	0

Next we perform pivotal elimination. Take $\frac{1}{2}r_1$, $-2r_1 + r_2$, $-2r_1 + r_3$, $100r_1 + r_4$, and relabel the departing variable with the entering variable name. The resulting tableau is:

	x	y	u	v	w	z	
y	$\frac{1}{2}$	1	$\frac{1}{2}$	0	0	0	8
v	2	0	-1	1	0	0	8
w	1	0	-1	0	1	0	2
	0	0	50	0	0	1	800

Since all the entries in the objective row are nonnegative, we have an optimal solution

$$x = 0, \qquad y = 8.$$

The slack variables are

$$u = 0, \qquad v = 8, \qquad w = 2,$$

and the optimal value of z is 800.

5. Referring to Exercise 4, we reformulate the dual using slack variables and obtain

Maximize $z' = 6y_1 + 10y_2$

subject to

$$
\begin{aligned}
2y_1 + 5y_2 + y_3 \quad\;\; &= 6 \\
3y_1 + 2y_2 \quad\quad + y_4 &= 5 \\
y_1 \geq 0, y_2 \geq 0, y_3 \geq 0, y_4 \geq 0.&
\end{aligned}
$$

Form the initial tableau.

	y_1	y_2	y_3	y_4	z'	
y_3	2	5	1	0	0	6
y_4	3	2	0	1	0	5
	-6	-10	0	0	1	0

Next we determine the pivotal column and pivotal row. The most negative entry in the objective row is -10, hence the y_2-column is the pivotal column. Thus the θ-ratios are $\left\{ \frac{6}{5}, \frac{5}{2} \right\}$ and $\frac{6}{5}$ is the smallest. Thus the pivotal row is the y_3-row. It follows that the pivot has the value 5. This information is marked in the next tableau.

	y_1	$\downarrow$ y_2	y_3	y_4	z'	
$\leftarrow y_3$	2	5	1	0	0	6
y_4	3	2	0	1	0	5
	-6	-10	0	0	1	0

Next we perform pivotal elimination. Take $\frac{1}{5}\mathbf{r}_1$, $-2\mathbf{r}_1 + \mathbf{r}_2$, $10\mathbf{r}_1 + \mathbf{r}_3$, and relabel the departing variable with the entering variable name. The resulting tableau is:

	y_1	y_2	y_3	y_4	z'	
y_2	$\frac{2}{5}$	1	$\frac{1}{5}$	0	0	$\frac{6}{5}$
y_4	$\frac{11}{5}$	0	$-\frac{2}{5}$	1	0	$\frac{13}{5}$
	-2	0	2	0	1	12

Since there is a negative entry in the objective row, we do not have an optimal solution. We repeat the process.

Determine the pivotal column and pivotal row. The most negative entry in the objective row is -2, hence the y_1-column is the pivotal column. Thus the θ-ratios are $\left\{ 3, \frac{13}{11} \right\}$ and $\frac{13}{11}$ is the smallest.

Thus the pivotal row is the y_4-row. It follows that the pivot has the value $\frac{11}{5}$. This information is marked in the next tableau.

		$\downarrow$				
		y_1	y_2	y_3	y_4	z'
y_2		$\frac{2}{5}$	1	$\frac{1}{5}$	0	0	$\frac{6}{5}$
$\leftarrow y_4$		$\frac{11}{5}$	0	$-\frac{2}{5}$	1	0	$\frac{13}{5}$
		-2	0	2	0	1	12

Next we perform pivotal elimination. Take $\frac{5}{11}\mathbf{r}_2$, $-\frac{2}{5}\mathbf{r}_2 + \mathbf{r}_1$, $2\mathbf{r}_2 + \mathbf{r}_3$, and relabel the departing variable with the entering variable name. The resulting tableau is:

	y_1	y_2	y_3	y_4	z'	
y_2	0	1	$\frac{3}{11}$	$-\frac{2}{11}$	0	$\frac{8}{11}$
y_1	1	0	$-\frac{2}{11}$	$\frac{5}{11}$	0	$\frac{13}{11}$
	0	0	$\frac{18}{11}$	$\frac{10}{11}$	1	$\frac{158}{11}$

Since all the entries in the objective row are nonnegative, we have an optimal solution

$$y_1 = \frac{13}{11}, \qquad y_2 = \frac{8}{11}.$$

The slack variables are

$$y_3 = 0, \qquad y_4 = 0,$$

and the optimal value of z is $\frac{158}{11}$. An optimal solution of the original problem appears in the objective row under the columns of the slack variables y_3 and y_4. The original problem has optimal solution

$$x_1 = \frac{18}{11}, \qquad x_2 = \frac{10}{11}.$$

Appendix A

Complex Numbers

Appendix A.1, p. A7

1. Let $c_1 = 3 + 4i$, $c_2 = 1 - 2i$, and $c_3 = -1 + i$.

   (a) $c_1 + c_2 = (3 + 4i) + (1 - 2i) = 4 + 2i$

   (b) $c_3 - c_1 = (-1 + i) - (3 + 4i) = -4 - 3i$

   (c) $c_1 c_2 = (3 + 4i)(1 - 2i) = 11 - 2i$

   (d) $c_2 \overline{c_3} = (1 - 2i)(-1 - i) = -3 + i$

   (e) $4c_3 + \overline{c_2} = (-4 + 4i) + (1 + 2i) = -3 + 6i$

   (f) $(-i)c_2 = (-i)(1 - 2i) = -2 - i$

   (g) $\overline{3c_1 - ic_2} = \overline{(9 + 12i) - (i - 2i^2)} = \overline{7 + 11i} = 7 - 11i$

   (h) $c_1 c_2 c_3 = (c_1 c_2)c_3 = (11 - 2i)(-1 + i) = -9 + 13i$

3.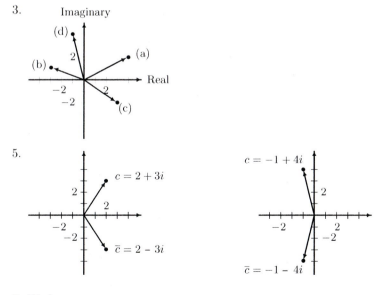

5.

7. We have

$$A^2 = \begin{bmatrix} -1 & 0 \\ 0 & -1 \end{bmatrix} = -I_2, \quad A^3 = A^2 A = -I_2 A = -A = \begin{bmatrix} 0 & -i \\ -i & 0 \end{bmatrix}, \quad A^4 = A^2 A^2 = I_2 = \begin{bmatrix} 1 & 0 \\ 0 & 1 \end{bmatrix}$$

and, in general, $A^{4n} = I_2$, $A^{4n+1} = A$, $A^{4n+2} = A^2 = -I_2$, and $A^{4n+3} = A^3 = -A$.

9. (a) The factors of 1, (1 and -1), are not roots. We obtain the roots by the quadratic formula:

$$x = \frac{-1 \pm i\sqrt{3}}{2}.$$

(b) Since -2 is a root (by trial and error),

$$x^3 + 2x^2 + x + 2 = (x + 2)(x^2 + 1) = 0,$$

so the other roots are $x = \pm i$.

(c) We find (by trial and error) that 1 and -1 are roots, so

$$x^5 + x^4 - x - 1 = (x + 1)(x - 1)(x^3 + x^2 + x + 1).$$

Now i and $-i$ are roots of $x^3 + x^2 + x + 1$ (by trial and error). Moreover,

$$x^3 + x^2 + x + 1 = (x + i)(x - i)(x + 1).$$

Hence, the roots are $1, -1, -1, \pm i$.

11. (a) We let $A = \begin{bmatrix} a & b \\ c & d \end{bmatrix}$. Then if $p(A) = O_2$, we have

$$\begin{bmatrix} a & b \\ c & d \end{bmatrix}^2 + I_2 = O_2 \implies \begin{bmatrix} a^2 + bc & ab + bd \\ ca + dc & cb + d^2 \end{bmatrix} = \begin{bmatrix} -1 & 0 \\ 0 & -1 \end{bmatrix}$$

so we must have

$$a^2 + bc = -1 \qquad ab + bd = \quad 0$$
$$ca + dc = \quad 0 \qquad cb + d^2 = -1.$$

Then $b(a + d) = 0$ and $c(a + d) = 0$. Thus $b = 0$ and $c = 0$ or $a = -d$. If $b = 0$ and $c = 0$, then $a^2 = -1$ and $d^2 = -1$, so $a = \pm i$ and $d = \pm i$. We then have as possible answers

$$A_1 = \begin{bmatrix} i & 0 \\ 0 & i \end{bmatrix}, \qquad A_2 = \begin{bmatrix} -i & 0 \\ 0 & -i \end{bmatrix}.$$

Note that other answers are possible.

(b) Substituting $A = \begin{bmatrix} 1 & 2 \\ -1 & -1 \end{bmatrix}$, we find that $p(A) = O_2$.

13. (a) Let $B = \begin{bmatrix} a & b \\ c & d \end{bmatrix}$. We want B to satisfy $B^2 = A$, or

$$B^2 = \begin{bmatrix} a & b \\ c & d \end{bmatrix} \begin{bmatrix} a & b \\ c & d \end{bmatrix} = \begin{bmatrix} a^2 + bc & ab + bd \\ ca + dc & cb + d^2 \end{bmatrix} = \begin{bmatrix} -1 & 0 \\ 0 & 0 \end{bmatrix}$$

so

$$a^2 + bc = -1 \qquad ab + bd = \quad 0$$
$$ca + dc = \quad 0 \qquad cb + d^2 = 0.$$

We can take $b = c = d = 0$ and $a = \pm i$. Thus we obtain as possible answers

$$\begin{bmatrix} i & 0 \\ 0 & 0 \end{bmatrix}, \qquad \begin{bmatrix} -i & 0 \\ 0 & 0 \end{bmatrix}.$$

(b) Letting $B = \begin{bmatrix} a & b \\ c & d \end{bmatrix}$, we must have

$$a^2 + bc = -2 \qquad ab + bd = 2$$
$$ca + dc = 2 \qquad cb + d^2 = -2.$$

Possible answers are

$$\begin{bmatrix} i & -1 \\ -i & i \end{bmatrix}, \qquad \begin{bmatrix} -i & i \\ i & -i \end{bmatrix}.$$

T.1. (a) $\operatorname{Re}(c_1 + c_2) = \operatorname{Re}((a_1 + a_2) + (b_1 + b_2)i) = a_1 + a_2 = \operatorname{Re}(c_1) + \operatorname{Re}(c_2)$.
$\operatorname{Im}(c_1 + c_2) = \operatorname{Im}((a_1 + a_2) + (b_1 + b_2)i) = b_1 + b_2 = \operatorname{Im}(c_1 + \operatorname{Im}(c_2)$.

(b) $\operatorname{Re}(kc) = \operatorname{Re}(ka + kbi) = ka = \operatorname{Re}(c)$.
$\operatorname{Im}(kc) = \operatorname{Im}(ka + kbi) = kb = k \operatorname{Im}(c)$.

(c) No.

(d) $\operatorname{Re}(c_1 c_2) = \operatorname{Re}((a_1 + b_1 i)(a_2 + b_2 i)) = \operatorname{Re}((a_1 a_2 - b_1 b_2) + (a_1 b_2 + a_2 b_1)i) = a_1 a_2 - b_1 b_2 \neq \operatorname{Re}(c_1) \operatorname{Re}(c_2)$.

T.3. (a) $\overline{a_{ii}} = a_{ii}$, hence a_{ii} is real. (See Property 4 in Section B1.)

(b) First, $\overline{A^T} = A$ implies that $A^T = \overline{A}$. Let $B = \dfrac{A + \overline{A}}{2}$. Then

$$\overline{B} = \overline{\left(\dfrac{A + \overline{A}}{2} \right)} = \dfrac{\overline{A} + \overline{\overline{A}}}{2} = \dfrac{\overline{A} + A}{2} = \dfrac{A + \overline{A}}{2} = B,$$

so B is a real matrix. Also,

$$B^T = \left(\dfrac{A + \overline{A}}{2} \right)^T = \dfrac{A^T + \overline{A}^T}{2} = \dfrac{A^T + \overline{A^T}}{2} = \dfrac{\overline{A} + A}{2} = \dfrac{A + \overline{A}}{2} = B$$

so B is symmetric.
Next, let $C = \dfrac{A - \overline{A}}{2i}$. Then

$$\overline{C} = \overline{\left(\dfrac{A - \overline{A}}{2i} \right)} = \dfrac{\overline{A} - \overline{\overline{A}}}{-2i} = \dfrac{A - \overline{A}}{2i} = C$$

so C is a real matrix. Also,

$$C^T = \left(\dfrac{A - \overline{A}}{2i} \right)^T = \dfrac{A^T - \overline{A}^T}{2i} = \dfrac{A^T - \overline{A^T}}{2i} = \dfrac{\overline{A} - A}{2i} = -\dfrac{A - \overline{A}}{2i} = -C$$

so C is also skew symmetric. Moreover, $A = B + iC$.

(c) If $A = A^T$ and $A = \overline{A}$, then $\overline{A^T} = \overline{A} = A$. Hence, A is Hermitian.

T.5. (a) Let

$$B = \dfrac{A + \overline{A^T}}{2} \quad \text{and} \quad C = \dfrac{A - \overline{A^T}}{2i}.$$

Then

$$\overline{B^T} = \overline{\left(\dfrac{A + \overline{A^T}}{2} \right)^T} = \dfrac{\overline{A^T} + \overline{(\overline{A^T})^T}}{2} = \dfrac{\overline{A^T} + A}{2} = \dfrac{A + \overline{A^T}}{2} = B$$

so B is Hermitian. Also,

$$\overline{C^T} = \overline{\left(\frac{A - \overline{A^T}}{2i}\right)^T} = \frac{\overline{A^T} - \overline{(\overline{A^T})^T}}{-2i} = \frac{A - \overline{A^T}}{2i} = C$$

so C is Hermitian. Moreover, $A = B + iC$.

(b) We have

$$\overline{A^T}A = \overline{(B^T + iC^T)}(B + iC) = (\overline{B^T} + \overline{iC^T})(B + iC)$$
$$= (B - iC)(B + iC)$$
$$= B^2 - iCB + iBC - i^2C^2$$
$$= (B^2 + C^2) + i(BC - CB).$$

Similarly,

$$A\overline{A^T} = (B + iC)\overline{(B^T + iC^T)^T} = (B + iC)(\overline{B^T} + \overline{iC^T})$$
$$= (B + iC)(B - iC)$$
$$= B^2 - iBC + iCB - i^2C^2$$
$$= (B^2 + C^2) + i(CB - BC).$$

Since $\overline{A^T}A = A\overline{A^T}$, we equate imaginary parts obtaining $BC - CB = CB - BC$, which implies that $BC = CB$. The steps are reversible, establishing the converse.

T.7. Let $A = B + iC$ be skew Hermitian. Then $\overline{A^T} = -A$ so $B^T - iC^T = -B - iC$. Then $B^T = -B$ and $C^T = C$. Thus, B is skew symmetric and C is symmetric. Conversely, if B is skew symmetric and C is symmetric, then $B^T = -B$ and $C^T = C$ so $B^T - iC^T = -B - iC$ or $\overline{A^T} = -A$. Hence, A is skew Hermitian.

Appendix A.2, p. A17

1. Form the augmented matrix and use row operations

(a) $\begin{bmatrix} 1 + 2i & -2 + i & \vdots & 1 - 3i \\ 2 + i & -1 + 2i & \vdots & -1 - i \end{bmatrix}_{\frac{1}{1+2i}\mathbf{r}_1} \longrightarrow \begin{bmatrix} 1 & i & \vdots & -1 - i \\ 2 + i & -1 + 2i & \vdots & -1 - i \end{bmatrix}_{-(2+i)\mathbf{r}_1 + \mathbf{r}_2}$

$\longrightarrow \begin{bmatrix} 1 & i & \vdots & -1 - i \\ 0 & 0 & \vdots & 2i \end{bmatrix}$

The system is inconsistent; there is no solution.

(b) $\begin{bmatrix} 1 + 2i & -2 + i & \vdots & 1 - 3i \\ 2 + i & -1 + 2i & \vdots & -1 - i \end{bmatrix}_{\frac{1}{2i}\mathbf{r}_1} \longrightarrow \begin{bmatrix} 1 & \frac{1}{2} + \frac{1}{2}i & \vdots & \frac{1}{2} - \frac{1}{2}i \\ 1 - i & 1 & \vdots & 1 - i \end{bmatrix}_{-(1-i)\mathbf{r}_1 + \mathbf{r}_2}$

$\longrightarrow \begin{bmatrix} 1 & \frac{1}{2} + \frac{1}{2}i & \vdots & \frac{1}{2} - \frac{1}{2}i \\ 0 & 0 & \vdots & 1 \end{bmatrix}$

The system is inconsistent; there is no solution.

(c) $\begin{bmatrix} 1+i & -1 & \vdots & -2+i \\ 2i & 1-i & \vdots & i \end{bmatrix}_{\frac{1}{1+i}\mathbf{r}_1} \longrightarrow \begin{bmatrix} 1 & -\frac{1}{2}+\frac{1}{2}i & \vdots & -\frac{1}{2}+\frac{3}{2}i \\ 2i & 1-i & \vdots & i \end{bmatrix}_{-2i\mathbf{r}_1+\mathbf{r}_2}$

$\longrightarrow \begin{bmatrix} 1 & -\frac{1}{2}+\frac{1}{2}i & \vdots & -\frac{1}{2}+\frac{3}{2}i \\ 0 & 2 & \vdots & 3+2i \end{bmatrix}_{\frac{1}{2}\mathbf{r}_2}$

$\longrightarrow \begin{bmatrix} 1 & -\frac{1}{2}+\frac{1}{2}i & \vdots & -\frac{1}{2}+\frac{3}{2}i \\ 0 & 1 & \vdots & \frac{3}{2}+i \end{bmatrix}_{\frac{1}{2}(1-i)\mathbf{r}_2+\mathbf{r}_1}$

$\longrightarrow \begin{bmatrix} 1 & 0 & \vdots & \frac{3}{4}+\frac{5}{4}i \\ 0 & 1 & \vdots & \frac{3}{2}+i \end{bmatrix}$

Thus the solution is $x_1 = \frac{3}{4}+\frac{5}{4}i$ and $x_2 = \frac{3}{2}+i$.

3. Form the augmented matrix and apply row operations to obtain row echelon form, then apply back substitution.

(a) $\begin{bmatrix} i & 1+i & 0 & \vdots & i \\ 1-i & 1 & -i & \vdots & 1 \\ 0 & i & 1 & \vdots & 1 \end{bmatrix}_{\substack{\frac{1}{i}\mathbf{r}_1 \\ -(1+i)\mathbf{r}_1+\mathbf{r}_2}} \longrightarrow \begin{bmatrix} 1 & 1-i & 0 & \vdots & 1 \\ 0 & 1+2i & -i & \vdots & i \\ 0 & i & 1 & \vdots & 1 \end{bmatrix}_{\substack{\frac{1}{1+2i}\mathbf{r}_2 \\ -1\mathbf{r}_2+\mathbf{r}_3}}$

$\longrightarrow \begin{bmatrix} 1 & 1-i & 0 & \vdots & 1 \\ 0 & 1 & -\frac{2}{5}-\frac{1}{5}i & \vdots & \frac{2}{5}+\frac{1}{5}i \\ 0 & 0 & \frac{4}{5}+\frac{2}{5}i & \vdots & \frac{6}{5}-\frac{2}{5}i \end{bmatrix}_{\frac{5}{4+2i}\mathbf{r}_3}$

$\longrightarrow \begin{bmatrix} 1 & 1-i & 0 & \vdots & 1 \\ 0 & 1 & -\frac{2}{5}-\frac{1}{5}i & \vdots & \frac{2}{5}+\frac{1}{5}i \\ 0 & 0 & 1 & \vdots & 1-i \end{bmatrix}$

Use back substitution; the solution is

$$x_3 = 1-i$$
$$x_2 = \frac{2}{5}+\frac{1}{5}i - \left(-\frac{2}{5}-\frac{1}{5}i\right)x_3 = 1$$
$$x_1 = 1 - (1-i)x_2 = i$$

(b) $\begin{bmatrix} 1 & i & 1-i & \vdots & 2+i \\ i & 0 & 1+i & \vdots & -1+i \\ 0 & 2i & -1 & \vdots & 2-i \end{bmatrix}_{-i\mathbf{r}_1+\mathbf{r}_2} \longrightarrow \begin{bmatrix} 1 & i & 1-i & \vdots & 2+i \\ 0 & 1 & 0 & \vdots & -i \\ 0 & 2i & -1 & \vdots & 2-i \end{bmatrix}_{-2i\mathbf{r}_2+\mathbf{r}_3}$

$\longrightarrow \begin{bmatrix} 1 & i & 1-i & \vdots & 2+i \\ 0 & 1 & 0 & \vdots & -i \\ 0 & 0 & -1 & \vdots & -i \end{bmatrix}_{-1\mathbf{r}_3}$

$\longrightarrow \begin{bmatrix} 1 & 1 & 1-i & \vdots & 2+i \\ 0 & 1 & 0 & \vdots & -i \\ 0 & 0 & 1 & \vdots & i \end{bmatrix}$

Use back substitution to get $x_3 = i$, $x_2 = -i$, $x_1 = (2+i) - (1-i)x_3 - x_2 = 0$.

5. (a) $\begin{bmatrix} i & 2 & \vdots & 1 & 0 \\ 1+i & -i & \vdots & 0 & 1 \end{bmatrix} \begin{array}{l} \frac{1}{i}\mathbf{r}_1 \\ \\ -(1+i)\mathbf{r}_1 + \mathbf{r}_2 \end{array}$ $\longrightarrow$ $\begin{bmatrix} 1 & -2i & \vdots & -i & 0 \\ 0 & -2+i & \vdots & -1+i & 1 \end{bmatrix} \begin{array}{l} \frac{1}{-2+i}\mathbf{r}_2 \\ \\ 2i\mathbf{r}_2 + \mathbf{r}_1 \end{array}$

$$\longrightarrow \begin{bmatrix} 1 & 0 & \vdots & \frac{2}{5}+\frac{1}{5}i & \frac{2}{5}-\frac{4}{5}i \\ 0 & 1 & \vdots & \frac{3}{5}-\frac{1}{5}i & -\frac{2}{5}-\frac{1}{5}i \end{bmatrix}$$

Thus the inverse is $\begin{bmatrix} \frac{2}{5}+\frac{1}{5}i & \frac{2}{5}-\frac{4}{5}i \\ \frac{3}{5}-\frac{1}{5}i & -\frac{2}{5}-\frac{1}{5}i \end{bmatrix} = \frac{1}{5}\begin{bmatrix} 2+i & 2-4i \\ 3-i & -2-i \end{bmatrix}.$

(b) $\begin{bmatrix} 2 & i & 3 & \vdots & 1 & 0 & 0 \\ 1+i & 0 & 1-i & \vdots & 0 & 1 & 0 \\ 2 & 1 & 2+i & \vdots & 0 & 0 & 1 \end{bmatrix} \begin{array}{l} \frac{1}{2}\mathbf{r}_1 \\ -(1+i)\mathbf{r}_1 + \mathbf{r}_2 \\ -2\mathbf{r}_1 + \mathbf{r}_3 \end{array}$

$$\longrightarrow \begin{bmatrix} 1 & \frac{1}{2}i & \frac{3}{2} & \vdots & \frac{1}{2} & 0 & 0 \\ 0 & \frac{1}{2}-\frac{1}{2}i & -\frac{1}{2}-\frac{5}{2}i & \vdots & -\frac{1}{2}-\frac{1}{2}i & 1 & 0 \\ 0 & 1-i & -1+i & \vdots & -1 & 0 & 1 \end{bmatrix} \begin{array}{l} \frac{2}{1-i}\mathbf{r}_2 \\ -\frac{1}{2}\mathbf{r}_2 + \mathbf{r}_1 \\ -(1-i)\mathbf{r}_2 + \mathbf{r}_3 \end{array}$$

$$\longrightarrow \begin{bmatrix} 1 & 0 & -i & \vdots & 0 & \frac{1}{2}-\frac{1}{2}i & 0 \\ 0 & 1 & 2-3i & \vdots & -i & 1+i & 0 \\ 0 & 0 & 6i & \vdots & i & -2 & 1 \end{bmatrix} \begin{array}{l} \frac{1}{6i}\mathbf{r}_3 \\ -(2-3i)\mathbf{r}_3 + \mathbf{r}_2 \\ i\mathbf{r}_3 + \mathbf{r}_1 \end{array}$$

$$\longrightarrow \begin{bmatrix} 1 & 0 & 0 & \vdots & \frac{1}{6}i & \frac{1}{6}-\frac{1}{2}i & \frac{1}{6} \\ 0 & 1 & 0 & \vdots & -\frac{1}{3}-\frac{1}{2}i & \frac{1}{3}i & \frac{1}{2}+\frac{1}{3}i \\ 0 & 0 & 1 & \vdots & \frac{1}{6} & \frac{1}{3}i & -\frac{1}{6}i \end{bmatrix}$$

Thus the inverse is $\begin{bmatrix} \frac{1}{6}i & \frac{1}{6}-\frac{1}{2}i & \frac{1}{6} \\ -\frac{1}{3}-\frac{1}{2}i & \frac{1}{3}i & \frac{1}{2}+\frac{1}{3}i \\ \frac{1}{6} & \frac{1}{3}i & -\frac{1}{6}i \end{bmatrix} = \frac{1}{6}\begin{bmatrix} i & 1-3i & 1 \\ -2-3i & 2i & 3+2i \\ 1 & 2i & -i \end{bmatrix}$

7. (a) We must determine if there are complex scalars c_1, c_2, c_3 such that

$$c_1\mathbf{v}_1 + c_2\mathbf{v}_2 + c_3\mathbf{v}_3 = (c_1(-1+i) + c_2 + c_3(-5+2i),$$
$$c_1(2) + c_2(1+i) + c_3(-1-3i), c_1 + c_2(i) + c_3(2-3i))$$
$$= (i, 0, 0).$$

Equating corresponding entries from both sides of the equation gives the linear system

$$\begin{array}{rl} (-1+i)c_1 + & c_2 + (-5+2i)c_3 = i \\ 2c_1 + (1+i)c_2 + (-1-3i)c_3 = 0 \\ c_1 + & ic_2 + (2-3i)c_3 = 0. \end{array}$$

Form the augmented matrix

$$\begin{bmatrix} -1+i & 1 & -5+2i & \vdots & i \\ 2 & 1+i & -1-3i & \vdots & 0 \\ 1 & i & 2-3i & \vdots & 0 \end{bmatrix}$$

and row reduce the matrix to obtain the reduced row echelon form

$$\begin{bmatrix} 1 & 0 & 0 & \vdots & -\frac{2}{5} - \frac{4}{5}i \\ 0 & 1 & 0 & \vdots & \frac{11}{10} + \frac{7}{10}i \\ 0 & 0 & 1 & \vdots & \frac{3}{10} + \frac{1}{10}i \end{bmatrix}.$$

Thus the system is consistent which implies that $\mathbf{v}$ belongs to W.

(b) Form the expression

$$c_1\mathbf{v}_1 + c_2\mathbf{v}_2 + c_3\mathbf{v}_3 = \mathbf{0}.$$

Expand, add the vectors, and equate corresponding entries from both sides of the equation. We obtain the homogeneous system

$$\begin{aligned}
(-1+i)c_1 + & & c_2 + (-5+2i)c_3 &= 0 \\
2c_1 + & (1+i)c_2 + & (-1-3i)c_3 &= 0 \\
c_1 + & ic_2 + & (2-3i)c_3 &= 0.
\end{aligned}$$

Form the coefficient matrix and apply row operations to obtain the reduced row echelon form which is I_3. Hence the only solution to the linear system is the trivial solution. Thus $\{\mathbf{v}_1, \mathbf{v}_2, \mathbf{v}_3\}$ is linearly independent.

9. (a) $\mathbf{u} = \begin{bmatrix} 1 - 3i \\ 1 + 3i \end{bmatrix}$, $\mathbf{v} = \begin{bmatrix} 2i \\ 6 \end{bmatrix}$

$$\mathbf{u} \cdot \mathbf{v} = \mathbf{u}^T\overline{\mathbf{v}} = (1-3i)(\overline{2i}) + (1+3i)(6) = 16i.$$

(b) $\mathbf{u} = \begin{bmatrix} 2 - 3i \\ 1 + 2i \\ 4 \end{bmatrix}$, $\mathbf{v} = \begin{bmatrix} 2i \\ 1 - i \\ 3 + 4i \end{bmatrix}$

$$\mathbf{u} \cdot \mathbf{v} = \mathbf{u}^T\overline{\mathbf{v}} = (2-3i)(\overline{2i}) + (1+2i)(\overline{1-i}) + 4(\overline{3+4i}) = 5 - 17i.$$

11. (a) No, since $\mathbf{u} \cdot \mathbf{v} = 9 - 2i$.

(b) No, since $\mathbf{u} \cdot \mathbf{v} = -2$.

(c) No, since $\mathbf{u} \cdot \mathbf{v} = 6$.

(d) Yes, since $\mathbf{u} \cdot \mathbf{v} = 0$.

13. (a) $A = \begin{bmatrix} 1 & 1 \\ -1 & 1 \end{bmatrix}$. Then

$$\det(\lambda I_2 - A) = \begin{vmatrix} \lambda - 1 & -1 \\ 1 & \lambda - 1 \end{vmatrix} = \lambda^2 - 2\lambda + 2.$$

The eigenvalues are $\lambda_1 = 1 + i$ and $\lambda_2 = 1 - i$. For $\lambda_1 = 1 + i$ we obtain

$$(\lambda_1 I_2 - A)\mathbf{x} = \begin{bmatrix} i & -1 \\ 1 & i \end{bmatrix} \begin{bmatrix} x_1 \\ x_2 \end{bmatrix} = \begin{bmatrix} 0 \\ 0 \end{bmatrix}.$$

The reduced row echelon form of the coefficient matrix is

$$\begin{bmatrix} 1 & i \\ 0 & 0 \end{bmatrix}.$$

Thus $x_1 = -ic$, $x_2 = c$, where c is any complex number. Choosing $c = 1$, we obtain the eigenvector

$$\mathbf{x}_1 = \begin{bmatrix} -i \\ 1 \end{bmatrix}$$

corresponding to eigenvalue $\lambda_1 = 1 + i$. For $\lambda_2 = 1 - i$ we have the system

$$(\lambda_2 I_2 - A)\mathbf{x} = \begin{bmatrix} -i & -1 \\ 1 & -i \end{bmatrix} \begin{bmatrix} x_1 \\ x_2 \end{bmatrix} = \begin{bmatrix} 0 \\ 0 \end{bmatrix}.$$

The reduced row echelon form of the coefficient matrix is

$$\begin{bmatrix} 1 & -i \\ 0 & 0 \end{bmatrix}.$$

Thus $x_1 = ic$, $x_2 = c$, where c is any complex number. Choosing $c = 1$, we obtain the eigenvector

$$\mathbf{x}_2 = \begin{bmatrix} i \\ 1 \end{bmatrix}$$

corresponding to eigenvalue $\lambda_2 = 1 - i$.

(b) $A = \begin{bmatrix} 1 & i \\ -i & 1 \end{bmatrix}$. Then

$$\det(\lambda I_2 - A) = \begin{vmatrix} \lambda - 1 & -i \\ i & \lambda - 1 \end{vmatrix} = \lambda(\lambda - 2).$$

The eigenvalues are $\lambda_1 = 0$ and $\lambda_2 = 2$. For $\lambda_1 = 0$ we have the system

$$(\lambda_1 I_2 - A)\mathbf{x} = \begin{bmatrix} -1 & -i \\ i & -1 \end{bmatrix} \begin{bmatrix} x_1 \\ x_2 \end{bmatrix} = \begin{bmatrix} 0 \\ 0 \end{bmatrix}.$$

The reduced echelon form of the coefficient matrix is

$$\begin{bmatrix} 1 & i \\ 0 & 0 \end{bmatrix}.$$

Thus $x_1 = -ic$, $x_2 = c$, where c is any complex number. Choosing $c = 1$ we have the eigenvector

$$\mathbf{x}_1 = \begin{bmatrix} -i \\ 1 \end{bmatrix}$$

corresponding to eigenvalue $\lambda_1 = 0$. For $\lambda_2 = 2$ we have the system

$$(\lambda_2 I_2 - A)\mathbf{x} = \begin{bmatrix} 1 & -i \\ i & 1 \end{bmatrix} \begin{bmatrix} x_1 \\ x_2 \end{bmatrix} = \begin{bmatrix} 0 \\ 0 \end{bmatrix}.$$

The reduced row echelon form of the coefficient matrix is

$$\begin{bmatrix} 1 & -i \\ 0 & 0 \end{bmatrix}.$$

Thus $x_1 = ic$, $x_2 = c$, where c is any complex number. Choosing $c = 1$ we have the eigenvector

$$\mathbf{x}_2 = \begin{bmatrix} i \\ 1 \end{bmatrix}$$

corresponding to eigenvalue $\lambda_2 = 2$.

(c) $A = \begin{bmatrix} 2 & 0 & 0 \\ 0 & 2 & i \\ 0 & -i & 2 \end{bmatrix}$. Then

$$\det(\lambda I_3 - A) = \begin{vmatrix} \lambda - 2 & 0 & 0 \\ 0 & \lambda - 2 & -i \\ 0 & i & \lambda - 2 \end{vmatrix} = (\lambda - 1)(\lambda - 2)(\lambda - 3).$$

The eigenvalues are $\lambda_1 = 1$, $\lambda_2 = 2$, $\lambda_3 = 3$. For $\lambda_1 = 1$, we have the system

$$(\lambda_1 I_3 - A)\mathbf{x} = \begin{bmatrix} -1 & 0 & 0 \\ 0 & -1 & -i \\ 0 & i & -1 \end{bmatrix} \begin{bmatrix} x_1 \\ x_2 \\ x_3 \end{bmatrix} = \begin{bmatrix} 0 \\ 0 \\ 0 \end{bmatrix}.$$

The reduced row echelon form of the coefficient matrix is

$$\begin{bmatrix} 1 & 0 & 0 \\ 0 & 1 & i \\ 0 & 0 & 0 \end{bmatrix}.$$

Thus $x_1 = 0$, $x_2 = -ic$, $x_3 = c$, where c is any complex number. Choosing $c = 1$ we obtain the eigenvector

$$\mathbf{x}_1 = \begin{bmatrix} 0 \\ -i \\ 1 \end{bmatrix}$$

corresponding to eigenvalue $\lambda_1 = 1$. For $\lambda_2 = 2$, we have the system

$$(\lambda_2 I_3 - A)\mathbf{x} = \begin{bmatrix} 0 & 0 & 0 \\ 0 & 0 & -i \\ 0 & i & 0 \end{bmatrix} \begin{bmatrix} x_1 \\ x_2 \\ x_3 \end{bmatrix} = \begin{bmatrix} 0 \\ 0 \\ 0 \end{bmatrix}.$$

The reduced row echelon form of the coefficient matrix is

$$\begin{bmatrix} 0 & 1 & 0 \\ 0 & 0 & 1 \\ 0 & 0 & 0 \end{bmatrix}.$$

Thus $x_1 = c$, $x_2 = 0$, $x_3 = 0$, where c is any complex number. Choosing $c = 1$ we obtain the eigenvector

$$\mathbf{x}_2 = \begin{bmatrix} 1 \\ 0 \\ 0 \end{bmatrix}$$

corresponding to eigenvalue $\lambda_2 = 2$. For $\lambda_3 = 3$ we have the system

$$(\lambda_3 I_3 - A)\mathbf{x} = \begin{bmatrix} 1 & 0 & 0 \\ 0 & 1 & -i \\ 0 & i & 1 \end{bmatrix} \begin{bmatrix} x_1 \\ x_2 \\ x_3 \end{bmatrix} = \begin{bmatrix} 0 \\ 0 \\ 0 \end{bmatrix}.$$

The reduced row echelon form of the coefficient matrix is

$$\begin{bmatrix} 1 & 0 & 0 \\ 0 & 1 & -i \\ 0 & 0 & 0 \end{bmatrix}.$$

Thus $x_1 = 0$, $x_2 = ic$, $x_3 = c$, where c is any complex number. Choosing $c = 1$ we obtain the eigenvector

$$\mathbf{x}_3 = \begin{bmatrix} 0 \\ i \\ 1 \end{bmatrix}$$

corresponding to eigenvalue $\lambda_3 = 3$.

T.1. (a) Let A and B be Hermitian and let k be a complex scalar. Then

$$(\overline{A + B})^T = (\overline{A} + \overline{B})^T = \overline{A}^T + \overline{B}^T = A + B,$$

so the sum of Hermitian matrices is again Hermitian. Next,

$$(\overline{kA})^T = \overline{kA}^T = \overline{k}A \neq kA,$$

so the set of Hermitian matrices is not closed under scalar multiplication and hence is not a complex subspace of C_{nn}.

(b) From (a), we have closure of addition and since the scalars are real here, $\overline{k} = k$, hence $(\overline{kA})^T = kA$. Thus, W is a real subspace of the real vector space of $n \times n$ complex matrices.

T.3. (a) Let A be Hermitian and suppose that $A\mathbf{x} = \lambda\mathbf{x}$, $\lambda \neq 0$. We show that $\lambda = \overline{\lambda}$. We have

$$(\overline{A\mathbf{x}})^T = \left(\overline{A}\,\overline{\mathbf{x}}\right)^T = \overline{\mathbf{x}}^T \overline{A} = \overline{\mathbf{x}}^T A.$$

Also, $(\overline{\lambda}\overline{\mathbf{x}})^T = \overline{\lambda}\overline{\mathbf{x}}^T$, so $\overline{\mathbf{x}}^T A = \overline{\lambda}\overline{\mathbf{x}}^T$. Multiplying both sides by $\mathbf{x}$ on the right, we obtain $\overline{\mathbf{x}}^T A\mathbf{x} = \overline{\lambda}\overline{\mathbf{x}}^T\mathbf{x}$. However, $\overline{\mathbf{x}}^T A\mathbf{x} = \overline{\mathbf{x}}^T \lambda\mathbf{x} = \lambda\overline{\mathbf{x}}^T\mathbf{x}$. Thus, $\lambda\overline{\mathbf{x}}^T\mathbf{x} = \lambda\overline{\mathbf{x}}^T\mathbf{x}$. Then $(\lambda - \overline{\lambda})\overline{\mathbf{x}}^T\mathbf{x} = 0$ and since $\overline{\mathbf{x}}^T\mathbf{x} > 0$, we have $\lambda = \overline{\lambda}$.

(b) $A^T = \begin{bmatrix} 2 & 0 & 0 \\ 0 & 2 & \overline{-i} \\ 0 & \overline{i} & 2 \end{bmatrix} = \begin{bmatrix} 2 & 0 & 0 \\ 0 & 2 & i \\ 0 & -i & 2 \end{bmatrix} = A.$

(c) No, see 11(b). An eigenvector $\mathbf{x}$ associated with a real eigenvalue λ of a complex matrix A is in general complex, because $A\mathbf{x}$ is in general complex. Thus $\lambda\mathbf{x}$ must also be complex.

T.5. Let A be a skew symmetric matrix, so that $\overline{A}^T = -A$, and let λ be an eigenvalue of A with corresponding eigenvector $\mathbf{x}$. We show that $\overline{\lambda} = -\lambda$. We have $A\mathbf{x} = \lambda\mathbf{x}$. Multiplying both sides of this equation by $\overline{\mathbf{x}}^T$ on the left we have $\overline{\mathbf{x}}^T A\mathbf{x} = \overline{\mathbf{x}}^T \lambda\mathbf{x}$. Taking the conjugate transpose of both sides yields

$$\overline{\mathbf{x}}^T \overline{A}^T \mathbf{x} = \overline{\lambda}\,\overline{\mathbf{x}}^T\mathbf{x}.$$

Therefore $-\overline{\mathbf{x}}^T A\mathbf{x} = \overline{\lambda}\,\overline{\mathbf{x}}^T\mathbf{x}$, or $-\lambda\overline{\mathbf{x}}^T\,\overline{\mathbf{x}} = \overline{\lambda}\overline{\mathbf{x}}^T\mathbf{x}$, so $(\lambda + \overline{\lambda})(\overline{\mathbf{x}}^T\mathbf{x}) = 0$. Since $\mathbf{x} \neq \mathbf{0}$, $\overline{\mathbf{x}}^T\mathbf{x} \neq 0$, so $\overline{\lambda} = -\lambda$. Hence, the real part of λ is zero.

Appendix B

Further Directions

Appendix B.1, p. A28

1. (b) $(\mathbf{v}, \mathbf{u}) = a_1 b_1 - a_2 b_1 - a_1 b_2 - 3 a_2 b_2 = (\mathbf{u}, \mathbf{v})$.

   (c) $(\mathbf{u} + \mathbf{v}, \mathbf{w}) = (a_1 + b_1)c_1 - (a_2 + b_2)c_1 - (a_1 + b_1)c_2 + 3(a_2 + b_2)c_2$
   $$= (a_1 c_1 - a_2 c_1 - a_1 c_2 + 3 a_2 c_2) + (b_1 c_1 - b_2 c_1 - b_1 c_2 + 3 b_2 c_2)$$
   $$= (\mathbf{u}, \mathbf{w}) + (\mathbf{v}, \mathbf{w}).$$

   (d) $(c\mathbf{u}, \mathbf{v}) = (ca_1)b_1 - (ca_2)b_1 - (ca_1)b_2 + 3(ca_2)b_2 = c(a_1 b_1 - a_2 b_1 - a_1 b_2 + 3 a_2 b_2) = c(\mathbf{u}, \mathbf{v})$.

3. $(\mathbf{u}, \mathbf{v}) = u_1 v_1 + 5 u_2 v_2$.

   (a) $(\mathbf{u}, \mathbf{u}) = u_1^2 + 5 u_2^2 \geq 0$. Moreover, $u_1^2 + 5 u_2^2 = 0$ if and only if $u_1 = u_2 = 0$ if and only if $\mathbf{u} = \mathbf{0}$.

   (b) $(\mathbf{v}, \mathbf{u}) = v_1 u_1 + 5 v_2 u_2 = u_1 v_1 + 5 u_2 v_2 = (\mathbf{u}, \mathbf{v})$.

   (c) $(\mathbf{u} + \mathbf{v}, \mathbf{w}) = (u_1 + v_1)w_1 + 5(u_2 + v_2)w_2 = u_1 w_1 + v_1 w_1 + 5 u_2 w_2 + 5 v_2 w_2 = (\mathbf{u}, \mathbf{w}) + (\mathbf{v}, \mathbf{w})$.

   (d) $(c\mathbf{u}, \mathbf{v}) = (cu_1)v_1 + 5(cu_2)v_2 = c(u_1 v_1 + 5 u_2 v_2) = c(\mathbf{u}, \mathbf{v})$.

5. (a) If $A = [a_{ij}]$, then

$$(A, A) = \text{Tr}(A^T A) = \sum_{j=1}^{n} \sum_{i=1}^{n} a_{ij}^2 \geq 0.$$

   Also, $(A, A) = 0$ if and only if $a_{ij} = 0$, that is, if and only if $A = O$.

   (b) If $B = [a_{ij}]$, then $(A, B) = \text{Tr}(B^T A)$ and $(B, A) = \text{Tr}(A^T B)$. Now,

$$\text{Tr}(B^T A) = \sum_{i=1}^{n} \sum_{k=1}^{n} b_{ik}^T a_{ki} = \sum_{i=1}^{n} \sum_{k=1}^{n} b_{ki} a_{ki},$$

   and

$$\text{Tr}(A^T B) = \sum_{i=1}^{n} \sum_{k=1}^{n} a_{ik}^T b_{ki} = \sum_{i=1}^{n} \sum_{k=1}^{n} a_{ki} b_{ki},$$

   so $(A, B) = (B, A)$.

   (c) If $C = [c_{ij}]$, then

$$(A + B, C) = \text{Tr}[C^T(A + B)] = \text{Tr}(C^T A + C^T B) = \text{Tr}(C^T A) + \text{Tr}(C^T B) = (A, C) + (B, C).$$

   (d) $(cA, B) = \text{Tr}[B^T(cA)] = c\,\text{Tr}(B^T A) = c(A, B)$.

7. $(\mathbf{u}, \mathbf{v}) = u_1v_1 - u_2v_1 - u_1v_2 + 3u_2v_2$, so if $\mathbf{u} = (1, 2)$ and $\mathbf{v} = (3, -1)$, we have

$$(\mathbf{u}, \mathbf{v}) = (1)(3) - (2)(3) - (1)(-1) + 3(2)(-1) = -8.$$

9. $(f, g) = \int_0^1 f(t)g(t)\, dt$, so if $f(t) = 1$ and $g(t) = 3 + 2t$, then

$$(f, g) = \int_0^1 (1)(3 + 2t)\, dt = \int_0^1 (3 + 2t)\, dt = \left[3t + t^2\right]\Big|_0^1 = 4.$$

11. $(A, B) = a_{11}b_{11} + a_{12}b_{12} + a_{21}b_{21} + a_{22}b_{22}$, so if

$$A = \begin{bmatrix} 1 & 2 \\ -1 & 3 \end{bmatrix} \quad \text{and} \quad B = \begin{bmatrix} 1 & 0 \\ 2 & -1 \end{bmatrix},$$

then

$$(A, B) = (1)(1) + 2(0) + (-1)(2) + 3(-1) = -4.$$

13. $(\mathbf{u}, \mathbf{v}) = u_1v_1 - u_2v_1 - u_1v_2 + 3u_2v_2$.

  (a) $\mathbf{u} = (1, 3)$: $\|\mathbf{u}\| = \sqrt{(1)(1) - (3)(1) - (1)(3) + 3(3)(3)} = \sqrt{22}$.
  (b) $\mathbf{u} = (-2, -4)$: $\|\mathbf{u}\| = \sqrt{(-2)(-2) - (-4)(-2) - (-2)(-4) + 3(-4)(-4)} = 6$.
  (c) $\mathbf{u} = (3, -1)$: $\|\mathbf{u}\| = \sqrt{(3)(3) - (-1)(3) - (3)(-1) + 3(-1)(-1)} = \sqrt{18}$.

15. (a) If $\mathbf{u} = t$, $\mathbf{v} = t^2$, then

$$\|\mathbf{u} - \mathbf{v}\| = \|t - t^2\| = \sqrt{\int_0^1 (t - t^2)^2\, dt}$$

$$= \sqrt{\int_0^1 (t^2 - 2t^3 + t^4)\, dt} = \sqrt{\left[\frac{t^3}{3} - \frac{2t^4}{4} + \frac{t^5}{5}\right]\Big|_0^1} = \sqrt{\frac{1}{30}}.$$

  (b) If $\mathbf{u} = e^t$, $\mathbf{v} = e^{-t}$, then

$$\|\mathbf{u} - \mathbf{v}\| = \|e^t - e^{-t}\| = \sqrt{\int_0^1 (e^t - e^{-t})^2\, dt} = \sqrt{\int_0^1 (e^{2t} - 2 + e^{-2t})\, dt}$$

$$= \sqrt{\left(\tfrac{1}{2}e^{2t} - 2t - \tfrac{1}{2}e^{-2t}\right)\Big|_0^1} = \sqrt{\tfrac{1}{2}(e^2 - e^{-2}) - 2}.$$

17. (a) $p(t) = t$, $q(t) = t - 1$, so

$$\cos\theta = \frac{(p(t), q(t))}{\|p(t)\|\, \|q(t)\|}.$$

We have

$$(p(t), q(t)) = \int_0^1 t(t - 1)\, dt = \int_0^1 (t^2 - t)\, dt = \left[\frac{t^3}{3} - \frac{t^2}{2}\right]\Big|_0^1 = -\frac{1}{6}.$$

Now,

$$\|p(t)\| = \sqrt{\int_0^1 t^2\, dt} = \sqrt{\frac{t^3}{3}\Big|_0^1} = \sqrt{\frac{1}{3}}$$

$$\|q(t)\| = \sqrt{\int_0^1 (t - 1)^2\, dt} = \sqrt{\int_0^1 (t^2 - 2t + 1)\, dt} = \sqrt{\left[\frac{t^3}{3} - t^2 + t\right]\Big|_0^1} = \sqrt{\frac{1}{3}}.$$

Hence

$$\cos\theta = \frac{-\frac{1}{6}}{\sqrt{\frac{1}{3}}\sqrt{\frac{1}{3}}} = \frac{-\frac{1}{6}}{\frac{1}{3}} = -\frac{1}{2}.$$

(b) $p(t) = \sin t$, $q(t) = \cos t$, so

$$\cos\theta = \frac{(p(t), q(t))}{\|p(t)\|\,\|q(t)\|}.$$

We have

$$(p(t), q(t)) = \int_0^1 \sin t \,\cos t \,dt = \tfrac{1}{2}\sin^2 t\big|_0^1 = \tfrac{1}{2}\sin^2 1$$

$$\|p(t)\| = \sqrt{\int_0^1 \sin^2 t \,dt} = \sqrt{\int_0^1 \left(\tfrac{1}{2} - \tfrac{1}{2}\cos 2t\right) dt} = \sqrt{\left(\tfrac{1}{2}t - \tfrac{1}{4}\sin^2 t\right)\big|_0^1} = \sqrt{\tfrac{1}{2} - \tfrac{1}{4}\sin(2)}$$

$$\|q(t)\| = \sqrt{\int_0^1 \cos^2 t \,dt} = \sqrt{\int_0^1 \left(\tfrac{1}{2} + \tfrac{1}{2}\cos 2t\right) dt} = \sqrt{\left(\tfrac{1}{2}t + \tfrac{1}{4}\sin 2t\right)\big|_0^1} = \sqrt{\tfrac{1}{2} + \tfrac{1}{4}\sin(2)}$$

Then

$$\cos\theta = \frac{\tfrac{1}{2}\sin^2 1}{\sqrt{\tfrac{1}{2} - \tfrac{1}{4}\sin(2)}\sqrt{\tfrac{1}{2} + \tfrac{1}{4}\sin(2)}} = \frac{\tfrac{1}{2}\sin^2(1)}{\sqrt{\tfrac{1}{4} - \tfrac{1}{16}\sin^2(2)}} = \frac{\tfrac{1}{2}\sin^2(1)}{\tfrac{1}{4}\sqrt{4 - \sin^2(2)}} = \frac{2\sin^2(1)}{\sqrt{4 - \sin^2(2)}}.$$

19. The polynomials are orthogonal provided $(p(t), q(t)) = \int_0^1 (3t+1)(at)\,dt = 0$. We have

$$\int_0^1 (3t+1)(at)\,dt = \int_0^1 (3at^2 + at)\,dt = \left[at^3 + \frac{a}{2}t^2\right]\Big|_0^1 = \frac{3}{2}a = 0$$

only if $a = 0$.

21. Let $B = \begin{bmatrix} b_{11} & b_{12} \\ b_{21} & b_{22} \end{bmatrix}$. Then

$$(A, B) = \operatorname{Tr}(B^T A) = \operatorname{Tr}\left(\begin{bmatrix} b_{11} & b_{21} \\ b_{12} & b_{22} \end{bmatrix}\begin{bmatrix} 1 & 2 \\ 3 & 4 \end{bmatrix}\right) = \operatorname{Tr}\left(\begin{bmatrix} b_{11} + 3b_{21} & 2b_{11} + 4b_{21} \\ b_{12} + 3b_{22} & 2b_{12} + 4b_{22} \end{bmatrix}\right).$$

This is one equation in four unknowns, the elements of B. Hence three of the elements of B can be chosen arbitrarily as long as at least one of them is not zero. For example, if we let $b_{21} = 1$, $b_{12} = 1$, and $b_{22} = 3$, then from the preceding relation we have that $b_{11} = -17$. Certainly there is more than one matrix B that is orthogonal to A.

23. (a) If $\mathbf{v}_1$ and $\mathbf{v}_2$ lie in W and c is a real number, then

$$((\mathbf{v}_1 + \mathbf{v}_2), \mathbf{u}_i) = (\mathbf{v}_1, \mathbf{u}_i) + (\mathbf{v}_2, \mathbf{u}_i) = 0 + 0 = 0 \quad \text{for } i = 1, 2.$$

Thus $\mathbf{v}_1 + \mathbf{v}_2$ lies in W. Also, $(c\mathbf{v}_1, \mathbf{u}_i) = c(\mathbf{v}_1, \mathbf{u}_i) = c0 = 0$ for $i = 1, 2$. Thus $c\mathbf{v}_1$ lies in W.

(b) A vector $\mathbf{w} = (a, b, c, d)$ is in W provided

$$(\mathbf{w}, \mathbf{u}_1) = 0 \quad \text{and} \quad (\mathbf{w}, \mathbf{u}_2) = 0.$$

Now

$$((a, b, c, d), (1, 0, 0, 1)) = a + d \quad \text{and} \quad ((a, b, c, d), (0, 1, 0, 1)) = b + d.$$

Thus $a = -d$ and $b = -d$. Hence a vector $\mathbf{w}$ in W has the form $(-d, -d, c, d)$. Writing

$$(-d, -d, c, d) = d(-1, -1, 0, 1) + c(0, 0, 1, 0),$$

we see that the vectors $(-1, -1, 0, 1)$ and $(0, 0, 1, 0)$ span W. Moreover, they are linearly independent since neither is a multiple of the other. Hence they form a basis for W.

25. (a) Let $S = \{\mathbf{u}_1, \mathbf{u}_2\} = \{t, 1\}$. Using the Gram–Schmidt process, we have:

$$\mathbf{v}_1 = \mathbf{u}_1 = t$$

$$\mathbf{v}_2 = \mathbf{u}_2 - \frac{(\mathbf{u}_2, \mathbf{v}_1)}{(\mathbf{v}_1, \mathbf{v}_1)}\mathbf{v}_1 = 1 - \frac{\int_0^1 (1)(t)\, dt}{\int_0^1 t^2\, dt}\, t = 1 - \frac{\frac{1}{2}}{\frac{1}{3}}\, t = 1 - \tfrac{3}{2}t.$$

The set $\{\mathbf{v}_1, \mathbf{v}_2\}$ is an orthogonal basis for W. To obtain an orthonormal basis for W we determine unit vectors in the same direction as the $\mathbf{v}_i$, $i = 1, 2$. Let

$$\mathbf{w}_1 = \frac{1}{\|\mathbf{v}_1\|}\mathbf{v}_1 = \frac{t}{\sqrt{\int_0^1 t^2\, dt}} = \frac{t}{\sqrt{\frac{1}{3}}} = \sqrt{3}\, t$$

$$\mathbf{w}_2 = \frac{1}{\|\mathbf{v}_2\|}\mathbf{v}_2 = \frac{1 - \frac{3}{2}t}{\sqrt{\int_0^1 \left[1 - \frac{3}{2}t\right]^2 dt}} = \frac{1 - \frac{3}{2}t}{\sqrt{\frac{1}{4}}} = 2 - 3t.$$

Then $\{\mathbf{w}_1, \mathbf{w}_2\}$ is an orthonormal basis for W.

(b) From the generalized Theorem 6.17, if $\mathbf{v}$ is any vector in P_2, then

$$\mathbf{v} = (\mathbf{v}, \mathbf{w}_1)\mathbf{w}_1 + (\mathbf{v}, \mathbf{w}_2)\mathbf{w}_2.$$

In this case $\mathbf{v} = 2t - 1$. We have

$$(2t - 1, \sqrt{3}\, t) = \int_0^1 (2t - 1)(\sqrt{3}\, t)\, dt = \int_0^1 (2\sqrt{3}\, t^2 - \sqrt{3}\, t)\, dt$$

$$= \left[\frac{2\sqrt{3}}{3}t^3 - \frac{\sqrt{3}}{2}t^2\right]\Bigg|_0^1 = \left(\frac{2}{3} - \frac{1}{2}\right)\sqrt{3} = \frac{1}{6}\sqrt{3},$$

$$(2t - 1, 2 - 3t) = \int_0^1 (2t - 1)(2 - 3t)\, dt = \int_0^1 (6t^2 + 7t - 2)\, dt$$

$$= \left[-2t^3 + \frac{7}{2}t^2 - 2t\right]\Bigg|_0^1 = -\frac{1}{2}.$$

Hence

$$2t - 1 = \frac{\sqrt{3}}{6}(\sqrt{3}\, t) - \frac{1}{2}(2 - 3t).$$

27. Let $S = \{\mathbf{u}_1, \mathbf{u}_2\} = \{t, \sin 2\pi t\}$. Using the Gram–Schmidt process we have:

$$\mathbf{v}_1 = \mathbf{u}_1 = t$$

$$\mathbf{v}_2 = \mathbf{u}_2 - \frac{(\mathbf{u}_2, \mathbf{v}_1)}{(\mathbf{v}_1, \mathbf{v}_1)}\mathbf{v}_1 = \sin 2\pi t - \frac{\int_0^1 t \sin 2\pi t\, dt}{\int_0^1 t^2\, dt}\, t = \sin 2\pi t - \frac{-\frac{1}{2\pi}}{\frac{1}{3}}\, t = \sin 2\pi t + \frac{3}{2\pi}\, t.$$

The integral in the numerator above was found using integration by parts; the denominator was computed in Exercise 5. The set $\{\mathbf{v}_1, \mathbf{v}_2\}$ is an orthogonal basis for W. To obtain an orthonormal basis for W we determine unit vectors in the same direction as $\mathbf{v}_i$, $i = 1, 2$. Let

$$\mathbf{w}_1 = \frac{1}{\|\mathbf{v}_1\|}\,\mathbf{v}_1 = \frac{t}{\sqrt{\int_0^1 t^2\,dt}} = \frac{t}{\sqrt{\frac{1}{3}}} = \sqrt{3}\,t$$

$$\mathbf{w}_2 = \frac{1}{\|\mathbf{v}_2\|}\,\mathbf{v}_2 = \frac{\sin 2\pi t + \frac{3}{2\pi}\,t}{\sqrt{\int_0^1 \left[\sin 2\pi t + \frac{3}{2\pi}\,t\right]^2 dt}} = \frac{\sin 2\pi t + \frac{3}{2\pi}\,t}{\sqrt{\int_0^1 \left[\sin^2 2\pi t + \frac{3}{\pi}\,t\sin 2\pi t + \frac{9}{4\pi^2}\,t^2\right] dt}}.$$

We integrate the terms in the preceding expression:

$$\int_0^1 \sin^2 2\pi t\,dt = \left[\frac{1}{2}\,t - \frac{1}{8\pi}\,\sin 4\pi t\right]\Big|_0^1 = \frac{1}{2}$$

$$\int_0^1 \frac{3}{\pi}\,t\sin 2\pi t\,dt = \frac{3}{\pi}\int_0^1 t\sin 2\pi t\,dt = \left(\frac{3}{\pi}\right)\left(-\frac{1}{2\pi}\right) = -\frac{3}{2\pi^2}$$

$$\int_0^1 \frac{9}{4\pi^2}\,t^2\,dt = \frac{9}{4\pi^2}\int_0^1 t^2\,dt = \left(\frac{9}{4\pi^2}\right)\left(\frac{1}{3}\right) = \frac{3}{4\pi^2}.$$

Then

$$\mathbf{w}_2 = \frac{\sin 2\pi t + \frac{3}{2\pi}\,t}{\sqrt{\frac{1}{2} - \frac{3}{2\pi^2} + \frac{3}{4\pi^2}}} = \frac{\sin 2\pi t + \frac{3}{2\pi}\,t}{\sqrt{\frac{1}{2} - \frac{3}{4\pi^2}}}.$$

Therefore $\{\mathbf{w}_1, \mathbf{w}_2\}$ is an orthonormal basis for W.

29. Let $p(t) = at^3 + bt^2 + ct + d$ be in $W^\perp$. Then $p(t)$ is orthogonal to $t - 1$ and t^2 and hence

$$(p(t), t - 1) = \int_0^1 (at^3 + bt^2 + ct + d)(t - 1)\,dt = -\tfrac{1}{20}a - \tfrac{1}{12}b - \tfrac{1}{6}c - \tfrac{1}{2}d = 0$$

$$(p(t), t^2) = \int_0^1 (at^3 + bt^2 + ct + d)(t^2)\,dt = \tfrac{1}{6}a + \tfrac{1}{5}b + \tfrac{1}{4}c + \tfrac{1}{3}d = 0.$$

As in Exercise 3 we form the coefficient matrix of this system and row reduce it:

$$A = \begin{bmatrix} -\frac{1}{20} & -\frac{1}{12} & -\frac{1}{6} & -\frac{1}{2} \\ \frac{1}{6} & \frac{1}{5} & \frac{1}{4} & \frac{1}{3} \end{bmatrix} \xrightarrow{\ \ } \underset{\text{steps omitted}}{\cdots} \xrightarrow{\ \ } \begin{bmatrix} 1 & 0 & -\frac{45}{14} & -\frac{130}{7} \\ 0 & 1 & \frac{55}{14} & \frac{120}{7} \end{bmatrix} = B.$$

Thus the solution is $a = \frac{45}{14}r + \frac{130}{7}s$, $b = -\frac{55}{14}r - \frac{120}{7}s$, $c = r$, $d = s$, where r and s are any real numbers. Therefore

$$at^3 + bt^2 + ct + d = \left(\tfrac{45}{14}r + \tfrac{130}{7}s\right)t^3 + \left(-\tfrac{55}{14}r - \tfrac{120}{7}s\right)t^2 + rt + s$$

$$= r\left(\tfrac{45}{14}t^3 - \tfrac{55}{14}t^2 + t\right) + s\left(\tfrac{130}{7}t^3 - \tfrac{120}{7}t^2 + 1\right)$$

Hence the set $S = \left\{\frac{45}{14}t^3 - \frac{55}{14}t^2 + t, \frac{130}{7}t^3 - \frac{120}{7}t^2 + 1\right\}$ is a basis for $W^\perp$.

31. Let

$$\mathbf{w}_1 = \frac{1}{\sqrt{2\pi}}, \qquad \mathbf{w}_2 = \frac{1}{\sqrt{\pi}}\cos t, \qquad \mathbf{w}_3 = \frac{1}{\sqrt{\pi}}\sin t.$$

Then by Example 11, $\{\mathbf{w}_1, \mathbf{w}_2, \mathbf{w}_3\}$ is an orthonormal basis for W. Hence, for any vector $\mathbf{v}$,

$$\text{proj}_W\,\mathbf{v} = (\mathbf{v}, \mathbf{w}_1)\mathbf{w}_1 + (\mathbf{v}, \mathbf{w}_2)\mathbf{w}_2 + (\mathbf{v}, \mathbf{w}_3)\mathbf{w}_3.$$

For $\mathbf{v} = t$, we find that

$$(t, \mathbf{w}_1) = \int_{-\pi}^{\pi} (t) \frac{1}{\sqrt{2\pi}}\, dt = 0$$

$$(t, \mathbf{w}_2) = \int_{-\pi}^{\pi} (t) \frac{1}{\sqrt{\pi}} \cos t\, dt = 0$$

$$(t, \mathbf{w}_3) = \int_{-\pi}^{\pi} (t) \frac{1}{\sqrt{\pi}} \sin t\, dt = 2\sqrt{\pi}.$$

Therefore

$$\text{proj}_W t = (0)\left(\frac{1}{\sqrt{2\pi}}\right) + (0)\left(\frac{1}{\sqrt{\pi}}\cos t\right) + (2\sqrt{\pi})\left(\frac{1}{\sqrt{\pi}}t\right) = 2\sin t.$$

33. The vector $\mathbf{w}$ is the projection of $\mathbf{v}$ onto the subspace W. Thus:

$$\mathbf{w} = (\mathbf{v}, \mathbf{w}_1)\mathbf{w}_1 + (\mathbf{v}, \mathbf{w}_2)\mathbf{w}_2 + (\mathbf{v}, \mathbf{w}_3)\mathbf{w}_3.$$

For the inner products, we find:

$$(\mathbf{v}, \mathbf{w}_1) = \int_{-\pi}^{\pi} (t - 1)\frac{1}{\sqrt{2\pi}}\, dt = -\sqrt{2\pi}$$

$$(\mathbf{v}, \mathbf{w}_2) = \int_{-\pi}^{\pi} (t - 1)\frac{1}{\sqrt{\pi}} \cos t\, dt = 0$$

$$(\mathbf{v}, \mathbf{w}_3) = \int_{-\pi}^{\pi} (t - 1)\frac{1}{\sqrt{\pi}} \sin t\, dt = 2\sqrt{\pi}.$$

Therefore

$$\mathbf{w} = -\sqrt{2\pi}\, \mathbf{w}_1 + (0)\mathbf{w}_2 + 2\sqrt{\pi}\, \mathbf{w}_3$$

$$= -\sqrt{2\pi}\left(\frac{1}{\sqrt{2\pi}}\right) + 2\sqrt{\pi}\left(\frac{1}{\sqrt{\pi}}\cos t\right) + 2\sqrt{\pi}\left(\frac{1}{\sqrt{\pi}}\sin t\right)$$

$$= -1 + 2\sin t.$$

Let $\mathbf{u} = \mathbf{v} - \mathbf{w} = (t - 1) - (-1 + 2\sin t) = t - 2\sin t$. Then $\mathbf{v} = \mathbf{w} + \mathbf{u}$, where $\mathbf{w}$ is in W and $\mathbf{u}$ is in $W^{\perp}$.

35. From Exercise 31, the projection of $\mathbf{v} = t$ onto W is $\mathbf{w} = 2\sin t$. Hence the distance from $\mathbf{v}$ to W is

$$\|\mathbf{v} - \text{proj}_W \mathbf{v}\| = \|t - 2\sin t\| = \sqrt{\int_{-\pi}^{\pi} (t - 2\sin t)^2\, dt} = \sqrt{\frac{2\pi^3}{3} - 4\pi}.$$

37. Let $\mathbf{v} = f(t) = t^2$. Then the Fourier polynomial of degree two for t^2 is

$$\left(t^2, \frac{1}{\sqrt{2\pi}}\right)\frac{1}{\sqrt{2\pi}} + \left(t^2, \frac{1}{\sqrt{\pi}}\cos t\right)\frac{1}{\sqrt{\pi}}\cos t + \left(t^2, \frac{1}{\sqrt{\pi}}\sin t\right)\frac{1}{\sqrt{\pi}}\sin t$$

$$+ \left(t^2, \frac{1}{\sqrt{\pi}}\cos 2t\right)\frac{1}{\sqrt{\pi}}\cos 2t + \left(t^2, \frac{1}{\sqrt{\pi}}\sin 2t\right)\frac{1}{\sqrt{\pi}}\sin 2t.$$

Now,

$$\left(t^2, \frac{1}{\sqrt{2\pi}}\right) = \int_\pi^\pi t^2 \frac{1}{\sqrt{2\pi}}\, dt = \frac{1}{\sqrt{2\pi}}\left(\frac{2}{3}\pi^3\right)$$

$$\left(t^2, \frac{1}{\sqrt{\pi}}\cos t\right) = \int_{-\pi}^\pi t^2 \frac{1}{\sqrt{\pi}}\cos t\, dt = \frac{1}{\sqrt{\pi}}(-4\pi)$$

$$\left(t^2, \frac{1}{\sqrt{\pi}}\sin t\right) = \int_{-\pi}^\pi t^2 \frac{1}{\sqrt{\pi}}\sin t\, dt = 0$$

$$\left(t^2, \frac{1}{\sqrt{\pi}}\cos 2t\right) = \int_{-\pi}^\pi t^2 \frac{1}{\sqrt{\pi}}\cos 2t\, dt = \frac{1}{\sqrt{\pi}}(\pi)$$

$$\left(t^2, \frac{1}{\sqrt{\pi}}\sin 2t\right) = \int_{-\pi}^\pi t^2 \frac{1}{\sqrt{\pi}}\sin 2t\, dt = 0$$

Thus

$$\text{proj}_W e^t = \frac{1}{\sqrt{2\pi}}\left(\frac{2}{3}\pi^3\right)\frac{1}{\sqrt{2\pi}} + \left(\frac{1}{\sqrt{\pi}}\right)(-4\pi)\frac{1}{\sqrt{\pi}}\cos t + \left(\frac{1}{\sqrt{\pi}}\right)(\pi)\frac{1}{\sqrt{\pi}}\cos 2t$$

$$= \frac{\pi^2}{3} - 4\cos t + \cos 2t$$

T.1. (a) $\mathbf{0} + \mathbf{0} = \mathbf{0}$ so $(\mathbf{0},\mathbf{0}) = (\mathbf{0},\mathbf{0}+\mathbf{0}) = (\mathbf{0},\mathbf{0}) + (\mathbf{0},\mathbf{0})$, and then $(\mathbf{0},\mathbf{0}) = 0$. Hence $\|\mathbf{0}\| = \sqrt{(\mathbf{0},\mathbf{0})} = \sqrt{0} = 0$.

(b) $(\mathbf{u},\mathbf{0}) = (\mathbf{u},\mathbf{0}+\mathbf{0}) = (\mathbf{u},\mathbf{0}) + (\mathbf{u},\mathbf{0})$ so $(\mathbf{u},\mathbf{0}) = 0$.

(c) If $(\mathbf{u},\mathbf{v}) = 0$ for all $\mathbf{v}$ in V, then $(\mathbf{u},\mathbf{u}) = 0$ so $\mathbf{u} = \mathbf{0}$.

(d) If $(\mathbf{u},\mathbf{w}) = (\mathbf{v},\mathbf{w})$ for all $\mathbf{w}$ in V, then $(\mathbf{u}-\mathbf{v},\mathbf{w}) = 0$ and so $\mathbf{u} = \mathbf{v}$.

(e) If $(\mathbf{w},\mathbf{u}) = (\mathbf{w},\mathbf{v})$ for all $\mathbf{w}$ in V, then $(\mathbf{w},\mathbf{u}-\mathbf{v}) = 0$ or $(\mathbf{u}-\mathbf{v},\mathbf{w}) = 0$ for all $\mathbf{w}$ in V. Then $\mathbf{u} = \mathbf{v}$.

T.3. Let $T = \{\mathbf{u}_1, \mathbf{u}_2, \ldots, \mathbf{u}_n\}$ be an orthonormal basis for an inner product space V. If

$$[\mathbf{v}]_T = \begin{bmatrix} a_1 \\ a_2 \\ \vdots \\ a_n \end{bmatrix},$$

then $\mathbf{v} = a_1\mathbf{u}_1 + a_2\mathbf{u}_2 + \cdots + a_n\mathbf{u}_n$. Since $(\mathbf{u}_i, \mathbf{u}_j) = 0$ if $i \neq j$ and 1 if $i = j$, we conclude that

$$\|\mathbf{v}\| = \sqrt{(\mathbf{v},\mathbf{v})} = \sqrt{a_1^2 + a_2^2 + \cdots + a_n^2}.$$

T.5. $\|\mathbf{u}+\mathbf{v}\|^2 = (\mathbf{u}+\mathbf{v}, \mathbf{u}+\mathbf{v}) = (\mathbf{u},\mathbf{u}) + 2(\mathbf{u},\mathbf{v}) + (\mathbf{v},\mathbf{v}) = \|\mathbf{u}\|^2 + 2(\mathbf{u},\mathbf{v}) + (\mathbf{v},\mathbf{v}) = \|\mathbf{u}\|^2 + 2(\mathbf{u},\mathbf{v}) + \|\mathbf{v}\|^2$, and $\|\mathbf{u}-\mathbf{v}\|^2 = (\mathbf{u},\mathbf{u}) - 2(\mathbf{u},\mathbf{v}) + (\mathbf{v},\mathbf{v}) = \|\mathbf{u}\|^2 - 2(\mathbf{u},\mathbf{v}) + \|\mathbf{v}\|^2$. Hence

$$\|\mathbf{u}+\mathbf{v}\|^2 + \|\mathbf{u}-\mathbf{v}\|^2 = 2\|\mathbf{u}\|^2 + 2\|\mathbf{v}\|^2.$$

T.7. $\|\mathbf{u}+\mathbf{v}\|^2 = (\mathbf{u}+\mathbf{v}, \mathbf{u}+\mathbf{v}) = (\mathbf{u},\mathbf{u}) + 2(\mathbf{u},\mathbf{v}) + (\mathbf{v},\mathbf{v}) = \|\mathbf{u}\|^2 + 2(\mathbf{u},\mathbf{v}) + \|\mathbf{v}\|^2$. Thus $\|\mathbf{u}+\mathbf{v}\|^2 = \|\mathbf{u}\|^2 + \|\mathbf{v}\|^2$ if and only if $(\mathbf{u},\mathbf{v}) = 0$.

T.9. Let $S = \{\mathbf{w}_1, \mathbf{w}_2, \ldots, \mathbf{w}_k\}$. If $\mathbf{u}$ is in span S, then

$$\mathbf{u} = c_1\mathbf{w}_1 + c_2\mathbf{w}_2 + \cdots + c_k\mathbf{w}_k.$$

Let $\mathbf{v}$ be orthogonal to $\mathbf{w}_1, \mathbf{w}_2, \ldots, \mathbf{w}_k$. Then

$$(\mathbf{v}, \mathbf{u}) = (\mathbf{v}, c_1\mathbf{w}_1 + c_2\mathbf{w}_2 + \cdots + c_k\mathbf{w}_k) = c_1(\mathbf{v}, \mathbf{w}_1) + c_2(\mathbf{v}, \mathbf{w}_2) + \cdots + c_k(\mathbf{v}, \mathbf{w}_k)$$
$$= c_1(0) + c_2(0) + \cdots + c_k(0) = 0.$$

Appendix B.2, p. A35

1. $L_1(x, y) = (x + y, x - y, 2x + y)$, $L_2(x, y, z) = (x + y + z, y + z, x + z)$.

   (a) $(L_2 \circ L_1)(-1, 1) = L_2(L_1(-1, 1)) = L_2(-1 + 1, -1 - 1, 2(-1) + 1)$
   $\qquad = L_2(0, -2, -1) = (0 - 2 - 1, -2 - 1, 0 - 1) = (-3, -3, -1)$.

   (b) $(L_2 \circ L_1)(x, y) = L_2(L_1(x, y)) = L_2(x + y, x - y, 2x + y)$
   $\qquad = ((x + y) + (x - y) + (2x + y), (x - y) + (2x + y), (x + y) + (2x + y))$
   $\qquad = (4x + y, 3x, 3x + 2y)$.

3. $L_1(at + b) = 2at - b$, $L_2(at + b) = t(at + b)$.

   (a) $(L_2 \circ L_1)(3t + 2) = L_2(L_1(3t + 2)) = L_2(6t - 2) = 6t^2 - 2t$.
   (b) $(L_2 \circ L_1)(at + b) = L_2(L_1(at + b)) = L_2(2at - b) = 2at^2 - bt$.

5. $L_1(x, y) = (x + y, x - 2y)$, $L_2(x, y) = (y, x - y)$.

   (a) $(L_2 \circ L_1)(1, 2) = L_2(L_1(1, 2)) = L_2(1 + 2, 1 - 4) = L_2(3, -3) = (-3, 3 + 3) = (-3, 6)$.
   (b) $(L_1 \circ L_2)(1, 2) = L_1(L_2(1, 2)) = L_1(2, 1 - 2) = L_1(2, -1) = (2 + (-1), 2 - 2(-1)) = (1, 4)$.
   (c) $(L_2 \circ L_1)(x, y) = L_2(L_1(x, y)) = L_2(x + y, x - 2y) = (x - 2y, x + y - (x - 2y)) = (x - 2y, 3y)$.
   (d) $(L_1 \circ L_2)(x, y) = L_1(L_2(x, y)) = L_1(y, x - y) = (y + (x - y), y - 2(x - y)) = (x, -2x + 3y)$.

7. $L_1(at + b) = 2at - b$, $L_2(at + b) = t(at + b)$,
   $S = \{t + 1, t - 1\}$, $T = \{t^2 + 1, t, t - 1\}$.

   From Exercise 3, $(L_2 \circ L_1)(at + b) = 2at^2 - bt$.

   (a) $(L_2 \circ L_1)(t + 1) = 2t^2 - t$, $(L_2 \circ L_1)(t - 1) = 2t^2 + t$. We now write $2t^2 - t$ and $2t^2 + t$ as a linear combination of the vectors in T:

   $$c_1(t^2 + 1) + c_2 t + c_3(t - 1) = 2t^2 - t$$
   $$d_1(t^2 + 1) + d_2 t + d_3(t - 1) = 2t^2 + t.$$

   Each of these equations leads to a linear system of three equations in three unknowns. Since the coefficient matrix is the same for both linear systems, we proceed as in Example 5 in Section 10.3 solving both systems at the same time by Gauss–Jordan reduction. Thus we transform the matrix

   $$\begin{bmatrix} 1 & 0 & 0 & \vdots & 2 & \vdots & 2 \\ 0 & 1 & 1 & \vdots & -1 & \vdots & 1 \\ 1 & 0 & -1 & \vdots & 0 & \vdots & 0 \end{bmatrix}$$

   consisting of the coefficient matrix and the right hand sides of each linear system to reduced row echelon form. We obtain

   $$\begin{bmatrix} 1 & 0 & 0 & \vdots & 2 & \vdots & 2 \\ 0 & 1 & 0 & \vdots & -3 & \vdots & -1 \\ 0 & 0 & 1 & \vdots & 2 & \vdots & 2 \end{bmatrix}.$$

Thus $c_1 = 2$, $c_2 = -3$, $c_3 = 2$, $d_1 = 2$, $d_2 = -1$, and $d_3 = 2$. Hence $B = \begin{bmatrix} 2 & 2 \\ -3 & -1 \\ 2 & 2 \end{bmatrix}$.

(b) $L_1(t+1) = 2t - 1$, $L_1(t-1) = 2t + 1$. We now write

$$c_1(t+1) + c_2(t-1) = 2t - 1$$
$$d_1(t+1) + d_2(t-1) = 2t + 1$$

We solve the two linear systems obtained from these equations by Gauss–Jordan reduction:

$$\begin{bmatrix} 1 & 1 & \vdots & 2 & \vdots & 2 \\ 1 & -1 & \vdots & -1 & \vdots & 1 \end{bmatrix} \xrightarrow[\text{steps omitted}]{\quad \cdots \quad} \begin{bmatrix} 1 & 0 & \vdots & \frac{1}{2} & \vdots & \frac{3}{2} \\ 0 & 1 & \vdots & \frac{3}{2} & \vdots & \frac{1}{2} \end{bmatrix} = B.$$

Thus $c_1 = \frac{1}{2}$, $c_2 = \frac{3}{2}$, $d_1 = \frac{3}{2}$, $d_2 = \frac{1}{2}$. Hence

$$A_1 = \begin{bmatrix} \frac{1}{2} & \frac{3}{2} \\ \frac{3}{2} & \frac{1}{2} \end{bmatrix}.$$

For L_2, we have $L_2(t+1) = t^2 + t$ and $L_2(t-1) = t^2 - t$. We now write

$$c_1(t^2 + 1) + c_2 t + c_3(t-1) = t^2 + t$$
$$d_1(t^2 + 1) + d_2 t + d_3(t-1) = t^2 - t.$$

We solve the two linear systems obtained from these equations by Gauss–Jordan reduction:

$$\begin{bmatrix} 1 & 0 & 0 & \vdots & 1 & \vdots & 1 \\ 0 & 1 & 1 & \vdots & 1 & \vdots & -1 \\ 1 & 0 & -1 & \vdots & 0 & \vdots & 0 \end{bmatrix} \xrightarrow[\text{steps omitted}]{\quad \cdots \quad} \begin{bmatrix} 1 & 0 & 0 & \vdots & 1 & \vdots & 1 \\ 0 & 1 & 0 & \vdots & 0 & \vdots & -2 \\ 0 & 0 & 1 & \vdots & 1 & \vdots & 1 \end{bmatrix}.$$

Thus $c_1 = 1$, $c_2 = 0$, $c_3 = 1$, $d_1 = 1$, $d_2 = -2$, and $d_3 = 1$. Hence

$$A_2 = \begin{bmatrix} 1 & 1 \\ 0 & -2 \\ 1 & 1 \end{bmatrix}.$$

We find that $B = A_2 A_1$.

9. (a) The matrix of $L_2 \circ L_1$ with respect to S and T is

$$A_2 A_1 = \begin{bmatrix} 0 & 1 \\ -2 & 3 \end{bmatrix} \begin{bmatrix} 1 & 2 \\ -1 & 3 \end{bmatrix} = \begin{bmatrix} -1 & 3 \\ -5 & 5 \end{bmatrix}.$$

(b) The matrix of $L_1 \circ L_2$ with respect to S and T is

$$A_1 A_2 = \begin{bmatrix} 1 & 2 \\ -1 & 3 \end{bmatrix} \begin{bmatrix} 0 & 1 \\ -2 & 3 \end{bmatrix} = \begin{bmatrix} -4 & 7 \\ -6 & 8 \end{bmatrix}.$$

11. $L(x, y) = (x + y, x - y, x + 2y)$ so $L: R^2 \to R^3$. We must determine whether L is one-to-one and onto. First, observe that if L is onto then $\dim(\text{range } L) = 3$. However, by Theorem 10.7 we must have

$$\dim R^2 = 2 = \dim(\ker L) + \dim(\text{range } L) \quad \text{and} \quad \dim(\ker L) \leq 2.$$

So L cannot be onto. Hence L is not invertible.

13. $L\left(\begin{bmatrix} x \\ y \\ z \end{bmatrix}\right) = \begin{bmatrix} 1 & 0 & 1 \\ 0 & 1 & 1 \\ 1 & 0 & 2 \end{bmatrix}\begin{bmatrix} x \\ y \\ z \end{bmatrix}.$

We determine whether L is one-to-one and onto. First, we see whether L is one-to-one by finding its kernel. Thus we seek all vectors

$$\begin{bmatrix} x \\ y \\ z \end{bmatrix}$$

such that

$$L\left(\begin{bmatrix} x \\ y \\ z \end{bmatrix}\right) = \begin{bmatrix} 0 \\ 0 \\ 0 \end{bmatrix}.$$

This leads to solving

$$\begin{bmatrix} 1 & 0 & 1 \\ 0 & 1 & 1 \\ 1 & 0 & 2 \end{bmatrix}\begin{bmatrix} x \\ y \\ z \end{bmatrix} = \begin{bmatrix} 0 \\ 0 \\ 0 \end{bmatrix},$$

a homogeneous system whose only solution is the trivial one. Hence L is one-to-one; $\dim(\ker L) = 0$. Next, to see if L is onto we use Theorem 10.7:

$$\dim R^3 = 3 = \dim(\ker L) + \dim(\operatorname{range} L)$$

so $3 = 0 + \dim(\operatorname{range} L)$. Since $\dim(\operatorname{range} L) = 3$ and range L is a subspace of R^3, it follows from Exercise T.7 in Section 6.4 that range $L = R^3$, so L is onto. Hence L is invertible.

We now find

$$L^{-1}\left(\begin{bmatrix} b_1 \\ b_2 \\ b_3 \end{bmatrix}\right).$$

Let

$$L^{-1}\left(\begin{bmatrix} b_1 \\ b_2 \\ b_3 \end{bmatrix}\right) = \begin{bmatrix} x \\ y \\ z \end{bmatrix}.$$

Applying L to both sides, we obtain

$$\begin{bmatrix} b_1 \\ b_2 \\ b_3 \end{bmatrix} = L\left(\begin{bmatrix} x \\ y \\ z \end{bmatrix}\right) = \begin{bmatrix} 1 & 0 & 1 \\ 0 & 1 & 1 \\ 1 & 0 & 2 \end{bmatrix}\begin{bmatrix} x \\ y \\ z \end{bmatrix} = \begin{bmatrix} x + z \\ y + z \\ x + 2z \end{bmatrix}.$$

Solving

$$\begin{aligned} x \quad\; + \; z &= b_1 \\ y + \; z &= b_2 \\ x \quad\; + 2z &= b_3 \end{aligned}$$

we find

$$x = 2b_1 - b_3, \qquad y = b_1 + b_2 - b_3, \qquad z = -b_1 + b_3.$$

Hence

$$L^{-1}\left(\begin{bmatrix} b_1 \\ b_2 \\ b_3 \end{bmatrix}\right) = \begin{bmatrix} 2b_1 - b_3 \\ b_1 + b_2 - b_3 \\ -b_1 + b_3 \end{bmatrix}.$$

Another way to find

$$L^{-1}\left(\begin{bmatrix} b_1 \\ b_2 \\ b_3 \end{bmatrix}\right)$$

is as follows. The matrix of L with respect to the natural basis for R^3 is the matrix defining L:

$$\begin{bmatrix} 1 & 0 & 1 \\ 0 & 1 & 1 \\ 1 & 0 & 2 \end{bmatrix},$$

so the matrix of L^{-1} is the inverse of this matrix:

$$\begin{bmatrix} 1 & 0 & 1 \\ 0 & 1 & 1 \\ 1 & 0 & 2 \end{bmatrix}^{-1} = \begin{bmatrix} 2 & 0 & -1 \\ 1 & 1 & -1 \\ -1 & 0 & 1 \end{bmatrix}.$$

Therefore

$$L^{-1}\left(\begin{bmatrix} b_1 \\ b_2 \\ b_3 \end{bmatrix}\right) = \begin{bmatrix} 2 & 0 & -1 \\ 1 & 1 & -1 \\ -1 & 0 & 1 \end{bmatrix}\begin{bmatrix} b_1 \\ b_2 \\ b_3 \end{bmatrix} = \begin{bmatrix} 2b_1 - b_3 \\ b_1 + b_2 - b_3 \\ -b_1 + b_3 \end{bmatrix}.$$

15. $L\left(\begin{bmatrix} x \\ y \\ z \end{bmatrix}\right) = \begin{bmatrix} 1 & 1 & 1 \\ 0 & 1 & 2 \\ -2 & -1 & 0 \end{bmatrix}\begin{bmatrix} x \\ y \\ z \end{bmatrix}.$

We determine whether L is one-to-one and onto. First, we see whether L is one-to-one by finding its kernel. Thus we seek all vectors

$$\begin{bmatrix} x \\ y \\ z \end{bmatrix}$$

such that

$$L\left(\begin{bmatrix} x \\ y \\ z \end{bmatrix}\right) = \begin{bmatrix} 0 \\ 0 \\ 0 \end{bmatrix}.$$

This leads to solving

$$\begin{bmatrix} 1 & 1 & 1 \\ 0 & 1 & 2 \\ -2 & -1 & 0 \end{bmatrix}\begin{bmatrix} x \\ y \\ z \end{bmatrix} = \begin{bmatrix} 0 \\ 0 \\ 0 \end{bmatrix},$$

a homogeneous system whose solution is

$$\begin{bmatrix} x \\ y \\ z \end{bmatrix} = \begin{bmatrix} r \\ -2r \\ r \end{bmatrix}.$$

Since this homogeneous system has nontrivial solutions, it follows that L is not one-to-one and thus is not invertible.

17. $L(at^2 + bt + c) = -at^2 + bt - c.$

We determine whether L is one-to-one and onto. To see whether L is one-to-one we find its kernel. we seek all vectors $at^2 + bt + c$ such that $L(at^2 + bt + c) = 0$. This leads to $-at^2 + bt - c = 0$, so $a = 0$, $b = 0$, $c = 0$. Hence L is one-to-one; $\dim(\ker L) = 0$. To see whether L is onto, we use Theorem 10.7:

$$\dim(P_2) = 3 = \dim(\ker L) + \dim(\operatorname{range} L)$$
$$3 = 0 + \dim(\operatorname{range} L).$$

Since $\dim(\operatorname{range} L) = 3 = \dim(P_2)$ and range L is a subspace of P_2, it follows by Exercise T.7 in Section 6.4 that range $L = P_2$, so L is onto. Therefore L is invertible. Now let

$$L^{-1}(dt^2 + et + f) = xt^2 + yt + z.$$

Applying L to both sides we obtain

$$dt^2 + et + f = L(xt^2 + yt + z) = -xt^2 + yt + z.$$

Equating coefficients of equal powers, we have

$$x = -d, \qquad y = e, \qquad z = f.$$

Hence

$$L^{-1}(dt^2 + et + f) = -dt^2 + et + f.$$

19. We have (by Theorem 10.7 in Section 10.2),

$$\dim R^4 = 4 = \operatorname{nullity} L + \operatorname{rank} L = \operatorname{nullity} L + 4$$

so nullity $L = 0$ and hence L is one-to-one. Moreover, $\dim(\operatorname{range} L) = \operatorname{rank} L = 4$, so L is onto. Hence L is invertible.

21. We have (by Theorem 10.7 in Section 10.2),

$$\dim(P_2) = 3 = \operatorname{nullity} L + \operatorname{rank} L$$
$$3 = 1 + \operatorname{rank} L$$

so rank $L = 2$. Since $\dim(\operatorname{range} L) = \operatorname{rank} L$, $\dim(\operatorname{range} L) \neq \dim(P_2)$, so range $L \neq P_2$ and hence L is not invertible.

23. Let S be the natural basis for R^3. The matrix representing L^{-1} is the inverse of the matrix representing L. Since S is the natural basis for R^3, the matrix representing L is the matrix whose jth column is $\left[L(\mathbf{e}_j)\right]_S$. Thus, the matrix representing L is

$$A = \begin{bmatrix} 1 & 0 & 1 \\ 2 & 1 & 1 \\ 3 & 1 & 0 \end{bmatrix}.$$

Hence, the matrix representing L^{-1} is

$$A^{-1} = \begin{bmatrix} 1 & 0 & 1 \\ 2 & 1 & 1 \\ 3 & 1 & 0 \end{bmatrix}^{-1} = \begin{bmatrix} \frac{1}{2} & -\frac{1}{2} & \frac{1}{2} \\ -\frac{3}{2} & \frac{3}{2} & -\frac{1}{2} \\ \frac{1}{2} & \frac{1}{2} & -\frac{1}{2} \end{bmatrix}.$$

25. The matrix of L^{-1} with respect to S is

$$A^{-1} = \begin{bmatrix} 2 & 0 & 4 \\ -1 & 1 & -2 \\ 2 & 3 & 3 \end{bmatrix}^{-1} = \begin{bmatrix} -\frac{9}{2} & -6 & 2 \\ \frac{1}{2} & 1 & 0 \\ \frac{5}{2} & 3 & -1 \end{bmatrix}.$$

T.1. Let $\mathbf{u}$ and $\mathbf{v}$ be in V_1 and let c be a scalar. We have

$$(L_2 \circ L_1)(\mathbf{u} + \mathbf{v}) = L_2(L_1(\mathbf{u} + \mathbf{v})) = L_2(L_1(\mathbf{u}) + L_1(\mathbf{v}))$$
$$= L_2(L_1(\mathbf{u})) + L_2(L_1(\mathbf{v}))$$
$$= (L_2 \circ L_1)(\mathbf{u}) + (L_2 \circ L_1)(\mathbf{v})$$

and

$$(L_2 \circ L_1)(c\mathbf{u}) = L_2(L_1(c\mathbf{u})) = L_2(cL_1(\mathbf{u})) = cL_2(L_1(\mathbf{u})) = c(L_2 \circ L_1)(\mathbf{u}).$$

T.3. We have

$$(L \circ O_V)(\mathbf{v}) = L(O_V(\mathbf{v})) = L(\mathbf{0}) = \mathbf{0} = O_V(\mathbf{v})$$

and

$$(O_V \circ L)(\mathbf{v}) = O_V(L(\mathbf{v})) = \mathbf{0} = O_V(\mathbf{v}).$$

T.5. Suppose that L_1 and L_2 are one-to-one and onto. We first show that $L_2 \circ L_1$ is also one-to-one and onto. First, one-to-one. Suppose that $(L_2 \circ L_1)(\mathbf{v}_1) = (L_2 \circ L_1)(\mathbf{v}_2)$. Then $L_2(L_1(\mathbf{v}_1)) = L_2(L_1(\mathbf{v}_2))$ so $L_1(\mathbf{v}_1) = L_1(\mathbf{v}_2)$, since L_2 is one-to-one. Hence $\mathbf{v}_1 = \mathbf{v}_2$ since L_1 is one-to-one. Next, let L_1 and L_2 be onto, and let $\mathbf{w}$ be any vector in V. Since L_2 is onto, there exists a vector $\mathbf{v}_1$ in V such that $L_2(\mathbf{v}_1) = \mathbf{w}$. Since L_1 is onto, there exists a vector $\mathbf{v}_2$ in V such that $L_1(\mathbf{v}_2) = \mathbf{v}_1$. Then we have

$$(L_2 \circ L_1)(\mathbf{v}_2) = L_2(L_1(\mathbf{v}_2)) = L_2(\mathbf{v}_1) = \mathbf{w},$$

and therefore $L_2 \circ L_1$ is onto. Hence $L_2 \circ L_1$ is invertible. Since

$$(L_2 \circ L_1) \circ (L_1^{-1} \circ L_2^{-1}) = I_V \quad \text{and} \quad (L_1^{-1} \circ L_2^{-1}) \circ (L_2 \circ L_1) = I_V,$$

we conclude that $(L_2 \circ L_1)^{-1} = L_1^{-1} \circ L_2^{-1}$.

T.7. Since L is one-to-one and onto, it is invertible. First, L is one-to-one. To show this, let $L(A) = L(B)$. Then $A^T = B^T$ so $(A^T)^T = (B^T)^T$ which implies that $A = B$. Also, if B is any element in M_{22}, then $L(B^T) = (B^T)^T = B$, so L is onto. We have $L^{-1}(A) = A^T$.

T.9. We show that (a) $\Longrightarrow$ (b) $\Longrightarrow$ (c) $\Longrightarrow$ (a).

(a) $\Longrightarrow$ (b): Suppose that L is invertible. Then L is one-to-one and onto, so $\dim(\text{range } L) = n = \text{rank } L$.

(b) $\Longrightarrow$ (c): If rank $L = n$, then $\dim(\ker L) = 0$ so nullity $L = 0$.

(c) $\Longrightarrow$ (a): If nullity $L = 0$, then rank $L = n$, which means that $\dim(\text{range } L) = n$. Hence L is one-to-one and onto and is then invertible.

T.11 We have

$$L(\mathbf{v}_1 + \mathbf{v}_2) = (\mathbf{v}_1 + \mathbf{v}_2, \mathbf{w}) = (\mathbf{v}_1, \mathbf{w}) + (\mathbf{v}_2, \mathbf{w}) = L(\mathbf{v}_1) + L(\mathbf{v}_2).$$

Also, $L(c\mathbf{v}) = (c\mathbf{v}, \mathbf{w}) = c(\mathbf{v}, \mathbf{w}) = cL(\mathbf{v})$.